Thomas L. Taylor

FISHERIES BIOENGINEERING SYMPOSIUM

The Fisheries Bioengineering Symposium was organized and sponsored by the Bioengineering Section of the American Fisheries Society. Generous financial contributions for publication of this volume were made by the following:

Fish and Wildlife Division
Bonneville Power Administration

Salmonid Enhancement Program
Fisheries and Oceans Canada

Recreational Fisheries Branch
British Columbia Ministry of Environment

Ontario Ministry of Natural Resources

CH_2M Hill
Engineering Consultants

Washington Department of Fisheries

Oregon Department of Fish and Wildlife

Michigan Department of Natural Resources

California Department of Fish and Game

Alaska Department of Fish and Game

National Marine Fisheries Service
U.S. Department of Commerce

Fish and Wildlife Service
U.S. Department of the Interior

Northwest Power Planning Council

AMERICAN FISHERIES SOCIETY
SYMPOSIUM 10

FISHERIES BIOENGINEERING SYMPOSIUM

JOHN COLT AND RAY J. WHITE

EDITORS

AMERICAN FISHERIES SOCIETY
BETHESDA, MARYLAND, USA
1991

The American Fisheries Society Symposium series is a registered serial. Suggested citation formats follow.

Entire book

Colt, J., and R. J. White, editors. 1991. Fisheries bioengineering symposium. American Fisheries Society Symposium 10.

Article within the book

Fleming, D. F., and J. B. Reynolds. 1991. Effects of spawning-run delay on spawning migration of Arctic grayling. American Fisheries Society Symposium 10:299–305.

Library of Congress Catalog Card Number: 91-71561

ISBN 0-913235-72-5 ISSN 0892-2284

Address orders to

American Fisheries Society
5410 Grosvenor Lane, Suite 110
Bethesda, Maryland 20814-2199, USA

CONTENTS

PERSPECTIVES

HABITATS

FISH HATCHERIES

Foreword

The 1988 Fisheries Bioengineering Symposium, held in Portland, Oregon, USA, was based on a theme of "Problems and Solutions in Developed River Systems." The emphasis was upon developed river systems both because development is more the norm than nondevelopment and because the seemingly worst problems—hence, the biggest challenges—for fish in rivers result from development. Further, one can reasonably predict that development will continue, so unless adequate solutions are found, river-based fisheries will experience progressively greater problems.

The symposium included a variety of regional perspectives and solutions already implemented in Japan, Canada, Norway, and the USA. Contributors emphasized U.S. river basins and focused on the nearby Columbia River and its tributaries, but also described the Tennessee and Sacramento–San Joaquin systems. Collectively these rivers span the fisheries spectra from anadromous to resident, from warmwater to coldwater, and from freshwater to esturarine.

Most of the symposium was devoted to solutions in two main and widely divergent topics: habitat improvement and restoration, and mitigation via hatcheries. Although these topics may seem strange bedfellows, they define the main alternatives available to fishery managers for correcting or mitigating fishery losses in developed rivers, and they are the professional concerns of bioengineering. In this regard, it was a primary goal of the steering committee to bring together and facilitate information exchange between biologists and engineers so that each group could gain a better appreciation for the other's perspectives, constraints, and strengths.

Many aspects of habitat improvement were addressed, and preferred solutions or alternatives were described by the experts. Whole sessions were devoted to planning and evaluating habitat improvement programs, to uses of side channels, to concerns relative to instream alternatives, and to instream flow technology.

Closely related to fish habitat management are the perplexing and costly problems of facilitating fish passage past dams and water-diversion structures. A bioengineering team approach is clearly needed to solve them. Symposium sessions were devoted to protective screens for juvenile fish, designs of fishways, passage of adult salmon, and the particularly important problem of providing passage for smolts at large, main-stem structures.

Hatcheries can be important tools for fisheries managers trying to rehabilitate fish stocks in developed river systems, and hatcheries are clearly important in commercial aquaculture. Hatchery technology varies with the fish species being cultured and not all hatchery programs are compatible with fishery management goals, but there is sufficient commonality to deal generically with many hatchery problems. The symposium addressed concerns in planning a hatchery, maintaining water quality, improving operation, and alternative hatchery production systems. Sessions on the disinfection of water supplies and the treatment of hatchery effluents were especially timely. Oxygen supplementation was discussed extensively, mainly as a means of increasing fish production with minimal capital and operating expenses.

A majority of all the papers presented at the symposium are included in this proceedings. The reader will find a rich harvest of ideas and solutions that have been applied to the problems listed herein.

GERALD BOUCK
Bonneville Power Administration
Portland, Oregon

Acknowledgments

The Fisheries Bioengineering Symposium held in Portland, Oregon, was a resounding success in terms of attendance, organization, informal discussion, and scientific information exchange. Many people went beyond the call of duty to put on and pull off the symposium. Dave Owsley and Harry Westers, cochairs of the symposium, provided able overall leadership and coordination. Dave Owsley was particularly active in finding funds for publication of the proceedings. John Colt's program of over 100 papers, reflected in the contents of this volume, was both focused on the two subareas of habitat and hatchery and comprehensive within those subdisciplines. Gerry Bouck coordinated all the local arrangements and was responsible for the excellent choice of accommodations and venue for the symposium. Ken Bates and his crew from Washington Department of Fisheries did a masterful job of handling registration and the many associated problems. Brian D'Aoust developed some innovative concepts for the commercial displays, and Wayne Daley completed logistic arrangements with his characteristic aplomb. Larry Visscher organized a mini-presentation of the symposium at Aquaculture '89 in Los Angeles. The other members of the Steering Committee all served their posts admirably: Bob Pearce (meals and social events), Bill Godby (publicity), Steve Rainey (video program), Jerry Bauer (audiovisual), John Ferguson (postsymposium tour), and Vic Kaczynski (finances). Joe Fuss and Howie Yoshida, as Secretary–Treasurers of the Bioengineering Section, also provided valuable service to the pre- and postsymposium finances.

Many members of the Steering Committee were also moderators of the technical sessions; additionally, Gary Logan, Steve Smith, Bob Piper, Ron Goede, Gary Boersen, and Charlie Smith helped out in this area and deserve recognition for jobs well done. The many speakers whose papers, for one reason or another, are not included in this volume are also thanked for their contributions to the symposium. Of particular acclaim was the outstanding keynote address by Dr. Howard Tanner on how change seekers can be effective in the management of public resources; his talk inspired many of us.

Coordination of the technical editing was handled in a timely and able manner by John Colt and Ray White. The American Fisheries Society's editorial staff, in particular Bob Kendall, Dan Guthrie, Beth Mitchell, and Sally Kendall, helped bring the quality of the manuscripts up to the high standard you see within this volume.

A very special thanks goes to the sponsors who contributed funds for publication of this volume. We also appreciate services provided by the staff of the Red Lion Columbia River Inn, who treated us very well. Finally, the Bioengineering Section would like to thank the many authors represented herein for their patience during the publication process and for their personal dedication and scientific excellence on the behalf of fishery resources.

Don D. MacKinlay
President, Bioengineering Section
American Fisheries Society

Symbols and Abbreviations

The Systeme International d'Unites (SI) is followed for metric units in this book. The SI symbols, English abbreviations, and other symbols listed below are used without further definition in the text. Also undefined are standard mathematical, statistical, chemical, and atomic (elemental) symbols and chemical acronyms given in standard dictionaries.

A	ampere
AC	alternating current
°C	degrees Celsius ([°Fahrenheit − 32]/1.8)
cm	centimeter (0.394 inch)
Co.	Company
Corp.	Corporation
cot	cotangent
d	day
DC	direct current
D.C.	District of Columbia
df	degrees of freedom
E	east
e	base of natural logarithms (2.71 . . .)
e.g.	(exempli gratia) for example
eq	equivalent
et al.	(et alia) and others
etc.	et cetera
°F	degrees Fahrenheit ([1.8 × °Celsius] + 32)
ft	foot (0.305 meter, 30.5 centimeters)
g	gram (0.0353 ounce)
h	hour
Hz	hertz
i.e.	(id est) that is
in	inch (2.54 centimeters)
Inc.	Incorporated
J	joule
k	kilo (10^3, as a prefix)
kg	kilogram (2.20 pounds)
km	kilometer (0.622 mile)
kWh	kilowatt-hour
L	liter (0.264 gallon, 1.06 quarts)
lb	pound (0.454 kilograms, 454 grams)
log	logarithm
Ltd.	Limited
M	mega (10^6, as a prefix); molar (as a suffix or by itself)
m	meter (as a suffix or by itself; 1.09 yards, 3.28 feet, 39.4 inches); milli (10^{-3}, as a prefix)
meq	milliequivalent
mi	mile (1.61 kilometers)
min	minute
mol	mole
N	north; normal; newton
N	sample size
NS	not significant
n	nano (10^{-9}, as a prefix)
o	ortho (as a chemical prefix)
P	probability
p	pico (10^{-12}, as a prefix)
p	para (as a chemical prefix)
pH	negative log of hydrogen ion activity
ppm	parts per million (per 10^6) milligrams per liter in the metric system
R	multivariate correlation or regression coefficient
r	univariate correlation or regression coefficient
rpm	revolutions per minute
S	siemens (for electrical conductance; = mho, 1/ohm); south
s	second
SD	standard deviation
SE	standard error
tonne	metric ton (1,000 kilograms, 2,200 pounds)
UK	United Kingdom
U.S.	United States (as adjective)
USA	United States of America
USSR	Union of Soviet Socialist Republics
V	volt
W	watt; west
μ	micro (10^{-6}, as a prefix)
Ω	ohm
°	degree (angular)
′	minute (angular)
%	percent (per hundred)
‰	parts per mille (per thousand, 10^3)

American Fisheries Society Symposium 10:1–11, 1991

PERSPECTIVES

Overview of Effects of Pacific Coast River Regulation on Salmonids and the Opportunities for Mitigation

J. H. MUNDIE
Department of Fisheries and Oceans, Biological Sciences Branch
Pacific Biological Station, Nanaimo, British Columbia V9R 5K6, Canada

Abstract.—Current awareness of the seriousness of losses of salmonid fishes associated with hydroelectric developments and with water abstraction from river systems has stimulated renewed commitments on the part of fishery agencies to mitigation of damage, restoration of degraded habitat, protection and promotion of wild stocks, and increased artificial production of salmonids. To achieve these aims the fish system, the fluvial system, and the system of human values and intentions must be integrated; in addition today's knowledge must be equal to the challenge. Examination of the present status of six aspects of the task shows that (1) facilitation of fish passage at dams is a very high priority but requires greater commitment, (2) hatchery production has generated unreasonable expectations and may be laying the basis for the demise of wild populations, (3) the practice of stocking fry has run ahead of its evaluation, (4) determination of instream flow requirements is bedeviled by spurious quantification, (5) drawdown requirements of impoundments seem incompatible with fishery objectives, and (6) stream habitat improvements give mixed results and may be of restricted application in terms of scale. Yet another limitation of fisheries' aspirations lies in political support. It is concluded that this is a time for stock-taking, improvement of current practices, and assessment of trends.

The story of dams and fish has been told frequently, at both the professional and popular level (e.g., Hedgpeth 1944; Netboy 1974, 1980; Schweibert 1977; Raymond 1988). Most attention has been paid to the impact of large dams on the Columbia River and its main tributaries, begun in 1933 with Rock Island Dam, and believed to be largely responsible for the steep decline of runs of spring and summer chinook salmon *Oncorhynchus tshawytscha* and steelhead *O. mykiss* between the 1960s and the early 1970s (Raymond 1988). The Central Valley Project in California brought similar destructiveness, but on a smaller scale. There the first large dam, the Shasta, was begun in 1938 and diverted water from the north-central part of the state to southern areas; other dams and a complex of diverted flows ensued. In consequence the runs of steelhead, coho salmon *O. kisutch* and chinook salmon declined by about 65% along the north coast between 1940 and 1969; in the Central Valley the chinook salmon spawning population dropped from 597,000 fish in 1953 to 332,000 in 1969—a 46% decline. These findings occasioned a report (CAC 1971) entitled "An Environmental Tragedy," which called for urgent positive measures to restore the runs. Today the report creates a feeling of déjà vu; the difference now perhaps is that we have a more united commitment to redress the harm done by dams. Yet another large dam that has been viewed with grave concern is one proposed for Moran Canyon of the Fraser River in British Columbia (Geen 1975).

Large-scale developments are not the sole cause of alarm. Following the U.S. Public Utility Regulatory Policies Act (1978), streams that formerly were considered too small to offer economically attractive hydroelectric development are now candidates. In consequence, several thousand proposed small hydropower projects—small by virtue of the power they could generate individually, not by virtue of their dimensions—now threaten salmon runs (Olson et al. 1985).

The general effects on salmonids of abstraction of water and regulation of flows have been well documented (e.g., Stalnaker 1980; Rochester et al. 1984; Raymond 1988). Storage of water (for hydropower or flood control), water removal (for irrigation and municipal and industrial uses), and rapid changes in water levels (for a variety of purposes) can inundate habitat, block migrations of anadromous species, impair downstream migrations of juveniles, impose stress on all stages of fish owing to changes in ambient temperatures and to high total gas pressures, alter the composition of gravel used for spawning and food production, strand fish, and change the species composition of fish communities. A recent survey of

TABLE 1.—Principal effects of flow regulation on fish populations following flow regulation. More than one effect may operate in a case history. (From Burt and Mundie 1986.)

Explanation	Number (%) of case histories
Reduced flows resulting in reduced habitat	29 (60)
Blockage of habitat	17 (35)
Sedimentation, deterioration of gravel quality	14 (29)
Fluctuating flows	9 (19)
Altered water temperature	8 (17)
Pollution	3 (6)
Difficulty in passage of downstream migrants	3 (6)
Lack of gravel recruitment	2 (4)
Inundation of fish habitat	2 (4)
Nitrogen supersaturation	1 (2)

81 miscellaneous case histories of the effects of regulated stream flow on salmonid populations of northwestern North America (Burt and Mundie 1986) brought out the frequency of occurrence of these effects (Table 1), and therefore of the kinds of problems that have to be overcome or mitigated. These data, however, do not convey the relative importance of effects in terms of fish losses. Clearly, blockage of fish from whole river systems and inundation of extensive areas of habitat rank high in severity of consequences. For example, habitat for salmon and steelhead in the Columbia River basin above Bonneville Dam has decreased from 18,800 river kilometers before 1850 to about 12,200 river kilometers in 1976, about a 35% loss (Northwest Power Planning Council 1987).

When bioengineering remedies are applied to the situations that limit fish production in developed rivers, three highly contrasting systems have to be meshed and harmonized—the fluvial system, the fish population system with its genetic content, and the "soft" system of human mixed motives, intentions, and values. Intentions and values, in this context, have been shaped by growing awareness of the scale and seriousness of losses of fish and habitat following dam construction, water abstraction, and multiple resource use. Values include not only more fish, however produced, but also satisfying fishing experience in aesthetically appealing environments. It is one of the limitations of democracy, however, that decisions on fish are not made only by those who value them most, but are weighed against other, conflicting, interests. This is often crucial to the outcome of programs aimed at conservation, mitigation, or enhancement.

The largest program directed at rebuilding fish populations that have been harmed by hydropower development is that of the Northwest Power Planning Council (1987) for the entire Columbia River basin; it is "quite possibly the most ambitious effort in the world to save a biological resource" (page 5). This takes a systems-wide approach, coordinates wild and hatchery production, controls harvest management, and has as an interim goal a twofold increase in runs of salmon and steelhead—from 2.5 to 5 million fish (Table 2). The program acknowledges genetic risks of actions, aims at using a variety of methods to increase production, and is prepared to change direction in the light of unfolding experience. It might be asked: irrespective of problems of scale, jurisdiction, and allocation of costs, do we know enough to achieve this, and to carry out similar but less ambitious programs elsewhere? To answer this question, it is instructive to highlight the state of knowledge and practice of a few selected issues that are fundamental to successful mitigation.

Mitigative Measures

Fish Passage at and between Dams

The future of salmonid stocks in several major river systems depends on our ability to reduce mortality of fish as they try to migrate downstream past dams. The average mortality of migrating smolts at a turbine lies between 10% and 15% but can rise to 30% (Ebel 1977). Average mortality of juvenile salmonids as they pass through a reservoir can amount to 14% (Rieman et al. 1988). The causes of mortality are the turbines themselves, predation by birds and fishes, delays in migration because of the impoundments, and to a lesser extent supersaturation of the water with atmospheric gases. Good progress has been made in the last decade on reducing these effects. Spillway deflectors go a long way toward solving the gas pressure problem, diversion screens can direct fish away from turbine intakes into a bypass system that returns them to the river below the dam, and fish may be collected and transported downstream in large barges and trucks, thereby bypassing several dams. In addition, water discharge may be budgeted so that there is sufficient flow in the impoundments to facilitate migrations. All these methods are helpful to varying degrees, but they are costly and there is still urgent need for improvements. Spilling of water and provision of sufficient flow through impoundments may be

TABLE 2.—Main emphasis of the Columbia River basin program. (Adapted from Northwest Power Planning Council 1987.)

Policy	Comment
Salmon and steelhead	
The area above Bonneville Dam is the accorded priority	Greatest losses of fish have been in upper Columbia and Snake river areas
Genetic risks will be assessed	Every effort must be made to maintain genetic diversity
Main-stem survival of fish will be improved expeditiously	This will be achieved by implementation of water budgets, interim spills, mechanical bypass systems, and transportation of fish
Increased production will be reached by a mix of methods	The main methods will be higher production at existing hatcheries, new hatcheries, stocking, habitat improvement, protection of habitat from new development, and management of mixed-stock fisheries
Harvest management will support rebuilding	There will be restraint on ocean and inriver harvests to accelerate rebuilding of runs
System integration will assure consistency	The plans of different institutions for production, passage, and harvest will be coordinated
Adaptive management will guide action and improve knowledge	Monitoring and evaluation will influence the development of the program
Resident fish	
Lake survival of resident fish will be improved	This will be done by setting limits to drawdown of certain reservoirs, by stocking reservoirs, and with hatcheries
Stream survival of resident fish be improved	Storage will be used to maintain suitable temperatures for fish habitat, water will be released at times to aid all stages of resident fish, rivers will be stocked and hatcheries will be built for raising rainbow trout *Oncorhynchus mykiss*, cutthroat trout *O. clarki*, bull trout *Salvelinus confluentus*, kokanee *O. nerka*, and white sturgeon *Acipenser transmontanus*

difficult or impossible in years of critically low water levels; there are still substantial losses in the transport of spring chinook salmon, especially if they are hatchery products infected with bacterial kidney disease (Raymond 1988); bypass sytems, both open and closed, need to be made more effective; and problems that are not understood arise in collecting and bypassing fall chinook salmon. Clearly, progress in this difficult area is highly important.

There remain also problems with upstream passage of adults. Low discharges of fresh water may delay the adults' departure from estuaries or upstream movements. Elevated temperatures may impose excessive demands on the fishes' energy reserves as they swim through impoundments (Brett 1957). Losses occur with passage at the dams themselves (Raymond 1988).

Hatcheries

Strictly speaking, hatcheries and spawning channels are not mitigative measures; they do not moderate the harmful effects of development. They are alternative ways of producing fish, and may compensate for losses. It is necessary, however, to consider hatcheries here because they are usually the main agent not only for maintaining and augmenting existing runs but for creating new ones where the original ones have been lost. For example, 54 hatcheries and 40 satellite rearing facilities produce 80% of the 2.5 million salmon and steelhead that return annually to the Columbia River (Bonneville Power Administration 1987). Some 79% of the funds of the Columbia River Fisheries Development Program are used to operate hatcheries, whereas 10% go to maintenance of irrigation screens and stream improvements (Delarm et al. 1987). It is estimated that over 314 million chinook salmon and 137 million coho salmon currently are released annually from all the hatcheries along the west coast of North America (Wahle and Pearson 1987).

The essential purpose of salmonid hatcheries is to reduce the exponential natural mortality associated with a life history strategy appropriate for environments that are highly heterogeneous in space and time (and anathemas to engineers because they imply uncontrollable variation). Hatcheries succeed in their aims up to the time of release of the fish, but at very substantial financial cost and with varying efficiency. Production costs at an average-size hatchery may exceed a quarter of a million dollars per annum; releases cost an average of US$4.72 per kilogram of fish produced, and the contribution of fish to the sport and commercial fisheries may range, for example for fall chinook salmon, from less than 2 to 12.9 fish per 1,000 released (Delarm et al. 1987). The cost of replacing lost natural production is therefore very high.

Of even greater concern than costs is the recent trend for hatchery releases to yield diminishing returns. The Oregon experience is well documented (Oregon Department of Fish and Wildlife 1981, 1985). Before the 1960s, coho salmon populations were predominantly naturally produced fish. By 1975, about 75% of coho salmon caught in the Oregon Production Index area were of hatchery origin; from 1970 to 1975, catches averaged 2.3 million fish, and in 1976 a record of 4.1 million occurred. This was followed by a low of 1.1 million in 1977, and stock size has remained low in subsequent years (Pacific Fishery Management Council 1988)—even though smolt releases rose from about 30 million in 1970 to about 60 million in 1981.

Unexpectedly poor returns were also obtained by the Salmonid Enhancement Program in British Columbia (Department of Fisheries and Oceans 1988). The goal of this program was the same as the interim goal of the Columbia River basin program—to double existing runs. In Phase I (1977–1983) Can$300 million were spent on 10 major hatcheries, 6 smaller ones, and numerous community projects. The objective was not reached, and by 1986 the program accounted for about 15% of the salmon caught in British Columbia (Department of Fisheries and Oceans 1988). Catches, by all gear types, of coho salmon in the south have declined in the 1980s from historical levels; catches of chinook salmon are the lowest in 25 years. Attempts to augment runs of chinook salmon by hatchery production in the interior of British Columbia have so far been unsuccessful. In contrast, there have been recent successes in increasing runs of chum salmon *Oncorhynchus keta* and sockeye salmon *O. nerka* (Department of Fisheries and Oceans 1989).

The explanation of the Oregon and British Columbia experiences is by no means clear. Various hypotheses have been examined: that there has been a decline in the quality of smolts released from hatcheries, bacterial kidney disease being the prime suspect; that reduction in genetic diversity has made the species more vulnerable to environmental changes; that the reductions in catch are attributable to a decline in wild fish that has not been offset by hatchery production; that the reductions are simply an example of natural variability in population numbers; that density-dependent mortality may have occurred in fresh water, in the estuary, or in the ocean where the carrying capacity has been exceeded. The most favored explanation (Oregon Department of Fish and Wildlife 1985) is cyclic changes in ocean conditions, particularly in the timing, stability, and magnitude of coastal upwelling (Nickelson 1986); associated with these changes are shifts in temperature, salinity, and food abundance that affect survival of smolts.

These possibilities, however, are as yet unproven. Some wild stocks of British Columbian coho salmon declined significantly from 1975 to 1985 (L. Lapi, Canada Department of Fisheries and Oceans, unpublished data). This finding indicates that decreasing survival is common to wild and hatchery fish and may therefore be attributed to environmental conditions rather than to factors within the hatcheries. Some notable declines, however, are hatchery-specific. The explanation of these may lie in fish quality.

The salient points emerge, then, that increased output of smolts from hatcheries does not necessarily bring proportional increases in returning adults, and the reasons for this are not understood.

If it is correct that cyclic changes in ocean conditions affect survival of smolts released from hatcheries, returns will improve in due course, at least for a period. Whether this occurs or not, the familiar difficulties and setbacks encountered in fish culture will, of course, continue. The main challenges are treatment and prevention of diseases, especially bacterial kidney disease, bacterial gill disease, and viral infectious hematopoietic necrosis (IHN); optimization of rearing density for survival; adjustment of the date of release and size of smolts to maximize returns; provision of fish for specific fisheries; and the maintenance of genetic diversity. The last is receiving overdue attention, and expositions of both the theory and practice of fish population genetics are now available (e.g., Kapuscinski and Jacobson 1987; Ryman and Utter 1987). One difficulty, however, is that the science of genetics has reached a stage of sophistication (as do all sciences) at which it is intelligible only to the specialist. Hatchery managers must therefore take the advice of geneticists on trust. A practical difficulty is that even a hatchery manager who tries to take a long-term view of genetic conservation may be compelled by shortage of brood stock to make do with what is available, which may not be sufficient to maintain genetic variation. Frequently there is a discrepancy between the theoretical ideal and the practical option.

Stocking

The practice of introducing fry or presmolts of hatchery origin into streams to supplement wild populations, to reestablish lost runs, or to establish runs where none existed is widely followed and is accepted as part of any program aimed at increasing stocks. If sufficient brood stock is available, a hatchery manager may incubate eggs to raise fry or fingerlings that are surplus to the hatchery production goals; these fish are used for stocking. Some managers aim to use half their annual egg take for this purpose. The fish may be given to public groups for dispersal under guidance. Most of this activity goes unevaluated, but when evaluation has been undertaken (reviewed by Smith et al. 1985), a wide range of survivals to the smolt stage has been found: for example, about 21% for spring chinook salmon fingerlings, less than 1 to 7% for coho salmon fry , and about 2% for steelhead fry (Smith et al. 1985; Hume and Parkinson 1987). Greater success seems to result when streams are stocked that have been devoid, historically, of the introduced species. Releases of early summer coho salmon fry into the Millstone River, British Columbia, gave a fry-to-smolt survival of 17.7% (Hurst and Blackman 1988). Adult returns averaging 18% were obtained following stocking of 5–10-g coho salmon fingerlings in September into the Quinsam River system, British Columbia (Van Tine 1986).

Nickelson et al. (1986) demonstrated that when releases of coho salmon presmolts were made into accessible coastal streams, more juveniles resulted but wild juveniles were displaced. Moreover, returns of adults to streams that received additions were not significantly different from adult returns to streams that had received none, and resulting densities of juveniles in the stocked streams were lower than in unstocked streams. The failure was attributed to the hatchery stock's early spawning time.

Some of the unintended consequences of stocking can be remedied by assessing densities of wild fish before stocking, by introducing fish of the same size as wild fish, and by obtaining brood stock from the target streams. Uncertainties and difficulties remain, however: for example, increases in the frequency of nonadaptive genes resulting from hatchery selection (Reisenbichler and McIntyre 1977), breakdowns of natural semi-isolated populations of the species, and assessments of the carrying capacity of habitats. The last is frequently an intractable problem in spite of attempts at predictive modeling. Moreover, in all these matters as in hatchery practice, biologists frequently are not able to follow even the currently accepted guidelines. They may be unable, for example, to obtain brood stock from the target stream. They therefore do the best they can.

Flow Requirements

The determination of instream flow needs of salmonid fishes, for example those living downstream of dams, has been and remains a highly important problem for fisheries managers. It is one of great economic significance, aggravated by intense conflict among diverse users of water. The recent upsurge of demands for small-scale hydropower developments gives further urgency to its resolution. The problem lies in evaluating the ecological consequences of controlled spatial and temporal changes in hydrologic regimes. This problem has been approached in ways that vary greatly in analytical detail, data requirements, and costs (reviewed by Loar and Sale 1981). Some approaches make use of historical flow records, some try to establish general relationships between discharge and habitat, and others try to predict usable species-specific habitat at different discharges. One aspect neglected in all these methods is the role, in high-gradient streams, of bank-full or dominant discharge in influencing the general physical, and consequently biological, nature of streams. For example, no predictive models can determine substrate conditions as a function of discharge regime.

Engineers have long been painfully aware of the consequences of bank-full stages in rivers that have erodible beds and transport their bed load. At this and higher discharge stages, a river reconditions its natural channel, flushes out sediment accumulated at low stages, and perhaps divides and coalesces by braiding (Abbett 1956). Dominant discharge has further major effects: it establishes the frequency of riffles and pools in the channel and the pattern of sinuosity or meanders. It is, then, an example in ecology of an extreme event acting as a master factor that "sets the stage" for years or decades. This is not unique in ecology; fire in forest ecosystems is another example of an extreme occasional event acting as a master factor. Fires, like floods, are not just devastation; they cycle, renew, and rejuvenate.

Fisheries biologists have been slow to recognize the full biological implications of dominant flows. For example, a 1972 review (Fraser 1972) of 354 papers on the effects of regulated flows on fish and other aquatic resources cites only one paper (Hynes 1970) dealing specifically with the role of high discharge in maintaining the general ecology of rivers. Yet dominant flows have a major influence on the sediment content of gravel beds, on production of invertebrates, on riffle: pool ratios, on undercut banks, and on the boulder layer of the substrate. All these features are fundamental to the ecology of salmon and trout. Today, the significance of dominant discharge is being acknowledged by some fishery biologists who are trying to define the magnitude and duration of flushing requirements for maintaining fish habitat (Reiser et al. 1987). Indeed, some biologists are recommending annual flushing flows that would restore habitat degraded by dams (e.g., Nelson et al. 1987; Reiser et al. 1989). This is equivalent to the practice of setting controlled fires to rejuvenate forests.

Dominant discharge is not the only physical variable neglected in current models that attempt to predict fish habitat at different discharges. Another is rate of change of discharge; this is most extreme with power peaking (reviewed by Cushman 1985). The neglect of biological variables, however, is equally apparent. Changes in flow affect production of fish-food organisms, competitive relations among fishes, and relations of fish to avian predators (Orth 1987). The influence of birds on salmonids is substantial. Common mergansers *Mergus merganser* may consume 24% to 65% of coho salmon smolt production in coastal streams (Wood 1987). Discharge greatly affects the availability of fry to birds. For example, there is an inverse relationship between the percentage of brook trout *Salvelinus fontinalis* in the diet of belted kingfishers *Megaceryle alcyon* and summer discharge (White 1938). Finally, most of today's instream flow models lack a time scale. The changes occurring below dams, especially those resulting from elimination of dominant discharge, are orderly and usually take several decades to attain ecological equilibrium (Petts 1980; Burt and Mundie 1986).

It is apparent from the above that no simple model can embrace the many changes relating discharge to fish numbers. Modelers who insist on seeking predictive relationships have not learned from the cautionary advice of biologists who argue that highly complex systems, especially those that alter *qualitatively* with changes in a physical variable (e.g., in species composition, or in unforeseen incursions of rooted vegetation), cannot be represented by a predictive model (Hedgpeth 1977; Fryer 1987).

Although developers understandably maintain that before water can be allocated for fish some precise quantitative defense must be offered by biologists, it seems that such formulae will not be forthcoming. In view of this it is better to follow a guideline based on professional judgment than to engage in spurious quantification. The guidelines seem to be that the historic pattern of annual flow should be followed as much as possible, for this is what the life history strategy of the fish is related to, and that flow should not be reduced by more than 25% to 30% of the mean monthly flows (Hazel 1976; Fraser 1979), and, on common sense grounds, not at all in periods of very low flows.

Drawdown Requirements

The impoundment of water frequently results in gross changes in water levels above and below dams. Examples are provided by Hungry Horse and Libby reservoirs in Montana. The level of the first is drawn down 26 m between December and April for purposes of power generation and flood control; in consequence, the surface area is reduced by half. Spring runoff usually refills the reservoir by summer. The fish population, which includes westslope cutthroat trout *O. c. lewisi* and bull trout, is unique in consisting almost entirely of native species (May and Weaver 1987). The spring drawdown of Libby Reservoir may be 52 m, which results in a 69% reduction in surface area. Seventeen species of fish, including kokanee occur in the reservoir (Chisholm and Fraley 1985).

Evidence exists (reviewed by Elder 1965) that changes in water levels of these magnitudes have profound effects on all aspects of lentic ecology. Species of fish that eat benthic invertebrates are likely to be impaired by the destruction of littoral benthos. Planktivorous species are affected by the removal of phytoplankton and zooplankton following gross abstraction of water, and by the loss in both surface area and volume of water in the impoundment. Studies are in progress to assess the impacts of these changes on game fishes and to develop mitigative operating procedures (Fraley et al. 1989). Here, however, we perhaps have an example of incompatibility of uses. It is difficult to see how a compromise may be reached that could satisfy the requirements of both industry and fisheries.

Stream Habitat Improvements

It is common knowledge that if wild fish are to be maintained, habitat must be protected from degradation caused by a wide variety of human activities. Where damage has occurred, mitigation by stream manipulation may have a role to play. The techniques are an interesting revival of ones developed in the 1930s (U.S. Bureau of Fisheries 1935). They are often attempted by community groups, but it must be said that, with a few exceptions such as seasonally adjusting the height of a weir or removing rubbish from streams, stream habitat improvement work is not suitable for amateurs. It requires experience and skill, and is most likely to be durable and to withstand changes in river morphology imposed by dominant flows when the work is done with heavy machinery.

Stream habitat modifications have been the subject of several reviews (e.g., Hall and Baker 1982; Wesche 1985). Among the commoner practices are bank protection and erosion control; gravel cleaning; placement of boulders, gabions, and logs; creation of pools for rearing; addition of large woody debris to make submerged cover; improvement of upstream access; side-channel development with either surface or ground water; and planting of riparian vegetation. Some applications are documented from Oregon (Everest and Sedell 1984; Bonneville Power Administration, Division of Fish and Wildlife 1987). In the Clackamas and Hood river basins, additions of logs and boulders have diversified habitat for spring chinook salmon, coho salmon, and summer and winter steelhead. Treatment extended over 5.9 km of stream. Side channels also were constructed to provide rearing area for coho salmon and protection for other species at high flows. Assessments showed that juvenile salmonids had increased both within the treated areas and between the treated areas and controls. A period of exceptionally high freshets offered an opportunity to examine the durability of more than 600 improvement structures installed over the previous 5 years. Approximately 90% were functioning as designed. Of those that failed, half were structures that had been installed by hand (Bonneville Power Administration, Division of Fish and Wildlife 1987). An off-channel pond made in the Clackamas system increased coho salmon smolt production by 102%. In this high-energy basin, the most effective habitat improvements were those applied to the stream edges, not to those applied to the entire cross section (Everest et al. 1987).

Not all stream habitat manipulations, however, bring such positive results. Habitat alterations for anadromous fish in the Clearwater and Salmon rivers of Idaho have had mixed results. Instream structures did not significantly increase the standing crop of chinook salmon and steelhead parr. This may be due, at least in part, to the widespread presence of granitic sand; densities of both species decrease as the percentage of sand increases. In contrast, off-channel development of connected ponds and side channels has shown promising results, and removal of barriers appears to be highly successful (Petrosky and Holubetz 1987). Great importance attaches to these mitigation projects, for the Clearwater and Salmon river drainages account for virtually all of Idaho's wild production of summer steelhead, spring and summer chinook salmon, and a remnant run of sockeye salmon. About 9,000 km of stream were formerly available to anadromous fishes in Idaho. With dam construction on the Snake River and the North Fork of the Clearwater River, 40% were lost.

Theoretically, provided water quality is acceptable, habitat models should be able to identify those aspects of damaged habitat that actually set limits to salmonid production, such as spawning, rearing, or overwintering habitat, and those mitigative measures that might usefully be applied. Existing models, however, are not helpful at answering such detailed questions, although they can direct attention to major variables (e.g., Wesche and Goertler 1987). Models are especially weak for anadromous fishes, and few have been tested in the field. The reasons why they cannot predict numbers of fish from habitat variables include the complexity of salmonid habitat, the interplay of habitat and fish behavior (e.g., fish may move to and from parts of a river system distant from the reach being studied), the climatically induced instability of streams, the mortality associated with passage at dams, and the exploitation of a fishery that may allow too few adults to escape to a river to breed (Milner et al. 1985).

In view of this, it is more reasonable to take a simple empirical approach, based on our understanding of the principles of stream and fish ecology, to attempt improvements. For example, a new appreciation of the role of large debris in streams—perhaps the most significant advance in stream ecology of the last decade—must influence decisions on this variable (Bisson et al. 1987).

Each decision should be in the direction of optimization of the environment for salmonids (Mundie 1974).

In Idaho, depression of upriver anadromous stocks has prevented full realization of the benefits of stream habitat management. As main-stem passage conditions improve and escapements of salmon and steelhead increase, the present low population levels should rise. The relative merits of various measures intended as improvements, however, should become apparent before this from partial responses. Once these are known, supplementation with hatchery fish may bring about adequate densities. Intensive studies are therefore being directed at defining population levels and smolt production for both treatment and control areas; posttreatment evaluations will measure physical changes and densities of fish in each reach (Petrosky and Holubetz 1987).

Clearly, the outcome of stream habitat work will be of great interest, although for some projects several years will have to elapse before a new equilibrium is reached with flow characteristics. Apart from the question of the efficacy of these measures, there must be concern over the limited scale, by virtue of costs and labor-intensiveness, on which it is possible to apply them. A long-term commitment seems inevitable.

Concluding Remarks

The world of fish, rivers, dams, hatcheries, and fisheries is a strange amalgam. It represents in microcosm some great issues of our time (Mundie 1977). What a person sees in it, as the prime concern, no doubt depends on his or her background. To some, it may be larger catches; to others, civil engineering, or high-technology in a modern hatchery, or aesthetically rewarding experiences on a river. Each user pursues self-interest, and by doing so accepts the resource as a utility—a utility, but not something that should exist in its own right (Evernden 1985), for such a view would take us back to the ethos of early native fishermen.

To satisfy these interests democratically, and to guarantee that they be perpetuated, fishery biologists and engineers have asserted a tireless empiricism. The outcome is an immense undertaking based on trial and error, with occasional injections of science that give it renewed impetus. Decisions are implemented that have far-reaching consequences; they are made at a rate that rules out careful assessment or reversal. Results are poorly documented and data are woefully imprecise. A sense of the distant future is not a constraint; we want more fish now. Our attempts at mitigation are enormously impressive but are far from ideal. We have certainly lost sight of Montaigne's advice (given in a very different context) of 400 years ago: "Let us give nature some chance to work; she understands her business better than we" (de Montaigne 1588).

Wild fish seem to be in a highly precarious situation. Many things seem to be against them and few for them. They are of high value to a relatively small number of people; more people want things that are incompatible with unspoilt habitat. Fishery agencies increasingly emphasize wild-salmonid policies, but a growing demand from agriculture and industry for water itself—perhaps soon to become of more concern than hydroelectric power or fish in some states (Lane 1977)—further endangers wild salmonids. The best chance of salmonids is, of course, in remote areas. Even, however, if decisions are made as to the desired ratio of wild to hatchery fish, the consequences of stocking and of fishing mixed hatchery and wild stocks will change the ratio over the years. In 1986 the landings of coho salmon at Oregon ports consisted of 12% wild fish and 88% fish of hatchery origin (Oregon Department of Fish and Wildlife 1987).

Essential to the survival of salmonid populations is the interplay of genetically induced variation of natural populations with environmental complexity and change. This variation, however, is being diluted as outplanting, fishing pressure and aquaculture, and an insidious blending, even hybridization, of stocks take place. There is an irrevocable trend in the direction of domestication. Before we plan more hatcheries, therefore, we should reevaluate; we should improve present practices and assess present trends to see where they are leading.

As was pointed out in the introduction, the success of mitigative measures lies not only with the manipulation of fish and rivers, but also with human motives and values. The last may be crucial. It is important to note that the political majority has always favored hydroelectricity over fish, for cheap energy has ramifications extending throughout the entire economy (Smith 1977). This being so, rivers and fish will always be threatened; democracy is so far not good at conservation. It follows that knowledge of biology and engineering is not enough. It must be backed by social and political support. This may not exist even at the level of working biologists, for fisheries staffs have been, and may continue to be, advised by political decision makers to temper their biological recommendations with what is politically acceptable. In consequence, the management goal of stock viability can be undermined (Fraidenburg and Lincoln 1985).

We must make every effort, then, to hold on to what is left of unspoilt habitat and wild runs. We must not forgo the option of natural production, for aquaculture may hold unpleasant surprises in the future (Hilborn 1987). To this end we should reexamine not only the democratic philosophy of multiple resource use of rivers, but also that of dual use by fisheries and hydroelectric power interests. Perhaps we may even be able to address the question: what are the long-term needs of the fish? For if even the present uneasy balance of wild and hatchery fish, of natural habitat and industrial development, is to continue, it will not be by our exercising yet more control and manipulation of fish but by our setting limits to the activities and demands of people.

Acknowledgments

D. F. Alderdice, R. A. Bams, D. W. Burt, and C. Shirvell of the Department of Fisheries and Oceans, Canada, kindly reviewed the manuscript. I am also indebted to the following for information and advice: G. F. Hartman, D. Marshall, W. E. McLean, E. A. Perry, and B. Riddell of the Department of Fisheries and Oceans; J. W. Buell of Buell Associates, Inc., Oregon; C. J. Cederholm, Washington Department of Natural Resources; W. J. Ebel, National Marine Fisheries Service; J. J. Fraley (who also reviewed the manuscript), Montana Department of Fish, Wildlife, and Parks; W. R. Heard, National Marine Fisheries Service; T. McMahon, Oregon State University; H. L. Raymond, formerly of the National Marine Fisheries Service; and M. F. Solazzi, Oregon Department of Fish and Wildlife. The staff of the library of the Department of Fisheries and Oceans, Pacific Biological Station, gave invaluable help, as did M. Sherry in word processing.

References

Abbett, R. W., editor. 1956. American civil engineering practice, volume 2, section 15: river engineering. Wiley, New York.

Bisson, P. A., and eight coauthors. 1987. Large woody debris in forested streams in the Pacific Northwest: past, present and future. University of Washington Institute of Forest Resources Contribution 57:143–190.

Bonneville Power Administration. 1987. Backgrounder, July. U.S. Department of Energy, Portland, Oregon.

Bonneville Power Administration, Division of Fish and Wildlife. 1987. Natural propagation and habitat improvement, volume 1, Oregon. U.S. Department of Energy, Portland, Oregon.

Brett, J. R. 1957. Salmon research and hydroelectric power development. Fisheries Research Board of Canada Bulletin 114.

Burt, D. W., and J. H. Mundie. 1986. Case histories of regulated stream flow and its effects on salmonid populations. Canadian Technical Report of Fisheries and Aquatic Sciences 1477.

CAC (Citizens' Advisory Committee on Salmon and Steelhead Trout). 1971. Report on California salmon and steelhead trout: an environmental tragedy. State of California, Sacramento.

Chisholm, I., and J. Fraley. 1985. Quantification of Libby Reservoir levels needed to maintain or enhance reservoir fisheries. U.S. Department of Energy, Bonneville Power Administration, Division of Fish and Wildlife, Contract DE-AI79-84BP 12660, Portland, Oregon.

Cushman, R. M. 1985. Review of ecological effects of rapidly varying flows downstream of hydroelectric facilities. North American Journal of Fisheries Management 5:330–339.

Delarm, M. R., E. Wold, and R. Z. Smith. 1987. Columbia River Fisheries Development Program annual report for 1986. NOAA (National Oceanic and Atmospheric Administration) Technical Memorandum NMFS (National Marine Fisheries Service) F/NWR-21, Seattle.

de Montaigne, M. E. 1588. Essays, book 3. Translated 1958 by J. M. Cohen. Penguin, New York. (Reprinted 1976.)

Department of Fisheries and Oceans. 1988. Salmonid enhancement program: 1986/87 update. Canada Department of Fisheries and Oceans, Vancouver.

Department of Fisheries and Oceans. 1989. Salmonid enhancement program: 1987/88 update. Canada Department of Fisheries and Oceans, Vancouver.

Ebel, W. J. 1977. Major passage problems. American Fisheries Society Special Publication 10:33–39.

Elder, H. Y. 1965. Biological effects of water utilization by hydro-electric schemes in relation to fisheries, with special reference to Scotland. Proceedings of the Royal Society of Edinburgh, Section B (Biological Sciences) 69:246–271.

Everest, F. H., G. H. Reeves, J. R. Sedell, D. B. Hohler, and T. Cain. 1987. The effects of habitat enhancement on steelhead trout and coho salmon smolt production, habitat utilization, and habitat availability in Fish Creek, Oregon, 1983–86. U.S. Department of Energy, Bonneville Power Administration, Division of Fish and Wildlife, Contract DE-A179-BP 16726, Portland, Oregon.

Everest, F. H., and J. R. Sedell. 1984. Natural propagation and habitat improvement. Volume 1: Oregon. Supplement A: Habitat enhancement evaluation of Fish and Wash creeks. U.S. Department of

Energy, Fish and Wildlife Division, Contract DE-A179-83BP 11968, Portland, Oregon.
Evernden, N. 1985. The natural alien. University of Toronto Press, Toronto.
Fraidenburg, M. E., and R. H. Lincoln. 1985. Wild chinook salmon management: an international conservation challenge. North American Journal of Fisheries Management 5:311–329.
Fraley, J., B. Marotz, J. Decker-Hess, W. Beattie, and R. Zubik. 1989. Mitigation, compensation, and future protection for fish populations affected by hydropower development in the upper Columbia system, Montana, U.S.A. Regulated Rivers Research and Management 3:3–18.
Fraser, J. C. 1972. Regulated stream discharge for fish and other aquatic resources—an annoted bibliography. FAO (Food and Agricultural Organization of the United Nations) Fisheries Technical Paper FIRI/T112.
Fraser, J. C. 1979. Emergency response: (How should administrators respond to a request for an opinion on abstraction within one month?). Pages 60–66 *in* D. Scott, editor. Proceedings of the seminar on the effects of water abstraction on fisheries. National Water Protection Committee of Acclimatisation Societies, Dunedin, New Zealand.
Fryer, G. 1987. Quantitative and qualitative: numbers and reality in the study of living organisms. Freshwater Biology 17:177–189.
Geen, G. H. 1975. Ecological consequences of the proposed Moran Dam on the Fraser River. Journal of the Fisheries Research Board of Canada 32:126–135.
Hall, J. D., and C. O. Baker. 1982. Rehabilitating and enhancing stream habitat: 1. Review and evaluation. U.S. Forest Service General Technical Report PNW-138.
Hazel, C. 1976. Assessment of effects of altered stream flow characteristics on fish and wildlife. Part B: California.U.S. Fish and Wildlife Service, Biological Services Program FWS/OBS-76/33.
Hedgpeth, J. W. 1944. The passing of the salmon. Scientific Monthly 59:370–378.
Hedgpeth, J. W. 1977. Models and muddles: some philosophical observations. Helgoländer wissenschaftliche Meeresuntersuchungen 30:92–104.
Hilborn, R. 1987. Living with uncertainty in resource management. North American Journal of Fisheries Management 7:1–5.
Hume, J. M., and E. A. Parkinson. 1987. Effect of stocking density on the survival, growth, and dispersal of steelhead trout fry (*Salmo gairdneri*). Canadian Journal of Fisheries and Aquatic Sciences 44:271–281.
Hurst, R. E., and B. G. Blackman. 1988. Coho colonization program: juvenile studies 1984 to 1986. Canadian Manuscript Report of Fisheries and Aquatic Sciences 1968.
Hynes, H. B. N. 1970. The ecology of flowing waters in relation to management. Journal of the Water Pollution Control Federation 42:418–424.
Kapuscinski, A. R., and L. D. Jacobson. 1987. Genetic guidelines for fisheries management. University of Minnesota, Minnesota Sea Grant College Program, Sea Grant Research Report 17, Duluth.
Lane, D. J. 1977. The role of fishery agencies in water-use planning. American Fisheries Society Special Publication 10:155–159.
Loar, J. M., and M. J. Sale. 1981. Analysis of environmental issues related to small-scale hydroelectric development: 5, instream flow needs for fishery resources. Oak Ridge National Laboratory, Environmental Sciences Division, Publication 1829, Oak Ridge, Tennessee.
May, B., and T. Weaver. 1987. Quantification of Hungry Horse Reservoir water levels needed to maintain or enhance reservoir fisheries. U.S. Department of Energy, Bonneville Power Administration, Division of Fish and Wildlife, Contract DE-A179-84 BP 12659, Portland, Oregon.
Milner, N. J., R. J. Hemsworth, and B. E. Jones. 1985. Habitat evaluation as a fisheries management tool. Journal of Fish Biology 27 (Supplement A):85–108.
Mundie, J. H. 1974. Optimization of the salmonid nursery stream. Journal of the Fisheries Research Board of Canada 31:1827–1837.
Mundie, J. H. 1977. Concluding remarks: the problem in its setting. Pages 299–36 *in* D. V. Ellis, editor. Pacific salmon: management for people.University of Victoria, Western Geographical Series 13, Victoria, British Columbia.
Nelson, R. W., J. R. Dwyer, and W. E. Greenberg. 1987. Regulated flushing in a gravel-bed river for channel habitat maintenance: a Trinity River fisheries case study. Environmental Management 11: 479–493.
Netboy, A. 1974. The salmon: their fight for survival. Houghton Mifflin, Boston.
Netboy, A. 1980. The Columbia River salmon and steelhead trout. University of Washington Press, Seattle.
Nickelson, T. E. 1986. Influence of upwelling, ocean temperature, and smolt abundance on marine survival of coho salmon (*Oncorhynchus kisutch*) in the Oregon Production Area. Canadian Journal of Fisheries and Aquatic Sciences 43:527–535.
Nickelson, T. E., M. F. Solazzi, and S. L. Johnson. 1986. Use of hatchery coho salmon (*Oncorhynchus kisutch*) presmolts to rebuild wild populations in Oregon coastal streams. Canadian Journal of Fisheries and Aquatic Sciences 43:2443–2449.
Northwest Power Planning Council. 1987. Columbia River basin fish and wildlife program. Northwest Power Planning Council, Portland, Oregon.
Olson, F. W., R. G. White, and R. H. Hamre, editors. 1985. Proceedings of the symposium on small hydropower and fisheries. American Fisheries Society, Bethesda, Maryland.
Oregon Department of Fish and Wildlife. 1981. Comprehensive plan for production and management of Oregon's anadromous salmon and trout. Technical draft. Oregon Department of Fish and Wildlife, Portland.
Oregon Department of Fish and Wildlife. 1985. Coho

salmon plan status report. Oregon Department of Fish and Wildlife, Portland.

Oregon Department of Fish and Wildlife. 1987. Stock assessment of anadromous salmonids. Annual progress report. Oregon Department of Fish and Wildlife, Portland.

Orth, D. J. 1987. Ecological considerations in the development and application of instream flow–habitat models. Regulated Rivers Research and Management 1:171–181.

Pacific Fishery Management Council. 1988. Review of 1987: ocean salmon fisheries. Pacific Fishery Management Council, Portland, Oregon.

Petrosky, C. E., and T. B. Holubetz. 1987. Evaluation and monitoring of Idaho habitat enhancement and anadromous fish natural propagation. U.S. Department of Energy, Bonneville Power Administration, Division of Fish and Wildlife, Contract DE-A179-84BP13381, Portland Oregon.

Petts, G. E. 1980. Long-term consequences of upstream impoundment. Environmental Conservation 7:325–332.

Raymond, H. L. 1988. Effects of hydroelectric development and fisheries enhancement on spring and summer chinook salmon and steelhead in the Columbia River Basin. North American Journal of Fisheries Management 8:1–24.

Reisenbichler, R. R., and J. D. McIntyre. 1977. Genetic differences in growth and survival of juvenile hatchery and wild steelhead trout, *Salmo gairdneri*. Journal of the Fisheries Research Board of Canada 34:123–128.

Reiser, D. W., R. P. Ramey, S. Beck, T. R. Lambert, and R. E. Geary. 1989. Flushing flow recommendations for maintenance of salmonid spawning gravels in a steep, regulated stream. Regulated Rivers Research and Management 3:267–275.

Reiser, D. W., M. P. Ramey, and T. R. Lambert. 1987. Considerations in assessing flushing flow needs in regulated stream systems. Pages 45–57 *in* J. F. Craig and J. B. Kemper, editors. Regulated streams: advances in ecology. Plenum, New York.

Rieman, B. E., R. C. Beamesderfer, S. Vigg, and T. P. Poe. 1988. Estimated total loss and mortality of juvenile salmonids to northern squawfish, walleye, and smallmouth bass. Pages 249–274 *in* T. P. Poe and B. E. Rieman, editors. Predation by resident fish on juvenile salmonids in John Day Reservoir, 1983–1986. Volume 1. U.S. Department of Energy, Bonneville Power Administration, Division of Fish and Wildlife, Contracts DE-AI79-82BP34796 and DE-AI79-82BP35097, Portland, Oregon.

Rochester, H., T. Lloyd, and M. Farr. 1984. Physical impacts of small-scale hydroelectric facilities and their effects on fish and wildlife. U.S. Fish and Wildlife Service, Biological Services Program FWS/OBS-84/19.

Ryman, N., and F. Utter, editors. 1987. Population genetics and fishery management. University of Washington Press, Seattle.

Schweibert, E., editor. 1977. Columbia River salmon and steelhead. American Fisheries Society Special Publication 10.

Smith, E. M., B. A. Miller, J. D. Rodgers, and M. A. Buckman. 1985. Outplanting anadromous salmonids: a literature survey. U.S. Department of Energy, Bonneville Power Administration, Contract DE-A179-85BP23109, Portland, Oregon.

Smith, H. 1977. Utility interests in compensation. American Fisheries Society Special Publication 10: 180–183.

Stalnaker, C. B. 1980. Effects on fisheries of abstractions and perturbations in streamflow. Pages 366–383 *in* J. H. Grover, editor. Allocation of fishery resources. Food and Agriculture Organization of the United Nations, Rome.

U.S. Bureau of Fisheries. 1935. Methods for the improvement of streams. U.S. Department of Commerce, Memorandum I-133, Washington, D.C.

Van Tine, J. 1986. Coho colonization of inaccessible headwater habitats in the Quinsam River watershed. Canadian Technical Report of Fisheries and Aquatic Sciences 1483:38–42.

Wahle, R. J., and R. E. Pearson. 1987. Listing of Pacific Coast spawning streams and hatcheries producing chinook and coho salmon with estimates on numbers of spawners and data on hatchery releases. NOAA (National Oceanic and Atmospheric Administration) Technical Memorandum NMFS (National Marine Fisheries Service) F/NWC-122, Seattle.

Wesche, T. A. 1985. Stream channel modifications and reclamation structures to enhance fish habitat. Pages 103–163 *in* J. A. Gore, editor. The restoration of rivers and streams: theories and experience. Butterworth, Boston.

Wesche, T. A., and C. M. Goertler. 1987. Modified habitat suitability index model for brown trout in southeastern Wyoming. North American Journal of Fisheries Management 7:232–237.

White, H. C. 1938. The feeding of kingfishers: food of nestlings and effect of water height. Journal of the Fisheries Research Board of Canada 4:48–52.

Wood, C. C. 1987. Predation of juvenile Pacific salmon by the common merganser (*Mergus merganser*) on eastern Vancouver Island. II: Predation of stream-resident juvenile salmon by merganser broods. Canadian Journal of Fisheries and Aquatic Sciences 44:950–959.

American Fisheries Society Symposium 10:12–18, 1991

Future of Salmon

WILLIAM J. MCNEIL

Cooperative Institute for Marine Resources Studies
Oregon State University, Newport, Oregon 97365, USA

Abstract.—World harvest of salmon (*Oncorhynchus* spp. and *Salmo salar*) still comes mostly from naturally produced fish, but the contribution from aquaculture is substantial and continues to grow rapidly. There is a strong likelihood that aquaculture production will exceed natural production before the year 2000. Two types of aquaculture production, "ranching" and "farming," are differentiated. Ranched salmon are released as juveniles into natural waters and harvested as maturing adults by sport and commercial fishers or at recapture facilities. Farmed salmon are held in captivity to harvest size. Ranching contributes a much larger biomass of salmon to today's world markets than farming, but the value of farmed salmon relative to ranched salmon is much greater than production levels indicate because species produced by farming are mostly of higher value than species produced by ranching. Furthermore, farming is growing at a much more rapid rate than ranching. Innovations in technology show promise for continued reductions in cost of producing ranched and farmed salmon. Ranching and farming applications are becoming more diversified because of innovations. Salmon are expected to become more price-competitive with other types of animal protein. Recognized advantages for human health make farmed and ranched salmon attractive commodities for international trade. Several environmental and social issues have emerged as a consequence of rapid growth of salmon aquaculture. Institutional barriers to salmon aquaculture typically result because environmental impacts have adverse implications for production and management of natural stocks of salmon. Other factors contributing to policies adverse to growth of aquaculture include concerns of commercial salmon fishers over rights to harvest stocks and competition for markets.

Recent developments in Prince William Sound, Alaska, highlight the rapid global transition from natural to artificial propagation of salmon. Stocks of pink salmon *Oncorhynchus gorbuscha* and chum salmon *O. keta* had been depressed for about three decades when a group of fishermen met in Cordova, Alaska, to consider implementation of a hatchery program to enhance their fishery. The Prince William Aquaculture Corporation was organized in 1974, and soon thereafter an inactive salmon cannery was converted into a hatchery. Four additional hatcheries were subsequently constructed along Prince William Sound—one by the corporation, one by the Valdez Fisheries Development Association, and two by the Alaska Department of Fish and Game. The five hatcheries make up one of the largest salmon hatchery systems in the world and can incubate about 800 million salmon eggs.

Response of the fishery to the recent hatchery initiative has exceeded most expectations. Figure 1 shows catches of Prince William Sound pink salmon since 1949. Recent catches, which include wild and hatchery pink salmon, exceed previous catches by a large margin, partly due to improved ocean survival and partly due to production of hatchery fish. The percentage of Alaska pink salmon harvested in Prince William Sound has trended upward in recent years (Figure 2), probably in response to hatchery production.

Salmon ranching, in which juveniles are released from hatcheries into natural waters, has become well established throughout much of the north Pacific basin and is developing in the north Atlantic and south Pacific oceans. Nearly 5×10^9 juvenile salmon (all species) are presently released from hatcheries into marine basins. Global production has been doubling in about a decade for the last several decades. Continued rapid growth is likely until biological or economic constraints limit expansion. The ultimate biological limitation will be the capacity of marine waters to grow salmon.

One analyst (Lowe 1988) partitioned the 1985 global harvest of salmon among natural production (63%), ranching (32%), and farming (5%). Salmon farming, in which fish are held captive until harvest, currently produces a much smaller biomass of salmon than ranching or natural production, but growth of farming is currently much more rapid than ranching. Farming places emphasis on high-value species, and the monetary value of farmed salmon is high even though the biomass is low relative to ranched and naturally produced salmon. Production of farmed salmon, insignificant in the 1970s, blossomed into an important

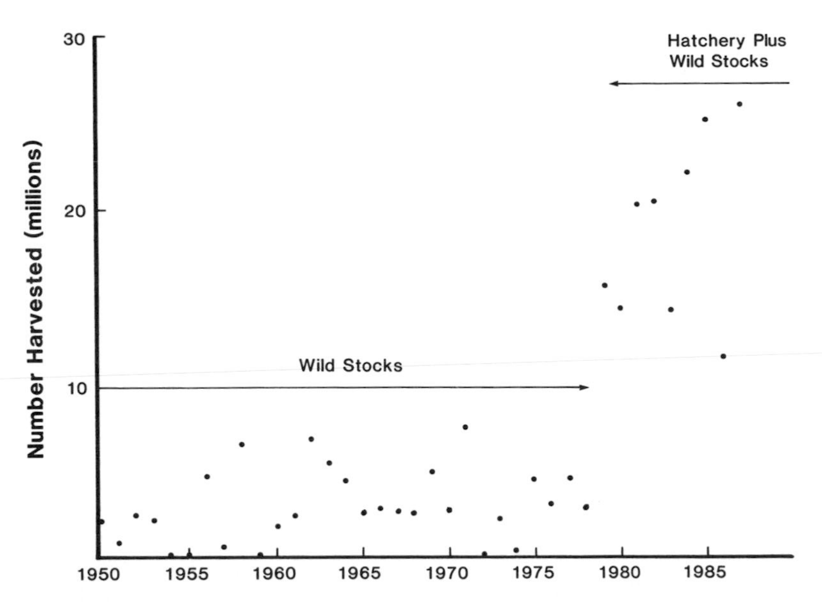

FIGURE 1.—Harvest of pink salmon in Prince William Sound, Alaska.

contributor to the world supply of salmon in the 1980s. Global production was only about 12,000 tonnes in 1981 but was forecast to reach 143,000 tonnes in 1990 (Lowe 1988). Other analysts, including White (1988), have more optimistic expectations for salmon farming, but even the more conservative forecast of a 12-fold growth in one decade is spectacular to say the least.

My purpose in writing this report is to offer some speculations about technological, environmental, and social issues affecting the future growth of salmon aquaculture. I expect global production of salmon from aquaculture to exceed that from natural stocks before the year 2000. Recent trends in growth of salmon aquaculture suggest that this expectation could be conservative. The transition from hunting natural stocks to aquaculture is well under way.

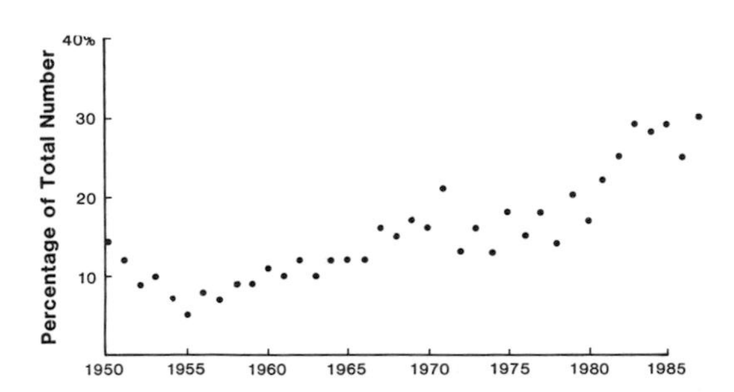

FIGURE 2.—Five-year moving average of percentage of total Alaska harvest of pink salmon landed in Prince William Sound.

Political opposition to salmon aquaculture has emerged, primarily from interests associated with traditional capture fisheries and environmentalists. Technological, environmental, and policy issues related to a transition from traditional fisheries to aquaculture merit continuing assessment because policies affecting salmon aquaculture are being debated and formulated in many countries in the northern and southern hemispheres.

Technology

Industrial-scale salmon ranching emerged in North America and Asia in the 1950s and 1960s. Mitigation for loss of salmon adversely affected by land- and water-use activities was a major stimulus for ranching in North America. Expansion of the supply of salmon for food was a major stimulus in Asia. North American technology emphasized production of species, such as coho salmon *O. kisutch*, chinook salmon *O. tshawytscha*, and steelhead *O. mykiss*, that require extended freshwater rearing. Asiatic technology emphasized production of species such as chum and pink salmon that do not require extended freshwater rearing.

Commercial salmon farming did not experience major growth until the late 1970s, but became well established by the mid-1980s. Hatchery technologies used for ranching found direct application for production of seed stock for farming. Floating cages in protected marine embayments became the primary technology for raising salmon in captivity to market size. The industry experienced initial rapid growth in northern Europe, but the focus for growth has recently shifted to British Columbia and Chile. My outlook for technological innovations for breeding, growing, releasing, and recapturing salmon is summarized below.

Breeding

Maturation and sex of progeny can now be manipulated with hormones (Ingram 1988). Hormones are also expected to find increased application to control smoltification and accelerate growth.

Techniques to produce sterile triploids (including hybrids) have potential applications in ranching and farming (Johnson et al. 1986; Seeb 1987). Triploids could reduce risk of transferring genetic material from hatchery to natural stocks in the event farmed or ranched salmon escape to reach natural spawning grounds. Sterile hybrid triploids have the potential for traits favoring early smoltification, rapid growth, delayed maturation, and disease resistance.

Selective breeding holds promise for applications in ranching as well as farming but objectives will differ. Selection for early or late-maturing ranched stocks could become an important method to minimize interbreeding between stray hatchery fish and natural stocks through temporal separation of spawning times. Selective breeding of farmed stocks will allow more efficient production of meat and provide increased resistance to disease and injury from handling and crowding.

Growth

Future programs for growing salmon are likely to feature use of hormones to regulate growth, procedures to maximize biomass production from limited space and water, prophylactic techniques to reduce incidence of and susceptibility to disease, and diets that make greater use of carcass waste from groundfish processing.

Growth-stimulating hormones (Donaldson et al. 1979) are likely to be used commonly to accelerate growth to the smolt stage and beyond, depending on restrictions to ensure clearance of drug metabolites before flesh is consumed by humans.

Greater biomass production is anticipated from available hatchery space and water through widespread use of technologies to heat, disinfect, oxygenate, and recirculate water to control waste metabolites. New and improved bacterins, vaccines, and other prophylactic products are anticipated. Production of "disease-free" stocks for transplantation of diploid, triploid, hybrid, and selectively bred stocks for ranching and farming is likely to involve hatcheries specializing in production of seed stock.

Diets for salmon show potential for improvement without a continued heavy dependence on quality herring meal, which is in short supply. Hydrolyzation and other means of preparing carcass waste from fish processing for conversion to salmon diets will become practical. Development of salmon diets from fish-processing wastes will benefit processors and aquaculturists alike. Salmon aquaculturists will enjoy a continuing supply of high-quality feed at a reasonable cost, and processors will generate income from sale of fish carcasses.

Land-based, pumped seawater systems for salmon aquaculture are being phased into operation at several locations in the north and south Pacific and north Atlantic oceans. These pump-ashore systems are in some instances used for combined ranching and farming operations and may someday incorporate other crops such as macroalgae and mollusks.

Release

Substantial gains in marine survival are possible through manipulations of size, time, location, and method of release (Gowan 1988). Achievement of a relatively modest reduction in marine mortality would have a major impact on fish returns. It is not unusual for natural marine mortality of salmon to vary around 95%. Should marine mortality of a hatchery stock average 95%, a 1% reduction to 94% produces a 20% gain in biomass of harvestable adults.

The ratio of biomass of adults harvested to smolts released is an important criterion to evaluate success of hatchery and release procedures. If a given biomass of small smolts results in a larger but less costly biomass of harvestable fish than the same biomass of large smolts, hatchery and release strategies should be programmed accordingly. Possible exceptions occur when brood fish are in short supply or when the number of smolts released is limited by government. In these cases it may be advantageous to release large smolts in order to maximize marine survival even though the ratio of biomass of adults to smolts may not necessarily be maximized.

Adult-to-smolt biomass ratios above 50:1 (i.e., 50 kg of adults harvested per 1 kg of smolts released) are common for hatchery pink and chum salmon in Alaska, Japan, and the USSR. Biomass ratios for Pacific Northwest hatchery chinook and coho salmon tend to be much less than 50:1, frequently in the range of 1:1 to 20:1. The challenge is to increase adult-to-smolt biomass ratios through optimization of size at release, release time, release method, and hatchery operations. Ratios greater than 50:1 are feasible for pink and chum salmon partly because of their small size at release (as small as 0.2 g in the case of pink salmon). Chinook salmon undergo smoltification at a relatively small size and may offer considerable potential for improved biomass ratios, but possibilities to improve biomass ratios with coho salmon and Atlantic salmon *Salmo salar* are likely to be more limited, partly because these fish are relatively large at smoltification.

Recapture

A ranched salmon will, under common law, remain a wild animal until recaptured (Hampson 1988). Most sovereignties are expected to allow sport and commercial fisheries on hatchery and natural stocks in marine waters, but restrictions on harvest rates will continue to be imposed with the primary objective of providing a sufficient number of spawners to reseed natural freshwater spawning and nursery habitats. Hatcheries are likely to receive adults in numbers surplus to the needs of restocking incubators and raceways. Sale of surplus fish and carcasses of hatchery spawners can generate revenues to defray costs of constructing and operating hatcheries.

State-of-the-art hatcheries designed for industrial-scale ranching feature facilities on or near the ocean and utilize pumped sea water for conditioning and imprinting smolts and recapturing adults. Operations in these saltwater facilities are becoming increasingly diversified to include both salmon farming and ranching. Several salmon species and hybrids, including sterile triploids, can potentially be ranched and farmed to spread marketing of fresh salmon over much of the year. Furthermore, pump-ashore systems as well as floating-cage systems can be located at points remote from freshwater hatcheries. Ranched smolts are imprinted to return as maturing adults to the saltwater release site.

Environmental Issues

Development of salmon aquaculture has stimulated debates over environmental impacts and interactions. McNeil (1979) and Walters (1988) have identified several pertinent issues, including those discussed below.

Food and Space

Competition for food and space in natural waters may involve single or multiple species of salmon and potentially nonsalmonids. Competition may initially occur in restricted coastal regions but may ultimately affect broader regions of major oceanic basins. The capacity of marine waters to grow salmon is undoubtedly finite, but natural variability is likely to make it exceedingly difficult to detect early signs of impaired growth and survival from density-dependent factors. There is also a possibility that increased numbers of ranched salmon will tend to displace nonsalmonids from marine ecosystems through competition for food and space. Sovereignties may someday seek "grazing rights" through international agreements to optimize economic benefits from ranched salmon.

Disease

Transplantation of salmon has associated risks for introductions of disease agents. Systems for production of "disease-free" stocks and methods to detect disease are expected to become more sophisticated and provide tools to reduce the risk of disease introduction. Regional hatcheries specializing in the provision of "certified" healthy seed stock for a variety of species and hybrids could become important components of a support infrastructure for ranching and farming.

Water Quality

Salmon aquaculture generates considerable quantities of nitrogenous by-products of metabolism and settleable organic solids. Effects of these wastes on water quality vary depending on local site characteristics, biomass of salmon, and pollution abatement practices employed by the aquaculture facility. One major advantage of land-based aquaculture facilities receiving pumped seawater is that management of wastes can be facilitated.

Genetics

As salmon aquaculture expands, the number of hatchery fish straying into natural spawning grounds will increase. Some will stray from ranching and some will escape from farming. Where industrial-scale ranches are in close proximity to natural streams, stray hatchery fish can outnumber natural spawners.

There is concern that interbreeding will negatively affect the genetic fitness of natural stocks. Depending on circumstances (e.g., genotypes and population sizes), interbreeding of hatchery with naturally produced stocks has the potential to reduce or increase fitness. Even though genetic impacts may not always be predictable, methods to reduce opportunities for interbreeding will undoubtedly receive increased attention. Some approaches (use of sterile triploids and selective breeding to separate times of spawning) have already been mentioned in this report. Other possibilities include use of imprinting chemicals at aquaculture facilites (Hasler et al. 1978).

Overharvest of Natural Stocks

Because of higher egg-to-adult survival, ranched stocks can support a higher harvest rate than natural stocks. Therefore, natural stocks are more vulnerable to overharvest in mixed-stock fisheries than hatchery stocks. The trend to manage ocean fisheries to allow escapements adequate for sustained high levels of productivity of natural stocks is expected to continue. Such a policy, if combined with policies to limit fishing power of capture fisheries and expand ranching, should provide increased catches for sport and commercial fishers. This is because the overall abundance of salmon available to capture fisheries will increase as the ratio of hatchery to naturally produced fish increases.

Public Policy

Seafoods have become widely recognized as healthful for humans, and per capita consumption is increasing in many countries (Nierentz and Josupiet 1987). Per capita consumption of seafoods in the USA remained near 5 kg for decades, but began to increase in the 1970s, reaching 6.7 kg in 1986 (Anonymous 1987a). Should per capita consumption continue to increase at the current rate of 2%/year, consumption of seafoods by Americans will double by 2020.

Global aquaculture is growing rapidly, whereas the growth of capture fisheries has slowed perceptibly since the 1960s. Aquaculture contributed only 11% of global harvest of fish and shellfish in 1983, but this percentage is expected to grow to 23% by the year 2000 (Anonymous 1987b).

Much, if not most, of the increased future demand for seafoods will be supplied from expanded aquaculture production. Salmon aquaculture is on the forefront of a global transition from hunting of natural stocks to aquaculture. An economic foundation exists for continued growth of salmon aquaculture. Some sovereignties will participate more actively than others due partly to differences in public policy as well as to environmental attributes. The role of policy is summarized below for several sovereignties that have encouraged development of salmon aquaculture.

Japan

Japan is the largest consumer of salmon. In the early 1960s, most salmon entering domestic markets were caught on the high seas by Japanese fishers. By the mid-1980s, the major source of supply had shifted to domestically ranched salmon.

The decline of Japanese high-seas capture fisheries was triggered by international policies. Treaties negotiated with Japan by the USSR, the USA, and Canada resulted in reduced harvest of salmon on the high seas. Japanese domestic policies favorable to aquaculture stimulated rapid growth of salmon ranching, and Japan has compensated for the loss of high-seas fisheries by releasing more than two billion juvenile salmon for ranching. The 1985 harvest of salmon produced by aquaculture (mostly ranching) in Japan accounted for 19% of the world tonnage (178,000 of 920,000 tonnes). Nasaka (1988) discussed the history and status of the successful salmon aquaculture industry in Japan.

Alaska

I introduced this report with an example from Alaska, where it appears that salmon fishers have benefited from policies favoring salmon ranching. The present institutional framework for salmon ranching in Alaska began to emerge in the early 1970s with creation by the legislature of a Fisheries Rehabilitation and Enhancement Division within the Alaska Department of Fish and Game. The Alaska constitution was amended in 1972 to permit aquaculture. Subsequent legislation provided a foundation for growth of hatchery programs operated by the state and by nonprofit corporations. Production of hatchery juveniles for ranching is approaching a scale comparable to Japan's.

Even though Alaska has encouraged ranching of salmon, policy barriers against farming remain. Political resistance to farming comes largely from traditional fishers who fear competition for markets. Meanwhile, Alaska's southern neighbor, British Columbia, has opted for policies favoring farming but not ranching.

British Columbia

A salmon farming "gold rush" has emerged in British Columbia (Fitzgerald 1987). Substantial private investments have been made in this developing industry and more are expected. Production of farmed salmon began in the late 1970s and was expected to reach 20,000 tonnes by 1990 (White 1988), which would likely exceed the combined harvest of coho and chinook salmon by British Columbia capture fisheries.

Ranching in British Columbia is conducted only by the federal government, which has investments in spawning channels and hatcheries (Bourne and Brett 1984). Walters (1988) pointed out that marine survival of hatchery-produced coho and chinook has trended downward. He recommended that further expansion of hatchery production be delayed until various hypotheses about factors causing reduced survival of hatchery salmon are evaluated.

Iceland

Iceland has prohibited commercial ocean fisheries on salmon since 1932. Natural stocks are harvested in rivers, primarily by sport fishers. Salmon aquaculture has become a growth industry. Atlantic salmon are grown to enhance natural stocks, for farming, and for ranching. Twenty-two companies produced salmon in 1984 and more than 50 had registered to engage in salmon aquaculture (Severson and McNeil 1985). Scarnecchia (1989) reviewed the history of salmon management and production in Iceland.

New Zealand

The New Zealand government has encouraged salmon aquaculture since the early 1970s as a means to enhance sport fisheries and promote economic growth. Private investments in facilities to ranch and farm salmon amounted to several million dollars by 1985 and were expected to increase rapidly. Commercial fisheries targeting salmon are banned, and disposal of salmon caught incidentally to marine commercial fish and shellfish catches has become a major policy issue (Young 1985).

Conclusions

Salmon aquaculture is expanding rapidly in temperate marine basins throughout the world. Production has been doubling in less than 3 years for farmed salmon and in about 10 years for ranched salmon. If present trends continue, aquaculture will provide more salmon for world markets than natural production by the year 2000.

Aquaculture techniques for breeding, growing, releasing, and recapturing salmon are improving and becoming more diversified. Industrial-scale systems for farming and ranching salmon have a potential to grow into a major international food commodity industry. Economic benefits to certain coastal states could someday rival those from intensive feedlot production of terrestrial farm animals.

As with any intensive production of a food commodity, environmental trade-offs are anticipated. Interactions between hatchery and natural stocks of salmon will increase in direct proportion to expanded production of farmed and ranched salmon. Strategies to minimize competition for food and space, transfer of disease, environmental effects, genetic modification, and overharvest of natural stocks in mixed-stock fisheries will require continued attention.

Policies that influence rate and direction of salmon aquaculture development vary greatly among sovereignties. Policies that impede growth are typically formulated in response to concerns over environmental impacts. Groups concerned about economic consequences adverse to their interests sometimes lobby for policies designed to impede growth of salmon aquaculture. However, the global outlook is for policies that facilitate growth of salmon aquaculture into an important international food industry.

References

Anonymous. 1987a. Price rises follow drop in supply. Fishing News International 26(6):6.

Anonymous. 1987b. Global output to double. Commercial Fish Farmer 11(6):9.

Bourne, N., and J. R. Brett. 1984. Aquaculture in British Columbia. Canadian Special Publication of Fisheries and Aquatic Sciences 75:25–41.

Donaldson, E. M., U. H. M. Fagerlund, D. A. Higgs, and J. R. McBride. 1979. Pages 455–597 *in* W. S. Hoar and D. J. Randall, editors. Fish physiology, volume 8. Academic Press, New York.

Fitzgerald, R. 1987. Salmon: Canada's aquaculture gold rush. Seafood Leader 7(2):144–152.

Gowan, R. 1988. Release strategies for coho and chinook salmon released into Coos Bay, Oregon. Pages 75–80 *in* W. J. McNeil, editor. Salmon production, management, and allocation. Oregon State University Press, Corvallis.

Hampson, A. 1988. The laws covering salmon ranching

from the salmon's point of view. Pages 19–24 *in* W. J. McNeil, editor. Salmon production, management, and allocation. Oregon State University Press, Corvallis.

Hasler, A. D., A. T. Scholz, and R. M. Horrall. 1978. Olfactory imprinting and homing in salmon. American Scientist 66:347–355.

Ingram, M. 1988. Farming rainbow trout in fresh water tanks and ponds. Pages 155–189 *in* L. Laird and T. Needham, editors. Salmon and trout farming. Ellis Horwood, Chichester, UK.

Johnson, O. W., W. W. Dickhoff, and F. M. Utter. 1986. Comparative growth and development of diploid and triploid salmon, *Oncorhynchus kisutch*. Aquaculture 57:329–336.

Lowe, M. 1988. Salmon ranching and farming net growing harvest. World Watch 1(1):28–32.

McNeil, W. 1979. Review of transplantation and artificial recruitment of anadromous species. Pages 547–554 *in* T. V. R. Pillay and W. A. Dill, editors. Advances in aquaculture. Fishing News Books, Farnham, UK.

Nasaka, Y. 1988. Salmonid programs and public policy in Japan. Pages 25–31 *in* W. J. McNeil, editor. Salmon production, management, and allocation. Oregon State University Press, Corvallis.

Nierentz, J. H., and H. Josupiet. 1987. World fish supply and demand projections. Infofish 87(3):27–30.

Scarnecchia, D. L. 1989. The history and development of Atlantic salmon management in Iceland. Fisheries (Bethesda) 14(2):14–21.

Seeb, J. E. 1987. Chromosome set manipulations in salmonids: survival and allozyme expression of triploid interspecific hybrids and gene–centromere mapping in genotypes. Doctoral dissertation. University of Washington, Seattle.

Severson, R. F., and W. J. McNeil. 1985. Impacts of ocean fisheries on natural and ranched stocks of Icelandic salmon. Oregon Aqua-Foods, Inc., Springfield, Oregon.

Walters, C. J. 1988. Mixed-stock fisheries and the sustainability of enhancement production for chinook and coho salmon. Pages 109–115 *in* W. J. McNeil, editor. Salmon production, management, and allocation. Oregon State University Press, Corvallis.

White, S. 1988. Anticipating the "blue revolution": the growth of the salmon farming industry and its public policy implications. Oregon State University Sea Grant Program, Anadromous Fish Law Memorandum 45:2–23, Corvallis.

Young, A. 1985. The optimal harvesting of salmon in New Zealand. Report of New Zealand Salmon Farmers Association to New Zealand Minister of Fisheries, Wellington.

American Fisheries Society Symposium 10:19–31, 1991

Bioengineering Problems in River Systems of the Central Valley, California

RANDALL L. BROWN

California Department of Water Resources
3251 S Street, Sacramento, California 95816, USA

Abstract.—California's Sacramento–San Joaquin River system drains the Central Valley and has been extensively developed to provide water to domestic, industrial, and agricultural users and to control floods. The system also provides essential habitat for several species of anadromous and resident fish, including chinook salmon *Oncorhynchus tshawytscha* and striped bass *Morone saxatilis*. In this article, I describe three situations (Red Bluff Diversion Dam, the upper Sacramento River, and Delta diversions) in which bioengineering has been used, or will be used, to develop measures designed to protect fishery resources at water project features. In general, mitigation measures have not always been effective, partly because of poor communication among biologists, planners, and design engineers. There are indications in California that project planners are now more aware of fishery concerns and that this awareness will result in more effective engineering solutions to complex biological problems.

The Sacramento and San Joaquin rivers (and their tributaries) drain about 40% of California's land area and provide valuable fishery and water-supply benefits. Over the past century or so, however, the uses of water by California's miners, loggers, farmers, cities, and industries have often adversely affected fish. In this paper I summarize some of the major problems and describe how biologists and engineers are working to solve them. Before describing the bioengineering problems, I provide some background on the climatic, geographical, demographic, and biological conditions that contribute to them.

Conflicts in water use occurred soon after gold miners began arriving in California in the 1850s. The miners fought over water supplies, cleared forests for mine timbers, filled streambeds with sediment from hydraulic mining, and polluted streams with toxic chemicals such as mercury and copper. Early settlers constructed levees to protect their farms and communities from flooding, built small water-storage projects and distribution systems for irrigation, and eliminated vast areas of swampland by constructing levees and draining lands behind them. These activities altered flow and streambed conditions in the Sacramento and San Joaquin systems.

Continued development after 1900 was accompanied by the need for a greater degree of flood protection and for year-round water supplies for California's towns and farms. In 1933 the Governor of California signed the Central Valley Project Act, which would redistribute water from the Sacramento and San Joaquin rivers to Central Valley and San Francisco Bay area residents. Although California was unable to finance the plan, the U.S. Bureau of Reclamation was authorized to build several of its key features. Friant Dam, completed in 1946, impounded the upper San Joaquin River, and the Friant–Kern Canal conveyed much of this water south. Shasta Dam, completed in 1944 on the upper Sacramento River, and Folsom Dam, completed in 1956 on the American River, provide water, power, and flood control. Distribution systems such as Folsom–South Canal, Contra Costa Canal, Delta–Mendota Canal, and Tehama–Colusa Canal were approved and constructed. A major project diversion, the Delta–Mendota Canal, carried water from the southern Delta to farmers in the northern San Joaquin Valley who had lost their supply from the San Joaquin River when Friant Dam was completed.

During the 1930s and 1940s, local agencies and private utilities were also constructing water- and power-supply projects on tributaries to the Sacramento and San Joaquin rivers. Projects such as Hetch Hetchy Dam on the Tuolumne River (City and County of San Francisco), Pardee Dam on the Mokelumne River (East Bay cities), Exchequer Dam on the Merced River (Merced Irrigation District), and Don Pedro Dam on the Stanislaus River (Modesto Irrigation District) were constructed. These and other projects often blocked access to upstream salmon spawning areas and left flows below the dams that were inadequate for fish.

In 1959 the State Water Project was approved

FIGURE 1.—Major features of state and federal water projects in California.

by California voters, authorizing construction of Oroville Dam on the Feather River, the California Aqueduct, and the North Bay and South Bay aqueducts (Figure 1). The aqueducts and downstream storage reservoirs are used to convey water from the Sacramento–San Joaquin Delta to consumers in the San Francisco Bay area, the San Joaquin Valley, and Southern California. Water is diverted from the Delta into the California Aqueduct by means of seven pumps with a total capacity of about 185 m^3/s. Four additional pumps now being installed will increase pumping capacity to about 300 m^3/s. As I discuss later, achieving full capacity will depend on finding environmentally acceptable ways to move the Sacramento River water across the Delta.

Unquestionably, there will be continued pressure on California's water supplies. The state's

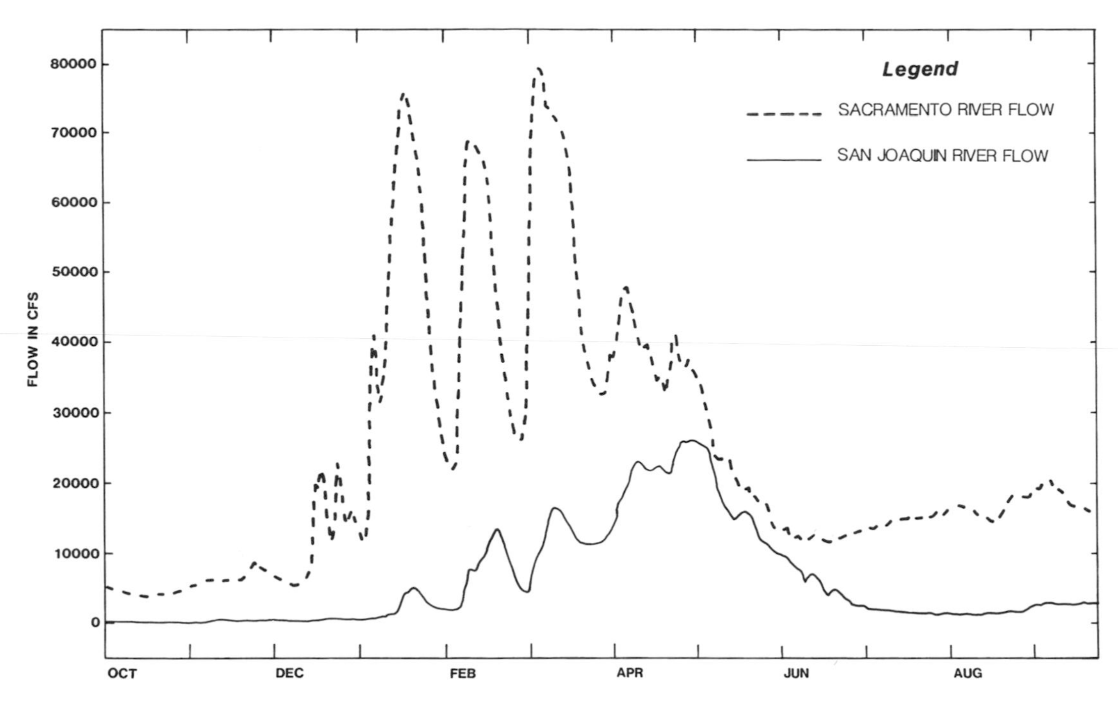

FIGURE 2.—Typical flow patterns in the lower Sacramento and San Joaquin rivers, California. CFS is cubic feet per second.

human population continues to expand at a high rate, recently passing the 28 million mark. Most of the people live in Southern California and the San Francisco Bay area—whereas most of the water originates in northern California and the Sierra Nevada mountain range.

Hydrology in the Sacramento–San Joaquin River basin is characterized by winter–spring precipitation and runoff followed by a dry summer and fall. Although every water year differs, flow during the 1978 water year (October 1977 through September 1978) was typical of present-day patterns in the Sacramento and San Joaquin rivers (Figure 2). The flows are measured at stations downstream of all major tributaries just before the rivers enter the Sacramento–San Joaquin Delta. Sacramento River flows generally are 5 to 10 times higher than those of the San Joaquin River due to the combination of natural differences in precipitation and water development. For the most part, dams have eliminated natural summer flows in the lower San Joaquin River (below Friant Dam), so the riverflow is predominantly municipal discharges and surface and subsurface agricultural drainage during summer. Fishery and water-quality releases from the recently completed New Melones Reservoir on the Stanislaus River have increased flows and helped improve water quality.

Water development, flood control, and hydroelectric projects plus riparian water uses have changed the timing and magnitude of flows in the tributaries and main stems of the Sacramento and San Joaquin rivers and into the estuary. Construction and operation of dams have blocked access to upstream spawning grounds for anadromous species, adversely affected downstream migration of juveniles, changed kilometers of stream habitat to lake habitat, and often changed downstream water temperatures. The dams and reservoirs, along with waste discharges, timber harvesting, and streambed and bank alteration, have created numerous bioengineering problems in the watershed and in the delta formed at the confluence of the two rivers.

I focus on three major problems: Red Bluff Diversion Dam, upper Sacramento River temperatures, and diversions from the Sacramento–San Joaquin Delta. Also, the discussion is limited to problems affecting populations of chinook salmon *Oncorhynchus tshawytscha* and striped bass *Morone saxatilis*. Organizationally, the report consists of a brief description of biology and stock status of chinook salmon and striped bass, followed by a review of the three problems and how they affect the species selected. Finally, I describe how fishery management agencies, water development agencies, and others are working to resolve the problems and restore fishery resources.

Fisheries: Biology and Status

Although both chinook salmon and striped bass are anadromous, their life histories have little in common. A knowledge of these life histories is essential to understanding and resolving bioengineering problems that often act to reduce stock size. The following is an introduction to the biology of these species—more complete reviews can be found in Skinner (1962).

Striped Bass

Striped bass were introduced into the Sacramento–San Joaquin estuary from the Atlantic coast of the USA in 1879. Although only a few fish were originally planted, the population expanded so rapidly that a commercial fishery was in place by the 1890s. The commercial fishery was stopped in 1935 by legislation sponsored by sportfishing interests. Striped bass continue to sustain an important recreational fishery in the Sacramento–San Joaquin system and in State Water Project reservoirs along the California Aqueduct. (Striped bass in project reservoirs result from entrainment by Delta pumps and transport into the California Aqueduct.)

Striped bass normally spawn in April and May in the Delta, in the lower San Joaquin River, and in the Sacramento River from Sacramento to Colusa. Females broadcast 0.5 to 1 million slightly buoyant eggs. After fertilization, the eggs drift downstream toward nursery areas in the Delta and Suisun Bay. The larvae feed first at about 5 d (6 mm), and generally reach 38 mm in length by late July. Striped bass remain in the upper estuary until adulthood. They enter the fishery at 3–4 years of age; a few females first spawn at about 4 years, but most at 5 years. Like Atlantic coast striped bass stocks, many adult striped bass from San Francisco Bay enter the ocean in the summer, although most do not migrate more than 30 km north or south from the Golden Gate.

Several striped bass life stages are vulnerable to engineering works. The eggs and early larvae measure 4–15 mm; thus, it is difficult to design fish screens that can protect fish and still keep head loss across the screen at an acceptable level. Adults are not readily attracted to fishways, and moving them around barriers presents difficulties.

The estimated number of adult striped bass has declined since the early 1970s (Figure 3). The number of juveniles of average size (38 mm) has declined even more during about the same period (Figure 3). The juvenile index has been positively correlated with spring outflows and negatively correlated with water diversion rates. Since 1977, however, given flows have resulted in smaller year-classes than did flows of the same magnitude during 1959–1976. Potential causes of the decline include entrainment of eggs, larvae, and juveniles; lack of food; decreased egg abundance resulting from fewer and younger spawners; overfishing; and pollution. Stevens et al. (1985) presented a more complete summary of the status of striped bass in the Sacramento–San Joaquin system.

Chinook Salmon

Chinook salmon are indigenous to the Sacramento–San Joaquin River system, which supports the southernmost populations of the species. Chinook salmon lay their eggs in the gravels of upriver riffles, and the young emerge after a few weeks. Most young chinook salmon in this system become smolts in their first spring and move rapidly to the ocean. Most adults return as 3-year-olds, although there are some 4- and a few 5-year-old spawners.

Historically, there have been four distinct runs to the drainage—fall, late-fall, winter, and spring races—characterized by the season of upriver migration through the estuary (Figure 4). Upstream or downstream migrants from some race are in the system at all times of the year.

The following information on status of Central Valley chinook salmon stocks is from California Department of Fish and Game (1987) and Vogel and Rectenwald (1987).

Late-fall run.—Late-fall-run fish are a relatively small component of Sacramento River spawning escapement. The size of this run has decreased during the past 20 years (Figure 5).

Winter run.—Historical patterns in the abundance of winter-run chinook salmon are not well known, but it is likely that the run was present before Shasta Dam was constructed. The initial years of Shasta operation resulted in large summer releases of cold water, which improved environmental conditions for winter-run fish in the main stem for several kilometers below the dam. Run size increased dramatically to the level where

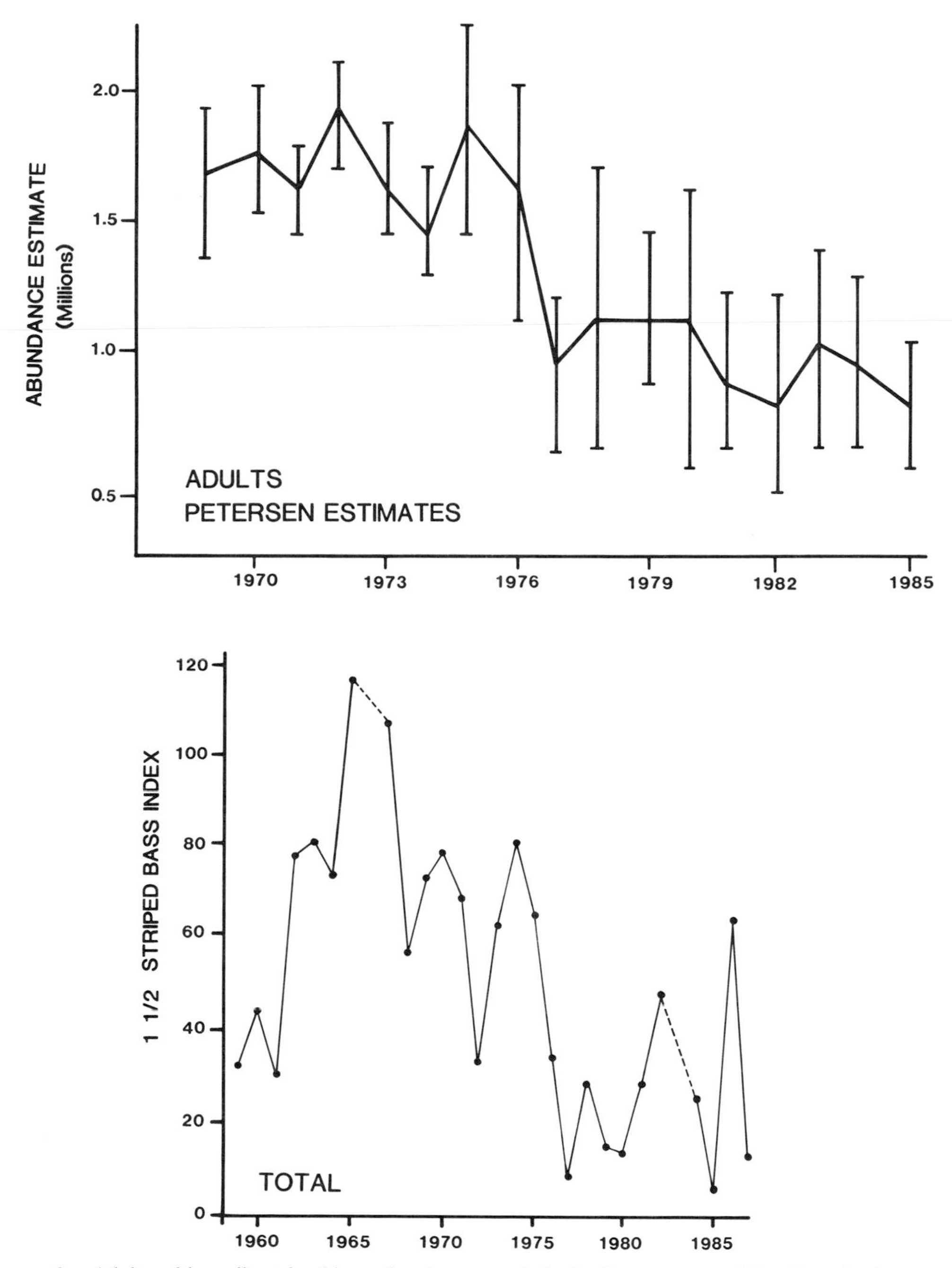

FIGURE 3.—Adult and juvenile striped bass abundance trends in the Sacramento and San Joaquin river system, California.

it approached that of the historically dominant fall run.

Spawning escapement has decreased in recent years to the point that the run is in danger of disappearing (Figure 5). The California–Nevada Chapter of the American Fisheries Society has petitioned the U.S. Department of Commerce to include the winter race on the federal list of threatened species. I summarize possible reasons for the decline of the winter run, and actions being

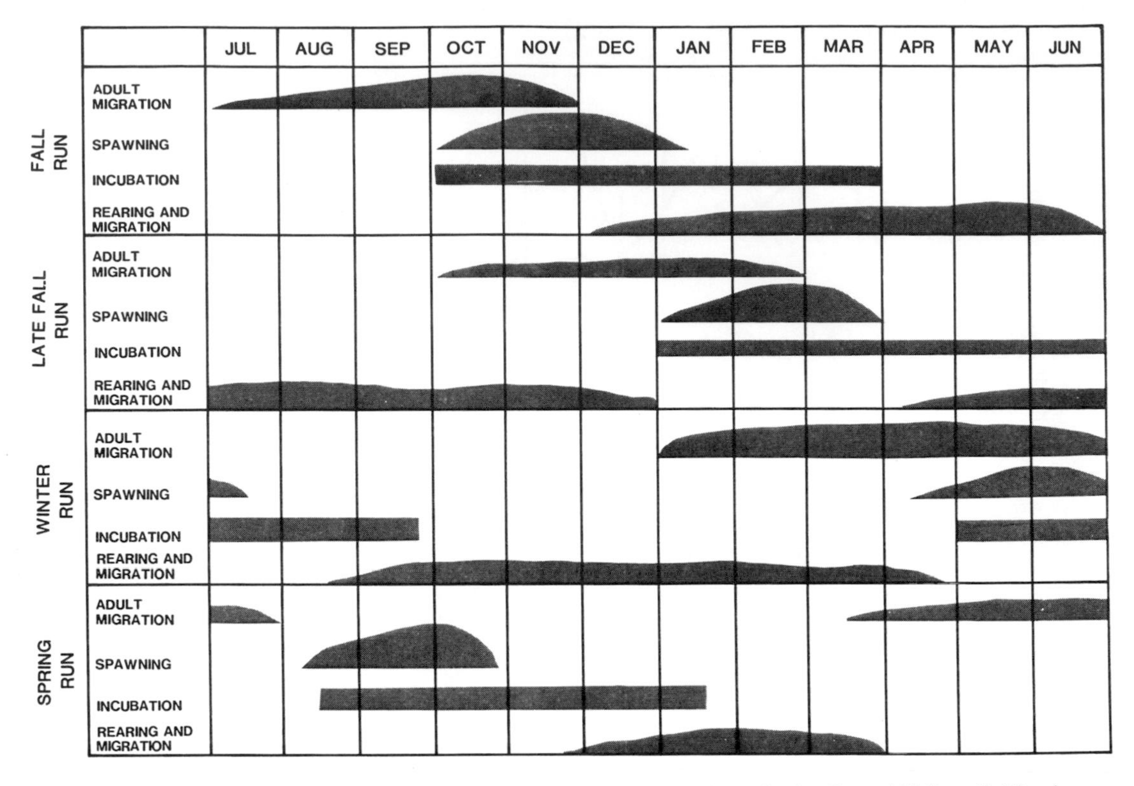

FIGURE 4.—Life history characteristics of four races of chinook salmon in the Central Valley, California.

taken to restore it, in later sections on the Red Bluff Diversion Dam and upper Sacramento River temperature.

Spring run.—The upriver spring spawning run has oscillated widely within the range of about 4,000–20,000 individuals over the past 20 years (Figure 5), with no apparent trend. Before the Shasta, Friant, and Oroville dams blocked spring-run spawners from their main spawning areas, the total run size was in the range of several hundred thousand fish. There is no longer a spring run in the San Joaquin River system. Since there is little suitable habitat for the run, and hatchery production is difficult, dramatic increases in run size are unlikely.

Fall run.—The fall run is now the principal contributor to the total chinook salmon spawning escapement in the Sacramento River and its major tributaries (American, Feather, and Yuba rivers). For the past 20 years or so, escapement has been remarkably stable at about 200,000 spawners—a run size that is less than that of the 1950s and early 1960s (Figure 5). (Reliable escapement data are available only since 1953.) The average ocean catch has been about 435,000 Sacramento fall-run fish over the last 34 years (Dettman et al. 1987); thus, the total catch plus escapement in an average year ranges between 600,000 and 700,000 fish. Hatchery production on Battle Creek (Coleman National Fish Hatchery for Shasta Dam mitigation), Feather River (Oroville Dam mitigation), and the American River (Folsom Dam mitigation) contributes significantly to the commercial and sport catches and to spawning escapements. Dettman and Kelley (1987) estimated that about half the recent ocean catch is from salmon reared in the American River and Feather River salmon hatcheries. Because of various problems (disease, water supply, and losses during outmigration), the Coleman hatchery has not contributed to the same extent as hatcheries on the Feather and American rivers.

Fall-run spawning escapements to San Joaquin River tributaries (there is no longer spawning in the main stem) have been much less stable than in the Sacramento River (Figure 6). The periodic appearance of fairly large runs produced in "wet" years does show that, given suitable environmental conditions, substantial salmon production can

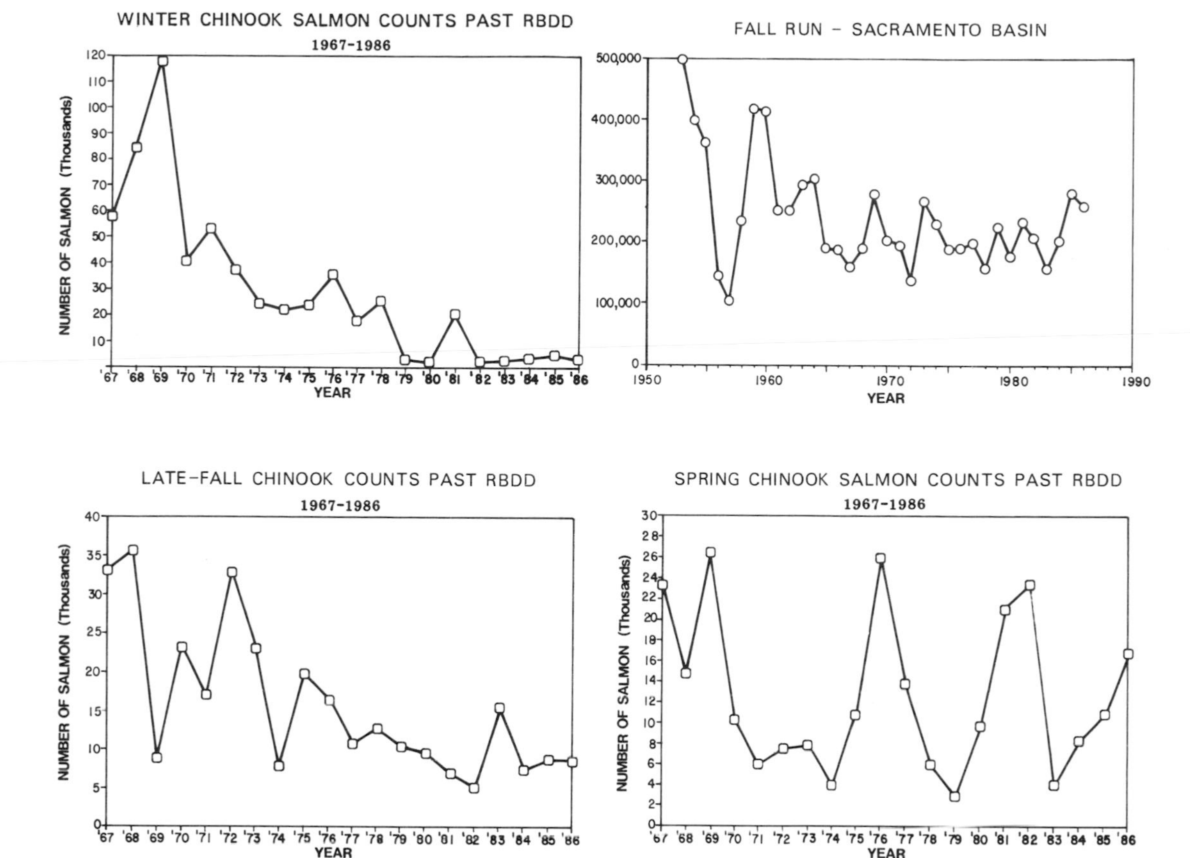

FIGURE 5.—Trends in spawning stock size of four races of chinook salmon in the Sacramento River, California. RBDD is Red Bluff Diversion Dam.

still occur from this system. There is little hatchery contribution to San Joaquin River stocks.

The stability of the fall run at relatively high stock size, in comparison with the other runs, is probably due to the Feather and American river hatcheries. This hatchery contribution is particularly important in that the estimated fraction of these fish harvested is in the range of 60–70% (Reisenbichler 1986).

Bioengineering Problems

The high level of development in the watershed and downstream areas, where four races of chinook salmon, striped bass, and numerous other anadromous and resident fish occur, has led to a variety of fishery problems, many of which may be solved through bioengineering. The problems I summarize only illustrate the types of problems encountered in the system. They are not necessarily the most important nor the most complex problems, but each is certainly important and complex. These types of problems can be expected in other areas of the world where large-scale water development has the potential to adversely affect fish resources.

Red Bluff Diversion Dam and Associated Facilities

Red Bluff Diversion Dam, an authorized feature of the federal Central Valley Project, is about 80 km below Shasta Dam on the main stem of the Sacramento River. Red Bluff Diversion Dam was constructed in the early 1960s (closed in 1967) to provide hydraulic head for a gravity-flow diversion into Tehama–Colusa Canal and Corning Canal. Corning Canal branches off Tehama–Colusa Canal about 5 km below the headworks. The dam and Tehama–Colusa Canal have the following principal fisheries-related features: fish ladders on both sides of the dam for upstream-migrating

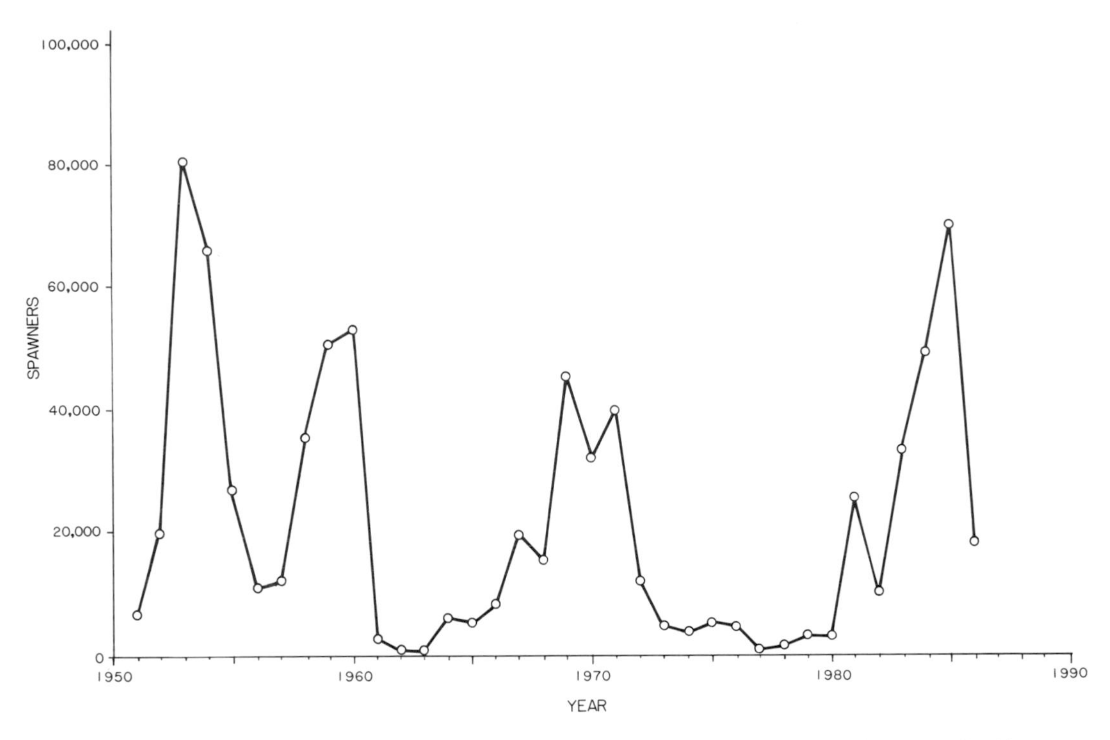

FIGURE 6.—Trends in spawning stock size of fall-run chinook salmon in the San Joaquin River system, California.

salmon; louver fish screens at the headworks of the canal; a dual-purpose spawning channel located in the first 5.1 km of the canal; and two single-purpose spawning channels located to the side and downstream of the dual-purpose spawning channel.

The general problem at Red Bluff is that none of the features built for fish protection worked as planned and designed by biologists and engineers. The main problems with each of the fish protection mitigation and enhancement facilities are discussed below.

Fish ladders.—The ladders are standard pool-and-weir ladders with staggered orifices, operated at a design flow of 2.4 m^3/s. A supplemental flow of 7.8 m^3/s is added through a diffuser to make a total flow of 10.2 m^3/s through each fishway. Each fishway has a counting structure at the upper end. There is a trap in the left-bank fishway to collect fish for the spawning channels. Fish not selected for transfer to the spawning channels are diverted back into the river above the dam.

The diversion dam has 11 gates, which can be opened or closed in response to changes in streamflow. However, gate operation causes complex circulation patterns below the dam, and the migrating adults have trouble finding the fishways. The problem is particularly severe when flows exceed about 280 m^3/s. Winter-run salmon, which pass the diversion dam from March through May, are especially vulnerable, since spring flows are often more than 280 m^3/s. The inability of fish to find the ladders causes delay in migration and may force some fish to spawn downstream of the dam in areas where summer water temperatures are too high for survival of winter-run eggs and alevins.

The U.S. Bureau of Reclamation has pledged to operate the gates to minimize delay and has funded a 5-year study by the U.S. Fish and Wildlife Service to identify methods of improving fish passage at Red Bluff Diversion Dam. It appears that a new fishway will be constructed on the east side of the dam.

Louver fish screens.—The louver screens at the head of Tehama–Colusa Canal were designed to prevent juvenile salmonids and other fish from entering the canal. The louver design, adapted from screens at the Bureau's Delta diversion, presents a behavioral barrier to minimize entrainment. Bypassed fish reenter the Sacramento River below the dam. The louvers are inefficient (about

70% for juvenile salmonids less than 41 mm long), and large numbers of salmon enter the canal. Some entrained fish are lost, others prey on fish reared in the spawning channels, and unquantified losses through the louvers have complicated evaluation of the spawning channel production because of an unknown contribution from upstream.

The louver screens are now being replaced by horizontal rotary-drain screens. The drain-screen structure will be 210 m long and will contain 32 drums, each 5.7 m in diameter and covered with a small-mesh stainless steel wire cloth. Maximum approach velocity is 100 cm/s. A new bypass system is also being constructed that will return the fish to the river near midchannel several meters below the dam. Construction of the replacement facility began in August 1988 with a 3-year completion schedule.

Spawning channels.—Two types of spawning channels were installed to compensate for the loss of about 3,000 spawners caused by inundation of spawning grounds above the dam. The U.S. Fish and Wildlife Service estimated that, properly operated, the channels would provide an additional annual enhancement of about 27,000 salmon spawners.

The canal that serves both for conveying waters and as a spawning channel (dual-purpose canal) is at the upstream end of Tehama–Colusa Canal. Gravel was placed in this 5.1-km channel, and a gravel cleaner was installed to maintain adequate intragravel flow for developing eggs and alevins. Although this dual-purpose canal has been used intermittently since 1972, it has been largely unsuccessful in providing either compensation or enhancement. Excessive algal growth, sedimentation, poorly graded gravel, and low efficiency of the gravel cleaner have caused poor survival of spawners placed in the channel. Problems at the headworks caused by debris and sediment have often prevented operators from providing intended velocities and flows down the channel.

The two single-purpose spawning channels have been somewhat more successful than the dual-purpose canal. Problems with algae and sediment have occurred, but not to the same extent. Since 1971 an average of about 3,000 spawners annually have been placed in the channels. Based on tag returns for 1973 through 1977, the channels contributed an average of about 11,500 fish to the ocean fishery and about 4,500 annually to the upper Sacramento River spawning escapement (U.S. Bureau of Reclamation 1985). The figures indicate that the channels may be meeting the mitigation obligation, but not the enhancement goals.

The U.S. Fish and Wildlife Service recently mothballed the Tehama–Colusa fish facilities. Agencies are now looking at the feasibility of a new salmon hatchery on the upper Sacramento River for propagating winter runs, and perhaps fall and spring runs.

In summary, the Red Bluff Diversion Dam and associated fish protective facilities created a variety of bioengineering problems for fishery scientists and engineers. (Not included in the discussion are some major concerns such as losses of downstream-migrating chinook salmon to Sacramento squawfish *Ptychocheilus grandis* and other predators at the dam.) The dam has had a major adverse impact on the four races of chinook salmon and on steelhead *Oncorhynchus mykiss*. The U.S. Bureau of Reclamation is working with fisheries agencies to resolve these problems while maintaining its ability to deliver water to farmers in the northern Sacramento Valley.

Upper Sacramento River Temperatures

Salmon spawning, incubation, rearing, and outmigration can be successful only if temperatures are in the range of 7.2–18.3°C. Incubation temperatures are particularly critical; optimum temperatures lie between about 7.2°C and 13.9°C. In the Sacramento River system, the need for these cool temperatures is complicated by the differing seasonal requirements of the four races and by California's hot summers and cool winters. For example, winter-run salmon spawn from April to June, and their eggs are in the gravel during the period of maximum water temperature. Historically, winter-run and spring-run fish spawned at high elevations where cooler waters could be found. Construction of high dams such as Shasta prevented access to these cooler waters.

Temperature problems on the upper Sacramento River are particularly severe during dry years, when Shasta Lake is drawn down. Although there are several outlet levels from Shasta, their flexibility in providing cold water is limited by the need to produce power from reservoir releases. In normal years, cold water from below the thermocline passes through the upper outlets, through the hydroelectric turbines, and out to the river, where it can exert a cooling influence downstream to below the Red Bluff Diversion Dam. The reach between Keswick Dam and Red Bluff Diversion Dam is critical habitat for spawning winter-run salmon. In dry years, or when the reservoir has been drawn down severely to provide water under contractual obligations, the upper outlets draw from the warm epilimnion. Withdrawal from the bottom outlet is possible, but hydroelectricity cannot be generated through this outlet.

As an emergency interim measure, in the summer of 1988 the U.S. Bureau of Reclamation bypassed the power generation and discharged cold water from the lower outlets to help protect winter-run salmon. After examining a number of potential long-term solutions, the Bureau of Reclamation has selected the multiple-level outlet as the best way to manage water temperature in the upper Sacramento River below Shasta Dam. The present engineer's estimate for construction cost is $50 million for a structure completed by about 1993. The federal and state governments are working on cost-sharing for this project.

Another temperature concern relates to long-term warming. Reuter and Mitchell (1987) reported that spring Sacramento River water temperatures are 2–3°C warmer at any given flow than they were 10 years ago. Reasons for this change are not known. The U.S. Bureau of Reclamation and others are developing a predictive temperature model for the Sacramento River that can be used to evaluate the effects on river temperatures of project operations and heat inputs such as irrigation return flows. Temperature management through reservoir operation and use of streamside vegetation may become even more important in maintaining salmonid populations.

Sacramento–San Joaquin Delta Diversions

The Sacramento–San Joaquin Delta is the freshwater portion of the upper estuary and provides critical habitat for about 45 species of resident and migrating fish. The Delta consists of 1,100 km of channels surrounding about 300,000 hectares of diked islands. The islands are intensively farmed with irrigation water supplied by about 1,800 small pumped or siphoned diversions from Delta channels.

The major unresolved bioengineering problems in the Delta are associated with operation of state and federal water-project pumps in the southern Delta. The problems with these diversions are twofold. First, the pumps entrain large numbers of fish and fish-food organisms. Second, the draft of water to the pumps can change flow patterns in the Delta and lower rivers, and the changes can adversely affect fish. These two problems are described briefly below.

Entrainment.—Intakes to the federal and state pumping plants are equipped with fish protective facilities. The primary screening system at each plant is a louvered behavioral barrier (Figure 7). At the state plant, the V-shaped primary louvers guide the small fish into a bypass leading to a secondary screening system. The primary louvers have a center divider wall that improves screening efficiency for small striped bass. The state's secondary screening system consists of two separate channels, one with a louver system and a second with a recently added perforated plate screen. The secondaries are intended to further concentrate the fish, and use of one or both depends on the number of primary channels in use. In practice, however, hydraulic problems have limited concurrent operation of the two secondary channels. This problem is being resolved by adding additional holding tanks. From the secondaries, the screened fish are bypassed to large tanks, where they are held pending return to the lower Sacramento River by trucks. Periodic counts are made to estimate numbers of fish (by species) in the holding tanks. The counts are used to determine hauling frequency and to provide a record of the number of fish salvaged.

The federal facility is similar to the state's except that it is smaller (127 m^3/s versus 184 m^3/s capacity), it does not have a V-shape, and the one secondary screen is a louver design. Operation of the two facilities is entirely different, however. The federal pumps essentially run continuously. The state, on the other hand, has a small (38 million m^3) regulation reservoir (Clifton Court Forebay) in front of the fish facilities and pumps. Radial gates are used to tidally pump water into the forebay, and the pumps are operated to minimize pumping head and to take advantage of off-peak power.

An extensive field evaluation of the efficiency of the state fish protective system was conducted by the California Department of Fish and Game (California Departments of Water Resources and Fish and Game 1973). Results of this evaluation have been used to develop operating criteria designed to maximize fish salvage, especially salvage of chinook salmon and striped bass. As might be expected, louver screen efficiency is quite low for fish less than 25 mm long and approaches 80–90% for fish larger than 38 mm when velocities are in the proper range. Table 1 provides estimates of the most common fish salvaged at the state facility during recent years. The large numbers of striped bass exposed to the

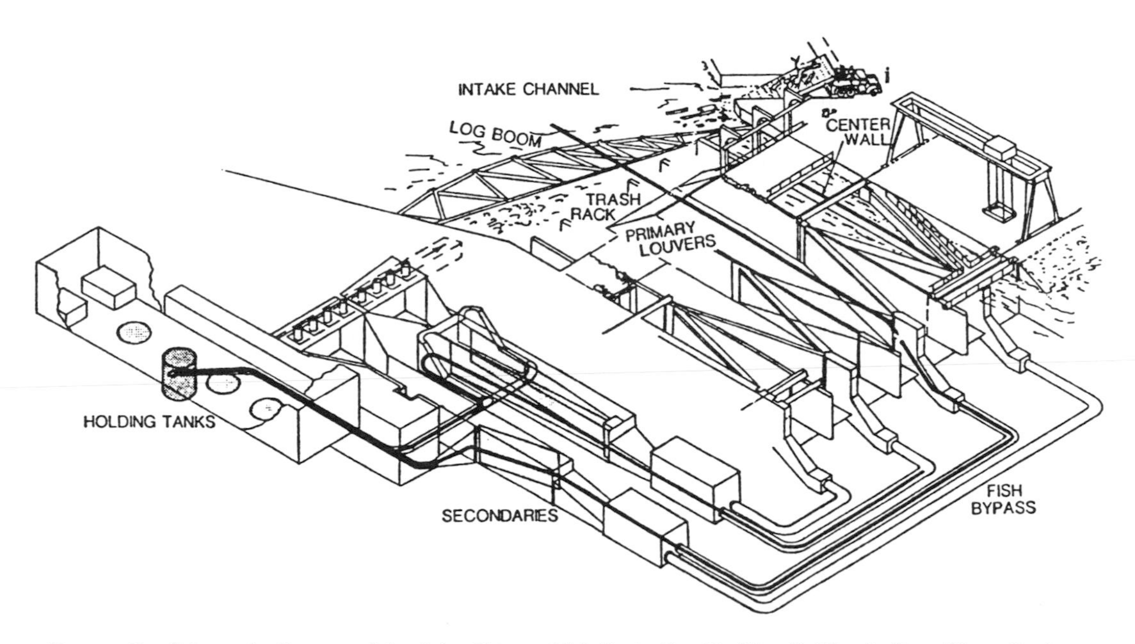

FIGURE 7.—Schematic diagram of the John Skinner Fish Protection Facility, California State Water Project.

pumps are of particular concern in that striped bass abundance has decreased dramatically during the past 13 years.

Another problem associated with the screens is the large numbers of fish lost incidentally to the salvage operations. There are losses to predators, losses through the screens (entrainment), losses in the holding tanks, and losses as the fish are returned to the Delta. Although exact numbers have not been agreed to, Fish and Game has estimated losses of small fish to predators in the state's Clifton Court Forebay to be in the range of 75–90%, depending on the size and species of fish. In 1986 the Departments of Fish and Game and Water Resources entered into an agreement that requires Water Resources to pay for striped bass, chinook salmon, and steelhead lost at the intake. The monies are used to fund projects that should replace the annual losses of the threespecies. A lump sum of $15 million is also included in this agreement to fund major restoration projects.

The final, and perhaps most significant, problem associated with the present Delta diversions is that they are in a dead-end location. Fish drawn to the screens cannot be directly bypassed to areas beyond the pumps' draft. There have been proposals to solve this dead-end problem by moving the intake to the Sacramento River as part of an isolated canal that would convey water around the Delta (the Peripheral Canal). The intake to this

TABLE 1.—Salvage of selected fish species at the State Water Project's Delta Fish Protective Facility, 1979–1988.

Year	Chinook salmon	Striped bass	White catfish[a]	American shad[b]	Threadfin shad[c]	Yellowfin goby[d]
1979	122,433	2,712,800	202,806	886,724	791,061	124,954
1980	154,764	1,327,670	700,203	552,889	1,351,560	87,110
1981	101,608	1,105,460	117,081	900,382	1,460,320	21,885
1982	278,421	976,493	563,780	848,862	663,404	69,828
1983	68,943	131,039	150,603	78,790	210,967	7,684
1984	145,037	6,376,160	811,200	423,483	386,790	1,010,490
1985	140,721	3,997,710	257,108	274,209	477,722	48,389
1986	435,231	13,854,700	427,930	560,975	534,825	532,689
1987	177,881	12,051,600	117,453	308,862	462,449	209,342
1988	151,907	13,038,100	126,390	255,775	357,088	74,191

[a]*Ictalurus catus*.
[b]*Alosa sapidissima*.
[c]*Dorosoma petenense*.
[d]*Acanthogobius flavimanus*.

400–600-m^3/s canal would be screened to prevent entrainment, and the bypassed fish would be returned to the Sacramento River (Odenweller and Brown 1982). Although the isolated canal continues to have considerable merit as means to reduce adverse impacts of Delta diversions, it is not politically feasible in the foreseeable future.

Delta flow patterns.—The state and federal pumps have a present maximum combined draft of about 340 m^3/s. (There is an additional internal consumption of 85–115 m^3/s by Delta agriculture during the irrigation season.) Flow to the pumps comes from the Sacramento and San Joaquin rivers. During most years, especially in the summer, San Joaquin River flows provide only a small part of the pumping demand. Water is drawn from the Sacramento River to the pumps by way of natural channels (Georgiana and Threemile sloughs) and the Bureau of Reclamation's Delta Cross Channel. If the draft exceeds the combined capacity of these channels, some Sacramento River water is drawn around the tip of Sherman Island and up the San Joaquin River to the pumps. In both instances water from the Sacramento River fills the Delta, and flow patterns and velocities are not the same as before pumping began in the southern Delta. These changes in Delta hydrology can confuse migrating fish, pulling them from normal migration pathways; they may also adversely affect primary and secondary productivity in the Delta.

This flow problem has been recognized since before the projects began operating in the 1950s and 1960s. Solutions proposed to resolve the flow problems while maintaining diversions at the present level (or even increasing them) include an isolated facility around the Delta, increased channel capacity in the Delta, and a larger channel leading from the Sacramento River to the interior Delta. For various political, social, economic, and technical reasons, none of the solutions has been adopted. The Department of Water Resources is now going through the environmental impact process to determine if there are environmentally acceptable means of diverting more water from the Delta while meeting the environmental, water quality, flood control, and other needs of the area and state.

Actions to Resolve the Fishery Problems

Several major bioengineering problems are apparent in the Sacramento–San Joaquin system. Fisheries concerns, however, are receiving additional emphasis in water development in California. Several interagency efforts are under way to identify bioengineering and other problems and to develop means of resolving them. Although these efforts mostly involve a core group of agencies and staff, they generally have distinct goals and objectives. Four significant efforts in this area are outlined below.

(1) Central Valley Fish and Wildlife Task Force. The task force consists of the California Departments of Fish and Game and Water Resources, the U.S. Bureau of Reclamation, the U.S. Fish and Wildlife Service, the U.S. Army Corps of Engineers, and the National Marine Fisheries Service. The group has input from local agencies. Its goal is to resolve present fishery and wildlife problems in the upper Sacramento River, such as those associated with Red Bluff Diversion Dam, using the existing agency budgets (or congressionally authorized increases) and staff. The Central Valley Fish and Wildlife Task Force is working to implement a number of projects developed through a legislatively sponsored study of the upper Sacramento River fisheries and riparian habitat resources. In addition to a temperature control device for Shasta Dam and improvements to fish protective features of Red Bluff Diversion Dam, projects include spawning gravel restoration, improvements to Coleman National Fish Hatchery, control of discharge of toxic metals, and improvements to fish protection at intakes of two major diversions in the upper river (Anderson–Cottonwood and Glenn–Colusa).

(2) Five-Agency Salmon Management Program. In response to a 3-year water right hearing process, the California Departments of Fish and Game and Water Resources, the U.S. Bureau of Reclamation, the U.S. Fish and Wildlife Service, and the National Marine Fisheries Service are working to categorize salmon problems and solutions in the entire drainage, including the Delta. This group was to report back to the hearing agency in early 1990. One important product of this program will be a mathematical operations model of the San Joaquin River system that should help engineers and biologists develop and assess solutions to fisheries problems in the basin.

(3) Delta Pumps Mitigation Advisory Committee. Water Resources and Fish and Game have convened a committee representing water contractors, fisheries, and environmental groups to recommend striped bass and chinook salmon enhancement projects to be funded by Department of Water Resources payments for fish lost at the

state's Delta pumps. Projects being supported by this process include gravel restoration, fish screens, hatchery improvements, purchase of hatchery fish, and flow augmentation.

(4) Interagency Ecological Studies Program. A slightly different mix of agencies (Fish and Game, Water Resources, U.S. Bureau of Reclamation, U.S. Fish and Wildlife Service, U.S. Geological Survey, and California State Water Resources Control Board) is evaluating the impacts of the state and federal water projects on Bay–Delta fish populations. Finding means of mitigating adverse impacts is an integral objective of this program.

The desired result of all these efforts, plus numerous others, is that the biological and engineering knowledge will be obtained to develop facilities that operate as planned and protect important fisheries. The need for clear communication among biological and engineering staff and management cannot be overemphasized. Fish protection and enhancement facilities must have a sound conceptual basis and must be constructed and operated within specification.

References

California Department of Fish and Game. 1987. The status of San Joaquin drainage chinook salmon stocks, habitat conditions, and natural production factors. California Department of Fish and Game, Region 4, Exhibit 15, submitted to the State Water Resources Control Board's Bay–Delta Hearing, Sacramento.

California Departments of Water Resources and Fish and Game. 1973. Evaluation testing program report for Delta Fish Protection Facility, State Water Project Facilities, California Aqueduct, North San Joaquin Division. California Department of Water Resources, Central District, Memorandum report, Sacramento.

Dettman, D. H., and D. W. Kelley. 1987. The roles of Feather and Nimbus salmon and steelhead hatcheries and natural reproduction in supporting fall run chinook salmon populations in the Sacramento River Basin. Final project report to California Department of Water Resources, Sacramento.

Dettman, D. H., D. W. Kelley, and W. T. Mitchell. 1987. The influence of flow on Central Valley salmon. Final project report to California Department of Water Resources, Sacramento.

Odenweller, D. B., and R. L. Brown, editors. 1982. Delta Fish Facility program report through June 30, 1982. Interagency Ecological Study Program for the Sacramento–San Joaquin Estuary, Technical Report 6 (FF/BIO-4ATR/82-6), Sacramento.

Reisenbichler, R. R. 1986. Use of spawner–recruit relations to evaluate the effect of degraded environment and increasing fishing on the abundance of fall-run chinook, *Oncorhynchus tshawytscha*, in several California streams. Doctoral dissertation. University of Washington, Seattle.

Reuter, J. E., and W. T. Mitchell. 1987. Spring water temperatures of the Sacramento River. Final project report to the California Department of Water Resources, Sacramento.

Skinner, J. E. 1962. An historical review of the fish and wildlife resources of the San Francisco Bay Area. California Department of Fish and Game, Water Project Board, Report 1, Sacramento.

Stevens, D. E., D. W. Kohlhorst, L. W. Miller, and D. W. Kelley. 1985. The decline of striped bass in the Sacramento–San Joaquin Estuary, California. Transactions of the American Fisheries Society 114:12–30.

U.S. Bureau of Reclamation. 1985. Fishery problems at Red Bluff Diversion Dam and Tehama–Colusa Canal fish facilities. Special report to the Central Valley Fish and Wildlife Management Study, Sacramento.

Vogel, D. A. and H. Rectenwald. 1987. Water quality and water quality needs for chinook salmon production in the upper Sacramento River. U.S. Fish and Wildlife Service, Exhibit 29, submitted to State Water Resources Control Board's Bay–Delta Hearing, Sacramento.

American Fisheries Society Symposium 10:32–41, 1991

An Investigation of Environmental Improvements for Fish Production in Developed Japanese Rivers

SHUNROKU NAKAMURA

Department of Civil Engineering/Regional Planning
Toyohashi University of Technology, Tempakucho Toyohashi, 441 Japan

NOBUHIKO MIZUNO

Department of Biology, Ehime University
Bunkyocho Matsuyama, 790 Japan

NOBUYUKI TAMAI

Department of Civil Engineering, University of Tokyo
Bunkyo-ku Tokyo, 113 Japan

RIKIZO ISHIDA

Laboratory for Fisheries Environment and Pisciculture
GE Japan Building, Kagurazaka 3-6, Shinjyuku-ku, Tokyo, 162 Japan

Abstract.—The present status of spawning beds, other aspects of fish habitat, and fish ladders in Japanese streams were investigated. In responding to questionnaires, 466 commercial fishery cooperatives reported many problems they perceived as limiting fish production in highly engineered rivers. Field observations in several streams revealed projects that had been designed to rehabilitate riffles, pools, spawning areas, and streambanks. Measures are proposed for further improvement of stream habitat for fish.

About 75% of Japan is mountainous terrain, through which many small, steep streams flow. The streams provide fisheries for mid- to high-priced food species as well as for recreational fishing. The fishing rights are owned only by cooperatives, which are organized by commercial fishermen and processors to jointly carry out economic activities, and which are obligated to produce and stock fish and also manage habitat in return for their fishing rights. Recreational anglers, whose numbers have been rapidly increasing, have to buy a temporary license from the local cooperatives. Most of Japan's human population, however, is concentrated on coastal alluvial plains. Here, summer and fall rains bring floods, whereas winter brings water shortages. To deal with the problems of large seasonal variation in flow, stream channels have been restructured by such agencies as the Ministry of Construction.

In recent decades, the stream environment for fish has been deteriorating due to pollution by industrial and household wastes and to channel reconstructions for flood prevention and water supply. During 1969–1984, the building of dams and revetments in rivers increased at a rather constant rate, and pollution accidents repeatedly occured (Figure 1). During the same period, although the total Japanese catch from all marine and freshwater fisheries increased, the inland catch underwent a general trend of increase only from 1971 to 1978, and thereafter decreased rather steadily (Figure 2). Despite ongoing rise in market prices for fish, the trend of increase in total sales value (unadjusted for inflation) for inland fisheries appears to have leveled off as of 1982, and fish-stocking costs have kept climbing (Figure 2); therefore, the economic return from inland fisheries must be declining. The objective of the present investigation was to find methods for improving environments for fisheries in Japan's flowing waters.

Methods

Via field observations and questionnaires sent to fishery cooperatives, we sought to analyze present conditions of spawning and rearing habitats, as well as of fish passage. From the results, recommendations were developed for improving stream habitats.

The National Federation of Inland Water Fisheries Cooperatives (NFIWFC) organized a study group funded by the Fishery Agency of Japan in 1985 and 1986. We, as the chairmen of four subcommittees of the study group, devised questionnaires at the request of NFIWFC and also analyzed the completed questionnaires returned

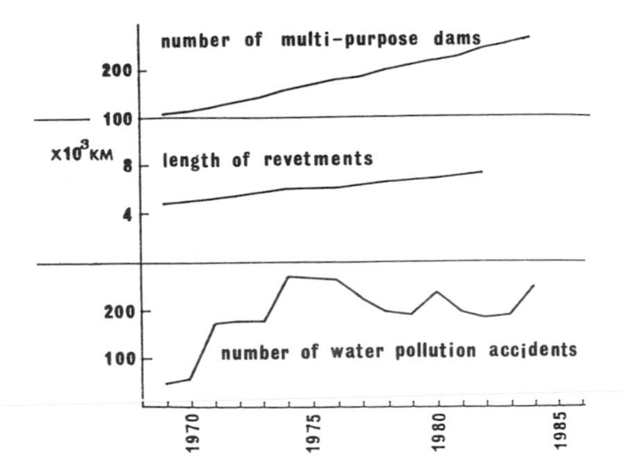

FIGURE 1.—Recent trends in dam construction, bank revetment construction, and water pollution accidents in Japanese streams. Sources: River Bureau (1967–1986, 1986) and Kawai (1986).

by the fishery cooperatives in 1985 (NFIWFC 1987). The questionnaires contained questions about the status of riffles, pools, spawning areas, and their management, plus questions on structures that had been built in channels and on riparian areas. Of 858 cooperatives provided with questionnaires, 466 responded.

We also made field observations on several streams that attracted our particular interests based on responses of the questionnaires, focusing on spawning and rearing habitat for fish. We sought information particularly on the effects of streambank structures, as well as on the success or failure of artificial alterations that had been designed to improve riffles and pools for these purposes.

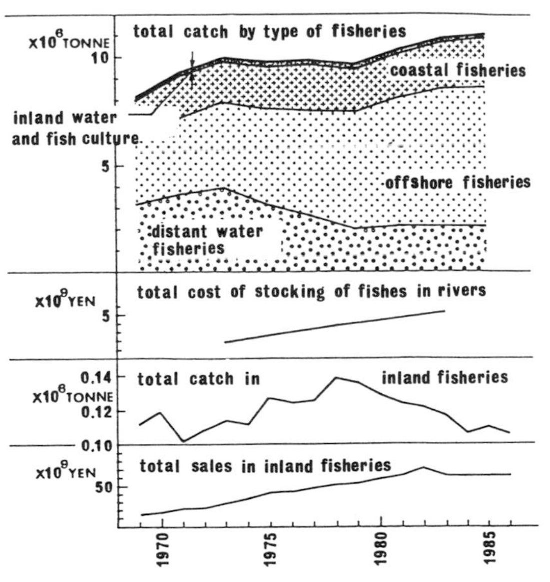

FIGURE 2.—Recent trends in Japanese fishery statistics. Sources: Ministry of Agriculture, Forestry and Fisheries (1980–1985, 1986) and Fishery Agency (1987).

TABLE 1.—Responses of Japanese fishery cooperatives to questions about stream pools that they administer.

Question	Response: Numerical	Response: Percentage
How many pools have vanished or become small in your district?	1,276	55.7
How many remain unchanged?	1,015	44.3
Total	2,291	100
What do you think is the cause of pool loss or change?		
Artificial change of channel route	51	2.8
Construction of revetments	304	17.0
Adjacent roadwork	120	6.7
Stream repair after floods	274	15.3
Other river works	117	6.5
Sedimentation after floods	366	20.4
Downcutting of stream bed	107	6.0
Construction of adjacent weirs	86	4.8
Decreased water flows	226	12.6
Other	142	7.9
Total	1,793	100
What was the average depth of past pools?		
<3 m	463	23.8
3–6 m	1,096	56.2
6–9 m	293	15.0
>9 m	97	5.0
Total	1,949	100
What is the average depth of present pools?		
<3 m	982	50.3
3–6 m	753	38.6
6–9 m	153	7.8
>9 m	64	3.3
Total	1,952	100
Why are pools necessary for fishes?		
Resting place		21.3
Hiding place		25.1
Refuge during floods		19.8
Source of food		11.3
Sustain adjacent stream bed		13.0
Stabilize adjacent stream bed		8.5
Other		1.0
(Total answers were 929)		
What do you think about trees near pools?		
Important for fishes		79.8
Unimportant for fishes		6.9
Other		13.8
(Total answers were 289)		

Information from the Fishery Cooperatives

Riffles and Pools as Habitat for Fish

Due primarily to human activities, more than half of the large pools that existed within memory of cooperative members have become smaller or have vanished (Table 1). The cooperatives believed that pools serve a variety of purposes for fish, and 80% of respondents felt that having trees near pools was important for fish (Table 1).

In response to the questionnaire's request to describe structural work in channels that had benefited pools and riffles as habitat for fish, the cooperatives listed 37 projects. Among pool–riffle alteration techniques involved, 18 of the projects included installation of large stones on the streambed; in 18 there was channelization of braided reaches and dredging of sediments from pools; in 5 cases concrete blocks were installed for fish; and other methods were used in 2 of the projects.

Revetment Effects and Their Amelioration

About 90% of responding cooperatives reported that some damage to fisheries had been caused by streambank revetments. About 90% of the damage was related to installation of concrete walls on one or both banks, and about 10% was associated with concrete structures covering the streambed and both banks. In recent years, bank revetment walls have almost always been made of concrete. Old-style walls, which the cooperatives regarded as less damaging, consisted of piled rock (riprap) or of stones held in cylindrical wire-mesh (gabion) containers. The fish populations considered impaired by revetments included all species important in Japanese inland fisheries. For example, with regard to concrete side-wall projects, the Japanese eel population was reported to have been damaged in 28% of all damages reported, ayu in 22%, yamame in 6.4%, carp in 5.5%, and Japanese dace in 5.2% (common and scientific names appear in Table 2). The population reductions were attributed mainly to disappearance of rapids and pools in 17.5% of all damages reported and to decrease in spawning area in 12.6%.

Sixteen projects of channel work intended to ameliorate detrimental effects of revetments on fish habitat were listed by the cooperatives. In some cases, fish blocks (Figure 3) had been incorporated in the lower parts of revetment walls. Also, large concrete blocks and log-framed rockwork had been installed on the streambed in front of walls.

Spawning Habitat

Target species of the 273 reported spawning habitat projects (Table 3) were primarily Japanese dace (31% of projects) ayu (26%), pale chub (6.3%), and salmon (5.5%). The main methods were construction of artificial spawning beds (46% of projects) and tillage (loosening) of streambed materials (25%). At least 75% of spawning bed construction projects and 68% of bed tillage projects were considered by the cooperatives to have had good-to-excellent results. For other methods, the success ratings were lower.

Fish Passage

Some difficulties were reported in about 62% of the 1,065 passage facilities (pool-and-weir ladders) installed for juvenile fish at weirs or dams of approximately 15–30 m height. The most prevalent general category of problems involved ladder entrances. Fish reportedly had difficulty finding entrances at 18% of ladders, and downward ero-

TABLE 2.—Japanese, English, and scientific names of fishes cited in this paper.

Japanese name	English name	Scientific names and others
Ayu	Ayu	*Plecoglossus altivelis* (Salmonidae)
Funa	Crucian carp	*Carassius* spp. (Cyprinidae)
Koi	Common carp	*Cyprinus carpio* (Cyprinidae)
Masu	Masu salmon	*Oncorhynchus masou masou* and *O. m. macrostomus* (Salmonidae)
Moroko	Moroko	*Gnathopogon* spp. and *Squalidus* spp. (Cyprinidae)
Oikawa	Pale chub	*Zacco platypus* (Cyprinidae)
Sake	Pacific salmon	*Oncorhynchus* spp., mainly chum salmon *O. keta* (Salmonidae)
Unagi	Japanese eel	*Anguilla japonica* (Anguillidae)
Ugui	Japanese dace	*Leuciscus* (*Tribolodon*) spp., mainly *L. (T.) hakonensis* (Cyprinidae)
Yamame	Yamame	Fluvial form of masu salmon

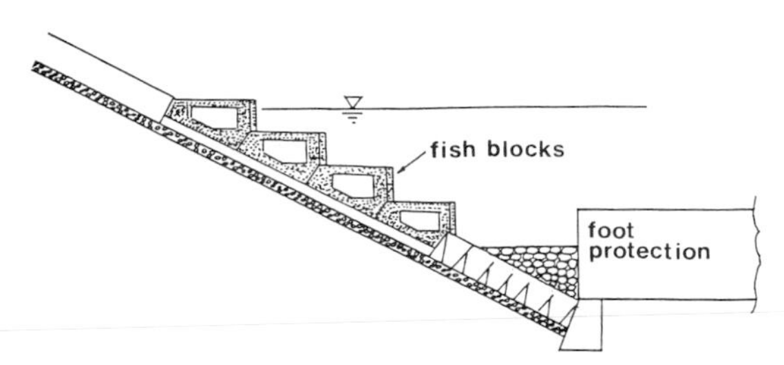

FIGURE 3.—Large, concrete, fish-block elements in a streambank revetment wall to provide spawning or hiding areas.

sion of the streambed had caused 17% of ladders to have entrances too high for fish. A remedy for the former problems is to provide an attraction flow, but few fish ladders had such a system. At 9.4% of the facilities, fish tended to become entrained in the inflow areas of dams' water-intake structures because ladder entrances were too close to the intakes.

TABLE 3.—Stream spawning habitat projects reported by fishing cooperatives.

Method	Target species	Number of reports in which results were judged					
		Excellent	Good	Fair	Poor	Unknown	Total
Streambed tillage	Ayu	3	14	4		9	30
	Japanese dace	1	14	2	1	4	22
	Pale chub		4	1		1	6
	Others	2	9				11
	All species	6	41	7	1	14	69
Nesting blocks	Ayu		1	1	1	2	5
	Japanese dace	1	3			3	7
	Crucian carp		2	2	1	1	6
	Common carp		2	1	1	1	5
	Japanese eel			2	1		3
	Others	1	2	1	3	4	11
	All species	2	10	7	7	11	37
Artificial spawning bed	Ayu	4	24			3	31
	Japanese dace	6	31	5		8	50
	Pale chub	1	8		1	2	12
	Vamame		5	1		1	7
	Others	2	13	3		7	25
	All species	13	81	9	1	21	125
Others	Ayu		4			2	6
	Japanese dace		2	1		3	6
	Crucian carp					3	3
	Yamame	2				3	5
	Japanese char	1	3	1		3	8
	Salmon		2	1			3
	Others	1	5	1		4	11
	All species	4	16	4		18	42
Totals	Ayu	7	43	5	1	16	72
	Japanese dace	8	50	8	1	18	85
	Pale chub	1	12	1	1	3	18
	Others	9	43	13	6	27	98
	All species	25	148	27	9	64	273

Another major category of problems involved flow discharge through ladders. Discharge was considered too low in 15% of the ladders and too high in 1%. In 11%, the period of use by fish was too short because discharge adjustment systems were lacking. Facilities for adjusting discharge have been installed on 23% of ladders.

The remaining problems related mainly to ladder size and proportions. Most ladders were 5–19 m long and 1–2 m wide (4% of stream width). Slopes ranged from 1:4 to 1:8. Slope was too steep or steps too high in 12.6% of ladders. Ladder width was thought to be too small in relation to stream channel width at 9.2% of the sites. Other problems existed in 7.1% of the ladders.

Field Observations and Recommendations

Pool and Riffle Improvements

Improvement of flow pattern.—When a flood prevention project or dam construction is completed in the upper reach of a stream, normal discharge in the downstream reach generally decreases and becomes less variable, so that the downstream riverbed sometimes becomes flattened by the stabilization of flow and the subsidiary channel reconstructions. The less-variable stream therefore becomes unfavorably shallow for fish. Observed remedial measures taken by local cooperatives include (a) channelizing braided channels into single, deeper channels, (b) dredging sediments from pools after channelization, (c) installing groups of large rocks on the streambed, (d) narrowing and deepening broad, shallow streambeds, and (e) moving channels to better locations (Figure 4). Some of these measures, because they decrease the area of wetted streambed, may reduce production of benthic algae, which could be disadvantageous for fish that feed on epilithic algae. However, the higher velocity of flow would prevent the attachment of fine sediment to rocks, so that the quality of algae may be improved. Thus the holding habitat and fishability of the stream for fish that feed on epilithic algae of high quality, such as the popular, high-priced ayu, will be improved.

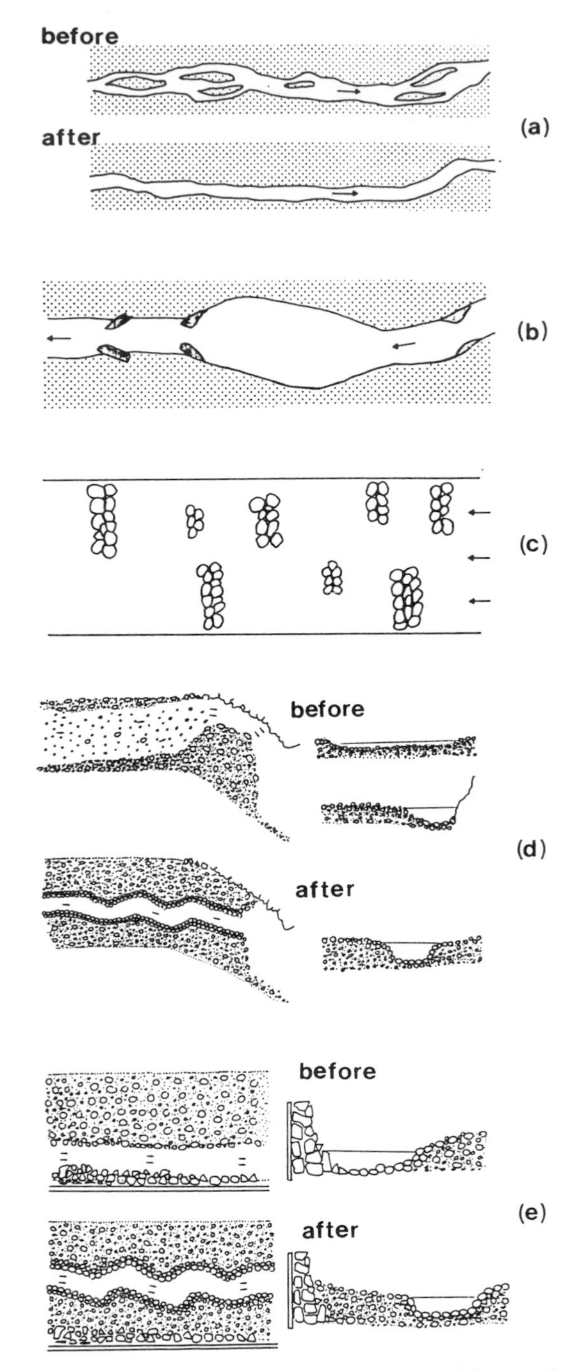

FIGURE 4.—Artificial alterations of channel form and flow patterns used in Japanese streams. Panels (a)–(e) are explained in text.

For success of these measures, several conditions must be met: there must be abundant supply of large stones nearby that are neither buried by fine sediment nor moved during high water, destructive high water must not occur at frequent intervals, and a large pool must exist nearby as holding habitat for ayu. Because, in many streams, these conditions do not usually exist, the above improvements often cannot be applied easily. For prolonged service, the stones can be placed in wire gabion baskets or in log frames near riffles so that they form sills.

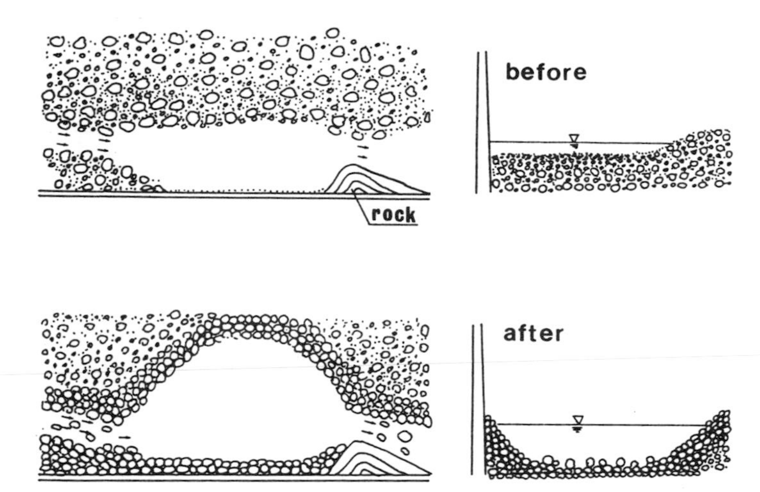

FIGURE 5.—Example of artificial pool enhancement in Japanese streams.

When channelization is done, it must be suited to frequent high-water flow. A deep, meandering channel pattern should be retained wherever possible, because a meandering flow pattern naturally forms riffles and pools over a long period. After channel alteration, pools are more readily lost than riffles, due to the movement of bed materials from riffles in high-water periods. Therefore, the manipulation of riffles should be done carefully so that pools are not damaged.

Construction of artificial pools.—Pools can be enlarged by excavation, then reinforced with rock to retain the shape (Figure 5), as was done by a local cooperative in Hiroshima prefecture. A key to the success of this project was to select a site where the pool formation was enhanced naturally by river characteristics.

In building channel side-wall revetments in streams, it is important that revetment on the outer, current-bearing side of a pool has low footings to prevent loss of pool depth (Figure 6; Mizuno 1980). This idea was adopted in the Saijyo River, which has a width of about 20–30 m, in Hiroshima prefecture. This will be more costly

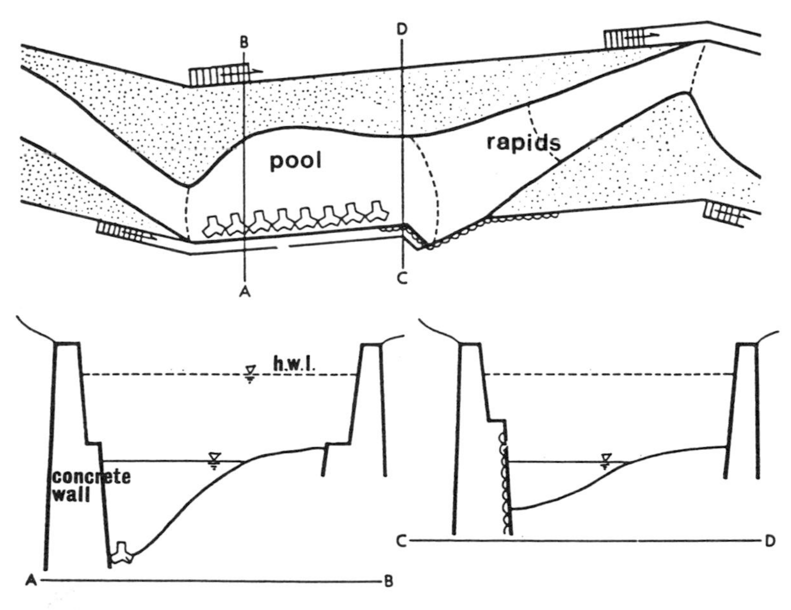

FIGURE 6.—Protection of a revetment toe area in a deep pool (vertical scale not exaggerated). Broken lines indicate high-water line (h.w.l.).

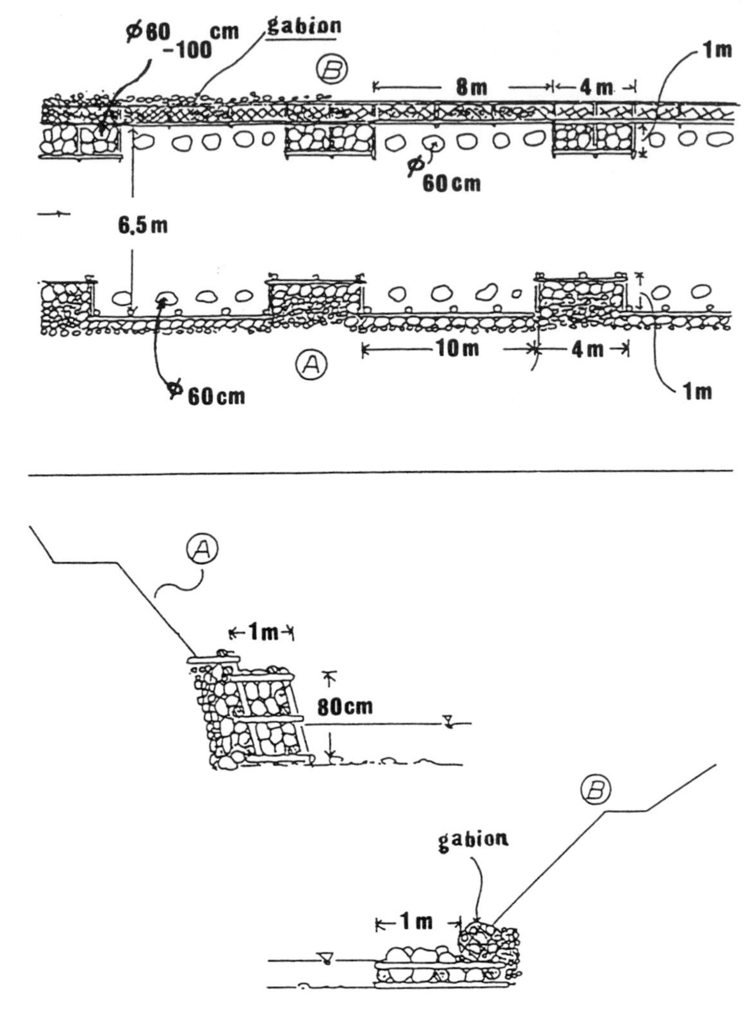

FIGURE 7.—Reinforcement of streambanks with log frames and rock structures (gabions).

than installing shallower footings, of course, but should be a requirement because preservation of the environment for fish is desirable.

Improvement of stream margins and banks.—To improve stream edges for fish, prefabricated concrete fish blocks can be incorporated in revetment walls (Figure 3). The open spaces within the blocks provide resting and hiding cover for fish. The large blocks with large (approximately 0.5-m^3) inner spaces installed in the revetment walls in the Ohta River, Hiroshima prefecture, gained a good reputation from the local cooperatives. Small blocks with small inner spaces often are occupied by a single species of fish and sometimes accumulations of sediment fill the spaces. Therefore, large blocks with large inner spaces are recommended. If one purpose of the installation is to provide refuge for fish during high waters, the blocks must be built up as high as the water will reach.

Also, various kinds of rock structures can be installed in ways that not only revet streambanks, but also provide fish habitat; these include rock-and-log frameworks or wire-mesh gabion baskets, as well as boulders placed on the streambed at the toe of the bank (Figure 7). The Nohgu River conservation project by Nagano prefectural government adopted this idea for a narrow (5–10-m wide) and straight stream. Reinforcing streambanks with artificial structures to protect human property has often damaged fisheries. Channel engineering to prevent disaster need not always conflict with fish habitat values, however. With proper knowledge and planning, it may often be

FIGURE 8.—Aerial view of an artificial spawning channel for ayu, built as a tributary to Lake Biwa.

possible for bank structures to satisfy both objectives, as the Nohgu River project proves.

Vegetation commonly occurs on streambanks. This benefits fish and can be extended onto revetments by filling the gaps in the facing blocks or rocks with soil. The effects of increased flow resistance caused by the vegetation should, however, be considered in advance. Slowly flowing parts of streams may be most suitable for vegetational enhancement.

Spawning Areas

Artificial spawning areas.—The largest Japanese project to create a fish-spawning area was the construction in 1981 of two spawning streams flowing into Lake Biwa for ayu (Figure 8 shows one, Ishida 1982). Since then, these artificial channels have almost annually produced more than half of the ayu larvae for Lake Biwa, except in years when exceptionally high numbers of larvae were supplied from natural streams tributary to the lake (NFIWFC 1987).

Construction and use of artificial spawning beds vary according to the species of fish. For example, spawning beds for Japanese dace are built, in Miyagi prefecture, by installing a bed of 2–4-cm gravel just downstream from an incline made of stones about 20 cm in diameter (Figure 9). Because the spawning beds are easily destroyed during high water, the period of construction of the bed is carefully chosen with respect to the flow condition and the spawning period of the fish. For fishes that spawn on aquatic plants, floating artificial spawning beds with living grass can be installed (Figure 10). The floating spawning beds are used mainly in some lakes in the central island of Japan.

Various species of fish can be benefited by preserving, improving, or building suitable spawning habitat. Because an artificial spawning area must be suited both to the characteristics of the species and to conditions of the stream, it is difficult to create artificial spawning areas for every important species in any one stream. Where engineering of stream channels for flood control, water supply, and other purposes cannot be

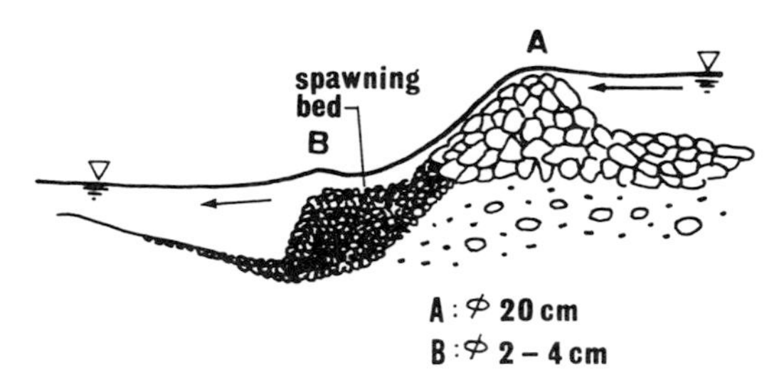

FIGURE 9.—Artificial spawning bed for Japanese dace.

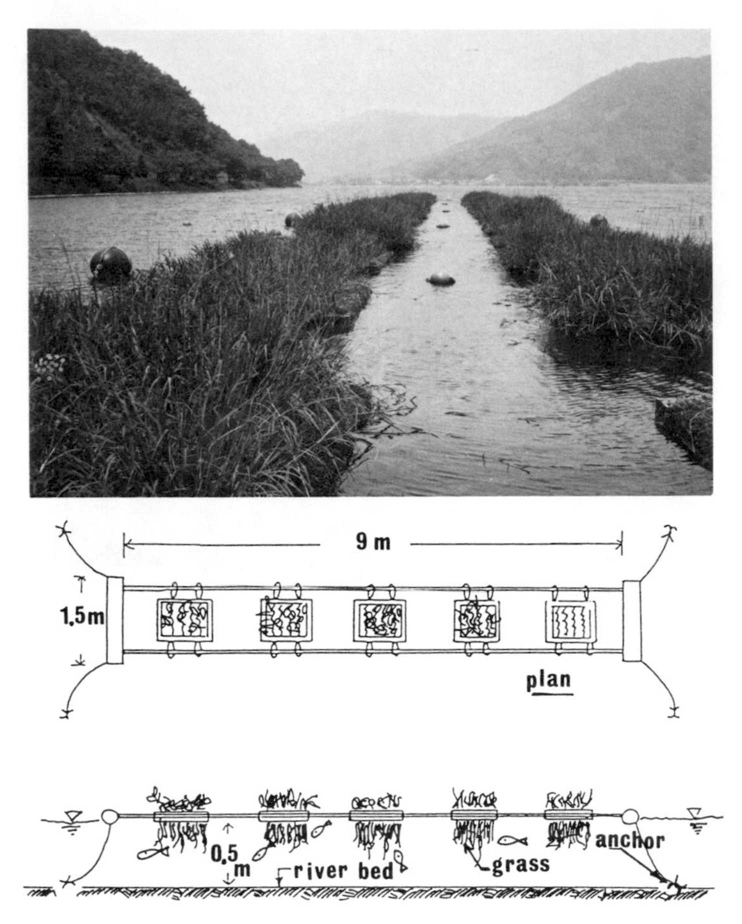

FIGURE 10.—Floating artificial spawning bed for moroko. The vegetation in the frames is live reeds or willows.

avoided, it may often be important to mitigate by artificial measures to assist the spawning of fish. In some cases, measures designed to improve riffles and pools as fish habitat may also improve conditions for spawning.

Fish Ladders

The dimensions, shape, and location of ladders should be designed for maximum effectiveness. Entrances to ladders should be such that every migrating fish can find and enter them easily. Discharge of water through ladders should be adequate for passage of fish at all times during the migration seasons of most species.

The design of new ladders to meet the above criteria will usually be easier than retrofitting existing ladders. Finding acceptable locations may often be difficult because ladders can be sources of resistance during high flows. However, the problem may be solved by installing a temporary ladder on a permanent fishway slope (Figure 11) where this is permissible, even though the idea is not adopted yet in Japanese rivers (Nakamura and Yotsukura 1987).

Special passage problems exist for young fish. One underlying reason has been that fishermen throughout a stream or river system often have not perceived their common interest and cooperated in resolving the fish-passage problem. Fishers of the upper reaches of stream systems tend to want sophisticated fish ladders, whereas those downstream tend to be less concerned about this.

In the past, many fish ladders were built because of high interest in ayu fishing. Most of these old fish ladders did not serve the intended purpose well because they were imitations of European designs suitable only for fish larger than ayu. Recently, the ayu fishery has had a lessening dependence on the migration of wild fish, as it has become increasingly based on artificial stocking

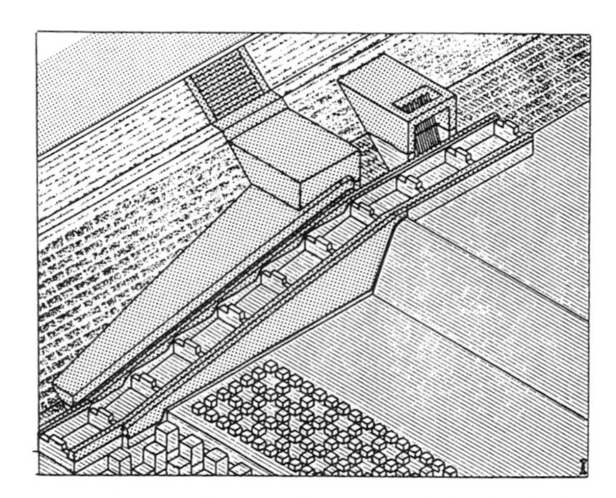

FIGURE 11.—A permanent fishway slope (left) with its temporary ladder (right), which is installed only during upstream migrations of fish.

(most of the recent rise in the cost of stocking streams [Figure 2] has been from an increase in number of ayu stocking sites). The trend toward management of ayu by stocking has decreased the demand for good fish ladders.

Even the stocked ayu migrate downstream in autumn to spawn, and many natually produced juveniles begin upstream migration in springtime. Moreover, almost all freshwater fishes, as well as shrimp and crabs, migrate at some time for some reason. Good fish ladders are therefore still strongly recommended for the maintenance and improvement of inland fisheries.

Acknowledgments

This paper is an extension of reports compiled by the chairmen of four subcommittees of a study group organized by NFIWFC and funded by the Fishery Agency of Japan in 1985 and 1986. Other members of the subcommittees include the following researchers: Yoshihisa Furuta, Fusao Goto, Kunio Watanabe, Wataru Tachikawa, Makoto Murata, Yasuyoshi Murakami, Shirou Tazaki, Takayoshi Yamazaki, Kenji Naka, Atsuyoshi Suzuki, Shigeo Koyama, Toshiaki Yada, Shigeru Satoh, Katsuji Shibuya, Toshio Satoh, Shinichi Wakabayashi, and Yuji Tsuno. We express our gratitude to NFIWFC and the Fishery Agency.

References

Fishery Agency (of Japan). 1987. Data book of inland fisheries in Japan 5. Ministry of Agriculture, Forestry and Fisheries, Tokyo.

Ishida, R. 1982. Artificial spawning streams flowing into Lake Biwa. Sakana 29, Tokyo.

Kawai, T. 1986. Fish program for the twenty-first century. Nosangyoson Bunka Kyokai, Tokyo.

Ministry of Agriculture, Forestry and Fisheries (of Japan). 1980–1985. Report on fisheries statistics of Japan 7, 8, and 10. Tokyo.

Ministry of Agriculture, Forestry and Fisheries (of Japan). 1986. Fisheries statistics of Japan. Tokyo.

Mizuno, N. 1980. Some problems and proposals on river improvement. Tansuigyo 6:1–7, Tokyo.

Nakamura, S., and N. Yotsukura. 1987. On the design of fish ladders for juvenile fish in Japan. Pages 499–508 *in* A. R. Kia and M. Albertson, editors. Design of hydraulic structures. Colorado State University, Fort Collins.

NFIWFC (National Federation of Inland Water Fisheries Cooperatives). 1987. Report of investigation of the river environment for fish production in Japanese rivers. NFIWFC, Tokyo.

River Bureau (of Japan). 1967–1986. Data of Japanese rivers. Ministry of Construction, Tokyo.

River Bureau (of Japan). 1986. Rule book of dam control. Ministry of Construction, Tokyo.

American Fisheries Society Symposium 10:43, 1991

HABITATS

Introduction

Part of the Fisheries Bioengineering Symposium was organized to deal with solutions to physical problems of fish in developed river systems—streams reshaped by humans during such landscape-altering activities as agriculture, urbiculture, road building, and damming—as well as in-channel manipulations intended to benefit fish. Authors were asked to address wild (naturally propagated) fish, which most fish in most streams are, but most of the information applies also to hatchery-produced fish that become part of stream communities.

Because this section concerns the environment of fish, we could aptly label it "fisheries *eco*engineering" rather than "fisheries bioengineering." If done with understanding, ecoengineering is applied ecology, like forestry and open-land agriculture. As such, it intervenes less directly in the target organisms' biology than does bioengineering by the standard definition of that term: the application of engineering principles and techniques to problems in medicine and biology (e.g., the design and production of artificial limbs and organs or the splicing of genes). Whether bio- or ecoengineering, the implied intent is to *benefit organisms,* namely fish—and ultimately, via fishing, us—by altering their physical environment in streams. Implied also, but not always, is an intent, in doing this, to *use organisms,* such as riparian plants and beavers, to create physical effects. This is emphasized in two papers (Beschta; Sedell and Beschta) and touched upon in a few others.

This section concentrates on salmonid fishes, though not exclusively. It deals with a variety of life stages, stream types, physical problems, and resource management outlooks. The problems considered for rather natural North American and European streams may not be as severe as posed by the intensively developed Japanese streams described by Nakamura et al. in the previous section. Nevertheless, among the papers is material geared not only to habitat management biologists and engineers, who design and carry out stream alterations and who evaluate the effects, but also to administrators and policy makers. It is not material for the stream improvement hobbyist.

As is fitting for practical symposia, many of the papers focus on techniques for manipulating stream conditions. Some present analyses of conditions to be altered or accomodated—or describe how to model or otherwise analyze them. Still others deal with the principles of design, of evaluating results, and of setting project objectives. Some authors present technical papers with detailed evidence and analyses. Others have written more general reviews or simply reports of methods that have worked (or not worked) to their satisfaction and that other professionals may wish to try (or avoid). Here and there, some promising speculative suggestions are offered. With the proviso that untested techniques should be used with professional caution, collegial pot-boiling and exchange of practical tips can be very useful. They are fully in keeping with the true spirit of a symposium, which, after all, means "a drinking together."

RAY J. WHITE
Edmonds, Washington

American Fisheries Society Symposium 10:44–52, 1991

Objectives Should Dictate Methods in Managing Stream Habitat for Fish

RAY J. WHITE[1]

Department of Biology, Montana State University, Bozeman, Montana 59717, USA

Abstract.—Objectives are desired future results of management. Effective planning of stream habitat management begins with broad, long-term objectives, such as "good fishing" and "health of stream and riparian ecosystems," which provide indefinite but fundamental guidance. Following from these is a continuum of intermediate and specific objectives, which are subordinate and contribute to attainment of the broad objectives. In habitat management, as in other fields, losing sight of (or never developing) broad objectives is common, often because of overorientation toward methods. Selection of methods should derive from objectives, not the other way around. Specific objectives, hence methods, while being consistent with broad objectives, should also be based on the fundamentals of ecology and physical science, on needs of the target species and of the ecosystem's full biotic community, on character of the specific stream and riparian area, and on values held by fishers. Fish abundance, angler satisfaction, and ecosystem health can probably be maximized and balanced with each other by promoting the physical and biological features and processes that occur and operate most often in nature—the stream attributes to which fish are adapted. As programs progress, objectives and methods can be improved. Objectives should be limited in number and mutually compatible, but a full range of objectives helps prioritize management phases and keeps managers aware of related parts of larger ecosystems; it is important to protect and restore floodplains and riparian zones before working on the channel. Channel stabilization is an objective that may often have been carried too far.

Confusion about objectives often reduces effectiveness of stream habitat management. Misplaced purpose or lack of purpose can plague professional managers (and amateurs), whether their training has been in biology, engineering, or other fields. It often stems from poor knowledge about basic ecology and about hydrology, geomorphology, and other physical sciences—in short, about streams and how fish and other organisms live in them. It also comes from ignorance of recreational or commercial fishery values, from inability to analyze habitat problems, and from the tendency to focus on techniques rather than on needs and results. Although know-how is important, knowing *why* is far more important. The literature of stream habitat improvement deals mainly with the hows, namely how to build structures. Such manuals may also have a few words about effects of structures on channel form and flow and may mention streambank vegetational management, but they rarely explain the supposed benefits for fish populations or for fishing.

Not only in stream habitat work, but in other branches of fishery management, the methods or technical operations themselves have all too often been regarded as objectives. Means have become ends, and programs have lurched off in unproductive and detrimental directions—tactics proceeding without strategy. Witness inward-looking fish propagation that produced domesticated fish unsuited for natural habitats, and witness a century of naive, wasteful, and sometimes destructive stocking with exotic fishes—in contrast to benefits and successes of some other fish propagation and stocking programs. Witness also many damaging stream engineerings, some done in the name of fish habitat improvement but with little or no knowledge of the channel features to which fish are naturally adapted.

The purpose of this paper is to emphasize the need for setting proper objectives in stream habitat management, especially objectives embodying a broad sense of purpose. More complete discussion of principles and of the complex analyses and strategies for ecological problem solving is to be found in such recent literature as that by Walters and Hilborn (1978), Orians et al. (1986), and Walters (1986).

What is a Management Objective?

A management objective is a desirable future result of a planned action. Management objectives (hereafter often termed simply objectives) include the same concepts often represented in goals,

[1]Present address: 320 12th Avenue North, Edmonds, Washington 98020, USA.

aims, targets, and the like. In recent business management literature, the term objective is advocated for all such concepts; this minimizes semantic confusion resulting from, for example, some people using objectives to mean simple, short-term results that are steps toward a larger goal, whereas others use the two terms in the reverse (Burke 1983).

The Continuum of Management Objectives

Within any area of management, objectives exist along a continuum of definiteness. At the indefinite end of the continuum are *broad* management objectives, expressions of desired future conditions (e.g., ecosystem health) that are supremely important but about which such vagueness exists as to make it impossible to say when the results have been achieved in the ultimate. At the other end of the scale are *specific* management objectives, narrow in scope and definite in method and schedule. Conception and design of successful management start with broad objectives and work downward through objectives that are more and more specific.

Broad Management Objectives

Broad objectives, the vague but important conditions, tend to involve long-term results, have philosophical implications, contain unstated complexity, and not be quantifiable. They are generally considered to be higher objectives. In the ecosystem health example, as in the health of any entity, be it an organism, a community, an ecosystem, or an institution, one can always wonder if things might be even healthier. Health of any system implies (1) operation near its inherent potential, (2) long-term stability, (3) resilience, and (4) possession of natural self-healing powers (J. C. Neess, unpublished talk, University of Wisconsin Extension Conference on Lake Protection and Management, Madison, 1974 [available from Ray J. White]).

Health of aquatic ecosystems is an objective toward which to strive in stream habitat work, but the main one is usually good fishing. In recreational fishing, the quality depends on human cultural tastes and satisfactions. Angling is a cultural activity; intangibles make its objectives hard to define and elusive to attain. As managers are often reminded, no matter how good fishing is, some folks will think it ought to be even better. Satisfactory angling means different things to different people. Generally, it involves fishing in natural, uncrowded settings and catching not only many fish per unit time, but also fish of desired species, sizes, appearance, and flavor. Besides catching large fish, many anglers also like having a diversity of sizes and perhaps of species to "fish over"; the adventure of not knowing what will be hooked next often makes fishing more fun. Chances to develop and test skills in meeting challenges of stream, fish, and gear also contribute to satisfaction. How many biologists, engineers, and others involved in stream habitat bioengineering adequately appreciate these recreational values?

Broad objectives tell us where we are going. It is a standard remark that if we don't know where we are going, almost any road will get us there. In the talk just cited above, John Neess suggested existence of a corollary in some resource management agencies that "knowing where we are going would impose unacceptable constraints on our choice of route."

Specific Management Objectives

Specific objectives, also called short-term objectives, are statements of definite, quantifiable results to be accomplished according to clear-cut procedures by a certain time. Some organizations distinguish between two kinds of specific objectives: results objectives, which state results to be achieved in 1–10 years but do not specify methods, and action objectives, which are shorter term and describe methods (day-to-day operations). An example of a results objective would be "achievement in overgrazed parts of Clear Creek, within 3 years, such restoration of streambank vegetative cover and channel form as to harbor 100 over-40-cm cutthroat trout per stream kilometer." Within this objective, an action objective might be that "during 1990, 5.2 km of stream will be fenced and special riparian grazing strategies will be started on pastures bordering 35 km of stream." Obviously, the results type of specific objectives are more intermediate on the continuum of definiteness than are the action objectives. It may be characteristic of intermediate objectives in habitat management that they deal largely with ecological principles, such as succession, diversity, interspersion, and needs for moderate perturbation (Odum et al. 1979). Succession is implicit in the above example. Within any program, intermediate and specific objectives must be compatible with the broad objectives and are usually considered essential for their attainment (Burke 1983).

Within the system of broad, intermediate, and

specific objectives involved in any major management program, possibilities for straying and perversion occur during the bustle of operations. The most common dangers are probably (1) becoming confused about which objectives are broad and which are specific—or about their priorities—and (2) losing sight of the broad objectives altogether. Also, resource-oriented objectives can be obfuscated by politically oriented objectives ranging from broad to specific. The tempering of resource-oriented objectives by reasonable political constraints is a common, essential process in democracy, and certain politically oriented intermediate objectives are often needed for accomplishing long-range resource objectives. But political interference in management is a problem. For example, legislatures sometimes allot funds with the stipulation that the agency use them for a specific management procedure in a specific stream before it is understood what needs to be done in that stream—or if anything needs to be done. This amounts to telling the managers not what should be accomplished, but how they should do it; the results are not likely to be compatible with resource needs.

Many people are uncomfortable with broad objectives and seek refuge in performing solely the technical details of specific objectives, but even dyed-in-the-wool technicians can be more effective if made aware of the broader objectives. One way to keep everyone on track is to have each program's and project's broad, intermediate, and specific objectives clearly and formally stated in writing, and to keep these written statements prominently displayed for all personnel involved, as well as for the public or other clients, to see and consider during all stages of planning and execution. Post the objectives at project sites and other workplaces. Mention them at meetings and in correspondence. State them in reports.

Objective-Oriented Management

Objectives are unlikely to be attained in any organization without leadership and commitment. Administrative systems focused on objectives and containing feedback communication can provide a framework for such leadership and help engender a sense of commitment among all personnel involved.

The management-by-objectives (MBO) approach (Drucker 1954 and subsequent publications; Odiorne 1965) includes a procedure of measuring results against objectives by continually reviewing objectives during all phases of a project, and by stating in progress reports the extent to which each objective has been reached. The reporting should also reveal how progress toward specific objectives is satisfying the broader objectives. Management by objective has not worked well in at least one major federal natural resource agency, but the observation has been made that it was not properly administered there (J. Addis, Wisconsin Department of Natural Resources, personal communication).

The foregoing descriptions of broad to specific objectives and of MBO may imply top-down objective setting but should not imply one-way communication during the process of objective setting. Without proper flow of information and opinion between all levels of personnel (and with other parties involved, including researchers and resource users), the dangers are that top-level administrators will be cut off from information about the resource and that field-level managers, researchers, and fishers will not be aware of the big picture and how they function in it. R. J. Behnke (Colorado State University, personal communication) has observed that in the USA, recent progress toward management focused on wild trout has been achieved more from the bottom up. The process may often have been impeded or perverted by upper-level administrators of state and federal agencies.

In any system for stream habitat management, not only should the role of objective setting be recognized, but also the need for fine-tuning, adding, and discarding objectives as programs proceed. Applicable here is Bailey's (1982) discussion of "cyclic incremental" approaches for readapting objectives and methods as better knowledge develops or conditions change during wildlife management programs; this constitutes the creative "muddling through" so often needed in solving complex ecologic problems. Stream managers who are forced, for example, by funding circumstances, to begin projects before they can gather as much information about the stream and its fishery as would be optimal, can, at the outset, choose the best objectives possible, then in later years improve objectives and procedures according to more detailed analysis of the stream and evaluation of work in progress.

Dangers in Not Recognizing Broad Objectives

In managing fish habitat, it is important for us to recognize and feel we are working toward broad objectives, of which specific habitat objectives (and indeed habitat management itself) are only parts. Failure to do this is probably common to most fields. In medicine, not seeing the forest of the whole patient's health for the trees of diseases and therapies is legendary (Stalker and Glymour 1985), and such a narrow perspective may have caused failure of railway companies that viewed themselves as running railroads, not as providing transportation. Likewise, stream habitat managers fail to properly serve the resource and society if they regard habitat therapies as their be-all and end-all, and if they lose (or never develop) sight of such overall objectives as good fishing, satisfactory recreation or food, and ecosystem health—the "whys" of management.

Pitfalls of working toward narrow, technique-oriented objectives without adequately appreciating broad objectives are illustrated by the State of Wisconsin's fishery management program in 1958–1960. I pick that place and time because I was involved in the situation; also, although Wisconsin's trout habitat management was then new and understandably groping for direction, a sound basis of conservation philosophy and outdoor recreational philosophy existed in that state, and forces were at work that would soon iron out many of the problems. Conditions were worse in many other states then, and similar problems still exist in some places.

Wisconsin's trout stream management at that time involved traditional restriction of angling harvest and stocking of hatchery-reared trout, as well as a recently launched habitat improvement program. There were also vigorous programs to protect streams from such abuses as damming, channelization, irrigation withdrawals, and pollution. Annual trout fishing seasons were set, and there were uniform statewide minimum size and daily creel limits. The size limit for trout was 6 inches (152 mm), the snout-to-tail-tip length of yearlings produced in hatcheries by the April opening of the fishing season. The trout propagation–stocking program had, in the mid-1940s, switched emphasis to put-and-take stocking of yearlings and away from planting fry and fingerling subyearlings, which had been thought of as seed for growth to harvestable size. (Since the late 1960s, subyearling stocking has been selectively reinstituted, where stream quality will sustain put-*grow*-and-take programs.)

The stream habitat effort was a post-World War II program derived from increasing general awareness of the importance of habitat to fish and wildlife and from urging by the state's growing cadre of college-trained aquatic and wildlife biologists and by sportsmen's groups. People who had been exposed to ecology in college or had worked in federal stream habitat programs of the Depression Era returned from the war and pressured the state to undertake habitat work.

In 1958–1960, the Wisconsin Fish Management Division was structured largely according to technical activities, each major activity having its own semiautonomous section. Such "organization by activity" tended to promote "management by activity" rather than management according to resource or recreational objectives. A fish-propagation system of 12 trout hatcheries was administered by an assistant superintendent for hatcheries. A habitat improvement unit, headed by another central administrator, had a habitat manager and crew in each of the five fishery administrative areas of the state. Rough-fish control was conducted by crews at four permanent stations and one seasonal camp; this program was warmwater-oriented and stemmed from problems with overabundance of common carp *Cyprinus carpio*. The fish propagation and habitat improvement programs dealt almost entirely with trout.

In overall trout stream management, problems deriving from organization by activity were exacerbated by the mind-set of most personnel, who were not oriented toward ecology and trout fishing. (The key to the manager's effectiveness lies in how he *thinks* about streams and the fisheries in them—his insights into the functioning of streams and the behavior of trout—in sum, his ecologic understanding. This understanding is affected by formal training, but also strongly by experience.) Few of the personnel were then biologists, fewer were biologists qualified in ecology, and perhaps most importantly, even fewer, whether biologist or not, were trout anglers.[2] Of 30 supervisory personnel responsible for hatchery and habitat operations, only 6 were biologists, and only 3 or 4, by being trout fishermen, could have genuinely

[2]The superintendent of the Fish Management Division was a biologist but not an angler. Among the administrator and 12 trout hatchery foremen of the propagation system, none was a biologist and only 1, a foreman, was a trout angler. The head of the habitat program was a biologist but not a trout fisherman, and of his 5 area habitat managers, none was a biologist and only 1 took part in the sport he was supposed to be benefiting. Of 10 area-level administrators in the direct hierarchy above hatchery and habitat operations, 4 were biologists, and only 1 or 2 were trout fishermen. Other professionals involved in trout stream management—5 area fish management biologists and about 15 multicounty district fish managers (some of whom were biologists)—had no line authority over hatchery and habitat operations, and most of these people did not greatly influence the quality of trout furnished by the hatcheries or of the habitat work done in streams.

appreciated the recreational objectives toward which they were—or should have been—working, although at least one of the non-trout-fishing habitat managers had a good feel for stream habitat anyway.

Good fishing was mentioned or implied now and then but was largely taken for granted and apparently was not really thought about much. In the agency's biennial report covering 1957 and 1958, the section on the Fish Management Division (Wisconsin Conservation Department 1959) was introduced as follows:

> . . .the objective of this division is fish management. And what is management? Management is giving Wisconsin's citizens and vacationists the opportunity to harvest the fishery resource in amounts and ways which will give optimum sustained yields year after year. To meet this objective here are the division's broad tasks: (1) provide fishing opportunities for anglers, (2) develop regulations for satisfactory harvest, (3) improve on or preserve the fish's habitat, (4) and maintain balanced fish populations.

The report went on to describe technical operations performed to accomplish the tasks, but it did not explain what would constitute optimum sustained yields, satisfactory harvest, or balanced fish populations.

In those days, habitat management personnel talked of watershed programs, erosion control, and streambank stabilization; hence they were wrestling implicitly with concepts of biotic community health, but emphasis was on operational details of programs. Somehow, thinking did not gel with regard to broad objectives that might have truly guided trout stream management.

That this situation existed in a state containing 40% of the Midwest's trout stream resource (14,500 of 36,000 stream kilometers) was remarkable. That it existed in the state where Aldo Leopold had, until 10–12 years before, done his most insightful teaching and writing, was ironic. Leopold (1949) had eloquently and publicly espoused perception of outdoor recreational values, including those of trout fishing per se, as well as natural resource management objectives involving ecosystem integrity. He had also served as a member of the State Conservation Commission. Leopold had filled a trough of broad objectives under the the very nose of fishery management, but after more than a decade, the leadership had not begun to drink from it.

Because the system focused poorly on the objective of good trout fishing, it floundered. The trout management effort had long been dominated by fish culturists. Their hatcheries were run for the sake of fish production rather than good fishing. Their trout, bred and fed primarily for performance in the hatchery, were, when stocked out, inferior in size and coloration to anglers' expectations and were so ill-suited to life in streams that few survived long enough in the wild to attain satisfactory size and appearance. About 40% of yearlings stocked 2–4 weeks before the spring opening of the fishing season died before anglers could get at them (Brynildson and Christenson 1961). Less than 1% of domesticated strains of brook trout survived a year after stocking in streams (Mason et al. 1967).

With respect to stream habitat work, the situation was generally better but not without major problems. The fledgling program consisted largely of applying techniques found in the literature of federal agencies and of other states. Many of the methods were inappropriate for Wisconsin conditions; some were inappropriate to conditions anywhere. Sometimes this cookbook approach met needs, sometimes not. Some habitat managers became infatuated with certain methods of reshaping channels and making the water flow this way and that without knowing what it would do for trout or angling. In one creek, a habitat crew used a dragline to dig deep, wide, pondlike "pools" (derided as "sucker holes" by local anglers) until work was halted by a biologist who was a trout fisherman and understood streams, fish, and angling values. In some other streams, large, ugly, vertical walls of planking (sheet piling) were built to deflect flow or retard bank erosion; these often deepened channels but did not provide cover. There were also innovations, some fanciful and having minor or detrimental effect, but others based on knowledge of trout habitat-selection behavior. The latter substantially benefited trout populations and angling (Hunt 1971; White 1975a; Hall and Baker 1982; White 1986). The most notable of the beneficial devices was a design for submerged bank overhangs that simulated those formed in nature, hence the structure had natural appearance and was highly effective as cover for trout—the "bank cover-wing" developed by R. B. Heding (White and Brynildson 1967); the designer was the state's only habitat manager at that time who was also a trout angler.

To sum up the Wisconsin example, in 1958–1960 trout stream management was bogged down in specific, technical (activity-oriented) objectives and lacked broad objectives. The program churned away at operations but did not know where it was going; thus it was not effective in achieving recreational satisfaction. It was small wonder that overt tension existed between trout anglers and fishery administrators. (An upper-level administrator once warned his staff to be on guard at an upcoming meeting because "avid trout fishermen" would be present.) Already, biologists, on the basis of field investigations, had begun to cull unsuccessful methods out of the habitat program. Research, begun in the 1950s, had by the 1960s generated ongoing improvement in both the habitat and stocking efforts. During the 1960s and 1970s, philosophical and organizational changes led to better integration of stocking and habitat work with overall stream fishery management.

Keeping Objectives Few and Compatible

We should avoid having objectives that are too many and too conflicting. Whenever, within a program, two or more objectives of about the same specificity exist, they will conflict, if for no other reason than that they compete for the manager's time. Having too many objectives, especially if not hierarchically organized with priorities kept clear, leads to diffusion of effort, confusion, inconsistency (switching of emphasis between objectives in midproject), and decreased likelihood of satisfactory results.

The classic example is the private pond owner who, egged on by government bulletins, tries futilely to have simultaneous fishing, livestock watering, bird-life enhancement, hunting, lawn watering, swimming, fire control, scenic beauty, and other benefits. Stream and riparian management are also subject to the myth of multiple use. Concurrently maximizing timber harvest, livestock production, farm crops, watercraft navigation, fish, wildlife, and homesite development is impossible. However, fostering *natural* physical and biological processes of streams, riparian zones, and floodplains should result in such habitat diversity as to benefit at least a modest range of uses. A way to examine which of several objectives may be compatible or conflicting is to place them in a two- or three-way matrix. Then those that are compatible can be integrated in a plan, and the needs for choices between conflicting objectives can be recognized, and the choices can be made (J. Addis, personal communication).

Ecological Appropriateness of Intermediate Objectives

Confusion about objectives and methods can derive from not specifying the kinds of fishes to be targeted by management. Each species, strain, life history stage, or body size has special habitat needs. Objectives in managing stream habitat for anadromous salmonids are often not the right ones for stream-resident salmonids, and vice versa. Picture a biologist charged with improving a spring creek of the Rocky Mountains, and who, having been trained on the Pacific Coast, creates large, open pools and immense spawning riffles. The result: vast numbers of young but few of the lunkers that anglers seek. It might be just as far off the mark to provide cover for large trout where the objective is to produce salmon smolts. Large cover for hiding and security is, however, apparently needed in anadromous salmonid streams. F. H. Everest (U.S. Forest Service, personal communication) has, during stream diving, observed steelhead *Oncorhynchus mykiss* and Pacific salmon spawners in close association with submerged large rocks, logs, and undercut banks.

Besides target species, the needs of the full community of fishes and other organisms should be heeded. Habitat management solely for a few target fishes risks harming habitat for other parts of the community. Jones (1987) asserted that the main impact of humans on ecosystems is simplification, and that the destruction of ecosystem diversity is detrimental to many organisms. Much of agriculture, aquaculture, and forestry is monocultural and purposefully makes habitat hospitable for one kind of organism and inhospitable for most others.

Not only are monocultural tendencies to be controlled in habitat management, but there may be more subtle habitat-density-related problems to deal with. Wildlife habitat managers have been urged not to create too much interspersion of small forest patches and clearings for the sake of edge effect. Often forgotten in the enthusiasm for this useful form of habitat bioengineering is Leopold's (1933) original description of edge effect: edges benefit some kinds of wildlife having relatively low mobility and requiring a variety of vegetational types, but there was no evidence that edges attract animals adapted to mobile existence in unbroken expanses of grasslands or oceans. Also, some organisms require large areas of uninterrupted forest and are harmed by edge-dwelling animals when woodland is fragmented to make more edge (Robbins 1979; Brittingham and Temple 1983; Alverson et al. 1988). I mention the overemphasis of edge effect in wildlife management to point up the potential for *over*management in any form of habitat work.

Hazards of negative edge effect may be minor in stream management because edge effect may be a primary natural characteristic of streams. Unless damaged, streams are, by their very nature, complex linear bundles of edges, edges between fast and slow current (shear zones), between deep and shallow water, between areas of differing bed material, between bed and bank, and between aquatic and riparian–terrestrial zones. The water surface itself is an edge where processes of vast importance to fish take place. Moreover, the tendency toward monotypic effects may be less in stream management than in some other forms of resource management, especially if we draw upon natural features and processes of streams as guides. Diverse stream features constituting habitat complexity may derive from meandering in low-gradient streams that have balanced in- and outflow of sediment, from the stair-stepped clogging by boulders and woody debris in steep erosive streams, and undoubtedly from other stream processes. Therefore, if we promote meandering in gently sloping streams and stair-stepping in steep streams, we will maintain or enhance habitat diversity, and an appropriate community of organisms should be benefited.

It may be a useful working hypothesis (hence an objective) that greatest fish abundance, greatest recreational satisfaction, and greatest ecosystem health will be achieved by promoting the physical and biological features and processes of streams and riparian areas that occur most frequently in nature. These must be the characteristics to which stream-dwelling fish are well-adapted and in which they will thrive. It follows that we should avoid creating physical features not common in nature—because it is improbable that fish will be adapted to such features. The meandering form and the habitat features associated with it may be the most common situation in streams, and it may be this set of conditions to which stream fishes are most closely adapted.

Physical features of streams include not only course pattern (meandering, irregular, or braided), related intermediate-scale morphometry (pools, riffles, boulder cascades, glides, backwaters, sand flats, etc.), and finer-scale shapes (bank undercuts and bed texture), but also hydrologic regimens, thermal regimens, hydraulic characteristics, and spatial and temporal patterns of light intensity. Some of the important physical features are due to organisms. Streambank vegetation, particularly woody plants, have important effects in and beside channels by binding soils, creating shade, and (mainly as debris) providing ledges, sills, dams, and textures (large and small roughness elements). Woody material, alive and dead, furnishes intermediate and fine-scale spatial compartmentation that fish use as hiding cover from predators, as visual isolation from competitors, and as refuge from current.

Regional and stream-to-stream differences exist in geomorphometry and climate, hence in hydrologic regimen, vegetation, and resulting channel shape. These have caused regional and stream-to-stream differences in fish communities and life histories. To be effective, the habitat manager must understand these matters. Descriptions of the physical effects of woody debris in Pacific coast streams and of its importance to salmonid abundance have become voluminous (e.g., in Lisle 1981; Meehan et al. 1984; Sullivan et al. 1987). Woody debris may be similarly important to stream fish of other regions. Salmonids elsewhere select habitat having complex structure (Hunt 1971; White 1975b; Enk 1977; Fausch and White 1981), as do also warmwater stream fishes (Fraser and Sise 1980; Probst et al. 1984; Schlosser 1987; Todd and Rabini 1989). The complexity of habitat structure is often, even in streams of arid regions, due to root masses, fallen logs, and accumulations of smaller woody debris. Adaptation to the kinds of microhabitat created by live and dead woody vegetation is likely a primary characteristic of stream fishes, and this should be incorporated in the objectives and methods of habitat management.

Objectives and the Broader Ecosystem

In stream habitat management, narrow focus on specific objectives has often led to (or derived from) narrow focus on the stream channel itself. Often disregarded are basic, interrelated problems of adjacent riparian land and of the parts of floodplains beyond the riparian vegetational zone. It should be continually recognized that these are parts of a greater stream ecosystem.

Doing in-channel habitat work without first protecting against or remedying human abuses of riparian and floodplain areas is a misprioritization of objectives. Often, restoration of damaged riparian ecosystems (sometimes involving removal of human development from floodplains) leads to natural improvement of fish habitat in the channel. Artificially building habitat in channels to fully mitigate human changes in floodplains and riparian zones can be very expensive. Less expensive, token, in-channel attempts at repayment for loss of riparian benefits to streams are likely to be ineffective Band-Aids. Greatest result for lowest cost can often come from removing the abuse and allowing natural restoration of a stream–riparian interface that is healthy, well-vegetated, and intermediately stable.

The most prevalent abuse to alleviate is probably riparian livestock grazing. In a 1988 survey of U.S. and Canadian fishery agencies, streambank grazing damage was cited as the primary trout stream habitat problem not only in the West, but also in the Midwest and East (my unpublished data). Much need also exists for controlling damage from other agriculture, logging, mining, organization, and transportation-route development in floodplains.

A dilemma for habitat managers can occur when they are asked to prevent channels from migrating. Moderate lateral wandering of streams is natural and plays roles in creating habitat for fish (e.g., gravel supply). However, people often want channel migration halted to protect roads, buildings, cropland, and other developments that exist on the floodplain. In the process of building habitat structures, habitat managers sometimes want to stabilize the channel so that it will not move away from the improvements. Overstabilizing streams and preventing floods where they naturally occur counters the principle that moderate environmental perturbation enhances biotic production (Odum et al. 1979; White 1991, this volume); long-term reduction of fish abundance may result. The "soft architecture" advocated by Sedell and Beschta (1991, this volume), should be useful in this regard. Two much-needed objectives in stream habitat work are to keep dynamism in the dynamic equilibrium of stream morphometry and to let floodplains be floodplains.

Conclusions

Aldo Leopold (1949) wrote that "To promote perception is the only truly creative part of recreational engineering." He was expressing a broader, higher objective than many past managers, as "recreational resource engineers," were willing to consider. Leopold's assertion and challenge apply to present efforts to meld biology and engineering, embodied in the symposium of which this paper is a part, because much of the reason for fishery bioengineering is recreation. A basic function of such conferences is for us to increase our technical know-how, but I submit that we will become even more effective by increasing our perceptivity of broad issues and objectives as well. Then we will know better where we are going before we select the methods to be used in projects of habitat improvement or other fishery management.

Acknowledgments

Lewis A. Posekany and Lyle M. Christenson supplied historical information from Wisconsin. Robert L. Hunt, Lyle M. Christenson, Kenneth Bates, James A. Posewitz, Robert W. Wiley, Gerry D. Taylor, and Alan Martin kindly reviewed drafts and provided comments and suggestions.

References

Alverson, W. S., D. M. Waller, and S. L. Solheim. 1988. Forests to deer: edge effects in northern Wisconsin. Conservation Biology 2:348–358.

Bailey, J. A. 1982. Implications of "muddling through" for wildlife management. Wildlife Society Bulletin 10:363–369.

Brittingham, M. C., and S. A. Temple. 1983. Have cowbirds caused forest birds to decline? BioScience 33:31–35.

Brynildson, O. M., and L. M. Christenson. 1961. Survival, yield, growth and coefficient of condition of hatchery-reared trout stocked in Wisconsin waters. Wisconsin Conservation Department, Miscellaneous Research Report 3 (Fisheries), Madison.

Burke, W. W. 1983. General management. Pages 1–78 *in* W. K. Fallon, editor. American Management Associations management handbook, 2nd edition. American Management Associations Special Projects Division, New York.

Drucker, P. F. 1954. The practice of management. Harper, New York.

Enk, M. D. 1977. Instream overhead bank cover and trout abundance in two Michigan streams. Master's thesis. Michigan State University, East Lansing.

Fausch, K. D., and R. J. White. 1981. Competition between brook trout (*Salvelinus fontinalis*) and brown trout (*Salmo trutta*) for positions in a Michigan stream. Canadian Journal of Fisheries and Aquatic Science 38:1220–1227.

Fraser, D. F., and T. E. Sise. 1980. Observation on minnows in a patchy environment: a test of a theory of habitat distribution. Ecology 61:790–797.

Hall, J. D., and C. O. Baker. 1982. Rehabilitating and enhancing stream habitat: 1. Review and evaluation. U.S. Forest Service General Technical Report PNW-138.

Hunt, R. L. 1971. Responses of a brook trout population to habitat development in Lawrence Creek. Wisconsin Department of Natural Resources Technical Bulletin 48.

Jones, G. E. 1987. The conservation of ecosystems and species. Croom Helm, London.

Leopold, A. 1933. Game management. Scribner, New York.

Leopold, A. 1949. A Sand County almanac and sketches here and there. Oxford University Press, New York.

Lisle, T. E. 1981. Roughness elements: a key resource to improve anadromous fish habitat. Pages 93–98 *in* T. J. Hassler, editor. Propagation, enhancement, and rehabilitation of anadromous salmonid habitat in the Pacific Northwest. Humboldt State University, California Cooperative Fishery Research Unit, Arcata.

Mason, J. W., O. M. Brynildson, and P. E. Degurse. 1967. Comparative survival of wild and domestic strains of brook trout in streams. Transactions of the American Fisheries Society 96:313–319.

Meehan, W. R., T. R. Terrill, Jr., and T. A. Hanley, editors. 1984. Fish and wildlife relationships in old-growth forests. American Institute of Fishery Research Biologists, Morehead City, North Carolina.

Odiorne, G. S. 1965. Management by objectives—a system of managerial leadership. Pitman, Belmont, California.

Odum, E. P., J. T. Finn, and E. H. Franz. 1979. Perturbation theory and the subsidy-stress gradient. BioScience 29:349–352.

Orians, G. H., and eight coauthors. 1986. Ecological knowledge and environmental problem-solving—concepts and case studies. National Academy Press, Washington, D.C.

Probst, W. E., C. F. Rabini, W. G. Covington, and R. E. Marteney. 1984. Resource use by stream-dwelling rock bass and smallmouth bass. Transactions of the American Fisheries Society 113:283–294.

Robbins, C. S. 1979. Effect of forest fragmentation on bird communities. Pages 198–212 *in* R. M. DeGraaf and K. E. Evans, editors. Management of north central and northeastern forests for non-game birds. U.S. Forest Service General Technical Report NC-51.

Schlosser, I. J. 1987. The role of predation in age- and size-related habitat use by stream fishes. Ecology 68:651–659.

Sedell, J. R., and R. L. Beschta. 1991. Bringing back the "bio" in bioengineering. American Fisheries Society Symposium 10:160–175.

Stalker, D., and C. Glymour, editors. 1985. Examining holistic medicine. Prometheus Books, Buffalo, New York.

Sullivan, K., T. E. Lisle, C. A. Dolloff, G. E. Grant, and L. M. Reid. 1987. Stream channels: the link between forests and fishes. University of Washington Institute of Forest Resources Contribution 57:39–97.

Todd, B. L., and C. F. Rabini. 1989. Movement and habitat use by stream-dwelling smallmouth bass. Transactions of the American Fisheries Society 118:229–242.

Walters, C. J. 1986. Adaptive management of renewable natural resources. Macmillan, New York.

Walters, C. J., and R. Hilborn. 1978. Ecological optimization and adaptive management. Annual Review of Ecology and Systematics 9:157–188.

White, R. J. 1975a. In-stream management for wild trout. Pages 48–58 *in* W. King, editor. Wild trout management. Trout Unlimited, Vienna, Virginia.

White, R. J. 1975b. Trout population responses to stream flow and habitat management in Big Roche-a-Cri Creek, Wisconsin. Internationalen Vereinigung für theoretische und angewandte Limnologie Verhandlungen 19:2469–2477.

White, R. J. 1986. Physical and biological aspects of stream habitat management for fish: the primacy of hiding/security cover. Pages 241–265 *in* J. C. Miller, J. A. Arway, and R. F. Carline, editors. Proceedings of the fifth trout stream habitat improvement workshop, Lock Haven, Pennsylvania. Pennsylvania Fish Commission, Harrisburg.

White, R. J. 1991. Resisted lateral scour in streams—its special importance to salmonid habitat and management. American Fisheries Society Symposium 10: 200–203.

White, R. J., and O. M. Brynildson. 1967. Guidelines for management of trout stream habitat in Wisconsin. Wisconsin Department of Natural Resources Technical Bulletin 39.

Wisconsin Conservation Department. 1959. Twenty-sixth biennial report of the Wisconsin Conservation Department for the fiscal years ending June 30, 1957 and June 30, 1958. Wisconsin Conservation Commission, Madison.

American Fisheries Society Symposium 10:53–58, 1991

Stream Habitat Management for Fish in the Northwestern United States: The Role of Riparian Vegetation

ROBERT L. BESCHTA

Department of Forest Engineering, Oregon State University
Corvallis, Oregon 97331, USA

Abstract.—Historical development and land-use patterns along streams draining forest and range watersheds in the northwestern USA have had major effects on riparian vegetation, channel characteristics, and fish habitat. The functional attributes of riparian vegetation that have been altered include the dissipation of stream energy and channel stability, stream shade and temperature control, nutrient cycling, sediment deposition and storage, water storage and release, and others. Recent attempts at enhancing degraded fish habitat include many bioengineering projects that are adding structures of various sizes, materials, and configurations to stream channels. However, a higher priority for the long-term improvement of fish habitat is the implementation of management practices that will allow and encourage the continued functioning and succession of riparian vegetation.

The effects of human activities on streams and streamside areas in the northwestern USA have received increased attention and scrutiny in recent years. For example, in 1988 the State of Oregon modified forest practices rules for streamside areas because of fish and wildlife concerns. Research efforts, such as the Coastal Oregon Productivity Enhancement program (COPE), have also been initiated to further improve our understanding of natural and anthropomorphic factors influencing the characteristics of riparian areas for fish and wildlife habitat, timber production, and water quality (Skaugset et al. 1989). Similarly, the State of Washington has embarked on the Timber, Fisheries and Wildlife program (TFW) in an attempt to resolve riparian issues and related management concerns associated with forested watersheds. In eastern Oregon, representatives of environmental groups, ranchers, and other interested parties have formed the Oregon Watershed Improvement Coalition (OWIC) to improve communications between groups with differing backgrounds and values, and to develop solutions to pressing riparian and watershed management issues.

Management strategies for streamside areas on public and private lands have begun to shift in recent years (Adams et al. 1988; Lee and Gross 1988) and continued change is expected. Because of the importance of these areas for a wide variety of values (e.g., water quality and quantity, aesthetics, fisheries, wildlife, forage, timber, and recreation), the general public is increasingly involved in streamside regulation and policy decisions. Thus, streams and adjacent lands have become a focal point of intense interest, and sometimes of conflict, for a diversity of groups. Resolving the often disparate views represents a major challenge for the future management of stream and riparian ecosystems (U.S. Forest Service 1985; Department of Ecology 1986).

Historical Overview

During exploration and early settlement by people of European descent, the northwestern USA was generally considered to be endlessly endowed with natural resources. Historical records document the belief that vast supplies of agricultural and grazing lands, timber resources, and minerals awaited development. As these resources became increasingly utilized, many streams, rivers, and adjacent riparian ecosystems were significantly altered.

Perhaps the first major change brought to western streams began with the trapping of beavers. Although meticulous records of pelt production were maintained by trading companies, little was written about the effects on streams of depleted or eradicated beaver populations. Evidence from other geographic areas (Parker et al. 1985; Naiman et al. 1986, 1988) indicates that beavers, often referred to as "nature's engineers," must have had an important influence on the character and functioning of many Pacific Northwest streams and riparian areas. Their dams stored sediment and nutrients, promoted saturation of floodplain soils at high flow, and created significant fish and wildlife habitat. Over long periods of time, beaver dams may have caused continual deposition of fine sediments in alluvial valleys.

Although periodic breaching of some beaver dams during high flow may have caused local channel scour, the overall effect of beaver activity was to generally enhance wetland–riparian functions and values. Because beaver dams promoted the storage of water in alluvial and floodplain soils during high flow, the slow release of this detained moisture helped sustain summertime flow for many streams.

Reductions in beaver populations were soon followed by increased grazing and agricultural development. The high productivity of many streamside areas in eastern Oregon and Washington and other rangeland areas throughout the West, having resulted from thousands of years of sediment and nutrient accumulation, was quickly recognized and utilized. In many areas, grazing and agricultural development led to the eradication of beavers, removal of tree and shrub vegetation, and changes of drainage patterns. Channels were often cleaned and straightened in attempts to reduce "flooding." They were also modified for irrigation, roads, and other cultural developments. In still other instances, flow regimes were modified by upstream dams and water diversions. Over time, the diversity and complexity of many riparian plant communities decreased and their stream systems became increasingly simplified.

The alteration of streamside vegetation, in addition to any upslope land uses that may have increased surface runoff and sedimentation, affected the capacity of many stream channels for transporting water and sediment. As a consequence, low-gradient streams that traversed alluvial deposits often experienced episodes of accelerated gully erosion and channel downcutting. Once channel downcutting began, it typically continued until a relatively resistant layer of sediment or bedrock was encountered; the result was an incised channel with characteristically steep, erodible banks. In the southwestern USA, notable examples of deeply incised channels are typically referred to as arroyos. Varying degrees of channel incision also prevail in many rangeland streams in the northwestern USA. During channel incision, the incremental additions of sediment that had previously built floodplains during high flows and over long periods of time became major sources of sediment for downstream areas. Even after a stream stopped downcutting, it generally continued to erode laterally and to rework ancient alluvial deposits, causing continued sediment production and instability.

Another consequence of channel incision is that former floodplains may no longer store water during high flows (VanHaveren and Jackson 1986). This hydrologic alteration appears to have been an important factor influencing summertime flow regimes in many rangeland areas; incised streams are less likely to have continuous summer flow. With a drop in riparian water tables, shifts in vegetation species and communities materialized. Riparian plant communities that had existed along channels and across floodplains were usually replaced by upland assemblages.

In forested areas of the Pacific Northwest, stream systems have also been affected by development and resource use. During early periods of logging, the protection of streams or riparian areas was often of relatively little concern. In the Coast Range of Oregon and Washington, for example, splash dams were frequently used to transport logs downstream (Sedell and Luchessa 1981), a practice that was extremely disruptive to channel banks and aquatic habitat. Forestry operations that involved road and landing construction, yarding, site preparation, fire control, and other forest practices also degraded many riparian areas. In steep terrain, roading practices often increased the potential occurrence of mass soil movements and landslides (Swanson et al. 1987).

In the larger rivers of the Northwest, the removal of large woody debris was undertaken to improve navigation and transportation. In the mid-1900s, logjam removal was widely practiced in streams and rivers because they were thought to block anadromous fish migration. Harvesting of mature conifers along many streams has reduced or depleted future sources of large woody debris, hence the replenishment of instream large woody debris from second-growth conifer stands may require decades, in some cases longer, to accomplish. Recent research has increasingly indicated that large woody debris accumulations can enhance salmonid production in many situations (Bisson et al. 1987) and programs of debris removal by state and federal agencies have largely been halted. Whereas accumulations of logging slash in streams used to be a common byproduct of many harvesting operations, the enactment of forest practices rules in the early 1970s greatly reduced the input of these materials.

The type and extent of changes in streams and riparian areas vary greatly across the region, but the biological and physical diversity associated with many streams has generally been reduced. The processing of energy and nutrients in aquatic and riparian ecosystems has similarly been altered and simplified. As a result, the value of these systems for fish and wildlife, for production of forage and timber, and for influencing water quality and quantity, aesthetic values, and recreation opportunities has been altered. Managing and improving the multiple resources associated with aquatic and riparian areas represent an immediate challenge to the public, to managers of public lands, and to private landowners.

Stream and riparian systems that develop across landscapes do so within broad geologic and geomorphic constraints. Such is the case in many headwater areas, where a stream usually has sufficient energy (because of generally steep channel slopes) to transport sediment inputs from hillslopes and upstream sources. In mountainous areas, bedrock channels can resist the erosive forces of flowing water and thus often constitute a major portion of the streambed and channel margins. Along these topographically constrained channels, adjacent riparian areas are relatively narrow.

As water continues its downstream path, it generally encounters decreasing gradients. Low-gradient reaches typically occur in relatively wide valleys where the stream has deposited sediments and formed floodplains. Over geologic time scales, a stream or river may slowly shift or migrate across the valley. Such unconstrained reaches are notable because of the extent to which their flow regimes interact with the floodplain and its associated riparian vegetation. In turn, riparian plant communities influence the relative stability and morphology of the stream. From an aquatic and riparian biology perspective, these unconstrained reaches are often extremely productive. They typically support a larger biomass and greater diversity of aquatic organisms, wildlife species, and terrestrial vegetation than are found in steeper or more constrained reaches.

The Role of Riparian Vegetation

Riparian areas and wetlands can be generally defined as those areas that are saturated by groundwater or intermittently inundated by surface water at a frequency and duration sufficient to support a prevalence of vegetation typically adapted for life in saturated soil. Thus, riparian areas fill an important niche between aquatic and terrestrial systems. In addition, riparian vegetation functions in a variety of special ways.

Stream Energy

Rainfall and snowmelt waters that move across landscapes as surface or subsurface flow are ultimately concentrated into discrete stream channels. Where water enters a channel, it has a specific amount of potential energy because of its elevation above sea level. As the water flows downstream, potential energy is converted to kinetic energy (energy of motion). The vast majority of this energy is dissipated as turbulence and heat as a stream flows from riffle to pool, around bends, and past large roughness elements (Beschta and Platts 1986). Where the shear stress exerted by flowing water exceeds the shear strength of a bank or bed, channel erosion occurs. Subtle changes in the roughness of a channel can cause major changes in the ability of a stream to mobilize or transport sediment. Thus, a stream may have sufficient energy to transport sediment through one reach only to deposit it at the next reach downstream. Riparian vegetation generally increases the structural complexity of channel margins and increases flow resistance, thereby reducing the rate at which a stream's energy is expended. Similarly, the occurrence of large woody debris increases the hydraulic roughness of a channel and tends to locally influence the time-rate dissipation of potential energy by flowing water. This debris also provides important microhabitat for a wide range of fish and other aquatic organisms.

Channel Roughness

The resistance to flow or general "roughness" of a channel depends on many factors, including stream discharge, particle sizes of bed and bank materials, the occurrence of large roughness elements (e.g., boulders, bedrock, large woody debris), stream sinuosity, bank characteristics, and streamside vegetation. These features are of critical significance for low-gradient, unconstrained streams because of their importance to energy dissipation and channel stability.

In recent years, there has been considerable interest in managing fish habitat by simply altering the flow-resistance characteristics of forest and rangeland channels with the addition of structures and large roughness elements (gabions, boulders, boulder clusters, spur dikes, check dams, large logs). These structural additions often focus the dissipation of stream energy and can cause increased local variability in channel morphology. However, in unconstrained channels they can also cause rapid and often undesirable changes in channel location. The integrity of the streambanks and the inherent stability of a channel are strongly influenced by the density and diversity of streamside vegetation.

Shade and Temperature

Vegetation canopies provide shade to a stream and thus play a significant role in moderating stream temperatures (Beschta et al. 1987). In summer, plant leaves and stems reduce the amount of solar radiation impinging on the stream surface and thus help prevent the occurrence of high water temperatures. In winter, forest canopies may similarly moderate thermal energy losses from a stream and retard the formation of ice.

Litter and Nutrients

The leaves of coniferous and deciduous trees and various brush species provide periodic inputs of litter to the stream system. Forbs and grasses along the banks of streams can also be sources of nutrients. These organic materials represent important food sources for microbes, invertebrates, and other organisms. The retention and recycling of these nutrients is also important to the growth and productivity of riparian vegetation (Gregory et al. 1987).

Large Wood

In forested riparian areas of the northwestern USA, the occurrence of large wood in streams has become increasingly recognized as a significant component of stream habitat for fish (Bisson et al. 1987; Sedell et al. 1988). Woody debris physically alters local flow patterns and channel characteristics, provides cover, and represents a long-term food source for aquatic organisms. Large woody debris changes the dissipation of stream energy and generally creates local channel scour and deposition. Increased local diversity in channel characteristics is generally considered a beneficial aspect of habitat for fish and other organisms.

Root Systems and Bank Stability

Often overlooked in an assessment of riparian vegetation is the importance of root systems to the long-term stability of channel banks. Where roots occupy a major portion of a streambank, they bind soil particles and often provide important resistance to the erosive forces of flowing water (Platts et al. 1987). The woody roots of trees are particularly important along many streams. Where trees grow and remain for decades or centuries along a channel, their root systems provide long-term bank and channel stability. For many rangeland streams, the root systems of shrubs, grasses, and sedges can provide this same function.

Sediment Deposition and Storage

The channel banks and floodplains of many unconstrained stream reaches consist of alluvial sediment deposits. Deposition of waterborne sediment is a normal process in well-vegetated riparian areas when streams overtop their banks and spread sediment-laden waters across a floodplain. Where vegetation provides sufficient roughness to reduce flow velocities, or where backwater areas occur, the deposition of sediment accounts for the long-term accretion of alluvium on floodplains in unconstrained valleys. The stems of sedges, grasses, forbs, and other vegetation can greatly increase surface roughness and accelerate the deposition process. Sediment deposition normally occurs along well-vegetated streambanks and floodplains during high flow, but if riparian vegetation has been eliminated or significantly reduced, the opposite effect—accelerated bank erosion and an unstable channel—is all too common.

Water Storage and Release

Low-gradient stream reaches and associated alluvial deposits represent areas of infiltration and subsurface water storage during periods of overbank flow. The slow down-valley release of this stored water during the summer months can provide an important source of base flow for many rangeland stream systems. Without streamside protection by vegetation from the erosive energy of high flows, channel erosion and downcutting can occur rapidly. Once channel incision has taken place, local water tables will similarly drop, and the potential for moisture recharge and storage may be greatly reduced or eliminated. Particularly in relatively arid areas, the importance of riparian vegetation in maintaining stable channels and floodplain functions (e.g., dampening of peak flows, groundwater storage, and low flow maintenance) may far outweigh evapotranspiration losses attributed to this same vegetation.

Management Direction

One of the major ongoing bioengineering activities directed at streams in the northwestern USA is that of fish habitat management. Much of this effort involves the construction of structures in channels or the addition of various large roughness elements. The primary purpose of these structures is to improve fish habitat by increasing the amount of cover and of diversity in channel morphology. Channel stability is also considered a desirable outcome of such structural additions and it is common for the structures to be anchored with cable, metal rods, or other means. Thus, many fish habitat managers have largely adopted a mechanistic approach to a relatively complex ecological and geomorphic problem. They have seemingly opted for the short-term addition or replacement of stream structure while often ignoring the necessity of understanding and improving the long-term functional attributes of riparian plant communities. As important as structural additions to channels might seem for fish habitat improvement, they may be significantly less important (and in many cases not needed) than a program of long-term management directed specifically at protecting and improving riparian vegetation.

Riparian vegetation functions in a variety of ways to influence such factors as channel roughness and energy dissipation, water temperature regimes, nutrient cycling, large woody debris loading, bank stability, sediment deposition, and water storage. It is obvious that structures alone are unable to replicate the various functions of healthy riparian vegetation. Because of riparian vegetation's major impact on other resources (water quality, fish and wildlife habitat, aesthetics, etc.) and the relatively degraded condition of many riparian areas, the recovery of aquatic and riparian systems represents one of the most important challenges facing today's managers. Enlightened management policies and practices are needed that will again allow riparian vegetation to maximize its effect upon the hydrology and channel morphology of stream systems.

Where seed sources of important riparian species have been depleted or destroyed, reestablishment of these species should be a high priority. Where the full range of native species is present but the prevalence of functionally important species has been diminished, management practices that promote regeneration and growth of the important species should be implemented. In systems where beavers were a significant influence on stream and habitat conditions, perhaps the reintroduction of beavers should be actively pursued. With our assistance to their recurring life cycle, plants will function and continually adjust to varying flow and sediment regimes. These needs are particularly important along the margins of unconstrained stream reaches because of the close interaction between vegetation, channel hydraulics, and floodplain processes. The importance of streamside vegetation, relative to its capability for influencing a channel, tends to decrease with increasing distance from the channel. Thus, it is especially crucial that we maximize the effectiveness of riparian vegetation that is closest to the stream.

The roles of streamside vegetation are diverse and complicated, and a large number of vegetation communities and species typically exist in most undisturbed riparian areas. Improved understanding of the ecology of riparian vegetation and interactions with environmental conditions (especially soils and hydrology) is a particularly pressing research need. However, the important point is that we should be managing riparian systems toward their ecological potential. Management strategies should allow and encourage plant succession. When stream managers only add structural features to channels in attempts to improve fish habitat, it indicates that they have learned little from historical experiences or the present knowledge base.

We are currently in a period when fisheries enhancement strategies are undergoing rapid change. We can continue existing efforts of simply adding large roughness elements to many degraded streams in the hopes that such structures will provide the solution to aquatic habitat problems, or we can move towards management policies that also protect, reestablish, and encourage the functional attributes of riparian vegetation. Over a wide range of channel types and conditions, the latter course of action represents the most effective and practical approach for the long-term improvement of aquatic habitat.

References

Adams, P. W., R. L. Beschta, and H. A. Froehlich. 1988. Mountain logging near streams: opportunities and challenges. Pages 153–162 *in* International mountain logging and Pacific Northwest skyline symposium. Oregon State University, Department of Forest Engineering, Corvallis.

Beschta, R. L., R. E. Bilby, G. W. Brown, L. B. Holtby, and T. D. Hofstra. 1987. Stream temperature and aquatic habitat: fisheries and forestry interactions. University of Washington Institute of Forest Resources Contribution 57:191–232.

Beschta, R. L., and W. S. Platts. 1986. Morphological

features of small streams: significance and function. Water Resources Bulletin 22:369–379.

Bisson, P. A., and eight coauthors. 1987. Large woody debris in forested streams in the Pacific Northwest: past, present and future. University of Washington Institute of Forest Resources Contribution 57:143–190.

Department of Ecology. 1986. Wetland functions, rehabilitation and creation in the Pacific Northwest: the state of our understanding. Washington Department of Ecology, Publication 86-14, Olympia.

Gregory, S. V., G. A. Lamberti, D. C. Erman, K. V. Koski, M. L. Murphy, and J. R. Sedell. 1987 Influence of forest practices on aquatic production. University of Washington Institute of Forest Resources Contribution 57:233–255.

Lee, L. C., and F. E. Gross. 1988. Restoration, creation, and management of wetland and riparian ecosystems in the American West: a summary and synthesis of the symposium. Pages 201–219 *in* Restoration, creation and management of wetland and riparian ecosystems in the American West. Society of Wetland Scientists, Denver.

Naiman, R. J., C. A. Johnston, and J. C. Kelly. 1988. Alteration of North American streams by beaver. BioScience 38:753–762.

Naiman, R. J., J. M. Melillo, and J. E. Hobbie. 1986. Ecosystem alteration of boreal forest streams by beaver (*Castor canadensis*). Ecology 67:1254–1269.

Parker, J., F. J. Wood, B. H. Smith, and R. G. Elder. 1985. Erosional downcutting in lower order riparian ecosystems: Have historical changes been caused by removal of beaver? U.S. Forest Service General Technical Report RM-120:35–38.

Platts, W. S., and twelve coauthors. 1987. Methods for evaluating riparian habitats with applications to management. U.S. Forest Service General Technical Report INT-221.

Sedell, J. R., P. A. Bisson, F. J. Swanson, and S. V. Gregory. 1988. What we know about large trees that fall into streams and rivers. U.S. Forest Service General Technical Report PNW-GTR-229:47–81.

Sedell, J. R., and K. J. Luchessa. 1981. Using the historical record as an aid to salmonid habitat enhancement. Pages 210–223 *in* N. B. Armantrout, editor. Acquisition and utilization of aquatic habitat inventory information. American Fisheries Society, Western Division, Bethesda, Maryland.

Skaugset, A. E., C. G. Bacon, A. J. Hansen, and T. E. McMahon. 1989. COPE research on riparian zone management in the Oregon Coast Range. Pages 277–286 *in* W. W. Woessner and D. F. Potts, editors. Symposium proceedings on headwaters hydrology. American Water Resources Association, Bethesda, Maryland.

Swanson, F. J., and six coauthors. 1987. Mass failures and other processes of sediment production in Pacific Northwest forest landscapes. University of Washington Institute of Forest Resources Contribution 57:9–38.

U.S. Forest Service. 1985. Riparian ecosystems and their management: reconciling conflicting uses. U.S. Forest Service General Technical Report RM-120.

VanHaveren, B. P., and W. L. Jackson. 1986. Concepts in stream riparian rehabilitation. Transactions of the North American Wildlife and Natural Resources Conference 51:280–289.

American Fisheries Society Symposium 10:59–61, 1991

Establishing Volunteers in Natural Resource Restoration Programs

ALAN W. JOHNSON

Alan W. Johnson and Associates
1421 17th Street SE, Auburn, Washington, 98002, USA

WAYNE J. DALEY

Kramer, Chin, and Mayo
1917 First Avenue, Seattle, Washington 98101, USA

Abstract.—In the past 5–10 years, there has been a large increase in the involvement of volunteers in natural resource restoration programs. This is particularly evident in the Pacific Northwest, where public awareness has motivated volunteers to participate in a wide variety of programs. We urge governmental agencies or other groups wishing to establish volunteer programs not to view them as panaceas. To maximize potential success, volunteer programs must have realistic expectations and be properly planned and implemented, and the anticipated results must not be oversold. The success of a volunteer program depends on setting clear specific program objectives; understanding the true cost of the program; working within the limitations of the volunteers, the agency, and especially the science; and realizing that the bottom line is credibility. We feel the success of volunteer programs need not be strictly defined by the number of structures installed or fish returning to a project site. Success may be more appropriately defined by the public's willingness to become involved and ultimately, to accept responsibility for the stewardship of the resource.

In the past 5–10 years there has been a large increase in the involvement of volunteers in natural resource restoration programs. This is particularly evident in the Pacific Northwest, where public awareness has motivated volunteers to participate in a wide variety of programs. In Washington, the Adopt-a-Beach Foundation (1988) has identified over 60 agencies and organizations that have sponsored projects involving volunteers. Washington and Oregon have instigated salmon and trout enhancement programs that involve and depend on volunteer groups (Oregon Department of Fish and Wildlife 1985; Volunteer Fisheries Resource Program 1985). The Municipality of Metropolitan Seattle developed and annually budgets for a citizen grant program to encourage the development of environmental activism at local levels (Johnson 1986). Bellevue, Washington, has successfully involved volunteers in activities ranging from water quality monitoring to habitat restoration (Hubbard-Gray and Tilander 1988). Other programs such as Adopt-a-Stream and Adopt-a-Beach have also been successful in educating and involving the public in a variety of stream and beach restoration projects.

We are concerned that, with the increasing emphasis on volunteer programs, these programs might be viewed as a panacea. To maximize potential success, volunteer programs must have realistic expectations, be properly planned and implemented, and not be oversold. It would be impossible to provide an account of all experiences related to this topic, but we will review some basic principles for establishing volunteer programs. We will also identify issues that may reduce or limit the success of these programs.

Set Clear and Specific Program Objectives

Setting clear and specific program objectives is perhaps the most important element of a successful volunteer program. Specifically, what is the purpose of the program? The types of programs we have been involved with include the following.

- Education and public involvement programs designed to inform or educate the public on issues, to build constituency or advocacy groups, or to involve the public in the planning process. These programs can also provide the volunteer with hands-on experience or demonstrate an agency's willingness to involve the public in management of the resource.
- Programs that encourage an advocacy group to become actively involved. These vary from the implementation of instream projects to the development and review of management policies to lobbying local and regional politicians.
- Programs to develop and implement public information surveys, stream surveys, or water quality monitoring. These projects vary in length

from onetime events to ongoing monitoring over several years.

• Programs to identify, develop, and implement stream and fishery stock-restoration projects. Projects can vary in size from small site-specific projects accomplished by individuals to the development of programs focusing on entire stream systems.

From our experience, we feel that the main emphasis of volunteer programs should be on the education and involvement of the public in resource management. Because of the rapid rate at which urbanization is occurring in many areas, it is imperative that the public understand the complexity of natural resource management, the limitations and constraints of the resource, and most importantly, the future implications of the many decisions that are being made today. Volunteers can and have completed many projects, but we feel in many cases, and especially in urban areas, the amount of data collected or the number of structures built should be considered a bonus. It is only through the education and involvement of the general public that we will achieve long-term stewardship of our natural resources.

Understand and Match the Goals of the Agency and the Volunteers

It is important to understand early in the development of a volunteer program the expectations, limitations, and level of commitment of all parties involved. To ensure success, all parties must agree on both the direction and desired outcome of the program. Goals and expectations need to be realistic and aligned with the biology of the system.

Be Aware of Hidden Costs

It is a common perception that volunteer programs by definition are free. To quote the old adage, "There is no free lunch." When establishing a volunteer program, agencies need to consider the following.

• How committed is the agency to the program? Is there a demonstrated need for it? Is the community ready and willing to support the program? Before a program is established, sufficient agency commitment should exist to carry the program through to a predefined completion point. It can be very detrimental to an agency's credibility to establish a program and then suddenly to withdraw support because of lack of funds or, worse, lack of interest.

• Does the agency have the personnel available to develop and establish the program, possibly to locate and recruit participants, and to provide longterm leadership? Personnel may also be needed to act as a liaison between the agency and the volunteers and to organize and coordinate project activities.

• What funds are available to support the program? Costs include not only the funding of agency personnel to establish and operate the program, but also outside needs. Most volunteer groups have limited resources in terms of both dollars and available equipment. They may request or require funds before undertaking a desired project. Two common methods of funding volunteers are outright grants from the sponsoring agency and match funds. Match funds are particularly attractive for significantly increasing the limited budgets often available for volunteer programs.

Two types of costs are involved in establishing a volunteer program—program costs and project costs. Program costs are associated with the establishment and operation of a program. As the program becomes established and the volunteers assume more responsibility, these costs may decrease. Longterm educational projects or community events are examples of projects with which costs can decrease with time. Conversely, project costs, such as site-specific construction designs and logistics, tend to remain somewhat constant (i.e., per unit cost) from site to site.

Time is probably the most underbudgeted, least accounted for, and yet most costly agency expense in a volunteer program. Time costs include the following.

• Planning. Because these programs rely heavily on donations of time, energy, and equipment, it is essential to have a complete plan and marketing strategy that guide the activities of the program. The skills and resources needed to accomplish the program or project must be clearly identified. These include the number of volunteers, needed skill levels, types of equipment and materials, and most importantly, the amount of leadership required. Within the program, different levels of commitment must be accommodated; not all volunteers or agency personnel will work with the same enthusiasm. It is important, especially in the initial stages, that the program start with small projects and achieve favorable results. Then, depending on program objectives, operations can continue, expand, or draw to a predefined conclusion. Finally, to the extent possible, everyone interested in the results of the project (other agencies, groups, or adjacent land owners) should be made aware of and given the opportunity to participate in the planning process.

• Volunteer development and maintenance. If the program is new and does not include experienced volunteers, time will be required to recruit sufficient volunteers. Once volunteers are in place, they must be interacted with frequently. The amount of time required for effective "volunteer maintenance" is frequently underestimated and, consequently, underbudgeted. Maintenance includes providing general information, status reports, updates, and "extras" such as recognition and awards, newsletters, and meetings where speakers may be invited to discuss specific topics of interest to the volunteers.

• Training. Because the labor force available for these programs may be untrained and have limited or marginal equipment, safety issues, especially for construction projects, should be of high priority. Some projects, such as water quality monitoring or habitat surveys, often require training for specific methods and procedures. Quality control and quality assurance may also be needed in volunteer monitoring or construction programs. Because volunteers may be unfamiliar with methods or equipment, allowance should be made for one or more "false starts" as projects begin.

• Permits and access agreements. Sufficient time should be allowed for the inevitable delays associated with obtaining necessary permits and access agreements. In Seattle, for example, 60–120 d are often required to obtain required permits from the various agencies.

• Technical issues. Technical matters can range from preparation of necessary designs and plans needed for construction projects, to coordination with laboratories analyzing water quality samples, to arranging for the printing of brochures and posters. For specific projects, technical issues can include the logistics of coordinating and arranging for work crews, obtaining equipment and supplies (especially if the program is relying on donations), and management of project sites.

Do Not Oversell the Anticipated Results

Especially in the early stages of the program, care must to taken to not promise what technically cannot be delivered. Natural resource restoration at all levels is still an art; results generally cannot be predicted nor guaranteed with reliable accuracy. Combined with the many other variables that exist in nature, it is best to set and promote realistic and achievable goals.

Volunteer programs can and have accomplished many significant results. However, we feel the bottom line of these programs—that is, the real product being sold by agencies sponsoring or otherwise involved with these programs—is *agency credibility*. If sufficient resources, time, and commitment are not budgeted for establishing and maintaining volunteer programs, the end result can be very detrimental to an agency's public relations. One should always deal with real concerns, real issues, and most importantly, projects that people can relate to and understand.

In summary, we feel that to ensure the success of any volunteer program, several factors are very important. At the start, clear specific program goals and objectives must be set. The true costs (personnel, time, and funding) involved in establishing and maintaining a volunteer program must be understood. The limitations of the volunteer, the agency, and especially the science must be accomodated. The bottom line—credibility—must be appreciated.

With volunteer programs, success need not be strictly defined by the number of structures installed or fish returning to a project site. Success is also defined by the public's willingness to become involved and, ultimately, to accept responsibility for the stewardship of the resource. We feel, in many instances, this is the true measure of success in a volunteer program.

References

Adopt-a-Beach Foundation. 1988. Volunteer resource guide. A citizen's directory to volunteer opportunities in caring for Washington's outer coast, Puget Sound, and associated watersheds. Adopt-a-Beach Foundation, Seattle.

Hubbard-Gray, S., and S. Tilander. 1988. Stream team handbook: how residents and volunteers can protect and enhance our water quality, fish and wildlife. Storm and Surface Water Utility, Bellevue, Washington.

Johnson, A. 1986. A summary of Metro's community action grant program. Municipality of Metropolitan Seattle, Water Resource Section, Seattle.

Oregon Department of Fish and Wildlife. 1985. Guidelines for public involvement in cooperative salmon and trout enhancement projects. Oregon Department of Fish and Wildlife, Portland.

Volunteer Fisheries Resource Program. 1985. Progress—providing a greater resource. Washington Department of Fisheries, Olympia.

American Fisheries Society Symposium 10:62–67, 1991

Responses of Anadromous Salmonids to Habitat Modification: How Do We Measure Them?

Gordon H. Reeves, Fred H. Everest, and James R. Sedell
U.S. Forest Service, Pacific Northwest Experiment Station
3200 Jefferson Way, Corvallis, Oregon 97331, USA

Abstract.—Evaluation of responses of anadromous salmonids to habitat manipulation should be an integral part of habitat modification programs. However, responses of anadromous fish populations to habitat manipulations are seldom measured. The primary reasons given for this neglect are inadequate funds, personnel, and time. This paper examines ways in which biological responses to habitat manipulation can be evaluated at different stages in the life history of anadromous salmonids. Responses that can be measured are changes in numbers of adult fish, changes in numbers of juvenile fish, and changes in numbers of smolts leaving a stream or stream system. We assess the merits of each approach and conclude that changes in smolt numbers are the best way to evaluate the effect of habitat manipulation projects on anadromous salmonid populations. Evaluation programs should be developed on a basin or subbasin scale because reach or site scales provide an inadequate context for evaluating change. Evaluations should also consider the response of the entire salmonid community to changes in habitat rather than the response of a single or target species.

Modification of degraded or marginal habitat is a major component of many fishery management programs throughout the range of anadromous fish in western North America. Millions of dollars are spent annually on such efforts, the primary goal being to increase the numbers of wild anadromous fish. Few attempts, however, have been made to evaluate the response of fish populations to habitat manipulations. Hall and Baker (1982) and Reeves and Roelofs (1982) noted this shortcoming and concluded that without such studies it is impossible to assess the performance or cost-effectiveness of various habitat modification techniques. In an updated review, Reeves et al. (1991) found that this general lack of evaluation has continued to the present.

Several reasons have been offered to explain this absence of sound evaluation programs or efforts. White (1975) noted that such assessments are costly and time-consuming. Managers often lack adequate funding and personnel to initiate and carry out evaluations. In addition, there are certain instances in which legislation provides money to undertake restoration work but fails to mandate funding for evaluation (Hashagen 1984). Several years may also be required for optimum physical conditions to develop after restoration or enhancement, and fish populations may not respond for several more years (Hunt 1976). There may also be large natural fluctuations in populations of anadromous salmonids that confound the problem of detecting changes (Larkin 1974; White 1975; Hall and Knight 1981).

The purpose of this paper is twofold: (1) to examine ways in which responses of anadromous salmonids to habitat modification projects can be evaluated at different stages of their life cycle and (2) to assess the merits of each approach.

Methods for Evaluating Biological Responses

The goal of modifying freshwater habitats of anadromous salmonids is to increase the number of fish leaving the system as smolts, eventually to return as adults. Although increased adult returns is the ultimate goal, smolt production is more easily measured because of the multitude of factors beyond the control of the freshwater manager, such as oceanic conditions, commercial harvest, and migration obstacles, that influence adult survival. To achieve these goals, it is assumed that the factor(s) that limit the production at a particular life history stage can be correctly identified and measures implemented to remove the limitation. Everest et al. (1991, this volume) discuss in more detail the importance of identifying limiting factors and cite a procedure for identifying them.

The spatial scale for evaluation programs should be a basin or subbasin. This scale is necessary because anadromous salmonids generally use all parts of a stream system during the freshwater phases of their life history. In the past, most evaluation attempts have focused on a small reach of stream or on individual structures. Assessing the effects of habitat manipulation at these levels may lead to erroneous or misleading conclusions, however. For example, fish within a basin may simply redistribute themselves in response to habitat work, moving from unimproved to improved areas with no increase in the total population. Numbers may increase in the modified area, but it does not necessarily follow that the number of fish surviving to the next life history stage will correspondingly increase.

Ideally, a basin or subbasin approach to evaluating fish responses to habitat manipulation is best accomplished through analysis of paired treatment and control watersheds. This approach may seem overwhelming, however, in terms of the amount of time, money, and effort required to analyze habitats and fish populations in two basins. It may also be physically impossible, because even adjacent subbasins can be quite different in terms of geology, geomorphology, and biology. Alternatives to a control-basin design include before-and-after studies of fish habitats and populations within a basin and the use of migrant traps to isolate smolt production in a treatment reach within a basin.

Many habitat enhancement programs attempt to increase the numbers of a particular species or race of fish. Managers should be aware that manipulation of habitat can result in changes within the salmonid community. Benefits to a target species may come at the expense of another species in the community. For example, creation of pools to increase habitat for juvenile coho salmon *Oncorhynchus kisutch* is a common objective of many programs. A consequence of this action is often the loss of rearing habitat for age-0 steelhead *Oncorhynchus mykiss*. The objective of increasing coho salmon numbers might be accomplished, but age-0 steelhead numbers may show a corresponding decline. Managers may be unaware of such trends without monitoring the community. Inclusion of all species or races of salmonids in the design and implementation of any assessment program can be accomplished in most situations, often without large increases in costs in time, money, or personnel.

Returning Adults

The decline of anadromous salmonids in western North America has frequently been attributed to losses in the quantity and quality of spawning habitat. Reeves et al. (1991) cited numerous examples of efforts to improve the quantity and condition of spawning habitat. Much of the effort has been directed towards collecting and retaining new gravels by the addition of roughness elements such as gabions, logs, and boulder berms (e.g., Anderson et al. 1984; House 1984; Moreau 1984; House and Boehne 1985; Klassen and Northcote 1986), and by improving the quality of existing spawning habitat by removing fine sediment from stream substrates (e.g., West 1984).

Evaluations of efforts to improve spawning habitat have generally been made by comparing the number of juvenile fish in the treatment area with the number in an unimproved reach after emergence from the gravel (e.g., House and Boehne 1985, 1986) or simply by comparing the numbers of spawners using the area before and after treatment (e.g., West 1984). In most instances, the abundance of spawners in modified areas increased relative to unmodified sections (West 1984).

Evaluations of this type can be misleading because the results lack a basinwide context to accurately assess the effectiveness of the program. Spawning fish may have concentrated their activities in recently modified habitats within a basin without an actual increase in total numbers or an increase in the number or survival of developing eggs and alevins. Modified habitat may act as a magnet and attract fish because of the concentration of usable habitat in comparison with unmodified areas. It might therefore appear that numbers increased when in reality there was simply a redistribution of fish within the system.

A more meaningful evaluation would consist of quantitative counts of spawning salmonids in control and treatment basins over time. The basins should be located close to each other and be similar in physical and biological features. Composition of the salmonid community should also be comparable (e.g., contain winter steelhead, spring chinook salmon *O. tshawystcha*, and coho salmon).

If a control basin is not available or it is not feasible to use such a design, a comparison could be made of the number of spawners in the modified basin with general trends in other streams in the area. An inference could be made that habitat modification increased the number of spawners in the treatment basin if (1) there was an opposing trend in numbers, the treated area increasing while the overall trend was down; or (2) the rate of increase in numbers was greater in the altered basin than the overall trend, or conversely the rate of decline was slower in the treated basin than the overall trend. This approach may not be as sound as use of a control basin, but it is an improvement over comparison of treated and untreated reaches within the same stream.

Evaluation of adult returns should also consider the contribution of a stock or population to the commercial and sport fisheries. Whereas it is usually impossible to quantify the catch exactly, techniques are available for estimating the contribution. Examples of thess can be found in Everest et al. (1987) and House (in press). The estimate of catch should be added to the number of returning adults to obtain a more realistic estimate of project results.

Most evaluations of spawning habitat improvement have been short, usually 3 years or less. We conclude that a time period equal to two times the length of the life cycle of the species of interest is the minimum needed to fully assess the impact of any habitat modification effort. This period can vary from 6 to 12 years, depending on species and race of fish. An extended evaluation period is also necessary because of the tremendous natural variablity in anadromous fish populations (Hall and Knight 1981). Bottom et al. (1986) suggested that stocks of anadromous salmonids in Oregon and California are especially variable because of the large natural fluctuations in ocean conditions. Thus, evaluation programs in these areas may require a longer period to fully assess the effect of habitat modifications. An extended period of evaluation is also recommended because of the length of time it may take a population to respond to changes in habitat induced by modification efforts. An actual response may not be seen until completion of at least one generation of the species in question.

Juveniles

Modification of rearing habitat for juvenile salmonids is probably the most common type of project undertaken by fishery managers to increase populations of anadromous fish (Hall and Baker 1982; Reeves and Roelofs 1982; Reeves et al. 1991). The basic premise of such projects is that habitat alteration will increase the abundance of juvenile fish, which should in turn lead to increased numbers of smolts and adults.

The most common way in which these projects have been evaluated is similar to that discussed for evaluation of spawning habitat improvement. Abundance of fish in a treatment area is compared with the pre- and posttreatment number in a nearby control area (e.g., Moreau 1984; West 1984; House and Boehne 1985; Fontaine 1988). This approach has some attractive attributes in that it is relatively easy to conduct and there may be a relatively short response time—as little as a year.

However, there are problems associated with this approach that should be weighed against the usefulness of the information gained. The primary weakness of comparing numbers of juveniles in treated and untreated reaches is the same as that discussed for evaluation of spawning habitat. There may only be a redistribution of juveniles within the basin rather than a basinwide increase in numbers, growth rate, or survival. Without such basinwide contexts, the success of the modification program may be overestimated.

Programs to evaluate responses of juvenile salmonids to habitat modifications should follow the design used for determining the responses of adult fish to habitat alterations. Control and treatment basins are needed so fish numbers can be compared between treated and untreated basins. Again, the physical and biological features of the basins should be as similar as possible.

Estimating total numbers of fish in a basin has generally been done by sampling "representative" reaches of a stream and then extrapolating to the entire basin. Recent evidence has shown that such extrapolations can be inaccurate and misleading. For example, Everest et al. (1985) attempted to estimate the total numbers of juvenile anadromous salmonids in Fish Creek, a tributary of the Clackamas River in Oregon, by extrapolating fish numbers from five representative reaches. Estimates from extrapolation were 2 to 2.4 times greater, depending on species and age-class, than were estimates derived from a statistically valid sampling scheme (i.e., Hankin and Reeves 1988). In another instance, Bisson (in press) estimated the number of juvenile fish in a small stream in Washington with a series of representative reaches drawn from different portions of the basin. He found that estimated numbers of fish in the basin varied by several orders of magnitude, depending on choices of representative reaches. In addition, the factors thought to limit production were correspondingly variable and diverse.

Hankin and Reeves (1988) suggested a cost-effective sampling design for estimating total fish numbers in small streams based on visual estimation as an alternative to sampling representative reaches. This technique generates estimates of total fish numbers and has provisions for calculating confidence intervals of the estimate. Confidence intervals are important for assessing the quality of the estimate, and they permit statistical comparison among estimates.

There are limitations to relying on changes in juvenile numbers to assess the success of habitat modification efforts. There may be large natural fluctuations in site-specific density due to density-independent or density-dependent factors (Hall and Knight 1981). For example, Everest et al. (1989) observed annual variations of 220% in the numbers of age-0 steelhead in Fish Creek, Oregon. There may also be no direct relationship between the number of juvenile fish at one life history phase and the number found at the next phase, particularly if a limiting factor exists at the later phase. For example, Murphy et al. (1986) found that increased numbers of age-0 coho salmon in the fall in Alaska did not necessarily lead to increased numbers of age-1 fish the following spring.

Smolts

Because of problems associated with quantifying adult returns and the limitations on the ability to draw inferences from juvenile numbers, monitoring changes in smolt production in a basin may be the best alternative for measuring the impact of a habitat modification program. The smolt stage represents the culmination of freshwater life for anadromous fish; therefore, creation of habitat conditions that favor growth and survival of juveniles should result in greater production of smolts. Response in smolt production may occur relatively quickly compared with the response in adult returns. Changes in the number of fish reaching the smolt stage therefore reflect the degree to which a program has been successful in reducing or eliminating bottlenecks (i.e., factors that were limiting) at earlier stages.

Despite the potential usefulness of quantifying smolt production in relation to habitat manipulation programs, few such assessments have been done. The apparent lack of effort is probably due to a number of problems associated with quantifying smolts. First, initial investments in equipment are usually substantial and smolt trapping requires a continual effort and a large investment of funds and personnel. As discussed earlier, most projects may not be able to provide these necessary people and funds.

Monitoring of smolt production is further complicated by the need to rely solely on physical appearance to identify smolts. Rodgers et al. (1987) found that the physiological state of wild fish classified as smolts was consistent but that outward physical characteristics were highly variable, making it difficult to correctly identify all fish that were true smolts.

There may also be extensive movements from a basin of young salmonids that are not true smolts. Everest (1973) observed emigration of summer steelhead fry from tributaries of the Rogue River, Oregon. Similar presmolt migrations have been observed for coho salmon in Knowles Creek, Oregon (Rodgers 1986) and chinook salmon in Elk River, Oregon (Reeves et al., unpublished data). The contribution of these fish to future production is unknown at present, and as a result, the importance of nonsmolt migrants to the evaluation of habitat work is unclear.

Smolt numbers can exhibit large annual fluctuations. Everest et al. (1989) reported that annual fluctuations in smolt numbers varied with species, exceeding 95% for steelhead and 30% for coho salmon. Almost an order-of-magnitude difference has been observed in numbers of fall chinook salmon smolts in Elk River between 1985 and 1986 (Reeves et al., unpublished data). During the same period, the number of steelhead smolts remained relatively constant, varying between 7,000 and 8,000 fish annually. Problems associated with this variability can be partially offset by establishing a control basin with which to make comparisons and by maintaining the monitoring program for an extended period. Everest et al. (1991) recommend 10 years for monitoring.

Smolt numbers can be measured in a number of ways. The most common way is to establish a stationary weir or trap to intercept and capture fish as they move. Traps range from elaborate, semipermanent structures constructed of metal or wood to temporary traps made of net material. Whelan et al. (1989) recently designed a portable trap that can be fished either upstream for adults or downstream for smolts. Stationary traps require daily tending to remove debris and maintain fish-capture capabilities.

Floating traps offer an attractive alternative to stationary traps. Everest et al. (1985, 1986, 1987, 1989) employed a modified Humphrey trap, consisting of a rotating screen assembly suspended between two pontoons, to enumerate smolts leaving Fish Creek, Oregon. The rotating screen is powered by a paddle wheel and has scoops attached at various intervals. Fish are impinged on the leading edge of the screen by water velocity, picked up by the scoops, and delivered to a holding box.

More recently, a more passive floating trap has

been developed and modified by biologists of the Oregon Department of Fish and Wildlife and the U.S. Forest Service working from Corvallis, Oregon. This trap consists of a wire-mesh cone with a fiberglass helix on the inside. Flow turns the helix, which in turn rotates the cone, thereby trapping fish inside the cone. Rotation of the helix then moves the fish back into a holding box. These traps have been able to capture fish more effectively, particularly large steelhead smolts, and operate over a wider range of flows than do Humphrey traps (Reeves et al., personal observations).

Floating traps only sample a portion of the stream, so calibration curves must be developed to estimate the total number of smolts. Although this technique has generated a point estimate of smolt numbers, at present there is no method for constructing confidence intervals about these estimate. The authors of this paper, in conjunction with D. G. Hankin of Humboldt State University, plan to develop a procedure for calculating confidence intervals.

Another way to estimate smolt production is to follow the methodology of Hankin and Reeves (1988) for estimating fish populations on a basin scale. Timing is critical in making smolt estimates by this procedure; counts have to be made immediately prior to smolt migration from the system. The shortcoming of this approach is that it provides an estimate of the number of fish at a single point in time but does not account for fish that either left the system earlier or did not exhibit smolt characteristics at the time the estimate was made.

Conclusions

Each of the techniques discussed for evaluation of habitat modification programs has inherent strengths and weaknesses. The most meaningful criterion for evaluation is a change in the number of adult anadromous salmonids. However, the long response time before it is possible to observe a change in adult returns, coupled with the large degree of variability due to conditions beyond the control of the freshwater habitat manager, severely restricts the use of changes in adult populations as a viable measure of evaluation. Estimating the abundance of juveniles is probably the easist way to assess population responses and will show the shortest response time to habitat manipulations. This approach, however, begs the question of whether a project has resulted in an actual increase in smolt production or adult returns, or whether there was just a temporary increase in juveniles that will be negated later because the factor(s) limiting production was not correctly identified and addressed. Because of these problems, we believe that smolt production should be the standard for evaluating the biological response of anadromous salmonids to habitat manipulations. Estimating changes in smolt production avoids many of the problems associated with estimating population size at other life history stages while providing sound evidence of the response of anadromous fish populations to habitat manipulations.

References

Anderson, J. W., R. A. Ruediger, and W. F. Hudson, Jr. 1984. Design, placement and fish use of instream structures in Southwestern Oregon. Pages 165–180 *in* T. J. Hassler, editor. Proceedings: Pacific Northwest stream habitat management workshop. California Cooperative Fishery Research Unit, Humboldt State University, Arcata, California.

Bisson, P. A. In press. The importance of identifying limiting factors. *In* J. W. Buell, editor. Proceedings of the stream habitat enhancement evaluation workshop. U.S. Department of Energy, Bonneville Power Adminstration, Division of Fish and Wildlife, Portland, Oregon.

Bottom, D. L., T. E. Nickelson, and S. L. Johnson. 1986. Research and development of Oregon's coastal salmon stocks. Oregon Department of Fish and Wildlife, Job Final Report AFC-127, Portland.

Everest, F. H. 1973. Ecology and management of summer steelhead in the Rogue River fishery. Oregon State Game Commission, Research Report 7, Portland.

Everest, F. H., G. H. Reeves, J. R. Sedell, and D. B. Hohler. 1989. Changes in habitat and populations of steelhead trout, coho salmon, and chinook salmon in Fish Creek, Oregon, 1983–1987, as related to habitat improvement. U.S. Department of Energy, Bonneville Power Adminstration, Division of Fish and Wildlife, Portland, Oregon.

Everest, F. H., G. H. Reeves, J. R. Sedell, D. B. Hohler, and T. Cain. 1987. The effects of habitat enhancement on steelhead trout and coho salmon smolt production, habitat utilization, and habitat availability in Fish Creek, Oregon, 1983–86. U.S. Department of Energy, Bonneville Power Adminstration, Division of Fish and Wildlife, Portland, Oregon.

Everest, F. H., G. H. Reeves, J. R. Sedell, J. Wolfe, D. Hohler, and D. A. Heller. 1986. Abundance, behavior, and habitat utilization by coho salmon and steelhead trout in Fish Creek, Oregon as influenced by habitat enhancement. U.S. Department of Energy, Bonneville Power Adminstration, Division of Fish and Wildlife, Portland, Oregon.

Everest. F. H., J. R. Sedell, G. H. Reeves, and M. D. Bryant. 1991. Planning and evaluating habitat projects for anadromous salmonids. American Fisheries Society Symposium 10:68–77.

Everest, F. H., J. R. Sedell, G. H. Reeves, and J. Wolfe. 1985. Fisheries enhancement in the Fish Creek basin—an evaluation of in-channel and off-channel projects. U.S. Department of Energy, Bonneville Power Adminstration, Division of Fish and Wildlife, Portland, Oregon.

Fontaine, B. L. 1988. An evaluation of the effectiveness of instream structures for steehead trout rearing in Steamboat Creek basin. Master's thesis. Oregon State University, Corvallis.

Hall, J. D., and C. O. Baker. 1982. Rehabilitating and enhancing stream habitat: 1. Review and evaluation. U.S. Forest Service General Technical Report PNW-138.

Hall, J. D., and N. J. Knight. 1981. Natural variation in abundance of salmonid populations in streams and its implications for design of impact studies. U.S. Environmental Protection Agency, EPA-600/S3-81-021, Corvallis, Oregon.

Hankin, D. G., and G. H. Reeves. 1988. Estimating total fish abundance and total habitat area in small streams based on visual estimation methods. Canadian Journal of Fisheries and Aquatic Sciences 45:834–844.

Hashagen, K. A. 1984. The social and political ramification of stream rehabilitation: the California perspective. Pages 267–272. *in* T. J. Hassler, editor. Proceedings: Pacific Northwest stream habitat management workshop. California Cooperative Fishery Research Unit, Humboldt State University, Arcata, California.

House, R. A. 1984. Evaluation of improvement techniques for salmonid spawning. Pages 5–13 *in* T. J. Hassler, editor. Proceedings: Pacific Northwest stream habitat management workshop. California Cooperative Fishery Research Unit, Humboldt State University, Arcata, California.

House, R. A. In press. Using habitat, juvenile sampling, and adult returns as indicatiors of success in a stream rehabilitation project. *In* J. W. Buell, editor. Proceedings of the stream habitat enhancement evaluation workshop. U.S. Department of Energy, Bonneville Power Adminstration, Division of Fish and Wildlife, Portland, Oregon.

House, R. A., and P. L. Boehne. 1985. Evaluation of instream enhancement structures for salmonid spawning and rearing in a coastal Oregon stream. North American Journal of Fisheries Management 5:238–295.

House, R. A., and P. L. Boehne. 1986. Effects of instream structures on salmonid habitat and populations in Tobe Creek, Oregon. North American Journal of Fisheries Management 6:38–46.

Hunt, R. L. 1976. A long-term evaluation of trout habitat development and its relation to improving management-related research. Transactions of the American Fisheries Society 105:361–365.

Klassen, H. D., and T. G. Northcote. 1986. Stream bed configuration and stability following gabion weir placement to enhance salmonid production in a logged watershed subjected to debris torrents. Canadian Journal of Forest Research 16:197–203.

Larkin, P. A. 1974. Play it again Sam—an essay on salmon enhancement. Journal of the Fisheries Research Board of Canada 31:1433–1459.

Moreau, J. E. 1984. Anadromous salmonid habitat enhancement by boulder placement in Hurdygurdy Creek, California. Pages 97–116 *in* T. J. Hassler, editor. Proceedings: Pacific Northwest stream habitat management workshop. California Cooperative Fishery Research Unit, Humboldt State University, Arcata, California.

Murphy, M. L., J. Heifetz, S. C. Johnson, K. V. Koski, and J. F. Thedinga. 1986. Effects of clear-cut logging with and without buffer strips on juvenile salmonids in Alaskan streams. Canadian Journal of Fisheries and Aquatic Sciences 43:1521–1533.

Reeves, G. H., J. D. Hall, T. D. Roelofs, T. L. Hickman, and C. O. Baker. 1991. Rehabilitating and modifying stream habitat. American Fisheries Society Special Publication 19:519–557.

Reeves, G. H., and T. D. Roelofs. 1982. Rehabilitating and enhancing stream habitat: 2. Field applications. U.S. Forest Service General Technical Report PNW-140.

Rodgers, J. D. 1986. The winter distribution, movement, and smolt transformation of juvenile coho salmon in an Oregon coastal stream. Master's thesis. Oregon State University, Corvallis.

Rodgers, J. D., R. D. Ewing, and J. D. Hall. 1987. Physiological changes during seaward migration of wild juvenile coho salmon (*Oncorhynchus kisutch*). Canadian Journal of Fisheries and Aquatic Sciences 44:452–457.

West, J. R. 1984. Enhancement of salmon and steelhead spawning and rearing conditions in the Scott and Salmon Rivers, California. Pages 117–127 *in* T. J. Hassler, editor. Proceedings: Pacific Northwest stream habitat management workshop. California Cooperative Fishery Research Unit, Humboldt State University, Arcata, California.

White, R. J. 1975. In-stream management for wild trout. Pages 48–58 *in* W. King, editor. Wild trout management. Trout Unlimited, Vienna, Virginia.

Whelan, W. G., M. F. O'Connell, and R. N. Hefford. 1989. An improved trap design for counting migrating fish. North American Journal of Fisheries Management 9:245–248.

American Fisheries Society Symposium 10:68–77, 1991

Planning and Evaluating Habitat Projects for Anadromous Salmonids

FRED H. EVEREST,[1] JAMES R. SEDELL, AND GORDON H. REEVES
U.S. Forest Service, Pacific Northwest Research Station
3200 Jefferson Way, Corvallis, Oregon 97331, USA

MASON D. BRYANT
U.S. Forest Service, Pacific Northwest Research Station
Post Office Box 20909, Juneau, Alaska 99802, USA

Abstract.—Improvement of salmonid habitat is currently being emphasized by state, provincial, and federal agencies and private interest groups throughout much of the range of anadromous fish in western North America. Once the decision is made to improve habitat in a stream system, planning, implementing, and evaluating improvements must address the appropriate spatial and temporal scales, including (1) subbasin inventory information for all seasons of the year—a departure from the usual site or reach inventory normally done during summer, (2) a thorough analysis of factors limiting fish production in the subbasin during all seasons of the year, (3) identification of improvement techniques that address limiting factors, and (4) selection of sites for habitat projects in the basin. Items (1) and (2) are usually done by fishery biologists; items (3) and (4) are more interdisciplinary, thereby requiring skills of both biologists and hydraulic engineers. The evaluation of habitat projects encompasses physical, biological, and economic aspects that must be placed in an appropriate spatial and temporal context. Some aspects of improvement are more easily measured than others; for example, physical changes resulting from habitat improvement are easily quantified, but must be placed in a subbasin context to assess their overall contribution to the system. Net gains or losses in fish production can best be assessed by enumerating smolt outmigrants from the subbasin, but increased survival of juvenile salmonids—from which smolt production can be estimated—may be adequate. Assessing economic efficiency of projects is the final and perhaps most difficult step in evaluation because estimating the economic value of fisheries is difficult and imprecise with the current state of knowledge. In this paper, we propose a sequential process for planning and evaluating habitat improvement projects, and we discuss examples from northwest Oregon and southeast Alaska.

Natural freshwater habitats of western anadromous salmonids have undergone massive alterations since the early 1900s. Habitat changes caused by hydropower development, agriculture, grazing, mining, and forest harvest have resulted in absolute losses of habitat where human-made barriers have denied adult salmonids access to previously used spawning areas and have caused insidious losses in habitat quality through simplification of habitat structure and loss of water quantity and quality. Obvious habitat losses and declining fish runs have resulted in various laws and regulations leading to progress in protection of salmonid habitats during the past three decades. Consequently, the rate of habitat loss has declined. Continued efforts to protect remaining habitats of western anadromous salmonids are extremely important and remain a higher priority than habitat rehabilitation because protecting habitat costs less than rehabilitating it. Protection also has more predictable results because many rehabilitation techniques are mostly experimental and unproven or have never been adequately evaluated (Hall and Baker 1982; Reeves and Roelofs 1982; Reeves et al. 1991).

Declines in catches of economically important anadromous salmonids, coupled with obvious habitat losses, have led to intense efforts by fishery managers and legislators to restore dwindling stocks. Major efforts have been made to regulate fisheries and improve degraded habitats. Extensive habitat rehabilitation programs for anadromous salmonids are currently under way in the western United States and Canada, and an estimated $100 million will be spent by the two countries on this activity in the next decade. The science of habitat rehabilitation is embryonic, however, and many facets of the program lack sufficient information for successful use of available funding. Other problems, including inade-

[1]Present address: U.S. Forest Service, Forestry Sciences Laboratory, Post Office Box 20909, Juneau, Alaska 99802, USA.

quate planning and lack of experienced personnel for implementing habitat projects, have frequently resulted in unsuccessful attempts to rehabilitate degraded habitats.

The best long-term solution to habitat restoration is good watershed and riparian management. In the short term, however, properly planned and executed habitat restoration work can provide an interim solution until long-term management objectives are met. We suggest a sequential process for planning, implementing, and evaluating habitat projects to rapidly increase the knowledge base for future habitat work and maximize the potential for success of current programs. Such a procedure is described below and is followed with examples of habitat projects in Oregon and southeast Alaska.

Planning

Managers responsible for planning and implementing habitat rehabilitation programs for anadromous salmonids must recognize that the present state of the art is primitive and detailed information on successful techniques and their biological effects are mostly lacking. Habitat rehabilitation efforts for anadromous salmonids have been under way in the western United States for more than a decade, but few projects have been adequately planned and evaluated; the knowledge base consequently has expanded slowly. Managers with large budgets for restoring habitat but with inadequate knowledge to successfully implement their programs run a high risk of wasting financial resources and of program failure. To reduce the risk of failure now and in the future, specific steps must be taken in planning and evaluating habitat work. Detailed planning of programs and projects must precede implementation of all habitat work. The success or failure of habitat work depends heavily on adequate planning. Key aspects of planning are discussed below.

Program Planning

Two levels of planning must precede habitat work. The first level, program planning, establishes the goals, objectives, priorities, and criteria for a geographically broad-based effort. Program planning provides the context for accomplishing more specific project planning. Program planners, with their overview perspective, can orchestrate a broad program with multiple objectives, and can help project planners avoid unsuccessful techniques, duplicated effort, or projects not fitting the overall context. Some of the key items that program managers must address are discussed below.

Coordination of financial resources.—Program managers are usually responsible for distributing funds designated for habitat rehabilitation. Funds can be divided evenly among several administrative subunits, which scatters the restoration effort over a wide geographic area with small simultaneous efforts occurring in many subbasins. Another, and perhaps better, approach is for program managers to set priorities for restoration work in subbasins within their areas of responsibility, thereby concentrating financial resources in the highest priority subbasin. When work in that subbasin is completed, the effort can be shifted to the subbasin with the next highest priority. This approach requires more coordination of staff and support resources, but it avoids wasting funds on small, widely scattered, uncoordinated projects.

Selection of subbasins.—Much habitat improvement work is still experimental, and predicting the effects a project will have is difficult with present knowledge. With this in mind, managers should attempt habitat restoration in severely damaged habitat as a first priority and avoid enhancement efforts in healthy habitats where implementing a project could either increase or inadvertently decrease current production. Basins with the most degraded habitat, or with the most opportunity for improving access for adult salmonids, probably have the greatest chances for successful work. Areas with simplified habitat structure and unstable channels may require watershed restoration before habitat restoration is begun in the channels.

Selection of species.—Program managers responsible for habitat work in large geographic areas may have to contend with habitat needs of eight or more species and races of anadromous salmonids and two or three age-classes of fish within each group. Changes in stream environments that improve habitat for one species or race might reduce or degrade habitat for another. Even within a species, increasing habitat for one age-class might reduce habitat for another. These intraspecific and interspecific dynamics require careful consideration by program managers so that a program emphasizing one species or age-group does not do so at the expense of others. Selection of species to emphasize usually requires interagency coordination because areas of agency responsibility frequently overlap.

Selection of personnel.—Project planning and implementation require the best efforts of the most experienced field fishery biologists because habitat improvement is presently as much an art as a science. Because habitat improvement is a relatively recent facet of most fishery management programs, only a few biologists in the western United States have extensive experience with this kind of work. To achieve success, those responsible for habitat work must receive the best consulting advice available from experienced habitat biologists, geomorphologists, hydrologists, and hydraulic engineers. Program managers need to coordinate experienced people and use these people to train less-experienced biologists.

Coordination of evaluation programs.—The need to evaluate habitat improvement work can not be overemphasized. With present knowledge, predicting the biological effects (especially smolt production) resulting from habitat projects is impossible. Projects planned and implemented with the best intentions might produce no tangible biological benefits because of inadequate knowledge or faulty assumptions. The only way this knowledge gap can be filled is to thoroughly evaluate a certain percentage of projects of each type and quickly use that information to guide future aspects of the program. Program managers with their broad overview are best able to design and coordinate an evaluation program.

Project Planning

Project planning, a more specific activity than program planning, must provide a spatial and temporal context for evaluating and comparing proposed projects. The minimum spatial scale for this level of planning should be a stream subbasin of not less than 50 km^2. Smaller areas may be inadequate because juvenile and adult anadromous salmonids are so mobile and responsive to changes in environmental variables. Seasonal, annual, and perennial changes in streamflow and water temperature, for example, can result in major shifts in distribution of fish within a subbasin, potentially affecting the way habitat projects are perceived and temporally used by fish. Planning specific projects therefore requires knowledge of fish distribution, abundance, and habitat use at the subbasin level. The temporal scale for project planning is also an important consideration. Project planners should have data on basinwide habitat availability for all seasons, preferably for more than 1 year. Without such data, a meaningful analysis of factors limiting production of salmonids in a subbasin cannot be completed.

Project planners should adhere closely to a stepwise sequence when planning specific habitat projects, and they should coordinate closely with program managers to assure that specific projects are meeting program goals. A useful sequence for project planning is given below.

Preimprovement inventory.—Inventory techniques for assessing habitat availability and use by salmonids at the subbasin level are rare. We recommend use of the basinwide, statistically based technique recently developed by Hankin and Reeves (1988). This procedure provides the baseline information for the second step in project planning: limiting-factors analysis.

Analysis of limiting factors.—Isolating factors that limit production of salmonids in a basin is a critical step in project planning. Limiting-factors analyses, at a minimum, require basinwide data on habitat availability during all seasons and information on seasonal distribution and abundance of fish within the habitats. Data from more than 1 year are desirable because of natural variability in fish populations and environmental variables. Basinwide inventory data must be analyzed for each species of fish and each season to identify specific habitat variables most likely to limit freshwater production of the target species. Identifying limiting factors is not an exact science, but progress is being made. A diagnostic dichotomous key for identifying limiting factors from stream-inventory information has been developed by Reeves et al. (1989) for coho salmon *Oncorhynchus kisutch*, and similar keys for steelhead *O. mykiss* and chinook salmon *O. tshawytscha* are being prepared.

Site selection.—Several criteria are important when specific sites are selected within a subbasin for habitat restoration work. Easy access for crews and heavy equipment is needed to assure that riparian and watershed damage is minimized during the projects. Areas in the subbasin with the most degraded habitat conditions should have the highest priority for restoration. Work should be avoided in good-quality habitats. Habitat work should be confined to areas likely to receive adequate seeding on a perennial basis. Project planners, in consultation with hydrologists, geomorphologists, and hydraulic engineers, should select sites where channel stability and cross-sectional geometry favor successful habitat work. Plans to provide anadromous salmonids with upstream access over specific barriers should be discussed with state and provincial fish and wildlife agencies because anadromous stocks often displace resident stocks. Finally, if multiple projects are planned within a subbasin, they should be carefully coordinated to achieve the best results.

Selection of techniques and materials.—Once limiting factors have been identified and sites selected for habitat work, the next step is selecting habitat restoration techniques and materials compatible with the physical and biological characteristics of the stream and the identified needs. If stream channels are unstable, the initial work might best be directed at watershed restoration; this can be followed by work in the stream channel. Again, we emphasize that this should be done in consultation with experienced fishery biologists, hydrologists, geomorphologists, and hydraulic engineers.

The material used in habitat structures needs special consideration. The dynamics of stream systems result in frequent channel changes and failure of habitat structures. Habitat improvement structures therefore should be built from native materials whenever possible; for example, native boulders and logs should be used to construct weirs, deflectors, groins, and other habitat features and not exotic materials such as wire-basket gabions, timbers, tires, or concrete. If structures are disaggregated by high-flow events or left marooned by channel changes, the remaining native materials will not interfere with the aesthetic characteristics of the stream as nonnative materials often do.

Selection of implementation procedures.—Planning for installation of proposed habitat work is a critical step in habitat projects. Planning at this level includes selecting personnel, equipment, contractors, materials, and the timing of project construction. It is wise to choose the most experienced personnel and contractors to install habitat projects. Work should be planned for the time of year least likely to cause biological or environmental damage; for example, avoid doing work in active stream channels when salmonid eggs are incubating in the streambed.

Evaluation of projects.—Two types of evaluation activities are necessary at the project-planning level. First, the biological benefits resulting from the project must be estimated (for example, smolts produced) so that preliminary estimate of cost-effectiveness can be made. Estimates of cost-effectiveness and total biological benefits will help managers set priorities for several proposed projects. Second, project planners should select and plan for the type and intensity of evaluation that will be used to assess the actual benefits that accrue from the project.

Project Implementation

Once program and project planning have been completed, managers can begin to install projects. This next critical phase in a habitat restoration program can succeed only if plans are accurately translated into the desired physical changes in the habitat. Only the most experienced habitat biologists should be responsible for administrating planned projects. Communication among planners, project administrators, and contractors must be clear and frequent to assure that plans are accurately and completely followed during construction. Without a determined effort by the project administrator, the most comprehensive planning effort can be ruined during project implementation.

Project Evaluation

Evaluating projects to understand their benefits is just as important as constructing the projects themselves. Because knowledge of the effects of habitat improvement on anadromous salmonids is inadequate, the only way the knowledge base can be expanded is through quantitative evaluations of habitat work. Evaluations can be used to examine three aspects of habitat work: (1) physical changes in habitat resulting from projects, (2) changes in biological production, and (3) the cost-effectiveness of the work. Intensive evaluations simultaneously examining all three aspects of projects are the most meaningful. Evaluations must be conducted at the same spatial scale as project planning—that is, subbasins with a minimum size of 50 km^2. The minimum time for evaluations should span two life cycles of the species being evaluated because full benefits of habitat work might not be realized for some time, the life expectancy of most projects is unknown, and high variability in environmental conditions and salmonid populations usually precludes effective short-term studies. Each of the main aspects of evaluation is discussed briefly below.

Evaluating Physical Changes

The objective of habitat improvement work is to create habitat characteristics that result in increased production of desired fish species. Habitat work is designed to alleviate factors limiting fish production. Habitat work might target improvements in such physical variables as water temperature, habitat structure in the stream channel, or riparian vegetation and watershed improvement. Most work, however, will be aimed at habitat work in the stream channel. An effective evaluation of work conducted in the stream channel must quantify all habitat in a selected subbasin for at least 1 year, and preferably longer, before habitat manipulation begins. At present, this is best accomplished by using the habitat types described by Bisson et al. (1982) and the basin inventory procedures of Hankin and Reeves (1988). When pretreatment habitat has been described, the data can be compared with posttreatment habitat surveys to assess changes in habitat related to improvement activities. Inventory data collected at the subbasin scale allow managers to compare pretreatment and posttreatment habitat conditions within a sufficiently broad context to accurately assess the effects of the work.

Evaluating Biological Changes

The biological effects of habitat work must also be evaluated at the subbasin scale to provide adequate context for interpreting change. The effects of habitat improvement on fish community structure in streams need to be assessed rather than the changes in the population of a single species. Habitat work can affect populations of all salmonid and nonsalmonid species in streams in ways not readily predictable, so all species should be included in evaluating habitat work. For anadromous salmonids, smolts are the critical life history stage to evaluate. If habitat work for a desired species does not ultimately increase the output of smolts, then the effort should probably be considered a failure. When changes in smolt numbers are evaluated, pretreatment and posttreatment data for several years are desirable because smolt numbers naturally differ significantly from year to year. Wherever possible, isolating smolt production in the treatment area of a stream from untreated areas by strategic placement of smolt traps is desireable.

Evaluating Economic Benefits

A preliminary benefit–cost analysis based on estimated benefits of a proposed project is usually conducted in the project-planning phase of a habitat improvement effort. This must be followed by one or more actual benefit–cost analyses after the project is finished and has been evaluated. Actual project costs are easy to determine but benefits are not, because of the difficulty in evaluating sport and commercial fisheries for anadromous salmonids (Everest and Talhelm 1982). A sequential benefit–cost analysis can nevertheless be conducted by following procedures described by Everest and Talhelm (1982) to assess the benefit–cost ratio and present net worth of habitat projects.

Example 1: Project Planning, Implementation, and Evaluation in Oregon

A major habitat improvement project for anadromous salmonids was begun by the U.S. Forest Service in Fish Creek, an upriver tributary of the Clackamas River, in northwest Oregon in 1982 (Everest and Sedell 1984). The Bonneville Power Administration became a funding partner of the project in 1983. Fish Creek is a 171-km^2 watershed containing 16.7 km of stream habitat used by anadromous salmonids. Variation in annual flow ranges from 0.5 m^3/s in late summer to more than 100 m^3/s during winter freshets. Stream gradient averages 2 to 3% in most of the area used by anadromous fish. The terrain is steep and mountainous with bluffs in the lower canyons typical of the Columbia River Basalt formation. The valley bottoms are typically narrow with incised stream channels and narrow floodplains.

The primary anadromous species in the basin are steelhead and coho salmon; low numbers of spring-run chinook salmon are also present. Habitat surveys in the basin in 1959, 1965, and 1982 showed a major loss of pool habitat over the 23 years. Pool habitat totaled about 45% of wetted surface area in 1959 but only 12% in 1982. A major flood in 1964 reduced pool habitat to about 25%, and the remainder of the loss was related to timber harvest, road construction, and past logjam removal practices in the basin. Habitat conditions in 1982 were typified by long homogenous reaches of boulder-armored riffles punctuated by a few pools and glides.

The loss of pool habitat was believed to be most detrimental to production of juvenile coho salmon. Juvenile coho salmon prefer pools and quiet edge habitats with abundant large woody debris for rearing in summer and quiet backwaters with abundant cover in winter. Low availability of these habitat features in the basin was thought to limit coho salmon smolt production.

Planning

Basin planning for habitat improvement in Fish Creek was started in 1982 by personnel of the

TABLE 1.—Area of habitat types in Fish Creek used by coho salmon for rearing, and numbers, densities, and biomass of juvenile coho salmon by habitat type, September 1982.

Habitat type	Unit biomass of fish (g/m²)	Area in stream (m²)	Estimated number of fish by habitat	Estimated biomass of fish by habitat (g)	Number of fish/m²
Alcove	0.80	1,080	140	870	0.13
Riffle	0.05	70,350	1,040	3,380	0.01
Side channel	0.78	1,600	180	1,250	0.11
Pool	0.35	8,110	290	2,850	0.04
Beaver pond	6.34	190	260	1,200	1.37
Total or mean	0.12	81,330	1,910	9,550	0.02

Mount Hood National Forest, the Pacific Northwest Research Station, and the Oregon Department of Fish and Wildlife. A final interagency plan was approved in 1987, but frequent meetings of habitat managers and researchers helped guide the effort during the interim. Steelhead and coho salmon were identified as primary target species for improvement work, and because habitat improvement was largely unproven on large, high-energy streams like Fish Creek, the work began with testing of small prototype projects. This example describes the first prototype project designed specifically to increase coho salmon production in the basin.

A survey of habitat and fish numbers in the basin in 1982 indicated that densities of juvenile coho salmon were highest in beaver ponds on the floodplain of Fish Creek, but that this habitat type was least abundant in the basin (Table 1). Beaver ponds made up only 0.2% of the wetted surface area of habitat in the basin, but they contained 13.6% of the late-summer standing crop of juvenile coho salmon. Because beaver pond habitat represents ideal summer and winter rearing conditions for young coho salmon, and a disproportionate share of young fish are raised in beaver ponds in relation to the availability of the ponds, plans were made to increase the amount of this habitat type in the basin. The first project was to develop a large vernal pond on a flood terrace next to Fish Creek. The pond historically contained water in winter and spring but was dry in summer and fall. The pond was heavily used by beavers in the wet season and contained an abundance of large and small woody debris from beaver activity and salvage logging.

Implementation

The pond was developed in 1983 as a perennial coho salmon rearing area by building a gravity-fed pipeline from Fish Creek to the flood terrace about 200 m below the pipe intake (Everest and Sedell 1984). The 25-cm-diameter pipe was capable of delivering 35 L/s to the vernal pond. The physical development of the pond was a success. The water source maintained a nearly constant-level pond 90 m long and 60 m wide with a surface area of about 0.5 hectare. Depth ranged from about 0.2 to 1.25 m, and pond volume was about 3,600 m³. A second water source was developed by diverting a small tributary of Fish Creek into the north end of the pond. Additional developments included a fish ladder with a downstream-migrant trap at the pond outlet and placement of spawning gravel in the inlet stream between the pipe outfall and the pond. Completion of the pond increased beaver pond habitat used by anadromous salmonids to 6% of the total wetted surface area.

Evaluation

Coho salmon smolt production from the pond was evaluated each spring from 1985 through 1987. The pond was initially stocked in spring 1984 when 1,326 age-0 coho salmon were collected from Fish Creek and placed in the pond (Everest et al. 1985). An unknown number of naturally reproduced age-0 fish also entered the pond from coho salmon redds in the north inlet to the pond. Between April 15 and June 8, 1985, 493 coho salmon smolts left the pond (Everest et al. 1986). Mean length of migrants leaving the pond was significantly greater than that of coho salmon smolts leaving Fish Creek, indicating that the pond was understocked in its first year of production.

No age-0 coho salmon were stocked in the pond in 1985, but in January 1985 seven adult female coho salmon and five males were trapped downstream at North Fork Dam and transported to the pond. The fish spawned naturally in the inlets, and an unknown number of emergent larvae migrated downstream into the pond in spring 1985. The 1986 smolt migration resulting from the natural reproduction totaled 1,196 fish (Everest et al. 1987).

In 1986, the pond was stocked with 5,000 Clackamas stock age-0 coho salmon from Clackamas Fish Hatchery (Everest et al. 1987). In total 1,234 smolts left the pond between February 20 and June 5, 1987. Mean length of pond smolts was significantly larger than those produced in Fish Creek, indicating that the pond was probably still below carrying capacity.

The pond, even though seemingly not fully stocked, has made a significant contribution to coho salmon production in the Fish Creek basin. The pond increased coho salmon smolt production in the basin by 19, 102, and 49%, respectively, in 1985, 1986, and 1987.

A benefit–cost analysis indicated that the pond is a cost-effective project at the current level of smolt production. Construction costs were $24,330, and an annual maintenance cost of $100 per year is expected. Based on an average annual smolt production of 1,200 fish, benefits were calculated as follows:

- 1,200 smolts × 7.5% smolt-to-adult survival (Oregon Department of Fish and Wildlife 1981) = 90 adults;
- 90 adults × 7:1 catch:escapement ratio (Meyer 1982) = 79 adults harvested;
- 79 adults × 64% commercial harvest (Meyer 1982) = 51 adults in commercial harvest;
- 51 adults × 3.2 kg × $3.24/kg = $525 dockside commercial value annually;
- 79 adults × 36% sport harvest (Meyer 1982) = 28 adults in sport harvest;
- 28 adults × $107/adult (Meyer 1982) = $2,996 sport benefit annually;
- $525 commercial benefit + $2,996 sport benefit = $3,521 annual benefit.

The benefit–cost ratio of the project is 1.6:1, and 1.2:1, at discount rates of 4 and 7%, respectively, figured on a project life of 20 years (calculations per Everest and Talhelm 1982). Benefits began to accrue in the third year of the project when adults from the first year-class of smolts entered the fishery. The actual benefits will probably be higher because the pond has not yet been seeded to capacity.

Critique

Planning, implementing, and evaluating the preceding coho salmon habitat improvement project on Fish Creek is one of the most complete examples on the subject that we are aware of. Most of our present recommendations for planning, executing, and evaluating a habitat project were also used in developing the Fish Creek project. Planning was conducted by an interagency team for the entire Fish Creek subbasin used by anadromous fish. Selection of techniques and sites for habitat work was done by an interdisciplinary group of managers and scientists after a complete basin inventory of fish populations and habitats and an analysis of limiting factors. The project was implemented by an experienced habitat biologist working with an experienced contractor. Finally, the work is being thoroughly evaluated and includes a count of coho salmon smolts attributable to the project and an analysis of the economic viability of the project. Data from this project should be useful for estimating the benefits that are likely to result from similar projects in a broad geographic area of western Oregon and Washington.

Example 2: Coho Salmon Habitat Improvement and Monitoring in Alaska

Example 2 is more typical of a broad cross section of habitat rehabilitation efforts and their evaluations being done annually in western North America. The evaluation of most habitat work is carried out without adequate staff or financial resources and therefore lacks the systematic approach recommended in this paper. The results of such work are difficult to interpret and may be inconclusive or lead to erroneous conclusions about the success of habitat projects.

Large woody debris in southeast Alaska streams is an important component of stream habitat for juvenile coho salmon (Dolloff 1983; Bryant 1985; Murphy et al. 1985). But many streams with extensively logged riparian zones will have limited future sources of large wood for habitat for the next 80–100 years (Bryant 1980). Remaining trees in riparian zones of those streams may contribute most efficiently to habitat for juvenile coho salmon if the trees are artificially placed in streams, rather than being allowed to fall naturally. Artificial felling and placement should be done only in streams adequately seeded with coho salmon fry but with inadequate pool or off-channel habitat. This procedure was used in a test project begun in Kennel Creek, Alaska, in 1986.

Planning

Kennel Creek is on Chichagof Island (57°53′N, 35°10′W) about 110 km southwest of Juneau, Alaska. Its riparian zone was extensively logged in 1966, and periodic timber harvest activities have continued during the past 20 years. Few trees greater than 1 m in diameter at breast height are left along the streambank. The basin was selected for habitat improvement based on logging history and an analysis of aerial photographs of the area. Qualitative surveys of fish populations in 1977 showed that age-0 coho salmon were present throughout the main stem of Kennel Creek, and subsequent surveys in 1986 indicated that the stream was well seeded with age-0 coho salmon. Habitat was classified from aerial photographs by using the channel type classification system developed by Paustian et al. (1983) and was verified through on-the-ground surveys. Most of the main channel lacked large woody debris and consisted primarily of long, homogeneous riffles. Debris accumulations were infrequent throughout the lower 6 km of the stream.

After the basin and the treatment reach were selected, a more intensive survey was done to determine the type and placement of structures needed to increase coho salmon production in the system. The apparent limiting factor was identified as a lack of off-channel habitat along the main stem of the river. Placing whole trees, including attached rootwads, in the stream was the best method for increasing this type of habitat (Bryant 1985). Six sites along lower Kennel Creek were selected for this treatment by a team of fisheries biologists, hydrologists, and soil scientists.

A specific habitat objective was assigned to each site as follows: site 1—placing a large Sitka spruce *Picea sitchensis* into the channel to create a backwater pool with rootwad cover; site 2—placing a single Sitka spruce across the channel to create a debris dam and plunge pool; site 3—felling Sitka spruce and western hemlock *Tsuga hetrophylla* over an existing pool to form overhead cover and provide instream cover; site 4—using a spruce next to the channel to form a lateral scour pool with rootwad cover; site 5—placing two Sitka spruce in the channel along a riffle at a tributary junction to provide cover and increase the depth of an existing backwater pool; and site 6—placing two Sitka spruce to create a backwater pool with rootwad cover and to deflect flows into a side channel.

Because whole trees with attached rootwads were key elements of the structures, explosives were selected to uproot trees and directionally fell them into the stream. Each site was examined by an experienced blaster to determine the requirments to implement the project. At site 3, directional felling with saws could accomplish the objective. Site 4 was eliminated because adjacent trees prevented blasting of the selected tree.

Before the onsite surveys in 1986 were conducted, a total cost of the project, not including personnel, was estimated at $2,180 for equipment, supplies, and travel. Personnel time was estimated at three person-months for all personnel working on the project. These estimates did not include costs for evaluation and monitoring. Projected benefits were computed from estimates that the project would create an additional 0.4 hectares of habitat and provide 165 additional coho salmon to the commercial catch at $3.00/kg for an annual return of $1,862/year. In reality, these estimates may not be close.

Implementation

Work at all five sites was done in 5 d in July 1986 after all coho salmon larvae had emerged. Explosives were used at four of the five sites. At site 3, two trees were cut and directionally felled across the pool. Two primary considerations for how long the work would take were access to the stream and availability of personnel. Although age-0 coho salmon were present, they were not abundant near the area of activity.

The methods and problems encountered during this project, and during a similar project that used explosives, were described by Henke (1988). Problems were encountered with trees whose extensive root systems impeded the digging of holes for the charges and also with controlling the direction of the fall. Soil conditions were an important factor and affected the amount of explosive needed to fell trees.

Monitoring and Evaluation

After the site-selection surveys, stream maps showing general channel morphology, water depth, velocity characteristics, and location of existing wood debris were constructed for each site. In addition, channel cross sections for documenting water depth and channel shape were installed at each site. Similar measurements were made for control sites at reaches adjacent to the treatment sites. Immediately after the treatment, locations of new debris and channel changes were recorded on the maps. Each site was remeasured in 1987 and 1988. Periodic remeasurement of the physical habitat is planned for future years.

Salmonid populations in the treated reaches and adjacent control reaches were estimated by mark–recapture methods immediately before the treatment and shortly afterward. The following year, surveys by snorkeling were used to count populations in these areas. Yearly monitoring is not currently funded and is being done informally. No provision was made to monitor smolt production from the habitat provided by the structures.

Critique

Kennel Creek was selected for habitat work because the watershed had been logged and there was a general feeling that habitat rehabilitation was necessary. Like most U.S. Forest Service enhancement projects, structures were included in plans as "targets," and once embedded in the planning process, they became difficult to alter or delete without the loss of funds. For Kennel Creek, the targets were assigned based on a map and an inspection of aerial photographs. The ground was surveyed and the sites were selected after the targets were assigned to the basin. A complete basin survey of the stream was not done before the enhancement project was started.

Usually the most difficult part of an enhancement project is the monitoring and evaluating phase, especially with respect to fish responses to structural improvements. Because monitoring is expensive and time-consuming, it should be included in the initial planning and budget process. In the Kennel Creek project, montoring was included in the planning documentation but was not adequately identified in the budget process for the project. This and the transfer of the district biologist caused a gap in monitoring. Although incomplete, the data collected were sufficient to determine the relative use of the structures.

Although qualitative and anecdotal data indicated that the system was completely stocked by age-0 coho salmon, estimates of the density of rearing age-0 coho salmon for each habitat type in the basin were not made before the treatments. It will therefore be difficult to adequately assess the effect of the treatments on production of age-0 coho salmon. Because populations were counted in the area of the treatment before and after the trees were placed into the stream, relative use by coho salmon of the new habitat may be compared with other areas in the system to determine if the structures increased the amount of area available to juvenile salmonids. A more rigorous evaluation plan, including provisions to monitor smolt output with a smolt trap, would provide a more complete assessment of the effectiveness of the structures.

Evaluation of the physical changes in the stream caused by the structures was an important part of the project. The data collected before and after the treatment will provide an effective means of determining if the project accomplished the objectives defined during planning. With these data and the results from studies of habitat use by juvenile coho salmon elsewhere in southeast Alaska, the success, or lack thereof, of the project can be qualitatively assessed. In addition, periodic remeasurement of the physical features of the treated and control sections will provide a picture of the stream processes that change and will define the aquatic habitat of the system. Often, an unstable large structure may initially appear to be a failure, but weathering and repositioning by high flows may eventually make such structures effective components of coho salmon habitat.

In Kennel Creek, where complete assessment of the biological output of the enhancement project was not possible, the data collected on the physical aspects of the project will provide at least some indication of effectiveness. Through remeasuring and remapping of the area, the effect of the trees on the stream and changes in habitat resulting from the treatment will be relatively easy to determine. Accompanied by surveys of the salmonid population, these data will provide at least a yes or no answer to the question, "Is this method an effective means of creating new habitat for juvenile coho salmon?" It might not, however, be able to answer the question, "How many more coho smolts were produced as a result of the project?" Answers to the second question are needed if results of this work are to be extrapolated beyond the immediate project area.

Summary

An urgent need exists for managers of habitats used by anadromous salmonids to use a systematic, sequential process for planning, implementing, and evaluating habitat rehabilitation projects. More than $100 million will be spent on habitat rehabilitation and enhancement for anadromous salmonids over the next decade, and much of it could be wasted unless managers take a systematic, scientific approach to habitat work. We suggest that habitat managers follow a stepwise sequence for program planning, project planning, implementation, and project evaluation. Evaluation is a key element in all habitat programs because it is the only means of assessing the success of habitat projects and providing information to guide and improve future aspects of a habitat program.

References

Bisson, P. A., J. L. Neilson, R. A. Palmason, and L. E. Grove. 1982. A system of naming habitat types in small streams, with examples of habitat utilization by salmonids during low streamflow. Pages 62–73 *in* N. B. Armantrout, editor. Acquisition and utilization of aquatic habitat inventory information. American Fisheries Society, Western Division, Bethesda, Maryland.

Bryant, M. D. 1980. Evolution of large organic debris after timber harvest: Maybeso Creek, 1949 to 1978. U.S. Forest Service General Technical Report PNW-101.

Bryant, M. D. 1985. Changes thirty years after logging in large woody debris, and its use by salmonids. U.S. Forest Service General Technical Report RM-120:329–334.

Dolloff, C. A. 1983. The relationships of wood debris to juvenile salmonid production and microhabitat selection in small southeast Alaska streams. Doctoral dissertation. Montana State University, Bozeman.

Everest, F. H., G. H. Reeves, J. R. Sedell, D. B. Hohler, and T. Cain. 1987. The effects of habitat enhancement on steelhead trout and coho salmon production, habitat utilization, and habitat availability in Fish Creek, Oregon, 1983–1986. U.S. Department of Energy, Bonneville Power Administration, Division of Fish and Wildlife, Portland, Oregon.

Everest, F. H., G. H. Reeves, J. R. Sedell, J. Wolfe, D. Hohler, and D. A. Heller. 1986. Abundance, behavior, and habitat utilization by coho salmon and steelhead trout in Fish Creek, Oregon, as influenced by habitat enhancement. U.S. Department of Energy, Bonneville Power Administration, Division of Fish and Wildlife, Portland, Oregon.

Everest, F. H., and J. R. Sedell. 1984. Evaluation of fisheries enhancement projects on Fish Creek and Wash Creek, 1982 and 1983. U.S. Department of Energy, Bonneville Power Administration, Division of Fish and Wildlife, Portland, Oregon.

Everest, F. H., J. R. Sedell, G. H. Reeves, and J. Wolfe. 1985. Fisheries enhancement in the Fish Creek basin—an evaluation of in-channel and off-channel projects. U.S. Department of Energy, Bonneville Power Administration, Division of Fish and Wildlife, Portland, Oregon.

Everest, F. H., and D. R. Talhelm. 1982. Evaluating projects for improving fish and wildlife habitat on National Forests. U.S. Forest Service General Technical Report PNW-146.

Hall, J. D., and C. O. Baker. 1982. Rehabilitating and enhancing stream habitat: 1. Review and evaluation. U.S. Forest Service Station General Technical Report PNW-138.

Hankin, D. G., and G. H. Reeves. 1988. Estimating total fish abundance and total habitat area in small streams based on visual estimation methods. Canadian Journal of Fisheries and Aquatic Sciences 45:834-844.

Henke, V. L. 1988. Large woody debris additions for coho habitat enhancement. U.S. Forest Service, Alaska Region, Habitat Hotline 88-1, Juneau, Alaska.

Meyer, P. A. 1982. Net economic values for salmon and steelhead from the Columbia River system. NOAA (National Oceanic and Atmospheric Administration) Technical Memorandum NMFS (National Marine Fisheries Service) F/NWR-3, Washington, D.C.

Murphy, M. L., K. V. Koski, J. Heifetz, S. W. Johnson, D. Kirchofer, and J. F. Thedinga. 1985. Role of large organic debris as winter habitat for juvenile salmonids in Alaska streams. Proceedings of the Annual Conference Western Association of Fish and Wildlife Agencies 64(1984):252–262.

Oregon Department of Fish and Wildlife. 1981. Comprehensive plan for production and management of Oregon's anadromous salmon and trout, part 2. Oregon Department of Fish and Wildlife, Portland.

Paustian, S. J., D. A. Marion, and D. F. Kelliher. 1983. Stream channel classification using large scale aerial photography for southeast Alaska watershed management. Pages 670–677 *in* RRM (Renewable Resources Management) symposium on the application of remote sensing to resource management. American Society of Photogrammetry, Bethesda, Maryland.

Reeves, G. H., F. H. Everest, and T. E. Nickelson. 1989. Limiting factors analysis—coho salmon. U.S. Forest Service General Technical Report PNW-245.

Reeves, G. H., J. D. Hall, T. D. Roelofs, T. L. Hickman, and C. O. Baker. 1991. Rehabilitating and modifying stream habitats. American Fisheries Society Special Publication 19:519–557.

Reeves, G. H., and T. D. Roelofs. 1982. Rehabilitating and enhancing stream habitat: 2. Field evaluations. U.S. Forest Service General Technical Report PNW-140.

American Fisheries Society Symposium 10:78–87, 1991

Hydraulic Analysis and Modeling of Fish Habitat Structures

BRUCE A. HEINER[1]

Department of Civil and Environmental Engineering
Washington State University, Pullman, Washington 99164, USA

Abstract.—Hydraulic analysis of fish habitat structures has received little attention in engineering research or in practical application. The objective of my work was to calibrate scour and discharge equations for common habitat structures. These equations predict water depths and velocities, which can be used in physical habitat models to predict quantities of habitat created or lost when structures are installed in streams. In addition to habitat estimation, this research can aid in proper placement of structures in streams and in flood-potential analysis. Structures that were physically modeled include log weirs, flow deflectors, digger logs, and a new structure called a digger weir. One existing scour equation slightly underestimated scour at log weir models. The same equation also predicted scour at models of digger weirs. The coefficient of discharge for model log weirs correlated with log diameter. Field conditions of sloped log crests and poorly sealed logs contributed to a wide variation in discharge coefficients at prototype logs. The submerged flow equation predicted backwater depths at model deflectors with less than 15% error. Three deflector scour equations tested were based on sand bed channels and did not accurately predict gravel bed model scour. Model digger logs developed the largest scour hole when the base of the log was level with the channel bed.

Fish habitat structures, most of them designed and constructed by fish biologists, have been placed in streams for years (Bossou 1954; Saunders and Smith 1962). Hydraulic analysis of fish habitat structures is a little-researched topic and yet is potentially useful in fish habitat management. In the past 25 years, understanding of the hydraulic effects of instream structures has increased considerably, but most habitat enhancement design is still based on "hydraulic intuition" (Klingeman 1984). Only recently has hydraulic engineering been used in habitat enhancement design (Klingeman et al. 1984; Beschta and Platts 1986; Orsborn 1987). Characteristics of spawning gravels have been previously described (Klingeman 1981; Scrivener and Brownlee 1981; Tappel and Bjornn 1983). Some studies on scour and deposition at deflectors are available, although results are inconclusive and most tests were conducted in uniform sand bed streams or flumes (Lane 1955; Garde et al. 1961; Franco 1967; Gill 1972; Klingeman et al. 1984). However, little has been done to develop prediction equations for discharge, water depth, velocity, or volume of scour at fish habitat structures for planning and design purposes.

Creating structures beneficial to a fish population depends on identifying the true limiting factors in the stream or basin. These limiting factors may vary with fish species, age-class, and season of year (Mason 1976; Everest and Sedell 1984). Structures alter fish habitat by changing local depths and velocities, resulting in local scour or deposition at the streambed or bank(s). Fish habitat changes that may result include spawning gravel recruitment, creation of resting areas in feeding zones, increased interstitial space and reduced water velocities for overwintering, and increased cover, either as water depth or overhangs or visual isolation. Some structures are used to control water elevations, either to create side-channel habitat or to enhance fish passage.

In this paper I use physical modeling and prototype measurements to quantify scour and water depth caused by certain instream structures.

Methods

Literature regarding hydraulic analysis of fish habitat structures is limited, so I physically tested only a few general structures. They included log weirs, solid deflectors, digger logs, and a combination of digger log and log weir called a digger weir.

Log Weirs

Discharge (Q) over a weir is calculated with the general equation

$$Q = C_d L(2g)^{1/2} (H + V_a^2/2g)^{3/2}; \quad (1)$$

L = length of the weir, not including notch;

[1]Currently with Washington Department of Fisheries, Olympia, Washington 98504, USA.

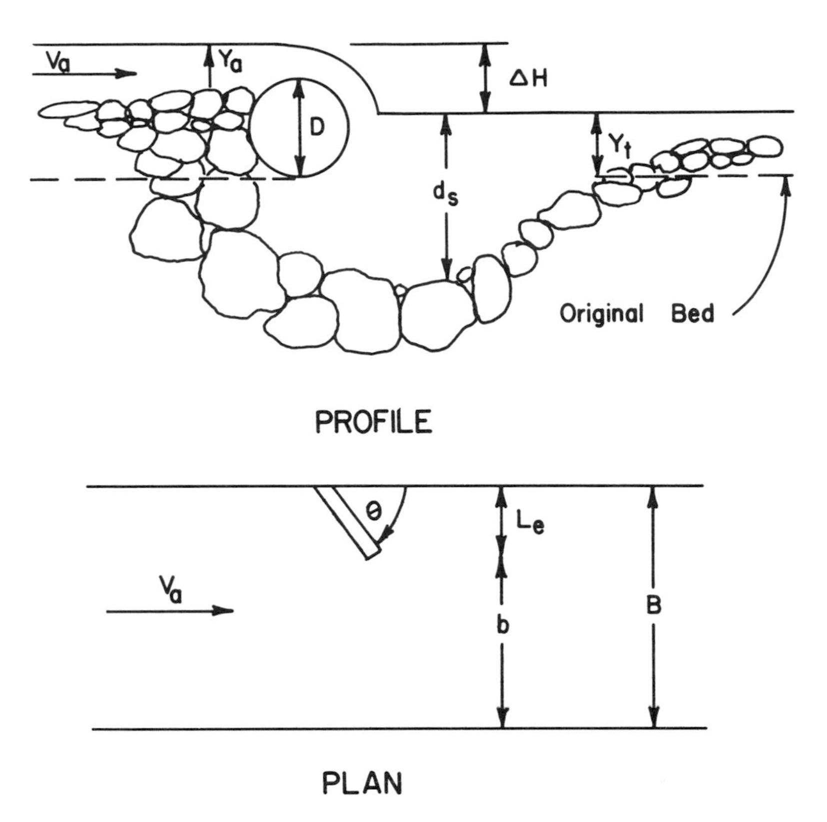

FIGURE 1.—Conceptual views of log weirs and deflectors showing log diameter (D), approach-flow depth (Y_a) and velocity (V_a), tailwater depth (Y_t), scour depth (d_s), head on weirs (ΔH), channel width (B), effective deflector length (L_e), deflector angle (Θ), and constricted channel width (b).

H = upstream water depth above the log crest;
$V_a^2/2g$ = approach-flow velocity head;
g = gravitational constant.

The coefficient of discharge (C_d) is a dimensionless value that allows model results to be applied to prototypes; for a well-sealed log, C_d depends on the log diameter (D), approach-flow velocity (V_a), approach-flow depth (Y_a), tailwater depth (Y_t), the angle of the log to the streambank (Θ), the geometry of any notch in the log, and the roughness of the log exterior (Figure 1). Discharge through a notch in a log can be estimated by using discharge coefficients from broad-crested weir tests (King 1954).

Log weir tests to determine discharge coefficients for equation (1) were performed in a steel flume 11 m long by 1.2 m wide by 1.2 m deep, with glass walls at the test section. A false wall was used to constrict the flume to a width of 0.6 m, and polyvinyl chloride (PVC) pipes bolted to a wood base were installed in the constricted section to model log weirs. Water was supplied by a recirculation system. Flow was measured with a rectangular sharp-crested weir at the flume outfall. Three diameters of pipe (0.32, 0.22, and 0.12 m) were tested with flows ranging from 0.01 to 0.27 m^3/s, and bark was also attached to each pipe to quantify roughness effects.

Log weirs, as well as deflectors, digger logs, and digger weirs, were tested in a gravel-bed flume. The flume was 7.3 m long by 0.87 m wide by 0.61 m deep, with a recirculating water supply. Flow rate was measured with a 60° triangular weir in an attached metal flume. A Plexiglas viewing window 1.2 m long was placed in the left wall 1.22 m from the downstream end of the flume. Metal rails mounted at the top of each side of the flume supported a platform on which a point gauge was mounted. Water, substrate, and structure elevations were measured with the point gauge.

The gravel used in the flume was narrowly graded to a mean diameter of 4 mm. Tests were performed with the gravel bed both horizontal and at a 2% slope. Subsurface flow was minimized by taping plastic sheets in the flume to form vertical barriers at the downstream end of the flume, at the test section, and 3.0 m from the upstream end of the flume. Flows in the flume ranged from 0.002 to 0.02 m^3/s.

Wood dowels and PVC pipe were used to model log weirs for scour research. Two diameters of pipe (0.06 and 0.03 m) and one diameter of wood dowel (0.04 m) were used in the mobile bed flume. All three log weir diameters were tested on the horizontal bed. Only the 0.04-m dowel was tested on the 2% slope. Test variables included discharge, elevation of the log crest relative to the streambed, and distance between two logs. Scour and deposition profiles below log weirs were measured with the point gage.

The variables influencing development of scour holes below log weirs are substrate gradation, D, V_a, Y_a, Y_t, height differential between upstream and downstream water surfaces (ΔH), and time (t) (Figure 1). Mason and Arumugam (1985) used data from many studies of spillways to calibrate scour equations for both prototypes and physical hydraulic models. I compared scour depths in my model to predictions from their model equation, which is

$$d_s = 3.27q^{0.60}\,\Delta H^{0.05}\,Y_t^{0.15}\,g^{0.30}\,D_m{}^{0.10}; \quad (2)$$

d_s = maximum scour depth measured from the tailwater surface;
q = flow rate per foot of channel width;
ΔH = change in head across the weir;
Y_t = tailwater depth measured from the original bed;
g = gravitational constant;
D_m = mean diameter of the substrate (D_{50}, the diameter at which 50% of the sample is smaller, by weight).

The particle sizes (0.001–0.028 m) and water surface differentials (0.325–2.15 m) of the models reviewed by Mason and Arumugam are near the range that would be encountered in the log weir design.

Deflectors

The variables affecting scour and deposition geometry around deflectors include deflector angle to the streambank (Θ), angle of the deflector crest to a horizontal plane (β), surface width of the stream (B), effective length of the deflector as measured perpendicular to the bank (L_e), substrate gradation, V_a, Y_a, Y_t, and time (Figure 1). Several deflector and groin scour equations were listed by Klingeman et al. (1984). I selected the equations by Ahmad (1951), Liu et al. (1961), and Gill (1972), as listed by Klingeman et al. (1984), to determine if they are applicable to a gravel-bed model.

Sections of ¾-inch plywood sheets were used to simulate deflectors in the mobile-bed flume. Deflector lengths of ⅙, ¼, and ⅓ channel widths (B) were modeled. The plywood was inserted well below the maximum scour depth; its crest was exposed at all flows; and it was held in place by the substrate. Only angles (Θ) of 90°, measured from the downstream wall, were tested. Other angles have been tested by Klingeman et al. (1984).

Deflectors were initially installed on one side of the flume and exposed to a range of flows. Two ¼-width deflectors also were installed concurrently on opposite sides of the flume at a 2% slope. One of the deflectors was sequentially moved farther upstream to document the effects of relative spacing on scour.

The submerged flow equation described by Fiuzat and Skogerboe (1983) was selected to predict the backwater created by flow deflectors. Their equation is

$$Q = Cg^{0.5}\,b(Y_1 - Y_3)n_1/(-\log_{10} S)n_2; \quad (3)$$

C, n_1, n_2 = empirical parameters dependent on b/B;
g = gravitational constant;
b = channel opening at the constriction;
Y_1 = flow depth at a distance b upstream of the constriction,
Y_3 = flow depth immediately downstream of the constriction, and
S = the submergence, which is Y_3/Y_1 (Figure 2).

Digger Logs

Logs cantilevered from banks and allowed to be undercut by the flow are called digger logs. Digger logs were experimentally modeled with the 0.06-m-diameter pipe on 0 and 2% slopes to determine general trends in their scour-producing capacity. Test variables included discharge, elevation of the digger log crest relative to the streambed, and horizontal distance between two diggers. Diggers were placed at 90° to the flow and spanned the entire channel.

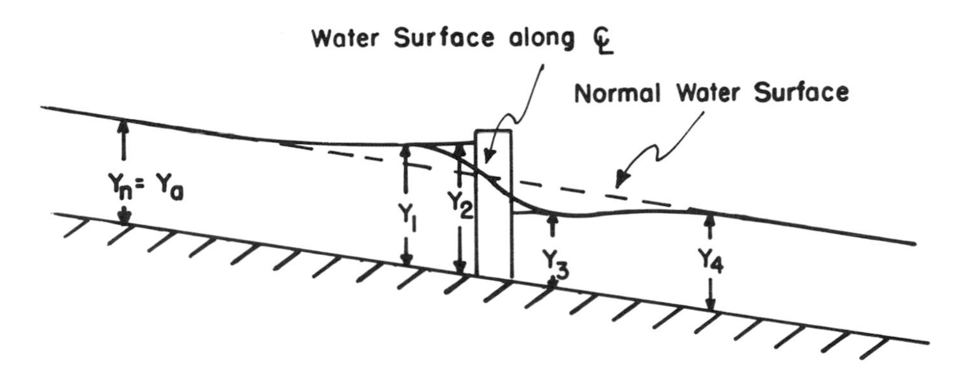

FIGURE 2.—Locations with respect to a deflector of various water depths used in the submerged flow equation of Fiuzat and Skogerboe (1983).

Digger Weirs

Several digger weir geometries were designed and briefly tested in the mobile-bed flume. The first design consisted of a 0.04-m diameter wood dowel with a square-shaped extension made of dowels attached to the upstream side (Figure 3). Flow plunged over the extension and was allowed to undercut the center of the main log. The second design had a triangular-shaped extension upstream, and tests were made of both 60° and 90° interior angles (α) of the triangle apex. The apex of the triangular extension was tested at positions lower than and level with the log crest. Discharges were 0.013 and 0.018 m^3/s. Formation of the deposition berm was noted, because the purpose of designing the structure was to provide reduced sediment in gravels during low flows. For comparison, a 0.04-m diameter straight weir was placed at the same location as the digger weir and tested at 0.018 m^3/s.

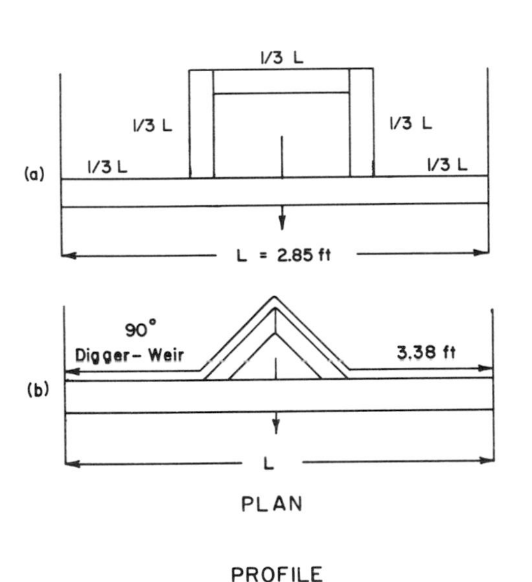

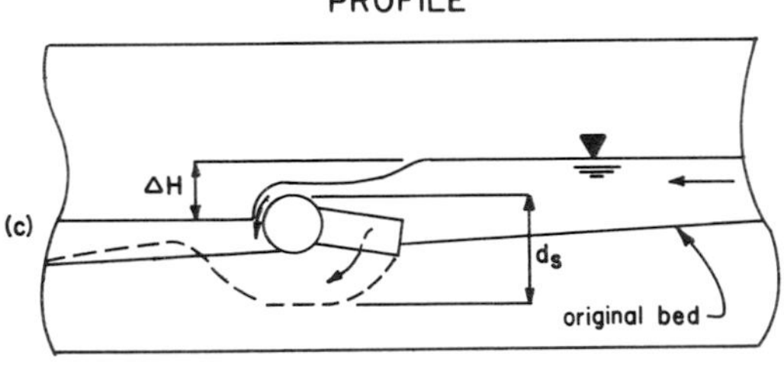

FIGURE 3.—Plan (a, b) and profile (c) views of square and triangular digger weirs. The dashed line indicates the final scour hole.

Results

Log Weir

Scour.—Scour depths predicted with Mason and Arumugam's equation generally matched measured scour depths in the model (Figure 4). Six of the poorest correlations were for low-profile weirs, where predicted scour depth was about double the measured depth. Aside from the low-profile weirs, the equation slightly underestimated measured scour.

Discharge.—The general form of the discharge equation for a rectangular weir appears suitable for describing flow over a log weir. Most of the exponents are close to the theoretical value of 3/2. With the exception of the 0.14-m pipe, the trend of discharge coefficients was to increase as pipe diameter decreased. Therefore, C_d was estimated as a function of pipe diameter to obtain the equation

$$C_d = 3.97\ D^{-0.106}, \tag{4}$$

for which r^2 was 0.934. The data from the 0.14-m pipe was not used when the equation for C_d was

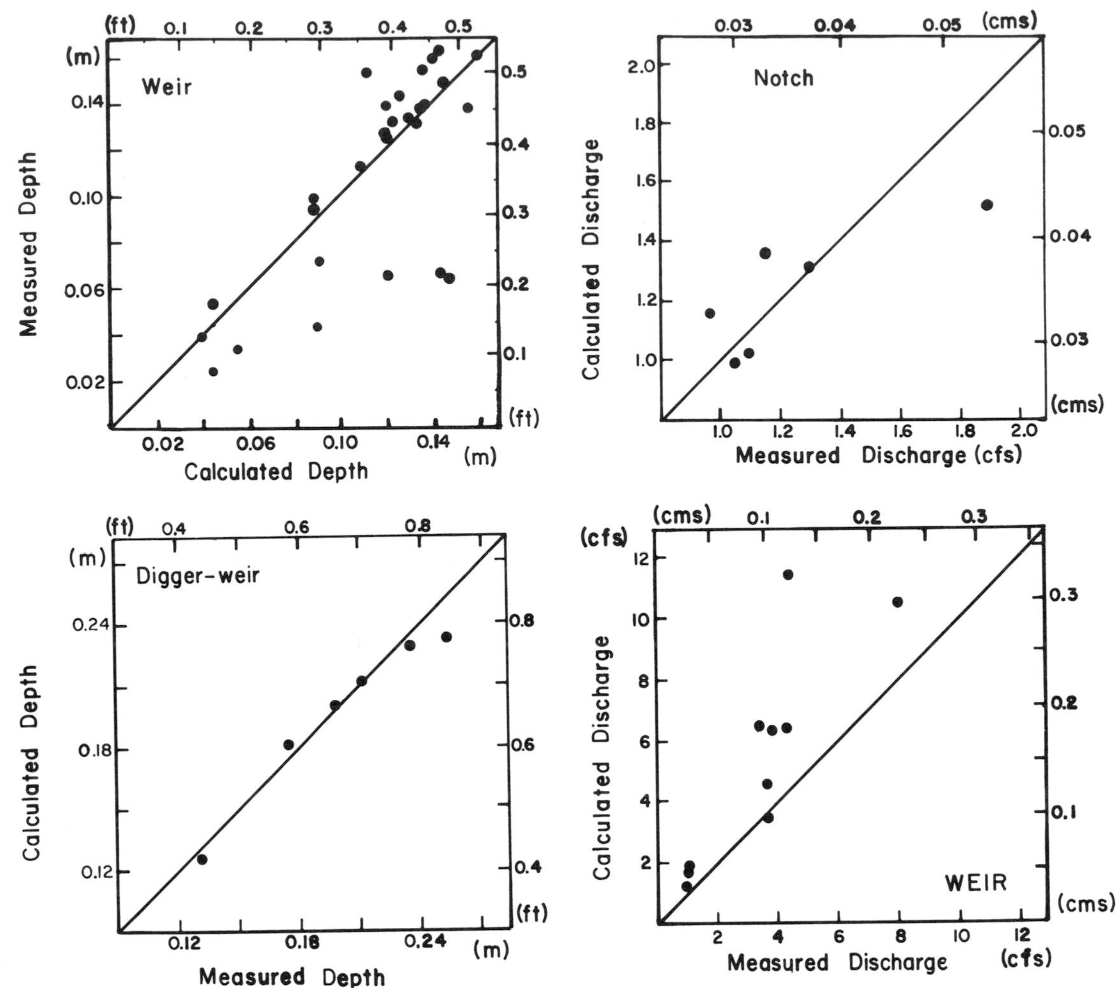

FIGURE 4.—Comparison of measured and predicted scour depths at model log weirs and digger weirs according to the equation by Mason and Arumugam (1985).

FIGURE 5.—Comparison of measured and predicted discharges at prototype log weirs and notches. Discharges are scaled in cubic meters (cms) or cubic feet (cfs) per second.

calculated. The final equation for log weir discharge (English units) is

$$Q = 3.97\ L\ D^{-0.106}\ H^{3/2}. \tag{5}$$

At low flows the velocity head term is usually insignificant and can be disregarded, but if $V_a^2/2g$ is 5–10% or more of the static head, it should be added to H.

Flow through the notches in log weirs can be estimated with an appropriate broad-crested weir discharge coefficient. Discharge coefficients for each notch geometry were found in King (1954). The broad-crested weir equation predicted notch flow at prototypes fairly well, but there was substantial variability between measured and calculated discharge at prototype log weirs (Figure 5). The discrepancy may have been caused by several prototype conditions, including (1) water flowing under the log, (2) sagging or bowed logs, (3) a sloped log crest instead of a level crest, or (4) flow measurement errors. The first three conditions are the most likely problems in this case. A sloped crest was observed at one of the prototypes.

Deflector

Scour.—The depth of scour at the 1/3-width deflectors was compared with predictions of scour from equations by Ahmad (1951), Liu et al. (1961), and Gill (1972) (Table 1). The depths predicted by

TABLE 1.—Comparison of measured and predicted scour depths (Y_s) as calculated with three scour equations (Ahmad 1951; Liu 1961; and Gill 1972) at 90° deflectors with the channel constriction ratio $M = 0.67$. Variables are flow rate per meter of channel width (q) and approach-flow depth (Y_a).

q (m³/[s · m])	Y_a (m)	Y_s (m) Ahmad	Liu	Gill	Measured
		2% Slope			
0.009	0.017	0.048	0.121	0.115	0.076
0.014	0.020	0.063	0.145	0.132	0.103
0.019	0.021	0.077	0.155	0.146	0.128
0.027		0.100			
		Zero slope			
0.017	0.054	0.073		0.168	0.094
0.027	0.064	0.099		0.197	0.148
0.032	0.072	0.113		0.215	0.172

Ahmad's equation were low, which is not surprising because no variables for channel contraction or substrate size were included. Equations by Gill and Liu et al. overestimated scour depth. Both equations are supposed to predict equilibrium scour depth, and because the flume tests lasted 30 min, equilibrium depth technically was not reached. Although these equations did not accurately predict scour at deflectors, they should indicate the general range of scour depths that may be expected. It should be noted that these equations do not account for channel curvature.

Discharge.—Equation (2) can be used to estimate either flow based on measured water depths or backwater depths based on measured or design flows. It was suggested that the original flow depth at the proposed constriction be used for Y_3 to predict backwater depth, Y_1, in project design (Figure 2). The equation estimated backwater stagnation depth at the deflector (Y_2) more accurately than it estimated discharge with my data. Percent error between estimated and recorded upstream depths at deflectors was 15% or less, generally 6% or less (Table 2).

Digger Logs

Digger log tests were of limited scope. When the 0.06-m digger log was placed at various elevations relative to the streambed, the maximum scour occurred when the base of the digger rested on the bed. When two diggers were placed side by side, the maximum scour occurred when there was a one-diameter gap between the logs.

Digger Weirs

The square digger weir formed a mound of gravel in the center of the square area because of the interference pattern of the flow over the sides of the upstream weir extension. Therefore, the square digger weir was not tested further.

The triangular digger weir concentrated the flow in the center of the channel and resulted in substantially deeper scour than a straight log weir.

TABLE 2.—Comparison of estimated (Y_{2e}) and measured (Y_{2m}) backwater depths at deflectors using the submerged flow equation at various discharges (Q) and channel constrictions (M). Y_3 is the flow depth immediately downstream of the constriction.

M	Q (m³/s)	Y_3 (m)	Y_{2e} (m)	Y_{2m} (m)	Error (%)
		Zero slope			
0.67	0.010	0.031	0.049	0.054	10
0.67	0.016	0.040	0.065	0.064	−2
0.67	0.019	0.050	0.076	0.072	−6
0.75	0.018	0.060	0.075	0.078	4
0.75	0.018	0.052	0.069	0.074	6
0.83	0.008	0.027	0.036	0.038	6
0.83	0.018	0.039	0.059	0.061	3
		2% Slope			
0.67	0.005	0.017	0.031	0.037	15
0.67	0.008	0.020	0.040	0.040	0
0.67	0.011	0.021	0.049	0.047	−4

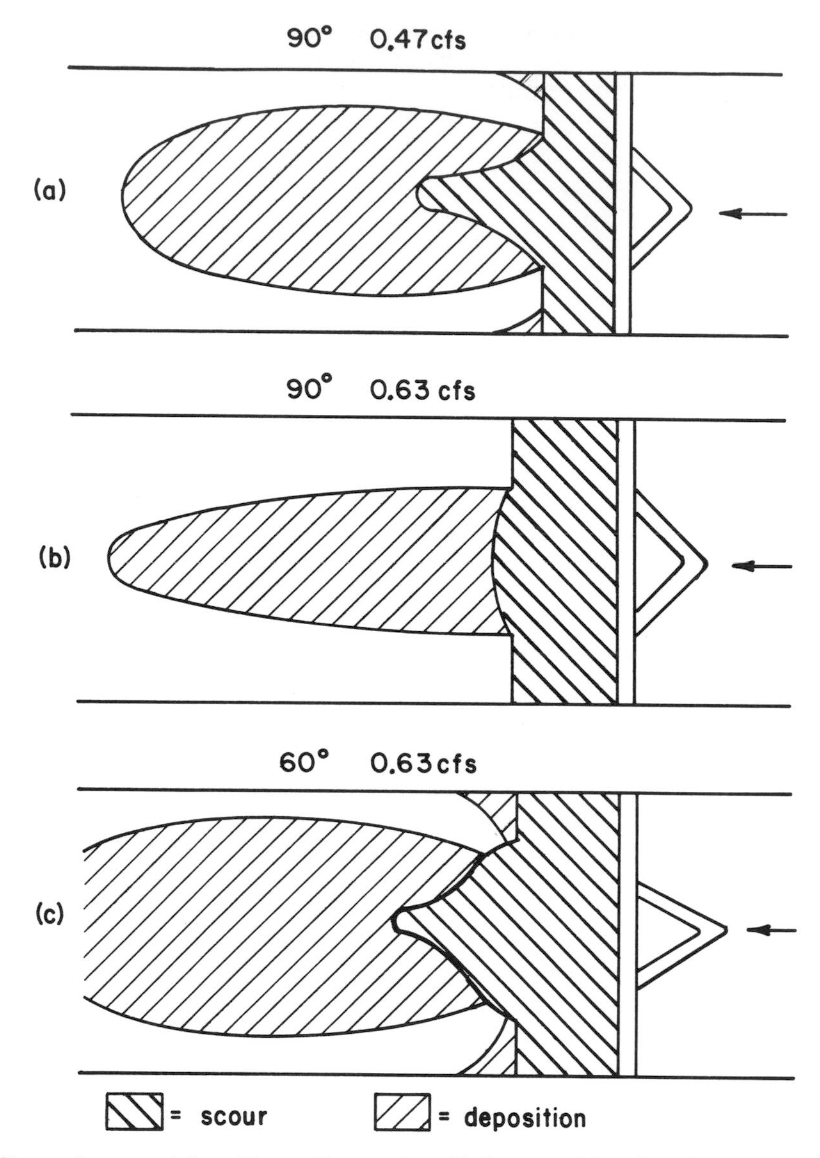

FIGURE 6.—Shape of scour and deposition at digger weirs with the apex of the triangular extension below the log crest (a, c) and level with the log crest (b). The interior angle of the triangle apex (degrees) and the test discharge (cubic feet per second) are indicated above each panel.

Scour depth was greater when the triangle apex was below the log crest than when it was level with the crest, because a greater percentage of the flow went through the digger section. Scour volume also increased when the apex was lowered.

When the apex of the triangle was lower than the log crest, a finger of scour projected downstream directly below the triangle (Figure 6). Two deep channels (thalwegs) were scoured out to the sides of the flume as the flow followed the solid boundaries of the sides of the triangle. A berm was deposited in the center of the channel, and the water flowing over it toward the sides of the channel was particularly noticeable at lower flows.

When the apex of the triangle was level with the log crest, the scour in the center of the channel was reduced. The thalwegs on each side were not as narrow or deep as previously, and the deposition berm was narrower. The deposition berm should provide good spawning conditions, particularly during low flows when the discharge is concentrated in the center and keeps the gravels relatively silt-free.

The scour equation by Mason and Arumugam was also used to calculate scour below the digger weir. It predicted maximum scour depth quite well when the coefficient was changed to 10.0 (Figure 3).

Discussion

The scour depth equation by Mason and Arumugam worked very well for approximating scour below log weirs in the mobile-bed flume. The key to applying their equation to prototypes is to accurately estimate the tailwater depth at design flows. Finding the head loss across the weir will require an estimate of the water depth above the weir from the log weir discharge equation. Also, substrate size has a significant effect on scour depth. I am not certain if the data analyzed by Mason and Arumugam involved uniformly sized substrate or if it included well-graded substrate. Armoring will occur in streams with graded substrate, resulting in less scour than in uniformly sized substrate. Tests with several different gradations of gravel will be necessary to determine the sensitivity of the equation to armoring. Ettema (1980) showed that at bridge piers where σ_g was greater than 4.0 ($\sigma_g = (D_{84}/D_{16})^{1/2}$), scour was only 20% of scour depths in streams with uniform sediments. The substrate in this study had a σ_g of 1.38.

One variable that was not considered was the effect of live-bed scour on the scour depth. Live-bed scour occurs when there is general movement of particles on the streambed, which results in less scour depth than the maximum theoretical depth. Most scour equations are designed to estimate the maximum scour depth for the purpose of protecting hydraulic structures. In estimating fish habitat, it is more important to predict the actual scour depth associated with flows in a stream during the critical life stage and season.

Scour at model deflectors was not accurately predicted with the equations by Ahmad (1951), Liu et al. (1961), and Gill (1972), although trends in scour were similar. It should be remembered that these equations were created with data from sand bed streams and flumes, whereas a narrowly graded gravel was used in this study. In fisheries applications the deflectors are usually not solid and have a variety of geometries, all of which will influence scour widths, lengths, and depths. Armoring will reduce scour depths at deflectors just as it does at other structures. A good description of the effects of deflector geometry, orientation, and spacing can be found in Klingeman et al. (1984) and should be referred to when deflectors are designed. The only data Klingeman et al. provided on scour depth at deflectors came from tests on deformable gabion groins, and no predictive equations for gabion scour were recommended. One problem with putting rock deflectors or groins in streams is that the scour hole at the tip of the groin can compromise the integrity of the groin. Scour equations will give the designer a better idea of how deep to place the toe of the groin to protect it from failure.

Digger logs have good potential for providing fish habitat and should be studied further. They have the advantage of providing both deep water cover and overhead cover. Digger logs are probably more susceptible to failure from rot because they are usually exposed at low flows. They will provide the most scour if placed in combination with other structures that direct the flow at the digger. When flow is directed at a structure near a streambank, care must be taken to protect the bank against excessive erosion if the bank is not well stabilized.

Triangular digger weirs appear to have potential for creating a deeper scour hole than log weirs create, and for providing cleaner gravels at low flows. The structures may be susceptible to rot in the center section of the main log because this section will be exposed at low flows. Care must also be taken to keep the triangular section high enough to prevent all the low flow from going through the triangular section. It can be anticipated that debris may collect on the digger portion of the weir. Debris may cause some problems, but it also adds cover diversity and may increase scour depth. The structure should be placed in areas with stable banks because the flow downstream is concentrated at the banks. Otherwise, the bank will need some protection. Either logs or boulders may be used to construct the triangular extension, but boulders are difficult to seal.

The submerged flow equation of Fiuzat and Skogerboe (1983) appears to be most useful as a backwater depth predictor in fish habitat calculations. Once a backwater depth has been calculated, a water surface profile computation could be used to calculate depths upstream to the next control. The scour at deflectors may introduce error into the results of the submerged flow equation, because the equation was derived for a rigid boundary channel section (Skogerboe and Hyatt

1967). If the deflector is porous, the backwater depth would be less than the equation predicts.

The consistency of discharge results for the model log weir (with the exception of the 0.14-m log) was encouraging. The log weir is consequently well suited, and is often used, as a water elevation control structure. Common applications would be at the entrances and exits of side channels, below culverts, and in other fish passage situations. The comparison of results from the derived equation (5) to measured discharge in the field was not as encouraging. Problems of poorly sealed logs and sagging logs introduced errors in discharge calculations. The use of larger diameter logs could help prevent the sagging problem. Ideally, logs installed to collect spawning gravels should not be sealed with geotextiles so intragravel flows are unrestricted. Realistically most logs are sealed to keep all the low flows from going subsurface. Logs used as elevation control structures should be well sealed.

If the log weir is notched, the broad-crested weir equation can be used with confidence. At low flows it is probably more important to know the flow through the notch than the flow over the rest of the log. A properly designed notch would *not* allow all the low flow to go through the notch. Instead, it would keep the entire log wet and prevent rapid deterioration.

The application of this research is in modeling the changes in fish habitat resulting from enhancement projects, and for designing safe structures. Many computer models of fish habitat are based on water depth and velocity, substrate size, and types of cover (Bovee and Milhous 1978; Fausch et al. 1988). Equations (1)–(5) can provide estimates of scour and the increased upstream depth created by each structure for the selected design flows. If the structure is used as a control point, water surface profile equations can be used to calculate the water depth upstream. The fish requirements for depth and velocity specified by the biologists would be used to evaluate the new habitat. Differences in amounts of habitat before and after construction could thus be estimated for the reach of stream being treated (Anderson 1985).

During project design the discharge equations could be used to determine proper elevations for log weirs and proper lengths of deflectors to achieve desired backwater elevations. The same equations could be used to analyze flood hazards. Results from modeling scour also indicate where protection may be needed for the upstream or downstream banks and where to protect the structure itself from scour. With continued research into hydraulics of habitat structures and fish habitat requirements, structures can be designed and installed that will fit the needs of the fish and fit the hydraulics of the stream.

Acknowledgments

This study was funded by the U.S. Forest Service and the American Water Resources Association. I thank John Orsborn and the faculty and staff at the Albrook Hydraulics Laboratory at Washington State University for direction and assistance in the project.

References

Ahmad, M. 1951. Spacing and projection of spurs for bank protection. Civil Engineering and Public Works Review (London) (March):172–174, (April): 256–258.

Anderson, J. W. 1985. A method for monitoring and evaluating salmonid habitat carrying capacity of natural and enhanced Oregon coastal streams. Proceedings of the Annual Conference of the Western Association of Fish and Wildlife Agencies 64(1984): 288–296.

Beschta, R. L., and W. S. Platts. 1986. Morphological features of small streams: significance and function. Pages 369–380 *in* W. L. Jackson, editor. Engineering considerations in small stream management. American Water Resources Association, Bethesda, Maryland.

Bossou, M. F. 1954. Relationship between trout populations and cover on a small stream. Journal of Wildlife Management 18:229–239.

Bovee, K. D., and R. T. Milhous. 1978. Hydraulic simulation in instream flow studies: theory and techniques. U.S. Fish and Wildlife Service Biological Services Program FWS/OBS-78/33.

Ettema, R. 1980. Scour at bridge piers. University of Auckland, School of Engineering, Report 216, Auckland, New Zealand.

Everest, F. H., and J. R. Sedell. 1984. Evaluating effectiveness of stream enhancement projects. Pages 246–256 *in* T. J. Hassler, editor. Proceedings: Pacific Northwest stream habitat management workshop. California Cooperative Fishery Research Unit, Humboldt State University, Arcata, California.

Fausch, K. D., C. L. Hawkes, and M. G. Parsons. 1988. Models that predict standing crop of stream fish from habitat variables: 1950–85. U.S. Forest Service General Technical Report PNW-213.

Fiuzat, A. A., and G. V. Skogerboe. 1983. Comparison of open channel constriction ratings. American Society of Civil Engineers, Journal of the Hydraulics Division 109:1589–1602.

Franco, J. J. 1967. Research for river regulation dike design. American Society of Civil Engineers, Journal of the Waterways and Harbors Division 93(WW3):71–88.

Garde, R. J., K. Subramanya, and K. D. Nambudripad. 1961. Study of scour around spur dikes. American Society of Civil Engineers, Journal of the Hydraulics Division 87(HY6[1]):23–37.

Gill, M. A. 1972. Erosion of sand beds around spur dikes. American Society of Civil Engineers, Journal of the Hydraulics Division 98(HY9):1587–1602.

King, H. W. 1954. Handbook of hydraulics, 4th edition (revised by E. F. Brater). McGraw-Hill, New York.

Klingeman, P. C. 1981. Short-term considerations on river gravel supply. State of Washington Water Research Center, Report 39, Seattle.

Klingeman, P. C. 1984. Evaluating hydrologic needs for design of stream habitat modification structures. Pages 191–213 *in* T. J. Hassler, editor. Proceedings: Pacific Northwest stream habitat management workshop. California Cooperative Fishery Research Unit, Humboldt State University, Arcata, California.

Klingeman, P. C., S. M. Kehe, and Y. A. Owusu. 1984. Streambank erosion protection and channel scour manipulation using rockfill dikes and gabions. Water Resources Research Institute, Final Technical Completion Report WWRI-98, Corvallis, Oregon.

Lane, E. W. 1955. Design of stable channels. Transactions of the American Society of Civil Engineers 120:1234–1279.

Liu, H. K., F. M. Chang, and M. M. Skinner. 1961. Effect of bridge constrictions on scour and backwater. Colorado State University, Report CER 60HKL22, Fort Collins.

Mason, J. C. 1976. Response of underyearling coho salmon to supplemental feeding in a natural stream. Journal of Wildlife Management 40:775–778.

Mason, P. J., and K. Arumugam. 1985. Free jet scour below dams and flip buckets. American Society of Civil Engineers, Journal of the Hydraulics Division 111:220–235.

Orsborn, J. F. 1987. Hydrology and hydraulics for fisheries improvement projects. Seminar sponsored by Siskiyou National Forest, U.S. Forest Service, Coos Bay, Oregon.

Saunders, B. W., and M. W. Smith. 1962. Physical alteration of stream habitat to improve brook trout production. Transactions of the American Fisheries Society 91:185–188.

Scrivener, J. C., and M. J. Brownlee. 1981. A preliminary analysis of Carnation Creek gravel quality data, 1973–1980. State of Washington Water Research Center, Report 39, Seattle.

Skogerboe, G. V., and M. L. Hyatt. 1967. Analysis of submergence in flow measuring flumes. American Society of Civil Engineers, Journal of the Hydraulics Division 93(HY4):183–200.

Tappel, P. D., and T. C. Bjornn. 1983. A new method of relating size of spawning gravel to salmonid embryo survival. North American Journal of Fisheries Management 3:123–135.

American Fisheries Society Symposium 10:88–103, 1991

A Method for Estimating Fishery Benefits for Galloway Dam and Reservoir[1]

JOHN L. MCKERN

U.S. Army Corps of Engineers, Walla Walla District
Building 602, City–County Airport, Walla Walla, Washington 99362, USA

Abstract.—To study the feasibility of constructing Galloway Dam on the Weiser River in Idaho, the U.S. Army Corps of Engineers developed a linear, multiplicative, computer model (FISHBEN) to estimate fishery benefits. The primary purpose of the dam would be to store water to augment flows on the Snake and Columbia rivers from Hells Canyon Dam to the Pacific Ocean to increase the downstream survival of juvenile salmon, *Oncorhynchus* spp., and steelhead *Oncorhynchus mykiss*. Fifty years of past flow data were used to predict future flow conditions. Juvenile fish populations and their survival through river reaches, reservoirs, dams, transportation, and in-river fish passage were predicted and used to estimate overall fish survival. Adult returns were predicted, and economic values based on potential harvest were used to estimate monetary fishery benefits. Fishery benefits derived by FISHBEN were compared with engineering and construction costs for 23 alternatives, and the alternative deemed best for national economic development as defined by the Corps of Engineer's principles and guidelines for planning studies was recommended.

Personnel of the U.S. Army Corps of Engineers developed a computer model (FISHBEN) to analyze the worthiness of structural and operational alternatives for augmenting river flows to increase the survival of juvenile salmon *Oncorhynchus* spp. and steelhead *Oncorhynchus mykiss* migrating down the Snake and Columbia rivers. FISHBEN was developed for a feasibility study for Galloway Dam and Reservoir, which was released to the public by the Corps of Engineers early in 1989.

The Northwest Power Planning Council included this study in its program (NPPC 1984, 1987) because flow augmentation would decrease losses to juvenile salmon and steelhead caused by alteration of river flows by federal dams in the Columbia basin, and Galloway Dam could provide water budget flows defined in the Council's fish and wildlife program. The project would consist of a new dam and storage reservoir in the Weiser River basin in southwest Idaho. This would provide a more cost-effective source of water for augmenting spring flows for juvenile salmon and steelhead migration than Brownlee and Dworshak reservoirs, which are currently used to provide water budget flows from the Snake River.

The flow area analyzed included the Snake River from Brownlee Reservoir to its mouth, the Clearwater River from Dworshak Dam to its mouth, and the Columbia River from Priest Rapids Dam to the tailrace of Bonneville Dam (Figure 1). Anadromous salmon and steelhead groups that occur above Bonneville Dam were analyzed for the area from Hells Canyon Dam on the Snake River and Grand Coulee Dam on the Columbia River (the upstream limits of anadromous fish distribution) to the Pacific Ocean. Salmon and steelhead production from hatcheries and natural spawning grounds were estimated for the main stem and tributary areas that would be affected by flow augmentation.

Principle partners in this study were the Corps of Engineers, the U.S. Fish and Wildlife Service, and the Idaho Department of Water Resources.

In this report, I summarize the FISHBEN analysis as it existed in the fall of 1988. For more information on FISHBEN as it has changed since that time, please write to Chief, Planning Division, Walla Walla District, U.S. Army Corps of

[1]The bioeconomic method reported in this paper is an evolving one. Here we see one of its stages that persisted for a significant time. The U.S. Army Corps of Engineers' FISHBEN model for examining economic justification for fish bypass systems in river-damming projects had, before 1988, been used not only in the Galloway project, but also in other Columbia River projects. Since the 1988 symposium, the model has been publicly reviewed in connection with its application in the Galloway project, particularly by fishery agencies. Further development of it has resulted, and some referees of this paper recommended that the changes be incorporated during the prepublication revision. This was not done, however, because the author felt it useful to present a record of the method's status in 1988. In the introduction, the author indicates where an updated version can be obtained.—Editors.

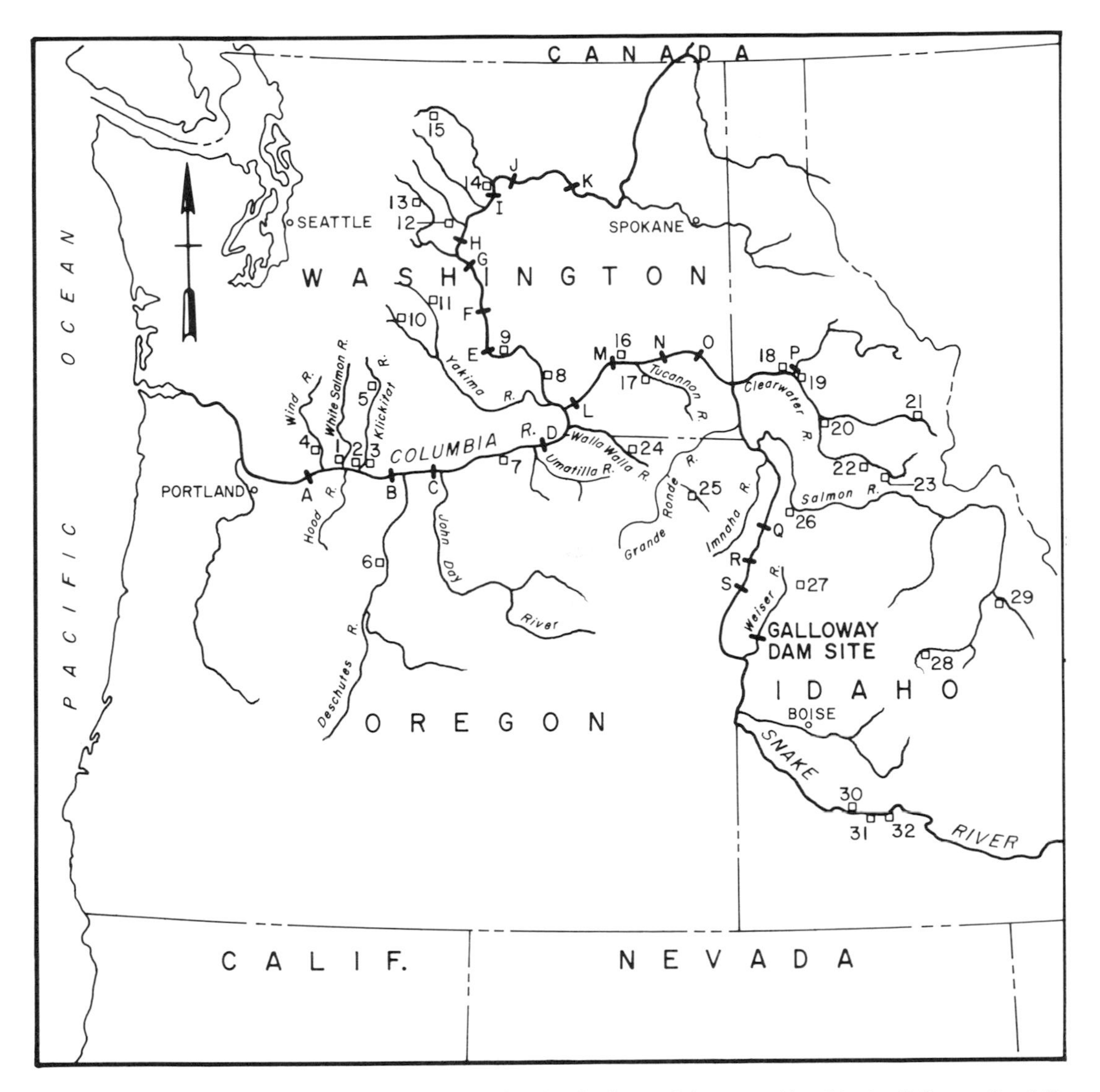

FIGURE 1.—Vicinity map showing rivers, tributaries, hatcheries, and dams considered in the Galloway feasibility study. Fish hatcheries are designated by numbers. 1, Little White Salmon; 2, Spring Creek; 3, Big White Salmon; 4, Carson; 5, Klickitat; 6, Round Butte; 7, Irrigon; 8, Ringold; 9, Priest Rapids; 10, Naches; 11, Yakima; 12, Leavenworth; 13, Entiat; 14, Wells; 15, Winthrop; 16, Lyons Ferry; 17, Tucannon; 18, Clearwater; 19, Dworshak; 20, Kooskia; 21, Powell; 22, Crooked River; 23, Red River; 24, Lookingglass; 25, Wallowa; 26, Rapid River; 27, McCall; 28, Sawtooth; 29, Pahsimeroi; 30, Niagra Springs; 31, Hagerman; 32, Magic Valley. Dams are designated by letters. A, Bonneville; B, The Dalles; C, John Day; D, McNary; E, Priest Rapids; F, Wanapum; G, Rock Island; H, Rocky Reach; I, Wells; J, Chief Joseph; K, Grand Coulee; L, Ice Harbor; M, Lower Monumental; N, Little Goose; O, Lower Granite; P, Dworshak; Q, Hells Canyon; R, Oxbow; S, Brownlee.

Engineers, Walla Walla Airport, Walla Walla, Washington 99362.

Methods

Computer model.—Because of the large volumes of information on flow, fisheries, and many factors influencing fish survival in the study area, a linear, multiplicative model was developed for use on a Harris 500 computer. Groups of juvenile salmon and steelhead entering the area analyzed with the model were multiplied by a survival rate for each condition they encountered thereafter until their return to the fishery (for the Galloway study, ocean, river, Indian, sport, and commer-

TABLE 1.—Monthly distribution of juvenile salmon and steelhead used in FISHBEN for the Galloway analysis.

Subgroup	Percent distribution by month[a]					
	Apr 1–14	Apr 15–May 14	May 15–Jun 30	Jul	Aug	Sep
Snake River steelhead	0.2	78.1	20.8	0.9	0.0	0.0
Columbia River hatchery steelhead	0.0	38.7	59.1	1.7	0.5	0.0
Columbia River wild steelhead	0.8	33.9	58.6	5.0	1.7	0.0
Snake River spring/summer chinook salmon	5.8	81.0	13.1	0.1	0.0	0.0
Columbia River hatchery spring chinook salmon	2.9	56.6	37.2	2.8	0.5	0.0
Columbia River wild spring chinook salmon	0.0	15.5	47.6	36.1	0.8	0.0
Columbia River hatchery summer chinook salmon	0.0	10.0	12.5	35.0	32.5	10.0
Columbia River wild summer chinook salmon	0.0	0.2	44.9	50.7	3.1	1.0
Hatchery fall chinook salmon	0.0	10.0	12.5	35.0	32.5	10.0
Wild fall chinook salmon	0.0	0.2	44.9	50.7	3.1	1.0
Hatchery coho salmon	0.0	42.2	52.3	0.5	5.0	0.0
Wild coho salmon	0.0	0.3	85.6	12.3	1.4	0.4
Wild sockeye salmon	0.0	39.4	56.7	3.9	0.0	0.0

[a]Zero distributions were used for all species for October–March.

cial fisheries were lumped together). At that point, returning adults were divided into escapement and potential harvest. Escapement was defined as the number of adult fish required to replenish the estimated juvenile fish population used in the model. Potential harvest, the return to the fishery minus escapement, was multiplied by a dollar value for each specific group of fish to estimate dollar benefits. Dollar benefits for all groups were summed to represent the gross fishery benefit for a given flow year. Fishery benefits varied with annual flow levels. Fifty years (1929–1978) of recorded annual flow data were used. The average of 50 years of fishery benefits based on these flow data was added to power generation, flood control, recreation, and employment benefits for comparison with the estimated project cost. Fishery losses, system power losses, construction costs, interest and amortization, operation and maintenance, and damage costs were used to figure average annual project cost. Because partial juvenile fish transportation was the accepted norm when the study was conducted, net benefits computed for that transportation option were compared with costs to determine benefit-to-cost ratios.

Flow data from the HYSSR model used by the Corps of Engineers' North Pacific Division were used to simulate projected flow conditions. Seventeen groups of juvenile anadromous fish were considered. Fish passage through up to four free-flowing river reaches and up to eight reservoirs and dams, and three options of collecting and transporting fish or leaving them to migrate in-river, were simulated. Over 30 alternatives of flow augmentation, including no action, were studied.

In comparing alternatives, preproject or base flow conditions were established for the 50-year record of flows. This information, juvenile fish numbers, survival rates, and one transport option were entered into the FISHBEN input file. On a year-by-year basis, augmentation flows for each test alternative for each of the 50 years were defined, and survival of smolts through the river reaches, reservoirs, and dams to Bonneville tailrace was estimated. Ocean survival and adult returns were estimated, partitioned for 1-, 2-, and 3-year rearing periods in the ocean for the various fish groups. Escapement required to perpetuate the runs was extracted, and economic values were estimated based on potential harvest and value per fish. This process was repeated with the other two transport options. The 50-year average value of potential fish harvest was used to compare alternatives.

For each flow year, a series of flow augmentation plans was systematically defined with different durations and timing relative to the timing of the juvenile fish outmigration. The economic value of each augmentation plan was estimated with FISHBEN, then the most favorable plan for each year was selected by the model. That plan was carried forward while each succeeding flow year was evaluated. This process allowed separate optimization of flow augmentation plans for each year based on potential harvest values and flow levels. For comparison, the average of 50 years of flow and augmentation analyses was used for each alternative.

Time parameters.—Recognizing that water budget managers may request flow augmentation ranging from short, high peaks to prolonged, small increases, augmentation durations of 5 to 31 d were simulated in 1-d increments. Separate simu-

TABLE 2.—Daily distributions of juvenile salmon and steelhead during the 31-d flow augmentation period used in FISHBEN for the Galloway analysis.

	Percent distribution by day[a,b]									
Date	Snake River steelhead	Snake River spring and summer chinook salmon	Columbia River hatchery steelhead	Columbia River wild steelhead	Columbia River hatchery spring chinook salmon	Columbia River wild spring chinook salmon	Columbia River hatchery summer chinook salmon	Hatchery coho salmon	Wild coho salmon	Wild sockeye salmon
Apr 15	1.4	1.3	0.0	0.3	1.0	0.0	0.6	0.0	0.0	0.0
16	1.1	1.6	0.0	0.3	1.1	0.0	0.5	0.0	0.0	0.0
17	1.2	1.4	0.0	0.2	0.9	0.0	0.5	0.0	0.0	0.0
18	1.0	2.8	0.0	0.3	1.2	0.0	0.5	0.0	0.0	0.0
19	1.3	3.2	0.1	0.4	1.8	0.0	0.4	0.0	0.0	0.0
20	1.7	3.8	0.1	0.5	1.9	0.0	0.3	0.0	0.0	0.0
21	1.5	4.1	0.2	0.6	1.5	0.1	0.5	0.0	0.0	0.1
22	2.0	3.7	0.3	1.5	2.6	0.0	0.4	0.0	0.0	0.3
23	2.4	3.8	0.3	1.3	2.8	0.1	0.5	0.0	0.0	0.5
24	2.4	4.0	0.5	1.7	2.0	0.2	0.6	0.0	0.0	1.2
25	2.6	3.8	0.4	1.3	2.3	0.1	0.8	0.0	0.0	1.1
26	1.9	4.3	0.5	1.2	1.9	0.3	1.1	0.0	0.0	1.9
27	4.0	4.1	0.6	1.3	1.8	0.3	0.7	0.0	0.0	0.8
28	2.8	4.6	0.4	1.0	2.1	0.4	0.9	0.0	0.0	1.0
29	3.1	4.7	0.6	0.6	1.6	0.5	0.8	0.0	0.0	1.4
30	3.6	3.2	0.3	0.7	1.3	0.3	1.6	0.0	0.0	3.3
May 1	4.6	3.4	0.4	1.0	1.1	0.6	2.0	0.0	0.0	3.1
2	4.1	2.0	0.8	1.1	1.2	0.7	2.5	0.0	0.0	3.0
3	3.9	1.8	0.6	1.3	1.0	0.3	1.8	0.1	0.0	1.8
4	3.6	2.4	1.0	1.1	1.8	0.6	1.8	0.1	0.0	1.7
5	3.2	1.8	0.9	1.0	2.2	0.5	2.5	0.2	0.0	2.5
6	2.3	2.7	2.4	1.3	1.8	0.3	2.5	0.4	0.0	0.0
7	2.5	2.5	3.5	1.5	2.8	0.4	2.5	1.2	0.1	2.2
8	2.5	1.9	3.8	1.1	1.9	0.5	2.7	3.3	0.0	1.7
9	2.2	1.5	3.2	1.3	1.8	0.6	3.2	3.7	0.1	0.9
10	2.2	1.3	3.4	1.4	2.3	0.4	2.3	4.5	0.0	1.0
11	3.0	1.1	2.6	1.3	2.2	0.8	2.3	5.6	0.0	1.3
12	2.8	0.9	1.9	1.3	2.3	0.7	3.5	6.0	0.1	2.5
13	2.5	1.0	4.0	2.1	2.1	0.3	2.5	6.3	0.0	2.5
14	2.3	1.1	2.8	2.1	1.6	0.7	2.6	5.4	0.0	1.6
15	2.4	1.2	3.1	1.6	2.7	0.3	2.4	5.5	0.0	1.8

[a]Fall chinook salmon distributions used were identical to Columbia River hatchery summer chinook salmon distributions.
[b]No distribution was made for Columbia River wild summer chinook salmon.

lations were run for each possible starting date. One 31-d duration, up to 27 possible 5-d durations, and a reducing range of durations in between were compared.

Because their outmigrations peak between April 15 and May 15, Snake River yearling chinook salmon *O. tshawytscha* and steelhead would benefit most from the proposed augmentation. Flows for each year were simulated based on the percentage of these juvenile fish passing during each month (Table 1), and daily passage percentages (Table 2) were used for the April 15 to May 15 augmentation period.

For each flow year, the duration and starting date that produced the greatest overall juvenile fish survival were identified to optimize the shape and timing of augmentation for given flow conditions, augmentation volumes, and smolt distributions.

Flow system.—The flow system analyzed with FISHBEN included discharges from Brownlee, Dworshak, and Priest Rapids dams, inflow from tributaries downstream of these dams, and flows

TABLE 3.—Hydraulic capacities and minimum flows required for spill at lower Snake and lower Columbia River dams used in FISHBEN for the Galloway analysis.

Dam	Powerhouse capacity (ft^3/s)	Minimum flow for spill (ft^3/s)
Lower Granite	133,373	120,000
Little Goose	136,141	122,500
Lower Monumental	136,145	122,500
Ice Harbor	101,816	91,600
McNary	211,884	190,700
John Day	330,847	297,800
The Dalles	360,847	324,800
Bonneville	291,082	262,000

TABLE 4.—Projected numbers of juvenile salmon and steelhead (in thousands) used in FISHBEN for the Galloway analysis.[a]

Species or race	Snake River segment					Total Snake River
	Hells Canyon	Salmon River	Grande Ronde River	Clearwater River	Lower Monumental Reservoir	
Hatchery						
Spring chinook salmon	1,330	4,730	1,340	4,980		12,380
Summer chinook salmon	1,370					1,370
Fall chinook salmon	500	1,490	500	7,672		10,162
Steelhead	1,130	5,610	1,660	6,750	470	15,620
Coho salmon						
Wild						
Spring chinook salmon		3,280	1,000	2,010	10	6,300
Summer chinook salmon		3,100				3,100
Fall chinook salmon	400			400		800
Steelhead		2,960	2,000	1,460		6,420
Coho salmon				100		100
Sockeye salmon		100				100

[a]Sources: Buechler (1982a, 1982b); IDFG (1984); Anonymous (1986a); C. Ross, Washington Department of Fisheries (personal communication); S. Pettit, Idaho Department of Fish and Game (personal communication); K. Witty, Oregon Department of Fish and Wildlife (personal communication); M. Dell, Grant County Public Utility District (personal communication).

at the eight lower Snake and lower Columbia River dams. Various combinations of flows from these projects and areas were analyzed, including not providing the water budget, providing the water budget by combinations of flows from existing projects, and providing the water budget and additional flow augmentation from existing projects or from Galloway Dam.

Monthly flow values from the HYSSR model were added to or subtracted from the base flow according to the alternative being studied. Positive changes were converted to volumes and identified as augmentation. Negative changes were subtracted from potential augmentation volumes. The augmentation volume for each alternative was spread over the augmentation duration, converted to flow, then added to the base April 15–May 15 flows for Brownlee, Dworshak, and the eight downstream dams. Flows for nonaugmentation periods were not changed. Water budget volumes for Priest Rapids Dam were shaped only at the 31-d duration for this study.

Dam flow routing.—From Lower Granite through Bonneville Dam, flows were split between the spillway and powerhouse according to powerhouse hydraulic capacity and spill criteria (Table 3). Flows exceeding 90% of hydraulic capacity were treated as spill because 10% downtime was allowed for turbine maintenance and repair. FISHBEN was structured to allow for other spill criteria, but altering spill levels was not considered appropriate for this study.

Projected fish numbers.—To ascertain the benefits of flow augmentation at the time when Galloway Dam and Reservoir could be constructed, juvenile fish numbers were projected to the year 2000. Four species of anadromous fish were treated in FISHBEN: steelhead, chinook salmon, coho salmon *O. kisutch*, and sockeye salmon *O. nerka*. Chinook salmon were split into spring-, summer-, and fall-run groups. Spring and summer chinook salmon were combined for the Snake River because they migrate together as yearlings and are indistinguishable. For the Columbia River, spring and summer chinook salmon were separated because spring chinook salmon migrate as yearlings and summer chinook salmon migrate as subyearling fish. Spring and summer chinook salmon and steelhead were divided into Snake and Columbia River subgroups because of differing economic values between Washington and Idaho fisheries. All groups except sockeye salmon were separated into wild and hatchery subgroups (no hatchery sockeye salmon were reported to occur in the Columbia Basin).

Projections of juvenile fish populations (Table 4) were based on current numbers (Anonymous 1986a) adjusted according to the assumptions (Buechler 1982a, 1982b) that (1) natural smolt production would increase from low levels observed in the late 1970s to above-average levels observed in the 1960s; (2) the anadromous fish hatcheries of the Lower Snake River Fish and Wildlife Compensation Plan would be in full production; (3) Idaho Power Company's mitigation for anadromous fish losses caused by Brownlee, Oxbow, and Hells Canyon dams would be fully implemented; (4) planned expansion of several

TABLE 4.—Extended.

Species or race	Columbia River segment					Total Columbia River
	Mid-Columbia River	McNary Reservoir	John Day Reservoir	The Dalles Reservoir	Bonneville Reservoir	
Hatchery						
Spring chinook salmon	4,410	660		1,500	5,500	12,070
Summer chinook salmon	1,850					1,850
Fall chinook salmon	7,700	4,290	3,550		23,358	38,898
Steelhead	632	614	410	1,075	1,358	4,089
Coho salmon	1,000				2,410	3,410
Wild						
Spring chinook salmon	5,000	500	450	300	750	7,000
Summer chinook salmon	1,500					1,500
Fall chinook salmon		20,000	600	100	300	21,000
Steelhead	600	400	500	500		2,000
Coho salmon			500	500		1,000
Sockeye salmon	4,900					4,900

state-owned anadromous fish hatcheries in the study area would be completed; and (5) planned production of Dworshak and Kooskia National Fish Hatcheries would be met. It was also assumed that habitat restoration, passage improvement, and hatchery production increases for anadromous fish in the Council's Fish and Wildlife Program (NPPC 1982, 1984, 1987) would be implemented.

Juvenile fish migration and survival.—Juvenile fish migration and survival through the system without flow augmentation was compared with survival for each flow augmentation alternative. The difference in initial survivability between hatchery and wild fish (Bell 1986) was accounted for by subtracting 5% release mortality for all hatchery fish. Thereafter, hatchery and wild fish were subjected to the same survival rates for events occurring throughout the outmigration, in the ocean, and upon return as adults.

To analyze the benefits of flow augmentation, a method had to be devised to estimate the effect of changes in flow volumes on juvenile fish survival. Other researchers and managers (Raymond 1974, 1979; Sims and Ossiander 1981; Brown 1982; Ceballos 1982) developed second-order polynomial curves based on percent survival with given flows. However, conditions in the river and at dams have changed considerably since those curves were developed, and for FISHBEN it was assumed fish passage conditions would be optimized by the year 2000. Therefore, based on available information (Bayha 1974; Bell et al. 1976; Sims et al. 1978, 1983; Ebel et al. 1979; J. B. Athearn, U.S. Army Corps of Engineers, personal communication) an optimized, third-order polynomial curve (Figure 2) relating percent survival to percent optimum flow (Table 5) was devised. Maximum survival was assumed to be 98%. Based on comments received from reviewing agencies in 1985, it was assumed there would be no negative effects above 100% optimum flow.

Based on the assumption that survival through a river reach or reservoir was related to the distance traveled by fish, and therefore to the duration of their exposure to mortality factors, a set of survival multipliers (Table 6) was established for reservoirs and river reaches in the study area. Survival through free-flowing reaches was assumed to be twice as high as through reservoirs of the same length. Percent survival established by the flow–survival curve was adjusted by the appropriate survival multiplier to account for the time of exposure to mortality in each segment.

Use of the optimized flow–survival curve and survival multipliers was the key to predicting the effects of flow augmentation with FISHBEN. Percent survival for the base condition was compared with percent survival for each alternative tested as measured from the flow–survival curve for each free-flowing river reach or reservoir through which a group of fish passed. It was assumed that the rate of survival at dams remained constant as flow changed up or down, so the measure of effectiveness of flow augmentation was the decrease or increase in survival in the river reaches and reservoirs as flow changed up or down on the flow–survival curve.

Dam passage and survival rates.—Juvenile fish could pass dams through spillways, powerhouse turbine units, ice and trash sluiceways, fish ladders, or navigation locks. For this study, no

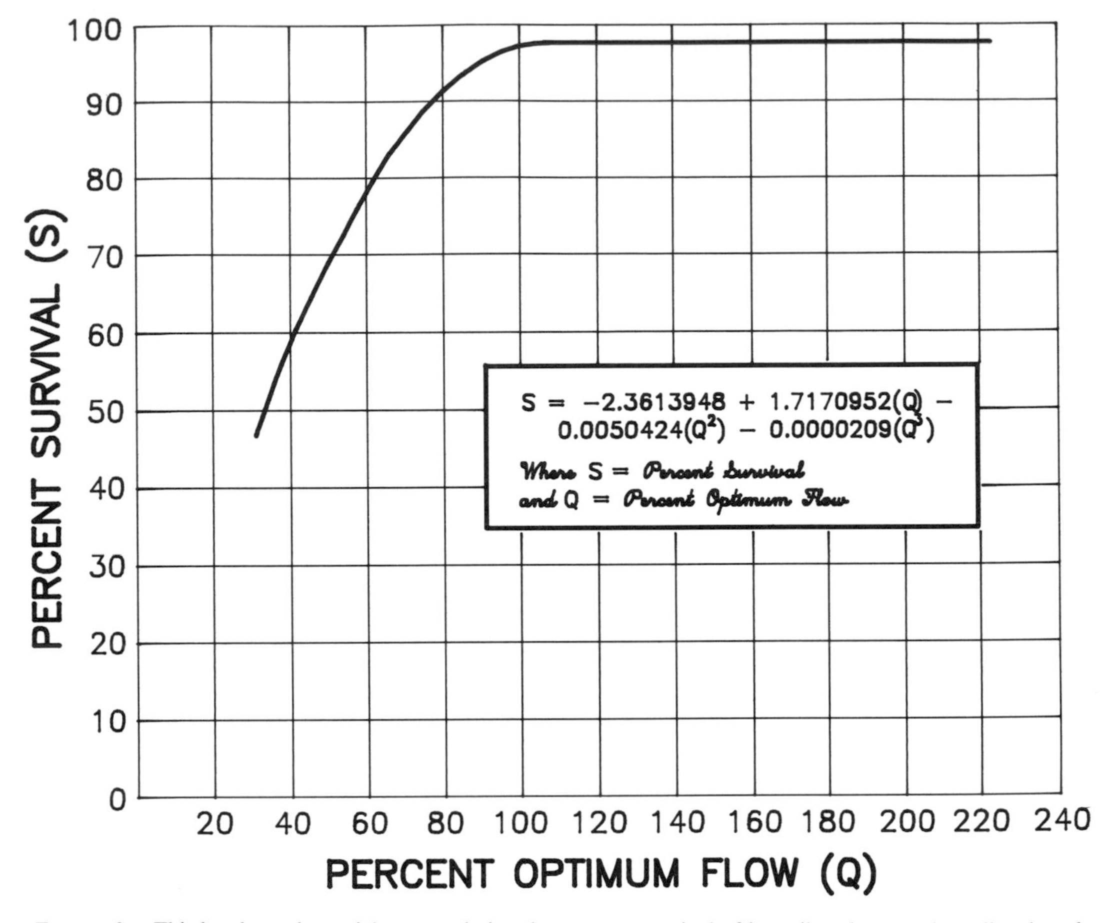

FIGURE 2.—Third-order polynomial curve relating the percent survival of juvenile salmon and steelhead to the percent optimum flow used in FISHBEN to relate flow levels to downstream fish survival for the Galloway feasibility study.

juveniles were assumed to pass through adult fish ladders or navigation locks. For powerhouse and spillway passage, a direct relationship between percent flow and percent passage was assumed (Johnson et al. 1983; Anonymous 1986a). Fish approaching each dam were divided between the powerhouse and spillway according to the percent flow through each structure.

TABLE 5.—Optimum flows for various river reaches and reservoirs for survival of juvenile salmon and steelhead used in the FISHBEN analysis. River miles (RM) are measured from the confluence of the Snake and Columbia rivers.

Site	Species	Optimum flow (ft^3/s)	Source
Hells Canyon Dam	All anadromous fish	27,000	Bayha (1974)
Snake River (RM 188–168)	All anadromous fish	60,000	Estimated
Snake River (RM 168–136)	All anadromous fish	70,000	Estimated
Lower Clearwater River	All anadromous fish	47,000	Estimated
Lower Granite Dam	Yearling chinook salmon and steelhead	140,000	NPPC (1984); Sims et al. (1983)
Priest Rapids Dam	All anadromous fish	150,000	NPPC (1984)
McNary Dam	All anadromous fish	330,000	NPPC (1982)
The Dalles Dam	All anadromous fish	330,000	Ceballos (1982); Brown (1982)
Bonneville Dam	All anadromous fish	330,000	Ceballos (1982); Brown (1982)

TABLE 6.—Multipliers used in FISHBEN to adjust fish survival according to river flow for the length of each river reach or reservoir.

River reach or reservoir	Survival multiplier
Hells Canyon Dam to Snake River mile 188	0.716
Snake River miles 188–168	0.239
Snake River miles 168–136	0.359
Lower Clearwater River	0.493
Mid-Columbia River	0.634
Lower Granite Reservoir	1.033
Little Goose Reservoir	0.955
Lower Monumental Reservoir	0.749
Ice Harbor Reservoir	0.826
McNary Reservoir	1.110
John Day Reservoir	1.962
The Dalles Reservoir	0.620
Bonneville Reservoir	1.214

Based on projected fish passage improvement programs (NPPC 1987), it was assumed that all dams were equipped with state-of-the-art traveling fish screens, balanced-flow vertical barrier screens, 12-in orifices, and bypass–collection systems (Krcma et al. 1982, 1983, 1984, 1985). Passage of fish approaching the powerhouse at dams with an ice and trash sluiceway was further subdivided between the sluiceway and turbine. The most recent bypass efficiency information (Nichols and Ransom 1982; Johnson et al. 1983; Krcma et al. 1983, 1984, 1985; J. Williams, U.S. Army Corps of Engineers, personal communication) was used to estimate state-of-the-art efficiency rates to be used in FISHBEN (Table 7). For projects without ice and trash sluiceways, bypass efficiency represented traveling-screen guiding efficiency. Where sluiceways were available as bypasses, it was assumed that screen-guiding efficiency was reduced 10% because sluiceways skimmed off fish higher in the water column (Krcma et al. 1982). For such projects, bypass efficiency was a combination of sluiceway and traveling-screen guiding efficiency.

Fish transportation and transport survival.—Under oversight of an interagency Fish Transportation Oversight Team, the Corps of Engineers could transport all or none of the juvenile salmon and steelhead collected at Lower Granite, Little Goose, and McNary dams to the tailrace of Bonneville Dam. Three rates were used in FISHBEN: (1) all collected fish were transported; (2) all fish collected at Lower Granite Dam and 20% of the yearling chinook salmon smolts and 80% of other smolts collected at Little Goose and McNary Dams were transported (Basham et al. 1983), or (3) no fish were transported (all were bypassed at the three collector dams). The second rate represented normal transport operations in the early 1980s, whereas the other rates represented maximum and minimum possible transportation.

Transport survival was estimated from the time fish entered the transport vehicle until they were released below Bonneville Dam (Ebel et al. 1973, 1974, 1975; Park et al. 1978, 1979, 1981, 1982, 1983; Ebel 1980; Park 1980; DeLarm et al. 1984; Athearn, personal communication). Direct transport mortality ranged from 0.1 to 0.5% (Koski et al. 1989) with no differentiation between truck and barge transport (Matthews et al. 1985). Indirect or delayed mortality was incorporated in the ocean survival rate.

Passage, transportation, and survival rates are summarized in Table 8. Figure 3 shows the interrelationship of transportation and passage downstream without transportation.

Adult returns.—Adult returns were determined by multiplying the number of smolts reaching Bonneville tailrace by ocean survival rates (Table 9). Ocean survival rates developed for FISHBEN were based on Meyer's (1982) optimum catch-to-escapement ratios adjusted to fit the subgroups analyzed with FISHBEN. Meyer (1982) did not separate wild from hatchery fish, divide runs by river segment, or consider the effects of transportation. Ocean survival rates used in FISHBEN reflected optimum production and dam passage conditions and were made comparable with Meyer's (1982) optimum ratios.

TABLE 7.—Fish bypass efficiency rates for dams studied in the Galloway analysis.

Species	Lower Granite Dam	Little Goose Dam	Lower Monumental Dam	Ice Harbor Dam	McNary Dam	John Day Dam	The Dalles Dam	Bonneville Dam
Steelhead	90.0	90.0	80.0	80.0	90.0	85.0	80.0	78.0
Spring and summer chinook salmon	87.0	87.0	78.0	78.0	87.0	72.0	67.0	76.0
Fall chinook salmon	60.0	60.0	40.0	40.0	60.0	35.0	30.0	54.0
Coho salmon	87.0	87.0	78.0	78.0	87.0	72.0	67.0	76.0
Sockeye salmon	70.0	70.0	55.0	55.0	70.0	47.0	42.0	60.0

TABLE 8.—Passage and survival rates for juvenile salmon and steelhead passing through spillways, turbines, bypasses, or sluiceways used for dams, and for transport methods used in FISHBEN for the Galloway analysis.

Rate	Lower Granite Dam	Little Goose Dam	Lower Monumental Dam	Ice Harbor Dam	McNary Dam	John Day Dam	The Dalles Dam	Bonneville Dam
Spillway passage				Equal to % flow spilled				
Spillway survival	0.98	0.98	0.98	0.98	0.98	0.98	0.98	0.98
Powerhouse passage				Equal to % flow through the powerhouse				
Sluiceway passage	0.00	0.00	0.00	0.51	0.00	0.00	0.45	0.00
Sluiceway survival	0.00	0.00	0.00	0.98	0.00	0.00	0.99	0.00
Turbine survival	0.85	0.85	0.85	0.85	0.92	0.87	0.85	0.93
Bypass survival	0.99	0.99	0.98	0.98	0.99	0.98	0.99	0.98
Transport survival	0.995	0.995	0.000	0.000	0.995	0.000	0.000	0.000

In keeping with Park's (1985) findings, 10% delayed mortality (steelhead, Columbia River summer chinook salmon, fall chinook salmon, coho salmon, and sockeye salmon) or 20% delayed mortality (spring chinook salmon and Snake River summer chinook salmon) was assumed for transported fish and incorporated in the ocean survival rates (Table 9). Columbia River summer chinook salmon migrate as subyearling smolts, as do fall chinook salmon, so it was assumed that decreased ocean survival caused by their small size would offset the increased number from the shorter duration of freshwater rearing. Columbia River summer chinook salmon and upriver bright fall chinook salmon were assumed to survive ocean rearing at half the rate of spring chinook salmon, whereas downriver fall chinook salmon were assumed to survive at a lower rate.

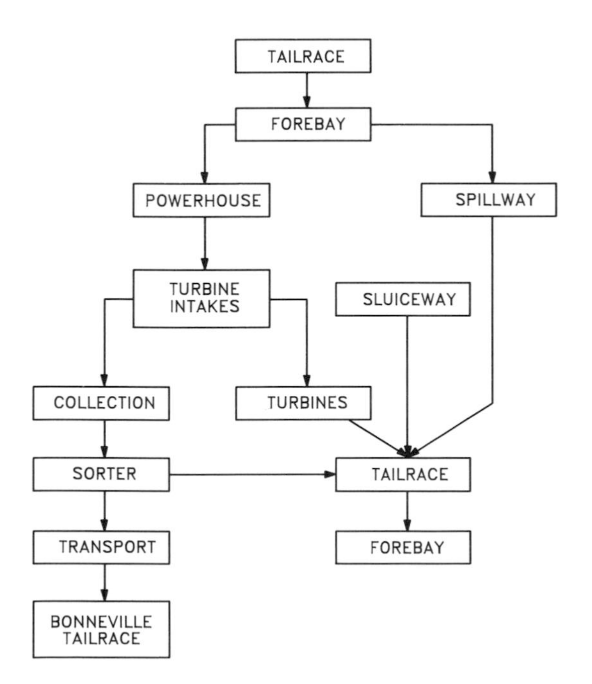

FIGURE 3.—Typical routing of fish passing downstream through rivers, reservoirs, and dams as analyzed in FISHBEN for the Galloway feasibility study.

As part of a sensitivity analysis, ocean survival rates used in FISHBEN were compared with smolt-to-adult-return rates used by Idaho Department of Fish and Game (IDFG 1984) and in the Lower Snake River Compensation Plan (USACE 1975) by converting the latter rates to ocean survival rates and the former to smolt-to-adult-return rates (Table 10). Because of the optimization in FISHBEN, upper river, lower river, and ocean catch were added to the other estimates to bring them in line with Meyer's (1982) optimum catch-to-escapement ratios. Proposed hatcheries, and additional lower river and ocean catch, were assumed to bring USACE (1975) and IDFG (1984) figures in line with Meyer's (1982) ratios. For comparison as ocean survival rates, smolt survival to Bonneville tailrace was estimated, then that number was divided into the adult return that had been estimated by the smolt-to-adult-return rate. Conversely, ocean survival rates were converted to smolt-to-adult-return rates by dividing the total adult return estimate by the number of smolts entered at the beginning of the model process for each species.

Because it was assumed for FISHBEN that fish survival conditions would be optimized by the year 2000, it was also assumed that adult fish would return at rates two to five times higher than estimated by USACE (1975) or the IDFG (1984).

Escapement in FISHBEN was defined as enough adult fish returning past Bonneville Dam to replenish the year 2000 smolt population (Table 11). Fecundity ranges (Scott and Crossman 1973) were adjusted according to information from hatchery and fishery managers for each river

TABLE 9.—Ocean survival rates for salmon and steelhead used in FISHBEN compared as catch-to-escapement ratios (CERs) with Meyer's (1982) CERs.

Species and group	Ocean survival		CER by transport mode			Meyer's CER
	Without transport	With transport	None	Partial	Full	
Hatchery steelhead						
Snake River	0.0600	0.0540	11.2:1	24.4:1	24.6:1	
Columbia River	0.0600	0.0540	28.7:1	30.3:1	30.8:1	
Wild steelhead						
Snake River	0.1200	0.1080	1.3:1	3.9:1	3.9:1	
Columbia River	0.1200	0.1080	3.2:1	3.7:1	3.8:1	
All steelhead combined			4.0:1	7.5:1	7.6:1	4:1
Hatchery spring chinook salmon						
Snake River	0.0800	0.0640	18.1:1	35.8:1	37.0:1	
Columbia River	0.0800	0.0640	44.4:1	45.1:1	47.7:1	
Wild spring chinook salmon						
Snake River	0.1600	0.1280	2.6:1	6.0:1	6.2:1	
Columbia River	0.1600	0.1280	5.8:1	6.0:1	7.0:1	
Snake River summer chinook salmon						
Hatchery	0.0800	0.0640	18.3:1	36.3:1	37.5:1	
Wild	0.1600	0.1280	2.6:1	6.0:1	6.2:1	
All yearling chinook salmon combined			7.2:1	10.2:1	10.9:1	8:1
Columbia River summer chinook salmon						
Hatchery	0.0400	0.0360	8.3:1	9.7:1	15.1:1	
Wild	0.0800	0.0720	6.3:1	6.8:1	9.2:1	
Subyearling summer chinook salmon			6.7:1	7.5:1	10.5:1	8:1
Fall chinook salmon						
Hatchery	0.0250	0.0225	11.1:1	12.3:1	12.6:1	
Wild	0.0500	0.0450	4.0:1	5.6:1	6.0:1	
Subyearling fall chinook salmon			6.2:1	7.7:1	8.1:1	8:1
Coho salmon						
Hatchery	0.0800	0.0720	29.2:1	30.9:1	31.3:1	
Wild	0.1600	0.1440	4.1:1	4.3:1	4.3:1	
Combined coho salmon			9.8:1	10.3:1	10.4:1	14:1
Wild sockeye salmon	0.0900	0.0810	2.1:1	2.8:1	3.0:1	2:1

segment. D. W. Chapman and T. C. Bjornn (Don Chapman Consultants and U.S. Fish and Wildlife Service, respectively, personal communications), as well and hatchery and fishery managers, provided egg-to-smolt and smolt-per-redd (or female) information. The number of females was divided by 0.4 (CBFTC 1974) to determine the number of adults needed. Survival to spawning was established by dividing the number of adults by survival rates through dams, reservoirs, and river reaches traversed in reaching hatcheries or spawning grounds. A loss of 2.5% was used for each dam and reservoir traversed (Liscom 1982; USACE 1988). Fish migrating up tributaries to hatcheries or spawning grounds were assigned an additional 2.5% mortality, and subgroups such as spring chinook salmon and summer steelhead that oversummer or overwinter in fresh water before spawning were assigned 5% more mortality.

Production models developed by the U.S. Fish and Wildlife Service (Buechler 1982a, 1982b) used to estimate the number of spawners required were wild spawners = (wild production/smolts per redd)/(ratio of returning adults to females/upstream survival); and hatchery spawners = (hatchery production/smolts per female)/(ratio of returning adults to females/upstream survival).

Adult fish return distributions were used to account for the fact that most anadromous salmon and steelhead spend 1 to 3 years at sea before reaching adulthood. Each group of smolts was apportioned as shown in Table 12, and returns were estimated according to that apportionment. Thus, each adult return comprised returns from two or three smolt outmigrations.

Economic values of salmon and steelhead.—Detailed analyses of the economic values of salmon and steelhead were described in the Galloway feasibility report. Current weighted average values, considering enhancement versus restoration values, sport fishery values, commercial fishery values, average versus marginal values,

TABLE 10.—Comparison of ocean survival rates used in FISHBEN with smolt-to-adult-return rates used by IDFG (1984) and in the Lower Snake River Fish and Wildlife Compensation Plan (LSRFWCP: USACE 1975).

	Comparison as ocean survival rates			Comparison as smolt-to-adult-return rates		
Species and group	FISHBEN	IDFG	LSRFWCP	FISHBEN	IDFG	LSRFWCP
Hatchery fish						
Snake River steelhead	0.0600	0.0114	0.1430	0.0210	0.0050	0.0040
Columbia River steelhead	0.0600			0.0418		
Snake River spring chinook salmon	0.0800	0.0570	0.0620	0.0262	0.0087	0.0080
Columbia River spring chinook salmon	0.0800			0.0543		
Snake River summer chinook salmon	0.0800	0.0565	0.0614	0.0264	0.0087	0.0080
Columbia River summer chinook salmon	0.0400			0.0090		
Fall chinook salmon	0.0250		0.0117	0.0100		0.0020
Coho salmon	0.0800			0.0872		
Wild fish						
Snake River steelhead	0.1200	0.0305		0.0420	0.0107	
Columbia River steelhead	0.1200			0.0968		
Snake River spring chinook salmon	0.1600	0.0807		0.0740	0.0160	
Columbia River spring chinook salmon	0.1600			0.0855		
Snake River summer chinook salmon	0.1600	0.1130		0.0528	0.0160	
Columbia River summer chinook salmon	0.0800			0.0300		
Fall chinook salmon	0.0500			0.0189		
Coho salmon	0.1600			0.0931		
Sockeye salmon	0.0900			0.0429		

TABLE 11.—Calculation of spawning escapement required to sustain year 2000 smolt production used in FISHBEN for the Galloway analysis.

Species and subgroup	Year 2000 smolt production (millions)	Smolts per redd or female	Number of females required	Number of adults required	Survival to spawning (%)	Escapement required at Bonneville Dam
Snake River						
Wild spring chinook salmon	6.30	250	25,200	63,000	76.3	82,536
Hatchery spring chinook salmon	11.76	2,500	4,704	11,761	81.3	14,461
Wild steelhead	6.42	200	32,100	80,250	76.3	105,136
Hatchery steelhead	14.84	2,000	7,420	18,549	81.3	22,807
Wild summer chinook salmon	3.10	250	12,400	31,000	76.3	40,613
Hatchery summer chinook salmon	1.30	2,500	521	1,302	81.3	1,600
Wild fall chinook salmon	0.80	750	1,067	2,667	81.3	3,279
Hatchery fall chinook salmon	9.65	3,000	3,218	8,405	85.9	9,365
Wild sockeye salmon	0.10	250	400	1,000	71.3	1,402
Wild coho salmon	0.10	140	714	1,786	76.2	2,339
Mid-Columbia River (McNary Reservoir and above)						
Wild spring chinook salmon	5.50	250	22,000	55,000	76.3	72,056
Hatchery spring chinook salmon	4.82	2,500	1,927	4,817	81.3	5,922
Wild steelhead	1.00	200	5,000	12,500	76.3	16,376
Hatchery steelhead	1.18	2,000	592	1,480	81.3	1,819
Wild summer chinook salmon	1.50	750	2,000	5,000	81.3	6,148
Hatchery summer chinook salmon	1.76	3,000	586	1,465	85.9	1,705
Wild fall chinook salmon	20.00	750	26,667	66,667	87.9	75,878
Hatchery fall chinook salmon	11.39	3,000	3,797	9,492	87.9	10,804
Wild sockeye salmon	4.90	250	19,600	49,000	71.3	68,695
Hatchery coho salmon	0.95	1,400	679	1,696	85.9	1,975
Lower Columbia and tributaries (McNary Dam to Bonneville tailrace)						
Wild spring chinook salmon	1.50	250	6,000	15,000	80.2	16,633
Hatchery spring chinook salmon	7.28	2,500	2,911	7,277	92.7	7,852
Wild steelhead	1.00	160	6,250	15,625	90.2	17,326
Hatchery steelhead	2.71	2,000	1,354	3,386	92.7	3,653
Wild fall chinook salmon	1.00	900	1,111	2,778	95.0	2,924
Hatchery fall chinook salmon	25.56	3,600	7,101	17,752	97.5	18,207
Wild coho salmon	1.00	140	1,143	17,857	95.0	18,797
Hatchery coho salmon	2.29	1,400	1,635	4,088	97.5	4,193

TABLE 12.—Percentages of adult salmon and steelhead returning to the Columbia River after 1, 2, or 3 years in the Pacific Ocean.

Species	1 ocean year	2 ocean years	3 ocean years
Spring chinook salmon[a]	11.3	46.2	42.5
Summer chinook salmon[a]	11.3	46.2	42.5
Fall chinook salmon[b]	15.0	45.0	40.0
Steelhead[c]	35.7	59.7	4.6
Coho salmon[d]	44.2	55.8	
Sockeye salmon[e]	20.0	80.0	

[a]J. Coon, Pacific Fishery Management Council, personal communication.
[b]W. Ebel, National Marine Fisheries Service, personal communication.
[c]G. James, Confederated Tribes of the Umatilla Indian Reservation, personal communication.
[d]USACE 1975.
[e]M. Dell, Grant County Public Utility District, personal communication.

and market and nonmarket values were used in FISHBEN as shown in Table 13.

Results

FISHBEN was used to estimate the economic value of salmon and steelhead for pre-water budget (no special flow regulations for juvenile salmon and steelhead migrations), no project (continue current flow regulations for juvenile salmon and steelhead), and 21 alternative conditions (Table 14). Gross fishery values estimated by FISHBEN, net fishery benefits, total project benefits, total project costs, and benefit-to-cost ratios were developed (Table 15).

Under the principles and guidelines of the Corps of Engineers, alternative TW3J2, construction of Galloway Dam with a reservoir capacity of 900,000 acre-feet, appeared to be the best alternative from a national standpoint. Selection of the preferred alternative was scheduled to follow public review of the plan in 1989. Comparison of alternative TW3J2 with preproject conditions is shown in Table 16.

Most benefits estimated with FISHBEN (73%) occurred between Hells Canyon and Lower Granite dams, primarily because transportation of juvenile salmon and steelhead from Lower Granite, Little Goose, and McNary dams removed them from the benefits of flow augmentation below these dams. The benefit on the middle Snake River (44.4%) indicated that augmentation, as modeled, greatly increased juvenile fish survival in the free-flowing reaches of that area. Eliminating water budget releases from Dworshak Dam would significantly reduce (by 18.4%) survival of fish through the lower Clearwater River.

Discussion

Compared with previous Corps of Engineers methods of estimating fishery benefits, FISHBEN is a thorough and comprehensive method for analyzing various ways of providing augmented flows to improve survival of juvenile salmon and steelhead in the Columbia River basin. FISHBEN was structured so that input could be updated as new data and information become available. Extensive sensitivity analysis was conducted, based on current fish numbers and survival rates. The sensitivity analysis indicated FISHBEN provides reasonable estimates for projecting fishery benefits for feasibility study purposes.

FISHBEN was developed in cooperation with the fishery agencies and Columbia basin tribes. It was used extensively in analyzing the benefits of improving juvenile fish bypass systems for lower

TABLE 13.—Weighted average fish values (1987 price level) used in FISHBEN for the Galloway analysis.

Species	Sport (S) or commercial (C)	Value per fish	Share of catch	Weighted value	Average value
Spring chinook salmon	S	$135.00	57%	$ 76.95	$ 95.06
Spring chinook salmon	C	42.12	43%	18.11	
Fall chinook salmon	S	101.00	20%	20.20	48.91
Fall chinook salmon	C	35.89	80%	28.71	
Snake River steelhead	S	83.00	82%	68.06	71.99
Snake River steelhead	C	21.81	18%	3.93	
Columbia River steelhead	S	136.00	82%	111.52	115.45
Columbia River steelhead	C	21.81	18%	3.93	
Coho salmon	S	101.00	36%	36.36	42.11
Coho salmon	C	8.98	64%	5.75	
Sockeye salmon	S				
Sockeye salmon	C	9.02	100%	9.02	9.02

TABLE 14.—Alternatives examined with FISHBEN to estimate fishery benefits for the Galloway study.

Alternative	Description
Preproject	Do nothing—continue current operations
TW3A	Pre-water budget condition
TW3C1	Use 900,000 acre-feet storage at Galloway Dam to replace water budget use of storage Brownlee and Dworshak dams; 85,000 ft^3/s target flow at Lower Granite Dam
TW3C2	Same as TW3C1 except 120,000 ft^3/s target flow at Lower Granite
TW3D1	Use 900,000 acre-feet of storage at Galloway plus Brownlee and Dworshak storage as needed; 85,000 ft^3/s target at Lower Granite
TW3D2	Same as TW3D1 except 120,000 ft^3/s target flow at Lower Granite
TW3E1	Use Brownlee and Dworshak without constraints to meet 85,000 ft^3/s target flows at Lower Granite
TW3E2	Same as TW3E1 except 120,000 ft^3/s target flow at Lower Granite
TW3F1	Use Brownlee and Dworshak storage first, Galloway last; 85,000 ft^3/s target flow at Lower Granite
TW3F2	Same as TW3F1 except 120,000 ft^3/s target flow at Lower Granite
TW3G1	Same as TW3D1 except flood storage of Galloway combined with flood storage of Brownlee
TW3G2	Same as TW3D2 except flood storage of Galloway combined with flood storage of Brownlee
TW3H1	Use Brownlee under current constraints, Dworshak unconstrained to meet 85,000 ft^3/s target flow at Lower Granite
TW3H2	Same as TW3H1 except 120,000 ft^3/s target flow at Lower Granite
TW3I1	Use Galloway as first source, Brownlee with current constraints, Dworshak without constraints; 85,000 ft^3/s target flow at Lower Granite
TW3I2	Same as TW3I1 except 120,000 ft^3/s target flow at Lower Granite
TW3J2	Same as TW3C2 except Galloway flood storage added to Brownlee flood storage
TW3C4	Same as TW3C2 except 4.5-MW power plant at Galloway
TW3C2S	Same as TW3C2 except 600,000 acre-feet storage used for Galloway
TW3C2L	Same as TW3C2 except 1,220,000 acre-feet storage used for Galloway
HM86GB	40-MW power plant, no flood control or water budget at Galloway; Brownlee, Dworshak, and Priest Rapids operated under pre-water budget criteria
HM86GC	Same as HM86GB except fixed flood control operation at Galloway
HM86GF	40-MW power plant at Galloway added to preproject; Brownlee, Dworshak proved water budget; Priest Rapids operated under pre-water budget criteria

Snake River dams and McNary Dam. Fishery information gathered for FISHBEN was used extensively by the North Pacific Division of the Corps of Engineers for their FISHPASS model.

It is recommended that this model be maintained and updated for use by the Corps of Engineers and fishery agencies in evaluating the economic benefits of fish facility improvements.

TABLE 15.—Fishery benefits estimated by FISHBEN for the Galloway feasibility study, and comparison of total project benefits with total project cost.

Alternative	Gross fishery value	Net fishery benefit	Total project benefit	Total project cost	Benefit-to-cost ratio
Preproject	$402,447,000				
TW3A (pre-water budget)	395,086,000	$ −7,361,000	$12,580,000	$ 7,361,000	1.71
TW3C1	402,400,000	−47,000	9,903,000	12,814,000	0.77
TW3C2	407,357,000	4,910,000	15,231,000	12,767,000	1.19
TW3D1	405,967,000	3,520,000	3,876,000	15,376,000	0.25
TW3D2	412,304,000	9,857,000	10,213,000	14,888,000	0.69
TW3E1	408,130,000	5,683,000	5,683,000	11,538,000	0.49
TW3E2	431,559,000	29,112,000	29,112,000	36,631,000	0.79
TW3F1	406,110,000	3,663,000	3,663,000	16,532,000	0.24
TW3F2	413,806,000	11,359,000	11,715,000	15,559,000	0.75
TW3G1	404,878,000	2,431,000	7,913,000	12,767,000	0.62
TW3G2	411,705,000	9,258,000	12,350,000	12,767,000	0.97
TW3H1	406,346,000	3,899,000	3,899,000	12,185,000	0.32
TW3H2	424,210,000	21,763,000	21,763,000	21,457,000	1.01
TW3I1	406,965,000	4,518,000	4,874,000	24,077,000	0.20
TW3I2	430,122,000	27,675,000	28,031,000	33,419,000	0.84
TW3J2	407,386,000	4,939,000	16,809,000	12,767,000	1.32
TW3C4	407,357,000	4,910,000	15,271,000	12,767,000	1.20
TW3C2S	405,780,000	3,333,000	13,379,000	10,981,000	1.22
TW3C2L	408,690,000	6,243,000	16,847,000	14,501,000	1.16
HM86GB	394,277,000	−8,170,000	19,631,000	20,937,000	0.94
HM86GC	392,262,000	−10,185,000	19,834,000	22,952,000	0.86
HM86GF	401,284,000	−1,163,000	7,772,000	13,930,000	0.56

TABLE 16.—Average annual fishery results obtained with FISHBEN for the preferred alternative (TW3J2) in the Galloway feasibility study.

Subgroup	Initial smolts	Estimated smolts to Bonneville tailrace	Increment above preproject	Adult return	Increment above preproject	Value of increment
Snake River						
Hatchery steelhead	15,620,000	10,893,259	180,805	589,896	9,791	$ 705,000
Wild steelhead	6,420,000	4,829,734	84,600	522,546	9,137	657,000
Hatchery spring chinook salmon	12,380,000	8,330,600	132,972	541,444	8,735	830,000
Wild spring chinook salmon	6,300,000	4,507,440	61,232	585,982	8,076	768,000
Hatchery summer chinook salmon	1,370,000	939,240	21,217	61,037	1,385	132,000
Wild summer chinook salmon	3,100,000	2,237,144	50,536	290,766	6,598	628,000
Columbia River						
Hatchery steelhead	4,089,000	2,921,837	13,706	172,362	829	95,000
Wild steelhead	2,000,000	1,359,355	5,132	158,175	630	72,000
Hatchery spring chinook salmon	12,070,000	8,074,166	41,150	638,233	3,332	317,000
Wild spring chinook salmon	7,000,000	4,025,030	34,523	626,414	5,590	531,000
Hatchery summer chinook salmon	1,850,000	466,911	2,623	18,266	105	10,000
Wild summer chinook salmon	1,500,000	613,682	1,013	48,284	80	7,000
Snake and Columbia Rivers						
Hatchery fall chinook salmon	49,060,000	20,906,040	74,970	513,636	1,849	90,000
Wild fall chinook salmon	21,800,000	11,437,877	15,466	542,926	746	37,000
Hatchery coho salmon	3,410,000	2,508,444	11,277	197,701	907	38,000
Wild coho salmon	1,100,000	701,388	1,232	111,353	195	8,000
Wild sockeye salmon	4,900,000	3,139,316	13,671	268,019	1,273	11,000
Total	153,969,000	87,891,463	746,125	5,877,040	59,258	$4,939,000

Acknowledgments

FISHBEN was first designed in the early 1980s by Mark Lindgren and Nick Iadanza of the Corps of Engineers with the assistance of Roy Heberger of the U.S. Fish and Wildlife Service. Since then, Cliff Fitzsimmons, the Corps' project manager for the Galloway feasibility study, has been the principal developer of the model. Biological input was provided by Corps biologists Teri Barila, Doug Arndt, and the author. Tim Bartish, Corps limnologist, was the main operator of the model and analyzer of input and results. Economic analyses were provided by Paul Fredericks and Ed Woodruff, Corps economists. Doug Arndt and Bolyvong Tannovan of the North Pacific Division office of the Corps reviewed and coordinated FISHBEN with FISHPASS. Valuable input and review were provided by the U.S. Fish and Wildlife Service, Idaho Department of Water Resources, Grant County Public Utility District, Idaho Department of Fish and Game, participating Indian tribes, and other agencies.

References

Anonymous. 1986a. Fish Passage Center weekly report for January 8. Columbia Basin Fishery Agencies and Tribes, Portland, Oregon.

Anonymous. 1986b. Detailed fishery operating plan with 1986 operating criteria. Prepared jointly by the Columbia River Basin Tribes and state and federal fish and wildlife agencies. Columbia Basin Fish and Wildlife Authority, Portland, Oregon.

Basham, L. R., M. R. DeLarm, S. W. Pettit, J. B. Athearn, and J. V. Barker. 1983. Fish transportation oversight team annual report. Fiscal year 1982, transport operations on the Snake and Columbia Rivers. NOAA (National Oceanic and Atmospheric Administration) Technical Memorandum NMFS (National Marine Fisheries Service) F/NWR-5, Portland, Oregon.

Bayha, K. 1974. Anatomy of a river. Pacific Northwest River Basins Commission, Vancouver, Washington.

Bell, M. C. 1986. Fisheries handbook of engineering requirements and criteria. U.S. Army Corps of Engineers, North Pacific Division, Fisheries Engineering Research Program, Portland, Oregon.

Bell, M. C., Z. E. Parkhurst, R. G. Porter, and M. Stevens. 1976. Effects of power peaking on the survival of juvenile fish at lower Columbia and Snake River dams. Report to U.S. Army Corps of Engineers, North Pacific Division, Portland, Oregon.

Brown, B. J. 1982. Benefits and costs of juvenile migrant bypass facilities. National Marine Fisheries Service, working paper, Portland, Oregon.

Buechler, D. G. 1982a. Report of March 15 to U.S. Army Corps of Engineers, Walla Walla, Washington.

Buechler, D. G. 1982b. Report of April 21 to U.S. Army Corps of Engineers, Walla Walla, Washington.

CBFTC (Columbia Basin Fishery Technical Committee). 1974. Lower Snake River hatchery subcommittee site selection report. Report to U.S. Army Corps of Engineers, Walla Walla, Washington.

Ceballos, J. R. 1982. Impact of incremental levels of riverflow on Snake and Columbia river juvenile fish survival. National Marine Fisheries Service, working paper, Portland, Oregon.

DeLarm, M. R., L. R. Basham, S. W. Pettit, J. B. Athearn, and J. V. Barker. 1984. Fish transportation oversight team annual report. Fiscal year 1983, transport operations on the Snake and Columbia rivers. NOAA (National Oceanic and Atmospheric Administration) Technical Memorandum NMFS (National Marine Fisheries Service) F/NWR-7, Portland, Oregon.

Ebel, W. J. 1980. Transportation of chinook salmon, *Oncorhynchus tshawytscha*, and steelhead, *Salmo gairdneri*, smolts in the Columbia River and effects on adult returns. U.S. National Marine Fisheries Service Fishery Bulletin 78:491–505.

Ebel, W. J., R. F. Krcma, and H. L. Raymond. 1973. Evaluation of fish protective facilities at Little Goose Dam and review of other studies related to protection of juvenile salmonids in the Columbia and Snake rivers, 1973. National Marine Fisheries Service, Northwest and Alaska Fisheries Center, Seattle.

Ebel, W. J., H. L. Raymond, G. E. Monan, W. E. Farr, and G. K. Tanonaka. 1975. Effect of atmospheric gas supersaturation caused by dams on salmon and steelhead trout of the Snake and Columbia Rivers. National Marine Fisheries Service, Northwest and Alaska Fisheries Center, Seattle.

Ebel, W. J., and six coauthors. 1974. Evaluation of fish protective facilities at Little Goose Dam and review of other studies related to protection of juvenile salmonids in the Columbia and Snake rivers, 1974. National Marine Fisheries Service, Northwest and Alaska Fisheries Center, Seattle.

Ebel, W. J., G. K. Tonanoka, G. E. Monan, H. L. Raymond, and D. L. Park. 1979. Status report, 1978; the Snake River salmon and steelhead crisis; its relation to dams and the national energy shortage. National Marine Fisheries Service, Northwest and Alaska Fisheries Center, Seattle.

IDFG (Idaho Department of Fish and Game). 1984. Draft Idaho anadromous fish management plan, 1984–1990. IDFG, Boise.

Johnson, L., C. Noyes, and R. McClure. 1983. Hydroacoustic evaluation of the efficiencies of the ice and trash sluiceway and spillway at Ice Harbor Dam for passing downstream migrating juvenile salmon and steelhead. BioSonics, Seattle.

Koski, C. H., S. W. Pettit, J. L. McKern. 1989. Fish transportation oversight team annual report—FY 1988. Transport operations on the Snake and Columbia rivers. National Marine Fisheries Service, Portland, Oregon.

Krcma, R. F., D. DeHart, M. Gessel, C. Long, and C. W. Sims. 1982. Evaluation of submersible traveling screens, passage of juvenile salmonids through the ice–trash sluiceway, and cycling of gatewell–orifice operations at the Bonneville first powerhouse. National Marine Fisheries Service, Northwest and Alaska Fisheries Center, Seattle.

Krcma, R. F., M. H. Gessel, W. D. Muir, C. S. McCutcheon, L. G. Gilbreath, and B. H. Monk. 1984. Evaluation of the juvenile collection and bypass system at Bonneville Dam—1983. Report to U.S. Army Corps of Engineers by National Marine Fisheries Service, Northwest and Alaska Fisheries Center, Seattle.

Krcma, R. F., M. H. Gessel, and F. J. Ossiander. 1983. Research at McNary Dam to develop and implement fingerling protection system for John Day Dam. National Marine Fisheries Service, Northwest and Alaska Fisheries Center, Seattle.

Krcma, R. F., G. A. Swan, and F. J. Ossiander. 1985. Fish guiding and orifice passage efficiency tests with subyearling chinook salmon, McNary Dam. National Marine Fisheries Service, Northwest and Alaska Fisheries Center, Seattle.

Liscom, K. L., and L. C. Stuehrenberg. 1982. Radio tracking studies of "upriver bright" fall chinook salmon between Bonneville and McNary Dams, 1982. National Marine Fisheries Service, Northwest and Alaska Fisheries Center, Seattle.

Matthews, G. M., D. L. Park, T. E. Ruehle, and J. R. Harmon. 1985. Evaluation of transportation of juvenile salmonids and related research on the Columbia and Snake rivers. National Marine Fisheries Service, Northwest and Alaska Fisheries Center, Seattle.

Meyer, P. A. 1982. Net economic values for salmon and steelhead from the Columbia River system. NOAA (National Oceanic and Atmospheric Administration) Technical Memorandum NMFS (National Marine Fisheries Service) F/NWR-3, Portland, Oregon.

Nichols, D. W., and B. H. Ransom. 1982. Continued evaluation of The Dalles Dam trash sluiceway as a juvenile salmonid bypass system during 1981. Report to U.S. Army Corps of Engineers, Portland, Oregon.

NPPC (Northwest Power Planning Council). 1982. Columbia River basin fish and wildlife program. NPPC, Portland, Oregon.

NPPC (Northwest Power Planning Council). 1984. Columbia River basin fish and wildlife program. NPPC, Portland, Oregon.

NPPC (Northwest Power Planning Council). 1987. Columbia River basin fish and wildlife program. NPPC, Portland, Oregon.

Park, D. L. 1980. Transportation of chinook salmon and steelhead smolts 1968–80 and its impact on adult returns to the Snake River. National Marine Fisheries Service, Northwest and Alaska Fisheries Center, Seattle.

Park, D. L. 1985. A review of smolt transportation to bypass dams on the Snake and Columbia rivers. Pages 2-1 to 2-66 *in* Comprehensive report of juvenile salmonid transportation. U.S. Army Corps of Engineers, Walla Walla, Washington.

Park, D. L., and eight coauthors. 1979. Transportation activities and related research at Lower Granite, Little Goose, and McNary dams. National Marine Fisheries Service, Northwest and Alaska Fisheries Center, Seattle.

Park, D. L., T. E. Ruehle, J. R. Harmon, and B. H. Monk. 1980. Transportation research on the Columbia and Snake rivers. National Marine Fisheries Service, Northwest and Alaska Fisheries Center, Seattle.

Park, D. L., and six coauthors. 1981. Transportation research on the Columbia and Snake rivers. National Marine Fisheries Service, Northwest and Alaska Fisheries Center, Seattle.

Park, D. L., and six coauthors. 1982. Transport operations and research on the Snake and Columbia rivers. National Marine Fisheries Service, Northwest and Alaska Fisheries Center, Seattle.

Park, D. L., and six coauthors. 1983. Evaluation of transportation and related research on Columbia and Snake rivers. National Marine Fisheries Service, Northwest and Alaska Fisheries Center, Seattle.

Park, D. L., J. R. Smith, E. Slatick, G. M. Matthews, L. R. Basham, and G. A. Swan. 1978. Evaluation of fish protective facilities at Little Goose and Lower Granite dams and review of mass transportation activities. National Marine Fisheries Service, Northwest and Alaska Fisheries Center, Seattle.

Raymond, H. L. 1974. Snake River runs of salmon and steelhead trout: trends in abundance of adults and downstream survival of juveniles. Report to U.S. Army Corps of Engineers by National Marine Fisheries Service, Northwest and Alaska Fisheries Center, Seattle.

Raymond, H. L. 1979. Effects of dams and impoundments on migrations of juvenile chinook salmon and steelhead from the Snake River, 1966 to 1975. Transactions of the American Fisheries Society 108:505–529.

Scott, W. B., and E. J. Crossman. 1973. Freshwater fishes of Canada. Fisheries Research Board of Canada Bulletin 184.

Sims, C. W., W. W. Bentley, and R. C. Johnson. 1978. Effects of power peaking operation on juvenile salmon and steelhead trout migrations. National Marine Fisheries Service, Northwest and Alaska Fisheries Center, Seattle.

Sims, C. W., A. E. Giorgi, R. C. Johnson, and D. A. Brege. 1983. Migrational characteristics of juvenile salmon and steelhead in the Columbia River basin. National Marine Fisheries Service, Northwest and Alaska Fisheries Center, Seattle.

Sims, C. W., and F. J. Ossiander. 1981. Migrations of juvenile chinook salmon and steelhead trout in the Snake River from 1973 to 1979. National Marine Fisheries Service, Northwest and Alaska Fisheries Center, Seattle.

USACE (U.S. Army Corps of Engineers). 1975. Lower Snake River fish and wildlife compensation plan. USACE, Walla Walla, Washington.

USACE (U. S. Army Corps of Engineers). 1988. Annual fish passage report. USACE, Portland Oregon).

American Fisheries Society Symposium 10:104–108, 1991

The Beaded Channel: A Low-Cost Technique for Enhancing Winter Habitat of Coho Salmon

C. J. CEDERHOLM AND W. J. SCARLETT
Washington Department of Natural Resources
Olympia, Washington 98504, USA

Abstract.—Tributaries of the Clearwater River, Washington, support substantial populations of overwintering juvenile coho salmon *Oncorhynchus kisutch*. In 1986, a 2-year study was initiated to further explore opportunities for enhancing the survival and growth of overwintering juvenile coho salmon in wall-base channel streams. In this study, six 7-m-long by 4-m-wide by 1–2.5-m-deep ponds were blasted into the mud substrate of an ephemeral wall-base channel, which came to be known as Swamp Creek. The serial arrangement of the small ponds that were created along the stream channel, termed a beaded channel, resulted in a significant increase in overwinter survival and growth of juvenile coho salmon. Survival increased from zero before construction to 43% the first winter (1986–1987) and 70% the second winter (1987–1988) after construction. Mean overwinter increases in length and weight of immigrants changed from nothing (due to zero survival) before construction to 20 mm and 5.6 g after construction. The total cost of the Swamp Creek beaded channel was $3,340 (personnel plus materials).

In Pacific Coast streams, juvenile coho salmon *Oncorhynchus kisutch* migrate into small tributaries in spring and fall presumably to escape high flows in the main river channels (Skeesick 1970; Peterson 1982b). This behavior has been documented along the Oregon (Skeesick 1970), Washington (Cederholm and Scarlett 1982; Peterson 1982a, 1982b), and British Columbia coasts (Bustard and Narver 1975; Brown 1985). Some of these tributaries have been termed "wall-base channels" by Peterson and Reid (1984). Wall-base channels are developed along abandoned river meanders, and tend to have mud bottoms, which can be highly productive sites. Peterson and Reid (1984) estimated that 20–25% of the total number of coho salmon smolts from the Clearwater River basin spent the winter in wall-base channel tributaries. Brown (1985) estimated 23% of the coho salmon smolts produced from Carnation Creek, British Columbia, came from similar tributaries.

Peterson (1982a, 1982b) found that coho salmon juveniles had good growth in a shallow wall-base channel pond but poorer overwinter survival than fish in a deep wall-base channel pond. Therefore, an opportunity may exist for winter habitat enhancement if one could combine the productivity of a shallow pond and the cover aspects of a deep pond on a wall-base channel. Further, work by Zarnowitz and Raedeke (1984) and Peterson (1985), in the same river basin, indicated that many coho salmon immigrants were dying of heavy predation and stranding while wintering in some wall-base channels. With knowledge of these mortality factors and the recommendation that greater pond depth may increase coho overwintering survival in wall-base channels, Cederholm et al. (1988) created a shallow pond with some 2–3-m-deep holes on a wall-base channel tributary. The result was significantly improved survival and growth of overwintering coho salmon, when compared with preconstruction conditions. The question still remained, however, as to what other pond configurations could be used in wall-base channel habitat enhancement. Therefore, this study was designed to look at the effectiveness of a beaded channel arrangement.

Study Area

This work was done between 1986 and 1988 in the Clearwater River basin, located on the western Olympic Peninsula about 120 km southwest of Port Angeles, Washington (Figure 1). The Clearwater basin receives over 350 cm of rain annually. Heavy winter rains result in main-river discharges three orders of magnitude higher than summer discharges, whereas flows in wall-base channels remain low.

Swamp Creek, a wall-base channel of the Clearwater River, was selected for this work. Swamp Creek is about 10 km upstream of the confluence of the Clearwater and Queets rivers (Figure 1) and has a catchment area of about 18 hectares. The flow of Swamp Creek is derived from a small tributary and a series of springs in the adjacent hillside. Before we began our work, coho salmon juveniles attempted to overwinter in Swamp Creek, but the stream dried up each winter, killing

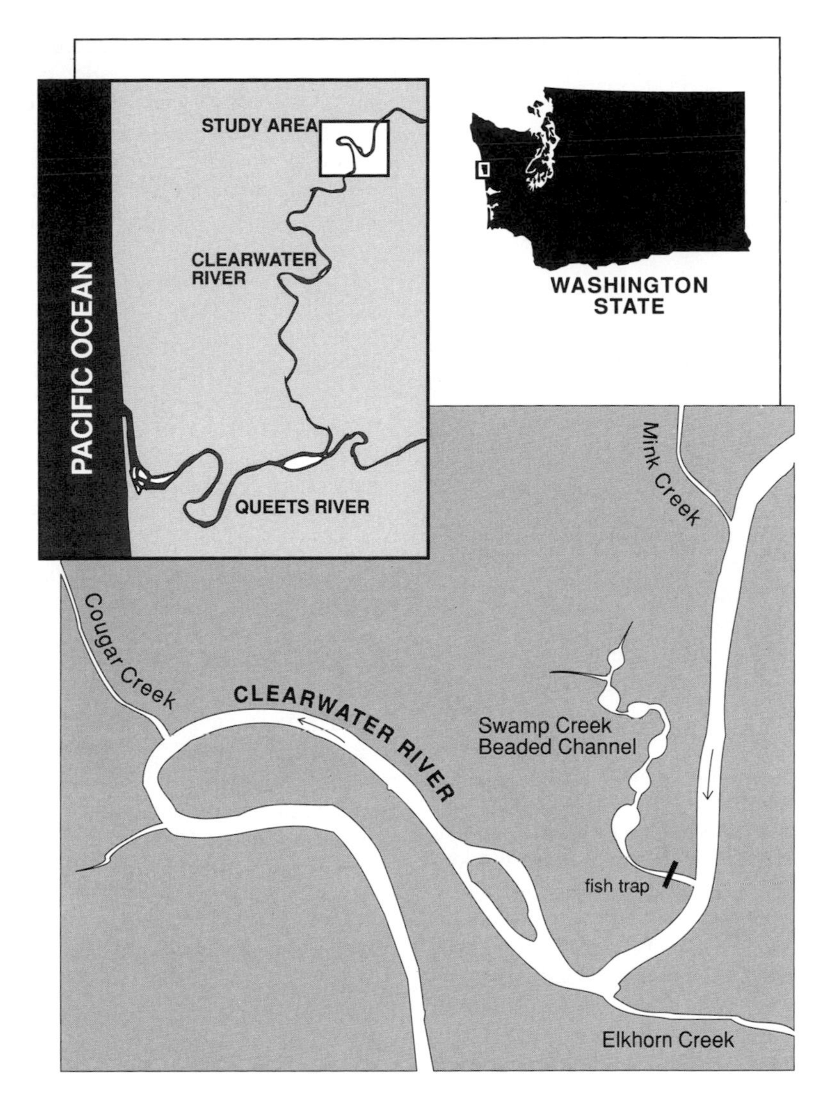

FIGURE 1.—Location of the Swamp Creek beaded channel.

all the immigrants. We thought that fish could survive the winter if we could create ponds of water by blasting holes below the winter water table, thereby providing refuge habitat.

Methods

In the summer of 1986, three 7-m-long by 4-m-wide by 1–2.5-m-deep holes were blasted into the mud substrate along Swamp Creek channel, and an additional three holes were added in the summer of 1987. The resulting holes filled with water and formed a beaded configuration of small ponds. We named the site Swamp Creek beaded channel.

The holes were excavated by exploding charges of ANFO, commonly called "fertilizer" because of its resemblance to commercial fertilizer. The formulation is 94% ammonium nitrate (AN) and 6% fuel oil (FO). It is classified as a blasting agent because it is not cap-sensitive and must be detonated by a high-velocity explosive.

To minimize excess excavation during blasting, care was taken when the hand-cored holes were loaded with the ANFO charges not to place the explosive deeper than the mud layer that sat on top of underlying gravel. This ensured against pond leakage. Each hole was filled with ANFO, and the charges were connected with primacord, a detonation wire, and a battery-operated trigger.

The six small ponds created at Swamp Creek

TABLE 1.—Overwinter survival rates of juvenile coho salmon in the Swamp Creek beaded channel, 1986–1987 and 1987–1988.

Year	Number of marked fish		Survival rate
	Upstream migrants	Downstream migrants	
1986–1987	332	144	0.43
1987–1988	294	207	0.70
Mean			0.57

had a combined wintertime water surface area of 180 m^2. These ponds averaged 1 m in depth, with maximum depths of 2.5 m.

To monitor fish movement, a two-way fish trap (which captured juvenile coho salmon moving either up- or downstream) was installed downstream of the lowermost pond (Figure 1). Movement of coho salmon to and from Swamp Creek beaded channel was monitored for 2 years following excavation. All immigrant coho salmon captured in the trap were measured to the nearest 1 mm (fork length), and subsamples (10%) were weighed to the nearest 0.1 g. During the two fall seasons, about 20–25% of the captured immigrants were freeze-branded with branding irons cooled with a slurry of dry ice and acetone. Annual overwinter survival and growth rates of coho salmon were estimated from these marked fish.

Annual overwinter survival rate of coho salmon using Swamp Creek beaded channel was calculated by dividing the number of branded smolts by the number of branded immigrants (Table 1). Because the trap was occasionally inundated during the winter rainy periods, the total number of immigrants actually using the site was estimated by dividing the total smolt yield of the site by the calculated survival rate of branded fish (Table 2).

Results and Discussion

Over the 2-year period of this study, 331 (1987) and 707 (1988) age-1 smolts were produced from Swamp Creek beaded channel. Remember that smolts did not survive the winter in Swamp Creek prior to our habitat enhancement of the site. While in residence in the beaded channel, juvenile coho salmon were evenly distributed in the ponds (beads), based on visual observation. We noticed after the first year of study, when the number of ponds was increased from three to six, the survival and number of smolts produced increased proportionately.

The overwinter survival of branded juvenile coho entering Swamp Creek beaded channel increased from zero before enhancement to 43% in 1986–1987 when we had three ponds, to 70% in 1987–1988 when we had six ponds (Table 1). The improved survival of overwintering fish in 1987–1988 compared with 1986–1987 might have been the result of doubling the number of ponds. Similar good survival rates were found by Cederholm et al. (1988) in another wall-base channel enhancement project on a nearby tributary of the Clearwater River.

Between October 1986 and June 1988, 1,487 age-0 fall immigrant juvenile coho salmon were captured as they moved upstream into the fish trap. The corrected estimate of the total number of immigrants is 1,780 (Table 2). As was also reported by Peterson (1982b), immigration of juvenile coho salmon coincided with rising autumn streamflows in the Clearwater River.

The average incremental increases in length (20 mm) and weight (5.6 g) (Table 3) were about half those reported by Cederholm et al. (1988) for a similar wall-base channel enhancement project. Cederholm et al. (1988) created a 5,000-m^2 pond with 85% of its area in shallows less than 1 m deep, compared with Swamp Creek beaded channel's six small ponds with a combined area of 180 m^2 and relatively little shallow area. Also, the riffles that connect the six small ponds at Swamp Creek beaded channel dry up during low-flow periods in winter, a condition not conducive to establishment of many aquatic invertebrates. Peterson (1982a), in his studies of natural Clearwater River riverine ponds (wall-base channels), found that shallows provide favorable growing conditions for overwintering juvenile coho salmon due to the detritus and associated invertebrates that accumulate there. The relatively small size of the Swamp Creek ponds and the lack of shallows for food production could have imposed space limitations and intense competition for a limited food supply on the overwintering coho salmon, which may explain the relatively low growth in Swamp Creek beaded channel. This is supported by the smaller densities of overwintering coho

TABLE 2.—Captured and estimated numbers of juvenile coho salmon immigrating past the trap at Swamp Creek beaded channel.

Year	Captured	Estimated
1986	523	770
1987	964	1,010
Total	1,487	1,780

TABLE 3.—Mean lengths and weights of juvenile coho salmon during the movement periods of 1986–1988 in Swamp Creek beaded channel.

	Upstream migrants		Downstream migrants		Growth increments	
Year	Mean length (mm)	Mean weight (g)	Mean length (mm)	Mean weight (g)	Length (mm)	Weight (g)
1986–1987	84	6.2	103	11.4	19	5.2
1987–1988	85	6.6	107	12.7	22	6.1
Mean					20	5.6

salmon ($1.1/m^2$) found by Cederholm et al. (1988) in their pond than occurred in Swamp Creek beaded channel ($5.6/m^2$) during our highest density year.

The total cost of the Swamp Creek beaded channel was $3,340 (personnel plus materials), compared with $50,000 to excavate a large pond with heavy equipment in 1983 (Peterson 1985), and $9,480 to excavated a pond with dynamite and install a small dam and fishladder in this same river basin (Cederholm et al. 1988). When one compares these initial costs with the mean annual number of smolts each project produced (Table 4), one can see that the beaded channel approach to wall-base channel enhancement is an effective and economical technique.

Explosives have been used to enhance the summer habitats of coho salmon in Oregon (Anderson 1973; Anderson and Miyajima 1975). Cederholm et al. (1988) successfully used dynamite to excavate deep areas in a wall-base channel pond in the Clearwater River, Washington. However, Brown (1985), in his exhaustive investigation of wall-base channel habitat in the Carnation Creek drainage, cautioned that when applying this blasting technique to muck-bottom ephemeral streams, one still runs the risk of insufficient streamflow in the spring to allow smolts to escape. We agree with this conclusion. However, considering the frequent spring rains and the significant number of overwintering coho salmon that are using wall-base channels, we feel that enhancement efforts using this blasting technique can frequently be beneficial in the Pacific Northwest, as demonstrated in this project.

TABLE 4.—Initial costs and mean annual number of coho salmon smolts produced from three wall-base channel habitat enhancement projects.

Project	Initial cost	Number of smolts
Peterson (1985)	$50,000	1,300
Cederholm et al. (1988)	9,481	1,968
Swamp Creek beaded channel	3,340	503

Acknowledgments

The Washington State Departments of Fisheries (WDF) and Natural Resources (DNR) funded this cooperative project. Support and encouragement were provided by Duane Phinney (WDF) and Jerry Kammenga (DNR) throughout the project. We extend thanks to Harry Bell of I.T.T. Rayonier–Northwest Forest Resources for allowing us to construct the beaded channel on company land. Phil Peterson, Larry Cowan, Greg Johnson, David King, Steve Jenks, and Dick Allen of the Washington Department of Fisheries reviewed a draft of this paper.

References

Anderson, J. W. 1973. Evaluation of an excavated fish pool in Vincent Creek. U.S. Bureau of Land Management, Technical Note, Filing Code 6763, Coos Bay, Oregon.

Anderson, J. W., and I. Miyajima. 1975. Analysis of the Vincent Creek fish rearing project. U.S. Bureau of Land Management, Technical Note, T/N 274, Coos Bay, Oregon.

Brown, T. G. 1985. The role of abandoned stream channels as overwintering habitat for juvenile salmonids. Master's thesis. University of British Columbia, Vancouver.

Bustard, D. R., and D. W. Narver. 1975. Aspects of the winter ecology of juvenile coho salmon (*Oncorhynchus kisutch*) and steelhead trout (*Salmo gairdneri*). Journal of the Fisheries Research Board of Canada 32:667–680.

Cederholm, C. J., and W. J. Scarlett. 1982. Seasonal immigration of juvenile salmonids into four small tributaries of the Clearwater River, Washington, 1977–1981. Pages 98–110 *in* E. L. Brannon and E. O. Salo, editors. Salmon and trout migratory behavior symposium. University of Washington Press, Seattle.

Cederholm, C. J., W. J. Scarlett, and N. P. Peterson. 1988. Low-cost enhancement technique for winter habitat of juvenile coho salmon. North American Journal of Fisheries Management 8:438–441.

Peterson, N. P. 1982a. Population characteristics of juvenile coho salmon (*Oncorhynchus kisutch*) overwintering in riverine ponds. Canadian Journal of Fisheries and Aquatic Sciences 39:1303–1307.

Peterson, N. P. 1982b. Immigration of juvenile coho

salmon (*Oncorhynchus kisutch*) into riverine ponds. Canadian Journal of Fisheries and Aquatic Sciences 39:1308–1310.

Peterson, N. P. 1985. Riverine pond enhancement project October 1982–December 1983. Washington Department of Fisheries, Progress Report 233, Olympia.

Peterson, N. P., and L. M. Reid. 1984. Wall-base channels: their evolution, distribution, and use by juvenile coho salmon in the Clearwater River, Washington. Pages 215–225 *in* J. M. Walton and D. B. Houston, editors. Olympic wild fish conference. Peninsula College, Port Angeles, Washington.

Skeesick, D. G. 1970. The fall immigration of juvenile coho salmon into a small tributary. Research Report of the Fish Commission of Oregon 2:90–95.

Zarnowitz, J. E., and K. J. Raedeke. 1984. Winter predation on coho fingerlings by birds and mammals in relation to pond characteristics. Report (Contract 1480) to Washington Department of Fisheries, Olympia.

American Fisheries Society Symposium 10:109–124, 1991

Construction, Operation, and Evaluation of Groundwater-Fed Side Channels for Chum Salmon in British Columbia

R. Gregory Bonnell

Department of Fisheries and Oceans, Pacific Region, Salmonid Enhancement Program
555 West Hastings Street, Suite 400, Vancouver, British Columbia V6B 5G3, Canada

Abstract.—Since 1978, more than 40 groundwater-fed side channels have been built in British Columbia, totaling over 100,000 m^2 of new or improved salmonid spawning and rearing area. The technique involves grading down, deepening, and widening of intermittent or relic side channels on river floodplains to intercept subsurface flow. The constructed channels are rock-armored and protected from floods by dykes or local landforms. The channels are unmanned and depend on volitional entry of spawners and self-regulation of fish density. Methods of site selection, design, and construction are discussed. Data collected from 1978 to 1987 from 24 channels showed an annual production of emergent chum salmon *Oncorhynchus keta* of over 290 fry/m^2 of developed spawning area, and a mean survival to emigration of over 16% of potential egg deposition. Yearling production of coho salmon *O. kisutch* was as high as 30 g/m^2 in one channel in 1 year. Increasing coho salmon populations did not coincide with declining chum salmon production. A conservative benefit/cost ratio calculated for the technique was 1.7. Channels constructed with introduced graded spawning gravel substrates (e.g., 0% fines of 9.5 mm or less in diameter) did not result in greater survival or annual production of chum salmon fry than those constructed with existing gravel substrate (containing up to 15% fines of 1.6 mm and smaller), although construction costs for graded gravel averaged $4.00 (Canadian)/$m^2$ higher. Production and survival of chum salmon remained high for more than 4 years following construction.

The use of groundwater- or seepage-fed areas for spawning and rearing by chum salmon *Oncorhynchus keta* and other salmonids in rivers along the northwestern coast of North America is well known. Adults returning to spawning areas in fresh water home with precision to locations influenced by groundwater. Groundwater flow often provides a more stable environment for incubation and rearing than does surface flow. In general, temperature, discharge, and hydraulic characteristics of groundwater-fed streams are more moderate and constant than those of streams fed by surface flow alone.

The Canadian Department of Fisheries and Oceans (DFO) recognized that many opportunities existed in British Columbia watersheds to improve or create groundwater-fed side-channel spawning areas to help stabilize or increase salmon and trout populations. The DFO began investigations in 1974 with the simple regrading of Bonsall Slough—a seepage-fed side channel of the Cowichan River on Vancouver Island. With the inception of the joint federal–provincial Salmonid Enhancement Program (SEP) in 1977, more funds became available for such work. Since then, more than 40 side channels have been constructed in British Columbia totaling over 100,000 m^2 of new or improved spawning and rearing habitat. The technique has been used successfully on 17 river systems including the Thompson, Fraser, Cowichan, Squamish, and Skeena rivers (Figure 1). A conservative estimate of overall annual production from these projects is 420,000 adult chum salmon, 20,000 adult coho salmon *Oncorhynchus kisutch*, and smaller numbers of other salmon and trout species.

Investigations of groundwater-fed side channels in British Columbia have concentrated on spawner recruitment and density, intragravel conditions, composition of gravel substrates, channel configuration, and protection measures, and on the effects of these on survival and production of chum salmon and other species. Lister et al. (1980) reported on the results from seven channels in British Columbia for the 1979 chum salmon brood (year of spawning) and provided the basis for many subsequent investigations. This paper describes site selection and design and construction of channels; it presents the results of chum salmon investigations following up those of Lister et al. (1980) and discusses economics and the effects of time.

Methods

The following section describes the general procedures involved in the development and assessment of groundwater-fed side channels in British Columbia. Some steps are optional, based

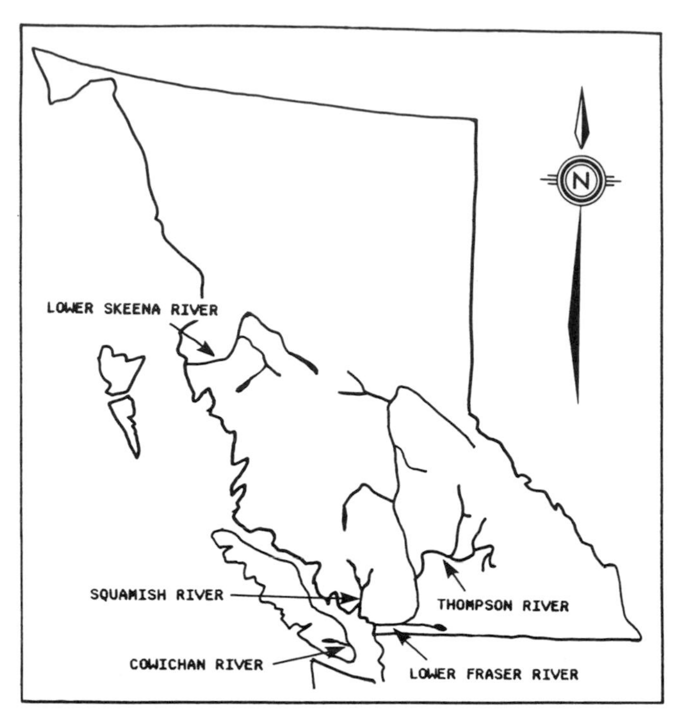

FIGURE 1.—Map of British Columbia showing five of the major watersheds in which groundwater-fed side channels have been constructed for chum salmon spawning.

on location, time, and cost. The descriptions of planning, design, construction, and maintenance are general. The analyses and interpretations presented are based on data collected from 24 channels by several DFO biologists and technicians during 1978–1987.

Planning

Reconnaissance and site selection.—Opportunities to develop side channels are often present on medium to large rivers, provided that the river bench lands chosen are composed of porous, gravel-type soils and there is an adequate supply of groundwater near the surface (Marshall 1986). Initially, only existing side channels with intermittent flow were chosen, but dry relic channels have also been successfully developed (Figures 2, 3). To reduce the amount of material to be excavated, old fluvial channels are selected that have cut through bench land. As many as 15 factors are reviewed in site selection (Table 1). A candidate project may be rejected on biological, engineering, or economic grounds. Major consideration is given to availability of spawner recruits or donor stock, presence of sufficient flow, and cost of excavation, flood protection (dyke construction), and bank armoring.

Survey.—Following selection of a potential site, a topographical survey is completed that includes geographical survey bench marks, several possible center lines for channel alignments, bank contours, high-ground elevations, water table elevations for both groundwater and adjacent river water, any potential obstructions, and all pertinent land features. The survey data are then used to create a plan drawing and to calculate the amount of material to be excavated. This drawing may be revised after construction to show the final alignment, cross sections, and other details. If the project is selected, the design begins. (Examples of engineering drawings that have proved useful are available on request from the author.)

Approvals and site inspections.—Reconnaissance and site selection usually result in a design concept. Once a design is developed, approvals are sought from property owners; federal and provincial fisheries agencies; provincial water management, municipal, or regional district officials; and others as appropriate. Site inspections with those involved have proved an invaluable

FIGURE 2.—A downstream section of Rotary Park channel after initial clearing of vegetation.

tool in the approval process; where convenient, they include visits to nearby completed channels to give an idea of the finished product.

Test pits.—After approvals are obtained, test pits are dug along or adjacent to the proposed channel alignment and stand pipes are installed to

FIGURE 3.—The same section as in Figure 2 after construction. Note rock bank armoring (foreground) and protective dyke (background).

TABLE 1.—Site-selection factors for assessing the feasibility of groundwater-fed side channel projects for chum salmon in British Columbia.

Factor	Criteria
Gradient	Approximately 0.002 to 0.005 along the center of the channel
Substrate	Clear, firm, gravel soil base with low proportion of fines
Water source	Unconfined aquifer with high percolation rate
Water table	Near ground surface with minimal fluctuation
Water quality	General suitability—dissolved oxygen, temperature, total gas pressure, water chemistry suitable for incubation and preferably for rearing
Spawner recruitment	Channels should be located within the upstream spawning limit, preferably with sufficient population size to recruit channel initially
Flood protection	Location to take advantage of existing protection such as high ground, dyke works, or road embankments. Low propensity for backwatering during floods
Existing habitat	Current value should be low or nil
Access	For heavy machinery during construction and for assessment and maintenance later
Sources of siltation	Low potential for siltation from road drainage, intermittent surface flow, or other sources
Availability of materials	Channels should be near sources of quarry rock for bank armoring and natural or graded gravels for substrate if necessary
Approvals	Property ownership (rights of way), interagency approvals, special approvals (e.g., pipeline crossing)
Maintenance	Low potential requirement; e.g., little or no beaver activity
Other	Public relations, interpretation, and local community involvement
Manageability	Potentially few enforcement problems; proposed adult production to be consistent with stock management plans

collect information on bed composition, potential discharge, water table elevation, and water quality—preferably over the course of a full year.

Design

Channel alignment and dimensions.—Channel alignment, length, width, and gradient are chosen from topographical data and estimates of potential discharge and bed composition. Earlier pondlike designs have been abandoned in favor of streamlike ones to increase gradient and velocity and decrease water depth. Channel cross sections are trapezoidal; widths and gradients are chosen to achieve spawning depths of 0.25–0.5 m. These are sometimes determined in advance, but most often are set during construction. A typical channel might be 6 m wide at the streambed with side slopes of 2 to 1 and a gradient of 0 to 0.005—a drop of 0 to 5 in 1,000. Channel length depends on the extent of groundwater percolation and amount of material to be excavated. The design may also include a narrow access channel to the main stream or routing of discharge water through a protected natural area to augment flow and increase its rearing potential.

Substrate and bank armoring.—Channels are constructed with either the existing gravel substrate, introduced graded gravel (screened and washed), or a combination of the two (Figure 4). If the native gravel is unacceptable, a nearby source of suitable river gravel or graded gravel is arranged. Similarly, a nearby source of angular quarry rock (0.2–0.6 m diameter) is sought for bank armoring. This rock often represents a major part of the overall construction cost but is worth the expense because it protects banks from being eroded by spawning and it provides rearing habitat for juvenile coho salmon (Figure 3). Proximity of these materials to the site is important in order to reduce cartage costs.

Flood protection.—High-water levels are determined from topographical information and local records. Sites are usually selected that require a minimum of additional flood protection. Where necessary to prevent the entry of floodwaters, a spur dyke or dam is constructed at the head of the

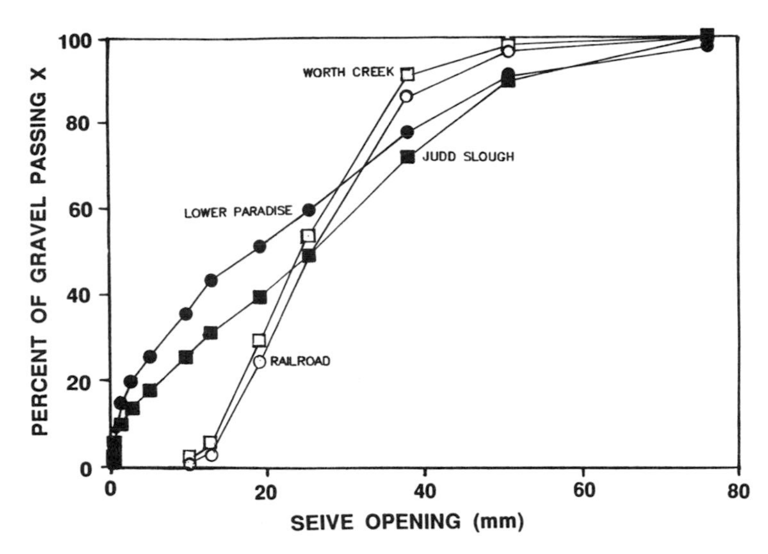

FIGURE 4.—Gravel composition of four groundwater-fed channels built for chum salmon spawning in British Columbia (data from Lister et al. 1980). Worth Creek and Railroad Creek channel had graded gravel. Judd Slough and Lower Paradise channel had existing gravel.

channel with the excavated material. Channels are often located to take advantage of existing roads, railways, or dykes.

Construction

Construction is timed so as to have the least adverse effect on salmonid populations. Work within the channel is done only during the summer when no spawning fish and few rearing fish are present. During construction, precautions are taken to minimize or eliminate the negative effects of siltation on downstream habitat by providing additional settling, diverting heavily silted water, operating machinery from dry land, and other means.

Construction commonly involves the following activities.

- Grading down of the channel, beginning at the downstream end, deep enough to intercept a large portion of the subsurface flow (e.g., to achieve a final water depth to 0.3 m). An initial excavation 1 m in width is often made to confirm the available flow.
- Lengthening and widening of the channel.
- Installation of timber or rock weirs above or below the bed surface along the channel to control hydraulic gradient and to maintain minimum water depth for spawning.
- Installation of timber cribbing spanning the channel at its outlet flush with the bed surface to allow upstream and downstream enumeration fences to be attached.
- Lining of both banks with quarry rock along the developed length to a height of about 1.0 m and a thickness of 0.5 m.
- Construction of flood protection measures as required with excavated material.

Maintenance

Minor routine maintenance procedures include redistributing gravel that has shifted as a result of spawning; cleaning and loosening bed material to help remove settled organic particles and to discourage rooted vegetation; and repairing weirs and fences. More ambitious maintenance projects are occasionally undertaken to rehabilitate or redesign channels. These may include replacing graded gravel with gravel containing a higher percentage of fine material; deepening, narrowing, and regrading channels; installing or removing weirs; and diverting channels to improve downstream flow in small streams.

Assessment

Site considerations (e.g., remoteness) and constraints on staff and funding limit the extent of assessment possible. In recent years, three or four channels representative of designs or locations have been chosen to evaluate over several years. Additional work is carried out as funds and manpower allow.

FIGURE 5.—The downstream part of British Columbia Railway channel showing a carcass fence used to count adult salmon.

Adult populations and sex ratios were estimated by adjusted Petersen mark–recapture estimates (Ricker 1975), by total carcass counts (usually with a fence to prevent loss of carcasses; Figure 5), or by frequent foot surveys of spawners throughout the spawning period. Fecundities were determined from channel-specific regressions of egg number on female length or were estimated from the literature (Beacham 1982). Egg retention was assumed to be zero. Potential egg deposition was calculated as the number of females using a channel multiplied by the mean fecundity.

Redds in channels were sampled to coincide with the embryo development stages between eyed (yolk vascularization; Velsen 1980) and hatch, determined by monitoring accumulated thermal units. Hydraulic redd sampling apparatus was used to estimate embryo survival (McNeil 1964). Sampling was typically carried out in a nonrandom fashion by choosing individual redds evenly throughout the channel rather than by choosing random locations based on a grid pattern. At least 100 embryos and larvae were taken per sample, and a minimum of 25 to 30 samples were taken per channel each year. Mean embryo survival in a channel at this stage was estimated as the grand mean of sample results.

To measure intragravel dissolved oxygen (DO), water samples of 200 to 500 mL were drawn from redds just prior to hydraulic sampling from about 0.15 m below the bed surface with a stainless steel probe. Samples were transferred immediately to a glass container, and DO was measured with an electronic oxygen meter and probe.

Migrating fry (newly emerged juveniles) were counted at fence traps installed at the downstream ends of developed spawning areas. Survival of fry was estimated as the number of fry trapped, divided by the potential egg deposition. Densities of spawners and females, egg deposition, and fry production were calculated for the total developed spawning area of the channel in question. Investigation into the utilization of channels by coho salmon involved downstream migrant trapping, baited minnow trapping, electroshocking, and mark–recapture techniques (M. Foy, DFO unpublished internal reports 1985, 1987).

Economic Analysis

In 1983, an independent bioengineering study team analyzed the economics of groundwater-fed channel developments in British Columbia (D.B. Lister and Associates et al., DFO unpublished consultant report 1983). Production that would have occurred without the improvement was subtracted from production observed after construction. The benefit–cost model used was the SEP Production Model—a tool used by DFO planners to evaluate potential projects. Benefits were based on the incremental production mentioned above, on assumptions for catch/escapement ratios, and on commercial, sport, and Indian food fishery values. Benefits were discounted by 10% per annum to account for decreasing future monetary value. Costs included capital and operating costs, administrative costs, contributions of time and materials from nongovernmental groups at market value, harvesting and processing costs, as well as the cost of labor in the fishing industry. The overall assumptions of the model greatly reduce the likelihood of achieving a benefit/cost ratio higher than 3.0, and ensure excellent benefits from projects with ratios greater than 1.0.

For their base case, the consultants assumed an average project life of 20 years under a regular cycle of maintenance involving gravel cleaning and channel repair. Two further evaluations were made under assumptions of no maintenance and on the different probabilities of flooding. For the latter case, engineering surveys determined the minimum severity and probability of flooding that would eliminate fish production benefits for the remainder of the project life and would require an expenditure equal to the original investment to restore net economic benefits.

Longevity

To examine the longevity of channels, data for spawning density, fry survival, and annual fry production were examined from four channels that had been in operation for 4 years or longer. Time scale was standardized to the number of brood years following construction; year 1 was the fall immediately following construction.

Results and Discussion

Spawning Density and Fecundity

The spawning densities observed for all channels ranged from 0.03 to 2.8 females/m^2 with a mean of 0.8 (SD = 0.6) and a mode of 0.6 female/m^2 (Table 2). The mean proportion of females observed in the spawning populations was 0.46 (SD = 0.11); the mean proportion in existing gravel channels (0.49) was slightly but not significantly higher than that in graded gravel channels (0.44). The mean fecundity (from observations and literature sources) was 3,240 eggs/female (SD = 242), with a range of 2,936 to 3,524 eggs/female.

TABLE 2.—Mean chum salmon spawning density, potential egg deposition, and survival and annual fry production in groundwater-fed side channels in British Columbia for brood years 1978–1987.

Factor	Channel type: Existing gravel substrates	Graded gravel substrates	Both
Female spawners/m^2			
Mean	0.8	0.8	0.8
SD	0.6	0.6	0.6
N	19	30	49
Potential egg deposition/m^2			
Mean	2,558	2,355	2,431
SD	1,934	1,640	1,759
N	18	30	48
Embryo survival (%) (from redd sampling)			
Mean	72.3	51.8	61.9
SE	20.5	23.9	24.5
N	33	34	67
Fry survival (%)			
Mean	17.4	15.8	16.3
SE	11.4	12.5	12.2
N	9	21	30
Annual fry production (fry/m^2)			
Mean	337	280	297
SD	167	177	176
N	9	21	30

Survival and Production

Embryo survival and DO.—Results from 1982 sampling showed mean embryo survival at 10 of our channels was significantly correlated with mean intragravel DO (r = 0.685, $P \leq 0.05$), although r^2 was only 0.48 (Figure 6). Embryo survival of 50% in individual redds generally occurred at DO levels between 3.0 and 6.5 mg/L (M. Foy and B. Snyder, DFO unpublished internal report 1983).

In groundwater, DO depends on depth of source, water temperature, and chemical and biochemical oxygen demands resulting from oxidization of dissolved metals (e.g., iron) and from decay of organic matter including salmon carcasses (Servizi et al. 1971; Hansen 1975). Suggestions for minimum DO range from 5.0 to 8.0 mg/L and higher (Alderdice et al. 1958; Servizi et al. 1971; Reiser and Bjornn 1979; Sigma Resource Consultants, DFO unpublished consultant report 1983; Sowden and Power 1985). When intragravel DO was 5.3 mg/L or less, Sowden and Power

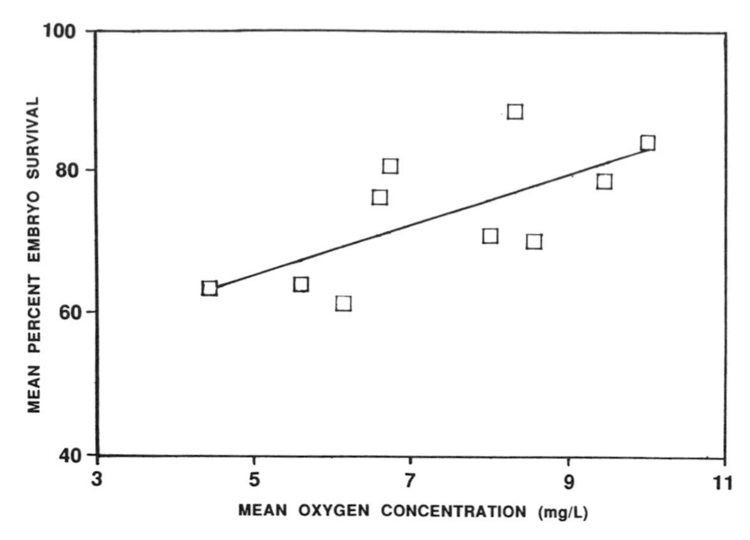

FIGURE 6.—Chum salmon embryo survival as a function of intragravel dissolved oxygen concentration at hatch, from Brood 1981 sampling, all data (correlation; $r = 0.685$, $P \leq 0.05$).

(1985) found little or no survival of embryos of rainbow trout *O. mykiss* in redds in a natural groundwater stream.

The moderate survivals observed at low temperatures in 1982—e.g., 63.3% survival at mean DO of 4.4 mg/L for Lower Paradise channel (Table 3)—may be partly explained by low oxygen consumption at low temperatures. Rombough (1986) developed models to predict incipient limiting DO levels (P_c) and levels at which metabolic rate is half the routine rate (P_{50}) for steelhead *O. mykiss* at different developmental stages and temperatures. All mean intragravel DO levels from the 1982 sampling exceeded the steelhead P_{50} level (2.6–2.8 mg/L) and half of them exceeded the steelhead P_c level (7.4–7.9 mg/L).

Embryo survival versus fry survival.—For pooled data from graded gravel channels, there was no significant correlation between embryo survival and subsequent fry survival. When pooled data from existing gravel channels were used, embryo survival was significantly correlated with subsequent fry survival ($r = 0.777$, $P \leq 0.025$, $N = 7$) (Figure 7). The data for two individual channels also showed significant correlations: Worth Creek (graded gravel), $r = 0.981$, $P \leq 0.01$, $N = 5$; Mamquam (existing gravel), $r = 0.953$, $P < 0.05$, $N = 4$ (Figure 8).

TABLE 3.—Mean intragravel and surface dissolved oxygen measurements at hatch in groundwater-fed side channels in 1982 (brood 1981). (From M. Foy and B. Snyder, DFO unpublished internal report 1983).

Channel	Intragravel dissolved oxygen (mg/L)			Surface dissolved oxygen (mg/L)	Temperature (°C)	Survival at hatch (redd sampling) (%)		
	Mean	SD	Number of samples			Mean	SD	Number of samples
Existing gravel								
Street[a]	6.1	1.8	22	8.4	7.0	61.2	39.5	24
Chilqua[a]	6.6	0.3	5	8.8	6.0	76.3	22.2	29
Upper Paradise[a]	5.6	2.2	29	8.3		64.0	37.0	30
Lower Paradise	4.4	1.5	25	8.6	6.3	63.3	28.3	25
Judd	6.7	1.2	29	10.6	6.0	80.7	21.8	29
Westholme	8.6	2.1	17		7.0	70.1	33.0	31
Graded gravel								
Billy Harris	9.5	1.8	33	11.1	7.0	78.8	29.7	33
Ed Leon	8.0	1.5	30	8.5	7.0	70.9	37.6	30
Railroad	8.3	1.9	21	9.5		88.6	10.8	20
Worth	10.0	1.0	25	10.0	5.9	84.3	22.4	24

[a]Undeveloped channels at the time of sampling.

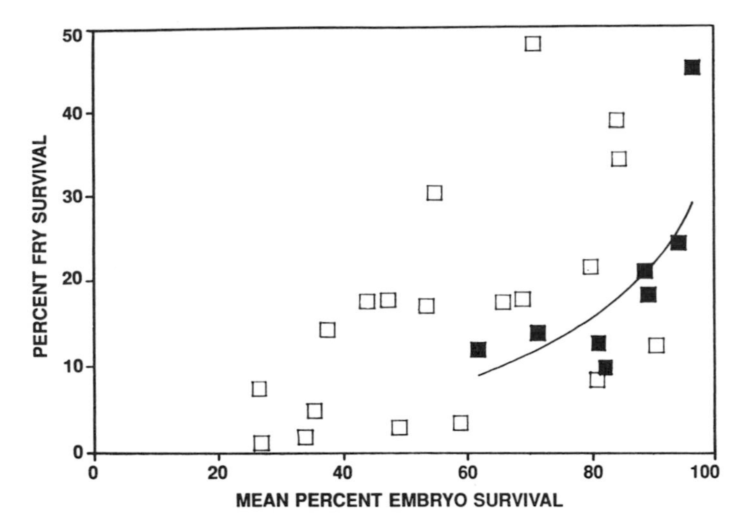

FIGURE 7.—Observed chum salmon fry survival as a function of earlier mean embryo survival. The data are estimates from downstream migrant trapping and redd sampling in existing (solid squares) and graded (open squares) gravel channels. The curve is the correlation for existing gravel channels ($r = 0.777$, $P \leq 0.025$).

Hydraulic sampling was usually carried out simply to obtain gross estimates of embryo survival. I made these comparisons to see whether embryo survival methods were predictive of overall survival to emergent fry. If so, assessment costs could be reduced by avoiding lengthy downstream trapping programs. The results suggest that for existing gravel channels, this may be possible, although the correlations came from small sample numbers.

Several factors likely contributed to the scatter observed in the egg survival versus fry survival plots. The loss of unfertilized eggs that dissolved prior to sampling would have caused the embryo survival estimate to be high. The loss of live embryos and larvae that moved beyond the sampled area (0.09 m^2) due to the larger void spaces in graded gravel could have biased the embryo survival estimate downward. Selective (rather than random) sampling of redds may not have given

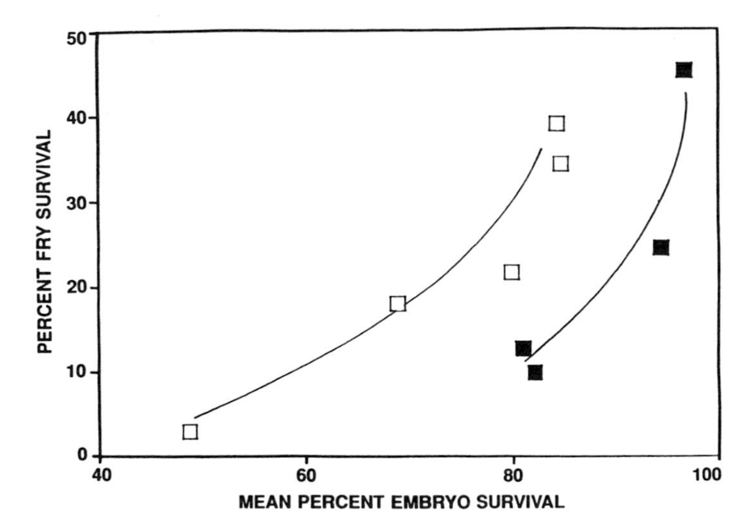

FIGURE 8.—Observed chum salmon fry survival as a function of earlier mean embryo survival. The data are estimates from downstream migrant trapping and redd sampling. The curves are the correlations for Mamquam channel (solid squares, $r = 0.953$, $P \leq 0.05$) and Worth Creek channel (open squares, $r = 0.981$, $P \leq 0.01$).

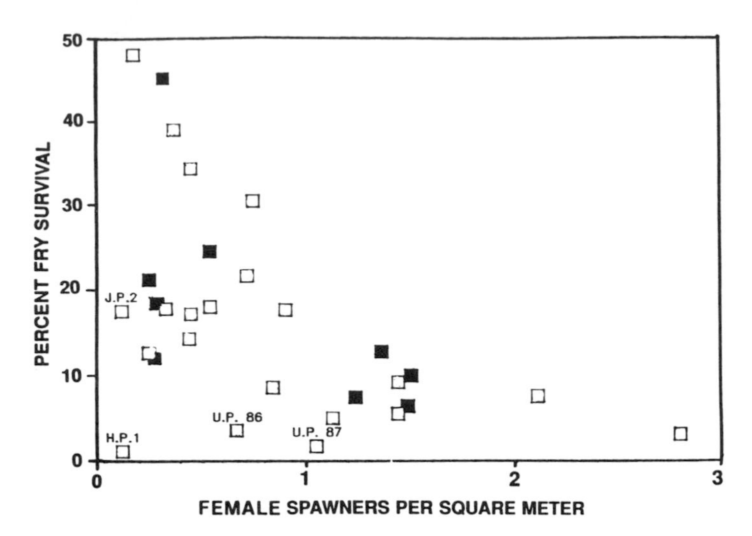

FIGURE 9.—Chum salmon fry survival as a function of female spawner density in channels with existing (solid squares) and graded (open squares) gravel. J.P.2 is Judd Pond; H.P.1 is Hopdale Pond; U.P. 86 and 87 are Upper Paradise Channel for brood years 1986 and 1987.

representative estimates of embryo survival. Differences in posthatch survival might also have prevented correlations.

Fry emergence.—Some studies indicate that a large percentage (>20–25%) of fine material (<6.4 mm in diameter) in the substrate can hinder fry emergence (Phillips et al. 1975; Reiser and Bjornn 1979; Harshbarger and Porter 1982). This effect, if pronounced, would lessen or eliminate the significance of a correlation between embryo and fry survival. From the 1986 hydraulic sampling of Mamquam channel (existing gravel) following fry emigration, only 0.5% of the total fry output was estimated to remain in the gravel (M. Foy, DFO unpublished internal memorandum, 1987). This compares with 4–9% of dead larvae remaining in gravel containing 2–11% fines (≤0.83 mm) in Sashin Creek, Alaska (McNeil 1969). The gravel composition of Mamquam channel was not analysed but was probably similar to that of two other existing gravel channels in the same alluvial plain—Lower Paradise (8.7% ≤ 1.1 mm) and Judd Slough (5.9% ≤ 1.1 mm) (Figure 4).

Fry survival.—For all channels at all spawning densities, fry survival averaged 16.3% (Table 2). Mean survival from existing gravel channels was slightly but not significantly higher than that from graded gravel channels. At low spawning densities (≤0.5 female/m^2), survival averaged 23% and ranged from 12 to 48% (Figure 9). Survival decreased with increasing spawning density to less than 10% at over 1.3 females/m^2 and to less than 5% at 2.1 females/m^2. At any point at or below a spawning density of 0.7 female/m^2 (67% of the data), survival was more than twice the 7.9% average found by Lister et al. (1980) for six natural spawning streams and well above the 9.0% suggested as standard survival for natural production in British Columbia (Enhancement Opportunities Subcommittee for the SEP, DFO unpublished discussion document 1983).

The relationship between fry survival and female spawner density shown in Figure 9, while not significant, is made clearer if four data points are disregarded. The results from Judd Pond number 2 (1979 brood year) and Hopdale Pond number 1 (1979 brood year) are unusual because these earlier projects were of pondlike rather than streamlike design. Although their local upwelling flow rates may be similar to those in streamlike channels, they have less horizontal velocity and may lack intragravel exchange as a result. The data from Upper Paradise channel for brood years 1986 and 1987 were collected when that channel had undergone a sharp decline in production—likely due to an accumulation of fine organic matter in a layer beneath the spawning gravel, which may have impeded upwelling flow and increased the intragravel biochemical oxygen demand.

Fry production.—For all channels, annual production of emergent fry averaged 297 fry/m^2; the means were 337 fry/m^2 for existing gravel channels and 280 fry/m^2 for graded gravel channels

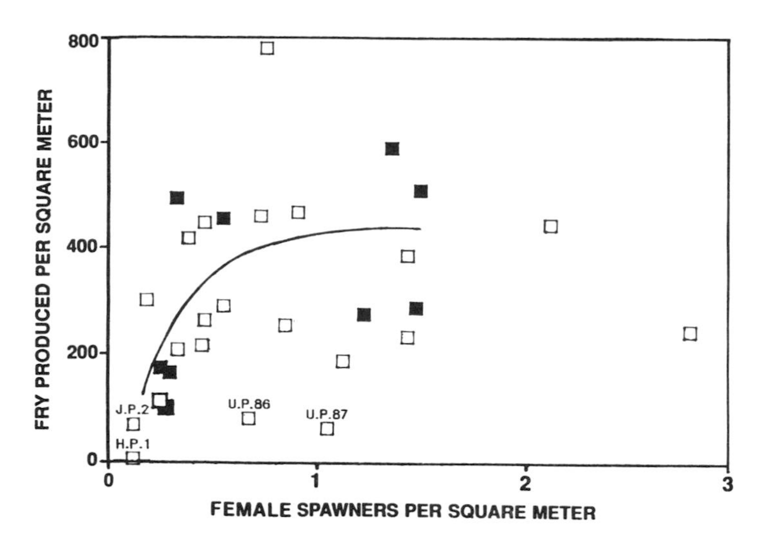

FIGURE 10.—Annual chum salmon fry production as a function of female spawner density in channels with existing (solid squares) and graded (open squares) gravel. The curve was drawn by eye. Abbreviations are defined in Figure 9.

(Table 2). Mean annual production for existing gravel channels was not significantly different from that for graded gravel channels. Fry production in groundwater channels increased with spawning density to roughly 400–500 fry/m^2 at about 0.8–1.0 female/m^2 (2,600–3,200 eggs/m^2) and remained fairly high even at densities in excess of 2.5 females/m^2 (8,000 eggs/m^2) (Figure 10). As with Figure 9, the relationship in Figure 10 is made somewhat clearer by the exclusion of the four data points mentioned.

The maximum production from groundwater channels is similar to that observed for pink and chum fry over 25 years at Sashin Creek (McNeil 1969)—a stable surface-fed creek in Alaska—which approached 500 fry/m^2 at a spawning density of about 1.0 female/m^2 (2,000–3,000 eggs/m^2). However, production from Sashin Creek declined rapidly to about 100 fry/m^2 at 2.2 females/m^2 (5,500 eggs/m^2) compared with levels of 443 fry/m^2 at 2.1 females/m^2 (5,900 eggs/m^2) for Railroad channel and 243 fry/m^2 at 2.8 females/m^2 (8,200 eggs/m^2) for Worth Creek channel.

The plot of fry production versus female spawning density for chum salmon in groundwater channels is similar to that for sockeye salmon in Fulton River and Pinkut Creek (West and Mason 1987)—two controlled-flow streams in northern British Columbia. Although production in those sttreams was generally higher than production in groundwater channels, maximum levels were approached at densities of 3,000 eggs/m^2 (roughly 1.0 female/m^2) or more, the relationship becoming asymptotic at higher densities. The authors advocated a ceiling on density of 3,000–4,000 eggs/m^2 (about 1.0–1.3 females/m^2). For chum salmon in groundwater-fed channels, 82% of the results were below this ceiling and corresponded to a mean fry survival of 28.4% and a mean annual fry production of 271 fry/m^2.

Effects of Coho Salmon

Almost all of our groundwater-fed channels have experienced some colonization by coho salmon. Juvenile coho salmon can be produced from adults spawning in channels or can migrate into the channels from elsewhere in the summer and fall. They can rear in channels until their seaward migration as yearling smolts the following spring, or they can leave the channel at various times to colonize other parts of the river system. Late summer stock densities of coho salmon have reached 7.0 fish/m^2 (M. Foy, DFO unpublished internal report 1985). Annual coho salmon production from Upper Paradise and Mamquam channels has been as high as 3.1 and 3.4 yearlings/m^2, respectively, with concurrent fry yields of 3.7 and 2.1 fry/m^2. The biomass of yearlings migrating in one year exceeded 30 g/m^2 at Upper Paradise channel (M. Foy, personal communication).

Examination of stomach contents of rearing coho salmon at Upper Paradise channel showed that chum salmon carcass remains, and (later)

TABLE 4.—Benefit–cost results for groundwater-fed side channels built for chum salmon spawning in British Columbia (from D.B. Lister and Associates et al., DFO unpublished consultant report 1983).

Channel	Base case		Base case without maintenance		Base case including probability of flooding	
	Benefit/cost ratio	Net benefits[a]	Benefit/cost ratio	Net benefits[a]	Benefit/cost ratio	Net benefits[a]
Existing gravel						
Westholme	1.03	9,000	0.94	−20,000		
Peach	1.95	174,000	1.89	117,000		
Lower Paradise	1.35	68,000	1.24	40,000	0.60	−77,000
Judd Slough	2.42	887,000	2.20	372,000		
Graded gravel						
Hopedale Pond	0.34	−345,000				
Billy Harris	2.19	961,000	2.11	812,000		
Ed Leon	1.84	570,000				
Upper Paradise	2.04	174,000	1.90	134,000	1.45	75,000
Worth	1.27	42,000				
Railroad	1.05	9,000				
Overall	1.66	2,549,000				

[a]Net benefits are in 1987 Canadian dollars.

larvae and fry, were the primary food items from January through May 1985. Approximately 23% of the output of chum salmon fry was estimated to have been lost to predation. The long-term effects of competition and predation are uncertain. Although the numbers of chum salmon produced annually declined in two of the three channels that have been monitored for 4 years or more, these declines did not correlate with increases in abundance of coho salmon fry and yearlings.

The design and operation of groundwater channels can be tailored to benefit coho salmon. Channels that dried partially or entirely during the summer months produced very few coho salmon yearlings. During electroshocking at Upper Paradise channel in the fall of 1984, the majority of juveniles were found in the interstices of the rock-bank armoring (M. Foy, personal communication). Following installation of angular rock on the bank of Worth Creek channel in 1986, the number of coho yearlings migrating from the channel increased to over 1,700 in 1987 from a previous 5-year average of 91 (M. Sheng, personal communication and DFO unpublished internal memorandum 1987).

Economic Analysis

The overall benefit/cost ratio for the base case (D.B. Lister and Associates et al., DFO unpublished consultant report 1983) was 1.7 (range, 0.34–2.42) (Table 4). Without any maintenance of the channels, the ratio was predicted to remain high but with a reduced net benefit. When the probability of flooding was taken into account, the ratio was reduced, falling below 1.0 for a poorly protected channel but staying greater than 1.0 for a well-protected channel.

Since that analysis, data have been gathered that suggest current benefit/cost ratios may be even higher. Among the assumptions of the 1983 analysis were annual chum salmon fry outputs of 400/m^2 for graded gravel channels and 250/m^2 for native gravel channels, with annual coho salmon yearling outputs of 1.0 per lineal meter of stream or about 0.2 yearling/m^2. Data now show that chum salmon fry yields have most often been 250–800 fry/m^2 per year for both types of substrates (62% of data), and that coho salmon densities have been over 1.0 yearling/m^2 in many cases.

In addition, since the first channels were constructed (1978–1982), increased efficiency in construction and better use of opportunities have resulted in reduced costs. Translated to 1987 Canadian dollars, the overall cost of channels built before 1983 was \$25.90/m^2 compared with \$24.30/m^2 for those built in 1983 and later (Table 5). This is true even though all channels built in more costly remote locations have been constructed since 1983. The mean development cost for existing gravel channels was \$24.56/m^2 compared with \$28.99/m^2 for graded gravel channels.

TABLE 5.—Development costs for groundwater-fed side channels built for chum salmon spawning in British Columbia.

Channel	Year constructed	Area in m²	Total cost of construction in 1987 dollars[a,b]	Cost/m² in 1987 dollars[b]
		Existing gravel		
Westholme	1978	2,930	191,400	65.32
Judd Slough	1978	5,360	48,900	9.12
Lower Paradise	1979	1,940	24,200	12.47
Peach Creek	1982	2,540	23,200	9.13
Mamquam	1983	2,000	52,800	26.40
Seabird	1984	2,000	24,600	12.30
B.C. Rail	1985	2,340	36,900	15.77
Deadman[c]	1985	690	36,600	53.04
Kitwancool[c]	1985	1,360	48,300	35.51
McNab[c]	1986	1,380	33,800	24.49
Moodies	1986	3,420	46,500	13.60
Smokehouse	1986	1,370	24,000	17.52
Mean		2,300	49,300	24.56
		Graded gravel		
Billy Harris	1979	7,490	159,200	21.26
Railroad	1979	770	32,000	41.56
Worth Creek	1979	850	17,300	20.35
Ed Leon	1980	5,800	203,700	35.12
Upper Paradise	1982	2,630	50,600	19.24
Chilqua	1984	960	22,300	23.22
Kitsumkalum[c]	1984	1,100	46,400	42.18
Mean		2,800	75,900	28.99

[a]Construction costs include development costs, internal administration costs, person-year costs, and costs to other agencies or groups in joint ventures.
[b]Costs are all in Canadian funds translated to 1987 dollars.
[c]Projects in remote locations—either in northern British Columbia or where access is difficult.

Longevity

Spawning density.—For the four channels examined, spawning densities of over 0.5 female/m² occurred as early as 1 or 2 years after construction (Figure 11), although adult returns from channel-produced fry did not contribute to spawning populations until years 4 or 5. The return rate of adults from channels and their contribution to subsequent spawning populations have not yet been studied in detail due to the technical difficulty, expense, and statistical problems associated with marking the fry and assessing total returns to the river systems. No obvious relationship was observed between annual fry outputs and subsequent channel spawning densities. It is likely that each year the channels are used by some river-produced fish and that some fish originating from channels spawn elsewhere in the river. The factors controlling spawning density are probably related primarily to ocean survival and harvest rates.

Fry survival.—In both graded and existing gravel channels, fry survival appeared to decline over time (Figure 12). However, the observed decline may be the result of high spawning densities rather than of a gradual deterioration in gravel quality, predation, or other effects. In general, fry survival estimates remained above those observed in natural spawning areas (Lister et al. 1980)—over 8% beyond year 4.

Fry survival is generally not a good indicator of channel longevity because of complicating factors such as errors in survival estimation and spawning density. Factors influencing variability in estimates of survival include uncertainty in spawner counts, degree of egg retention, variations in fecundity, predation, and problems associated with fry trapping. The survival estimates presented here are considered conservative, because they assume zero egg retention and use only the number of fry counted from traps. The numbers of fry produced per square meter per year is probably a better measure of channel success over time.

Fry production.—Of the channels examined, annual production in only one (Upper Paradise) declined to less than 100 fry/m² after the fourth

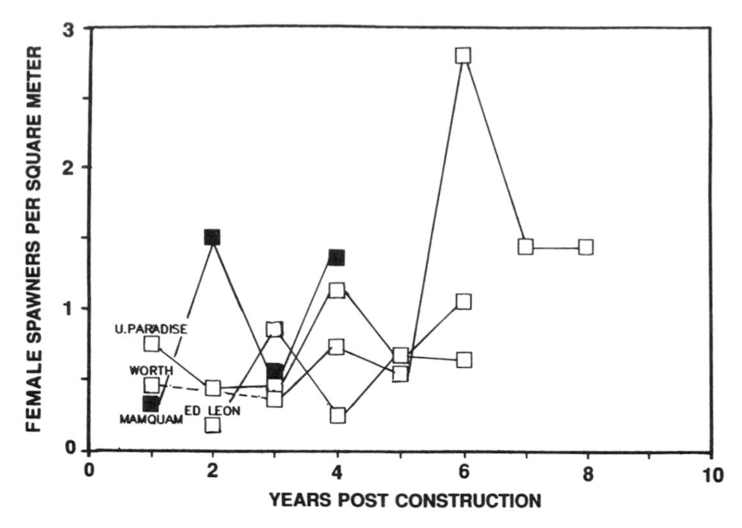

FIGURE 11.—Density of female chum salmon spawners during postconstruction years in four groundwater-fed channels (solid squares denote existing gravel channels; open squares denote graded gravel channels). Data for year 1 represent female densities during the fall and winter immediately following summer construction.

year (Figure 13). Mamquam and Worth Creek channels maintained annual production levels of over 250 fry/m^2.

Fry production tended to be constant over a range of spawning densities. The majority of results (66%) were between 200 and 525 fry/m^2 annually (Figure 10). A decline in production may be the best indicator of the deterioration of a channel. Estimates of annual fry production are less subject to error than estimates of survival. In groundwater-fed channels, trapping is not affected by floods or large debris typical in surface-fed streams. Water velocity tends to be low and stable in groundwater-fed channels (about 0.15 m/s) during downstream migration (M. Foy, personal communication). Apart from occasional periods of backwatering in some channels, trap counts have been very reliable.

The reasons for declining annual production, such as that observed for Upper Paradise, are not clear. One factor might be the gradual deterioration of the intragravel environment due to the settling of fine organic particles (from carcasses, dead eggs, leaves, or other debris) into a layer that

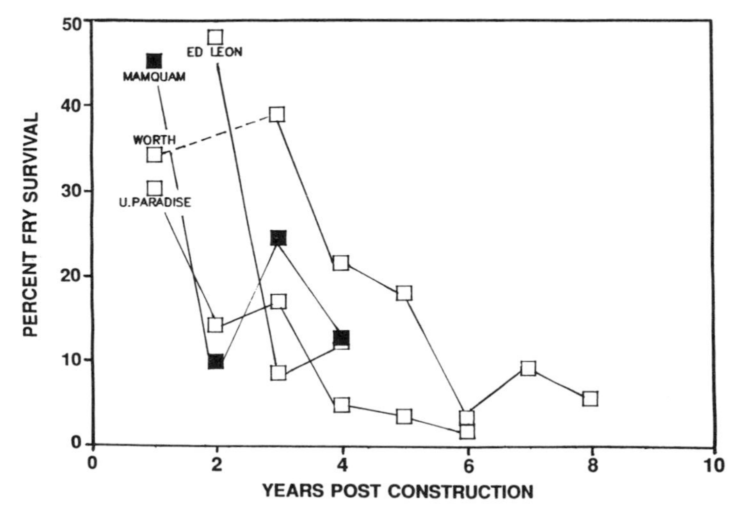

FIGURE 12.—Chum salmon fry survival during postconstruction years in four groundwater-fed channels (solid squares denote existing gravel channels; open squares denote graded gravel channels).

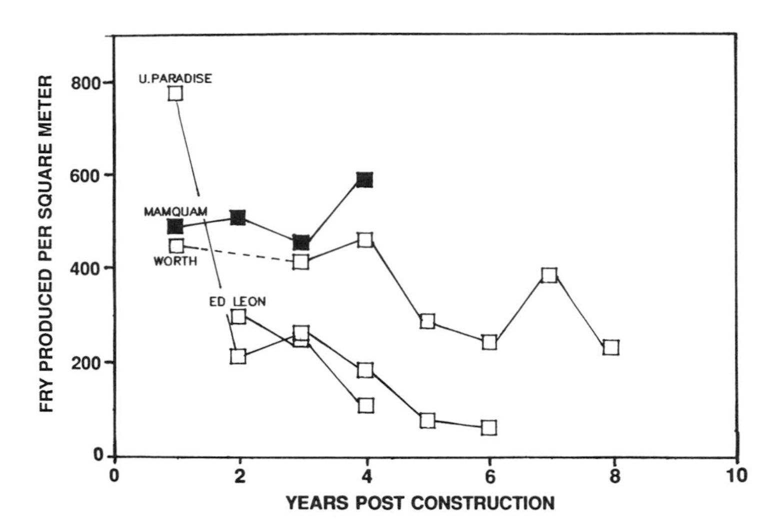

FIGURE 13.—Annual chum salmon fry production during postconstruction years in four groundwater-fed channels (solid squares denote existing gravel channels; open squares denote graded gravel channels).

limits water exchange and lowers dissolved oxygen levels. During rebuilding of Upper Paradise channel in 1988, it was found that a dense layer of organic material had accumulated in the gravel at a depth of about 0.3 m below the stream bed (R. Brown, DFO, personal communication). In general, a decline in annual chum salmon fry production from groundwater-fed channels in British Columbia to less than 100 fry/m^2 is cause for concern and may indicate the need for rehabilitation measures.

In the 1983 economic analysis, the consultants suggested that deterioration might proceed 4–8 years before productivity would be reduced to that of natural streams (150 fry/m^2) (D.B. Lister and Associates et al., DFO unpublished consultant report 1983). The results from the channels examined here suggest that fry survival can remain above natural levels beyond year 4, and annual fry production can remain high well beyond that.

Summary and Conclusions

(1) Planning, design, and construction of groundwater-fed side channels in British Columbia are done by a multidisciplinary bioengineering team to avoid problems, to further the understanding of the technique, and to improve the potential for success.

(2) Streamlike designs, use of existing gravel, dyke building, and armoring of banks with large rock are features now routinely used to increase longevity of channels and to lower maintenance costs.

(3) Mean chum salmon fry survival was 16.3% overall and 20.1% at spawning densities of less than 1.0 female/m^2 (75% of the data). Mean annual production of chum salmon fry from channels was 297 fry/m^2 overall and 275 fry/m^2 at spawning densities of less than 1.0 female/m^2.

(4) Lister et al. (1980) noted that any advantages of graded gravel substrates were not clear from their survival data. In this review, there was no significant difference between graded gravel channels and existing gravel channels in either fry survival or annual fry production for chum salmon. Indeed, survival and production from existing gravel channels were slightly higher, whereas development costs were lower.

(5) Use of groundwater channels by coho salmon can be substantial, producing yearling stock densities of over of 3.0/m^2, standing crops of 30 g/m^2, plus considerable numbers of fry. Although coho salmon prey on chum salmon fry, there is no clear relationship between declining annual production of chum salmon fry and increasing coho salmon numbers.

(6) A conservative benefit/cost ratio calculated from 1978 to 1983 data was 1.7, assuming a 20-year life span for groundwater-fed side channels. Recent information suggests this ratio may have increased due to improved design and efficiency of construction.

(7) Spawning densities of chum salmon of 0.4

female/m^2 (1,300 eggs/m^2) or more (resulting in annual fry production levels over 200 fry/m^2) were achieved before adults from channel-produced fry contributed to the spawning populations. Annual fry production generally remained high (>180 fry/m^2) 4 years following construction.

Acknowledgments

G. A. Logan directed and encouraged this project. D. E. Marshall and R. J. Finnigan have directed the development and assessment of groundwater-fed side channels in British Columbia, and they gave much useful advice. M. P. Foy, M. D. Sheng, R. F. Brown, and others generously provided advice and allowed me to use their unpublished data and reports. R. J. White, R. L. Kendall, and two anonymous referees made comments that greatly improved the manuscript. A. Ho and her staff patiently typed several drafts of the manuscript and B. Bowler assisted in the preparation of the figures.

References

Alderdice, D., W. P. Wickett, and J. R. Brett. 1958. Some effects of temporary exposure to low dissolved oxygen levels on Pacific salmon eggs. Journal of the Fisheries Research Board of Canada 15:229–249.

Beacham, T. D. 1982. Fecundity of coho salmon (*Oncorhynchus kisutch*) and chum salmon (*O. keta*) in the northeast Pacific Ocean. Canadian Journal of Zoology 60:1463–1469.

Hansen, E. A. 1975. Some effects of groundwater on brown trout redds. Transactions of the American Fisheries Society 104:100–110.

Harshbarger, T. J., and P. E. Porter. 1982. Embryo survival and fry emergence from two methods of planting brown trout eggs. North American Journal of Fisheries Management 2:84–89.

Lister, D. B., D. E. Marshall, and D. G. Hickey. 1980. Chum salmon survival and production at seven improved groundwater-fed spawning areas. Canadian Manuscript Report of Fisheries and Aquatic Sciences 1595.

Marshall, D. E. 1986. Development and assessment of groundwater-fed side channels. Canadian Technical Report of Fisheries and Aquatic Sciences 1483:106–109.

McNeil, W. J. 1964. A method for measuring mortality of pink salmon eggs and larvae. U.S. Fish and Wildlife Service Fishery Bulletin 63:575–588.

McNeil, W. J. 1969. Survival of pink and chum salmon eggs and alevins. Pages 101–117 *in* T. G. Northcote, editor. Salmon and trout in streams. H. R. MacMillan Lectures in Fisheries, University of British Columbia, Vancouver.

Phillips, R. W., R. L. Lautz, E. W. Claire, and J. R. Moring. 1975. Some effects of gravel mixtures on emergence of coho salmon and steelhead trout fry. Transactions of the American Fisheries Society 104:461–466.

Reiser, D. W., and T. C. Bjornn. 1979. Habitat requirements of anadromous salmonids. U.S. Forest Service General Technical Report PNW-96.

Ricker, W. E. 1975. Computation and interpretation of biological statistics of fish populations. Fisheries Research Board of Canada Bulletin 191.

Rombough, P. J. 1986. Mathematical models for predicting the dissolved oxygen requirements of steelhead trout (*Salmo gairdneri*) embryos and alevins in hatchery incubators. Aquaculture 59:119–137.

Servizi, J. A., D. W. Martens, and R. W. Gordon. 1971. Toxicity and oxygen demand of decaying bark. Journal of the Water Pollution Control Federation 43:278–292.

Sowden, T. K., and G. Power. 1985. Prediction of rainbow trout embryo survival in relation to groundwater seepage and particle size of spawning substrates. Transactions of the American Fisheries Society 114:804–812.

Velsen, F. P. J. 1980. Embryonic development in eggs of sockeye salmon, *Oncorhynchus nerka*. Canadian Special Publication of Fisheries and Aquatic Sciences 49.

West, C. J., and J. C. Mason. 1987. Evaluation of sockeye salmon (*Oncorhynchus nerka*) production from the Babine Lake Development Project. Canadian Special Publication of Fisheries and Aquatic Sciences 96:176–190.

American Fisheries Society Symposium 10:125–131, 1991

Physical Characteristics and Intragravel Survival of Chum Salmon in Developed and Natural Groundwater Channels in Washington

LARRY COWAN

Washington Department of Fisheries
115 General Administration Building, Olympia, Washington 98504, USA

Abstract.—Biological and physical evaluation of five groundwater-fed side channels (three artificially developed, two natural) on the East Fork Satsop River showed that recruitment of chum salmon spawners (*Oncorhynchus keta*) was positively correlated with streamflow discharge of channels. Straying was an important factor affecting initial colonization and seasonal densities of chum salmon spawners in the Satsop groundwater channels. Egg-to-fry survival ranged from 21 to 55%, substantially higher than the 6–31% range reported in other coastal streams. Correction of poor spawning conditions in a natural channel (Maple Glen) may more than double the egg-to-fry survival there. Female chum salmon spawner densities ranged from 0.07 to 0.24 female/m^2, which was below optimum (0.5 female/m^2) for groundwater channels. It is postulated that increased adult escapements into the Satsop system would provide a fivefold increase in fry production per square meter without diminishing egg-to-fry survival rates in all channels, except Maple Glen. Total dissolved gas ranged from 100 to 104% saturation, although no detrimental effects were noted in fry of the Satsop channels.

The high-quality groundwater (dissolved oxygen, 9.0–11.8 mg/L; total gas, 100–104% saturation) in parts of the Satsop River system led the Washington State Department of Fisheries (WDF) to develop four groundwater-fed side channels (Schafer, Kelsey, Simpson, Creamer) for spawning of chum salmon *Oncorhynchus keta*. Habitat development techniques used included channel excavation and diking, gravel placement and cleaning, log installation to control gradient and water surface elevation, and channel bank revegetation. My study was undertaken to identify relationships between physical conditions and productivity of chum salmon fry in the channels. Better understanding of these relationships should enable improvements in channel design and subsequently greater chum fry production.

Study Areas

The East Fork Satsop River is tributary to the Grays Harbor Pacific estuary system. Figure 1 shows the location of the five study sites in this report.

The source of flow in these channels was groundwater percolating through porous soils. Discharge was stable but fluctuated seasonally with river stage, increasing as the water table rose with seasonal runoff and diminishing as the water table receded in summer drought. In some channels there were also occasional brief increases in flow (freshets) because of surface runoff during rainstorms.

Schafer Park

Schafer Park (Schafer), a 440 × 9.5-m groundwater channel constructed in 1982, enters the East Fork Satsop River at river kilometer 21.1. Development included diking at the upstream end with rock riprap and a surface water diversion via a valved culvert installed through the dike. An existing intermittent side channel was excavated to a level that provided year-round groundwater flows, and seven log weirs were installed to control the channel gradient (0.003) and water depth (0.3 m). Round, washed, evenly graded gravel ranging in diameter from 1.0 to 7.5 cm was added to a depth of 0.5 m. Development created 2,750 m^2 of spawning habitat. The surface water diversion was opened to attract 1984 brood spawners and closed for the 1985 and 1986 brood years. The gravel was mechanically cleaned (Bates and Johnson 1986) in summer 1986.

Schafer Park Tributary 3

Schafer Park tributary 3 (Schafer #3) is a 610-m^2 (140 × 4-m) natural groundwater-fed channel that enters the left bank of Schafer 220 m from the confluence of Schafer and the East Fork Satsop River. No improvements were made to this channel; it was evaluated separately as a natural site. Visual observations determined good gravel quality. The channel gradient was 0.004 and the mean depth was 0.18 m. Chum salmon had spawned naturally here for 20 years or more.

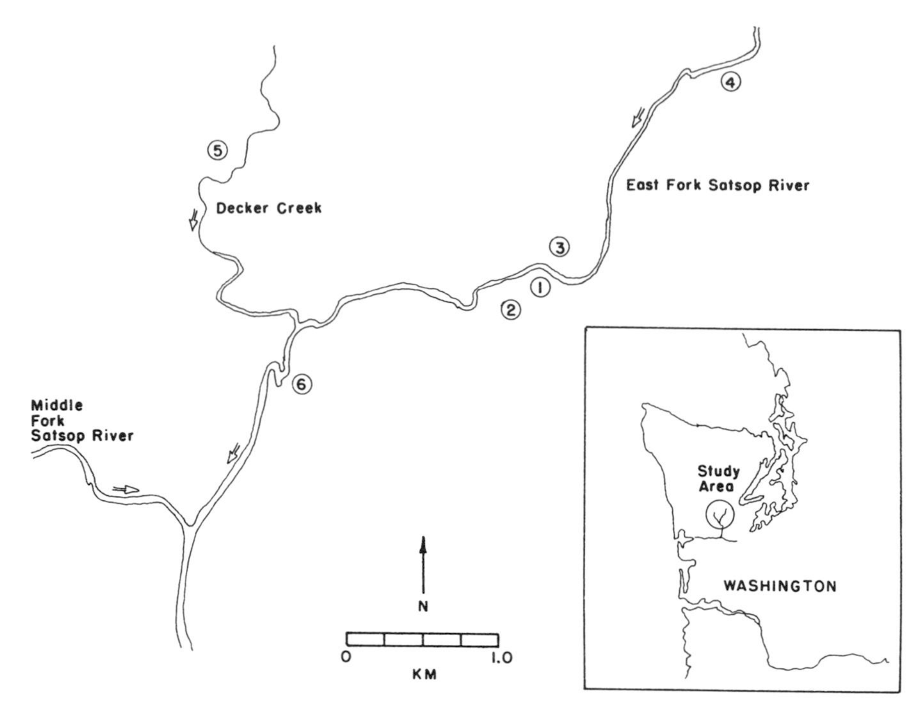

FIGURE 1.—Map of the East Fork Satsop River showing the Satsop groundwater channel sites: 1—Schafer, 2—Schafer #3, 3—Kelsey, 4—Simpson, 5—Maple Glen, 6—Creamer.

Kelsey

Kelsey is a 1,754-m^2 (160 × 12-m) groundwater channel, constructed in 1982 that enters the East Fork Satsop River at river kilometer 21.2. Development involved upgrading an existing groundwater-fed area by excavating the mud substrate and placing graded gravel. The streambed gradient is nearly flat (0.0005), and mean depth is 0.45 m. At the uppermost end of the channel is a "spring hole" about 25 m in diameter, which was left unchanged. It contained gravel ranging from 1.5 to 0.1 cm in diameter. Chum salmon spawning occurred at this site before development but was confined to the spring hole area.

Simpson

Simpson is a 680-m^2 (160 × 5-m) groundwater channel excavated at the site of an existing dry swale in 1983. It enters the East Fork Satsop River at river kilometer 24.1. An earth berm, built from the excavated material, provides flood protection at the upstream end of the channel. A small section of the channel was provided with graded gravel; the remaining native bed material was mechanically cleaned (Bates and Johnson 1986) in 1984. The channel gradient is 0.004 and mean depth is 0.18 m.

Maple Glen

Maple Glen is a 2,170-m^2 (530 × 5-m) natural groundwater-fed channel that enters Decker Creek at river kilometer 2.25. Decker Creek is tributary to the East Fork Satsop River, entering at river kilometer 19.3. Development was limited to periodic beaver dam removal. Chum salmon spawning occurred naturally at this site, but gravel quality was poor. The channel gradient was 0.005, and mean depth was 0.17 m.

Maple Glen is a natural channel, large areas of which are contaminated by fines and organic sediment because they are occasionally impounded by beavers. Poor spawning conditions in Maple Glen channel tend to concentrate spawners into favored gravel patches on which spawner densities are significantly higher than those calculated for the entire channel.

Methods

Productivity of the developed channels (Schafer, Kelsey, Simpson, and Creamer) and two

TABLE 1.—Mean annual production statistics for chum salmon in Satsop groundwater-fed spawning channels, 1984–1986.

Site	Egg-to-fry survival	Fry per m^2	Annual spawners	Females per m^2	Fry production
Schafer	0.31	88	634	0.11	241,927
Kelsey	0.38	109	423	0.12	190,387
Schafer #3	0.47	241	246	0.20	146,776
Maple Glen	0.17	65	672	0.15	141,861
Simpson	0.45	175	211	0.16	119,341

natural undeveloped channels (Schafer #3 and Maple Glen) was assessed for three brood years (1984–1986) by WDF staff biologist David King. He used weekly counts of chum salmon spawners conducted by foot survey, to estimate spawner recruitment into each channel by standard WDF methods (Symons and Waldichuk 1984); he then estimated total egg deposition for each brood year by multiplying recruitment by 0.5 female/spawner and by 2,500 eggs/female. Swim-up chum salmon (fry) production was assessed by means of downstream migrant traps that operated continuously during the fry migration season. Trap box and wing walls had 6.35 × 6.35-mm mesh screen (Blankenship and Tivel 1980). King and Young (1986) documented chum salmon productivity of these channels for the 1984 brood year. For this report, I also used King's unpublished evaluation data for brood years 1985 and 1986. I compared wetted area calculations for each channel, shown in King and Young (1986), with those in Althauser (1985) and found some substantial differences that could not be explained by discussions with King. King then remeasured the sites in question and found that some of his previous measurements were in error, as well as one in Althauser (1985). I then used the most accurate measurements to calculate fry/m^2 and female/m^2 for this report, which resulted in more accurate estimates of these parameters than previously reported.

Chum salmon spawner densities were calculated for brood years 1984 through 1986 by dividing the wetted channel area by the total number of female spawners (total spawners × 0.50) for each brood year. The mean female spawner per square meter for each channel was then computed by using all three brood years (Table 1). For this report, 0.5 female/m^2 was used as the optimum spawner density, defined as the density that maximizes the use of available spawning area without significantly sacrificing egg-to-fry survival rate.

Physical data for the channels, including channel dimensions, wetted area, flow volume, groundwater inflow rate, and gravel quality, were presented by Althauser (1985). He measured flows, considered typical spawning season base flows, near the outlet of each channel during the fall of 1984. I calculated a simple linear regression equation for all channels combined, using these flows as the independent variable and adult chum spawner recruitment for three brood years 1984–1986 as the dependent variable.

Althauser (1985) created an index of groundwater inflow in the Satsop channels by measuring the increase in flow between channel cross-section stations and calculating the drainage rate (inflow per unit length of stream, m^3/[s·m]). An estimate of drainage per unit area of stream, termed "approximate apparent velocity" (AAV), was then derived by dividing approximate drainage rate by the average stream width. This was multiplied by 3.6 to convert it to velocity in mm/h.

Groundwater is often supersaturated with dissolved gases. Gas bubble disease is caused by supersaturation of gases in water; bubbles form within the tissues of aquatic organisms, resulting in death when emboli block vital circulatory pathways. Total dissolved gas analysis was conducted on all channels with a Weiss saturometer and a Hach digital titration kit to determine dissolved oxygen content.

Results and Discussion

Spawner Recruitment as a Function of Flow

Streamflow discharge and adult chum salmon spawner recruitment were positively correlated (r^2 = 0.56; $P < 0.005$: Figure 2). I infer that discharge flow volume is a dominant factor affecting the number of spawners attracted into each channel.

I did not try to include surface runoff freshets or flow increases caused by short-term river-stage fluctuations in analysis of spawner recruitment as a function of flow. These flow fluctuations may influence spawner recruitment by increasing flow, but the irregularity and infrequency of these

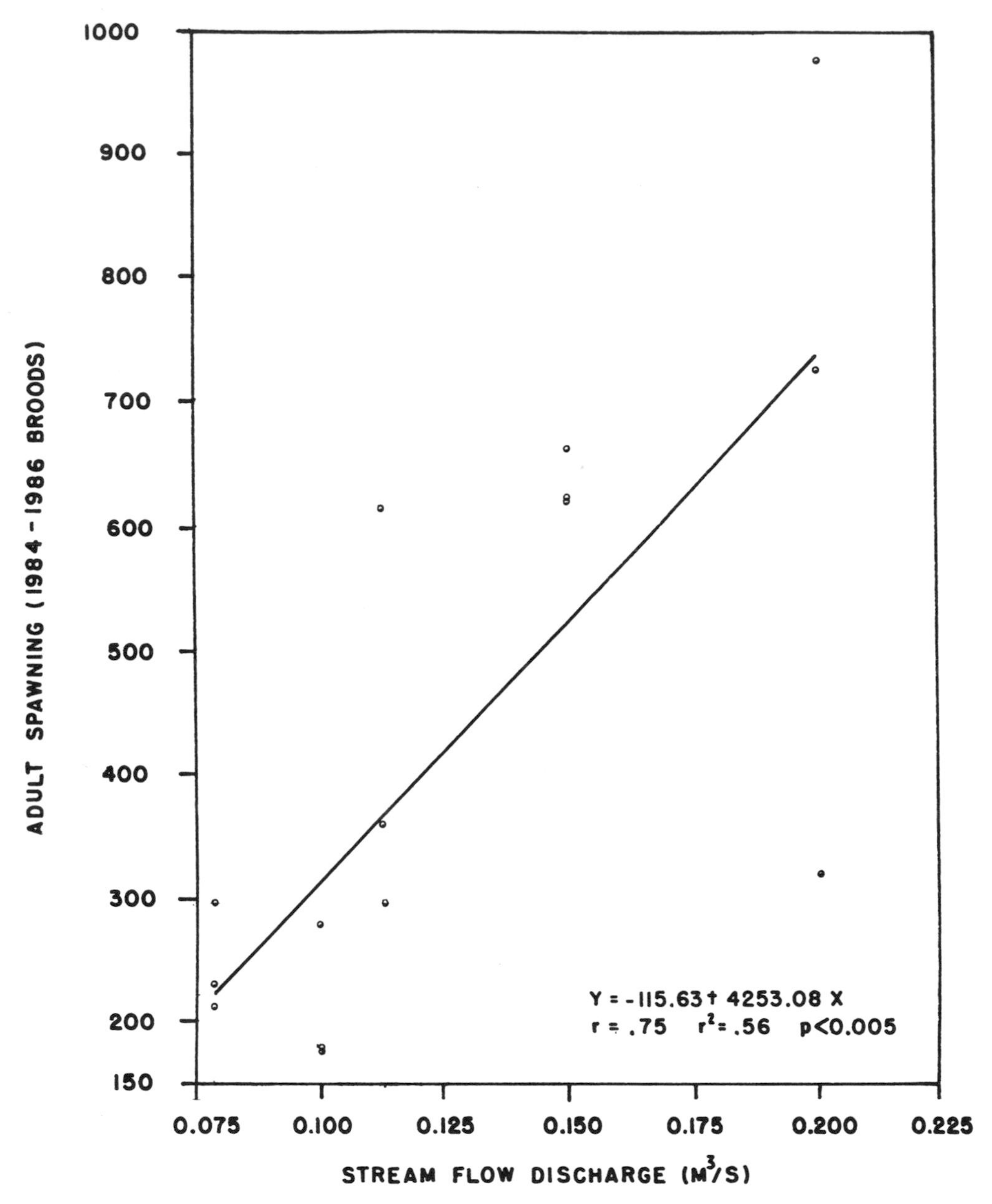

FIGURE 2.—Chum salmon spawner recruitment (*Y*) plotted against flow (*X*) for the Satsop groundwater channels. Channels with higher flows recruited larger numbers of chum salmon spawners.

events probably limited their influence on mean spawner recruitment over the three brood years addressed in this report.

Straying by chum salmon spawners is an important factor affecting initial colonization and may influence spawner densities in groundwater channels. Simpson had a mean chum salmon spawner density of 0.16 female/m^2, which was exceeded only by the density in Schafer #3 (0.20 female/m^2), an unimproved channel that has existed for 20 years or more (Tables 1, 2). Simpson was a recently (1983) excavated channel with no previous chum fry production, hence could have had no homing chum salmon adults. These data weaken the supposition that chum salmon spawner densities at the Satsop groundwater channels depend on returning progeny. The effects of imprinting and subsequent homing behav-

TABLE 2.—Summary of chum salmon production data at Satsop groundwater-fed spawning channels.

Site[a]	Brood year	Egg-to-fry survival (%)	Fry produced	Fry per m^2	Spawners	Females per m^2
Schafer	1984	23	177,751	65	623	0.11
(2,750 m^2)	1985	21	174,355	63	661	0.12
	1986	48	373,675	136	619	0.11
Kelsey	1984	26	199,597	114	613	0.17
(1,754 m^2)	1985	51	229,198	131	359	0.10
	1986	38	142,366	81	296	0.08
Schafer #3	1984	54	199,228	327	296	0.24
(610 m^2)	1985	33	93,724	154	229	0.19
	1986	55	147,376	242	213	0.17
Maple Glen	1984	20	182,285	84	724	0.17
(2,170 m^2)	1985	15	179,942	83	975	0.22
	1986	16	63,357	29	317	0.07
Simpson	1984	46	159,395	234	280	0.21
(680 m^2)	1985	37	81,269	119	176	0.13
	1986	53	117,359	173	178	0.13

[a]Schafer, Kelsey, and Simpson are developed groundwater-fed spawning channels. Schafer #3 and Maple Glen are natural groundwater-fed spawning channels.

ior on spawner densities in the Satsop channels are unclear from the current Satsop data.

Egg-to-Fry Survival as a Function of Groundwater Inflow

Groundwater inflow contributes to intragravel flow and is the main water supply in groundwater channels. Intragravel flow has been shown to relate closely to intragravel survival, fitness, and timing of hatching in chum salmon (Wickett 1954; McNeil 1966; Koski 1975). Phillips and Campbell (1963) and Sowden and Power (1985) found direct dependence of embryonic survival on apparent velocity. Coble (1961) and Turnpenny and Williams (1980) also found that embryo survival in streambeds was directly related to apparent velocity. In groundwater channels this relationship may be influenced by low dissolved oxygen (<7 mg/L), high total dissolved gas levels (>103%) or other water-quality problems. Dissolved oxygen in the Satsop channels was near saturation and should not have interfered with the relationship between groundwater inflow and survival (Table 3). The potential influence of total gas levels on embryonic survival is discussed later.

A simple linear regression equation was calculated for Schafer, Kelsey, Schafer #3, and Simpson with egg-to-fry survival (Table 2) as the dependent variable and AAV as the independent variable for each brood year separately and for all brood years combined (1984–1986). Percent egg-to-fry values were normalized by arcsine transformation, and in all cases analysis-of-variance testing failed to support a relationship (Figure 3). Maple Glen was treated as an outlier because of its atypical spawning conditions.

Maple Glen had a high rate of groundwater inflow, exceeded only by Simpson. The mean egg-to-fry survival (1984–1986 brood years) was 40% for Schafer, Kelsey, Simpson, and Schafer #3, all with good spawning gravel conditions. In contrast Maple Glen had a mean survival of 17%. I postulate that fry production in Maple Glen channel is limited by its poor gravel quality, and that by enhancing its gravel, mean fry production can be improved from 17% to the 40% rate measured at the other channels.

TABLE 3.—Dissolved-gas analysis in Satsop groundwater-fed channels, February 6, 1987.

Site	Dissolved oxygen (mg/L)	Total gas (% saturated)
Upper Creamer Slough	9	101.6
Lower Creamer Slough	10.9	101
Upper Simpson Slough	9.7	103
Lower Simpson Slough	10.4	104
Upper Maple Glen	9.6	101
Lower Maple Glen	10.5	101
Upper Kelsey Slough	9.7	102
Lower Kelsey Slough	9.9	103.6
Upper Schafer Slough (main channel)	11.2	100
Lower Schafer Slough (main channel)	11.8	100
Schafer #3 (upper channel)	9.6	100.3
Schafer #3	10.1	101.6
Satsop River	12.3	100.5

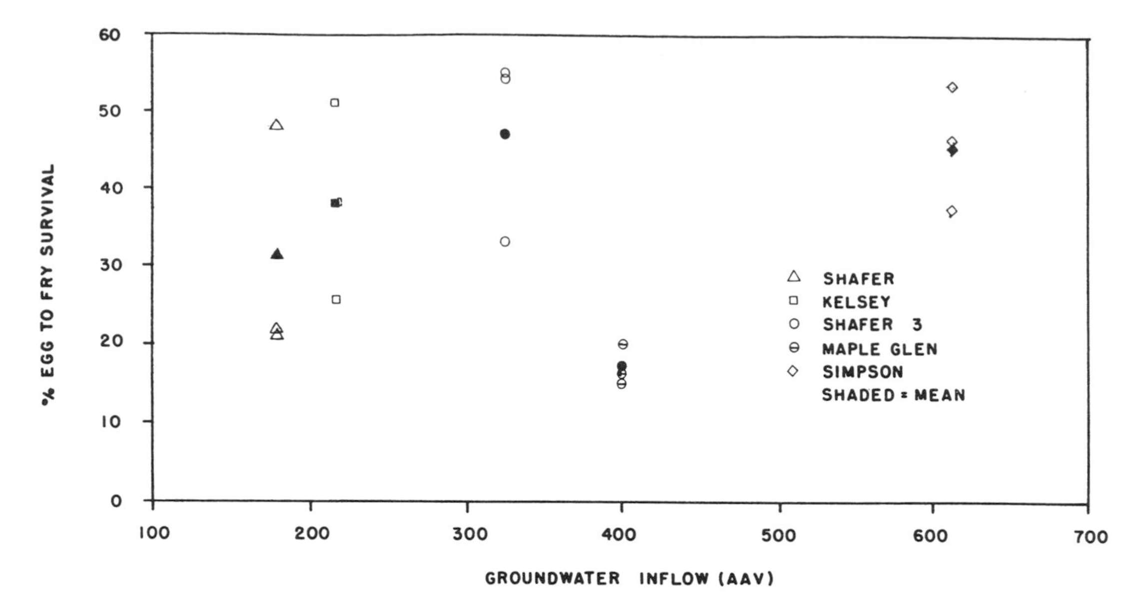

FIGURE 3.—Relationship between egg-to-fry survival and groundwater inflow (approximate apparent velocity, AAV). Each open data point represents a brood year, 1984–1986,; solid points are 3-year means.

Spawner Densities

With the possible exception of Maple Glen, the Satsop channels were underutilized by spawners during the study period. The highest spawner density (female/m^2) was 0.24 in Schafer #3 and the lowest was 0.11 in Schafer (Table 1). These densities are below the optimum 0.5 female/m^2 demonstrated by Lister et al. (1980) for groundwater channels in British Columbia and by Schroder (1974) for the Big Beef Creek chum salmon spawning channel in Washington. Mean egg-to-fry survival was 40% for all channels combined, except Maple Glen, during the 1984–1986 study period. Assuming optimum spawning densities, and 40% egg-to-fry survival, I estimate that 500 fry/m^2 could be produced by the Satsop channels. Production from all channels combined would yield 3,982,000 fry annually, a fivefold increase over the 840,292 mean annual fry production measured. The above production estimates assume enhancement of spawning conditions at Maple Glen.

Total Dissolved Gas

Total dissolved gas was slightly supersaturated in most channels; the range was from 100% saturation in Schafer to 104% in Lower Simpson (Table 3). Wood (1979) indicated that a gas saturation level of 103–104% "appears to be detrimental or cause death in yolk sac fry and newly 'buttoned-up fry'." Simpson (104%) and Kelsey (103.6%) were the only channels in which measured total gas levels exceeded 103%. Although no abnormal conditions were seen in swim-up fry during assessment of either of these channels, detrimental effects would be most prevalent in yolk sac fry while in the gravel.

The relatively high egg-to-fry survivals measured in Simpson (54%) and Kelsey (38%) limit the possibility that an undetected mortality occurs during incubation due to high total gas levels. Hydrostatic pressure was found to influence the occurrence of gas bubble disease by Knittel et al. (1980), with each 10 cm of water depth counteracting one percentage unit of supersaturation at an atmospheric pressure of 760 mm Hg. Water depth in the channels averaged 30 cm, which may have normalized the effects of supersaturation in all channels. Further easing of lethal threshold levels of total gas concentrations may be due to the reduced stress levels provided by optimal natural incubation conditions in the groundwater channels. This would not have been the case in the hatchery incubation conditions, under which Wood (1979) made his observations that saturations of 103–104% were detrimental.

This paper expands our knowledge and understanding of chum salmon production in groundwater channels. Its value is in describing physical and biological characteristics of the Satsop channels, as well as presenting evidence of some mechanisms affecting production of chum salmon fry in groundwater channels. Some of these are the effects of streamflow discharge on spawner recruitment and the tendency of chum salmon spawners to stray into newly created spawning areas. This information should stimulate further study to verify the existence of these mechanisms at other groundwater sites.

The predictions of increased fry production, based on attainment of optimum female spawner densities, need to be verified. Present production is only 20% of the predicted optimum. Optimum productivity may not be achieved because of uncontrolled factors such as nonrandom selection of spawning sites or unevenness of specific groundwater upwelling patterns; however, substantial potential for increase in fry production is evident.

The excellent egg-to-fry survival rates of 21–55% in the Satsop channels emphasize the value to chum salmon of groundwater spawning channels relative to natural, nongroundwater channels. McNeil (1966) concluded that mortality of pink salmon *O. gorbuscha* and chum salmon from egg deposition to fry migration is seldom less than 75% and commonly exceeds 90% in small coastal streams. Hunter (1959) concluded after a 10-year study on Hook Nose Creek that 69–94% of the pink and chum salmon eggs potentially available for deposition were lost before emergence of fry. The reliable survival in well-protected groundwater channels should moderate adverse effects by floods on total fry production of the river system and other catastrophes that cause fluctuation in egg-to-fry survival in the main river channel.

Managers may use this information to add emphasis and credibility to habitat and floodplain protection of river bench lands that have existing or potential groundwater sites. This information should also be used to improve design and operation of groundwater enhancement projects.

References

Althauser, D. R. 1985. Groundwater-fed spawning channel development in the Pacific northwest. Master's thesis. University of Washington, Seattle.

Bates, K., and R. Johnson. 1986. A prototype machine for the removal of sediment from gravel streambeds. Washington Department of Fisheries Technical Report 96.

Blankenship, L., and R. Tivel. 1980. Puget sound wild stock coho trapping and tagging. Washington Department of Fisheries, Progress Report 111, Olympia.

Coble, D. W. 1961. Influence of water exchange and dissolved oxygen in redds on the survival of steelhead trout embryos. Transactions of the American Fisheries Society 90:469–474.

Hunter, J. G. 1959. Survival and production of pink and chum salmon in a coastal stream. Journal of the Fisheries Research Board of Canada 16:835–886.

King, D. D., and R. Young. 1986. An evaluation of four groundwater-fed side channels of the east fork Satsop River—spring 1985 outmigrants. Washington Department of Fisheries Technical Report 90.

Knittel, M. D., G. A. Chapman, and R. R. Garton. 1980. Effects of hydrostatic pressure on steelhead survival in air-supersaturated water. Transactions of the American Fisheries Society 109:755–759.

Koski, K. V. 1975. The survival and fitness of two stocks of chum salmon from egg deposition to emergence in a controlled stream environment at Big Beef Creek. Doctoral dissertation. University of Washington, Seattle.

Lister, D. B., D. E. Marshall, and D. G. Hickey. 1980. Chum salmon survival and production at seven improved groundwater-fed spawning areas. Canadian Manuscript Report of Fisheries and Aquatic Sciences 1595.

McNeil, W. J. 1966. Effect of the spawning bed environment on reproduction of pink and chum salmon. U.S. Fish and Wildlife Service Fishery Bulletin 65:495–523.

Phillips, R. W., and H. J. Campbell. 1963. The embryonic survival of coho salmon and steelhead trout as influenced by some environmental conditions in gravel beds. Pacific Marine Fisheries Commission Annual Report 14:60–73.

Schroder, S. L. 1974. Assessment of production of chum salmon fry from the Big Beef creek spawning channel. University of Washington, College of Fisheries, Fisheries Research Institute, Annual Report 7411, Seattle.

Sowden, T. K., and G. Power. 1985. Prediction of rainbow trout embryo survival in relation to groundwater seepage and particle size of spawning substrates. Transactions of the American Fisheries Society 114:804–812.

Symons, P. E. K., and M. Waldichuk, editors. 1984. Proceedings of the workshop on stream indexing for salmon escapement estimation. Canadian Technical Report of Fisheries and Aquatic Sciences 1326.

Turnpenny, A. W. H., and R. Williams. 1980. Effects of sedimentation on the gravels of an industrial river system. Journal of Fish Biology 17:681–693.

Wickett, W. P. 1954. The oxygen supply to salmon eggs in spawning beds. Journal of the Fisheries Research Board of Canada 11:933–953.

Wood, W. W. 1979. Diseases of Pacific salmon: their prevention and treatment. Washington Department of Fisheries, Hatchery Division, Disease Manual, Olympia.

American Fisheries Society Symposium 10:132–135, 1991

Some Environmental Requirements of Atlantic Salmon

TOR G. HEGGBERGET
Norwegian Institute for Nature Research, Tungasletta 2, N-7004 Trondheim, Norway

Abstract.—Homing of presmolts and adults of Atlantic salmon *Salmo salar* was demonstrated within the Alta River by hydroacoustic tagging and analysis of growth differences. In spite of small nonsystematic differences between spawning habitats used by sympatric populations of Atlantic salmon and brown trout *Salmo trutta*, no interspecific superimposition of redds of the two species was observed. Peak spawning by brown trout takes place earlier than that of Atlantic salmon, which helps to segregate the species. Time of optimal emergence of fry is regulated by differences in time of spawning. Atlantic salmon inhabiting streams that are cold in winter spawn early, whereas Atlantic salmon in warm winter streams spawn late. Enhancement programs involving fish transplantations should take into account the thermal regimes of the donor and target streams. As a consequence of interspecific competition, presmolt Atlantic salmon are found at higher water velocities, farther from the banks, and in deeper water than other sympatric salmonids. In a Norwegian river partially dammed by weirs, brown trout became the dominant species in the weir basins, where water velocity was reduced, but the proportion of Atlantic salmon remained unchanged in the areas of the river between the weirs, where water velocity was unchanged. Different competitiveness by salmonid species under varying physical conditions should be considered when the effects of human activities in salmon streams are analyzed.

Atlantic salmon *Salmo salar* are distributed around the Atlantic Ocean between 40°N and 70°N. Atlantic salmon spawn and live their first years in fresh water. Life history characteristics are similar for Atlantic salmon and most species of Pacific salmon *Oncorhynchus* spp.

Environmental changes caused by humans take place in most salmon streams, thereby affecting the aquatic habitat and fish present. Salmon population enhancement programs, including habitat manipulations and fish transplantations, together with rapidly increasing numbers of escaped farmed fish, may also influence certain populations of Atlantic salmon. Knowledge of environmental requirements of salmon during the freshwater stage is essential for understanding the consequences of environmental and biological changes in streams.

The present review analyzes some environmental requirements of different stages of Atlantic salmon in relation to human activities. The data are based on some recent studies in Norwegian streams (Heggberget 1988), and special emphasis is put on spawning, incubation, and early life history.

Prespawning Migrations

Adult Atlantic salmon enter streams between April and November, with a peak in most Norwegian streams in June and July. However, little detailed information is available on the migration pattern after salmon have entered the river and before spawning occurs. In the River Vefsna, northern Norway, changes in water temperature and flow are the most important influences on the upstream migrations of adult Atlantic salmon (Jensen et al. 1986).

In the River Alta (70°N, 7,400 km^2) hydroacoustic tagging of wild, upstream-migrating salmon showed a unidirectional upstream migration pattern (Heggberget et al. 1988). The fish entered the river at high tide (50% or more) and mean time until they established a holding position was 14 d. All tagged salmon remained in the holding area until spawning. The consistent pattern of migration without backtracking suggests that the salmon select and are able to locate special parts of the river. Experiments carried out about 2 months before spawning in this river showed that 71% of salmon displaced 6 km downstream from the site of capture returned to the donor area (within ±2 km of capture site). Of the downstream-displaced salmon, 57% returned to the pool where they were initially captured. Salmon displaced 7 km upstream from the initial capture site showed a low degree of return to the donor site (one of seven fish).

The results from the hydroacoustic tagging of Atlantic salmon show that the fish are able to relocate formerly occupied areas within a river, a behavioral mechanism necessary for local homing.

To test the hypothesis that Atlantic salmon spawners return to their nursery areas in the

TABLE 1.—Mean total lengths of presmolt Atlantic salmon and mean age of smolts in three sections of the Alta River, 1981–1986. Data in parentheses are half the 95% confidence intervals.

River section	Presmolt length (mm)			Smolt age (years)
	Age 1	Age 2	Age 3	
Upper	73 (0.8)	101 (2.6)	123 (1.9)	3.85 (0.05)
Middle	61 (1.1)	86 (1.4)	105 (3.2)	4.31 (0.04)
Lower	68 (1.1)	92 (2.0)	109 (3.2)	4.16 (0.08)

stream, growth differences among presmolt and adult salmon from different sections of the river were used as biological tags (Heggberget et al. 1986b). The relationship between presmolt growth rate and age at smolting is based on the well-known life history characteristic that faster growing juveniles smolt earlier (Elson 1962). Another assumption for using growth differences as biological tags is that fast-growing individuals of presmolt Atlantic salmon originating from different parts of the River Alta do not aggregate as spawners in one of the sections. The results showed a predominance of multi–sea-winter fish in the upper section of the River Alta. However, comparisons of the smolt ages of fish with similar sea ages caught in the different sections of the River Alta showed that fish from the upper section had a lower smolt age in all sea-age groups than fish from areas farther down the river.

The results (Table 1) showed that presmolt Atlantic salmon in the upper part of the river had significantly faster growth than fish in the middle and lower section. Analysis of scales from adult Atlantic salmon caught in the different parts of the river showed that Atlantic salmon caught in the upper section had smolted at younger ages and at greater lengths than Atlantic salmon caught farther down.

The hydroacoustic studies and the analysis of growth differences support the hypothesis that Atlantic salmon home in the River Alta. Local homing may have two important consequences. First, it provides a basis for the evolution of local populations. Second, it allows for the establishment and continuation of an appropriate density pattern of Atlantic salmon spawners within a river.

Habitat Selection by Spawners

The extent of spawning activity by Atlantic salmon in several Norwegian streams was assessed from light aircraft (Heggberget et al. 1986a). According to White (1942), Stuart (1953), and Berg (1964), ideal spawning sites occur where the tail end of a pool merges into a riffle and where the water flow is fairly fast. More recent findings (Heggberget et al. 1986b) showed that the spawning sites of Atlantic salmon and brown trout *Salmo trutta* were not limited to the lower tails of the pools; spawning redds were frequently found in the pools, often close to the riverbanks (1–2 m) in shallow water (10–20 cm).

Atlantic salmon and brown trout normally are sympatric in Norwegian streams. Spawning sites used by the two species in sympatric populations show great intra- and interspecific variability and overlap when related to water depth, water velocity, substrate, and area of the redds (Table 2). Atlantic salmon redds in the River Eira were found where water velocity was about 40 cm/s, about 9 m from the riverbanks in depths of about 50 cm water depth, and where the gravel was about 10 cm in diameter. The differences between redds made by salmon and trout were usually small and unsystematic (Table 2). Despite this overlap, no hybrids between the two species were found (Heggberget et al. 1988). Interspecific superimposition of redds of salmonid species is known to occur, especially where one of the species is introduced outside its native range. The main mechanism separating the two species seems to be temporal segregation: brown trout normally spawn earlier than Atlantic salmon. It is concluded that a long period of sympatry between Atlantic salmon and brown trout in Norwegian streams has permitted the evolution of sufficient isolating mechanisms to prevent superimposition and hybridization.

Time of Spawning and Incubation of Eggs

Recent studies in Norwegian streams (Heggberget 1988) have shown a close relation between the timing of spawning and thermal regime of the streams. In streams with high water temperatures during incubation, the salmon spawn later than in streams that are cold during winter.

Studies of several species of Pacific salmon support this relationship between timing of spawning and temperature during incubation. Burger et al. (1985) observed that spawning by chinook salmon *Oncorhynchus tshawytscha* peaked during August in lake-fed, tributaries that are warmer in winter, whereas spawning took place during July in tributaries not influenced by lakes. Similarly, Sheridan (1962) observed that pink salmon *O. gorbuscha* spawned early in cold

TABLE 2.—Physical characteristics of Atlantic salmon and brown trout redd sites in the River Eira (data from Heggberget 1988). CI is confidence interval. Sample sizes were 234–236 Atlantic salmon redds and 125 brown trout redds.

Characteristic and species	Mean	SD	One-half 95% CI
Water velocity (cm/s)			
Atlantic salmon	38.6	18.2	2.4
Brown trout	27.4	13.6	2.4
Distance from banks (m)			
Atlantic salmon	8.9	5.0	0.7
Brown trout	10.5	5.7	1.0
Water depth (cm)			
Atlantic salmon	49.3	16.6	2.2
Brown trout	50.0	15.5	2.8
Area (m^2)			
Atlantic salmon	4.8	4.8	0.6
Brown trout	2.3	3.3	0.6
Substratum (cm)			
Atlantic salmon	9.2	3.8	0.5
Brown trout	8.1	2.7	0.5

streams in Alaska and later in warmer streams; the ova thus received about the same number of temperature units by the beginning of spring and fry in these streams emerged about the same time.

Genetic differences between stocks of Atlantic salmon have been observed (Ryman and Ståhl 1981; Thorpe and Mitchell 1981). No evidence of stock adaptation to temperature was found in a study of embryonic development of Norwegian Atlantic salmon (Wallace and Heggberget 1988), indicating that timing of emergence is regulated by time of spawning rather than by different temperature responses during incubation. This seems to be a general adaptation occurring both in Atlantic salmon and in Pacific salmon.

For salmonids living under temperate and cold conditions and having a restricted growing season, there appears to be a strong selection pressure on the timing of spawning, hatching, and emergence so that fry have optimal fitness. Offspring from individuals breeding at any time of the year other than the ideal will normally have limited possibilities for survival. A practical implication of this is that enhancement programs involving fish transplantations should take into account the thermal regime of the donor streams.

Habitat Selection of Presmolt Atlantic Salmon and Brown Trout

Allen (1969) suggested that limitations on salmonid production in streams could be imposed by territorial behavior. It has been shown that presmolt Atlantic salmon and brown trout are territorial and show agonistic behavior, both inter- and intraspecifically (Kalleberg 1958; Karlstrøm 1977).

TABLE 3.—Physical characteristics of sites occupied by presmolt Atlantic salmon and brown trout in the Gaula River. Values are mean (one-half 95% confidence intervals). All differences between species were significant (analysis of variance, $P > 0.001$).

Species	Distance from riverbank (m)	Water depth (cm)	Water velocity (cm/s)
Atlantic salmon ($N = 723$)	3.6 (0.2)	24.6 (0.9)	26.7 (1.2)
Brown trout ($N = 209$)	2.0 (0.3)	19.6 (1.2)	12.1 (1.8)

In spite of considerable habitat overlap between juvenile Atlantic salmon and brown trout, significant differences between the species with respect to water velocity, water depth and distance from the riverbanks can be observed (Table 3). This is a result of interspecific competition, and it implies that small ecological differences are magnified in competing species (Brian 1956), which has been demonstrated in many salmonid species (Kalleberg 1958; Gibson 1966).

The fact that Atlantic salmon are more competitive than trout at high water velocities is important when the water velocity of streams is changed by human activities. Water velocity is often reduced because of hydropower development, water abstraction for different purposes, or the establishment of weirs in streams. This reduction will most severely affect the fish species that is most competitive at high water velocities. For example, in the River Skjoma the proportion of salmon in the weir basins was estimated at about 17%, whereas, the proportion in the fast-flowing areas between the weirs constituted about 71%.

Atlantic salmon production and the severity of competitive interactions probably depend very much on the habitat. Several studies have shown that Atlantic salmon densities appear to be greater than those of brown trout in the main stem of rivers, whereas brown trout have greater production in the tributaries (Lindroth 1955; Power 1973). In North America, the only native salmonid species living sympatrically with Atlantic salmon throughout its range is the brook trout *Salvelinus fontinalis*. These two species are frequently the dominant fish species. Although there is considerable habitat overlap between the two species, brook trout usually associate more with pools than do presmolt Atlantic salmon (Keenleyside 1962; Gibson 1966).

Different competitiveness of fish species under varying physical conditions should therefore be considered when the effects of human activities in streams are analyzed.

References

Allen, K. R. 1969. Limitations on production in salmonid populations in streams. Pages 3–18 *in* T. G. Northcote, editor. Salmon and trout in streams. H. R. MacMillan Lectures in Fisheries, University of British Columbia, Vancouver.

Berg, M. 1964. Nord-norske lakseelver. J. G. Tanum Forlag, Oslo. (In Norwegian with English summary.)

Brian, M. W. 1956. Segregation of the species of the ant genus *Myrmica*. Journal of Animal Ecology 25:319–337.

Burger, C. V., R. L. Wilmot, and D. B. Wangaard. 1985. Comparison of spawning areas and times for two runs of chinook salmon (*Oncorhynchus tshawytscha*) in the Kenai River, Alaska. Canadian Journal of Fisheries and Aquatic Sciences 42:693–700.

Elson, P. F. 1962. Predator–prey relationship between fish-eating birds and Atlantic salmon. Fisheries Research Board of Canada Bulletin 133.

Gibson, R. J. 1966. Some factors influencing the freshwater distribution of brook trout and young Atlantic salmon. Journal of the Fisheries Research Board of Canada 23:1977–1982.

Heggberget, T. G. 1988. Reproduction in Atlantic salmon (*Salmo salar*). Aspects of spawning, incubation, early life history and population structure. Doctoral dissertation. University of Trondheim, Trondheim, Norway.

Heggberget, T. G., L. P. Hansen, and T. F. Næsje. 1988. Within-river spawning migration of Atlantic salmon (*Salmo salar*). Canadian Journal of Fisheries and Aquatic Sciences 45:1691–1698.

Heggberget, T. G., T. Haukebø, and B. Veie-Rossvoll. 1986a. An aerial method of assessing spawning activity of Atlantic salmon, *Salmo salar* L., and brown trout, *Salmo trutta* L., in Norwegian streams. Journal of Fish Biology 28:335–342.

Heggberget, T. G., R. A. Lund, N. Ryman, and G. Ståhl. 1986b. Growth and genetic variation of Atlantic salmon (*Salmo salar*) from different sections of the River Alta, north Norway. Canadian Journal of Fisheries and Aquatic Sciences 43:1828–1835.

Jensen, A. J., T. G. Heggberget, and B. O. Johnsen. 1986. Upstream migration of adult Atlantic salmon, *Salmo salar* L., in the River Vefsna, northern Norway. Journal of Fish Biology 29:459–465.

Kalleberg, H. 1958. Observations in a stream tank of territoriality and competition in juvenile salmon and trout (*Salmo salar* L. and *Salmo trutta* L.). Institute of Freshwater Research Drottningholm Report 39:55–59.

Karlstrøm, Ø. 1977. Habitat selection and population densities of salmon and trout parr in Swedish rivers. Information from Institute of Freshwater Research Drottningholm, Stockholm. (Mimeograph; in Swedish with English summary.)

Keenleyside, M. H. A. 1962. Skin-diving observations of Atlantic salmon and brook trout in the Miramichi River, New Brunswick. Journal of the Fisheries Research Board of Canada 19:625–634.

Lindroth, A. 1955. Distribution, territorial behaviour and movements of sea trout fry in the River Indalsälven. Institute of Freshwater Research Drottningholm Report 36:104–119.

Power, G. 1973. Estimates of age, growth, standing group and production of salmonids in some north Norwegian rivers and streams. Institute of Freshwater Research Drottningholm Report 53:78–111.

Ryman, N., and G. Ståhl. 1981. Genetic perspective of the identification and conservation of Scandinavian stocks of fish. Canadian Journal of Fisheries and Aquatic Sciences 38:1562–1575.

Sheridan, W. L. 1962. Relation of stream temperatures to timing of pink salmon escapements in southeast Alaska. Pages 87–102 *in* N. J. Wilimovsky, editor. Symposium on pink salmon. H. R. MacMillan Lectures in Fisheries, University of British Columbia, Vancouver.

Stuart, T. A. 1953. Spawning migration, reproduction, and young stages of the Loch trout (*Salmo trutta*). Scottish Home Department, Freshwater and Salmon Fisheries Research Report 5, Edinburgh.

Thorpe, J. E., and K. A. Mitchell. 1981. Stocks of Atlantic salmon (*Salmo salar*) in Britain and Ireland: discreteness and current management. Canadian Journal of Fisheries and Aquatic Sciences 38:1576–1590.

Wallace, J. E., and T. G. Heggberget. 1988. Incubation of eggs of Atlantic salmon (*Salmo salar*) from different Norwegian streams at temperatures below 1°C. Canadian Journal of Fisheries and Aquatic Sciences 45:193–196.

White, H. C. 1942. Atlantic salmon redds and artificial spawning beds. Journal of the Fisheries Research Board of Canada 6:37–44.

American Fisheries Society Symposium 10:136–149, 1991

Restructuring Streams for Anadromous Salmonids

NEIL B. ARMANTROUT

U.S. Department of the Interior, Bureau of Land Management
Post Office Box 10226, Eugene, Oregon, 97440 USA

Abstract.—The production capability of anadromous fish streams in the Bureau of Land Management's Eugene District, Oregon, is well below potential as a result of the loss of large woody debris. District managers placed 395 structures made of wood, rock, and gabions to restructure selected streams to restore part of the production potential. The fate and success of structures of different designs and materials are discussed in the paper. All types of structures improved aquatic habitat by changing channel characteristics, substrate composition, and availability of habitat features used by juvenile salmonids.

At the time settlement began in western Oregon in the last century, the stream channels were dominated by large woody debris. These large fallen trees, predominantly Douglas fir *Pseudotsuga menziesii* and western red cedar *Thuja plicata*, remained in the channels for extended periods, trapping and retaining sediments. Movement of water interacting with the woody materials and channel substrates created pools, cover, and other fish habitats. Vegetation and deposits helped moderate stream flows. Occasional disasters, such as floods, fires, and major land movements disrupted stream channels, but woody material was usually present in riparian areas or headwaters to restore stream habitats (Bryant 1982; Sedell and Swanson 1982).

Fishes established in western Oregon streams have adapted to the habitats provided by the interaction of the large woody debris, stream channels, and seasonal flow patterns. The dominant species are the anadromous salmonids, including coho salmon *Oncorhynchus kisutch*, chinook salmon *O. tschawytscha*, chum salmon *O. keta*, steelhead *O. mykiss*, and cutthroat trout *O. clarki*. These salmonids have similar life histories but different habitat preferences. Use of habitat varies seasonally and habitat selection is influenced by stage of reproduction, size of fish, and environmental conditions such as flow and temperature (Murphy et al. 1982; Everest et al. 1985).

Early programs of woody debris removal reduced instream structure in larger streams throughout western Oregon. Land-use activities, particularily timber harvest, removed much of the instream structure from all but a few streams (Sedell and Luchessa 1982). With the removal of large trees from riparian areas and adjacent slopes, there were few trees to fall into the channel to replace the lost structure.

Without the retentive action of woody debris, sediments, including the gravels needed for spawning, washed out of the basins during high flows. Channels downcut until a firmer substrate was reached, lowering water storage capabilities of alluvial areas. The depth and quality of pools declined. Instream cover and off-channel nursery areas largely disappeared (Marzolf 1978; Bilby 1984). The altered streams had lower diversity and less habitat that could be used by salmonids, resulting in a decline in potential fish runs (House and Boehne 1987).

The role of habitat loss in the decline of fish runs led to an increase in efforts to restore anadromous fish habitat. A variety of techniques has been used (Paquet 1986; Patterson 1986; House et al. 1990). Most techniques are adapted from other disciplines or are attempts to duplicate natural habitat. Few thorough evaluations of these projects have been done, but success has been quite variable (Hall and Baker 1982; Reeves and Roeloffs 1982; BPA 1985; Osborn and Anderson 1986).

The Bureau of Land Management's (BLM) Eugene District implemented a series of habitat improvements from 1969 through 1973. Using observations on these earlier projects and projects by other biologists, the District in 1983 initiated a program of anadromous fish habitat restoration. Various instream structures were tried. This paper summarizes project design and results, and offers some general comments on aquatic habitat restoration.

Methods

Project work was in tributaries of the Siuslaw River, western Lane County, Oregon. All work was on lands managed by the BLM's Eugene District. These lands are managed primarily for

TABLE 1.—Dimensions of project streams in Oregon.

Stream	Order	Flow (m^3/s)	Total basin area (km^2)	Average width (m)	Number of structures	Project length (m)
Fish	4–5	0.03–14	21	5	64	2,000
Greenleaf	5	0.05–55	31	7	30	200
North Fork Whittaker	4	0.003–6	5.2	2	54	1,000
North Fork Leopold	4	0.001–3	5.2	1	113	1,500
Grenshaw	4	0.003–7	6.5	3	27	1,000
Saleratus	4	0.003–5	7.8	3	27	1,000
Big Canyon	5	0.006–8	5.2	5	5	100
Whittaker	6	0.1–56	25.9	12	5	600
Esmond	6	0.1–70	41.4	10	17	1,000
Nelson	5	0.0001–8	36.3	4	7	100
Bounds	4	0.003–6	5.6	4	53	1,000

timber production under a plan approved in 1983. This plan provides for protection of riparian areas on third-order and larger streams (Boehne and House 1983), and for some protection of smaller headwater streams to sustain water quality. Habitat management plans for fisheries were prepared for each subbasin under the timber management plan and included land management recommendations to protect fisheries as well as project work. Work was done in 11 streams, listed in Table 1.

Prior to the initial project work, streams were inventoried according to an unpublished intensive inventory system developed by BLM for western Oregon. In this system, the stream is divided into individual habitats, and each habitat is then analyzed for channel characteristics, substrate, cover, and adjacent land and vegetation characteristics. Pool typing relied on a modified version of a system developed by Bisson et al. (1982). Inventory information was used to identify habitat-deficient areas and to develop site-specific habitat improvement proposals. Reinventory following implementation of projects was used for comparison of habitat before and after project work.

Design of projects was influenced by the size of stream, access, availability of materials, and costs (Armantrout 1990). Because of limited funding and availability of volunteers, emphasis was given to projects that could be done by hand with on-site materials. Most projects placed structures in the stream. Excavation was restricted to blasting to improve passage over structures or at velocity barriers.

In one stream, Greenleaf Creek, projects were to correct localized problems, primarily bank erosion. In the other 10 streams, series of projects were undertaken along one or more stream reaches. Designs, as much as possible, incorporated existing channel features such as boulders, woody material, pools, and cover.

Blasting.—All blasting was done to create pools in bedrock to facilitate passage upstream by migrating adult salmonids. Fourteen pools were blasted by the U.S. Marine Corps Reserves as part of training exercises. Shape charges placed on the bedrock were used to create fractures. Explosives were packed into those fractures and discharged to make pools. Additional charges were sometimes detonated under the edges of the pools to increase the size. Final pools at low water were up to 75 cm deep and 4 m × 8 m in size.

Four other pools were blasted by techniques developed by BLM's Coos Bay District (Anderson and Miyajima 1975; Ruediger 1990). Holes up to 1 m deep were drilled in rows up to 60 cm apart in the shape of the desired pool, charged with two-part explosives, and detonated. Fractured rock and debris from the blasting were left in place in the stream (Figure 1).

Boulders.—Boulder weirs were made of both small and large boulders. Large boulders, 60 to 150 cm in diameter, were quarried and delivered to the site, then placed in the stream by tire-mounted front-end loaders. The size of boulders used was limited by what the equipment could carry. Weirs were in diagonal lines placed either at an angle up to 30° with the channel line or in upstream or downstream arches spanning the channel. Boulders were not anchored. The ends of the weirs were higher than the middle to protect against erosion around the ends and to concentrate flows at midchannel during low-flow periods. The boulders were placed in natural depressions or fractures in the underlying bedrock when

FIGURE 1.—Greenleaf Creek, blasted pool with fractured rock still in place.

present. Several individual boulder clusters were also created (Figure 2).

Small-boulder weirs utilized on-site rocks up to 50 cm in diameter that could be moved and placed by hand; usually they were taken from the stream channel. Rocks were piled up to 60 cm in height in a series or in coordination with woody structures (Figures 3, 4). Weir alignment and shape varied, conforming to the channel.

Gabions.—Fifty gabions were built in four streams, all using on-site rock to fill the wire baskets. Initially, gabions were made in a V-shape with the point downstream. After viewing gabions built by other biologists, particularly those of John Anderson, BLM, Coos Bay (e.g., Anderson and Cameron 1980), we changed designs to primarily straight lines at a 30° diagonal across the creek. In addition, we used gabions extending only partially across the channel and shaped into a downstream arch. The designs conformed more

FIGURE 2.—North Fork Whittaker Creek, large-boulder weir. Boulders are 70 to 120 cm in diameter.

FIGURE 3.—North Fork Whittaker, small-boulder weir.

FIGURE 4.—Saleratus Creek, small boulder and alder structure.

closely to the natural and desired flow and deposition patterns. Five gabions were built by a volunteer group that used different design criteria.

Gabions ranged in height from 50 cm to 100 cm, depending upon the flow in the stream. The gabions were anchored into the streambank and, by rebar and cable, to the channel bottom. They were placed in series with other gabions or in conjunction with other structures.

Woody materials.—Large woody debris remains in riparian areas along some streams. Some of this debris was incorporated into instream structures. Logs used in projects ranged from 3 to 10 m in length, and up to 1 m in diameter; root wads were up to 5 m high. Most of the debris was culled from past timber operations. The large woody debris was anchored in place by cables fastened to standing live trees.

In the project area, red alder *Alnus rubra* is the dominant tree species in second-growth riparian communities. Alder trees were selectively cut in the riparian area at least 15 m away from the stream and placed in the channel. Logs were up to 20 m long, the length being determined by the size of the channel. Diameter of the logs ranged up to 75 cm.

Most of the work was by volunteers, who were encouraged to experiment with different structural designs. The three most commonly used were termed the fence post, log cabin, and logjam.

The fence post design was patterned on trash catchers. An alder log was placed across the channel and wedged upstream behind standing live trees or large boulders. Short pieces of alder, less than 10 cm in diameter and up to 125 cm in length, were placed upright behind the alder log at an angle of 30° to the channel so that water pressure would push the uprights against the log. The uprights were wired to the log. Brush, alder tops, or small Douglas firs were packed behind the uprights to trap debris and sediments (Figures 5, 6).

The log cabin design resembled a split rail fence in appearance and was built to protect eroding banks. Several logs were stacked atop one another. For extended structures, one set of logs overlapped another in the same way as the cor-

FIGURE 5.—Bounds Creek, alder fence post structure with thinned Douglas fir packed behind.

FIGURE 6.—Fish Creek, alder fence post structure after two years; some beaver augmentation on ends.

ners of log cabins. The logs were wired or cabled together and anchored by cable at one or both ends to live trees or posts on the bank. Thinned Douglas firs and brush were placed behind the structures (Figure 7).

Logjams were created by placing several alder logs in a pile across the stream and cabling the ends to live trees or posts on the bank. Small Douglas firs, brush, or small boulders were often added to the structure (Figure 8).

In Douglas fir forest management, trees are precommercially thinned while less than 25 years old in order to achieve proper spacing. We began using these thinned trees as structural material, first to provide bulk to the fence post structures, then later as woody structural material. The thinned trees range up to 20 m tall, with a butt diameter up to 30 cm. They were used to create the same type of structure as the alders. In addition, they were used to provide bank protection against erosion and to create cover. For both bank protection and fish cover, the butt ends of the thinned trees were anchored with wire to live trees or posts on the bank and the top end was allowed to hang into the stream.

FIGURE 7.—Fish Creek log cabin alder structure for bank protection.

Evaluation.—Each structure was observed several times during the year. The condition of the structure was noted, along with general observations of its effect on the stream. Observations of earlier structures led to modifications of later structure designs. Corrections and repairs to structures were made as needed. A detailed evaluation of all structures was done in 1988 and is summarized in Table 2.

Two streams, Fish Creek and North Fork Whittaker Creek, were completely reinventoried in 1988. In other streams, reinventory was limited to portions of the stream where juvenile fish sampling was done as part of the study on structure utilization.

Spawning observations were begun in Fish and Greenleaf creeks in designated index areas in project and nonproject areas in 1983 prior to instream project work. By 1986, spawning observations were being made of all anadromous species in most of the habitat managed by BLM in the Siuslaw Basin. Separate index areas were established for project reaches in order to compare use of the improvements with other habitat throughout the basin. One or more counts of live and dead fish and redds were made during the peak spawning period of each anadromous species.

Very limited juvenile salmonid sampling was done prior to project construction, although visual estimates were made in each habitat during habitat inventories. Following project implementation, sampling was conducted in several project reaches to assess structure utilization by juvenile salmonids. Population measurements were made by two-pass electrofishing (Robson and Regier 1968). As a result of drought conditions in 1987 and 1988, winter flows were low enough to permit electrofishing studies of winter use of structures and habitat in unimproved areas.

Results

The status of the 396 structures in 1988 is summarized in Table 2. Of the 396 structures, 85% were completely or partially intact and in place and improving aquatic habitat. Eight percent were completely or partially intact and in place but were judged not to be improving aquatic habitat. Seven percent of the structures were substantially or totally gone. None of the structures was judged to have degraded the aquatic habitat or to have created additional problems. An improvement in aquatic habitat is defined as increasing pool or

FIGURE 8.—Grenshaw Creek, logjam alder and thinned Douglas fir structure.

spawning areas, increasing deposition in the channel, or increasing cover or off-channel nursery areas. Whether the contribution is significant or moderate is my subjective judgment based on knowledge of the streams and local conditions.

Channel Changes

North Fork Whittaker Creek is a small fourth-order stream that was partially channelized and straightened when a road was constructed adjacent to it. The increased gradient and loss of instream structure produced constant scouring at high flows. Initial inventories were completed in February 1986, and project work was implemented in the summer of the same year. Fifty-five structures were placed along approximately 1 km of stream, including four large-boulder weirs, 23 small-boulder weirs, and 28 alder–thinned Douglas fir structures.

Table 3 summarizes changes in the channel in the improved section. Flows were considerably reduced during the reinventory in 1988 due to drought, or comparable values, particularily for pools and glides, would have shown an even greater difference. Pool habitat increased from 12.9% of the stream length to 49.4% and glides from 1.4% to 14.7%. Riffle habitat declined by almost half, and cascades and rapids were nearly eliminated.

The most pronounced change in substrate was the reduction in cobble (7.5–15 cm) and rubble (15–30 cm), which dropped from 18.2% and 19.5%, respectively, to 5.7% and 8.7%. Sand increased from 5.9% to 22.1%, and silt from 0.9% to 8.3%, reflecting the increase in pools and glides. Types and causes of pools, cover, and habitat dimensions also showed substantial changes.

Fish Creek is a fourth- and fifth-order stream along the 8.5 km managed by BLM. Initial inven-

TABLE 2.—Current status of habitat improvement projects in Oregon.

Type of structure	Number of structures by category: Intact, major improvement	Intact, moderate improvement	Partial, adding habitat	Intact or partial, neutral	Missing	Total
Gabion	22	16	7	5	0	50
Large-boulder weir	29	3	0	2	0	34
Small-boulder weir	54	40	48	12	0	154
Alder fence post	7	1	34	4	18	64
Alder log cabin	3	2	1	0	0	6
Othe alder–fir	16	9	14	5	8	52
Large woody debris	4	2	1	0	1	8
Blasted pools	5	2	7	3	1	18
Other	4	5	0	0	1	10

TABLE 3.—Summary of changes in stream characteristics associated with improvement projects on North Fork Whittaker and Fish creeks.

Characteristics	North Fork Whittaker Before (0.17 m^3/s)	North Fork Whittaker After (0.01 m^3/s)	Fish Creek, main project area Before	Fish Creek, main project area After
Substrate (%)				
Bedrock	25.4	22.4	33	20
Small boulder	5.5	2.4	3.2	1.3
Cobble	18.2	5.7	6.6	5.7
Rubble	19.5	8.7	9.4	16
Large gravel	12.1	12.1	21.2	19.2
Small gravel	10.1	12.9	9.3	8.8
Sand	5.9	22.1	5.7	10.8
Silt	0.9	8.3	6.4	13.1
Habitat (%)				
Pool	12.9	49.4	39.7	47.8
Riffle	62.3	34.3	39.5	33
Rapid	17.8	0.1	3.8	5
Cascade	2.3	0	1.1	0.9
Runs	3.2	0	0	2.6
Glide	1.4	14.7	15.7	10.4
Mean length (m)	9.1	7.0	11.9	12.3
Pool type				
Dam	2 (3%)	22 (31%)	7 (6%)	21 (20%)
Lateral scour	5 (33%)	14 (22%)	24 (23%)	34 (33%)
Plunge	5 (33%)	14 (22%)	28 (27%)	31 (30%)
Trench		13 (18%)	19 (18%)	4 (3%)
Secondary		2 (2%)	5 (4%)	9 (8%)
Eddy		2 (2%)	3 (3%)	2 (1%)
Backwater	3 (20%)	1 (1%)	14 (3%)	
Alcove		1 (1%)		
Pool cause				
Enhancement project	0	23 (24%)	0	49 (33%)
Large woody debris	2 (11%)	22 (23%)	33 (23%)	37 (25%)
Boulder	1 (5%)	21 (22%)	14 (10%)	7 (4%)
Bedrock fracture	0	13 (13%)	0	2 (1%)
Stream bend	4 (23%)	7 (7%)	30 (21%)	31 (21%)
Beaver dam	0	4 (4%)	9 (6%)	10 (6%)
Falls	3 (17%)	1 (1%)	0	8 (5%)
Channel changes	4 (23%)	0	15 (10%)	0
Debris torrent	3 (17%)	0	0	0
Other	0	2 (2%)	35 (25%)	1 (1%)
Pool cover type (%)				
Plunge	55	8	1	13
Woody debris	19	20	38	32
Boulder	1	21	16	16
Undercut bank	13	19	12	16
Pool dimension				
Average depth/width	22%	12%	7%	6%
Average pool volume (m^3)	30.75	9.42	26.21	37.67
Total pool volume (m^3)	461.4	651.6	2,673.0	3,842.0

tories were conducted in 1983, and projects were completed during the summers of 1984, 1986, and 1987. Some structures were placed along the lower 6 km, but most were concentrated in a 2-km section. Projects included 34 gabions, 3 large-boulder weirs, 1 small-boulder weir, 18 alder–thinned Douglas fir structures, 2 large woody debris structures, and 5 blasted pools. The stream was reinventoried in the spring and summer of 1988. The results are summarized in Table 3.

In the 2-km section where most of the structures are located, pool habitat increased from 40% to 48%; riffles decreased by 5.5 percentage points and glides by 5.3 points. The majority of projects were in bedrock areas, primarily glide and riffle habitat. The percentage of substrate as exposed bedrock decreased from 33% to 20%; rubble, sand, and silt showed the greatest percentage increase.

Structure Evaluation

Four of the 50 gabions were submerged by ponding from downstream beaver activity, and 6 sustained damage, 3 by large woody debris from upstream mass movements and a logjam breakup. The damaged gabions were partially repaired. Two were built on unstable substrates, resulting in some collapse and rotation. The sixth was damaged by slumping and rotation caused by faulty design. All remain in place and functioning.

Seven gabions in Greenleaf Creek were designed as deflectors, six of them in a series, to narrow a shallow channel and create meander. The deflectors were built at 30° angles to the bank and one-third of the way across the channel. The channel narrowed as a result of extensive deposits behind the gabions as planned, but the gabions were neither built far enough into the channel nor spaced properly to create meander, although the technique looks very promising.

Thirty-four large-boulder weirs were placed in five streams, and only two did not remain in place. One was built partially of smaller boulders from the site when large boulders from the quarry ran out; the smaller boulders shifted downstream with the first high flow. Two boulders in the second weir shifted after a fallen tree and an upstream beaver dam on top of another boulder weir redirected the main current against them. The shifted boulders remain in the channel downstream and create habitat.

In Grenshaw Creek seven boulder weirs were built close to each other to see if interactions would increase the amount and diversity of habitat. Results are shown in Figure 9. There was an increase in the number and diversity of habitats and in the number and depth of pools and the availability of cover. The numbers of juvenile salmonids also increased. A similar design has been used with 17 boulder weirs in Esmond Creek. The weirs have not been in place long enough to evaluate their impact, but the initial response is promising.

Of the 154 small-boulder weirs, only 12 were nonfunctional. Over a third were judged to create major habitat improvement, primarily through creation of pools and increased deposition of salmonid spawning gravel. The most common problem was a breach where currents were strongest. Even breached weirs still retained deposits on the sides of the channel behind them, thereby narrowing and deepening flows.

Weirs built upright like a fence were least likely to remain intact. Unless quickly filled in behind by deposition, they were breached. The second-most common cause of breaching was undercutting by eddy action at the downstream face of the weir, causing the weir to collapse foward. Weirs survived best when they had a sloping downstream face that provided support against currents and reduced downstream eddy action or when they were built in a series.

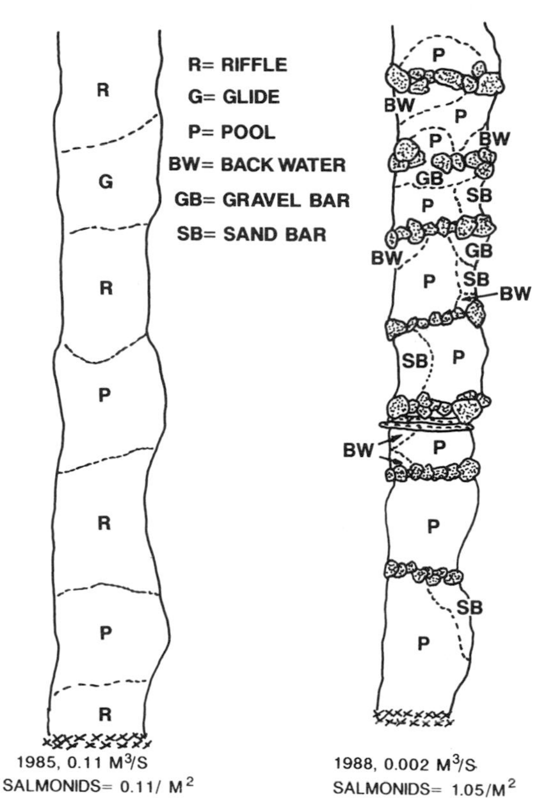

FIGURE 9.—Comparison of habitats in a 90-m reach of Grenshaw Creek before and after placement of boulder weirs.

The 122 alder–thinned Douglas fir structures had a variety of designs, all of which were successful to some degree. The major cause of failure was insufficient structural strength. The force of the current during peak flows broke the structures unless they were well armored upstream by deposits. Even when broken, the structure materials usually remained along or in the channel, providing some habitat improvement.

Only 8 of the 64 of the fence post structures remained intact, and 18 were substantially gone. One common problem was rotation of the upright posts. The posts were built slanting upstream so water pressure would push them against the spanning log and stream bottom. When the substrate was firm, the posts remained stable. However, when substrates were eroded and the posts were no longer supported by the channel bottom, the posts rotated downstream and became nonfunctional for retaining debris and deposits.

All six of the log cabin structures built to protect exposed banks remained in place, although one was inundated by the pond created by a downstream beaver dam. One structure was built too far from the bank and not backed by sufficient brush, and eddy action caused some additional erosion at high flows.

The oldest alder and Douglas fir structures were in place 3 years when evaluated, so longevity of the structures is not known. Alder trees that fell naturally into district streams have been observed for 10 years and remain stable and functioning, so longevity of these structures could exceed 10 years. Of the structures built, those imbedded in the stream and remaining wet all year had the greatest integrity, whereas those exposed for extended periods dried out, became weaker and brittle, and were much likelier to fail.

Six structures were created with large woody debris, three from root wads and three from short, large logs. One of the root wads, plus one large log not included in this summary, were lost when the anchoring cables were stolen and the debris floated away during subsequent high flows. One root wad was partially isolated by a channel change. The remaining four debris structures were in place and functioning when surveyed in 1988.

All 18 pools blasted were intended to improve fish passage. All were successful in this regard, although their final appearance was not as expected. One pool was covered by a large tree that fell into it, creating a logjam that flooded a second blasted pool upstream. Seven pools were partially filled in by deposits. The large rock fragments created when the pools were blasted remained in the channel, where they slowed currents and caused sediment deposition. The result was a series of short rapidlike areas, which were successful in breaking up velocity barriers.

Thinned Douglas fir trees proved moderately successful in reducing erosion along exposed banks. The technique was developed in eastern Oregon with juniper trees (Sheeter and Claire 1981). The Douglas firs successfully created instream cover when the butt ends were anchored on the bank and the trees were thrown into the stream. Needles were lost within months, but debris was trapped in brushy parts of the trees, creating good cover.

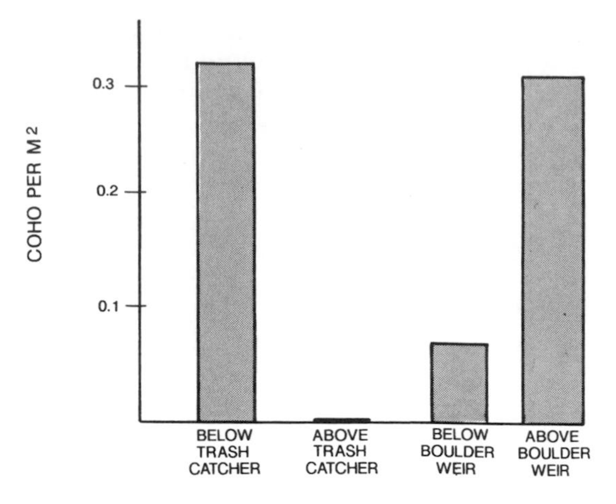

FIGURE 10.—Coho salmon densities above and below structures on North Fork Whittaker Creek.

Fish

Because of funding and time limitations, measurements of existing fish communities were limited. Visual estimates, made concurrently with habitat inventories and structure evaluations, indicated increases in juvenile salmonids in project locations, particularily in association with cover. In summer, the fish particularily used pools behind structures.

Low winter flows in 1987 and 1988 allowed electrofishing during cold periods. Preliminary results reported by Armantrout (1988) indicated that juvenile fish used structures for overwintering. During the summer, juvenile salmonids are highly visible in pools, glides, and other quieter waters or shallow riffles. Project areas, where pools and other habitats increased, appeared to have increased juvenile populations. In winter, fish seek cover. Coho salmon, steelhead, and cutthroat trout used undercut banks, root masses, and woody debris in pool areas behind all types of structures for winter habitat; they also used the alder and thinned Douglas fir structures as cover (Figure 10). Only a few steelhead used boulder weirs and gabions as cover.

Spawning counts of adult salmon and steelhead were begun on selected index areas in 1983 and have now been extended to most suitable habitat in the Siuslaw River basin managed by BLM. Most project work was done in 1985 or later, so there has not been sufficient time since completion of the projects to document changes in the size of returning runs in project areas because most of the fish have a 3–5-year life cycle. By

TABLE 4.—Fish Creek redd counts.

Season	Chinook salmon	Coho salmon	Steelhead
Index area 1 (reference)[a]			
1983–1984	0	6	6
1984–1985	0	3	4
1985–1986	10	45	10
1986–1987	0	16	6
1987–1988	0	11	6
1988–1989	0	6	1
Alder index area (reference)[a]			
1983–1984	2	12	3
1984–1985	1	No count	4
1985–1986	0	7	3
1986–1987	4	13	3
1987–1988	0	6	11
1988–1989	0	11	2
Project 1 (13 gabions built summer 1984)			
1983–1984	No count	1	3
1984–1985	3	0	16
1985–1986	0	21	15
1986–1987	1	22	17
1987–1988	0	12	7
1988–1989	5	5	3
Project 2 (7 gabions built summer 1984)			
1983–1984	No count	0	2
1984–1985	0	2	7
1985–1986	2	15	11
1986–1987	6	15	6
1987–1988	0	9	1
1988–1989	15	6	4

[a]Index-1 and alder-index areas are downstream from the project areas and have good to excellent habitats.

1988, numbers of fish using project areas were greater than the numbers using the areas prior to project work, in part because of a shift in habitat selection, and in part because of increases in spawning adults in the first years after projects were completed (Table 4). During the 1987–1988 and 1988–1989 spawning seasons, there was a sharp decline in escapement of coho salmon and steelhead, which, together with low flows due to drought and an increase in illegal harvest, made assessment of project use by fish difficult.

Beavers

Beavers built dams on top of 22 structures; dams ranged in height from a few inches to over 4 feet. In Fish Creek, 9 of the 34 gabions were topped by beaver dams, as well as two alder structures and one large-boulder weir. Eight structures were inundated by ponds created behind beaver dams built on top of structures.

Because of the lack of structure in the streams in the project area, and the extensive exposure of bedrock substrate, there is a high failure rate of beaver dams during peak flows. I surmise that the beavers build on our structures because the structures provide footing to secure their dams. Although no special study was made, we observed that a high percentage of the beaver dams in several streams, and all the beaver dams in two streams, were built on or in association with our project structures.

Discussion

Instream structure is integral to the creation and maintenance of aquatic habitat in the streams of western Oregon. The least costly and and most efficient way to develop and maintain good fish habitat is through management of riparian areas so that large trees can fall into streams. As a result of past management activities, few riparian areas retain the large conifers that provide the best habitat. It requires 100 years or more to grow conifers large enough to remain in place in stream channels for extended periods, although smaller conifers and some hardwoods will provide structure for shorter periods.

As a result of changes in management of riparian areas along District streams, including withdrawal from timber harvest of riparian areas along third-order and larger streams, existing riparian areas with larger trees will continue to be sources of woody debris. Riparian vegetation is recovering along streams where trees have been harvested, but it will take 50–100 years before trees become large enough to remain in place after falling into the stream channel.

Until natural sources recover sufficiently to provide structure, placement of structures in streams is one method for restoring habitat lost with removal of natural woody debris. The projects discussed in this paper represent several types of structures that can be used in restoring streams. The structures were designed to mimic natural structures in entrapping sediments, altering the types and distribution of habitats, and adding complexity to channels. The restructuring creates habitat used by anadromous salmonids, the target species in Oregon. All types of structures we tried improved aquatic habitat. As shown by Table 3, the physical conditions in the streams were changed, so there were increased pool areas and depths, sediment deposition, nursery areas, complexity, and cover, as well as reduced areas of exposed bedrock and faster flows.

Large-boulder weirs were the most successful of the structures used. Twenty-nine of the 34 weirs remained fully intact and made significant contributions to instream habitat. Boulders from the two weirs that partially broke up remained in the stream and continued to provide cover for fish. Because of their size, the boulders could be used in larger streams where the other types of structures we used were less suitable. Placement is very flexible and can be closely coordinated with channel conditions and other structures. Because of the differential size and shape of the boulders and the flexibility of their placement, they produced a greater complexity of habitats than the other types of structures, and they did so at moderate costs. Using two rubber-tired front-end loaders, we were able to build up to 175 linear meters of weirs of large boulders a day at a cost of $21 to $30 per linear meter. Capabilities of our equipment limited the size of the boulders we could use; in Esmond Creek, a sixth-order stream, we would have tried boulders 2 m in diameter or larger if equipment had permitted. The larger boulders would have created more pool habitat and increased deposition, and would have been more stable. Boulders from weirs that do break remain in the stream channel and continue to provide habitat.

Small-boulder weirs had variable success. Later weirs were more successful than earlier efforts because we learned how to build to reduce the effects of currents and downstream eddies. All small-boulder weirs incorporated on-site rock; thus when weirs did breach, the only result was localized redistribution of the rocks. The weirs were all made by hand by volunteers. Two volunteers could make up to 10 weirs per day, each up to 7 m long. If the value of the labor is not added, projects using volunteer labor cost less than $1 per linear meter. Channel changes due to the small-boulder weirs were similar to those for the large-boulder weirs, but on a smaller scale. Pools, in particular, were smaller and shallower. The most common changes were increased deposition behind the weirs and a narrowing and deepening of the channel. Individual small-boulder weirs were less successful than such weirs in a series with other boulder weirs or other types of structures. Interactions among the structures not only increased the complexity and quality of habitats, but also helped to control flow patterns and reduce current energies during high flows when most damage to structures occurs.

All blasting we did was to improve passage by breaking up high-velocity flows over bedrock or to create pools to facilitate fish jumps over barriers. All pools we blasted ultimately achieved the desired objective except for the two affected by a fallen tree. Six of the 18 pools we blasted became at least partially filled in, but even though no pool remained, the area blasted altered flow patterns to make it easier for migrating salmonids to swim through. The 10 pools that remain are used for rearing by resident and juvenile anadromous salmonids. Blasting was used to create pool habitat in bedrock areas in the District in 1969 through 1973. During the drought in 1987 and 1988, we observed that several of these blasted pools had little or no flow over the bedrock between pools, and the pools, lacking any surface or groundwater flow, became stagnant, causing loss of fish. Pool blasting appears to produce best results where good surface flows are maintained and when done in conjunction with habitat features that ensure adequate flows.

Large woody material has been used in several successful habitat restoration projects (House and Boehne 1987). Two of our structures were lost when anchoring cables were stolen. All structures were placed on the side of the channel to create cover and scour pools, to protect exposed banks, and to improve habitat through creation of scour pools and deposition of gravels in quiet water areas behind the structures. Since the large woody debris was on-site and placed without heavy equipment, actual costs were limited to hand equipment and anchoring cables and wire.

The most diverse results were obtained with the alder and thinned Douglas fir structures. More of these structures failed than others, but we also produced some of the best habitat with them. When remaining intact, the structures trapped sediments and created a diversity of pool and off-channel habitat. In addition, these structures were used much more often than boulders or gabions as cover by juvenile salmonids in all seasons.

The alder and thinned Douglas fir structures are easy to build and use on-site materials. A crew of six volunteers could build four or more structures in 1 d. With contributed labor, costs were $10 to $25 per structure. Longevity of the structures is a question, because alder dries quickly and becomes brittle when exposed to air continuously for several months. Observations made in the district on alder and bigleaf maple *Acer macrophyllum* that fell naturally into streams suggest that logs that remain submerged remain intact and provide stream structure for at least 10 years. Our structures that became exposed for extended periods during the year did become dry and spongy, and broke easily, but structures that have remained submerged for 3 years show little indication of weakening.

The main problem was making structures stout enough to withstand peak winter flows. This problem was greatest during the first storm events after the structures were built, and lessened as debris and alluvium accumulated behind the structures. As with other structures, it was important to anchor the ends to prevent erosion around them during high flows.

All 50 gabions were still in place after periods up to 5 years, although some sustained damage from woody debris or erosion. Gabions have been used successfully by others (Anderson and Cameron 1980; Anderson 1981). The gabions were quite successful in retaining sediments, creating or maintaining downstream plunge pools and undercut bank cover, and controlling the direction of flow. We used volunteers and temporary employees to build the gabions with on-site rock at a cost of $30 to $45 per linear meter.

Gabions were the most expensive of the structures we tried but were adaptable to a range of stream sizes and habitat needs. As Anderson has shown, the gabions can be used in a variety of designs and for a range of habitat needs. I used the gabions most often in larger streams where their bulk resisted the force of currents well, and in inaccessible locations where heavy equipment could not be used.

One problem encountered was vandalism. As mentioned, structures were lost when anchoring cables were stolen. In addition, several alder structures and some logs used to protect the ends of structures were cut up as firewood and hauled away, exposing structures to damage or loss.

We have insufficient information to document overall changes in fish use of project streams as a result of project work. Spawning surveys in the past three seasons provide a good record of adult anadromous salmonid use of the project areas compared with fish use elsewhere in the project streams and the Siuslaw River basin. However, we lack a similar level of preproject information for the streams. As shown in Table 4, we had spawning activity on two project sites comparable with two downstream index areas with good to excellent habitat. For the projects overall, we observed an increase in spawning activity compared with adjacent unimproved areas through the 1986–1987 spawning season (Bureau of Land Management, Eugene District, unpublished data). Major declines in the escapement of coho salmon and steelhead, and extremely low flows during chinook salmon spawning seasons, in the 1987–1988 and 1988–1989 spawning periods make assessment of the impact of projects on spawning difficult.

Our data on juvenile salmonid populations in project areas are more limited than spawning data. All population assessments prior to project work were visual estimates made during habitat inventories. Estimates following project work showed increased use of projects, particularily pool areas (unpublished data). Bob House in the adjacent BLM Salem District found an increase in juvenile use of habitat projects compared with use prior to project implementation (House and Boehne 1985; R. A. House, personal communication). Populations in our project areas, based on visual estimates, were similar to those in nonproject areas with good to excellent habitat, and greater than in most nonproject habitats.

We have found that juvenile salmonids and adult cutthroat trout use the alder and thinned Douglas fir structures for cover during both summer and winter. Gabions and boulder weirs are seldom used directly for cover, except in North Fork Leopold Creek where other types of cover are essentially absent, and juvenile steelhead and cutthroat trout use the small-boulder weirs for cover. All pools behind and below all structure types were used by fish in summer. Winter use was much more limited, but occurred in quiet waters, in pools behind structures, or below alder structures where suitable cover was available. Figure 10 shows some of the winter habitat preferences in North Fork Whittaker Creek. Comparisons were also made with areas where no project work was done, and they showed a decided preference by wintering juvenile salmonids for improvement structures (unpublished data).

Conclusions

All structure types we used improved fish habitat in our streams. The structural features provided many of the same functions as large woody debris, including retention of sediments, creation of habitat diversity, and increased cover. Comparison of stream reaches before and after project work documented changes in channel characteristics considered beneficial to anadromous salmonids, the target species.

Decisions on the type of structures to be used were influenced by the size of a stream, access to stream reaches, availability of materials, and cost.

Some structures were built alone, but I found the best results were from a series of structures along a stream reach, particularily when I increased the number of structures and placed them closer together. The interaction among structures and the better control of flow patterns increased the number and diversity of habitats. Some of the simplest designs worked as well as or better than more expensive options. Of major importance was an understanding of stream flow patterns, including peak winter flows, in reference to placement and design of structures. Success also improved when I used existing structures and channel patterns to place new structures.

Our evaluations were based primarily on changes in the physical characteristics of the channel. Because of budget and time constraints, we were unable to do preproject assessment of adult and juvenile salmon and trout populations. Fish population monitoring has been primarily for spawning adults. Comparison of spawning activity in project areas with activity elsewhere in the Siuslaw River basin could demonstrate the effectiveness of projects in attracting fish; however, the sharp decline in runs the past two seasons greatly reduces the effectiveness of these comparisons. As a general statement, we see a trend to increased use of the structures. Habitat use studies of juvenile salmonids, summarized in part by Armantrout (1988), have shown an increase in juveniles, particularily in winter, in project areas compared with nonproject areas. The availability of pools and cover appears to be the principal attractant for the juveniles. Increased numbers depend on adequate seeding by adults, plus an increase in survival of juveniles, although we are unable to document this. Monitoring of structures and fish populations will be continued.

References

Anderson, J. W. 1981. Anadromous fish projects 1981 USDI–BLM Coos Bay District. Pages 109–114 *in* T. J. Hassler, editor. Proceedings: propagation, enhancement, and rehabilitation of anadromous salmonid populations and habitat in the Pacific Northwest symposium. Humboldt State University, Arcata, California.

Anderson, J. W., and J. J. Cameron. 1980. The use of gabions to improve aquatic habitat. U.S. Bureau of Land Management, Technical Note 342.

Anderson, J. W., and L. Miyajima. 1975. Vincent Creek fish rearing pool project. U.S. Bureau of Land Management, Technical Note 274.

Armantrout, N. B. 1989. Use of various types of habitat improvements by salmonids in winter. Pages 142–147 *in* B. G. Shepherd, editor. Proceedings, 1988 Northeast Pacific chinook and coho workshop. American Fisheries Society, North Pacific International Chapter. (Available from B.C. Ministry of Environment, Penticton.)

Armantrout, N. B. 1990. Cost-effective labor and equipment. Section 8 *in* R. House, J. Anderson, P. Boehne, and J. Suther, editors. Stream rehabilitation manual. American Fisheries Society, Oregon Chapter, Corvallis.

Bilby, R. E. 1984. Removal of woody debris may affect stream channel stability. Journal of Forestry 82: 609–613.

Bisson, P. A., J. L. Nielsen, R. A. Palmason, and L. E. Grove. 1982. A system of naming habitat types in small streams, with examples of habitat utilization by salmonids during low streamflow. Pages 62–73 *in* N. B. Armantrout, editor. Acquisition and utilization of aquatic habitat inventory information. American Fisheries Society, Western Division, Bethesda, Maryland.

Boehne, P., and R. A. House. 1983. Stream ordering: a tool for land managers to classify western Oregon streams. U.S. Bureau of Land Management, OSO T/N:OR-1.

BPA (Bonneville Power Administration). 1985. Natural propagation and habitat enhancement. Volume 1—Oregon. U.S. Department of Energy and BPA Division of Fisheries and Wildlife, DOE/BP-726, Portland, Oregon.

Bryant, M. D. 1982. Organic debris in salmonid habitat in southeast Alaska: measurement and effects. Pages 259–265 *in* N. B. Armantrout, editor. Acquisition and utilization of aquatic habitat inventory information. American Fisheries Society, Western Division, Bethesda, Maryland.

Everest, F. H., and seven coauthors. 1985. Salmonids. Pages 199–210 *in* E. R. Browne, editor. Management of wildlife and fish habitats in forests of western Oregon and Washington. U.S. Forest Service, Pacific Northwest Region, R6-F&WL 192-1985, Portland, Oregon.

Hall, J. D., and C. O. Baker. 1982. Rehabilitating and enhancing stream habitat. 1. Review and evaluation. U.S. Forest Service General Technical Report PNW-138.

House, R. A., J. Anderson, P. Boehne, and J. Suther, editors. 1990. Stream rehabilitation manual. American Fisheries Society, Oregon Chapter, Corvallis.

House, R. A., and P. L. Boehne. 1985. Evaluation of instream enhancement structures for salmonid spawning and rearing in a coastal Oregon stream. North American Journal of Fisheries Management 5:283–295.

House, R. A., and P. L. Boehne. 1987. The effect of stream cleaning on salmonid habitat and populations in a coastal Oregon drainage. Western Journal of Applied Forestry 2(3):84–87.

Marzolf, G. R. 1978. The potential effects of clearing and snagging on stream ecosystems. U.S. Fish and Wildlife Service, Biological Servies Program FWS/OBS-78/14.

Murphy, M. L., J. F. Thedinga, K. V. Koski, and G. B.

Grette. 1982. A stream ecosystem in an old-growth forest in SE Alaska: part V. Seasonal changes in habitat utilization by juvenile salmonids. Pages 89–98 *in* W. R. Meehan, T. R. Merrell, Jr., and R. A. Hanley, editors. Fish and wildlife relationships in old growth forests. American Institute of Fisheries Biologists, Morehead City, North Carolina.

Orsborn, J. F., and J. W. Anderson. 1986. Stream improvements and fish response: a bio-engineering assessment. Water Resources Bulletin 22:381–388.

Paquet, G. 1986. Guidelines for the improvement and restoration of fish habitat in small streams. Ministère de la Chasse et de la Pêche, Service des Etudes Ecologiques, Quebec City, Quebec.

Patterson, J. H., editor. 1986. Proceedings of the workshop on habitat improvements. Canadian Technical Report of Fisheries and Aquatic Sciences 1483.

Robson, D. S., and H. A. Regier. 1968. Estimation of population number and mortality rates. IBP (International Biological Programme) Handbook 3:124–158.

Ruediger, R. 1990. Blasting considerations for fisheries enhancement projects. Section 11 *in* R. House, J. Anderson, P. Boehne, and J. Suther, editors. Stream rehabilitation manual. American Fisheries Society, Oregon Chapter, Corvallis.

Sedell, J. R., and K. J. Luchessa. 1982. Using the historical record as an aid to salmonid habitat enhancement. Pages 210–223 *in* N. B. Armantrout, editor. Acquisition and utilization of aquatic habitat inventory information. American Fisheries Society, Western Division, Bethesda, Maryland.

Sedell, J. R., and F. J. Swanson. 1982. Ecological characteristics of streams in old-growth forests of the Pacific Northwest. Pages 9–16 *in* W. R. Meehan, T. R. Merrell, Jr., and T. A. Hanley, editors. Fish and wildlife relationships in old growth forests. American Institute of Fisheries Research Biologists, Morehead City, North Carolina.

Sheeter, G. R., and E. W. Claire. 1981. Use of juniper trees to stabilize eroding stream banks in the South Fork John Day River. U.S. Bureau of Land Management, Technical Note T/N-OR-1, Portland, Oregon.

American Fisheries Society Symposium 10:150–159, 1991

Habitat and Channel Changes after Rehabilitation of Two Coastal Streams in Oregon

ROBERT HOUSE, VAL CRISPIN, AND JOAN M. SUTHER
U.S. Bureau of Land Management, Salem District Office
1717 Fabry Road SE, Salem, Oregon 97306, USA

Abstract.—Two stream reaches of the upper Nestucca River basin, severely degraded by a dam failure, roads, and stream cleaning, were structurally modified in 1986. In the upper Nestucca River, which drains 51.3 km^2, 77 structures that fully spanned the stream and 120 that partially spanned it, as well as 9 off-channel areas, were constructed along 8.5 km of stream. In Elk Creek, a drainage of 26.6 km^2, 37 fully spanning and 55 partially spanning structures, including the creation of 7 off-channel areas, were constructed along 1.9 km of stream. Follow-up evaluations in 1987, after two storms caused flows greater than would be expected to recur with a 1-year interval, showed substantial changes in habitat and channel conditions favoring anadromous salmonid use. In the upper Nestucca and Elk Creek, respectively, stream surface area increased 14% and 57%, water volume increased over 60% on each stream, and gravel substrate increased 44- and 50-fold. Pool habitat, the most important rearing component for salmonids, increased significantly between pre- and postenhancement evaluations. Pool area increased approximately two- and fivefold, and pool volume increased from 25 to 48% and from 14 to 46%, respectively, in the upper Nestucca River and Elk Creek.

Past human activities in Oregon watersheds have altered the physical characteristics of stream channels. Historical accounts (Sedell et al. 1982; Baker et al. 1986) indicate presettlement streams were filled with large woody material, which decreased stream velocity, promoted development of pools, trapped spawning gravels, and created off-channel habitat. In western Oregon, fires, floods, and forest management practices, particularly the removal of woody debris from stream channels (House and Boehne 1987), have contributed to the formation of homogeneous stream reaches. Streams dominated by riffles are common in coastal Oregon today. These conditions will continue to limit stream habitat diversity unless riparian areas are protected and instream habitat is rehabilitated to improve salmonid production. Stream rehabilitation projects should be designed to restore the quality and quantity of fish habitat. Evaluations of such projects report increased habitat diversity (Ward and Slaney 1979; House and Boehne 1985, 1986; Everest et al. 1987). Rehabilitation of northwestern coastal streams is a relatively recent but widespread practice, and few evaluations have been undertaken. The primary objective of this study is to quantify changes in stream habitat by examining certain physical characteristics of stream channels prior to and following the installation of stream rehabilitation structures.

Study Area

The upper Nestucca River basin, with its major tributary system, Elk Creek, is in Tillamook County, Oregon (Figure 1). The basin receives 200–300 cm of precipitation annually, primarily from 1 October through March 31 (Franklin and Dyrness 1973). Elevation ranges from 340 to 1,065 m. The riparian overstory vegetation is dominated by red alder *Alnus rubra,* with substantial amounts of Douglas fir *Psuedostuga menziesii,* western red cedar *Thuja plicata,* and bigleaf maple *Acer macrophyllum.* Major understory riparian species are salmonberry *Rubus spectabilis,* thimbleberry *Rubus parviflorus*, and nettle *Urtica* spp. The basin is characterized by tuffaceous sediments interbedded with basaltic breccia and flow rock (Baker et al. 1986). Soils are commonly silty with gravel or (less often) clay topsoil.

The upper Nestucca River and Elk Creek support populations of spring and fall chinook salmon *Oncorhynchus tshawytscha,* coho salmon *O. kisutch,* summer and winter steelhead *O. mykiss,* sea-run and resident cutthroat trout *O. clarki,* Pacific lamprey *Lampetra tridentata,* and sculpins *Cottus* spp.

Past fire, floods, road building, logging, and stream-cleaning activities have contributed to stream channelization, loss of spawning gravel, loss of off-channel and pool habitats, and reductions in the number of large conifers in riparian

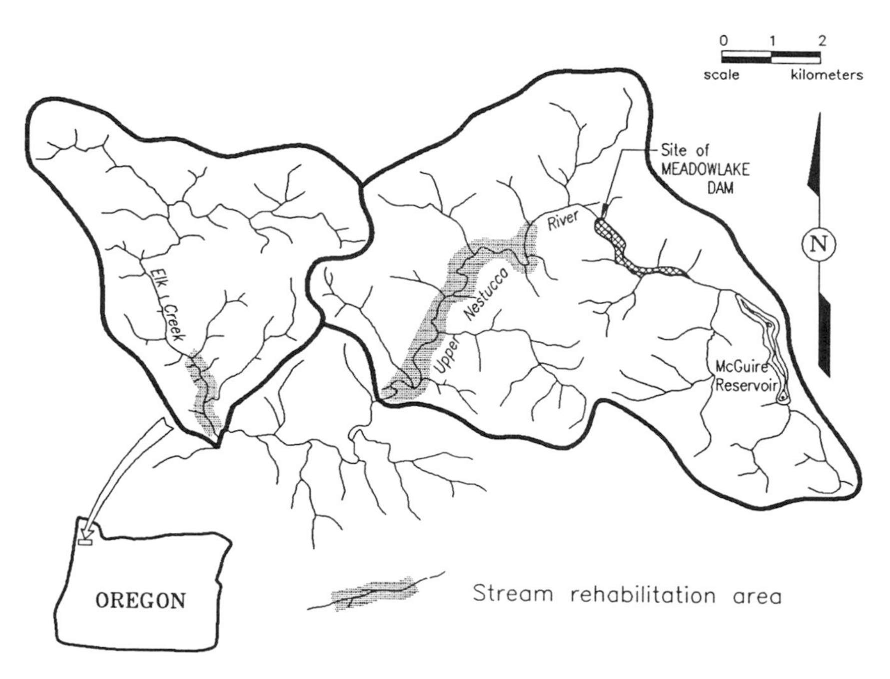

FIGURE 1.—Upper Nestucca River basin, Oregon, showing study site locations.

areas. Forest fires repeatedly burned portions of the basin from 1919 to the 1950s. In the aftermath of the fires, increases in erosion, runoff, stream sediment, and stream temperature certainly occurred (Baker et al. 1986). Floods occurred in 1945, 1950, 1955, 1964, 1965, and 1972.

Access roads, completed around 1960, preceded intensive timber harvesting in the basin. The close proximity of roads to waterways reduced riparian and floodplain habitat, confined the stream channel, and contributed sediment from cut-slope and fill failures; consequently, spawning habitat was reduced in quality and quantity (Baker et al. 1986). Over 50% of the timber in the basin has been logged in the past 20 years and many riparian zones lack large coniferous species.

Upper Nestucca River

The upper Nestucca River, which includes 14.4 km of stream, drains 51.3 km^2. Streamflow ranges from 0.11 to 0.22 m^3/s in late summer to peak winter flows of 28.3 to 84.9 m^3/s. Two storm events during the winter of 1986–1987 produced flows exceeding 39.6 m^3/s.

In November of 1962, the main channel was scoured severely by a major flood and debris torrent caused by the failure of Meadow Lake Dam (Baker et al. 1986). Habitat changes resulting from the flood probably reduced the capability of the upper Nestucca River to produce anadromous salmonids. Visual surveys conducted in 1948 and 1954 (ODFW: Oregon Department of Fish and Wildlife, unpublished data) indicated that patches of good-quality spawning areas occurred throughout the upper Nestucca River prior to dam failure. These observations were supported by aerial photographs taken in July 1962, prior to dam failure, which revealed the presence of large woody debris (LWD) and off-channel habitat and extensive graveled areas. Additional surveys in 1962 (ODFW, unpublished data) conducted after dam failure, documented the nearly complete loss of spawning gravel and straightening of the channel.

Although accurate historical records of salmonid use of the upper Nestucca River are limited, it appears coho salmon use declined substantially after dam failure. The only count prior to the Meadow Lake Dam failure was made on 28–29 December 1951, when 21 coho salmon adults per kilometer were observed but, because of turbid water, 133 coho salmon/km were estimated in 7.5 km of the Upper Nestucca River (ODFW, unpublished data). This estimate was recorded in a year when coho salmon escapement numbers were around 2.9 times higher than the 10-year (1950–1960) average of 14/km (peak counts) in the Nestucca River drainage (Skeesick 1973). Later peak counts (1984–1986) in the same 7.5 km revealed an average of only 2 coho salmon/km (BLM: U.S. Bureau of Land Management, unpublished data) during a period when average north coast Oregon basin counts were 40% of the 1950–1960 levels (S. Jacobs, ODFW, personal communication). Thus, coho spawning declined approximately 75% in the upper Nestucca River after dam failure and other management activities.

Elk Creek

Elk Creek, which includes 4.5 km of stream up to an impassable falls, drains 26.6 km^2 of recently harvested or second-growth timberland. Flows range from summer lows of 0.03–0.14 m^3/s to winter flood events exceeding 68.0 m^3/s. In the winter of 1986–1987, two storms produced flows of 19.8 and 22.7 m^3/s (J. Fogg, BLM, personal communication). Stream-cleaning activities following the 1972 flood caused major changes in instream habitat conditions. A habitat survey on Elk Creek in 1958 (ODFW, unpublished data) estimated that 5% of the substrate was suitable for spawning, most of it associated with the 13 small logjams recorded in this 1.9-km reach. Use of the creek by salmonids, specifically coho salmon, seems to have decreased following the 1972 flood; adult counts declined from 150/1.6 km in 1970 to an average of 7/1.6 km in 1983–1986 (BLM, unpublished data).

Methods

Habitat Surveys

Habitat assessment from 1984 through 1987 was based on habitat types described by Bisson et al. (1982). Our method separates by habitat type and then quantifies physical aspects of these types by length, width, average and maximum depth, and dominant substrate and cover types. Length and width of each microhabitat unit were measured to the nearest 30 cm with a hip-chain. Average depth (nearest 3 cm) was derived from a series of measurements taken within each microhabitat unit. Dominant substrate in each habitat type was determined visually by size class, and major components of stream cover were recorded (e.g., boulder edge, deep pool, undercut banks). Large, stable woody material (greater than 15 cm in diameter and 3 m in length) was also recorded. Length (to nearest 30 cm) and diameter (to nearest 3 cm) of logs and rootwads were measured. After the first year, individual structures were assessed to quantify the dimensions of newly created habitats and amounts of deposited sediment. To determine the physical factors limiting coho production in the study streams, the technique developed by Reeves et al. (1990) was used to identify potential seasonal limiting factors. This technique, which relies on habitat and smolt factors to estimate the survival rate between life history stages, determines the amount of habitat needed at a given life history stage to support or produce the potential summer population.

Upper Nestucca River.—Pre- and posttreatment surveys of the upper Nestucca River were conducted in the summers of 1984 and 1987, respectively. A review of aerial photographs taken in July 1962 (prior to Meadow Lake Dam failure) provided information on channel characteristics, including stream width, off-channel habitat, and the presence of large woody debris.

Elk Creek.—Pre- and posttreatment surveys of Elk Creek were conducted in the summers of 1985 and 1987, respectively (Figure 2). Surveys conducted in 1970 and 1980 were used to assess habitat conditions. Although earlier survey meth-

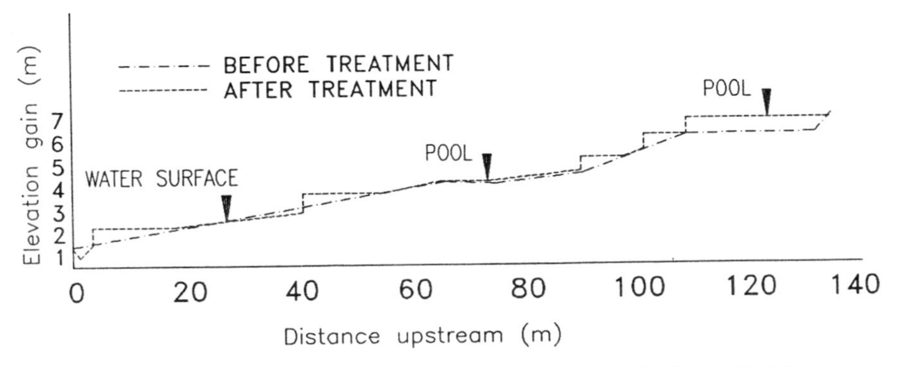

FIGURE 2.—Stream profile before and after treatment in a 140-m section of Elk Creek. (The two labeled pools were present prior to treatment.)

FIGURE 3.—Typical series of fully spanning structures installed on the upper Nestucca River.

ods differed from those conducted in 1985 and 1987, all were designed to determine habitat complexity based on a variety of physical characteristics. In the 1970 and 1980 surveys, variables measured were stream width, spawning gravel, maximum pool depth, pool area, and pool/riffle ratio.

Treatment

Boulders, logs, and rootwads obtained from the floodplain, the channel, and off-site sources, were installed with a track-mounted hydraulic excavator in August and September 1986. Structures were installed mainly on straight, wide-riffle and shallow-glide sections of the main channel or in newly created off-channel (secondary channels) areas. Two or more structures that fully spanned the stream were constructed at each site. Structures were installed with no rigid design standards in an attempt to mimic existing natural structures. Treatment included fully spanning structures in diagonal, semicircular, and downstream V- and T-shaped configurations, as well as partially spanning (⅓ to ½ of channel width) deflectors and cover structures, consisting of rootwads, boulder clusters, and combinations of logs and boulders (Figure 3). Deflectors and cover structures were usually installed upstream from fully spanning structures to act as both fish cover and scour agents to maintain pool depth. Fully spanning structures averaged 1–1.5 m in height. Manual labor was required to stabilize large log structures. Logs were connected to boulders or bedrock with steel cable (1.3 cm in diameter), and cable ends were secured with polyester adhesive resin in 1.5-cm holes drilled in boulders or bedrock.

Upper Nestucca River.—A total of 3,121 m (37%) of 8,458 m of the Upper Nestucca River were treated with rehabilitation structures at a cost of US$55,000. An aggregate channel length of 1,342 m (16%) was not considered due to gradient or access constraints, which left 3,995 m (47%) of untreated stream for comparison. Altogether 91 fully spanning structures and 106 partially spanning and midchannel structures were installed at 17 sites, and nine secondary channels with a total length of 1,047 m were created in the upper Nestucca River (Table 1).

Elk Creek.—A total of 853 m (45%) of 1,910 m of Elk Creek were treated by rehabilitation structures at a cost of US$25,000. An aggregate of 645 m (34%) was not considered due to gradient or access constraints, which left 412 m (21%) of untreated stream for comparison and controls. Altogether, 46 fully spanning structures and 46 partially spanning and midchannel structures were installed at 19 sites, and seven secondary channels were created with a total length of 636 m in Elk Creek (Table 1).

Results

Evaluation of the treatment measures showed major changes in channel characteristics, habitat types, substrate composition, water volume and quantity of large woody debris (LWD).

Channel Characteristics

The installation of fully spanning structures, which widened and deepened the channel, increased substantially the wetted surface area and the water volume capacity of both systems. For-

TABLE 1.—Types and numbers of structures installed in upper Nestucca River and Elk Creek in 1986.

	Upper Nestucca River		Elk Creek	
Structure type	Main channel	Secondary channel	Main channel	Secondary channel
Fully spanning	77	14	37	9
Partially spanning	27	15	7	1
Midchannel	44	20	21	17

TABLE 2.—Pool habitat created by fully spanning structures installed in the upper Nestucca River and Elk Creek.

Pool type	Average surface area (m^2) per structure	
	Nestucca River	Elk Creek
Dammed	314	175
Plunge	34	8
Backwater	5	8
Scour	7	2
Secondary channel	22	1
Total	382	193

mer glide and riffle areas were converted into large dammed pools (Table 2). These structures also aggraded the bed and increased upstream water levels, thereby creating new off-channel areas and flooding vegetated stream margins.

Upper Nestucca River.—In the upper Nestucca River, after only 37% of the total channel length was treated, water surface area increased 14.3%, water volume increased 65% (Table 3), and average stream width increased 0.7 m.

Elk Creek.—In Elk Creek, surface area measured under similar flow regimes varied from 10,480 to 11,471 m^2 in preproject years (1970, 1980, and 1985) but increased after project construction in 1987 (Table 4). Increase in surface area was due to the creation of dammed pools in the main channel and the reopening of off-channel habitat. Main channel average width increased 39%, from 5.9 m in 1985 to 8.2 m in 1987. Average stream depth increased 43%, from 23 cm to 33 cm, resulting in a 130% gain in water volume after project construction. The 37 fully spanning main channel structures resulted in a series of stairstep drops totaling 25 m in elevation and accounting for 41% of the total 61-m drop through the study reach (Figure 2). The 35 dam pools created by fully spanning structures now contain more water (3,202 m^3) than all habitats combined prior to rehabilitation.

Habitat Types

Potential summer habitats (pool and off-channel areas) and winter habitats (dam and backwater pools and off-channel areas) for coho salmon showed the most substantial changes after treatment, increasing 57% and 14-fold respectively, in the upper Nestucca River, and 6-fold and 15-fold, respectively in Elk Creek. The amount of pool habitat created by individual fully spanning structures varied between the larger upper Nestucca River and the smaller Elk Creek (Table 2). Fully spanning structures created mainly dammed pools in both streams, total pool habitat averaging 382 m^2 per structure in the Nestucca River and 193 m^2 in Elk Creek. The relative costs for creating summer pool habitat varied from \$2.00/m^2 in the Nestucca River to \$3.50/m^2 in Elk Creek.

Upper Nestucca River.—Following treatment, total pool habitat in the upper Nestucca River doubled in surface area, and dammed, plunge, secondary channel, and trench pools increased

TABLE 3.—Habitat changes in the upper Nestucca River before (1984) and after (1987) the project.

Habitat type	Surface area (m^2)		Water volume (m^3)	
	1984	1987	1984	1987
Riffles				
Low-gradient	20,169	17,880	2,823	3,934
High-gradient	6,126	6,198	977	1,266
Rapids	1,931	1,120	311	230
Cascades, falls	1,188	687	192	148
Secondary channel	563	2,175	51	261
Total	29,977	28,060	4,354	5,839
Glides	21,208	10,950	4,242	3,504
Pools				
Dammed	131	24,744	76	11,877
Plunge	998	2,827	509	1,329
Lateral and under scour	11,233	2,655	5,617	1,301
Trench	3,463	6,828	1,732	3,482
Backwater	1,352	862	297	155
Secondary channel	346	1,591	69	350
Total	17,523	39,507	8,300	18,494
Grand total	68,708	78,517	16,896	27,837

TABLE 4.—Habitat changes on lower Elk Creek before (1985) and after (1987) the project.

Habitat types	Surface area (m^2)		Water volume (m^3)	
	1985	1987	1985	1987
Riffles				
High-gradient	2,754	3,154	578	883
Low-gradient	4,207	4,183	841	962
Rapids, cascades, falls	841	227	168	69
Secondary channel	120	1,734	12	191
Total	7,922	9,298	1,599	2,105
Glides	1,942	323	544	149
Pools				
Dammed	191	6,536	65	3,202
Plunge	871	282	272	133
Lateral and under scour	154	221	46	106
Trench	85	144	65	71
Backwater	150	505	24	121
Secondary channel	156	729	25	190
Total	1,607	8,417	497	3,823
Grand total	11,471	18,038	2,640	6,077

considerably, while lateral scour and backwater pools diminished in size (Table 3). In contrast, glide habitat was reduced by 50% and main channel riffle habitat by 13%. Pool depth showed little overall change, declining from 0.41 m to 0.39 m. However, the average depth of riffle habitat increased 35% from 0.17 m to 0.23 m. Although secondary channel habitat increased over 400% in surface area after treatment, the 1987 secondary channel habitat quantity still appears to be 50% below levels inferred from the 1962 aerial photographs. The most significant change was in dammed pool habitat; the total number of pools increased from 2 to 55 and area increased from 131 m^2 to 24,744 m^2.

Elk Creek.—Total pool habitat in Elk Creek increased fivefold in surface area following treatment. There were large increases in dammed, backwater, and secondary channel pools. Plunge pools decreased in numbers due to the rearranging of boulders into larger structures that prevented downstream scour. Main channel riffle habitat remained constant (Table 4).

Pool habitat showed substantial changes over the past 17 years in Elk Creek. In 1970, prior to the major flood of 1972 and subsequent stream cleaning, pool habitat made up 25% of the total available habitat (Figure 4). By 1985, pool habitat within the main channel had declined and accounted for only 14% (1,456 m^2) of the available habitat. However, following treatment, main channel pool habitat increased to 47% (7,688 m^2) of the available habitat (Table 4). Pool habitat following treatment increased in numbers (85 to 112) and in average size (19 to 75 m^2). Pool depths increased markedly after treatment. Maximum pool depths increased 52% from 1970 and 18% from 1985 measurements, while mean depths increased from 33 cm to 48 cm between 1985 and 1987, respectively. Dammed pool habitat showed the greatest change after treatment, increasing 34-fold. The amount of riffle habitat did not change markedly following treatment, but the overall percentage decreased from 67% to 42% (Table 4). Glide habitat decreased 83%, having been changed to dam pool habitat by structures.

Substrate Composition

Prior to treatment, most gravel in both the upper Nestucca River and Elk Creek was found in pockets, intermixed with larger material in low-gradient riffles. A few larger patches of gravel were located in glides and at the tail of pools. After treatment, large expanses of gravel and fines were deposited just behind and for considerable distances upstream from the fully spanning structures in both streams.

Upper Nestucca River.—Boulders, cobble, and rubble substrate decreased in surface area within the treated reaches of the upper Nestucca River. Conversely, smaller substrates increased dramatically; large gravel increased 30-fold, small gravel 126-fold, and sand 3-fold. Potential spawning gravel (8–16 cm, located at pool tails and low-gradient riffles and glides), which occupied less than 1% (400 m^2) of the substrate prior to treat-

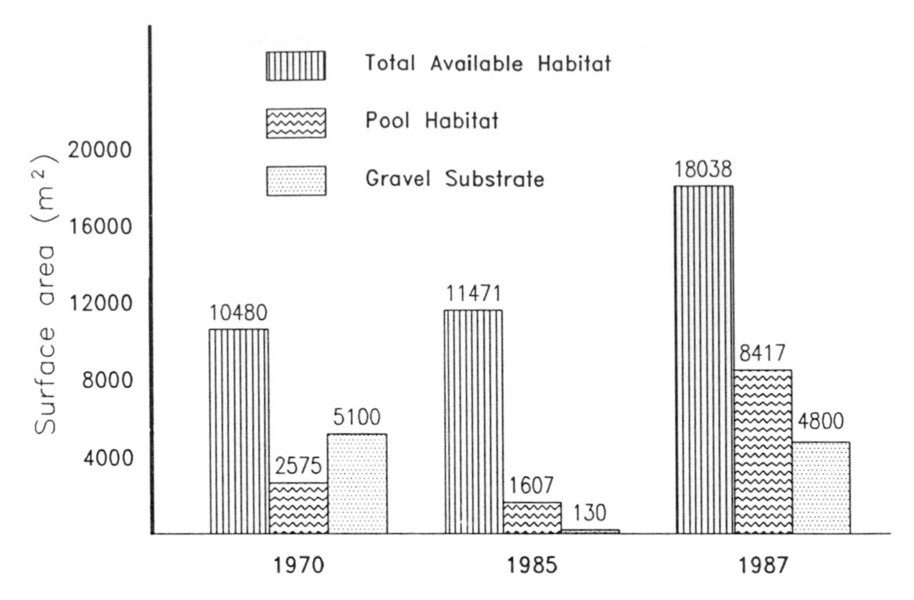

FIGURE 4.—Comparison of total available habitat, pool habitat, and gravel substrate in 1,910 m of Elk Creek for 1970, 1985, and 1987.

ment, increased to 9% (7,400 m^2) after treatment. Bedrock decreased slightly, due to deposition of smaller substrate behind structures installed on bedrock.

Elk Creek.—The amount of gravel substrate in Elk Creek changed substantially between survey years (Figure 4). In 1958 and 1970, prior to a major flood and intensive stream cleaning, about 5,100 m^2 of gravel (of which only about 10% was regarded as usable spawning gravel) was recorded in the study reach of Elk Creek. By 1985, however, only 2% of this gravel (130 m^2) remained in the reach. After treatment, total gravel amounted to 27% (4,800 m^2) of the substrate, and 1,600 m^2 (9%) were estimated to have usable spawning gravel. Structures increased large gravel 37-fold and small gravel and sand 7-fold. Gravel increased by 4,850 m^2; gravel trapped immediately behind the fully spanning main channel structures amounted to 1,450 m^2, an average of 59 m^2 per structure. An additional 3,170 m^2 of gravel were deposited just below and considerable distances upstream from these structures. Newly deposited gravel was found at varying depths of up to 1 m. The amount of larger substrate (cobble, rubble, and boulders) and bedrock (9,500 m^2) was similar before and after treatment.

Large Woody Debris

The amount and size of LWD increased greatly after installation of wood structures in the upper Nestucca River and Elk Creek (Table 5). Altogether, 134 pieces of LWD with a combined length of 1,123 m were added to the upper Nestucca, and 78 pieces of LWD with a combined length of 338 m were added to Elk Creek. Cabling increased the stability of newly added wood structures.

Upper Nestucca River.—Interpretation of 1962 aerial photographs of the upper Nestucca River prior to the failure of Meadow Lake Dam but after initial road construction indicated densities of large single pieces or small logjams (complexes of LWD greater than 8 m in length and 0.60 m in diameter) of 0.8 structures per 100 m of stream. After dam failure, continued road building, stream cleaning, and two major floods, the densities of LWD influencing the channel decreased 50% to approximately 0.4 structure per 100 m of stream in

TABLE 5.—Comparison of large woody debris before and after stream enhancement work on upper Nestucca River and Elk Creek.

Large woody debris	Upper Nestucca River		Elk Creek	
	1984	1987	1985	1987
Number of pieces	262	423	76	181
Average diameter (m)	0.46	0.64	0.30	0.66
Total length (m)	1,101	2,192	265	697
Average length (m)	4.2	5.2	3.4	3.9
Pieces per 100 m	3.1	5.0	4.0	9.5
Pieces >8 m long per 100 m	0.4	1.0		

1984. This density would have been even lower if the upper Nestucca River did not still flow through a predominantly older-growth coniferous riparian zone that continues to contribute LWD to the stream channel. With the addition of the fully spanning log and boulder structures (which hydraulically act like LWD) during rehabilitation, the density of large main channel structures (greater than 8 m in length) increased to 1.0 per 100 m of stream. This compares favorably with the quantity of structures that occurred naturally in the upper Nestucca River in 1962. Smaller pieces of LWD (3 m in length and 15 cm in diameter) also increased in total numbers and average length (Table 5) as winter flows deposited them behind newly added large structures.

Elk Creek.—In Elk Creek the number of pieces, average diameter, total length, and density of LWD more than doubled after treatment. In addition to LWD added during rehabilitation, an additional 27 pieces, totaling 94 m in length, collected on or were deposited behind newly installed structures. On Elk Creek, the smaller LWD component increased in number but remained the same in average length.

Discussion

In Pacific Northwest streams, LWD is probably the most important factor governing stream diversity (Swanson et al. 1976; Lisle 1981; Bryant 1982; Beschta and Platts 1986). In undisturbed valley-wall channels of the South Fork River, Washington, Sedell et al. (1982, 1984) found that five (1982) and four (1984) pieces of LWD per 100 m of stream, respectively, either dammed the channel or directly influenced it by creating pool habitat. In Oregon, a relatively undisturbed section of Tobe Creek (a smaller stream than the upper Nestucca River or Elk Creek) had 18 pieces of LWD per 100 m of stream that were large enough to influence the channel (House and Boehne 1986). Prior to treatment, three pieces of LWD per 100 m were located in the upper Nestucca River channel. Of those, only 0.4 piece per 100 m of stream was large enough to directly influence the channel. After treatment, the installation of structures increased the number of interventions that dammed or directly influenced the channel to 2.5 pieces per 100 m of stream. This is still only half the number of pieces influencing streams in undisturbed old-growth coniferous riparian zones (Sedell et al. 1982).

The reintroduction of structure increased pool and off-channel habitat, which should improve conditions for anadromous salmonids, especially coho salmon.

The fully spanning, main channel structures installed in the upper Nestucca River and Elk Creek averaged 1 m in height and acted like LWD. Only input from old-growth coniferous riparian zones (where trees exceed 200 years in age) can provide structure similar in size and function. Debris inputs need to be of sufficient height and length to provide stable structures capable of creating large, deep pools, which are necessary components for overwintering presmolt salmonids in coastal streams.

Prior to treatment, surveys indicated that overwintering habitat and spawning habitat were potentially limiting coho production in the upper Nestucca River and Elk Creek. Pre- and postanalysis by the technique of Reeves et al. (1990) substantiated that overwintering habitat (off-channel ponds and deep pools) was and is limiting. The restructuring of both streams increased all seasonally available habitats; however, the most significant changes occurred in winter habitat and spawning habitats. Based on these preliminary results, the rehabilitation of the upper Nestucca River and Elk Creek appears to be successful in increasing coho salmon habitat.

To quantify possible changes in seasonal carrying capacities for coho salmon, ODFW and BLM have conducted limited summer and winter sampling of coho salmon juveniles. This sampling was not intensive or designed for statistical rigor, but preliminary results indicate summer densities of juveniles in Elk Creek pools have increased from 0.93 fish/m^2 (average of 1980, 1985, and 1986 sampling) to 2.08 fish/m^2 in 1987 (BLM, unpublished data). In the upper Nestucca River, cursory sampling in control and treated pools in the main channel and off-channel areas in 1987 revealed that summer densities of juvenile coho salmon averaged 0.34 fish/m^2 in control sites and 0.82 fish/m^2 in treated sites (M. Solazzi, ODFW, personal communication). Limited sampling is scheduled to continue for several years to provide long-term monitoring results for these projects.

Other studies have documented the reduction of coho salmon rearing habitat by removal of LWD (Tschaplinski and Hartman 1983; Bisson and Sedell 1984; Koski et al. 1984; House and Boehne 1987) and the increase in summer numbers of rearing coho salmon juveniles after channel restructuring (House and Boehne 1985, 1986). As yet, few studies have quantified losses of coho salmon production due to limited overwintering habitat. Mason (1976) found that supplemental feeding of coho salmon juveniles increased summer rearing densities six- to sevenfold; however, overwintering capacity negated this increase, and smolt output remained unchanged. Also, Rodgers (1986) found only 9% of the summer juvenile coho salmon left as smolts in Knowles Creek, a system lacking good overwintering habitat.

Based on better habitat conditions and higher coho salmon spawner densities, conditions in the upper Nestucca River prior to Meadow Lake Dam failure were more favorable for coho salmon. The reintroduction of large structures has reestablished preferred coho salmon habitat and should increase future smolt production and adult escapement. Results of this short-term evaluation are further evidence that coastal streams need structural input sufficient in size and quantity, derived naturally from old-growth riparian zones, to reestablish and maintain instream habitat conditions necessary to produce optimum numbers of anadromous salmonids.

Acknowledgments

We appreciate the manuscript reviews provided by P. Boehne and A. Oakley.

References

Baker C., V. Crispin, B. House, and C. Kunkel. 1986. Nestucca River basin anadromous salmonid habitat overview. U.S. Forest Service, Siuslaw National Forest, Hebo Ranger District, Corvallis, Oregon.

Beschta, R. L., and W. S. Platts. 1986. Morphological features of small streams: significance and function. Water Resources Bulletin 22:369–379.

Bisson, P. A., J. L. Nielsen, R. A. Palmson, and L. E. Grove. 1982. A system of naming habitat types in small streams with examples of habitat utilization by salmonids during low streamflow. Pages 62–73 *in* N. B. Armantrout, editor. Acquisition and utilization of aquatic habitat inventory information. American Fisheries Society, Western Division, Bethesda, Maryland.

Bisson, P. A., and J. R. Sedell. 1984. Salmonid populations in streams in clearcut versus old-growth forests of western Washington. Pages 121–129 *in* W. R. Meehan, T. R. Merrell, Jr., and T. A. Hanley, editors. Fish and wildlife relationships in old-growth forests. American Institute of Fishery Research Biologists, Morehead City, North Carolina.

Bryant, M. D. 1982. Organic debris in salmonid habitat in southeast Alaska: measurement and effects. Pages 259–265 *in* N. B. Armantrout, editor. Acquisition and utilization of aquatic habitat inventory information. American Fisheries Society, Western Division, Bethesda, Maryland.

Everest, F. H., G. H. Reeves, J. R. Sedell, D. B. Hohler, and T. Cain. 1987. The effects of habitat enhancement on steelhead trout and coho salmon smolt production, habitat utilization, and habitat availability in Fish Creek, Oregon, 1983–1986. Annual Report 1986 to Bonneville Power Administration, Portland, Oregon.

Franklin, J. F., and C. T. Dyrness. 1973. Natural vegetation of Oregon and Washington. U.S. Forest Service General Technical Report PNW-88.

House, R. A., and P. L. Boehne. 1985. Evaluation of instream enhancement structures for salmonid spawning and rearing in a coastal Oregon stream. North American Journal of Fisheries Management 5:283–295.

House, R. A., and P. L. Boehne. 1986. Effects of instream structures on salmonid habitat and populations in Tobe Creek, Oregon. North American Journal of Fisheries Management 6:38–46.

House, R. A., and P. L. Boehne. 1987. The effect of stream cleaning on salmonid habitat and populations in a coastal Oregon drainage. Western Journal of Applied Forestry 2(3):84–87.

Koski, K. V., J. Heifetz, S. Johnson, M. Murphy, and J. Thedinga. 1984. Evaluation of buffer strips for protection of salmonid rearing habitat and implications for enhancement. Pages 138–155 *in* T. J. Hassler, editor. Pacific Northwest stream habitat management workshop. Humboldt State University, Arcata, California.

Lisle, T. E. 1981. Roughness elements: a key resource to improve anadromous fish habitat. Pages 93–98 *in* T. J. Hassler, editor. Propagation, enhancement and rehabilitation of anadromous salmonid populations and habitat in the Pacific Northwest. Humboldt State University, Arcata, California.

Mason, J. C. 1976. Response of underyearling coho salmon to supplemental feeding in a natural stream. Journal of Wildlife Management 40:775–788.

Reeves, G. H., F. H. Everest, and T. E. Nickelson. 1990. Coho salmon limiting factors analysis. Section 2 *in* R. House, J. Anderson, P. Boehne, and J. Suther, editors. Stream rehabilitation manual. American Fisheries Society, Oregon Chapter, Corvallis.

Rodgers, J. D. 1986. The winter distribution, movement and smolt transportation of juvenile coho salmon in an Oregon coastal stream. Master's thesis. Oregon State University, Corvallis.

Sedell, J. R., P. A. Bisson, J. A. June, and R. W. Speaker. 1982. Ecology and habitat requirements of fish populations in South Fork Hoh River, Olympic National Park. Ecological research in National Parks of the Pacific Northwest. Pages 35–42 *in* J. W. Matthews, editor. Second conference on scientific research in National Parks, San Francisco, California 1979. Oregon State University, Forest Research Laboratory, Corvallis.

Sedell, J. R., and K. J. Luchessa. 1982. Using the historical record as an aid to salmonid habitat enhancement. Pages 210–223 *in* N. B. Armantrout, editor. Acquisition and utilization of aquatic habitat inventory information. American Fisheries Society, Western Division, Bethesda, Maryland.

Sedell, J. R., J. E. Yuska, and R. W. Speaker. 1984. Habitats and salmonid distribution in pristine, sediment-rich river valley systems: S. Fork Hoh and Queets rivers, Olympic National Park. Pages 33–46 *in* W. R. Meehan, T. R. Merrell, Jr., and T. A. Hanley, editors. Fish and wildlife relationships in old-growth forests. American Institute of Fishery Research Biologists, Morehead City, North Carolina.

Skeesick, D. G. 1973. Spawning fish surveys in coastal watersheds, 1972. Fish Commission of Oregon, Management and Research Division, Coastal Rivers Investigation Information Report 73-3, Portland.

Swanson, F. J., G. W. Lienkaemper, and J. R. Sedell. 1976. History, physical effects, and management implications of large organic debris in western Oregon streams. U.S. Forest Service General Technical Report PNW-56.

Tschaplinski, P. J., and G. F. Hartman. 1983. Winter distribution of juvenile coho salmon (*Oncorhynchus kisutch*) before and after logging in Carnation Creek, British Columbia, and some implications for overwintering survival. Canadian Journal of Fisheries and Aquatic Sciences 40:452–461.

Ward, B. R., and P. A. Slaney. 1979. Evaluation of instream enhancement structures for the production of juvenile steelhead trout and coho salmon in the Keogh River: progress 1977 and 1978. British Columbia Ministry of the Environment, Fish and Wildlife Branch, Fisheries Technical Circular 45, Vancouver.

American Fisheries Society Symposium 10:160–175, 1991

Bringing Back the "Bio" in Bioengineering

JAMES R. SEDELL
U.S. Forest Service, Pacific Northwest Research Station
3200 SW Jefferson Way, Corvallis, Oregon 97331, USA

ROBERT L. BESCHTA
Department of Forest Engineering
Oregon State University, Corvallis, Oregon 97331, USA

Abstract.—The current emphasis in fish habitat management is directed primarily towards the physical modification of streams and channels. However, added physical structures cannot replicate the important interactions between dynamic streams and associated riparian vegetation. Management of streamside vegetation was an early enhancement method and was used over a century ago. The management of streamside vegetation provides a basis for an interactive approach to fisheries enhancement and leads to a more productive, diverse, and stable biotic community. Channel roughness values index the hydraulic importance of snags and large wood in addition to their role as habitat and cover. Bioengineering approaches to fish habitat management must complement the natural dynamics of functional riparian communities that are well adapted and connected to the fluvial system.

> After the fullest investigation and examination of these and other streams, I have become satisfied that the destruction of the trees bordering on these streams and the changed condition of the banks produced thereby, has resulted in the destruction of the natural harbors or hiding places of the trout, that this is the main cause of the depletion, and that until these harbors are restored, it will be useless to hope for any practical benefit from restocking them. (Van Cleef 1885)

Van Cleef's remarks, made over 100 years ago, still strike a resonant chord. In the interval between those remarks and today, bioengineering has taken a circuitous path. Streams have been cleaned and flows have been contained in a single channel by state fish wardens and the U.S. Army Corps of Engineers working together to keep rivers manageable and fish from being stranded in side channels at low flows. From the 1940s to 1960s, timber harvesting inundated Pacific Northwest streams with logging slash that usually formed jams. From that period until the mid-1970s, 90% of the state fisheries and U.S. Forest Service budgets for habitat improvement was devoted to cleaning streams of debris. Intensive agriculture since the late 1940s stimulated massive channelization to "square-up" fields for plowing and planting and to move water quickly off the land. Streamside areas in agricultural and rangeland areas often experienced a loss of overstory vegetation that provided shade and other values to stream and riparian areas. Today managers are attempting to mitigate the effects of past practices and to restore damaged habitat with a variety of engineered and other types of instream habitat improvement structures. Some structures have provided local benefits, but in other instances they either failed or provided limited habitat benefits—limited to a type of fish, age-class, or season of the year.

Van Cleef's (1885) suggestions for restoring trout habitat are also applicable today.

> First—prohibit the further destruction of either tree or brush upon or near the bank of the stream.
>
> Second—where the soil is wet and suitable, protect the pools by an abundant growth of alders or other bushes.
>
> Third—plant trees on the banks wherever feasible, especially where their roots will protect the surface of the ground, and at the same time permit the working away of the soil underneath, so that large hollows may be formed as hiding places for the fish.
>
> Fourth—in each year, after the spring freshets are over, protect every pool as far as practicable by placing stumps, or trees or bushes in them, so that fishing with nets will be impossible. And also that the trout may be provided with artificial harbors until the natural ones are again restored. (Pages 54–55)

Van Cleef captured the dynamic nature of streamside vegetation interacting with streams. His comments ring as true in the Catskills of New York as in the Cascades of the Pacific Northwest. Snags in rivers have long been recognized by fishermen and biologists as important summer and winter rearing habitats.

Keown et al. (1977) compiled an extensive literature survey of streambank protection methods, covering approximately 1,500 references. Of these, fewer than 15% pertained to the use of vegetation. It is readily apparent that vegetation has not been ranked nearly as high as structural devices (e.g., riprap, concrete, dikes, fences, asphalt, gabions, matting, and bulkheads) in providing bank stability to many waterways and rivers. While significant amounts of research and project funds have been invested in structural approaches to streambank protection and stability, the crucial role of streamside vegetation seems to have been forgotten.

We believe that any attempt to improve the quality of stream biota must use a broadly based, multipurpose approach, because various organisms are adapted to different environmental conditions. Stream management must optimize for a number of environmental variables, including flow characteristics, water temperature, habitat diversity, food availability, and water quality, all of which interact to determine the quality of the stream biota. Management of streamside vegetation provides the basis for such an interactive approach and leads to a more productive, diverse, and stable biotic community.

Streamside vegetation is much more than an agent of physical change to a stream. Quality of energy inputs, nutrient regulation, algal and macrophytic production, and the structure and function of biotic communities are largely controlled by streamside vegetation (Meehan et al. 1977; Wharton and Brinson 1978).

In this paper we identify the physical role that streamside vegetation plays in the channel and on the floodplain for both small streams and large rivers. We also discuss how streamside vegetation allows streams to function in ways that enhancement structures cannot replicate. We conclude with a discussion of the biological significance of vegetation from the perspectives of structure and cover.

Streamside Vegetation and Channel Dynamics

Streambank vegetation affects channel morphology and hydraulic characteristics by (1) reducing the effective size (width and depth) of a channel, (2) increasing hydraulic resistance, and (3) increasing the resistance of banks to erosion (Nunnally 1978). Because of their greater resistance, vegetated channels tend to be narrower and deeper than nonvegetated channels. Vegetation provides stability to streambanks by (1) binding soils, (2) reducing water velocities at the soil surface, (3) inducing deposition of sediment, and (4) acting as a buffer against transported debris (Parsons 1963; Nunnally and Keller 1979).

Channels and Vegetation from a Geomorphic Perspective

The geomorphic features of streams and river channels tend to change systematically in a downstream direction (i.e., with increasing stream order). Streams in headwater reaches are characterized by V-shaped valleys, relatively steep gradients, coarse bed sediments, and little floodplain development. Bedrock channels may be common. Downstream reaches have lower gradients, finer sediments, and broader alluvial floodplains. As a result, the function and type of stream biota tend to respond to the varying downstream influence of vegetation and the changing physical characteristics of a stream system (Beschta and Platts 1986).

Fluvial geomorphology texts (e.g., Leopold et al. 1964; Gregory 1977; Schumm 1977; Richards 1982) provide relatively limited insights regarding the importance of vegetation to channels. The current emphasis on the quantitative analysis of physical processes affecting fluvial systems led Hickin (1984, page 111) to conclude that "The physical science of fluvial geomorphology is flawed because it ignores processes that are not easily quantifiable and physically or statistically manipulable. The influence of vegetation on river behavior and fluvial geomorphology is a set of these processes. Vegetation may exert significant control over fluvial processes and morphology." Hickin (1984) identified five important mechanisms whereby vegetation influences rivers in British Columbia: flow resistance, bank strengthening, bar sedimentation, formation of logjams, and concave bank deposition. Although the first two of these mechanisms have been acknowledged in the literature, such is not the case for the other three. Hickin (pages 123–124) continued, "there is a clear need . . . for studies designed to isolate the influence and quantity and, particularly, quality of vegetation on channel morphology and on lateral migration rates." We agree with Hickin's (1984) conclusions.

The interconnection between soil, hydrology, and vegetation is recognized in the jurisdictional delineation of wetlands and riparian areas, but we have not yet made the conceptual leap of applying this view to many fish habitat management programs. Past practices have tended to simplify, straighten, and "improve" channels with a view towards increased flow capacity, but the roles of vegetation in providing bank integrity and stability, in storing nutrients, in processing contaminants and sediments, and in diffusing and dissipating stream energy along the channel instead of focusing it at "hardened" locations have apparently been disregarded or ignored. Even where increased structural diversity of channels may be desired, ecosystem processes that will ultimately provide and sustain such diversity are seldom identified and promoted.

Channel Roughness

Energy from water moving down a channel can be used to do work (e.g., transport sediment), but the vast majority is used to overcome frictional resistance along the bed and banks and is eventually dissipated as heat. The relative effectiveness of the water–channel interface in dissipating potential energy is often characterized by a roughness coefficient, such as Manning's n. Ideally, such a coefficient would remain relatively constant for a given boundary condition, regardless of channel slope, size of channel, or depth of flow. Although Manning's n (an integrative index of a channel's resistance to flow) is most appropriately applied to channels of uniform dimensions, it has been extensively used by practicing hydrologists and others to hydraulically characterize the flow resistance for a broad range of natural and altered channels. The channel roughness concept has seldom been used to promote understanding of the long-term effects of vegetation on aquatic habitat, however.

Tabular values of Manning's roughness coefficients associated with commonly encountered stream conditions are indicated in Table 1. Barnes (1967) provided photographic examples of stream reaches representing a range of channel sizes with associated high-flow roughness values. A given channel at flood stage tends to have a smaller value of Manning's n than the same channel at a lower stage (Jarrett 1984). However, the effectiveness of bank vegetation in influencing overall channel resistance is generally most important at high flows, when floodwaters are at or above bankfull stage and when bank or channel erosion is most likely. Where water flows across a floodplain, the channel and floodplain components of a channel's roughness can be estimated separately for velocity and discharge calculations, and then combined (Figure 1; Table 1).

Cowan (1956) presented a procedure that identifies several factors that constitute and influence Manning's n (Table 2). It is clear from the tabular values of n_4 that vegetation plays a major role in the hydraulic roughness of stream channels. Furthermore, vegetation influences the relative surface irregularity of channels (n_1), the variation in channel shape and size (n_2), and the degree of meandering (m); further, at least in some channels, vegetation is a source of obstruction as in the case of large woody debris (n_3). Because vegetation has a complex array of shapes, sizes, growth forms, and mechanical strength properties, its effects on channel roughness are similarly di-

TABLE 1.—Manning's n associated with indicated boundary conditions.

Boundary condition	n[a]	Source[b]
Top width at flood stage generally <30 m		
Rigid channels (smooth or shot concrete, planed timber, vitrified clay, brick, firm gravel)	0.011–0.017	1, 4
Earth canals		
In best condition	0.017	1
In good condition	0.020	1
In fair condition, some growth	0.025	1
In poor condition, considerable moss growth	0.035	1
Mountain streams with rocky beds, rivers with variable sections and some vegetation along banks	0.040–0.050	1
Alluvial channels, sand bed, no vegetation		
Lower Regime (Froude number <1)	0.014–0.035	1
Upper Regime (Froude number >1)	0.011–0.020	1
Natural channels, full stage		
Straight, no pools, clean	0.029	4
Straight, no pools, weeds and stones	0.035	4
Winding, pools and shallows, clean	0.039	4
Winding, pools and shallows, weeds and stones	0.042	4
Natural channels, low stage		
Winding, pools and shallows, clean	0.047	4
Winding, pools and shallows, large stones	0.052	4
Natural channels		
Sluggish, deep pools, weedy	0.065	4
Sluggish and very weedy	0.112	4
Natural stream, sinuous, snag-choked, sandy	0.150	5
Top width at flood stage generally >30 m		
Regular section with no boulders or brush	0.025–0.060	3
Irregular and rough section	0.035–0.100	3
Souris River, USA		
Channel	0.033–0.058	6
Overbank areas	0.060–0.120	6
Tamish River, Yugoslavia		
Channel	0.035	2
Floodplain under grass	0.073	2
Floodplain under planted trees	0.094	2
Floodplain under natural forest	0.104	2

[a]Values indicated are averages; variations of 20% or more should be expected, especially in natural channels.

[b](1) Albertson and Simons (1964), (2) Bruk and Volf (1967), (3) Chow (1959), (4) Linsley et al. (1982), (5) Ramser (1929), (6) Shields and Nunnally (1984).

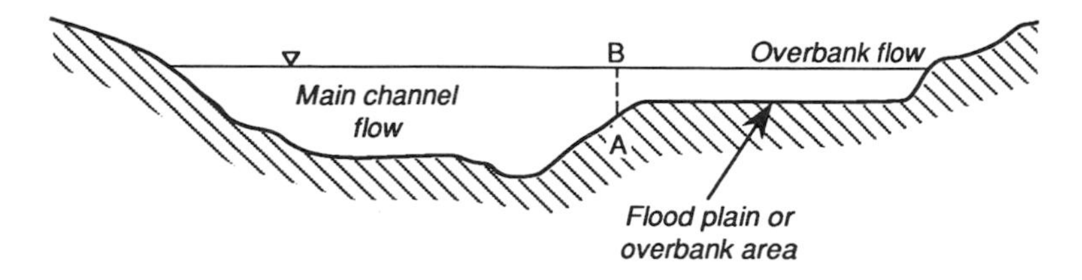

FIGURE 1.—Channel and floodplain components of a river system. (From Albertson and Simons 1964.)

verse. Thus, vegetation (both directly and indirectly) is potentially the most important single factor influencing the hydraulic roughness of many streams and rivers. As a result, it undoubtedly has major short- and long-term consequences for channel characteristics and morphology, streambank and channel stability, ecosystem functioning, and the biological productivity of natural and engineered channels.

Relatively few studies have evaluated changes in Manning's *n* associated with changes in streamside vegetation. However, where such evaluations have been conducted (Figure 2), it is plain that vegetation is a major modifier of overall channel roughness. In Mississippi, the roughness of a channelized stream increased more than twofold as streamside vegetation grew following cleaning (Wilson 1973). In Arizona, the removal of mesquite *Prosopis juliflora* and saltcedar *Tamarix pentandra* reduced the flood-flow roughness by 70% (Burkham 1976). Similarly, the removal of large woody debris from river bars in Pacific Northwest rivers caused more than a 20% reduction in channel roughness (Shields and Nunnally 1984). Even though short-term changes in roughness can obviously be dramatic, there is a lack of studies of long-term changes in channel roughness caused by more pervasive management practices or impacts.

Floodplains represent special landforms associated with riverine landscapes where streamside vegetation naturally assisted the long-term deposition of fine sediments. The effect of vegetation on floodplain roughness is illustrated in Figure 3. Where floodplain forests have been converted to agricultural and other uses, roughness coefficients are generally one-third or less of their original values. Thus, the removal *or* reestablishment of riparian forests and associated understory vegetation can greatly influence channel and floodplain stability and subsequently affect aquatic habitat.

Long-term modifications of river channels not only influence the dynamics and fragmentation of floodplain forests (Décamps et al. 1988), they also affect floodplain functions. For example, streambank dikes can sever the connectedness between a stream and the overbank areas that are important for the temporary storage of floodwaters; and revetments may cause accelerated channel downcutting or accelerated bank erosion farther downstream. In both instances, physical changes in the channel will probably have concurrent effects upon instream biological communities.

Channels that are hydraulically rough are most likely to be channels characterized by considerable trapping and retention of organic matter during high flows (Speaker et al. 1984, 1988). This material is a major food source for aquatic organisms. The importance of in-channel trapping and retention mechanisms tends to decrease with increasing stream size. With larger streams, floodplains become the major location of organic matter trapping and retention during high flows.

Large organic debris inputs to a stream from the riparian environment may also significantly affect

TABLE 2.—Manning's *n* based on Cowan's (1956) method.[a]

Channel category	Range in channel conditions	n_i or m	Range in values for n_i or m
Bed material	Earth to coarse gravel	n_0	0.020–0.028
Degree of surface irregularity	Smooth to severe	n_1	0.000–0.020
Variations in channel cross section	Gradual to frequently alternating	n_2	0.000–0.015
Relative effect of obstructions	Negligible to severe	n_3	0.000–0.060
Effects of vegetation	Minor to very large	n_4	0.005–0.100
Degree of meandering	Minor to severe	m	1.0–1.3

[a] $n = m(n_0 + n_1 + n_2 + n_3 + n_4)$ for channels where the hydraulic radius (i.e., area/wetted perimeter) is less than 5 m.

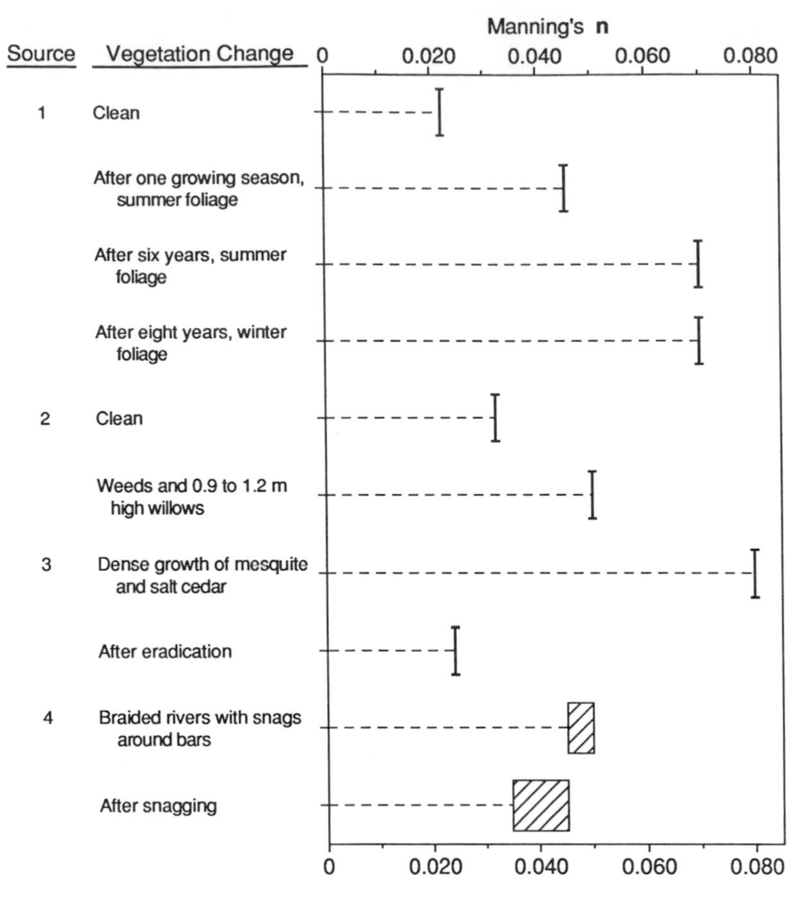

FIGURE 2.—Manning's *n* associated with changes in streamside vegetation and woody debris.

channel structure and roughness. For example, in sand-bottom streams, deep holes are commonly scoured near fallen trees and other large woody debris, creating relatively permanent pools in an otherwise uniform and unstable habitat (Hickman 1975; Mendelson 1975). In high-gradient, bedrock streams, habitat diversity is enhanced by pools and waterfalls that are formed in sediment captured behind debris dams (Bilby and Likens 1980; Bisson et al. 1987).

Streambank Integrity and Stability

Vegetation is generally effective at protecting banks and riparian soils from the erosive forces of moving water. Aboveground plant parts (stems of trees, shrubs, grasses, and forbs) tend to dampen turbulence and slow velocities at the water–channel interface (Klingeman and Bradley 1976). Root systems and organic matter help bind soil particles and alluvial sediments. Thus bank and bed sediments can remain in place even though they experience considerable shear stress during high flow. Shear stresses are highly variable across a given cross section, but relatively high shear stresses are common along the outside banks of meandering channels (Figure 4).

Zimmerman et al. (1967) found significant differences in channel dimensions for a stream passing through meadow and woodland. The dense, fibrous network of grass roots in the meadow stabilized banks and maintained a narrow channel even for extreme flood events. Although tree roots provided long-term stability at a given location, channel form was wider, shallower, and

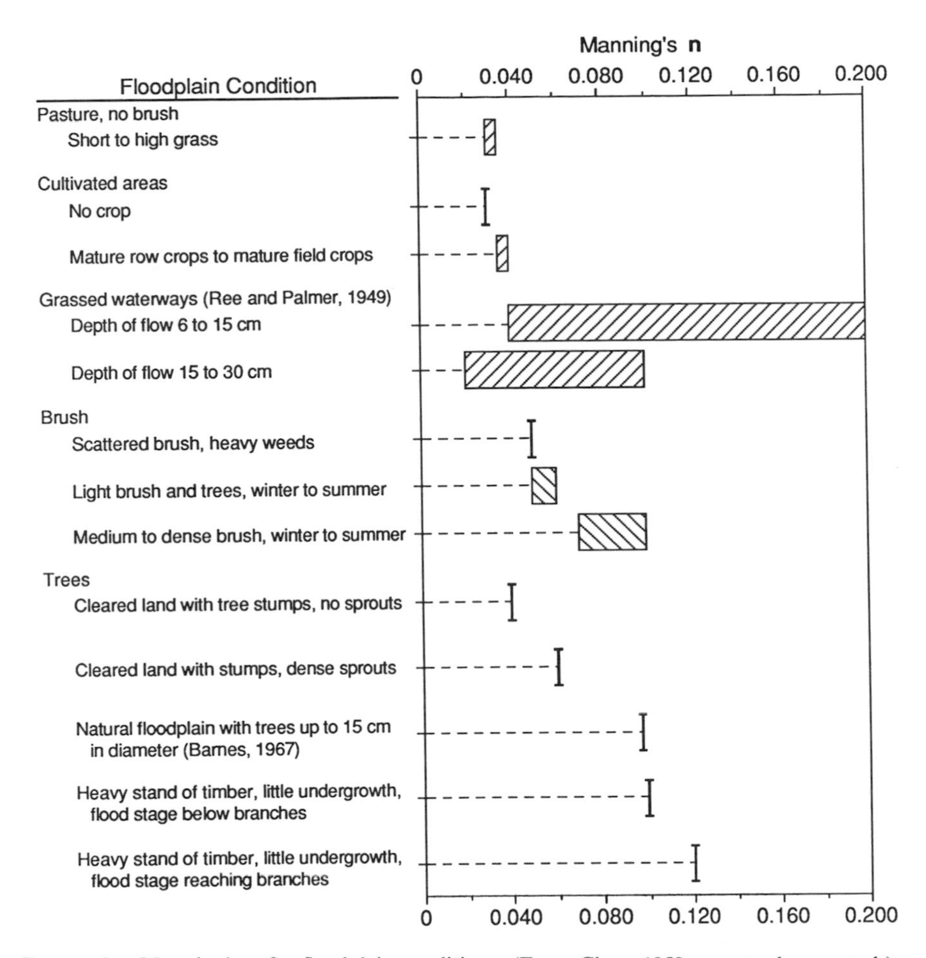

FIGURE 3.—Manning's *n* for floodplain conditions. (From Chow 1959 except where noted.)

more variable than in the meadow reach. Thus, at least for small headwater streams, the effects of vegetation can sufficiently dominate channel morphology such that general hydraulic geometry relations are obscured.

Hickin (1984) indicated that riverbanks that are well bound by roots offer far greater resistance to lateral erosion than relatively unvegetated banks of rivers of western Canada. For similar hydraulic conditions and bank characteristics, a river migrating through a cleared or cultivated floodplain would erode at nearly twice the rate of one reworking a naturally forested floodplain. For a glacial meltwater river, the effects of roots upon bank erosion rates had an even more pronounced effect (Smith 1976). In root-free silt, a bank erosion rate of 1,600 mm/h was measured, in contrast to a rate of 0.2 mm/h (nearly five orders of magnitude difference) with roots present.

Plant regeneration, growth, and succession provide the mechanism by which riparian communities can continually respond to natural changes in streamflow, sediment loads, and local shifts in channel morphology. Structures alone seldom have that capability.

Sediment Transport

Few field studies have been conducted on the effect of streamside vegetation on bedload transport. Mosley (1981) found that organic debris accumulations in a 3–5-m-wide forest stream significantly influenced the bedload transport and represented "a major constraint on the application of physical laws and theories explaining sediment movement in, and the morphology of, this stream." In southeast Alaska, Helmers (1966) constructed log-debris jams and concluded that they had important effects on streambed stability and sediment movement.

On floodplains, vegetation (including canopies

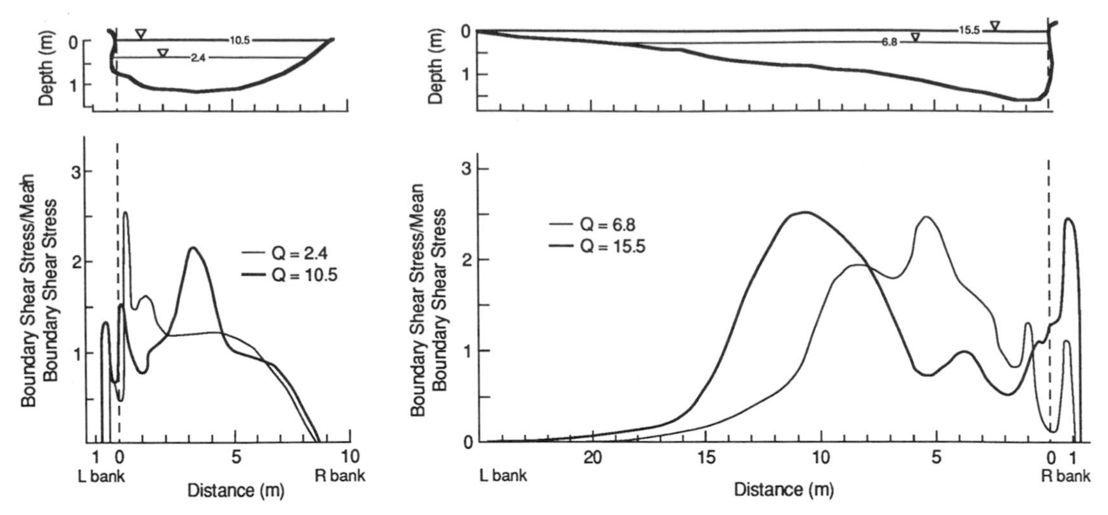

FIGURE 4.—Distribution of the ratio of boundary shear stress to mean boundary shear stress for channel bends of the River Severn, Wales. (From Bathurst 1979.) Q = discharge (m^3/s); L and R are left and right, respectively.

and litter) protects the soil from direct rainfall impact, reduces the velocity of overbank flows, and decreases the erosion of soil. Grasses and other low-growing plant species are particularly effective at protecting soils during high flows. The amount of roughness associated with grasses depends on species, stage of growth, and degree of submergence (Wilson 1967). For many grasses, it is common for the roughness to remain relatively high until the grass is submerged and then to decrease asymptotically with increasing water depth (Ree and Palmer 1949). If grass stems become bent or "laid down" with increasing flow, roughness and sediment trapping decrease, but the plants can still mechanically protect the soil.

Trees also retard the velocity of overbank flow (Figure 5). The application of flow resistance theory to heavily wooded floodplain conditions led Petryk and Bosmajian (1975) to conclude, for situations with high tree density, that flow resistance is due almost entirely to the tree trunks. Thus, the total shear force on the floodplain soil surface is negligible in comparison with the total drag force on the vegetation.

Li and Shen's (1973) theoretical assessment of flow hydraulics and sediment transport associated with the spacing, size, and relative numbers of vertical cylinders (representing boles of standing trees) illustrated several principles related to floodplain forests. Average boundary shear stress increased as flow increased across a floodplain, but the stress was inversely related to the diameter of stems. Thus, for a given flow, the average boundary shear stress should decrease as trees grow larger. Stems grouped into staggered patterns best retarded flow. There was an exponential increase in relative sediment transport rate as cylinders were removed (i.e., as trees were harvested). This relationship (Figure 6) suggests that leaving at least some trees on a formerly forested floodplain can greatly reduce the sediment-transporting capability of floodwaters.

Distribution of Fishes in Streams and Rivers

Numerous studies have demonstrated that fish species and communities vary along the continuum from headwater streams to large rivers. Excellent reviews were given by Karr et al. (1983) and Matthews and Heins (1987). Although some species exhibit longitudinal zonation suggesting adaptation to habitat conditions related to stream size, others are broadly distributed and can be found from small streams to large rivers. Distributions of stream fishes may also vary over time and with changing environmental conditions (temperature or hydrology). Species requiring specific habitat conditions may undergo extensive movements to maintain that association in unstable or fluctuating environments. Populations may be relatively stationary when their habitats are not constantly available or the species adapts to a wide range of conditions. These distribution patterns clearly indicate that management of lotic fish populations may transcend boundaries of stream reaches. Hence, preservation of fish community

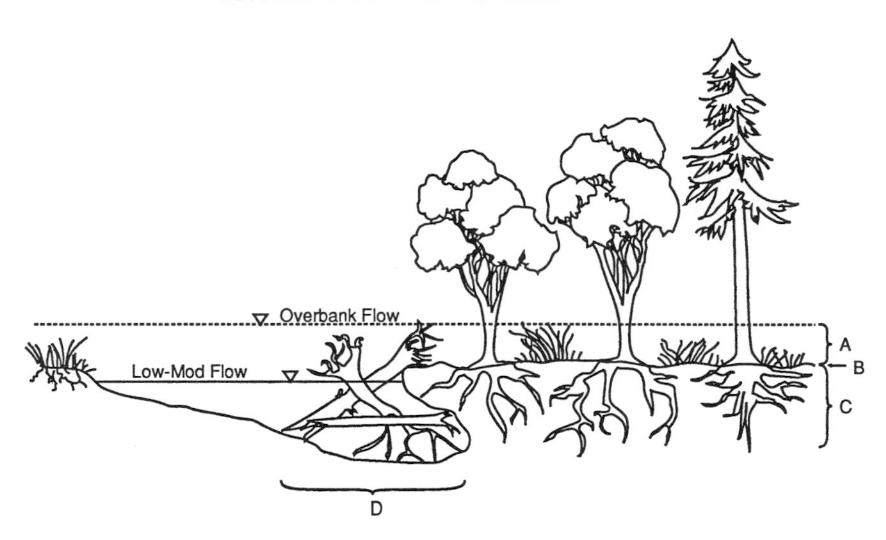

FIGURE 5.—The role of vegetation in providing boundary roughness and shear strength at channel margins for a floodplain stream system: (A) aboveground portions of plants are a major component of flow resistance during overbank flow; (B) accumulated organic matter and litter, and bases of plants and roots at the surface, provide resistance to flow; (C) live roots increase strength properties of soil, thus reducing rates of bank erosion; (D) large woody debris functions as large roughness elements that alter local channel geometry and represent effective trapping sites for transported organic debris.

integrity requires an integrative perspective of the entire stream basin.

Fish species in streams and rivers are associated to various degrees with distinct habitat types. These habitats form primarily as a result of natural fluvial processes, and their characteristic physical and chemical attributes vary considerably with discharge. Like their general distribution patterns, the habitats fish use may change with age, sex, reproductive state, geographic area, and fluctuating environmental conditions. Pools, riffles, and glides are the primary habitat divisions in small to medium-sized streams. In addition to these main channel habitats, larger streams and rivers have a diverse array of other important habitat types. Side-channels, extra channels, sloughs, and backwater lakes are continually created and destroyed by natural fluvial processes.

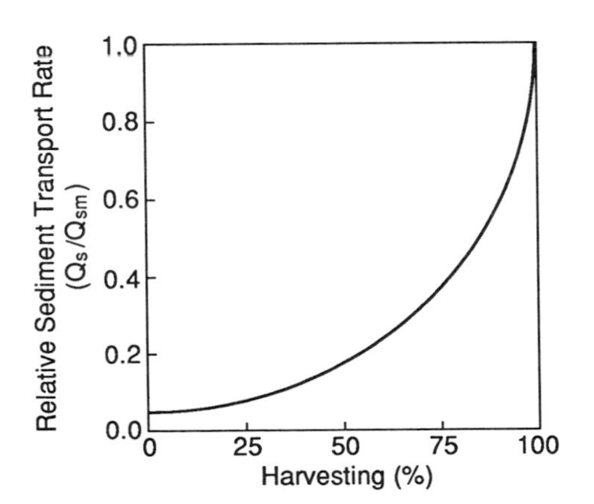

FIGURE 6.—Example of relationship between sediment yields and forest harvesting based on the theoretical evaluation by Li and Shen (1973). (Q_s = sediment discharge, Q_{sm} = maximum sediment discharge when there is no vegetation.)

Fish communities are very responsive to local site conditions as well as to reach characteristics. This complex interaction with structural and hydrologic variability is very difficult to capture in an engineered approach to fish habitat. The vegetation is a much more sensitive agent of habitat structure because it too responds to local geomorphic and hydrologic settings. Almost universally, channels with living vegetation on the banks that provides organic debris to the stream are chosen by biologists when they seek a representative reach within an ecoregion.

Streamside Vegetation and Cover

Over the past 20 years, numerous investigations have documented the importance of woody debris as instream structure and cover structures for salmonids; useful reviews of many of these studies were given by Harmon et al. (1986) and Bisson et al. (1987). Woody debris and snags in streams act to (1) create and maintain pools, (2) cause local reductions in stream velocities that serve as foraging sites for fish feeding on drifting food items, (3) form eddies where food organisms are concentrated, (4) supply protection from predators, (5) provide shelter during winter high flows, and (6) trap and store organic inputs from streamside forests, enabling them to be biologically processed (a key link in the food chain of salmonids of the Pacific Northwest and Alaska).

Cover in lotic environments may be divided into two rather ill-defined classes: instream and streamside cover. Instream cover includes undercut banks, tree roots, large rocks, logs, brush, and aquatic vascular plants. Streamside cover refers to vegetation type, canopy density, and other characteristics of the riparian environment. Instream cover features tend to attract and concentrate fish. They provide spawning sites, protection from current or predators, and hiding places from which predators can ambush prey. They support important food resources for fish and cause changes in stream morphology that increase habitat diversity (Marzolf 1978; Karr et al. 1983). Furthermore, they may influence local channel morphology, sediment transport, and the extent of scour and deposition.

The importance of cover has been most intensively investigated in coldwater streams. For example, the relative amounts of various forms of cover may largely account for spatial variation in the density of brook trout *Salvelinus fontinalis* (Hunt 1971) and the standing crops of steelhead *Oncorhynchus mykiss* and cutthroat trout *Oncorhynchus clarki* (Bisson et al. 1987). In warmwater streams, instream cover has similar but perhaps broader influences on fish communities. In the Missouri River, the standing crop of all fish was 25% higher, and that of catchable-size fish 51% higher, in a section with snags than in a comparable section without instream cover (Hickman 1975). Some striking differences in community structure were also observed. In a section of a small Illinois stream with instream cover of logs and brush, fish biomass was five to nine times higher than in an adjacent section that was cleared of all cover features (Karr and Dudley 1981; Angermeier and Karr 1984). Fish species diversity may also be higher in sections of streams with instream cover (Sheldon 1968).

For streams and rivers with fine-grained substrates, instream cover structures are particularly important to fish because they lead to increased habitat diversity (e.g., by promoting scouring of pools) and provide substrate for as much as 90% of the macroinvertebrate biomass (Marzolf 1978; Wharton and Brinson 1978; Benke et al. 1985). Wood debris also provides a major substrate for invertebrate populations in high-gradient, bedrock streams (Triska and Cromack 1979).

An important function of streamside vegetation in headwater reaches is to lend stability to these small-stream ecosystems. By stabilizing streambanks, intercepting eroding sediment from the adjacent land surface, moderating water temperatures, and limiting growths of algae and aquatic macrophytes, streamside vegetation is largely responsible for maintaining the physical integrity of small-stream channels over a wide range of environmental conditions. The importance of streamside vegetation is best illustrated for modified streams where forested reaches provide critical refuges for fishes during severe environmental conditions (Karr and Gorman 1975). Fish communities reside intact in forested portions of the streams throughout the year, whereas only some species of a fish community persist in areas with little or no woody riparian vegetation.

Habitat Modifications of Streamside Vegetation

Small-Stream Environments

As suggested by previously described fish–habitat associations, reductions in habitat complexity (pool–riffle frequency, size, type, etc.) typically result in a significant loss of fish species richness (North et al. 1974; King and Carlander 1976; Griswold et al. 1978). Elimination of deep-pool habitats, for example, commonly results in loss of game-fish populations.

Loss of biotic integrity in small streams is linked to modifications of specific habitat variables. The absence of instream brush piles, for example, may account for lower fish biomass in small-stream reaches where the source of these cover structures (streamside vegetation), as well as the environment favoring their retention, has been destroyed by channelization (King and Carlander 1976). Fish food supplies are also affected because brush piles support dense aquatic macroinvertebrate populations and removal of streamside vegetation reduces inputs of terrestrial insects (Lynch et al. 1977; Angermeier and Karr 1984).

Large-River Environments

Modifications of the physical environment may have an even more significant impact on fish communities in large rivers than in small streams. For example, although 30 years had elapsed since the Chariton River channel in Missouri was cleaned of debris and straightened, the standing crop of fish remained 83% less than in an adjacent unmodified section (Congdon 1971). The straightened reach also had eight fewer fish species.

Snag removal contributed to a serious decline in catfish fisheries in the Missouri River (Funk and Robinson 1974). Fish samples from modified sections of the Olentangy River, Ohio (Griswold et al. 1978), and the Luxapalila River, Alabama–Mississippi (Arner et al. 1976), indicate that, in addition to consistently supporting lower fish standing crops, snagged and straightened sections of large rivers have markedly different fish community structure than unmodified sections. The decline in game-fish populations is particularly striking. Moreover, many fish that are captured in straightened sections are actually transients en route to more complex habitats (Hansen 1971; Arner et al. 1976).

Removal of snags also results in a severe reduction in fish food resources (Hansen 1971; Arner et al. 1976; Benke et al. 1985), since most of the aquatic invertebrate production in large, unstable-bottom rivers is associated with these structures. Benke et al. (1985) estimated that snags represent only 4% of the area habitat of the Satilla River, Georgia, but that over 78% of the drifting invertebrates originate from snags. Several species of game fishes obtain at least 60% of their food from the snag habitat in the Satilla River.

Many biologists and engineers are not aware of the historical frequency of snags per kilometer in large rivers (Sedell and Froggett 1984). Table 3 provides a conservative estimate of frequency of snags that were removed from several rivers in virtually every corner of the United States. Even the largest river in North America, the Mississippi River, had over 600 snags/km removed from its channel between St. Louis and New Orleans. Table 3 indicates the extent to which fish habitat in large rivers depended on large snags and how lumbering and navigation concerns converted riparian forests to brush, and diverse and productive streams to navigable aquatic highways. Over time, individual snags tended to aggregate at the outside bend of meanders and at channel cutoffs. These aggregations resulted in deep and complex scour pools (Wallace and Benke 1984). Interestingly, the frequency of snags in large rivers is similar to the frequency of large woody debris and downed trees in intermediate-sized streams (Harmon et al. 1986).

The common thread between warmwater fishes and coldwater fishes is that higher biomasses of fish generally occur in areas with structurally complex habitats. Woody debris manipulations have improved the rate of survival (Gard 1961; Hunt 1971), growth, and density (Burgess and Bider 1980; House and Boehne 1985, 1986) of salmonids. Most techniques have involved introducing debris to provide resting areas, overhead cover, and new pools.

Challenges in Bioengineering

Habitat structure is a strong determinant of biotic conditions in stream and river systems. Throughout the United States, habitat diversity of many streams and rivers has largely been lost. Even in the Pacific Northwest, where human influence has been of major significance for the past 60 to 80 years, lack of habitat diversity is a major problem. Pools without large woody debris lack structural complexity and are generally too few to provide adequate low- and high-flow rearing habitat for salmonids. Habitat losses are particularly severe in urban and agricultural areas of the Pacific Northwest, but they occur over a broad spectrum of land uses and vegetation types. A complex habitat seems to provide two things: rearing space for small fish, including a diversity of substrates for food organisms; and hiding places (cover) from which large fish can prey on smaller species.

Much habitat complexity is created during periods of high flow when streams and rivers interact with streambanks and floodplains. This interaction provides a diversity of edges and rearing areas. As long as a river is allowed to locally erode and to shift its course, riparian plant communities will continue to be rejuvenated along the banks and on the floodplain. If the channel becomes fixed by embankments or levees, it typically becomes functionally disconnected from the streamside vegetation and the floodplain. Straightened channels, unless experiencing increased sediment loads from upstream areas, will generally undergo streambed erosion. And as the channel drops in elevation, the adjoining floodplain water table will also drop. This lowering will change the soil moisture gradient and thus alter the character of riparian vegetation.

The "bio" in the bioengineering of fish habitat pertains to practices that (1) encourage and promote the growth of important riparian species, (2)

TABLE 3.—Summary of snags pulled from rivers in the USA for navigation improvement from 1867 to 1912. Most rivers in the USA lost important amounts of fish habitat by the year 1910. (From Sedell et al. 1982.)

Rivers by region	Drainage area (km^2)	Kilometers snagged	Snags per kilometer
Southeast region			
Pamunkey River, Virginia	2,740	48	86
North Landing River, North Carolina and Virginia		27	394
Pamlico and Tar rivers, North Carolina	12,500	78	480
Contentnia Creek, North Carolina	1,870	112	151
Black River, North Carolina	3,960	112	172
Edisto River, South Carolina	8,700	120	307
Savannah River to Augusta, Georgia	27,100	397	123
Oconee River, Georgia	11,300	158	398
Noxubee River, Alabama and Mississippi	3,070	110	1,306
Pearl River, Mississippi	18,700	722	408
Tombigbee River, Mississippi	26,900	770	372
Guyandot River, West Virginia	3,140	130	62
Cumberland River above Nashville, Tennessee	45,800	573	135
Choctawhatchee River, Florida and Alabama	11,200	339	524
Oklawaha River, Florida	7,070	99	112
Caloosahatchee River, Florida		35	452
Central region			
Grand River, Michigan	13,400	61	33
Minnesota River, Minnesota	43,300	384	71
Red River, North Dakota and Minnesota	171,000	512	16
Red Lake River, North Dakota and Minnesota	13,500	240	6
Wabash River, Illinois and Indiana	83,900	77	102
Missouri River, Missouri	542,000	2,800	9
Arkansas River, Arkansas	411,000	1,920	100
White River, Arkansas	71,700	480	124
Cache River, Arkansas	2,660	157	217
St. Francis and L'Anguille rivers, Arkansas	29,000	352	81
Southwest region			
Guadalupe River, Texas	25,900	83	848
West coast region			
Sacramento River, California	60,200	368	91
Chehalis River, Washington	4,600	24	202
Willamette River above Albany, Oregon	11,500	88	61

provide woody vegetation along stream channels that help maintain important riparian functions and groundwater levels, (3) allow plant succession to occur relatively unhindered, and (4) locate reaches within a basin that not only are important for fish but are where the stream has its greatest interaction with streamside vegetation. Areas where flowing water interacts strongly with riparian vegetation and adjacent terrain are typically the areas that have the most diverse and productive habitats. These are also areas that are difficult places in which to add habitat structures other than large woody debris. Vegetation is less costly than other habitat structures and provides a more complex and functional habitat.

Unconstrained sections of a stream or river (where floodplains are wide) are generally lower-gradient reaches and areas where downed trees accumulate in the channel and along channel margins. These aggregations of trees or debris jams increase channel roughness, create multiple channels, and help divert water onto the floodplain. In these areas, channel morphology and location are dynamic, not fixed, and are hydrologically and morphologically interactive with the floodplain. Debris jams form, are abandoned by the channel on the floodplain, create backwaters and side channels, or are broken-up and floated downstream by floods. Yet, the key fish habitat features are predictable and consistent because of the woody riparian vegetation. Specific habitat features are not always in the same location but continue to be found along the stream from year to year. The maintenance of natural process and not the stabilizing of a channel at a fixed point is the most important aspect of maintaining fish habitat.

Maintaining or improving riparian vegetation along streams is vital to long-term maintenance and complexity of fish habitat in most streams.

Silvicultural practices that provide for (1) large woody debris, (2) erosion control, and (3) nonpoint nutrient removal represent important mechanisms for achieving physical habitat goals. We will illustrate three silvicultural practices with examples from different regions of the United States. However, the basic principles will hold anywhere; only the tree species will change.

Recruitment of large woody debris to streams is difficult in many areas because of competition from smaller trees and shrubs. For example, west of the Cascade mountains in the Pacific Northwest, red alder *Alnus rubra* dominates the riparian zone, with a scattering of conifer trees, because of past natural and human disturbances. Although red alder satisfies some fish habitat needs, it is not as desirable as conifers for creating complex fish habitat because it does not attain the size and strength of conifers and is less resistant to decay. A mixture of tree types is more desirable. To attain this situation, conifers must be established by underplanting in riparian areas dominated by red alder or brush species, or by releasing suppressed and intermediate-sized conifers through manipulation of the alder overstory (e.g., girdling or complete felling of overstory). The release of understory trees already present on a site may cost less time, money, and planning than understory planting. Because suppressed understory trees are often large compared with a typical seedling, they may reach sizes useful for production of large woody debris and snags more quickly than underplanted seedlings. Field trials using these techniques are currently under way in coastal Oregon (C. Bacon, Oregon State University, personal communication).

Agriculture and riparian vegetation for erosion control can also be compatible along floodplain systems. For centuries, native American and Spanish American farmers of the arid Southwest have managed riparian vegetation adjacent to their agricultural fields (Nabhan 1985). They planted, pruned, and encouraged tree species for flood erosion control, soil fertility renewal, buffered field microclimates, and fuel wood production.

Living fencerows were constructed by weaving brush between the trunks of lines of cottonwood *Populus fremontii*, willow *Salix goodingii*, and mesquite *Prosopis juliflora* adjacent to their floodplain fields. The fencerows were planted during the winter months when farmers cut branches of cottonwood and willow and planted the pruned, trimmed branches at field margins on the edge of the most recently developed stream channel. These cuttings sprouted, sent down roots that stabilized streambanks, and grew into large trees. Between their trunks, brush of seep willow *Baccharis salicifolia*, burrobush *Hymenuclea* spp., and mesquite were woven to form a horizontal, permeable barrier up to 2 m high. This woven fence slowed later floodwaters without channelizing the primary streambed the way in which concrete or riprap channel banks would. As a result, when summer or winter floods covered a floodplain terrace, channels were less likely to become entrenched and erosion was less pervasive than would occur on a barren floodplain (Nabhan 1985).

This technique of living fence rows is readily adapted to projects where the objective is to maintain or improve the complexity of fish habitat. Furthermore, the living fence posts could be clustered and staggered across the floodplains. Li and Shen's (1973) theoretical assessment of flow hydraulics and sediment transport associated with spacing, size, and relative numbers of vertical cylinders (surrogate tree stems) indicated that stems grouped into staggered patterns were the most effective at retarding flow.

Streamside vegetation also maintains and enhances water quality in streams draining agricultural lands. In the Midwest and on the southeastern coastal plain, woody riparian vegetation not only physically stabilizes banks and falls into streams to create complex fish habitat, but it also filters nutrients and maintains water quality on agricultural watersheds. The removal of nutrients such as nitrogen and phosphorus occurs via several mechanisms (Brinson et al. 1984; Lowrance et al. 1984; Petterjohn and Correll 1984): by surface filtration of sediments, by incorporation of N and P into living woody plants, and by the nitrification–denitrification process below ground and at the soil surface. Soils of the riparian ecosystem are ideal sites for denitrification: they receive large inputs of organic matter from forest litter; they are seasonally waterlogged; and large amounts of nitrate arrives in subsurface flow. Most of the nitrogen goes to the atmosphere as gas via denitrification and only a small amount is incorporated into growing trees. In Maryland, a 15-m-wide buffer strip of trees is required between agricultural land and the Chesapeake Bay and adjacent tributaries. This width of a filter strip can remove more than 75% of the groundwater nitrogen and more than 40% of the phosphorus before either gets into the adjacent stream or water body.

The point of this discussion is to indicate that a bioengineering approach utilizing woody vegetation along streams can result in multiple benefits to fish habitat and water quality that extend beyond habitat complexity.

Another problem with fisheries programs that focus on installing enhancement structures is the general difficulty in finding suitable construction sites. From a logistic standpoint, not all stream distance can be improved with enhancement devices even if unlimited funding is available. For example, one river basin in western Oregon has 850 km of streams producing anadromous fish, yet less than 100 km are feasible for structural habitat management. Nearly 90% of the stream kilometers cannot be structurally modified because of access problems and other constraints. In some areas, enhancement structures may well be the only fish improvements possible, but the majority of stream kilometers must be managed for vegetation diversity and plant succession over a wide range of riparian areas to maintain and improve fish production.

Based on current and projected trends, a variety of enhancement structures will be constructed in streams and at an accelerated rate during the 1990s. These structures, even if they perform as desired, are often only short-term solutions and are typically placed without any long-term plan. For example, a comprehensive plan might call for reestablishment of woody vegetation and accelerated growth of trees to provide future snags and wood to the stream. Along some streams, shrubs may play a dominant role in maintaining bank integrity and habitat. Thus, vegetation management must be a greater-than-equal partner in fish habitat improvement programs. In many instances, structures may only represent stopgap measures until suitable vegetation diversity and size return. Managing streamside vegetation is managing fish habitat. Yet management prescriptions designed to make streamside vegetation ecologically and biologically diverse are generally lacking. Ecological diversity can provide the structural diversity of dead wood, different canopy heights, and a varied vegetative community. However, maintaining biological diversity and habitat for fish species in managed watersheds may involve much more than setting aside strips of riparian vegetation. For example, additional research and administrative studies are needed to understand how various tree species, root systems, debris sizes, etc., function through time with channels of different widths, gradients, and valley form. Such information is necessary to aid decisions regarding where, what kind, and how many trees (or shrubs) are to be selectively planted or maintained in riparian areas to optimize fish habitat requirements over the long term. This is obviously a major challenge.

The current direction in fish habitat management is oriented primarily towards the physical modification of streams and channels. Although structural additions and alterations may improve habitat for streams lacking in physical diversity, they cannot replicate the important interactions between a dynamic physical system (water, sediment, and an array of energy transfers) and the associated riparian vegetation. For a wide variety of resource values, including fisheries, reestablishing the functional attributes of streamside vegetation is crucial to improving the bank characteristics and riparian conditions for streams and rivers throughout the United States.

Engineers have much to offer stream biologists by helping define roughness patterns associated with a multitude of channel and vegetation combinations. These patterns have important implications regarding fish community diversity and production. The channel roughness concept does not focus on habitat units such as individual pools or riffles. Instead, channel roughness indexes the overall hydraulic character of a stream reach and is determined by the summation of all the various channel and vegetation features that provide resistance to flow. The concept of channel roughness provides a useful means of integrating the diverse associations of conditions and features encountered in stream and river systems. Similarly, optimal fish habitat cannot be attained by simply concentrating management on a particular channel unit or structural feature (e.g., creating a bigger pool or adding large woody debris). It can only be accomplished by helping an entire stream reach operate in a productive and self-sustaining capacity. Thus, to be successful, bioengineering approaches to fish habitat management must complement the natural dynamics of structurally functional riparian communities that are well adapted and connected to the fluvial system.

References

Albertson, M. L., and D. B. Simons. 1964. Fluid mechanics. Pages 1–49 *in* V. T. Chow, editor. Handbook of applied hydrology. McGraw-Hill, New York.

Angermeier, P. L., and J. R. Karr. 1984. Relationships between woody debris and fish habitat in a small warmwater stream. Transactions of the American Fisheries Society 113:716–726.

Arner, D. H., H. R. Robinette, J. E. Frasier, and M. Gray. 1976. Effects of channelization of the Luxapalila River on fish, aquatic invertebrates, water quality, and furbearers. U.S. Fish and Wildlife Service Biological Services Program FWS/OBS-76-08.

Barnes, H. H., Jr. 1967. Roughness characteristics of natural channels. U.S. Geological Survey Water-Supply Paper 1849.

Bathurst, J. C. 1979. Distribution of boundary shear stress in rivers. Pages 95–116 *in* D. D. Rhodes and G. P. Williams, editors. Adjustment of fluvial systems. Kendall/Hunt, Dubuque, Iowa.

Benke, A. C., R. L. Henry III, D. M. Gillespie, and R. J. Hunter. 1985. Importance of snag habitat for animal production in southeastern streams. Fisheries (Bethesda) 10(5):8–13.

Beschta, R. L., and W. S. Platts. 1986. Morphological features of small streams: significance and function. Water Resources Bulletin 22:369–379.

Bilby, R. E., and G. E. Likens. 1980. Importance of organic debris dams in the structure and function of stream ecosystems. Ecology 61:1107–1113.

Bisson, P. A., and eight coauthors. 1987. Large woody debris in forested streams: past, present, and future. University of Washington Institute of Forest Resources Contribution 57:143–190.

Brinson, M. M., H. D. Bradshaw, and E. S. Kane. 1984. Nutrient assimilative capacity of an alluvial floodplain swamp. Journal of Applied Ecology 21:1041–1057.

Bruk, S., and Z. Volf. 1967. Determination of roughness coefficients for very irregular rivers with large floodplains. Proceedings of the Twelfth Congress of the International Association for Hydraulic Research, International Association for Hydraulic Research 1:95–99, Delft, Netherlands.

Burgess, S. A., and J. R. Bider. 1980. Effects of stream habitat improvements for invertebrates, trout populations, and mink activity. Journal of Wildlife Management 44:871–880.

Burkham, D. E. 1976. Hydraulic effects of changes in bottom-land vegetation on three major floods, Gila River in southeastern Arizona. U.S. Geological Survey Professional Paper 655-J.

Chow, V. T. 1959. Open channel hydraulics. McGraw-Hill, New York.

Congdon, J. C. 1971. Fish populations of channelized and unchannelized sections of the Chariton River, Missouri. Pages 52–62 *in* E. Schneberger and J. L. Funk, editors. Stream channelization: a symposium. American Fisheries Society, North Central Division, Special Publication 2, Bethesda, Maryland.

Cowan, W. L. 1956. Estimating hydraulic roughness coefficients. Agricultural Engineering 37:473–475.

Décamps, H., M. Fortuné, F. Gazalle, and G. Pantou. 1988. Historical influence of man on the riparian dynamics of a fluvial landscape. Landscape Ecology 1:163–173.

Funk, J. L., and J. W. Robinson. 1974. Changes in the channel of the lower Missouri River and effects on fish and wildlife. Missouri Department of Conservation, Aquatic Series 11, Jefferson City.

Gard, R. 1961. Creation of trout habitat by constructing small dams. Journal of Wildlife Management 25: 384–390.

Gregory, K. J., editor. 1977. River channel changes. Wiley, New York.

Griswold, B. L., C. Edwards, L. Woods, and E. Weber. 1978. Some effects of stream channelization on fish populations, macroinvertebrates, and fishing in Ohio and Indiana. U.S. Fish and Wildlife Service Biological Services Program FWS/OBS-77/46.

Hansen, D. R. 1971. Stream channelization effects on fishes and bottom fauna in the Little Sioux River, Iowa. Pages 29–51 *in* E. Schneberger and J. L. Funk, editors. Stream channelization: a symposium. American Fisheries Society, North Central Division, Special Publication 2, Bethesda, Maryland.

Harmon, M. E., and 12 coauthors. 1986. Ecology of coarse woody debris in temperate ecosystems. Advances in Ecological Research 15:133–302.

Helmers, A. E. 1966. Some effects of log jams and flooding in a salmon spawning stream. U.S. Forest Service Research Note NOR-14.

Hickin, E. J. 1984. Vegetation and river channel dynamics. Canadian Geographer 28:111–126.

Hickman, G. D. 1975. Value of instream cover to the fish populations of Middle Fabius River, Missouri. Missouri Department of Conservation, Aquatic Series 14, Jefferson City.

House, R. A., and P. L. Boehne. 1985. Evaluation of instream enhancement structures for salmonid spawning and rearing in a coastal Oregon stream. North American Journal of Fisheries Management 5:283–295.

House, R. A., and P. L. Boehne. 1986. Effects of instream structure on salmonid habitat and populations in Tobe Creek, Oregon. North American Journal of Fisheries Management 6:38–46.

Hunt, R. L. 1971. Responses of a brook trout population to habitat development in Lawrence Creek. Wisconsin Department of Natural Resources Technical Bulletin 48.

Jarrett, R. D. 1984. Hydraulics of high-gradient streams. ASCE (American Society of Civil Engineers) Journal of the Hydraulics Division 110:1519–1539.

Karr, J. R., and D. R. Dudley. 1981. Ecological perspectives on water quality goals. Environmental Management 5:55–68.

Karr, J. R., and O. T. Gorman. 1975. Effects of land treatment on the environment. U.S. Environmental Protection Agency, Report EPA-905/9-75-007, Chicago.

Karr, J. R., L. A. Toth, and G. D. Garman. 1983. Habitat preservation for midwest stream fishes: principles and guidelines. U.S. Environmental Protection Agency, EPA-600/3-83-006, Corvallis, Oregon.

Keown, M. P., N. R. Oswalt, E. B. Perry, and E. A.

Dardeau, Jr. 1977. Literature survey and preliminary evaluation of streambank protection methods. U.S. Army Corps of Engineers, Waterway Experiment Station, Technical Report H-77-9, Vicksburg, Mississippi.
King, L. R., and K. F. Carlander. 1976. A study of the effects of stream channelization and bank stabilization on warmwater sport fish in Iowa. U.S. Fish and Wildlife Service Biological Program FWS/OBS-76-13.
Klingeman, P. C., and J. B. Bradley. 1976. Willamette River basin streambank stabilization by natural means. Oregon State University, Water Resources Research Institute, Corvallis.
Leopold, L. B., M. G. Wolman, and J. P. Miller. 1964. Fluvial processes in geomorphology. Freeman, San Francisco.
Li, R., and H. W. Shen. 1973. Effect of tall vegetations on flow and sediment. ASCE (American Society of Civil Engineers) Journal of the Hydraulics Division 99:793–814.
Linsley, R. K., M. A. Kohler, and J. L. H. Paulus. 1982. Hydrology for engineers. McGraw-Hill, New York.
Lowrance, R., R. Todd, J. Fail, Jr., O. Hendrickson, Jr., R. Leonard, and L. Asmussen. 1984. Riparian forests as nutrient filters in agricultural watersheds. BioScience 34:374–377.
Lynch, J. A., E. S. Corbett, and R. Hoopes. 1977. Implications of forest management practices on the aquatic environment. Fisheries (Bethesda) 2(2):16–22.
Marzolf, G. R. 1978. The potential effects of clearing and snagging on stream ecosystems. U.S. Fish and Wildlife Service Biological Services Program FWS/OBS-78/14.
Matthews, W. J., and D. C. Heins. 1987. Community and evolutionary ecology of North American stream fishes. University of Oklahoma Press, Norman.
Meehan, W. R., F. J. Swanson, and J. R. Sedell. 1977. Influences of riparian vegetation on aquatic ecosystems with particular reference to salmonid fishes and their food supply. U.S. Forest Service General Technical Report RM-43:137–145.
Mendelson, J. 1975. Feeding relationships among species of *Notropis* (Pisces: Cyprinidae) in a Wisconsin stream. Ecological Monographs 45:199–230.
Mosley, M. P. 1981. The influence of organic debris on channel morphology and bedload transport in a New Zealand stream. Earth Surface Processes and Landforms 6:571–579.
Nabhan, G. P. 1985. Riparian vegetation and indigenous southwestern agriculture: control of erosion, pests, and microclimate. U.S. Forest Service General Technical Report RM-120:232–236.
North, R. M., A. S. Johnson, H. O. Hillestad, P. A. R. Maxwell, and R. C. Parker. 1974. Survey of economic–ecologic impacts of small watershed development. University of Georgia, Institute of Natural Resources, Technical Completion Report ERC 0974, Athens.
Nunnally, W. R. 1978. Stream renovation: an alternative to channelization. Environmental Management 2:403–411.
Nunnally, W. R., and E. Keller. 1979. Use of fluvial processes to minimize adverse effects on stream channelization. University of North Carolina, Water Resources Research Institute, Report 144, Raleigh.
Parsons, D. A. 1963. Vegetative control of streambank erosion. U.S. Department of Agriculture Miscellaneous Publication 970:130–136.
Petryk, S., and G. Bosmajian III. 1975. Analysis of flow through vegetation. ASCE (American Society of Civil Engineers) Journal of the Hydraulics Division 101:871–884.
Petterjohn, W. T., and D. L. Correll. 1984. Nutrient dynamics in a agricultural watershed: observations on the role of the riparian forest. Ecology 65:1466–1475.
Pickles, G. W. 1931. Run-off investigations in central Illinois. University of Illinois, Engineering Experiment Station, Bulletin 232, Urbana.
Ramser, C. E. 1929. Flow of water in drainage channels. U.S. Department of Agriculture Technical Bulletin 129.
Ree, W. O., and V. J. Palmer. 1949. Flow of water in channels protected by vegetative linings. U.S. Department of Agriculture Technical Bulletin 967.
Richards, K. 1982. Rivers: form and process in alluvial channels. Methuen, New York.
Schumm, S. A. 1977. The fluvial system. Wiley, New York.
Sedell, J. R., F. H. Everest, and F. J. Swanson. 1982. Fish habitat and streamside management: past and present. Pages 244–255 *in* Proceedings of the Society of American Foresters Annual Meeting. Society of American Foresters, Bethesda, Maryland.
Sedell, J. R., and J. L. Froggett. 1984. Importance of streamside forests to large rivers: the isolation of the Willamette River, Oregon, U.S.A., from its floodplain by snagging and streamside forest removal. Internationale Vereinigung für theoretische und angewandte Limnologie Verhandlungen 22: 1828–1834.
Sheldon, A. L. 1968. Species diversity and longitudinal succession in stream fishes. Ecology 49:193–198.
Shields, F. D., and N. R. Nunnally. 1984. Environmental aspects of clearing and snagging. ASCE (American Society of Civil Engineers) Journal of the Environmental Engineering Division 110:152–165.
Smith, D. G. 1976. Effect of vegetation on lateral migration of anastomosed channels of a glacier meltwater river. Bulletin of the Geological Society of America 87:857–860.
Speaker, R. W., K. J. Luchessa, and J. F. Franklin. 1988. The use of plastic strips to measure leaf retention by riparian vegetation in a coastal Oregon stream. American Midland Naturalist 120:22–31.
Speaker, R., K. Moore, and S. Gregory. 1984. Analysis of the process of retention of organic matter in stream ecosystems. Internationale Vereinigung für

theoretische und angewandte Limnologie Verhandlungen 22:1835–1841.

Triska, F. J., and K. Cromack, Jr. 1979. The role of wood debris in forests and streams. Pages 171–190 *in* R. H. Waring, editor. Forests: fresh perspectives from ecosystem analysis. Oregon State University, Corvallis.

Van Cleef, J. S. 1885. How to restore our trout streams. Transactions of the American Fisheries Society 14:50–55.

Wallace, J. B., and A. C. Benke. 1984. Quantification of wood habitat in subtropical Coastal Plain streams. Canadian Journal of Fisheries and Aquatic Sciences 41:1643–1652.

Wharton, C. H., and M. M. Brinson. 1978. Characteristics of southeastern river systems. U.S. Forest Service General Technical Report WO-12:32–40.

Wilson, K. V. 1973. Changes in floodflow characteristics of a rectified channel caused by vegetation, Jackson, Mississippi. Journal of Research of the U.S. Geological Survey 1:621–625.

Wilson, L. G. 1967. Sediment removal from flood water by grass filtration. Transactions of the ASAE (American Society of Agricultural Engineers) 10: 35–37.

Zimmerman, R. C., J. C. Goodlett, and G. H. Comer. 1967. The influence of vegetation on channel form of small streams. Bulletin of the International Association of Scientific Hydrology 75:255–275.

American Fisheries Society Symposium 10:176–190, 1991

Rehabilitation of Estuarine Fish Habitat at Campbell River, British Columbia

C. D. LEVINGS AND J. S. MACDONALD

Department of Fisheries and Oceans
Biological Sciences Branch, West Vancouver Laboratory
4160 Marine Drive, West Vancouver, British Columbia V7V 1N6, Canada

Abstract.—This study, at the Campbell River estuary in British Columbia, involved a detailed evaluation of the use of artificial islands by juvenile salmonids and their epibenthic prey species. The islands were built in 1982 at an intertidal portion of the estuary used for log storage since 1904 and rehabilitated as part of an estuary management strategy. Wild salmonid fry, particularly chinook salmon *Oncorhynchus tshawytscha*, were caught in the created habitat in numbers equal to or greater than in the reference habitats. Hatchery fish were less abundant in the islands. The abundance of several invertebrates increased with time at the created habitat sites. By the third or fourth year after island construction, they had reached densities similar to those at reference site (e.g., *Manayunkia* spp., oligochaetes, chironomids, and other dipterans). However, the abundance of some amphipods (e.g., *Eogammarus* sp.) did not match that observed at the reference site. Invertebrate communities located lower in the intertidal zone merged with reference communities much faster than those higher on the beaches. The study was not designed to determine if the artificial islands were performing all the ecological functions of natural habitats. However, after 5 years they were providing detrital production, fish habitat, and food items that are frequently eaten by juvenile salmonids. This project therefore resulted in a net gain of the capability of the Campbell River estuary to support fish production.

Interest in restoration methodology for salmonid habitat has been growing in recent years, and some available publications explain and evaluate freshwater restoration techniques for spawning and rearing areas (e.g., SEP 1980). Estuarine habitats have been degraded in southwestern British Columbia and wild salmonids, especially chinook salmon *Oncorhynchus tshawytscha*, are known to rear in estuaries (e.g., Levings et al. 1986). Estuarine restoration is therefore being considered as a technique for modifying the capacity of coastal habitats to produce salmon and, where necessary, to compensate habitat losses due to industrial developments.

As pointed out by Race (1985) very few long-term research projects have evaluated any estuarine restoration projects, especially on the Pacific coast. Our study, at the Campbell River estuary in British Columbia, involved a detailed 5-year evaluation of the use of artificial islands (built in 1982) by juvenile salmonids and their epibenthic prey species. A full description of the techniques used for constructing these islands has been presented elsewhere (Brownlee et al. 1984) along with results of fish and invertebrate studies that were obtained in the first year of evaluation. This paper presents the results of the 5-year program. Except for work by Shreffler et al. (1988) in the Puyallup estuary in Puget Sound, projects that have been evaluated on the Pacific coast have not considered fish and invertebrates (Good 1987). The short-term results from the Puyallup estuary project showed that juvenile salmon utilized a developed habitat, in this instance a wetland created by removal of landfill and planting of vegetation (Shreffler et al. 1988). Wilcox (1986) found that bird use of restored estuarine habitat in California was probably related to succession of the invertebrate communities used as food by waterfowl.

The objectives of our study were to answer several questions relating to the fish and invertebrate use of the islands. Did juvenile salmonids occupy island habitats and were levels of abundance and patterns of use comparable to reference habitats within the estuary? Did macrobenthic invertebrates used as fish food colonize all intertidal elevations of the new islands, and did community type, abundance, and species composition differ from a reference habitat? What role did vegetation play in development of the invertebrate communities?

Study Area and Island Construction

The Campbell River estuary is on the northeast coast of Vancouver Island and connects with Discovery Passage (Figure 1). Most of the estuarine habitat is located behind Tyee Spit (Figure 1) and comprises about 73 hectares of intertidal

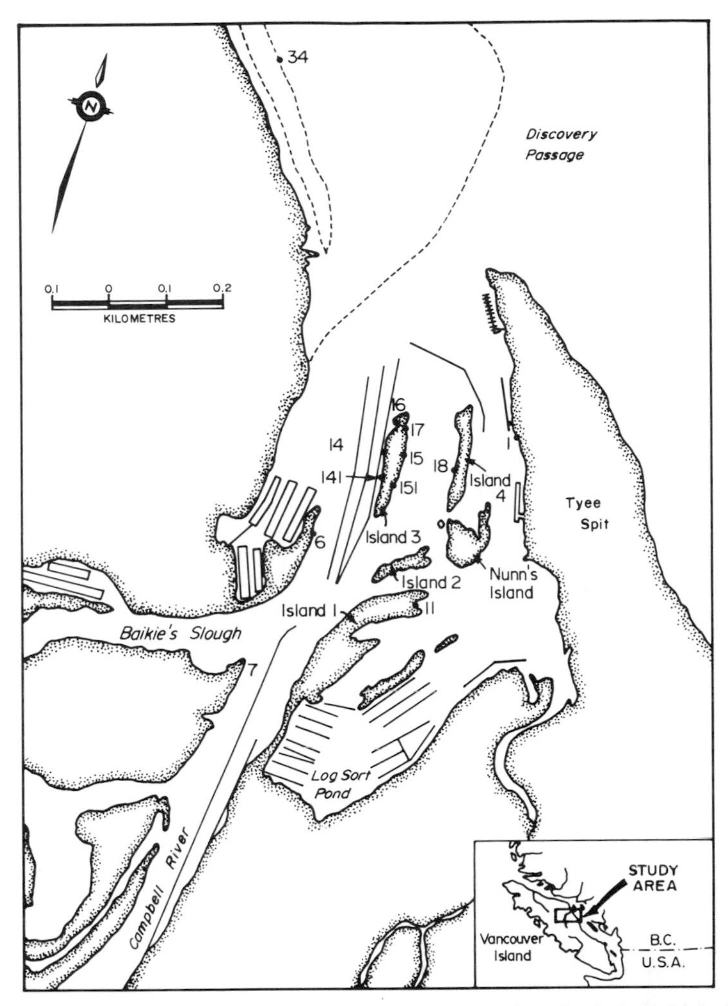

FIGURE 1.—Map of the study area showing the four constructed islands (Islands 1, 2, 3, 4). Nunn's Island was used as a reference location for collection of invertebrate and vegetation samples. Fish counts were made from seine catches at eight sites indicated on Islands 1, 3, and 4 and from three reference sites indicated at other locations in the estuary.

lands and channels. The tidal range in the estuary is about 5 m. The river is regulated by a hydroelectric dam built in 1947 and discharges are partially controlled. Natural accretion in the estuary is very reduced because the suspended sediment load of the river is low. Natural habitats in the estuary consist of gravel beaches, marshes dominated by sedges (especially *Carex lyngbyei*), and river channels with riparian vegetation (Kennedy 1982). The estuary has been extensively modified by bank protection, marina construction, log storage, and gravel removal. A more complete description of past industrial activity in the estuary has been published by Bell and Thompson (1977).

In 1981, an area within the estuary used for storage and sorting of logs since 1904 was reduced from 32.8 hectares to 6.8 hectares by B.C. Forest Products Ltd. (now Fletcher Challenge Canada). Wood debris and wastes from log storage activities were removed. Four artificial islands were then constructed within about 3.2 hectares (including side slopes) of the middle sector of the estuary with gravel dredged from the construction of a new dryland log-sorting facility (Figure 1). Dredging of terrestrial and riparian habitat was also required to construct a log pond (7.8 hectares) for temporary storage of logs before their removal by towboat. The artificial islands were constructed with embayments to maximize shoreline perimeter. Except for the south end of Island 3, island habitats were totally submerged at high tide. The embayments provided low-tide refuge for fish because they did not dewater except on low spring tides. Vegetation salvaged from construction activity was planted in specific patterns on the islands in February and March 1982. Approximately 23,000 cores of *Carex lyngbyei*, *Juncus balticus*, and other wetland vegetation were transplanted (Brownlee et al. 1984). Close to half of the area of the islands was left unplanted to see if adjacent vegetation would colonize the vacant areas.

Each year the Quinsam River hatchery, about 8 km upstream of the estuary on a tributary of the Campbell River, released 2–3 million reared salmonids as fry or smolts. The hatchery released 71,565 chum salmon fry *Oncorhynchus keta* in 1982, but none in other years. Several million fry of pink salmon *O. gorbuscha* (range, 2.8–7.2 million per year, 1982–1985) were released from the hatchery or reared in pens in the estuary. Coho salmon *O. kisutch* smolt releases from the hatchery (including outplanted fish, i.e., smolts produced from fry released in the Quinsam River watershed in September of the previous year; assumed survival rate of 25%) ranged from approximately 0.37 million in 1985 to 3.93 million in 1984. Steelhead *O. mykiss* releases (smolts) ranged from about 20,500 in 1986 to 76,700 in 1982. Chinook salmon from both production and experimental releases ranged from about 0.76 million smolts in 1982 to 3.13 million smolts in 1986.

Reference stations used to compare fish catches and invertebrate communities with those at the artificial islands were chosen primarily because of their proximity to the developed sites. Nunn's Island in the center of the estuary (Figure 1) was sampled as a reference habitat for invertebrates and vegetation. Three stations were used as reference sites for fish abundance (Figure 1): station 1 (MR: north side of channel separating Nunn's Island from Tyee Spit, a site partially disturbed by an adjacent seaplane ramp and having gravel substrate with fringes of sedge in the high intertidal zone); station 6 (BULK: west side of Campbell River across from Island 3, a gravel–cobble beach on artificial fill placed about 30 years ago to construct a parking lot and supporting riparian vegetation of willows *Salix* spp. above high tide); and station 7—(NBM: mouth of Baikie's Slough at the junction with Campbell River, having eelgrass *Zostera marina* at lower elevations and sedges and riparian vegetation at higher elevations).

Methods

Vegetation.—Detailed data on vegetation patterns and their change through time are being analysed by Neil Dawe (Canadian Wildlife Service, Qualicum Beach, British Columbia). However, the quadrat sampling for invertebrates in our program included incidental observations on percent cover of the prominent species, *Carex lyngbyei*, *Juncus arcticus*, and the algae *Enteromorpha* sp. and *Fucus* sp., at six sites on each of the four constructed islands and on the Nunn's Island reference site (Riley et al. 1987). Estimates were made during May, June, and July of 1982 to 1986.

Invertebrate sampling.—Annual benthic samples in planted areas on the tops of islands were obtained and analyzed by contractors (Riley et al. 1987) to B.C. Forest Products Ltd. working under our direction. The benthic sampling in May 1983 was conducted by a contractor to the Department of Fisheries and Oceans (Anderson and Galbraith 1985) and included embayment sites at lower elevations not sampled in the annual work.

Macroinvertebrates at higher elevations were sampled by quadrat at low tide in May, June, and July 1982–1986. Elevations of sites on Islands 1, 2, and 4 were about the same, approximately 3.0 m above chart datum. Quadrats on Island 3 were about 0.5 m lower. Three replicate samples were obtained by randomly placing a wooden quadrat (0.06 m^2) at sites within (vegetated) and outside of (unvegetated) planted areas, so that six samples were obtained on each constructed island and on Nunn's Island in each of the three months. Detailed information on station locations is presented elsewhere (Riley et al. 1987). Sediments were mainly gravel, sand, and cobble mixed with a thin layer ($<$0.5 cm) of silt.

In July 1982, quadrat samples (0.06 m^2; three replicates) were obtained at seven additional sites in the embayments of Islands 1 and 3 (elevation about 1.0 m over chart datum). These sites were approximately 2.0 m lower than the tops of islands where vegetation was planted and were characterized by finer sediments, mainly silt and mud (Riley et al. 1987). In May 1983, the sites in the embayments on Islands 1 and 3 were resampled, and three new sites were added in the embayments on Islands 2 and 4 and at lower elevations in an eelgrass bed at the mouth of Baikie's Slough (Figure 1). Methods used were identical to those used in the regular surveys, and details on sampling are described elsewhere (Anderson and Galbraith 1985).

Fish sampling.—Juvenile salmonids were sampled with a 15 × 1.5-m beach seine, described in detail in Brown et al. (1987b), at eight sites located on Islands 1, 3, and 4 and at three reference sites within the estuary (Figure 1). Sampling was conducted approximately biweekly in March–December 1982, January–December 1983, and March–September 1984, 1985, and 1986. Each year, Island 1 (one site) was sampled with a minimum of 15–20 hauls, Island 3 (four sites) with 84–100 hauls, and Island 4 (two sites) with 28–50 hauls. Each of the reference sites was sampled at least 28 times per year. At some sites, seining was conducted more than once per trip so that sample sizes for each sampling trip varied from 4 to 18 seine hauls. At sites 18, 7, 6, and 1, duplicate seine hauls were obtained with an outboard motor boat. One end of the seine was held on shore and the boat was used to pull the other end out and to retrieve the net in a semicircle against the beach. All other stations were narrow embayments that could be sampled entirely by hand with one pass of the net.

For the purposes of abundance estimates, the number of hatchery chinook salmon in each haul was computed by adding the number of marked fish (adipose fin removed) to the number identified as unmarked hatchery fish (CHINM and CHINH in Brown et al. 1987b).

To help characterize physiochemical conditions in the estuary, surface salinity and temperature measurements were usually obtained at low tide during trips for fish sampling. A Beckman RS5-3 portable salinometer was used.

Statistical analysis.—Separate discriminant functions analysis (DFA) was used to examine spatial and temporal changes in vegetation cover, invertebrate densities, and fish abundance among the sample sites. A group in the data matrix consisted of pooled samples taken from an island or reference site during each year. In testing spatial and temporal differences among groups in this manner, DFA was being used in much the same manner as a MANOVA (multivariate analysis of variance). Both tests share the same underlying mathematical method: comparison of pooled within-group variation with pooled among-group variation (Porebski 1966). We did not compare groups with a predetermined probability level because our intent was to use DFA as an exploratory rather than a hypothesis-testing tool. Used in this manner, DFA is robust in the analysis of data that violate the assumptions of multivariate normality and equality of covariance matrices. The plots and histograms that resulted from the analyses provided a more effective method of describing group separation and spatial orientation than formal multivariate tests of significance (Green 1974).

By considering the weighting coefficients (or loadings) of variables (species of fish or vegetation) on the first two or three discriminant functions, we were also able to identify variables that exhibited a great degree of difference among groups (Green 1979).

Anderson and Galbraith (1985) used principal components analysis, a related multivariate technique, to analyze benthic invertebrate data from the quadrat samples obtained in May 1983. Details on procedures are found in Orloci (1967). Raw data were transformed ($\log_{10}[X + 1]$), converted to a covariance matrix, and then ordinated or projected onto a small number of statistically independent axes. The first axis-array samples account for the maximum variance, and successive axes account for decreasing amounts of variance. Usually the first two or three axes account for most of the original sample variance. Depending on the variability of the data, interpretable patterns emerge when data are plotted either as aspects of the sampling environment that correspond to the order of samples along the axes or as discrete groupings of samples.

Results

Temperature and Salinity

Surface salinity at Islands 1, 3, and 4 ranged from 0‰ to approximately 9‰ during the five field seasons. The seasonal fluctuations at reference sites 6 and 7 were very similar to those observed at the island sites. However, surface water at station 1, closer to Discovery Passage, was more saline, ranging up to about 15‰ in July 1982 and May 1986.

Surface temperatures at island and reference sites ranged from winter lows of approximately 4°C to about 21°C in August. Maximum summer temperatures in the island habitats, especially at Islands 1 and 4, were usually higher than those at Island 3 or the reference locations. The shallow embayments at low tide were subjected to diurnal heating effects, especially at Islands 1 and 4. Temperatures at Island 3 and the reference locations were cooled by Campbell River and Discovery Passage water, respectively.

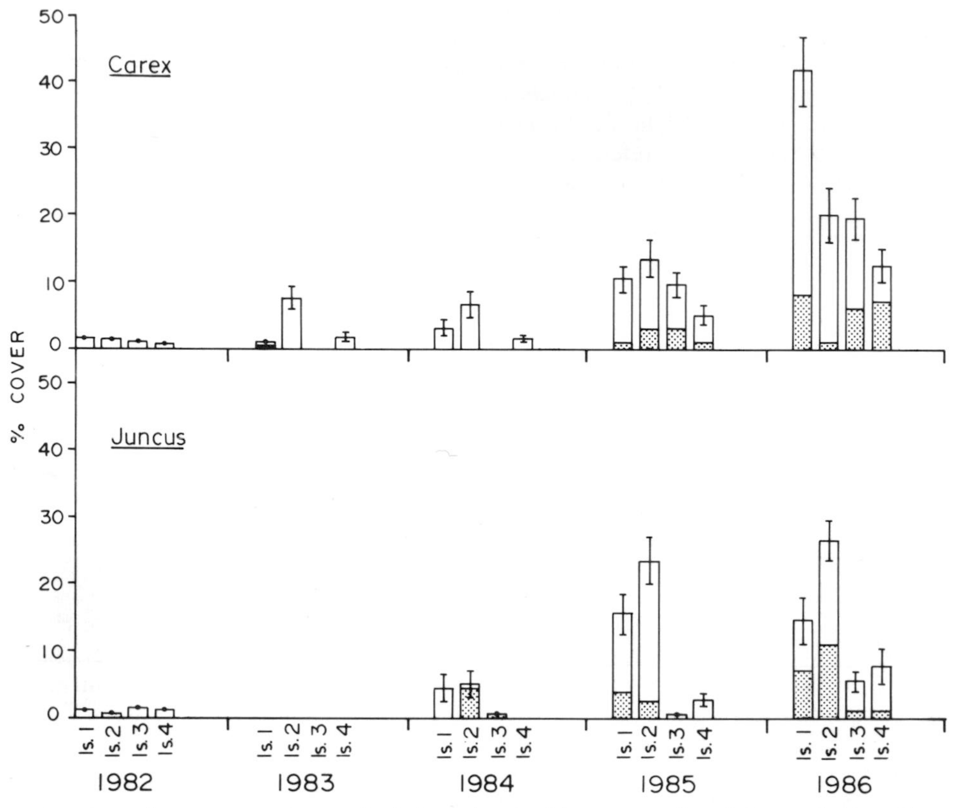

FIGURE 2.—Annual mean cover (percent) of *Carex lyngbyei* (upper panel) and *Juncus balticus* (lower panel) observed in benthic quadrat sampling (0.06 m^2) at artificial islands, 1982–1986. Shading indicates the proportion of the cover estimate that was contributed by samples from unplanted (naturally established) areas. Vertical bars are ± 1 SE of the mean. Cover estimates are minimum values and indicate temporal trend only because the quadrat used for sampling covered insufficient area.

Vegetation

As estimated from percent cover in quadrats, there was a significant increase in areal coverage of vegetation from 1982 to 1986 ($P < 0.05$: DFA). *Carex lyngbyei* increased most significantly on Island 1, from less than 5% in 1982 to more than 40% in 1986 (Figure 2). *Juncus arcticus* increased on Island 2 from 0% in 1982 to about 25% in 1986. These are minimum estimates because, as indicated above, vegetation observations were only made incidentally to benthic invertebrate sampling. The quadrat used for invertebrate sampling covered an insufficient area (0.06 m^2) for adequately estimating vegetation cover. N. Dawe (Canadian Wildlife Service, personal communication) provided the following detailed information on cover and biomass from 1986 sampling, which used standard botanical methods: Island 1—90% cover by *Carex lyngbyei* and *Eleocharis palustris*; mean aboveground biomass of *C. lyngbyei*, 380 g dry weight/m^2; Island 2—75% cover by *C. lyngbyei*, *Potentilla pacifica*, and *Juncus arcticus*; mean aboveground biomass of *C. lyngbyei*, 292 g dry weight/m^2; Island 4—50% cover by *C. lyngbyei*, *E. palustris*, and *J. arcticus*; mean aboveground biomass *C. lyngbyei* 244 g dry weight/m^2. Vegetation transplanted on the islands in 1982 has grown and expanded onto the barren areas intentionally left between planted plots (Figure 2). Vegetation cover on Nunn's Island remained relatively constant (close to 100%) between 1982 and 1986 (Dawe, personal communication).

Invertebrate Communities

Benthic organisms in quadrats from the island sampling included representatives from most of the common estuarine taxa, as follows: nematodes, oligochaetes, polychaetes (six taxa identified), harpacticoids, isopods (two taxa), gammarid amphipods (seven taxa), ostracods, acarids, cu-

TABLE 1.—Occurrence—presence (+) or absence (−)—of benthic invertebrates sampled by quadrats on island tops and embayments, 1982–1986.

Taxon	Nunn's Island		Tops of Islands 1 to 4		Embayments of Islands 1, 3	
	1982	1986	1982	1986	1982	1983
Nematoda	+	+	+	+	+	+
Oligochaeta	+	+	+	+	+	+
Polychaeta (unknown)	+	+	+	+	−	−
Ampharetidae (unknown)	+	+	−	+	+	+
Arenicola marina	−	+	−	+	−	−
Hobsonia florida	−	−	−	−	−	+
Manayunkia aestuarina	+	+	+	+	+	+
Nereis limnicola	+	+	+	+	+	+
Corophium spinicorne	+	+	+	+	+	+
Corophium juveniles	+	+	+	+	+	+
Eogammarus sp.	+	+	+	+	+	+
Eogammarus confervicolus	+	+	+	+	+	+
Eogammarus o'clairi	+	+	−	+	+	−
Lagunogammarus setosus	−	−	−	−	+	−
Pontogeneia spp.	+	+	−	−	−	−
Gnorimosphaeroma insulare	+	+	+	+	+	+
Harpacticoida	+	+	+	+	+	+
Ostracoda	+	+	+	+	−	−
Ceratopogonidae larvae	−	+	−	+	−	−
Chironomid larvae	+	+	+	+	+	−
Diptera larvae			−	+	−	−
Diptera pupae	+	−	+	+	+	−
Diptera adults	+	+	+	+	−	−
Dolichopodidae larvae	+	+	+	+	−	−
Ephydridae larvae	−	−	+	−	−	−
Heleidae larvae	+	−	+	−	−	−
Simulidae pupae	−	−	−	+	−	−
Tabanidae larvae	+	−	−	−	−	−
Tipulidae pupae	−	+	−	+	−	−
Muscidae larvae	−	−	−	−	+	−
Acari	+	−	+	+	+	−
Coleoptera larvae	+	−	−	−	−	−
Collembola	+	−	−	−	−	−
Macoma sp.	−	−	−	−	−	+

maceans, mysids, insects (larvae, pupae, and adults: 15 taxa, mostly chironomids), and one bivalve mollusc. In addition, a few incidental invertebrates that could not be classified as benthic organisms were obtained (cladocerans, calanoid and cyclopoid copepods, hyperiid amphipods, lepidopteran adults, and coleopterans). Detailed tabulations of the data are presented elsewhere (Anderson and Galbraith 1985; Riley et al. 1987).

In May 1982, approximately 3 months after the islands were built, there was a fair degree of faunal similarity between Nunn's Island (reference location) and the artificial islands. Eighteen of 26 taxa that were recorded on Nunn's or the new islands were in common (Table 1). The same general pattern was observed in 1986. The fauna at lower elevations in both 1982 and 1983 was quite different from that at higher-elevation sites for both the artificial islands and Nunn's Island; only 14 out of 27 taxa were in common. This was mainly attributable to the lack of insects at the lower elevations.

Difficulties were encountered in identifying *Corophium* spp. and *Eogammarus* spp. to species, particularly in the juvenile stages. The majority of specimens were identified as *C. spinicorne* and *E. confervicolus*, which is consistent with other invertebrate surveys in the Campbell River estuary (e.g., Raymond et al. 1985). Less than 5% of the samples of *Eogammarus* sp. were *E. o'clairi*.

Invertebrate Abundance—Early Colonization of Embayments

Invertebrates appeared to colonize lower elevation habitats rapidly. Gammarid amphipods, specifically *E. confervicolus* and *Lagunogammarus setosus*, and mysids *Neomysis mercedis* were observed by qualitative sampling in the channels on Island 3 within a few days of its construction.

By July 1982, *Corophium spinicorne*, *E. confer-*

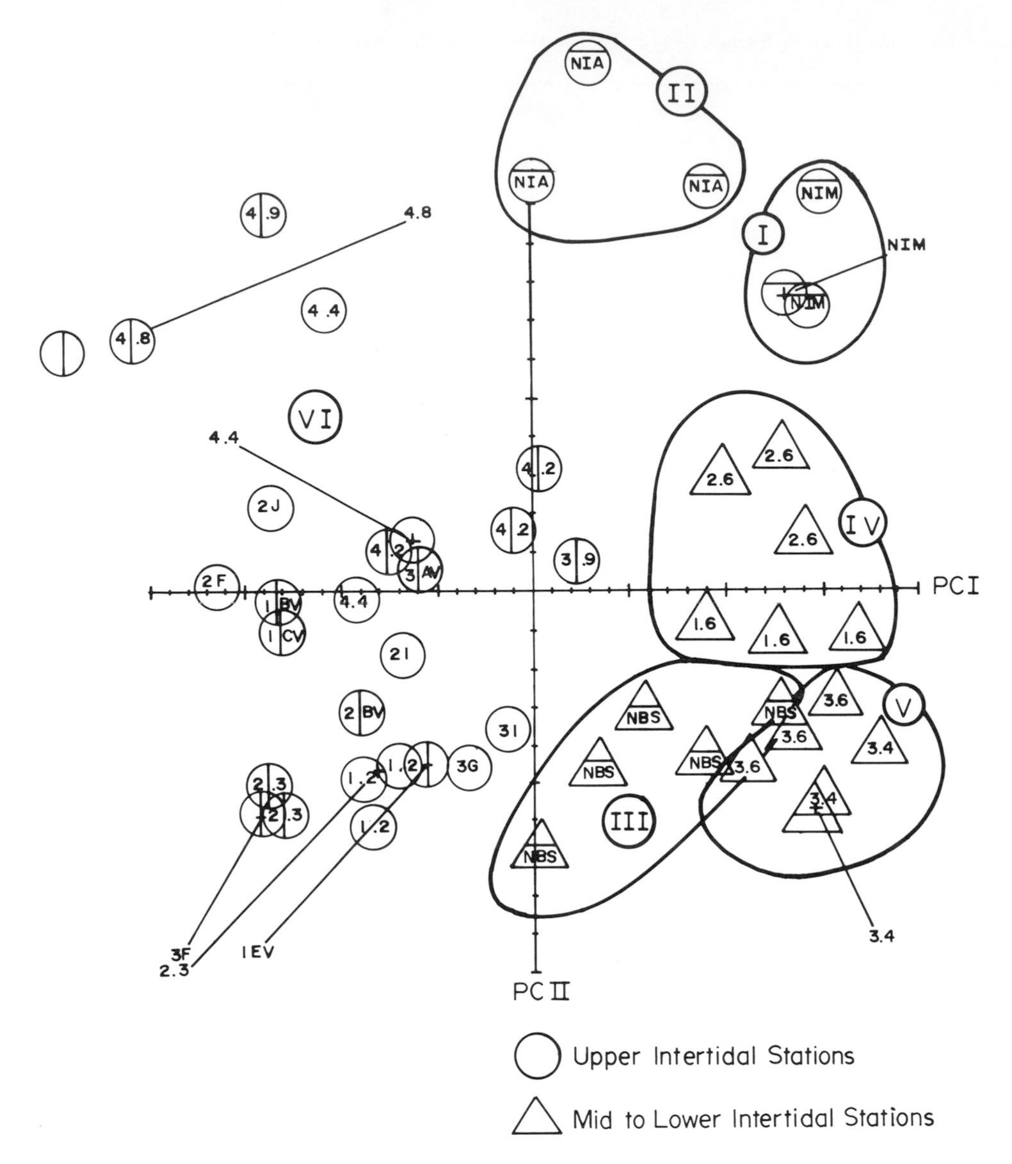

FIGURE 3.—Principal component analysis of benthic invertebrate data from island and reference stations sampled in May 1983 (from Anderson and Galbraith 1985). The first two components account for 42.1% of the variance in the data set.

vicolus, and *L. setosus* were more abundant in the embayments of Island 3 than in those of Island 1 (mean abundance: *C. spinicorne* 22,624 versus 4,320/m^2, *E. confervicolus* 19,616 versus 48/m^2). In May 1983 mean abundance of *E. confervicolus* and *C. spinicorne* in the embayments was 3,952 and 3,776/m^2 versus 12,720 and 26,624/m^2 at Islands 1 and 3, respectively.

There was a strong indication that the invertebrate assemblages sampled in the embayments on the experimental islands in May 1983 were more similar to low-elevation reference habitats than those at higher elevations on the islands. When results from a principal component analysis were plotted, the samples could be separated into six interpretable groups (Figure 3). There was considerable overlap between low-elevation, reference location assemblages (group III) and low-elevation, Island 3 assemblages (group V). Group III was dominated by *Corophium* spp. (75.2%), whereas prominent taxa in group V were *Corophium* spp. (26.5%), tubificids (17.3%), and *E. con-*

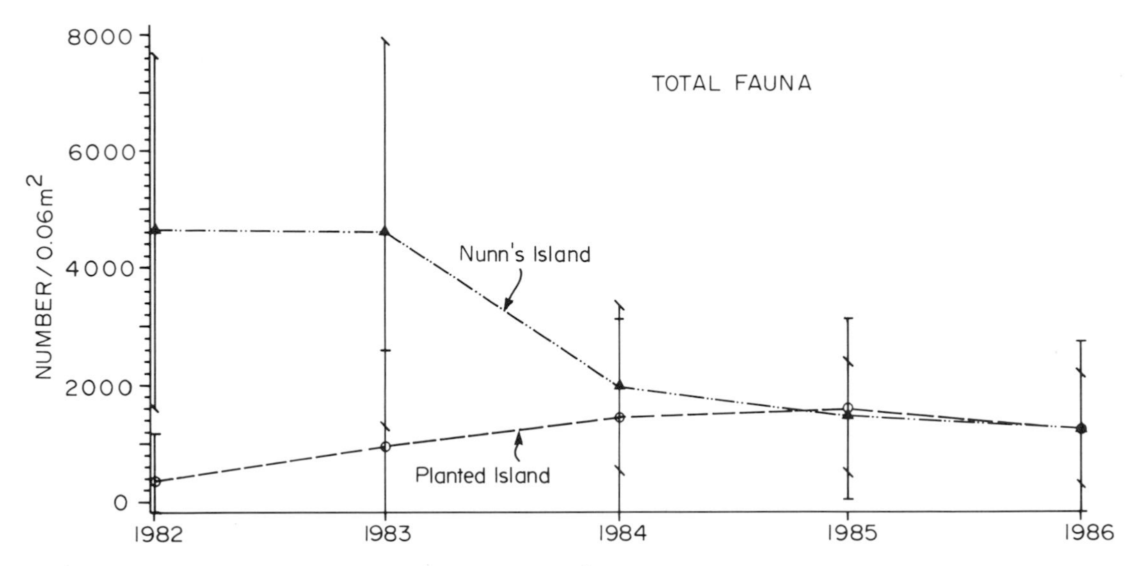

FIGURE 4.—Yearly changes in mean number (per 0.06 m^2) of all invertebrate taxa at the Nunn's Island reference site (dotted line) and at the experimental island sites (dashed lines). Vertical bars are ±1 SE of the mean.

fervicolus (17.0%). Most of the upper intertidal stations on the artificial islands were in group VI, separated on the left-hand side of the first principal component axis. Group VI was dominated by annelid worms, specifically enchytraeids (29.0%), the sabellid polychaete *Manayunkia aestuarina* (23.1%), and naidid oligochaetes (9.3%). The reference habitat groups (groups I and II) from marsh sites on Nunn's Island were characterized by *M. aestuarina* (30.3%), naidid oligochaetes (21.3%), and the isopod *Gnorimosphaeroma insulare* (14.2%).

Temporal Trends of Colonization at Higher Elevations

Of the invertebrate taxa identified in samples between 1982 and 1986 (Riley et al. 1987), only six occurred in large enough numbers for inclusion in the analysis of spatial and temporal variation (DFA). Of these, amphipods (*Eogammarus* sp. and *Corophium* sp.) and insect larvae (chironomids and other diperans) were frequently eaten by juvenile salmonids (Kask et al. 1986). Two taxa of abundant annelid worms (oligochaetes and *Manayunkia* sp.) were rarely consumed by salmon but are probably important food for the juvenile flatfish and sculpins found in the estuarine islands.

The abundance of invertebrates at Nunn's Island (reference location) declined during the course of the experiment ($P < 0.05$: DFA; Figure 4). The decline of *Manayunkia* sp., *Eogammarus* sp., oligochaetes, and dipterans other than chironomids was most pronounced between 1982 and 1984 (Figure 4). During the same period, the abundance of invertebrates on the experimental islands (e.g., *Manayunkia* sp., chironomids, and other dipterans and oligochaetes) increased in number until, by the third or fourth year of the experiment, they were comparable to abundances at the reference island (Figure 5). However, some of the invertebrates on the experimental islands did not achieve the abundance observed on the reference island by the end of the experiment. Despite their colonization and establishment on Island 3 by 1983, abundances of *Corophium* sp. and *Eogammarus* sp. usually were greater at the reference location than on the experimental islands throughout the experiment (Figure 5).

The degree to which a site was initially planted with vegetation had little effect on the abundance of most invertebrate taxa with the exception of the insect larvae, which rapidly colonized and became established on the vegetated sites. As vegetation expanded onto barren areas, insect abundance increased at unplanted sample sites. The abundance of dipteran larvae was spatially and temporally more variable than other invertebrate taxa.

Except for 1982 and 1983, maximum abundance of combined invertebrate taxa on the islands occurred in May of each year, and abundance was

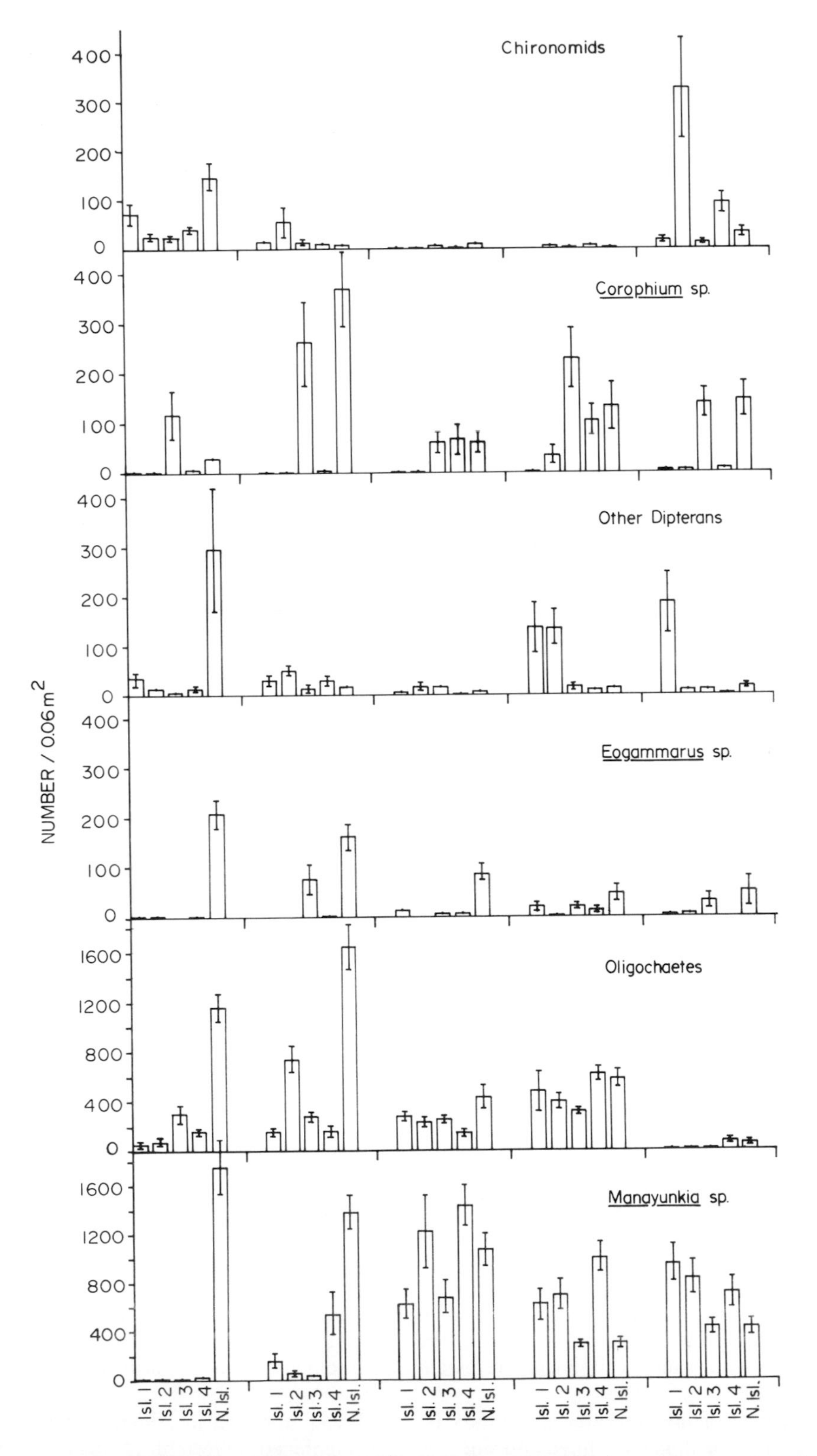

FIGURE 5.—Annual mean abundance (number/0.06 m^2) of six taxa of benthic invertebrates at Islands 1 to 4 and at Nunn's Island, 1982 (left-most group of bars) to 1986 (right-most group) sampled on tops of islands. Vertical bars are ±1 SE of the mean.

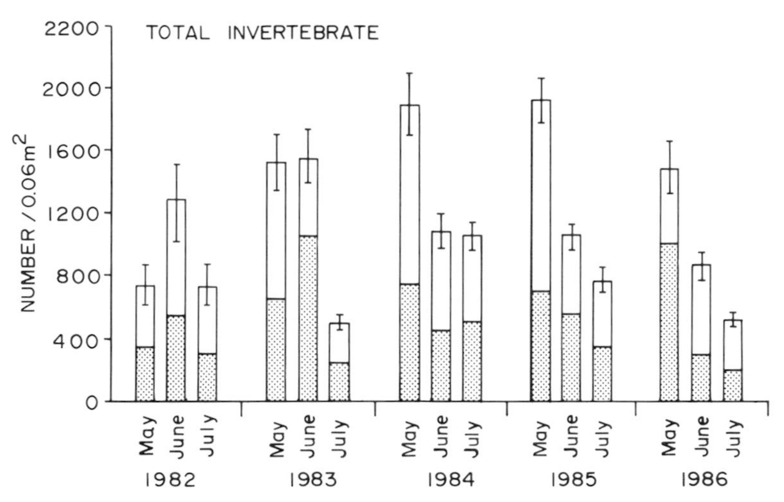

FIGURE 6.—Temporal changes in all invertebrate taxa (number/0.06 m^2) based on combined data from all islands. The shaded portion of each vertical bar indicates the proportion of the mean densities of fauna sampled from unplanted areas of the islands.

usually lower in June and July. Total invertebrate fauna decreased from 1984 to 1986; the maximum animal density occurred in May 1984 (Figure 6).

Fish Catches

Over the 5 years of sampling, juvenile salmonids originating from hatchery and wild stocks used the islands. Chinook salmon, coho salmon, chum salmon, pink salmon, steelhead, and cutthroat trout *O. clarki* were caught (references to complete catch reports are in Brown et al. 1987b).

Based on their loadings on the first two discriminant functions (DF), hatchery and wild chinook salmon exhibited greater spatial and temporal variation in abundance within the estuary than other salmonids ($P < 0.05$). Wild chinook salmon used BULK (reference site 6), NBM (reference site 7), and Island 3 (experimental sites) to a greater extent than other locations ($P < 0.05$: DF I—Figures 7, 8). BULK was used consistently throughout the study period, whereas Island 3 utilization increased and NBM utilization declined as the study progressed (1985–1986).

Peak catches of wild chinook salmon at island locations were recorded from May 3 to May 29 and at reference sites from May 3 to May 24, except for 1986 when the maximum occurred June 25–27. In each of the five field seasons, maximum catches were consistently recorded in the three tidal channels at the seaward end of Island 3 (stations 15, 16, 17). Wild chinook salmon ranged from 33 to 90 mm, except in 1983 when wild chinook salmon up to 180 mm were caught in November.

Catches of hatchery chinook salmon (62–189 mm long) at reference locations were as high as 200 fish/100 m^2 at NBM in 1985 (Figures 7, 8—DF II). However, with the exception of Island 1 in 1985, the islands were infrequently used by hatchery-produced fish. Timing of peak catches of hatchery fish at reference sites was closely related to hatchery release patterns. Increased catches of hatchery chinook salmon in 1985 and 1986 correlated with increased releases from the hatchery during those years (Figure 7). Except for 1984, when large catches of marked chinook salmon were recorded in April, maxima were recorded over the period May 4 to June 12. Timing of peak catches of hatchery chinook salmon in the islands was later (June 3 to July 10). In 1983, peak abundance occurred at island stations during November 7–9, indicating that some hatchery chinook salmon were rearing in the estuary even in late fall.

Annual mean abundance of coho salmon smolts (60 to 215 mm long) caught in the islands never exceeded 1.5 fish/100 m^2, compared with over five fish/100 m^2 at reference sites in 1984 and 1986 (Figure 7).

Abundance of chum salmon fluctuated widely among sites and years and was particularly low in 1983 and 1984. The largest catches of chum salmon were at MR (reference site 1) and Island 3 (experimental site); at Islands 1 and 4, abundance was much lower. Chum salmon fry were caught in island habitats, but larger fish (up to or over 100 mm) were taken at station 1 (MR). These fish were probably migrating to sea from other river sys-

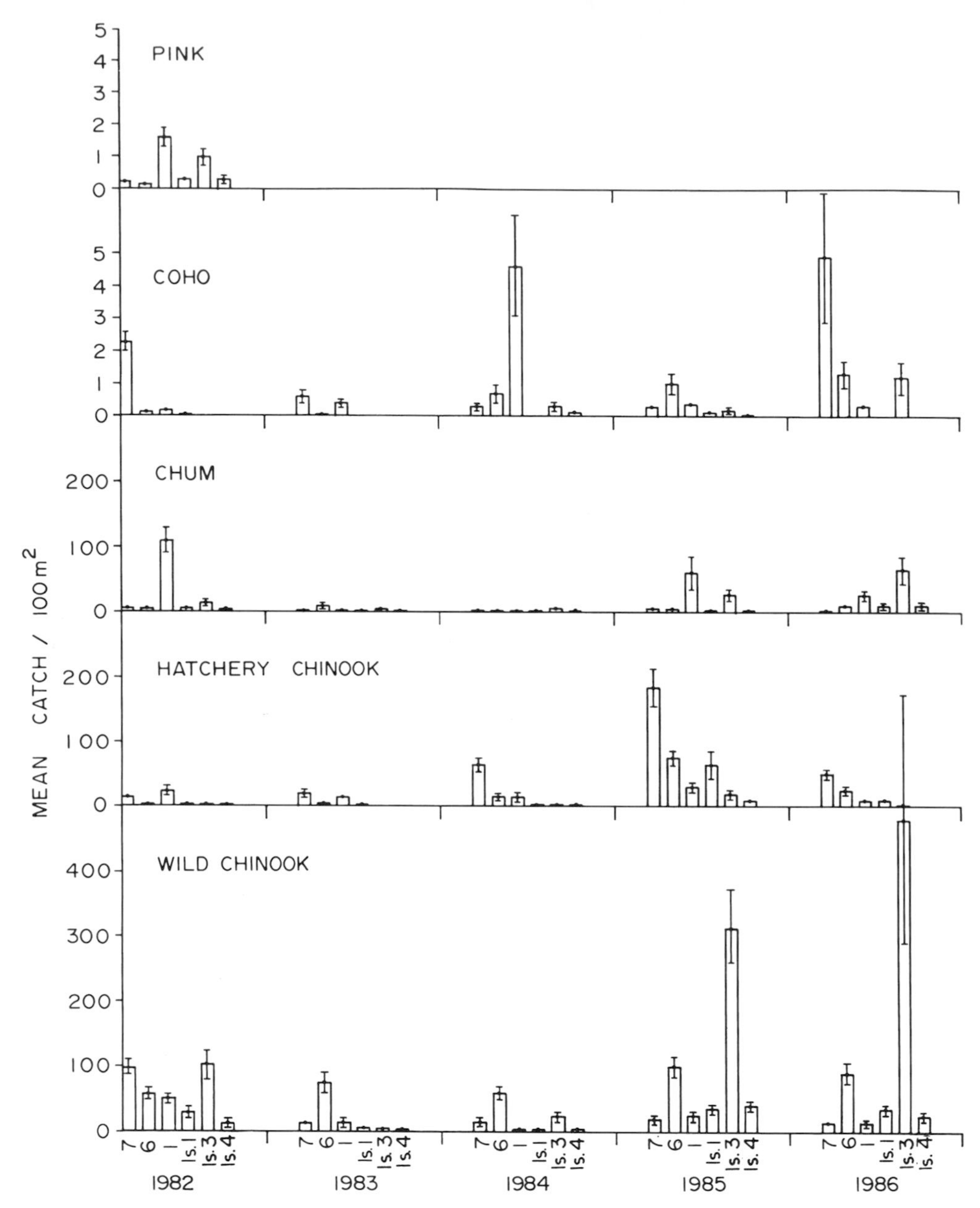

FIGURE 7.—Annual mean abundance (number/100 m^2) and standard deviation for juvenile pink, chum, coho, hatchery chinook, and wild chinook salmon caught at island and reference sites, 1982–1986. Note that coho and pink salmon catch data are on different scales than the other salmon species.

tems. Catches at MR between 1982 and 1986 declined, but abundance at the artificial islands increased over this period (Figure 7). At the islands, maximum catches were recorded in the four tidal channels at the seaward end of Island 3 (Figure 1—sites 14–17), except for 1986 when site 151 at the south end of Island 3 was heavily utilized (up to 100 fish/m^2). Peak catches of chum salmon at island and reference locations occurred during the period May 5–15 in each of the 5 years except in 1983, when maximum catches occurred in the islands March 28–30.

Despite the large number of pink salmon fry produced by the hatchery each year, this species was only taken in the estuary in 1982 (Figure 7). Abundance estimates for pink salmon were unre-

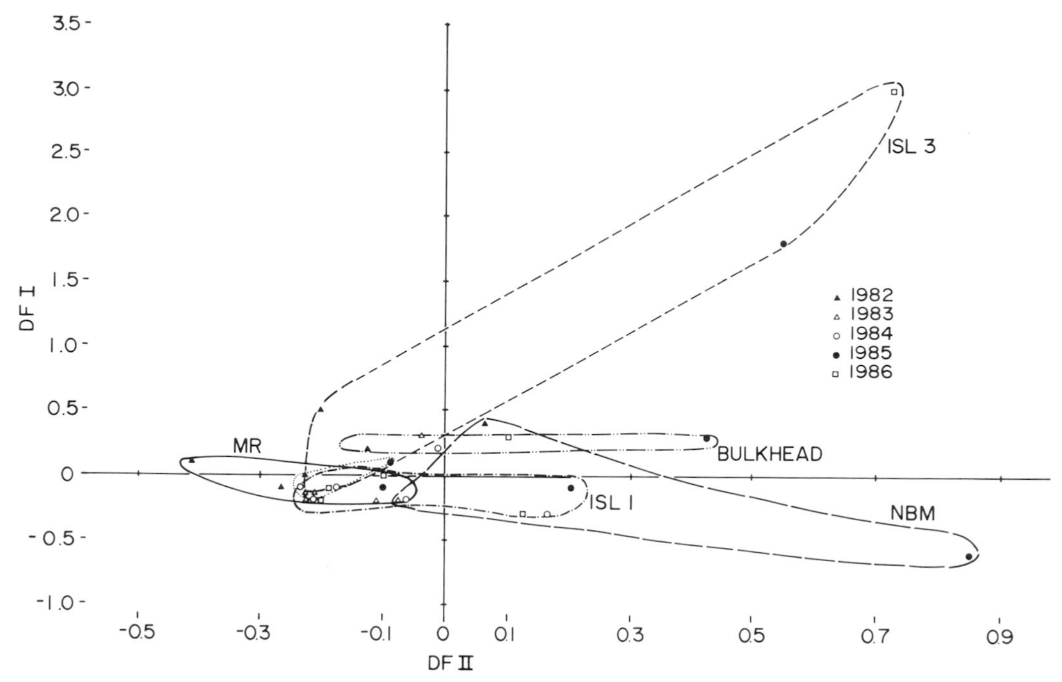

FIGURE 8.—Discriminant function analysis of salmonid catch data from island and reference sites, 1982–1986. Forty-four percent of the total variation was described by DF I and 23% was described by DF II.

liable because they passed through the estuary very quickly.

Discussion

Structure and Composition of Fish and Invertebrate Communities on Artificial Islands

Our data indicate that juvenile salmonids used the shorelines of the constructed islands, especially Island 3. Juvenile chinook and chum salmon were the dominant salmonids on island shorelines, but smaller numbers of other salmonids were taken during the study (Figure 7). Detailed data on the abundance of salmonids in the area before the islands were constructed, when this part of the estuary was being used for log storage, are not available. However, Raymond et al. (1985) sampled in April–June 1980 at three sites on the east side of the log storage area (their stations 9, 11, 22), and recorded mean catches of 18 chum salmon and 16 chinook salmon per 100 m^2. Over the 5 years of our study, maximum mean catches at Island 4, close to the stations of Raymond et al., were approximately 10 chum salmon and 35 chinook salmon per 100 m^2 (1985). These data indicate there may have been a decrease in chum salmon fry use and an increase in chinook salmon fry use of this sector of the estuary after log removal. There has been an increase in the number of spawning chinook salmon in the Campbell-Quinsam system over 1982–1986, so the number of wild fry in the estuary may be higher because of this.

The number of hatchery-reared juvenile salmon using the estuary has increased significantly since the Quinsam hatchery began operating in 1975. Releases of hatchery chinook salmon peaked in 1986: 3.13 million smolts left the hatchery that year. This represents an approximate threefold increase in hatchery chinook salmon using the estuary since 1980. Although our data indicated relatively little use of island habitats by the larger hatchery fish, these fish may influence the distribution of the smaller wild chinook salmon, possibly resulting in increased use of the shallow island habitats by the latter group. The vegetated island habitats also provide cover and possibly refuge from predation on the small wild fry. The provision of cover for fish has been widely recognized as an important ecological function of aquatic vegetation (Macdonald et al. 1987).

The speed of colonization of the artificial islands by invertebrates depended on elevation. In terms of community structure, the low-elevation communities appeared to merge with reference communities much faster than the high-elevation communities. Abundance of the latter assemblages on both reference and artificial island sites was variable, and on Nunn's Island invertebrate densities appeared to decrease over time, perhaps due to natural fluctuations or sampling error. Low-elevation sites on artificial islands created in North Carolina (Cammen 1976) were also colonized faster by macroinvertebrates and were considered more productive than higher sites. Increased submergence time at lower elevation sites provides more opportunity for colonization by motile epibenthic invertebrates. Sedentary macroinvertebrates such as the sabellid polychaete *Manayunkia aestuarina* must have colonized the islands via pelagic larval stages, a recognized life history strategy for this species (e.g., Light 1969) and other polychaetes. Larval dispersal to a restored marsh in California by a variety of sedentary polychaetes occurred within 6 months (Boyd 1982). The larger adult animals, such as gammarid amphipods, are able to move by swimming, whereas the smaller meiobenthos are probably passively moved by water currents (Bell and Sherman 1980). Meiobenthic animals sampled in a related study rapidly colonized the lower elevations of the islands in the Campbell River, and by spring 1982, meiofauna communities and abundance were similar to or higher than those of other sites in the estuary. Abundance in the new islands and reference sites reached seasonal peaks of approximately 5,000 and 1,100 organisms per m^2, respectively (Kask et al. 1986).

Provision of Fish Food

A food web analysis is one way of evaluating whether or not the islands are supplying usable food for juvenile salmonids. Such an analysis is fraught with difficulties because juvenile salmon are known to be opportunistic feeders (Hyatt 1979) and diets can change with time of day, fish size, tidal fluctuations, and season. However, several lines of direct evidence indicate that food organisms being produced on the islands can be used by juvenile salmonids. As well, diets of juvenile salmon caught in island habitats were similar to those of fish taken in natural habitats in other British Columbia estuaries (e.g., Healey 1982).

In 1982, juvenile chinook salmon caught in the islands were feeding on pelagic calanoids and cyclopoids, harpacticoids, epibenthic amphipods, juvenile and adult insects, and freshwater cladocerans (Kask et al. 1986). Of these, it is likely that the harpacticoids, insects, and amphipods were produced on the islands. Invertebrate production may have come from an upstream area, but it is likely that all invertebrates originating from island habitats were potentially available as fish food.

Considerations also must be given to microhabitat differences in food availability. Fish close to the deep channel of Campbell River (Islands 3 and 4) can access marine planktonic organisms in the salt wedge, as well as in the epibenthos, as indicated above. Those found in the shallow channels in the vicinity of Islands 1 and 2 can utilize epibenthic organisms and insects but have reduced access to marine zooplankton. Sheer zones in flows can concentrate prey organisms and also provide a low-velocity holding and feeding area for juvenile salmonids (Macdonald et al. 1987). The artificial islands, especially Island 3, create such zones in river and tidal flows and may increase the availability of plankton such as cladocerans, calanoids, and cyclopoids.

Levels of Production

Construction of the islands resulted in a net gain of about 3.2 hectares of intertidal marsh habitat. Five seasons after planting (1986), peak standing crop of *Carex lyngbyei* was approximately 302 g dry weight/m^2 (mean over samples from Islands 1, 2, and 4) (Dawe, personal communication). As a first approximation, this could result in an additional annual contribution of 9,600 kg dry weight of detrital material from marsh plants previously not available but known to be a significant component of estuarine food webs leading to juvenile salmon (Sibert et al. 1978).

The peak standing crop estimate in 1986 on the islands was comparable to 1976 estimates of *C. lyngbyei* production in the whole Campbell River estuary (225 g dry weight/m^2 per year; Kennedy 1982). This indicates that the brackish salinity regime (0 to 10‰), low wave and current energy, and substrate stability of the islands matched natural conditions for this plant. However, peak standing crop for *C. lyngbyei* at a low-elevation site on the reference island (Nunn's) in 1986 was 720 g dry weight/m^2 (Dawe, personal communication). This is considered a maximum value for the estuary and is comparable to biomass of the "tall *C. lyngbyei* ecotype" (777 g dry weight/m^2) recorded in the estuary in 1980 (Raymond et al. 1985). Therefore, further increase in detritus production from the islands might be expected.

There has been an increase in open-water habitat (7.8 hectares) due to the creation of the log pond. Phytoplankton and zooplankton are produced in this area (Brown et al. 1987a). In addition, the middle sector of the estuary previously used for log storage (32.8 hectares) is now almost completely available for aquatic biota. Aside from the intertidal marshes, approximately 22.8 hectares of mud and sand in the intertidal and shallow subtidal area can produce microbenthic algae such as *Enteromorpha* sp. and interstitial diatoms. *Enteromorpha* sp. was observed in most years in dense concentrations on the islands, trapped in marsh plants, as well as in areas between the islands (Riley et al. 1987). (In 1989, 7 years after log removal, we observed that eelgrass had also colonized the islands.)

Balanced against the gain in unvegetated intertidal production of detrital material is the loss of production from areas that were lower in elevation before they were covered by the artificial islands. However, the islands were built in sectors of the log-sorting area where algal productivity was reduced by the effects of shading from floating logs, scouring from log grounding, and covering by wood debris deposits (5–25 cm deep) (Raymond et al. 1985). Natural accretion has been reduced in the estuary because of the dam upriver. There has also been considerable gravel mining in the estuary (Bell and Thompson 1977). Therefore, there probably has been little natural decrease in intertidal volumes over the past 30 years. A recent production loss has occurred on the east side of the estuary, where about 15 hectares that were formerly primarily riparian or upland habitat are now part of the dry land sort facility or unproductive riprap shore.

This project therefore resulted in a net increase in approximately 18.8 hectares of usable fish habitat, and a net gain of habitat capability or productive capacity, if simple areal data are used as a proxy for the latter measurements. We feel this gain has significantly reduced the risk that estuarine capacity will be affected by log storage and its accompanying primary and secondary effects on estuarine fish habitats. It is difficult to measure the benefits of this net gain in terms of fish production, but related research at the Campbell River estuary has shown that in 2 of 3 years tested, survivorship of hatchery chinook salmon was higher when fish were exposed to estuarine conditions (Levings et al. 1989). Because of the problems in evaluating the complex and variable ecosystems supporting salmon, precise fisheries values cannot be determined for an economic analysis of the rehabilitation. Total net benefits for the forest company were positive even before these nonquantified benefits were considered (Hagen 1987). In order to secure the restored habitat for future fisheries values, a management plan for the entire estuary needs to be developed, and the rehabilitated sector of the estuary should be given permanent conservation status. The Campbell River estuary could also be used as a benchmark to evaluate future restoration during compensation projects in British Columbia.

Acknowledgments

This project would not have been possible without the administrative assistance of Mike Brownlee, Fisheries Branch, Vancouver (now with the Ministry of Forests, Province of British Columbia, Victoria), and several people within B.C. Forest Products Ltd. (now Fletcher Challenge Canada Ltd.), especially Bob Willington and Ted Mattice. We could not have assembled the data in this paper without the assistance of colleagues on the Campbell River team: T. J. Brown, M. S. Kotyk, C. D. McAllister, Bev Bravender, and a host of temporary technical assistants. Sally Leigh-Spencer was responsible for most of the invertebrate sampling and laboratory analyses, with financial assistance from B.C. Forest Products Ltd. The contract work by Edward Anderson Marine Sciences on invertebrate communities conducted in May 1983 was supported by Ocean Dumping Control Act research funding.

References

Anderson, E. M., and M. Galbraith. 1985. Use by juvenile chinook salmon of artificial habitat constructed from dredged material in the Campbell River estuary. Edward Anderson Marine Sciences, Sidney, British Columbia. Available from Department of Fisheries and Oceans, West Vancouver Laboratory, West Vancouver, British Columbia.

Bell, L. M., and J. M. Thompson. 1977. The Campbell River estuary: status of environmental knowledge to 1977. Department of Fisheries and Environment, Special Estuary Series 7, West Vancouver Laboratory, West Vancouver, British Columbia.

Bell, S. S., and K. M. Sherman. 1980. A field investigation of meiofaunal dispersal: tidal resuspension and implications. Marine Ecology Progress Series 3:245–249.

Boyd, M. 1982. Salt marsh faunas: colonization and monitoring. Pages 75–84 *in* M. Josselyn, editor. Wetland restoration and enhancement in California. University of California, Sea Grant College Program Report T-CSGCP-007, La Jolla.

Brown, T. J., C. D. McAllister, and B. A. Kask. 1987a.

Plankton samples in Campbell River and Discovery Passage in relation to juvenile chinook diets. Canadian Manuscript Report of Fisheries and Aquatic Sciences 1915.

Brown, T. J., C. D. McAllister, and M. S. Kotyk. 1987b. A summary of the salmonid catch data from Campbell River estuary and Discovery Passage for the years 1982 to 1986. Canadian Data Report of Fisheries and Aquatic Sciences 650.

Brownlee, M. J., E. R. Mattice, and C. D. Levings, compilers. 1984. The Campbell River estuary: a report on the design, construction, and preliminary follow-up study findings of the intertidal marsh islands created for purposes of estuarine rehabilitation. Canadian Manuscript Report of Fisheries and Aquatic Sciences 1789.

Cammen, L. M. 1976. Abundance and production of macroinvertebrates from natural and artificially established salt marshes in North Carolina. American Midland Naturalist 96:487–493.

Good, J. W. 1987. Mitigating estuarine development impacts in the Pacific Northwest: from concept to practice. Northwest Environmental Journal 3:93–112.

Green, R. H. 1974. Multivariate niche analysis with temporally varying environmental factors. Ecology 55:73–83.

Green, R. H. 1979. Sampling design and statistical methods for environmental biologists. Wiley, New York.

Hagen, M. E. 1987. Log-handling impacts on estuarine environments: an analysis of the Campbell River estuary, British Columbia. Master's project in Natural Resources Management, Report 42, Simon Fraser University, Burnaby, British Columbia.

Healey, M. C. 1982. Juvenile Pacific salmon in estuaries: the life support system. Pages 315–341 *in* V. S. Kennedy, editor. Estuarine comparisons. Academic Press, London.

Hyatt, K. D. 1979. Feeding strategy. Pages 71–119 *in* W. S. Hoar, D. J. Randall, and J. R. Brett, editors. Fish physiology, volume 8. Academic Press, New York.

Kask, B., C. D. McAllister, and T. J. Brown. 1986. Nearshore epibenthos of the Campbell River estuary and Discovery Passage, 1982, in relation to juvenile chinook diets. Canadian Technical Report of Fisheries and Aquatic Sciences 1449.

Kennedy, K. A. 1982. Plant communities and their standing crops on estuaries of the east coast of Vancouver Island. Master's thesis. University of British Columbia, Vancouver.

Levings, C. D., C. D. McAllister, and B. Chang. 1986. Differential use of the Campbell River estuary, British Columbia, by wild and hatchery-reared juvenile chinook salmon (*Oncorhynchus tshawytscha*). Canadian Journal of Fisheries and Aquatic Sciences 43:1386–1397.

Levings, C. D., C. D. McAllister, J. S. Macdonald, T. J. Brown, M. S. Kotyk, and B. A. Kask. 1989. Chinook salmon (*Oncorhynchus tshawytscha*) and estuarine habitat: a transfer experiment can help evaluate estuary dependency. Canadian Special Publication of Fisheries and Aquatic Sciences 105: 116–122.

Light, W. J. 1969. Extension of range of *Manayunkia aestuarina* (Polychaeta: Sabellidae) to British Columbia. Journal of the Fisheries Research Board Canada 26:3088–3091.

Macdonald, J. S., I. K. Birtwell, and G. K. Kruzynski. 1987. Food and habitat utilization by juvenile salmonids in the Campbell River estuary, B.C. Canadian Journal of Fisheries and Aquatic Sciences 44:1233–1246.

Orloci, L. 1967. Multivariate analysis in vegetation research, 2nd edition. Dr. W. Junk, The Hague, Netherlands.

Porebski, O. R. 1966. On the interrelated nature of the multivariate statistics used in discriminatory analysis. British Journal of Mathematical and Statistical Psychology 19:197–214.

Race, M. S. 1985. Critique of present wetlands mitigation policies in the United States based on an analysis of past restoration projects in San Francisco Bay. Environmental Management 9:71–82.

Raymond, B. A., M. M. Wayne, and J. A. Morrison. 1985. Vegetation, invertebrate distribution and fish utilization of the Campbell River estuary, British Columbia. Canadian Manuscript Report of Fisheries and Aquatic Sciences 1829.

Riley, P. M., B. A. Raymond, S. Leigh-Spencer, and M. S. Kotyk. 1987. Data report on benthic sampling on artificial islands at Campbell River estuary 1982–1986. Canadian Data Report of Fisheries and Aquatic Sciences 659.

SEP (Salmonid Enhancement Program). 1980. Stream enhancement guide. Department of Fisheries and Oceans, Vancouver.

Shreffler, D. K., R. M. Thom, C. A. Simenstad, J. R. Cordell, and E. O. Salo. 1988. Habitat utilization of a restored wetland system by juvenile salmonids. Pages 504–514 *in* Proceedings of the First Annual Meeting on Puget Sound Research. Puget Sound Water Quality Authority, Seattle.

Sibert, J. R., T. J. Brown, M. C. Healey, B. A. Kask, and R. J. Naiman. 1978. Detritus-based food-webs: exploitation by juvenile chum salmon (*Oncorhynchus keta*). Science (Washington, D.C.) 196:649–650.

Wilcox, C. G. 1986. Comparison of shorebird and waterfowl densities on restored and natural intertidal mudflats at Upper Newport Bay, California, U.S.A. Colonial Waterbirds 9:218–226.

American Fisheries Society Symposium 10:191–199, 1991

Cooperative Efforts to Protect Salmonid Habitat from Potential Effects of Highway Construction in British Columbia

FRED J. ANDREW
Andrew Consulting Ltd.
5024 Buxton Street
Burnaby, British Columbia V5H 1J6, Canada

Abstract.—The British Columbia Ministry of Transportation and Highways cooperated with federal and provincial agencies to protect salmonid habitat from potential effects of constructing the first phase of the Coquihalla Highway, which was built adjacent to two salmonid streams for about 54 km of the 118 km between Hope and Merritt, British Columbia. Biological and engineering consultants were employed by the Highway Ministry to work in collaboration with the fisheries agencies prior to construction to study potential environmental problems and to provide conceptual design of mitigative and compensatory measures. These measures included excavations to enhance off-channel rearing areas and instream rock placements to reduce adverse effects of stream channelization caused by river diversions and bank armoring. Some of the other factors that minimized adverse effects of construction activities included stipulation of clearing restrictions, riprapping at culverts and in drainage channels, reseeding of cut-and-fill slopes, riprapping of cut-and-fill slopes extending into the river, inclusion of specifications for environmental protection in the contract documents, use of steel-pile bridge abutments rather than concrete, and employment of environmental monitors to inspect construction activities and assist contractors in resolving environmental problems. An environmental coordinator employed by the Highway Ministry maintained a good liaison between the consultants, the regulatory agencies, and the highway design engineers during the planning and design stages, and it is suggested that he or another senior employee should also have been assigned full responsibility to coordinate environmental protection during the construction stage.

Construction of the first phase of the four-lane Coquihalla Highway through the mountainous terrain of the Cascade Range in British Columbia presented a serious threat to important fisheries resources of the Coquihalla and Coldwater rivers. The highway followed the river valleys and was built adjacent to these streams or their major tributaries for about 54 of the 118 km between Hope and Merritt (Figure 1). The Coquihalla River originates on the coastal slope of the Cascade Mountains and flows westward about 55 km to enter the Fraser River at Hope. The Coldwater River originates near the summit of the Cascades and flows easterly about 76 km to enter the Nicola River at Merritt. The Coquihalla River (average discharge, 33.0 m^3/s; once-in-200-year flood, 890 m^3/s) has a renowned summer run of steelhead *Oncorhynchus mykiss*; many of the fish spawn and rear in a 20-km reach where the highway was built immediately adjacent to the streams. Pink salmon *O. gorbuscha*, coho salmon *O. kisutch*, and chum salmon *O. keta* also spawn in the Coquihalla River but these fish use the portion immediately downstream from the area directly affected by highway construction. The Coldwater River (average discharge, 8.47 m^3/s; 200-year flood, 140 m^3/s: Water Survey of Canada 1985, 1986) supports steelhead, coho salmon, and chinook salmon *O. tshawytscha*, and a large proportion of these populations spawn and rear over a 21-km distance where the highway was built immediately adjacent to the river. Construction began in 1979 on the Coquihalla River and in 1983 on the Coldwater. The work was greatly accelerated in 1984 to complete the Hope–Merritt section in 1986.

The precipitous terrain and the proximity of two major pipelines—one for oil and one for natural gas—made it extremely difficult to avoid encroaching on the rivers. The project involved construction of 9 bridges over these rivers and 21 bridges or culverts on major tributaries. In addition, it was necessary to encroach upon or divert the rivers into new channels at about 65 locations to provide space for the highway.

This major undertaking, costing an estimated Can$400 million, was completed without any immediate impact on the fish-producing capability of Coquihalla and Coldwater rivers. Extensive biophysical investigations now under way may reveal possible future problems, but it is useful at this time to examine the administrative and technical

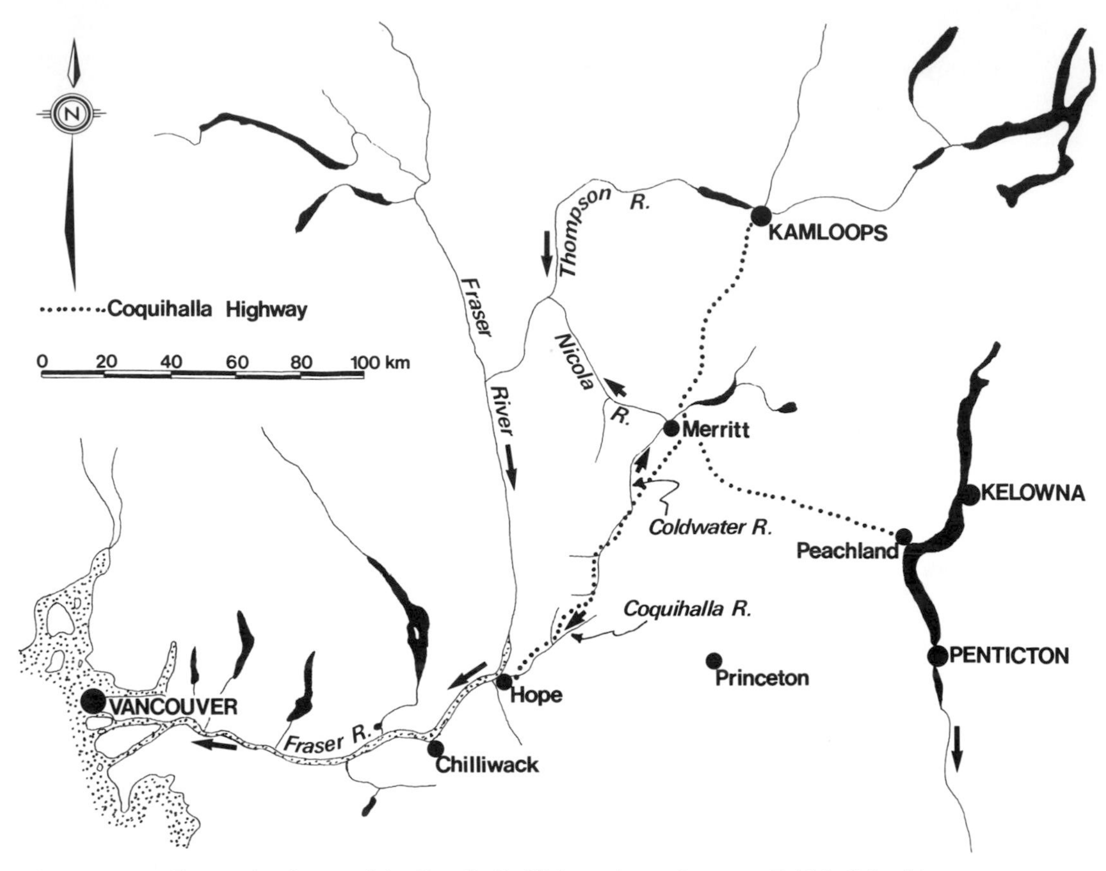

FIGURE 1.—Route of the Coquihalla Highway in southwestern British Columbia.

measures taken in efforts to protect the fish habitat and to compensate for potential adverse effects of construction activities.

Information in this report is primarily related to the Coldwater River portion of the project because my involvement with the Coquihalla River portion was only in reviewing the drawings and specifications, attending preconstruction meetings, and making limited inspections of construction activities.

Preliminary Investigations

Throughout the planning, design, and construction periods, the British Columbia Ministry of Transportation and Highways (MOTH) maintained a close working relationship with the appropriate agencies to ensure that every effort was made to protect the river resources. An environmental coordinator employed by MOTH was responsible for liaison with the fisheries agencies—the federal Department of Fisheries and Oceans and the provincial Ministry of Environment, Fisheries Branch. Being a staff member of the Design and Surveys Branch, and having some biological training, he was able to maintain a good relationship with the MOTH engineering staff and the habitat protection biologists employed by the agencies. Because the Highway Ministry does not employ staff biologists, consultants were engaged prior to the start of detailed engineering design to collaborate with the agencies in evaluating potential effects of the highway on fish production (Anonymous 1978; Miles et al. 1979; Harding et al. 1981). During this data-gathering phase, studies included biological and engineering investigations of the portions of the streams used by salmonids as well as an investigation of methods that could be used to mitigate and compensate for potential impacts. At every proposed river encroachment, consideration was given to methods of mitigating or compensating for potential adverse effects. These studies were extremely valuable because they brought the fisheries protection problems into focus early in the design process.

By thus raising the awareness of fisheries problems, these preliminary studies assisted the highway design engineers in avoiding harmful encroachments and in specifying effective mitigative and compensatory measures. The Highway Ministry agreed to fund a 3-year postconstruction evaluation of the fish-protective measures and, if necessary, to compensate for any proven loss of fish production, thus meeting the requirements of the Canada Department of Fisheries and Oceans policy, which states that a guiding principle of no net loss of fish production must be applied in the planning of development projects (Anonymous 1986).

Provision of Instream Refuges

Many potential adverse effects of road construction adjacent to fish streams have been documented (Villamere et al. 1983). In addition to destroying natural bank vegetation, which provides essential cover and a food supply for salmonids, bank armoring to protect bridge abutments and highway embankments tends to confine the stream and restrict the natural fluvial meandering process. Stream meanders create pools, riffles, and small, low-velocity side channels that provide prime fish habitat. The adverse effects of this channelization are especially severe when the river must be diverted into a new channel, which is usually shorter than the natural, meandering river channel. When a stream must be diverted into a new, shorter channel, bank armoring and increased bed slope may result in severe bed erosion, causing further narrowing of the channel. To minimize channelization and to provide habitat diversity with refuge and feeding areas for fish, large boulders, as recommended by Harding et al. (1981), were placed in riprapped portions of the stream. The boulders were approximately 1–1.5 m across their largest dimension, and they were arranged in three configurations to create scour holes and fish refuge areas: single boulders were placed along the toe of the riprapped banks, clusters of approximately four boulders on about 2–3-m centers were placed in midchannel or adjacent to riprapped banks, and spurs consisting of several boulders were placed immediately adjacent to riprapped banks in high-velocity areas. Approximately 11 boulder groups and 9 rock spurs were placed, mostly at bridge sites. At 11 other locations, fish habitat was created along riprapped banks by placing single boulders along the toe or by placing the riprap to create a particularly rough toe (M. Miles and Associates, Ltd. 1985). To test one possible method of compensating for the loss of natural bank vegetation in areas where banks were armored, fully branched coniferous trees were anchored along riprapped banks in two small experimental areas.

Design Considerations for Habitat Protection

For design and construction purposes, the highway was divided into sections about 5–20 km long, some sections being designed by Highway Ministry staff and others by engineering consultants. All construction was done by private contractors. The proposed highway alignment and design details were examined by the fisheries agencies during the initial route selection, during the preliminary engineering design, and again upon completion of the contract drawings, which had to be approved by the fisheries agencies before construction could commence. Any changes that the agencies considered necessary for fish protection were transmitted to the Ministry and the designated design consultants for inclusion in the drawings and contract documents.

Clearing was restricted to the minimum area necessary for constructing highway embankments. Thin fills were avoided on steep slopes to avoid sloughing. Where highway embankments extended onto the river floodplain, a riprap layer 1.0–2.0 m thick was placed on the bank up to the 200-year flood level to reduce erosion and consequent stream siltation. The riprap toe trench was 0.5–1.2 m deep. Sedimentation basins consisting of widened and deepened areas were provided in drainage channels where excessive amounts of silt were expected. The banks of newly constructed river channels were armored with riprap in areas where it was determined that excessive erosion could occur. Erosion at culvert entrances and exits was also controlled by riprap placement.

Bridge design was considered particularly important because the emergency measures taken to correct excessive scour and debris accumulation at bridge abutments during flood periods are frequently very harmful to resident fish and fish habitat. Bed scour at bridges often erodes and displaces valuable fish habitat during floods. Emergency dumping to protect against bridge damage often degrades fish habitat for considerable distances downstream because of sedimentation. Bridges were designed for a 200-year flood, designs provided adequate streambed width and freeboard for debris passage and included no instream piers. Bridge abutments were steel-pile structures, which greatly minimized instream excavations and the dewatering and stream contamination problems frequently associated with construction of concrete abutments. Bridge abutments were aligned parallel to the direction of flow, and the banks upstream and downstream were heavily riprapped, including the provision of a toe trench over the full length of riprapping.

Construction Considerations for Habitat Protection

An important element of the Highway Ministry's efforts to protect the fisheries resources was a section of the contract specifications entitled "Protection of the Environment," which had been prepared in collaboration with the fisheries agencies. It specified the arrangements required before work could be undertaken in streambeds and other environmentally sensitive areas, and it emphasized that precautions were to be taken to avoid stream siltation, to dispose of deleterious wastes, to install culverts and store petroleum products in appropriate ways, and other aspects of the work. At preconstruction meetings attended by representatives of the Highway Ministry as well as by fisheries and other regulatory agencies, the MOTH director of construction advised each contractor that the environmental stipulations described in the contract documents would be enforced and that an environmental monitor employed by MOTH would be assigned to each contract to inspect the work in progress and to assist the contractor in resolving environmental problems. The environmental monitors prepared biweekly reports on each contractor's work and these were circulated to the fisheries agencies. The MOTH resident engineer for each contract was responsible for instructing the contractor to correct any unsatisfactory conditions discovered by the environmental monitor or the fisheries agencies.

Where it was necessary to work within the active flood zone of designated fish streams, the contractors were required to obtain approval of the two fisheries agencies. Special restrictions were required on the Coldwater River because spawners, eggs, alevins, or newly emerged fry were in the stream during every month of the year. Therefore, instream work was generally restricted to the least biologically sensitive period—July 15–August 15—following the main period of steelhead fry emergence and prior to the commencement of chinook and coho salmon spawning.

The agencies often required that a berm or dike of washed gravel be placed in the stream to isolate the work areas from flowing water. This clean gravel barrier could be in the form of a dike surrounding a section of streambed or an island through which necessary excavations could be made. When there was excessive percolation through these clean gravel berms, a polyethylene sheet or a layer of fine-grained streambed material was placed on the upstream face of the berm.

No bank preparation, pile driving, or other work associated with bridge construction was permitted until the agencies approved a written submission from the contractor detailing proposed construction procedures and work schedule. Culverts on tributary streams were installed in the dry, either during a period when there was no discharge in the creek or after a clean granular berm had been built to divert the stream temporarily out of any area to be excavated or filled. All cut and embankment slopes were cleaned up immediately and they were reseeded as soon as weather permitted.

Special precautions were taken when it was necessary to divert the main river channels. To minimize siltation, the new channels were constructed in the dry, including placement of bank riprap where necessary on newly excavated side slopes. The river was then diverted into the new channel by building a dike of clean granular material across the existing river channel. Simultaneously, a fish salvage crew equipped with seines, dip nets, buckets, and electroshockers captured fish in the old channel and released them in the new channel. A riprap sill was built across one of the new channels into which the river was to be diverted. This sill was considered necessary to reduce the possibility of excessive bed scour caused by the greater slope in the diversion channel than occurred in the natural stream.

Compensatory Features

New habitat was created in an effort to compensate for unavoidable impacts on fish-producing capability. The new habitat consisted of spawning and rearing areas constructed adjacent to the rivers. Along the Coquihalla River, a combination spawning and rearing channel was developed by enhancing Karen Creek, a tributary stream. The gradient was reduced, the streambed was widened and graded, washed gravel was added, and fry refuge areas were created by the addition of overhead cover and instream boulders. Also, a heavy rock dike was built on the floodplain of the Coquihalla River to protect this

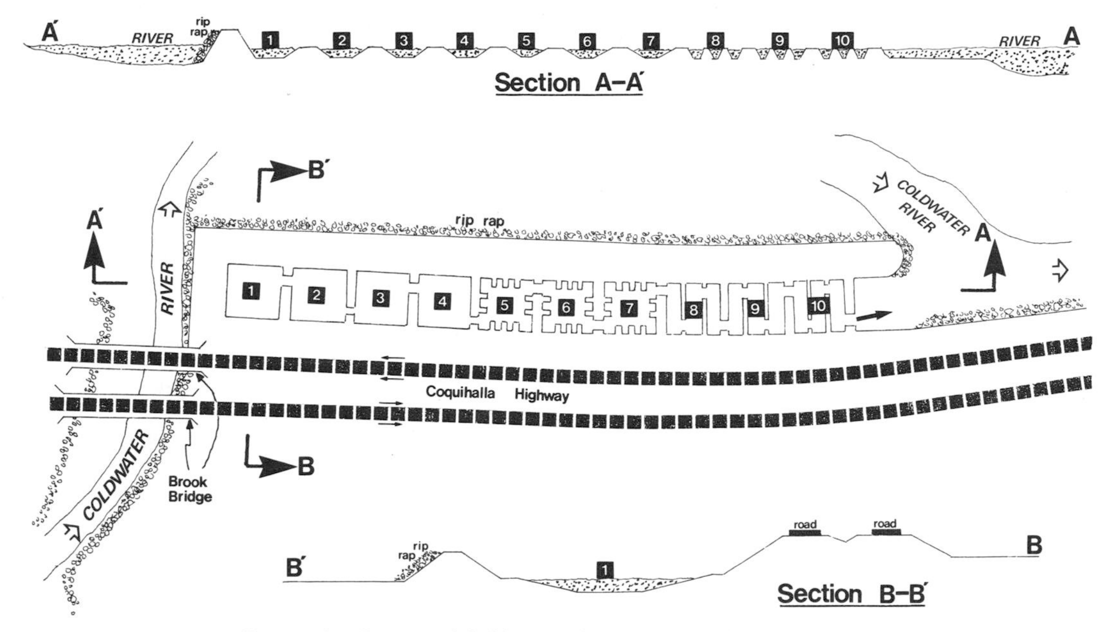

FIGURE 2.—Layout of Coldwater River compensation ponds 1–10.

enhanced tributary stream from severe Coquihalla River floods. Along the Coldwater River, spawning areas were abundant, but because it was expected that the already-limited rearing capacity could be adversely affected by diversions, encroachments, and riprapping, new rearing areas were built by deepening and improving fry access to existing groundwater-fed areas adjacent to the river and by excavating additional off-channel groundwater-fed ponds and channels. Four channels were excavated on the floodplain in accordance with design criteria suggested by Harding et al. (1981). The total channel length was about 600 m; widths were 0.5–12 m and maximum depths were 0.5–2 m, giving a total area of 1,400 m^2. Coniferous trees, some with root clusters, were placed in these channels to provide fry refuge. Most of the wetland areas that had been recommended in the preconstruction survey as potential sites for rearing enhancement had been destroyed by severe floods that occurred after the report was completed and by emergency measures taken by pipeline companies to protect the two pipelines and an access road from flood damage. An additional 10-pond channel and another large pond were therefore provided to develop additional rearing capacity in an area that could be protected from flood damage.

Design criteria for this additional habitat were obtained from the available literature and from an examination of the existing habitat and fish distribution in Coldwater River and adjacent beaver ponds (Rosenau et al. 1987). Trapping indicated a year-round high density of underyearling salmonids, mostly coho salmon, in beaver ponds adjacent to the stream, but very few juvenile salmonids overwintered in the river, where low flows, low temperatures, thick ice cover, and moving ice jams created a very severe environment (Swales et al. 1986). On the basis of this information, one of the sites recommended (Harding et al. 1981) for development of a small wetland rearing channel was used for excavation of 10 rearing ponds, each 30 m square with a maximum depth of 2 m (Figure 2). The riprap that was intended to be placed to protect the highway embankment was instead placed on the river side of the ponds for flood protection. The total length of this 10-pond channel was about 375 m and the surface area at low water was about 7,400 m^2 in ponds and 100 m^2 in riffles between the ponds.

Three pond configurations were used to test the effect of pond shape on fish productivity. The upstream four ponds had side slopes of 1:5 (vertical:horizontal), leaving a central area 10 m square and 2 m deep. The next three ponds had the same overall size and maximum depth as the first four, but the shoreline length was increased by providing 10 "fingers" of soil projecting into each pond. The side slopes in these ponds varied from 1:3 to 1:1.5. The lower three ponds were also 30 m square and 2 m deep. Each was divided into

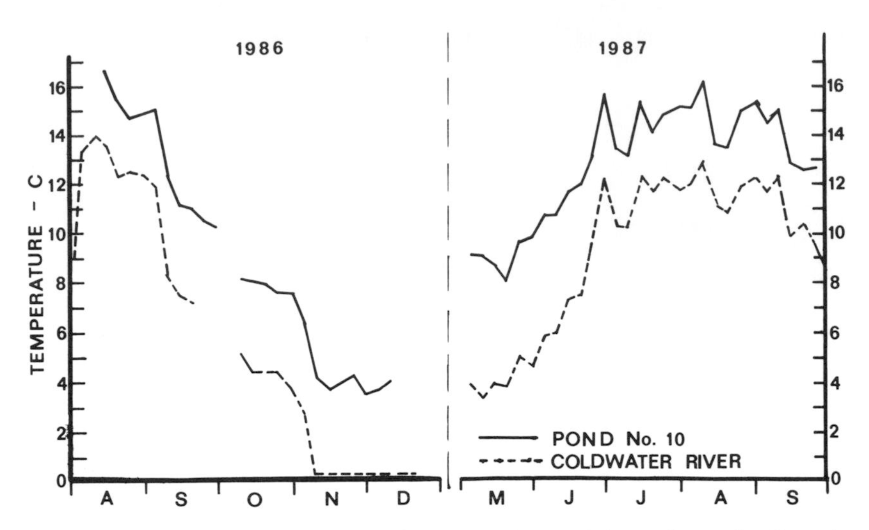

FIGURE 3.—Five-day average temperatures (°C) of Coldwater River and the outflows of a 10-pond compensation channel in 1986 and 1987.

three trapezoidal channels 9 m wide by 30 m long with 1:1.5 side slopes. A shallow gravel riffle about 3 m wide and 5 m long was provided at the outlet of each of the 10 ponds, and fully branched coniferous trees with their root clusters attached were placed in 7 of the ponds.

Immediately downstream from the 10-pond channel, a river oxbow channel had been cut off by one of the river diversions. To regain this area for fish rearing, a culvert (2 m in diameter by 60 m long) was placed under the highway to provide fish access, and the old river channel was deepened to provide a water depth of 1.0 m at low water. This pond was about 80 m long and it provided a surface area of about 1,050 m^2 (Beniston et al. 1988). The total surface area of the compensatory ponds and channels at low water was about 9,900 m^2.

Postconstruction Evaluation

Preliminary observations of and assumptions about the effect of highway construction on productivity of coho salmon suggested that the rearing habitat provided in the newly created off-channel streams and ponds would be sufficient to compensate for any probable loss of instream rearing habitat (Andrew 1987). It was realized, however, that predominantly stream-rearing species such as chinook salmon and steelhead might not benefit from the off-channel habitat to the same extent as coho salmon.

In this preliminary assessment, it was assumed that the major off-channel facility—the 10-pond channel off the Coldwater River—would be as productive for coho salmon rearing as existing beaver ponds. Temperature measurements indicated a much more favorable thermal regime for salmonid rearing in the off-channel habitat than in the river (Figure 3). Average temperatures were above the 7°C winter mode (Bustard and Narver 1975) for a much longer period each year and during summer months the highest 5-d average temperature of 16.7°C was well below the 25°C upper lethal level for coho salmon (Brett et al. 1952) but approached the 20°C level at which maximum cruising speed occurs (Brett 1958) and the 23°C level that produces the maximum rate of evacuation of stomach contents (Brett and Higgs 1970).

A much more thorough impact assessment is being made in the postconstruction evaluation program funded by MOTH, the third and final field year of which was 1988–1989. In this program, the fingerling densities of all rearing species were measured by trapping, netting, and electroshocking in altered portions of the stream, at mitigation structures (instream boulders), and in the off-river ponds and channels, and these densities are being compared with those measured in control areas assumed to represent preconstruction conditions. Preliminary results (Beniston et al. 1987, 1988) indicated an increase in rearing capabilities for chinook salmon, wild coho salmon, and steelhead fry and parr, and a major decrease in rearing capability for larger, hatchery coho salmon, based on low-flow summer sampling. Overwintering capability was estimated to be reduced for all salmonids except steelhead. Instream boulder placements appeared to increase the rearing capability for chinook salmon and wild and hatchery coho salmon. The experimental placement of coniferous trees on riprapped banks produced a fingerling density 3–4 times higher than in untreated bank areas. The rearing ponds and channels supported much lower coho salmon densities than natural off-channel habitat but it is likely that higher densities will occur in future years, after bank vegetation and pond biota develop. Because many years will be required to establish "natural" environments in the altered areas, it is much too early for a final assessment of the postconstruction rearing capability.

Conflict-Avoidance Factors

The environmental review process for the Coquihalla project was the same as for other projects in British Columbia, necessitating approval of all aspects of the work by the responsible federal and provincial regulatory agencies. The task of these agencies in setting environmental standards and ensuring compliance was simplified by the cooperative and progressive attitude shown by MOTH toward environmental protection. My personal observations while engaged over 4 years as a consultant on this project led me to believe that the following factors were important in reducing environmental conflicts and allowing construction to proceed with minimum impact on the fish-producing capability of Coquihalla and Coldwater rivers:

- the cooperative attitude of the Highway Ministry in funding fisheries studies and preparing an environmental design manual in collaboration with the fisheries agencies during the exploratory stage of route selection and conceptual highway design;
- the use of biological and engineering consultants rather than in-house staff to undertake preconstruction studies, thus avoiding confrontational attitudes that often develop between the in-house staffs of environmental agencies and development proponents;
- the efforts of the environmental coordinator in maintaining good liaison between fishery and highway personnel;
- consideration and conceptual design of mitigative and compensatory options at an early stage in planning of the highway project;
- selection by the design engineers of options that would minimize environmental conflicts (for example, the choice of steel-pile bridge foundations rather than reinforced concrete bridge abutments was very favorable for fish protection);
- review and approval by the fisheries agencies of all contract drawings and specifications prior to commencement of construction;
- agreement by the Highway Ministry to fund legitimate mitigation and compensation proposals and undertake a postconstruction assessment, with a commitment to rectify any deficiency in fish-producing capability;
- inclusion of environmental protection requirements in the contract documents, with emphasis drawn to their importance by the MOTH director of construction and by fisheries resource agencies at preconstruction meetings with each contractor;
- employment of environmental monitors to inspect construction activities and assist contractors in resolving environmental protection problems;
- frequent inspection of construction activities by environmental monitors, fisheries resource agency personnel, and environmental consultants;
- employment of independent consultants by the Highway Ministry to assist the fisheries agencies in reviewing contract drawings and inspecting construction activities from a fisheries protection perspective.

Problems Encountered

As might be expected, some aspects of the work did not proceed as smoothly as desired and in some cases there were unnecessary conflicts and minor environmental damage. Pointing out these difficulties and possible methods of avoiding them may provide a basis for improving habitat protection in future projects. It is important to note that the following seven problems could have been avoided if they had been anticipated in time.

(1) A serious problem occurred west of Hope, where the existing TransCanada Highway was being upgraded. The contractor stockpiled sand and crushed rock on the floodplain of a small stream, and some of this material washed onto a spawning area in the Fraser River during a very unusual flood, severely degrading the spawning ground quality. The stockpile of material was expected to be removed prior to the spawning period but construction delays prevented this. On several occasions, the environmental monitor and the local representative of the provincial Ministry of Environment, Fisheries Branch, had discussed

actual and potential environmental problems with the project engineer, but they felt that the problems were never satisfactorily resolved. Then, when the damage occurred to the Fraser River spawning area, it was decided to charge the contractor and MOTH with being in violation of the Canada Fisheries Act. This unsatisfactory outcome could have been avoided if a dispute settlement procedure had been prearranged or if the environmental coordinator or other senior Highway Ministry employee had been assigned responsibility for ensuring that construction problems reported by environmental monitors and fisheries agency personnel were promptly resolved to the satisfaction of all parties.

(2) There were many communications between the fisheries agencies, the Water Management Branch of the Provincial Government, and MOTH to establish environmental requirements prior to and during the construction period. Because of different needs of the several species administered by the two fisheries agencies, misunderstandings sometimes occurred regarding the necessity for and importance of specific environmental constraints. In order to avoid confusion, it would have been preferable if the agencies had agreed on the necessary constraints beforehand and a lead agency had been chosen to communicate with MOTH.

(3) Logging of merchantable timber on the right-of-way was done by private contractors under supervision of the provincial Ministry of Forests, which appeared to maintain a lower standard of environmental protection than MOTH. As a result of this lower standard and lack of supervision, and because of the lack of discrete logging boundaries, minor damage was done in some areas by removal of streambank vegetation, dragging of logs across streams, and other unsatisfactory practices. These problems were not anticipated but they could have been avoided if the logging activities had been treated the same as the highway construction activities, with preplanning and input by the fisheries agencies to establish environmental requirements, employment of environmental monitors, and other safeguards.

(4) A major effort was required to move and replace parts of the oil and gas pipelines in preparation for highway construction. This work, like the logging of merchantable timber, was done by a contractor not directly employed by MOTH. In at least two instances, the work resulted in major discharges of silt and sand into Coldwater River. These problems could have been avoided if arrangements had been made with the pipeline companies to follow the same environmental safeguards as MOTH, with preplanning, environmental specifications, and employment of environmental monitors.

(5) In an effort to avoid instream work during critical periods when fish populations were high, it was considered advisable to place the required riprap at proposed bridge locations prior to the start of pile driving for the bridge abutments. Riprapping involved excavation of a toe trench, which could not be done in flowing water without excessive stream sedimentation, and placement of large pieces of broken rock in the trench and on the bank. Cofferdamming was therefore required in some areas. It was expected that the rock pieces could later be moved slightly to permit pile driving to proceed, even if fish or eggs were in the stream. The pile-driving contractor, however, required that all of the rock at the pile locations be removed before pile driving could start. Unnecessary instream work could have been avoided if the riprapping had been postponed until completion of pile driving.

(6) Despite the many precautions taken, erosion of cut-and-fill slopes caused some inflow of silt and sand to the Coldwater River. Although there were no massive inflows, minor erosion on the many cut-and-fill slopes immediately adjacent to the river over approximately 21 km of river length could reduce productivity of the river due to siltation. In one area, an unexpected seepage from a cut slope of fine-grained soil caused localized sloughing, and the water flowing from the bank carried a stream of silt-laden water to valuable off-channel rearing areas and to the river. Serious localized erosion also occurred on cut-and-fill slopes due to joint leakage from corrugated culvert pipes that carried drainage from the highway surface to the bottom of highway embankments. A greater effort should have been made to avoid bank seepage by draining uphill areas. Erosion should have been better controlled by placing broken rock in bank seepage areas and by using more secure, leak-proof conduits for carrying highway surface drainage down the highway embankments.

(7) It is inevitable that further postconstruction problems will occur and they may not be corrected to the satisfaction of the fisheries agencies because no commitment was made by MOTH to study future problems affecting the fisheries resources. Furthermore, such problems will have to be resolved by the local highway maintenance

units, which claim to be chronically understaffed and underfunded. For instance, the above-noted problem of bank erosion at downpipes was not evident until after a moderate rainstorm, and by that date it was difficult to rectify the problem because the contractor, having fulfilled his obligations, was anxious to leave the area. Therefore, a minimum repair job was done with the hope that the local maintenance unit would correct any deficiencies that might appear later. It is evident that the fisheries resource would have been better protected if MOTH and the fisheries agencies had prearranged a specific program to study and correct potential erosion, sedimentation, and other operating problems for a few years after completion of construction. It would also have been advantageous to include a study of the practical operational and maintenance problems involved in obtaining satisfactory production from the mitigative and compensatory facilities in these postconstruction studies.

References

Andrew, F. J. 1987. Preliminary evaluation of compensatory ponds and channels on Coldwater River. Report of Andrew Consulting Ltd. to British Columbia Ministry of Transportation and Highways, Victoria, and Canada Department of Fisheries and Oceans, New Westminster, British Columbia.

Anonymous. 1978. Coquihalla (Hope–Merritt) highway preliminary environmental report. Report to British Columbia Ministry of Transportation and Highways, Design and Surveys Branch, Victoria.

Anonymous. 1986. Policy for the management of fish habitat. Canada Department of Fisheries and Oceans, Ottawa.

Beniston, R. J., W. E. Dunford, and D. B. Lister. 1987. Coldwater River juvenile salmonid monitoring study—year 1 (1986–87), volume 1. Report of D. B. Lister and Associates, Ltd., to British Columbia Ministry of Transportation and Highways, Victoria.

Beniston, R. J., W. E. Dunford, and D. B. Lister. 1988. Coldwater River juvenile salmonid monitoring study—year 2 (1987–88). Report of D. B. Lister and Associates, Ltd., to British Columbia Ministry of Transportation and Highways, Victoria.

Brett, J. R. 1952. Temperature tolerance in young Pacific salmon, genus *Oncorhynchus*. Journal of the Fisheries Research Board of Canada 9:265–323.

Brett, J. R., and D. A. Higgs. 1970. Effect of temperature on the rate of gastric digestion in fingerling sockeye salmon, *Oncorhynchus nerka*. Journal of the Fisheries Research Board of Canada 27:1767–1779.

Brett, J. R., M. Hollands, and D. F. Alderdice. 1958. The effect of temperature on the cruising speed of young sockeye and coho salmon. Journal of the Fisheries Research Board of Canada 15:587–605.

Bustard, D. R., and D. W. Narver. 1975. Aspects of the winter ecology of juvenile coho salmon (*Oncorhynchus kisutch*) and steelhead trout (*Salmo gairdneri*). Journal of the Fisheries Research Board of Canada 32:667–680.

Harding, E. A., R. Kellerhals, and M. Miles. 1981. Hydrology and fisheries study—Coldwater River. Kellerhals Engineering Services, Heriot Bay, Quadra Island, British Columbia.

M. Miles and Associates, Ltd. 1985. Hope to Merritt highway inventory of fisheries mitigation structures, volume 2, Coldwater River. Report to British Columbia Ministry of Transportation and Highways, Victoria.

Miles, M. J., E. A. Harding, T. Rollerson, and R. Kellerhals. 1979. Effects of the proposed Coquihalla Highway on the fluvial environment and associated fisheries resource. British Columbia Ministry of Highways and Public Works and British Columbia Ministry of Environment, Victoria.

Rosenau, M. L., F. J. Andrew, and L. Berg. 1987. Protection of Coldwater River salmonid habitat from potential effects of Coquihalla Highway construction. Report of Andrew Consulting Ltd. to British Columbia Ministry of Transportation and Highways, Victoria, and Canada Department of Fisheries and Oceans, New Westminster, British Columbia.

Swales, S., R. B. Lauzier, and C. D. Levings. 1986. Winter habitat preferences of juvenile salmonids in two interior rivers in British Columbia. Canadian Journal of Zoology 64:1506–1514.

Villamere, J., A. Stockwell, B. G. Dane, and J. S. Mathers. 1983. An annotated bibliography of literature pertaining to the protection of fish and fish habitat during road development. Hatfield Consultants Ltd. and Canada Department of Fisheries and Oceans, Vancouver.

Water Survey of Canada. 1985. Historical streamflow summary, British Columbia to 1984. Environment Canada, Inland Waters Directorate, Water Resources Branch, Ottawa.

Water Survey of Canada. 1986. Magnitude of floods—British Columbia and Yukon. Environment Canada, Inland Waters Directorate, Vancouver, British Columbia.

throat trout *O. c. lewisi* of 10–23 cm occupied sites in water only about 15–25 cm deep directly along current-bearing lateral surfaces of aquatic plant beds, rocks, and artificially built rock deflectors. All such surfaces were more or less parallel to current. No trout existed in plunge pools of much greater depth below log overpour structures that had been installed 3 years before. By contrast, in Wisconsin creeks of lower gradient than the Montana streams, large brown trout were abundant in plunge pools created by artificial overpour structures, as well as in lateral scour situations (my personal observations, 1960–1970). Such nonstatistical observations, made during fish population inventories for other purposes, bear testing in experiments specifically designed to measure and compare trout abundance in lateral scour situations and other habitats of streams.

In sympatry, young anadromous salmonids of different species and ages differentially occupy a variety of habitat types in small Pacific coast creeks; age-1 and older steelhead *Oncorhynchus mykiss* prefer plunge, trench, and lateral scour pools with wood debris and undercut banks (Bisson et al. 1982). Surely, in any well-adapted community of stream fishes, all habitat types will be used by fish of one species or another when of proper body size. It is my impression, however, that lateral scour pools having strong current beneath undercut banks or among large woody debris or large rocks constitute the most common general daytime resting–hiding habitat for trout of sizes that interest anglers.

Advantages for Salmonids

Lateral scour against banks made cohesive by vegetation tends to underwash the rootwork and fallen logs and limbs, as well as associated soil and rocks, creating undercut banks. The undercuts and the projections of objects offer concealment for fish. The submerged projections, in dissipating the stream's energy, create eddies of slow current close to the stream's fastest current. This situation enables drift feeders, such as salmonids, to occupy sites that have slow current and visual isolation from predators and competitors but that are close to swift current that brings into view the most food per unit time. Such sites of great current differential, (shear zones of energy dissipation, discussed by Beschta 1991, this volume), offer some kinds of salmonids the most potential for energy profit, hence for growth (Fausch 1984). As any skilled angler of trout streams knows, the biggest trout often lie where deep, swift current grinds along rough undercut banks.

Lateral Scour as Intermediate Perturbation

The advantages fish gain from lateral pool formation, undercut banks, and current-swept bankside rocks and wood debris are consistent with the concept that biotic productivity tends to be greatest at intermediate levels of environmental perturbation. The full concept is that biotic production is low when there is very little environmental disturbance, increases as perturbations occur, but decreases when perturbation becomes severe (frequent fire, flood, or various human actions on ecosystems; Odum et al. 1979).

Current veering against a streambank is a source of perturbation. If the bank is also disturbed by heavy grazing of livestock, or if (for whatever reason) its vegetation is made too weak to resist the force of high water, the bank will rapidly erode, undercuts will be destroyed, and the channel will become wide and shallow and will harbor far fewer large fish. If, on the other hand, a stream is so stabilized with rock, steel, or concrete that it can hardly change at all, and especially if it is thus locked in place for a great distance, the lack of perturbation will result in unnatural regimens of sediment and nutrient flow, that have various unfavorable consequences for the stream biota.

Between and far from either of these extremes is the current-washed streambank made resistant to erosion by a binding mantle of live vegetation and lodged debris, but not made so resistant that it cannot be eroded slowly and gracefully, taking on the sheltering shapes in which large fish find resting and hiding places close to the food-bearing swift currents. It will be best to let channels wander a bit—and keep the dynamism in the conservative dynamic equilibrium of stream systems.

Management Implications

It will usually be highly beneficial to incorporate resisted lateral scour in habitat management for salmonids. This can involve various methods to protect and promote meandering, bank overhangs, and the features of large woody debris. Such objectives can often be rapidly achieved by installing artificial structures to simulate the optimal natural conditions, but these are typically massive and expensive. (It is a common misconception that only a few logs and stones need be

FIGURE 2.—"Nature-hostile" engineering to stop streambank erosion and lateral migration of a channel. A heavy, fitted masonry wall reinforces the bank. Transverse log sills, placed radially along the meander curve, spill flow parallel to channel alignment, preventing laterally scouring flow. The objective is to protect a canyon road built for forest management. Fish habitat was not considered in the design. Weyereggerbach, Upper Austria, 1964.

installed or rearranged to significantly improve creek channels for large fish; many people are dismayed to learn the cost of proper in-channel habitat work.)

Lateral scour can be enhanced by accentuating the meandering channel form with current deflectors, as well as with oblique sills (White and Brynildson 1967). Current deflectors can be built to have resisted lateral scour along their current-bearing faces, in addition to the scour along the streambanks against which they deflect current.

Along the current-bearing banks, artificial overhangs ("bank covers" or "bank coverts") can be built to simulate the best of submerged undercuts. An error sometimes seen in construction of such coverts is that they are installed along streambanks that receive little or no current; thus they are far less attractive to salmonids and eventually fill up with sediments. This mistake stems from lack of appreciation of the lateral scour process and of fish behavior.

Also, oblique log (or, less favorably, rock) sills can be built to create lateral scour—and have natural appearance. The principle of flow departure from a transverse face has also been used to guide current and train channels so as to prevent lateral scour, as is sometimes done in stream engineering that lacks fish habitat objectives (Figure 2). Drop structures, such as low overpour dams or sills, can often be more beneficial for various kinds of larger salmonids if oriented diagonally to channel alignment, so as to send current against one side of the channel, creating lateral scour (as for the natural materials in Figure 1B). The traditional perpendicular orientation of drop structures forms only a plunge pool and directs the current straight downstream.

Fascination with our ability to achieve lateral scour by artificial means should not, however, blind us to situations in which nature could be left to do the work. It may often be most cost-effective in the long term to focus on protecting and restoring a healthy plant community on the streambanks, which in various ways can result in resisted lateral scour habitat for salmonids.

References

Beschta, R. L. 1991. Stream management and the role of riparian vegetation. American Fisheries Society Symposium 10:53–58.

Bisson, P. A., J. L. Nielsen, R. A. Palmason, and L. E. Grove. 1982. A system of naming habitat types in small streams, with examples of habitat utilization by salmonids during low streamflow. Pages 62–73 *in* N. B. Armantrout, editor. Acquisition and utilization of aquatic habitat inventory information. American Fisheries Society, Western Division, Bethesda, Maryland.

Fausch, K. D. 1984. Profitable stream positions for salmonids: relating specific growth rate to net energy gain. Canadian Journal of Zoology 62:441–451.

Odum, E. P., J. T. Finn, and E. H. Franz. 1979. Perturbation theory and the subsidy-stress gradient. BioScience 29:349–352.

White, R. J., and O. M. Brynildson. 1967. Guidelines for management of trout stream habitat in Wisconsin. Wisconsin Department of Natural Resources Technical Bulletin 39.

American Fisheries Society Symposium 10:204–212, 1991

Two-Pin Method for the Layout of Stream Habitat Enhancement Designs

DALE E. MILLER
Inter-Fluve, Inc., 211 North Grand, Bozeman, Montana 59715, USA

ROBERT J. O'BRIEN
Inter-Fluve, Inc., 1665 Grant Street, Denver, Colorado 80203, USA

GREGORY P. KOONCE
Inter-Fluve, Inc., Post Office Box 773, Hood River, Oregon 97031, USA

Abstract.—We present a new technique by which stream habitat enhancement measures designed by experienced personnel can be implemented by unskilled labor with minimum supervision. This method allows all elements of a detailed design to be triangulated to a pair of steel pins driven into the streambank near the enhancement site. An implementation document is prepared containing a descriptive text, a scaled plan map of existing and proposed features, a table of distances and elevations of existing and proposed features, and a table of construction material dimensions, quantities, and sources. The document can readily serve as construction specifications for contractors. The two-pin method was used to design and implement habitat enhancement measures for chinook salmon *Oncorhynchus tshawytscha*, coho salmon *O. kisutch*, and steelhead *O. mykiss* in 3.7 miles of Tarup Creek, Del Norte County, California.

Physical alteration of stream channels to improve fish habitat has become increasingly popular with resource managers (Duff 1982; Duff and Wydoski 1982). Documentation has shown positive results in many cases, but some efforts have been criticized for negatively affecting channel stability (Wesche 1985). In many of these cases the enhancement measures were inadequately designed; in others the original designs were incorrectly implemented.

The method we describe in this paper cannot assure functional designs, but it can allow accurate implementation of design details. Enhancement designs are often prepared after extensive site evaluations by specialists from different disciplines. Such an in-depth design process is not only justified, it is essential for project success (Brouha and Barnhart 1982; Wydoski and Duff 1982). But problems often arise when the design crew cannot devote the necessary time to personally implement construction specifications. Limited budget or, more often, limited personnel prevents full-time construction supervision. Regardless of the reason, supervision is rarely performed by the original design team. Uninformed or untrained personnel are often left to interpret a preponderance of costly design specifications prepared in an effort to make up for the lack of supervision. Nonetheless, specifications are commonly altered by contractors to fit their perception of the construction conditions encountered. The lack of design personnel on-site during construction results in many well-conceived designs being improperly implemented.

In view of these common problems, there is a need for an implementation method that addresses design criteria, reduces the amount of time required for construction supervision, is relatively inexpensive, is uncomplicated, and allows little room for interpretation or error. Ideally, the method would also require that the design team mentally build the proposed alterations during the design process to anticipate overall feasibility and construction difficulties. The two-pin method described in this paper facilitates the preparation of detailed enhancement designs and their accurate construction. Perhaps more importantly, the method encourages the participation of a multidisciplinary design team while making the most efficient use of project personnel and funds.

Procedure

The Two-Pin Method

The two-pin method consists of three basic steps. First, a site is selected for enhancement and the existing geomorphic conditions of the channel are described and mapped. In the second step, the actual design is conceptualized and then laid out on the ground. In the third step, locations and elevations are measured to prepare design specifications. This procedure is described in detail below.

The two-pin method is based on establishing a reference pin at the upstream and downstream ends of an enhancement site. These pins are placed on one side of the stream, with each pin located further upstream and downstream than the extent of the enhancement structure. By convention, pin 1 is downstream of pin 2.

Enhancement sites along the channel are referenced according to a curvilinear traverse line along one side of the stream. For example, site 12+50 is equivalent to a distance of 1,250 ft upstream from an assigned starting point. By convention, references begin at the downstream end of an enhancement section and progress upstream. Generally, the traverse line simply follows the course of the stream; bearings and azimuths are not recorded for meander curvature. Specific locations on the traverse line could be referenced to a nearby stadia or route survey (e.g., a road centerline) by recording distance and bearing from specific points along such surveys.

By using simple triangulation of intersecting horizontal distances from each reference pin, all significant existing channel features are located. With the pins in place, a fiberglass tape is stretched to the feature from each pin; their intersection defines a particular measuring point (Figure 1). By defining a number of points, a scaled plan map is drawn on site. For example, channel margins, existing in-stream boulders, and large woody debris are located on a plan map (Figure 2).

As an enhancement structure is conceptualized, important points are indicated on the ground with temporary wire flags. For example, each end of a proposed log weir is marked with a wire flag and then triangulated until these features can be accurately depicted on the map. Similarly, the centers of boulders are indicated and triangulated. Remaining design elements are defined in a similar fashion, producing a highly detailed plan schematic (Figure 2). Triangulation distances are recorded in a standardized table (Table 1).

The prescribed elevations for the top of log ends and boulder centers are then measured from each reference pin with a survey instrument; field notes are kept according to standard profile-leveling methods (Brinker et al. 1981; Brinker and Wolf 1984). Design elevations are recorded based on the convention that the elevation of pin 1 equals zero (Table 1).

During the site measurements, notes are also taken describing the condition of the channel and the type of habitat present and absent. The pre-

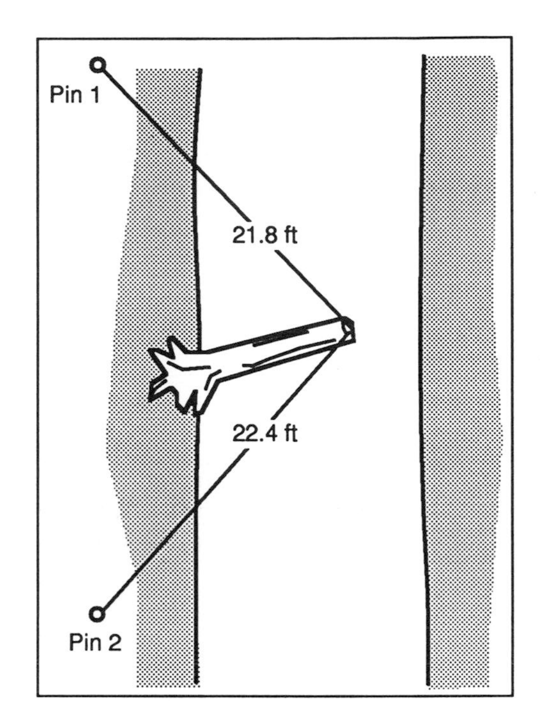

FIGURE 1.—Location of the instream end of a downed tree as determined by the intersection of the distances from two reference pins.

scribed enhancement measure and the anticipated effects are also noted, as is the availability of native material (such as logs or boulders) for construction. Based on these field notes, a brief textual description of the problem, solution, and expected results can be prepared in the office. A table of materials and availability is also developed. This information helps the construction crew conceptualize what has been designed (see Tables 2 and 3).

Construction

The enhancement design is followed during construction simply by reversing the processes of drafting the scaled plan map and dimension table. Fiberglass tapes are stretched from each reference pin to the distances noted in the structure location table; the intersection designates a structural feature, such as the bank end of a log or the center of a boulder. Temporary wire flags are placed at each intersection point until all important features are delineated. In this way, the location and orientation of a log weir can be identified; in addition, the log can be related to other elements of the design. Once the scaled plan map is verified, construction commences. The

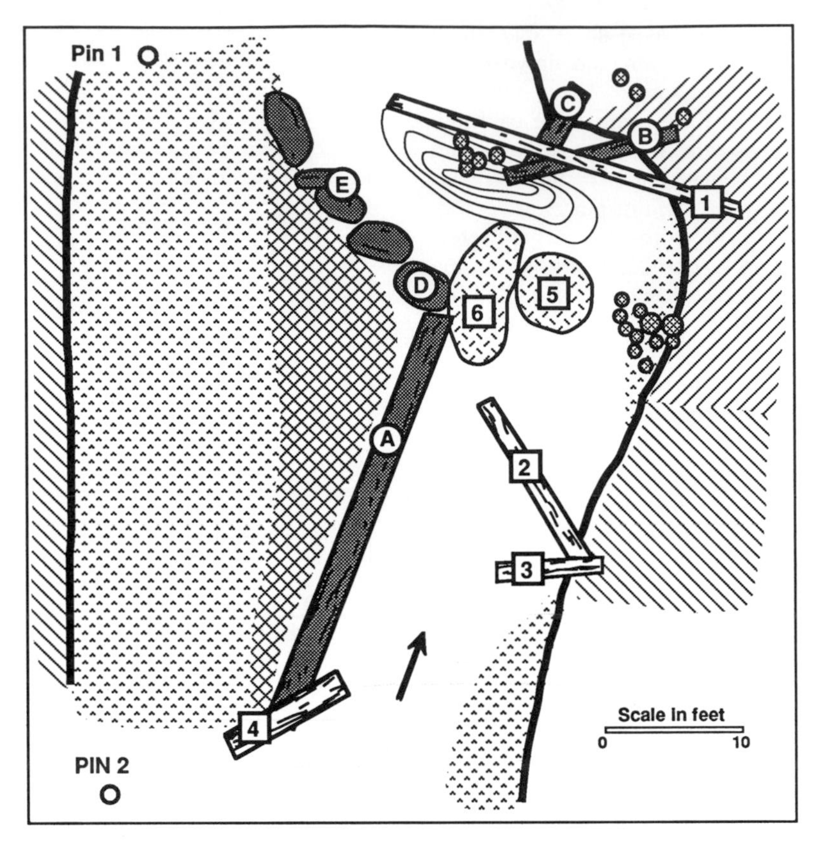

FIGURE 2.—Typical to-scale plan schematic depicting existing (numbered and nonshaded) and prescribed (lettered and shaded) features relative to reference pins at station 171+20 along Tarup Creek, California. Features correspond with those in Tables 1, 2, and 3. (Adapted from Inter-Fluve 1987.)

TABLE 1.—Measured distances and elevations of existing and prescribed structural components relative to the two reference pins at station 171+20 along Tarup Creek, California. Existing features are described by numbers; prescribed features by letters; all correspond with those in Figure 2. (Adapted from Inter-Fluve 1987.)

Item		Distance (ft)		Elevation (ft)[a]	
		Pin 1	Pin 2	Pin 1	Pin 2
Pin 1			57		−0.4
Pin 2		57		+0.4	
Log 1,	instream end	19	57		
Log 2,	instream end	37	41		
Log 3,	instream end	47	34		
Log 4,	instream end	50	20		
Boulder 5,	center	36	51		
Boulder 6,	center	31	48		
Log A,	upstream end, top	49	16	−3.5	−3.9
Log A,	downstream end, top	30	44	−4.9	−5.3
Log B,	bank end	41	67		
Log B,	instream end	29	56	−5.4	−5.8
Log C,	bank end	33	65		
Log C,	instream end	31	58	−5.7	−6.1
Boulder D,	center	28	45	−4.9	−5.3
Boulders E,	left bank edge	10	54		

[a]Elevation from pin: + above; − below.

TABLE 2.—Typical text describing the habitat limitations of the site, the prescribed measures, and the anticipated effects of these measures at station 171+20 along Tarup Creek, California. The station number represents the location of the enhancement site relative to other sites along the stream. Existing features are described by numbers; prescribed features by letters; all correspond with those in Figure 2. LB and RB refer to left and right bank, respectively. (Adapted from Inter-Fluve 1987.)

Site consists of a poorly developed pool on the RB associated with 2 large boulders in the channel and a clump of willows on the RB.

Prescription involves enhancing the pool by constricting the channel with a log and large boulders on the LB and providing cover by adding logs over the pool on the RB.

Place log A along LB. Excavate trench and place the bottom of log A approximately 1 foot below existing stream grade, with the upstream end approximately 1.5 feet higher than downstream end. Wedge the downstream end of log A against boulder 6; cable or pin log end to boulder. Cable the upstream end of log A to log 4 on LB. Remove boulder 5 and place as boulder D on the downstream LB side of log A. Place additional boulders E in a line from boulder D to the LB. The downstream end of log A and boulders E should all be of the same height. Place cobble fill between log A/boulders D, E and LB to height of log/boulders tapering up slope to match existing bar height on LB. Use material excavated from site of log A and boulders D and E, as well as material taken from pool.

Excavate pool approximately 1 to 2 feet below existing stream grade. Place material on LB side of log A and boulders D and E.

Place logs B and C to provide cover in the pool and to restrict flow between the clump of willows and the RB. Instream ends of logs B and C should overhang into pool. Cable bank ends of logs B and C to big leaf maple on RB.

Plant willow cuttings on disturbed areas of LB at a frequency of 2 by 2 feet.

During construction, do not disturb existing logs in the channel, especially logs 1, 2, 3, and 4. Do not disturb boulder 6. Protect the clumps of willows on the RB from foot traffic or other disturbances.

amount of excavation or fill for prescribed features is determined by measuring the appropriate depths with a hand-held Abney level, a tripod-based builders' level, or in simple cases, a string level.

Case Study

The California Conservation Corps (CCC), in cooperation with the California Department of Fish and Game and Simpson Timber Company, contracted with Inter-Fluve, Inc., to provide designs for fish habitat enhancement and watershed stabilization within a 3.7-mile portion of Tarup Creek, a tributary to the Klamath River in Del Norte County, California. The two-pin method was developed to design, establish on site, and provide construction assistance for 97 structures, 86 of which were located within the stream channel and 11 in the watershed. Enhancement was directed towards improving spawning and rearing habitat for a vestigial stock of winter-run chinook salmon *Oncorhynchus tshawytscha* as well as coho salmon *O. kisutch* and steelhead *O. mykiss*. Design philosophy focused on accentuating stream processes in selected areas to enhance preexisting desirable trends in channel conditions (after Lisle 1981).

Designs were prepared with two primary con-

TABLE 3.—Materials, with estimated dimensions and quantities, required for station 171+20 along Tarup Creek, California. The sources for individual items are indicated as available on-site, near to the site, or through a specified supplier. Lettered items correspond with those in Figure 2. (Adapted from Inter-Fluve 1987.)

Material	Size		Quantity	Source
	Diameter	Length		
Log A, redwood	2 ft	32 ft	1	On-site
Log B, redwood	1.5 ft	14 ft	1	Near-site
Log C, redwood	1.5 ft	7 ft	1	Near-site
Galvanized cable	3/8 in	35 ft	1	Supplier 3
Galvanized cable	1/2 in	20 ft	1	Supplier 3
Cable clamps	3/8 in		4	Supplier 3
Cable clamps	1/2 in		4	Supplier 3
Rock	2 ft			Near- and off-site
Willow cuttings	1/2 in	2 ft	100	Near- and off-site

siderations. First, all improvements were to be implemented by CCC hand-crew members with only 3 d per month of construction supervision by a member of the design team. Second, although some of the crew members were experienced in construction methods for habitat improvements, none of the crews or crew members had experience using this specific implementation method.

Initially, a curvilinear traverse line was established on one side of the stream; at intervals of 100 ft, each station was marked with numbered wooden stakes. This allowed identification of all stream segments as discrete locations. Generally, a field crew consisted of a fisheries biologist, a hydrologist, a fisheries technician, and two laborers.

After habitat enhancement measures were designed for a specific site on the traverse, the field crew placed the reference pins necessary for triangulation. Each design site was identified with an aluminum tag wired to each pin. For example, a tag attached to the upstream pin at design site 1+53 would read "STA 1+53 Pin 2." This information quickly identified each site and its reference pins. Where two or more sites were nearby, the tags prevented reference pins at one site from being confused with those of another.

After the pins were in place, the field crew operated as two teams. One team (professional and laborers) measured triangulation distances and developed the scaled plan map. The other team (professional and technician) surveyed elevations for certain design elements. Condensation of the field data (see Tables 1–3; Figure 2) resulted in a usable design and implementation document (Inter-Fluve, Inc. 1987).

Using this document, CCC work crews consisting of 10–15-person groups supervised by a crew leader followed a specific construction sequence.

(1) Once a specific site was located, the geomorphic description, design prescription, and details for construction were carefully read.

(2) The scaled plan map was oriented to fit the site, and reference pins were located so that existing objects could be triangulated.

(3) The work to be accomplished, the location of items to be placed, and features to be protected or left undisturbed were visually estimated and flagged if appropriate.

(4) The exact locations of design elements were triangulated as follows. From the structure location table for each work station, the distance from each reference pin to a specific element was determined. Two fiberglass measuring tapes were stretched to the appropriate distances, one from each reference pin, and oriented until the tapes intersected. This intersection indicated a specific location, such as the center of a boulder or the end of a log. If triangulation appeared to provide an erroneous result, the table numbers could be cross-referenced to distances measured on the scaled plan map.

(5) The triangulated location of each design element, such as the instream end of a log, was marked and identified with a temporary wire flag inserted into the ground.

(6) For design elements requiring specific elevations, a rod and level (or transit instrument) were used. The level was set up so that the entire site and either pin 1 or pin 2 was visible. The stadia rod was placed on one of the pins so that a backsight reading could be taken. From the structure location table for the appropriate work station, the elevation was read for the first component (for example: top, upstream end log A), either above or below the chosen reference pin (for example: −3.5 ft). This number was then subtracted from the backsight; the result was the proper foresight reading for the same component (top, upstream end log A) after it had been properly placed. Actual excavation, filling, or other placement was done and the site was checked until the measured elevation matched the predicted elevation. To illustrate, consider that the end of log B should be buried in the streambed so the top, instream end of the log is 5.4 ft below pin 1 (the structure location table reads −5.4 in the pin 1 column). A rod reading from pin 1 indicates a backsight of 8.3. Therefore, the finished elevation for the top, instream end of log B should be read on the rod as 13.7 (8.3 − [−5.4] = 13.7). If log B is 1.5 ft in diameter, the bottom of the hole before the log is placed should be excavated to a rod reading of 15.2 (13.7 + 1.5 = 15.2).

(7) The materials were placed and secured in the proper location according to the scale plan map and description. Triangulation of the completed structure confirmed its proper location, and elevations could be cross-checked, again using the structure location table.

Results

About two-thirds of the structures designed for Tarup Creek were constructed by the CCC during the winter of 1987 and the spring and summer of 1988, demonstrating the viability of the two-pin method in implementing enhancement designs. The total personnel-hours for design averaged 4.2,

2.2, and 3.0 h of professional, technical, and labor time, respectively, per structure. The fieldwork alone required 2.2, 1.1, and 3.0 h of professional, technical, and labor time, respectively, per structure. The time required for construction of each structure averaged 215 personnel-hours of manual labor, whereas construction supervision by a member of the design team amounted to 1.0 h per structure. About 8 h were spent orienting the CCC crew leaders to the two-pin method.

Discussion

Inherent Error in the Methodology

Under ideal conditions, the two-pin method allows little room for error between design intention and construction implementation. Aside from simple misinterpretation, any errors are the result of limits in the measurement system. Elevations can be established to a level of accuracy by extending measurement precision to a higher level of magnitude. For example, if 0.5 ft is an acceptable margin of error, measurements should be recorded in the structure location table to within 0.1 ft.

The plan location of enhancement features is based on intersecting distances from two fixed points. The intersection of two nonparallel lines defines only one position. The two-pin method is similar to that used in maritime navigation, where a fix is determined by the intersection of two distance circles. However, error can arise in prescribed feature locations due to the inherent shortcomings of the triangulation method. Because measurements occur twice, recorded during both design and construction, errors can be compounded. Therefore, the placement of reference pins relative to the plan location of given structural components must be done carefully to reduce the margin of error. From navigational standards, the ideal angle of intersection of the two lines is 90 degrees (Bowditch 1984). Accordingly, "fix error" increases when the angle decreases; if the angle is small, a slight error in measuring or plotting either line results in a relatively large error in position. Error increases proportionally to the cosecant of the angle of intersection (Figure 3). The increase in fix error becomes rapid after the angle has decreased to less than 30°; for example, with an angle of 30°, the error is about twice that of 90° (Bowditch 1984).

When measured from reference pins with the two-pin method, a small intersecting angle would occur in two scenarios: one, when a measured position is quite far from the reference points (Figure 4), and two, when a position does not fall roughly between the reference points (Figure 5). Thus, two important guidelines should be followed to reduce fix error.

(1) Reference pins should be placed so that the longest measurement across the channel does not exceed one-half of the distance between pins. In general, results are acceptable if pins are located

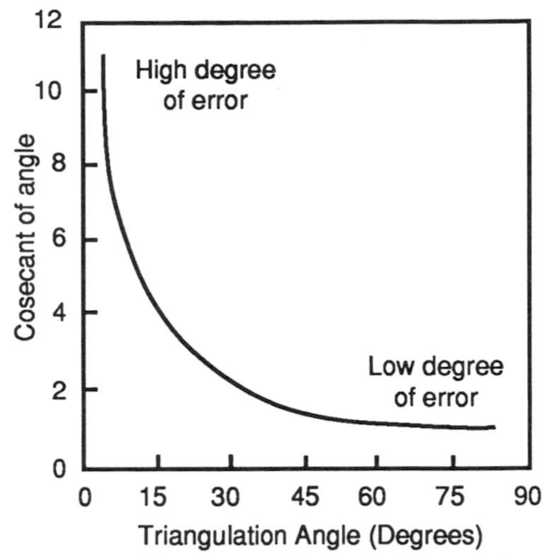

FIGURE 3.—The degree of fix error is inversely proportional to the angle formed by triangulation. (From Bowditch 1984.)

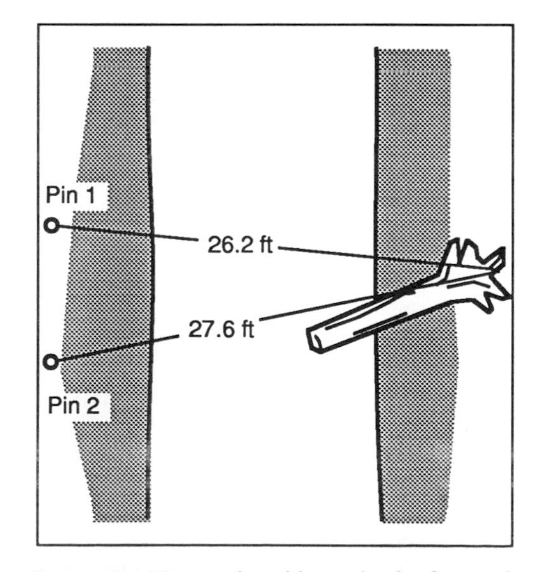

FIGURE 4.—Measured position quite far from reference pin results in an acute angle with a high potential for measurement error.

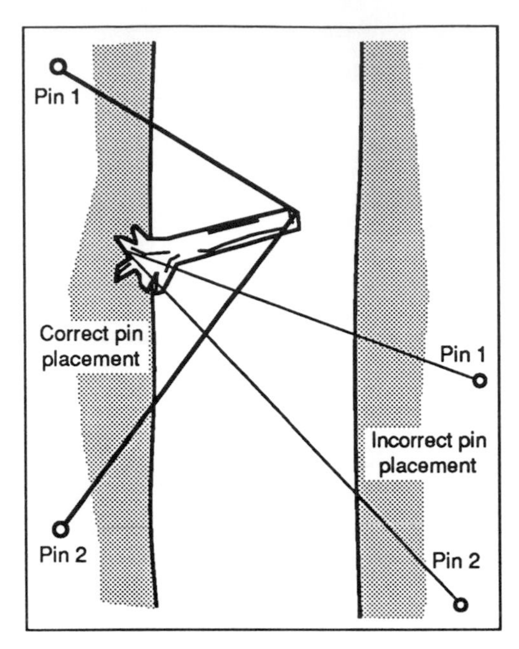

FIGURE 5.—Reference pins should be placed well upstream and downstream of the area to be measured. In this way, the intersecting angle approaches 90 degrees.

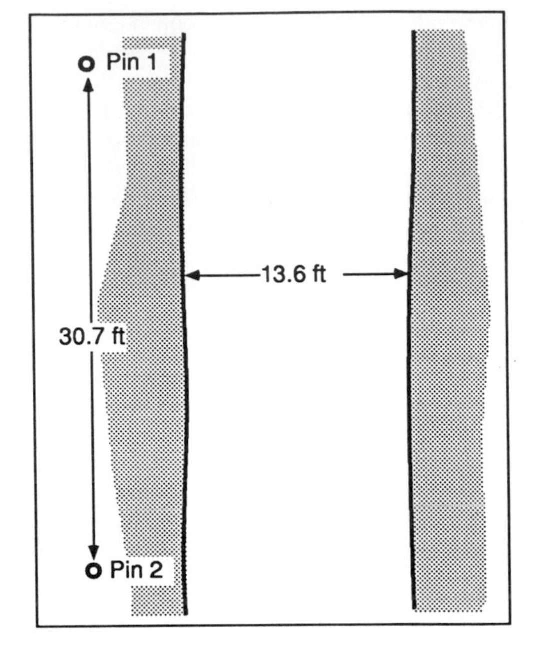

FIGURE 6.—Proper location of reference pins is slightly more than twice the channel width.

so the distance between them is slightly more than twice the channel width (Figure 6).

(2) Reference pins should always be placed upstream and downstream from the furthest extent of the prescribed enhancement structure. In this way, position lines from the downstream pin never extend downstream and those from the upstream pin never extend upstream (Figure 5). More than two pins may be used on very large sites or on those sites occurring on tight radius bends.

Comparison with Other Mapping Methods

More extensive methods have been used to create detailed plan maps of stream features. For example, a plane table and alidade are often used to develop plan and topographic maps (Compton 1961). Bisson et al. (1981) used a much more qualitative system of identifying habitat types and mapping these to scale. They reported that 100 m of stream could be mapped per day per person depending on channel complexity (they did not prepare enhancement designs in their study). Computer programs that plot topography have also been adapted from other disciplines for use on streams. For example, Ruediger and Engels (1981) used a U.S. Bureau of Land Management–U.S. Forest Service road design program to map the topography of 1,200 ft of stream with a two-person crew per day, not including computer data entry or corrections of wrongly entered data. However, the method requires access to the proper computer and software, and habitat must then be mapped onto the contour map following plotting. Another technique is the use of aerial photographs as the basis to map stream channels, although this method tends to be inexact unless data are field-checked.

The most common method of mapping stream conditions before or after habitat enhancement involves the use of channel cross sections. Typically, cross sections are spaced at regular intervals; average conditions between transects are approximated (Dunham and Collotzi 1975; Bryant 1981; Helm et al. 1981; House and Boehne 1985). Cross sections may be set up as parallel lines, regardless of channel curvature, or perpendicular to flow. Placing reference pins at each end of every cross section requires much effort when numerous transects are used. The information generated by one cross section usually contains only a portion of an item under evaluation. For example, several cross sections are required to accurately map an arc-shaped boulder weir. Furthermore, use of transects necessitates a benchmark system that measures the distance or angle between cross sections.

Generally, the two-pin method requires less effort, specialized equipment, or highly trained personnel than these other mapping methods. In some cases, the other methods require expensive tools during the preparation of design specifications, such as a computer program or a surveyor's theodolite and EDM (electro-optical distance measuring) equipment. Furthermore, in order to implement such specifications one must have both the equipment and the expertise to translate blueprint or engineering-style plan and profile drawings into on-site staking. The two-pin method tends to allow more cost-effective implementation of designs than these other mapping methods, primarily because it is less complicated and less time-consuming. The two-pin method requires no special surveying equipment or expertise to translate enhancement designs into blueprint specifications or to translate specifications into the finished product.

Management Implications

Stability Analysis

The two-pin method is a relatively precise, inexpensive means to evaluate the long-term stability of enhancement structures. Postconstruction documentation of selected locations of structural components can be compared with measurements following annual high-flow periods or even specific runoff events. For example, the plan and profile location of a group of boulders in an instream cluster can be recorded by measuring the fix and elevation at the center of each boulder. With such records, any movement of the boulders following runoff events can be readily quantified.

Habitat and Bedform Mapping

Mapping of fish habitat often involves sampling several variables (Binns 1982; Platts et al. 1983). These data are valuable, but there is often no rapid means of depicting the relative location of these measurements. Where sampling occurs over a short area, a plan schematic diagram can be developed with the two-pin method to depict the locations of sites where measurements of variables were made. For example, the location of benthic invertebrate, substrate, and water quality sampling sites can be illustrated on a plan map. Additionally, general geomorphic maps of stream channels can be developed by the two-pin method. Bed forms such as pools, riffles, and bars can be mapped based on definitions of features to be depicted (Bluck 1971).

Contract Specifications

Managers involved with the setup and coordination of stream habitat construction who prepare typical engineering-style blueprint design documents may find the two-pin method less expensive and more easily implemented by work crews. Similarly, in projects that stipulate background biological sampling, the two-pin method could be used to identify sample sites for contractors. In some cases, design details or sampling areas identified with this method could result in more specific proposals from bidders, thereby allowing more accurate comparison of contract line items.

Acknowledgments

David Muraki and Dan Ferreira of the California Conservation Corps, Carl Harral of the California Department of Fish and Game, and the Simpson Timber Company deserve thanks for extensive cooperation in preparing the Tarup Creek design document.

References

Binns, N. A. 1982. Habitat quality index procedures manual. Wyoming Game and Fish Department, Cheyenne.

Bisson, P. A., J. L. Nielsen, R. A. Palmason, and L. E. Grove. 1981. A system of naming habitat types in small streams, with examples of habitat utilization by salmonids during low streamflow. Pages 62–73 in N. B. Armantrout, editor. Acquisition and utilization of aquatic habitat inventory information. American Fisheries Society, Western Division, Bethesda, Maryland.

Bluck, B. J. 1971. Sedimentation in the meandering River Endrick. Scottish Journal of Geology 7:93–138.

Bowditch, N. 1984. American practical navigator, volumes 1 and 2. Defense Mapping Agency Hydrographic/Topographic Center, Washington, D.C.

Brinker, R. C., B. A. Barry, and R. Minnick. 1981. Noteforms for surveying measurements, 2nd edition. Landmark Enterprises, Rancho Cordova, California.

Brinker, R. C., and P. R. Wolf. 1984. Elementary surveying, 7th edition. Harper and Row, New York.

Brouha, P., and R. Barnhart. 1982. Progress of the Browns Creek fish habitat development project. *In* R. Wiley, editor. Proceedings of Rocky Mountain stream habitat management workshop. Wyoming Game and Fish Department, Laramie.

Bryant, M. D. 1981. Organic debris in salmonid habitat in southeast Alaska: measurement and effects.

Pages 259–265 in N. B. Armantrout, editor. Acquisition and utilization of aquatic habitat inventory information. American Fisheries Society, Western Division, Bethesda, Maryland.

Compton, R. R. 1961. Field manual of geology. Wiley, New York.

Duff, D. A. 1982. Historical perspective of stream habitat improvement in the Rocky Mountain area. *In* R. Wiley, editor. Proceedings of Rocky Mountain stream habitat management workshop. Wyoming Game and Fish Department, Laramie.

Duff, D. A., and R. S. Wydoski. 1982. Indexed bibliography on stream habitat improvement. U.S. Forest Service, Intermountain Region, Ogden, Utah.

Dunham, D. K., and A. Collotzi. 1975. The transect method of stream habitat inventory: guidelines and applications. U.S. Forest Service, Intermountain Region, Ogden, Utah.

Helm, W. T., J. C. Gosse, and J. Bich. 1981. Life history, microhabitat and habitat evaluation systems. Pages 150–153 in N. B. Armantrout, editor. Acquisition and utilization of aquatic habitat inventory information. American Fisheries Society, Western Division, Bethesda, Maryland.

House, R. A., and P. L. Boehne. 1985. Evaluation of instream enhancement structures for salmonid spawning and rearing in a coastal Oregon stream. North American Journal of Fisheries Management 5:283–295.

Inter-Fluve. 1987. Design specifications for fish habitat enhancement and watershed stabilization of Tarup Creek, a tributary of the Klamath River. Report, Contract 87-1878-47, to California Conservation Corps, Del Norte Center, Klamath, California.

Lisle, T. E. 1981. Roughness elements: a key resource to improve anadromous fish habitat. Pages 93–98 *in* T. J. Hassler, editor. Proceedings, symposium on propagation, enhancement, and rehabilitation of anadromous salmonid populations and habitat in the Pacific Northwest. Humboldt State University, Arcata, California.

Platts, W. S., W. F. Megahan, and G. W. Minshall. 1983. Methods for evaluating stream, riparian and biotic conditions. U.S. Forest Service General Technical Report INT-138.

Ruediger, R. A., and J. D. Engels. 1981. A technique for mapping stream channel topography and habitat using the RDS/PAL computer programs. Pages 162–169 *in* N. B. Armantrout, editor. Acquisition and utilization of aquatic habitat inventory information. American Fisheries Society, Western Division, Bethesda, Maryland.

Wesche, T. A. 1985. Stream channel modifications and reclamation structures to enhance fish habitat. Pages 103–106 *in* J. A. Gore, editor. The restoration of rivers and streams: theories and experience. Butterworth, Boston.

Wydoski, R. S., and D. A. Duff. 1982. A review of stream habitat improvement as a fishery management tool and its application to the Intermountain West. *In* R. Wiley, editor. Proceedings of Rocky Mountain stream habitat management workshop. Wyoming Game and Fish Department, Laramie.

American Fisheries Society Symposium 10:213–218, 1991

Vortex Mechanisms of Local Scour at Model Fishrocks

REG T. CULLEN[1]
Department of Civil and Environmental Engineering, Washington State University
Pullman, Washington 99164, USA

Abstract.—In a laboratory flume, geometrically different model fishrocks were used to induce scour. Fishrocks are defined as boulders placed in streams to improve fish habitat. Fishrocks create scoured areas in streams by increasing local water velocities near the substrate, which increase the local drag and lift forces that act on the substrate grains. Vortices act as the mechanism of local scour. The complexity of the local scour phenomemon arises from the hydraulic interaction among water flow, large roughness elements, vortices, and a deformable substrate.

Large roughness elements are solids whose dimensions are of the same order of magnitude as the depth of flow (Lisle 1981). These elements change the lotic environment by altering the local water velocity, which modifies the drag and lift forces acting upon substrate particles. Fishrocks, defined as boulders placed in streams to improve fish habitat, cause local changes in water speed and direction. This disturbed water is visible as a wake that manifests the complex flow field of turbulent, accelerated, and decelerated water. This wake contains many transient vortices that are either shed from or concentrated by the fishrock. A vortex is a swirling mass of water with concentric, helical, or spiral streamlines. Vortices are commonly found in stirred liquids.

The changed hydraulic forces of the water cause the streambed to deform into a more complex and diverse environment. The two most common bedforms found near fishrocks are gravel bars and scour pools. The streambed deformation at a fishrock is an anticipated hydraulic response to the placement of a large roughness element in a stream and should be predictable.

Fishrocks, and the stream bedforms they create, are among the types of habitat used by both juvenile and adult salmonids. Preferred summer microhabitat for juvenile salmonids consists of deep water in conjunction with submerged cover. Fishrocks can provide both deeper water and cover when a scour pool is formed that undercuts the rock. This cover is used to elude predators. Adult fish also use the scour pools for resting and hiding. Spawning adults appear to select spawning sites on the basis of the closeness of cover (Reiser and Bjornn 1979). Evaluation of structures that improve fish habitat has shown that fishrocks create salmonid habitat and can increase carrying capacities of streams for fish (Ward and Slaney 1979, 1981; Overton et al. 1981; Fontaine 1987).

Fish Response to Fishrocks

The evaluations of fish habitat improvement projects have shown a high variability in the change of the carrying capacity of a stream reach. This variability is due to differences in fish seeding levels, species and ages of fish, season of year, the location of the project in the stream, and current climatic conditions. Hall (1984) explained that gross variation in natural populations prevents the accurate evaluation of fish habitat improvement projects except on the basis of potential benefits to fish. He contended that complex sampling designs are needed to determine fish population response to stream enhancement structures and that 8 years may be needed for fish to fully respond to habitat improvement structures.

Ward and Slaney (1979, 1981) found that of seven stream-enhancement structures tested in the Keough River, British Columbia, fishrocks with additional log cover were most successful in increasing the stream's carrying capacity for steelhead *Oncorhynchus mykiss* and coho salmon *O. kisutch*. Overton et al. (1981) found rearing and cover habitat limiting for juvenile steelhead in many streams in the Six Rivers National Forest, California. The use of fishrocks in clusters in Red Cap Creek caused an absolute increase in steelhead parr biomass by 315%.

House and Boehne (1985) found gabions to be much more effective than fishrocks in increasing the carrying capacity for coho salmon. They tested one reach with fishrocks of diameters greater than 0.6 m and placed the fishrocks directly upon bedrock. They did find that the fishrocks were able to reduce exposed bedrock by trapping gravel.

[1]Present address: Trihey & Associates, 4180 Treat Boulevard, Suite O, Concord, California 94520, USA.

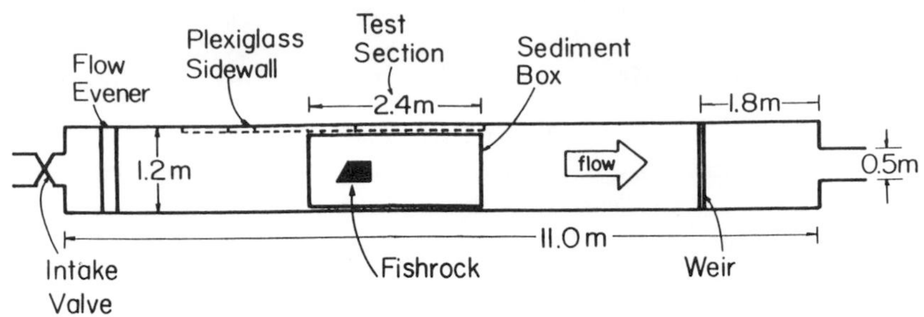

FIGURE 1.—Plan view of the flume used for the model fishrock study. Plexiglas sidewalls allow views into the movable bed region.

Fontaine (1987) studied habitat use by juvenile steelhead in Steamboat Creek, Oregon, in all four seasons of 1 year. She found that fishrocks in combination with log structures did not significantly increase the late-summer standing crop of steelhead. However, Fontaine did find that location of the structure with respect to the thalweg was important: structures at or near the thalweg were the most highly used. In winter months of higher flows, she found that stream enhancement structures with the highest concentration of fishrocks had the largest steelhead population.

The purpose of this work is to present the vortex mechanisms and hydraulics of local scour at fishrocks, which I believe can be useful tools for stream enhancement.

Methods

Flume studies.—Flume experiments have been very helpful in understanding fluid flow. For example, El-Samni (1949) used pressure taps in hemispheres to investigate the distribution of hydrodynamic forces. Adamek (1968) used strain gages in cubes to investigate lift and drag forces.

In August 1984, Fisher and Klingeman published the first article to combine local scour hydraulics with fish habitat enhancement and they originated the word "fishrock." They investigated the relationships between hydraulic variables, fishrock characteristics, and resultant scour pool quantities. Flume test durations averaged 22 h. Two model fishrocks were used during runs, and their results showed that the maximum scour occurred around the leading edge of a fishrock. Functional relationships were developed between seven variables in those clear-water scour experiments. The reader is directed to Klingeman (1984), and Klingeman et al. (1984) for further concepts in scour manipulation.

Laboratory flume.—A model stream reach with a deformable, movable bed was installed in a flume in the Albrook Hydraulics Laboratory at Washington State University. The movable bed section was 2.4 m long, 1.2 m wide, and 0.2 m deep. The entire flume was 11 m long, 1.2 m wide, and 1.2 m deep (Figure 1).

Discharge through the flume was controlled by an upstream gate valve and was measured at a downstream sharp-crested rectangular weir. The flow depth was controlled by an adjustable downstream weir. Average water velocities were calculated and point velocities were determined through the use of either an electromagnetic flow meter or a vaned anemometer. Discharge ranged from 0.01 to 0.25 m^3/s, and flow depth varied from 5 to 60 cm. Maximum pump discharge was 0.30 m^3/s.

The average diameter of the experimental substrate was 4.6 mm. Sieve analysis revealed that 0.6 mm was approximately one standard deviation. The experimental substrate can be classified as coarse sand to fine gravel. The maximum particle size was less than 9.5 mm.

Shapes of model fishrocks.—Four different model fishrocks were used in the flume. The width and height of each rock was 20 cm, whereas the lengths were 20, 30, 40, or 80 cm. The model fishrocks were constructed of plywood, filled with metal to give a specific gravity of 2.0, and sealed with an epoxy paint.

Results and Discussion

Hydraulics and Mechanism of Local Scour

A body immersed in static water is subject to the hydrostatic surface force of pressure (F_P) as well as the body forces of weight (W) and buoyancy (F_B) (Figure 2A). A body immersed in a flowing fluid is subject to the additional hydrodynamic forces of pressure (normal to the body surface) and viscous shear forces (tangential to the body surface) (Figure 2B). The normal and tangential hydrodynamic forces can be resolved

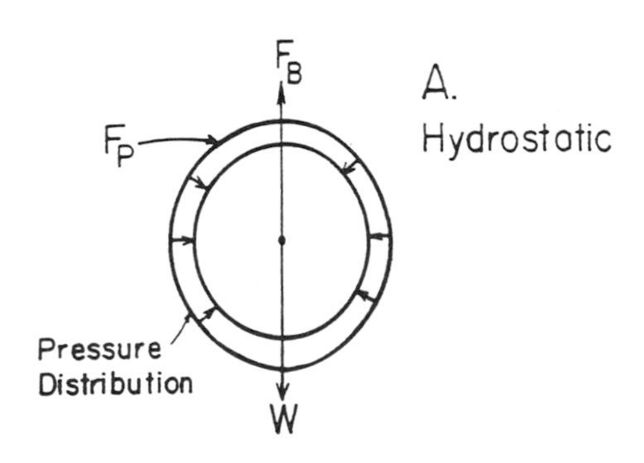

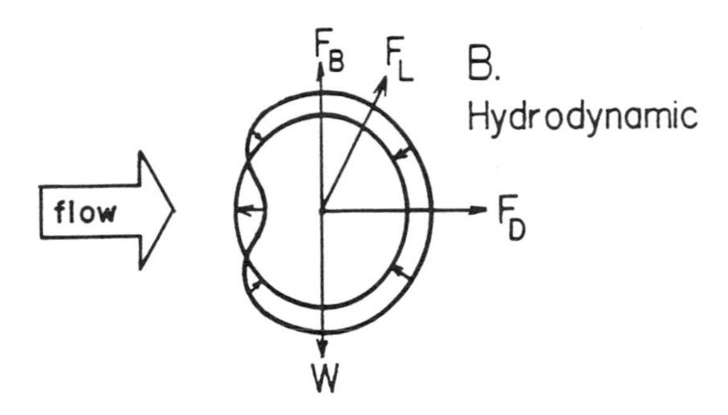

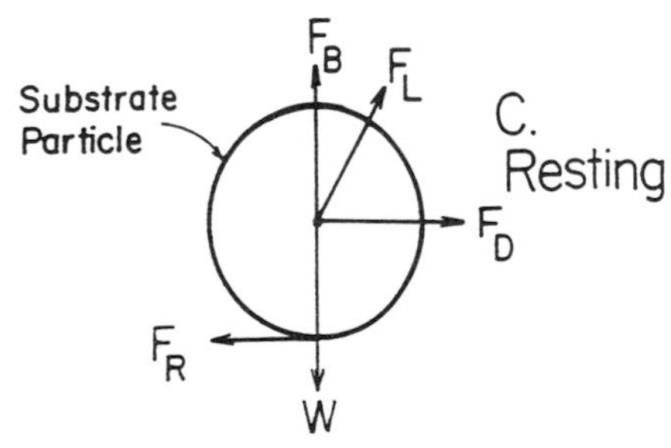

FIGURE 2.—**(A)** Hydrostatic forces acting upon a submerged particle. **(B)** Hydrodynamic forces and hydrostatic forces acting upon a submerged particle. **(C)** All forces acting upon a substrate particle. F denotes force, W weight. Force subscripts denote buoyancy (B), drag (D), lift (L), pressure (P), or frictional resistance (R).

into a single resultant force acting upon the submerged body. This resultant force can further be resolved into the two perpendicular component forces. The drag force (F_D) is the component parallel to the direction of the fluid velocity, and the lift force (F_L) is in the vertical direction (Daily and Harleman 1966; Roberson and Crowe 1985). If the immersed body is resting upon the streambed, there is a friction force (F_R) that acts opposite to the direction of flow (Figure 2C).

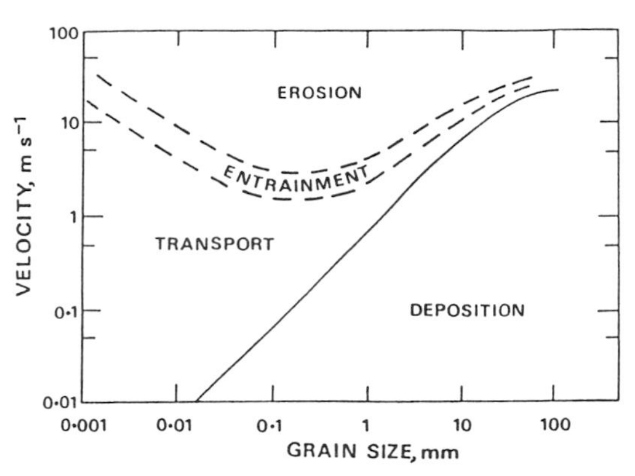

FIGURE 3.—Hjulstrom's diagram of substrate mobility. This diagram depicts general (not local) substrate activity for homogeneous grains.

A substrate particle will remain at rest as long as the active forces of drag, lift, and buoyancy are less than the resistive forces of weight and friction. A large roughness element causes a scour pool to develop by increasing the drag and lift forces on the substrate particles until the substrate moves (Bagnold 1966). It is important to note that both drag and lift are functions of the approach velocity raised to the second power. Hjulstrom (in Knighton 1984) described the relationships between the grain size of the substrate, the velocity of the water at the substrate, and the motion of substrate particles, as shown in Figure 3. Raudkivi and Ettema (1977) discussed the effects of substrate gradation on clear-water scour. They showed that the maximum depth of clear-water scour depends on the standard deviation of the substrate gradation.

The boundary layer of water near a fishrock has a decreased velocity that is theoretically zero at the rock surface. The deformed velocity gradient causes an adverse pressure gradient, which can cause the boundary layer to separate from the rest of the flow. This flow separation leads to the formation of eddies or vortices. A vortex is a swirling mass of water with either concentric, spiral, or helical streamlines. Vortex formation, as well as vortex concentration, occurs in the wake of a fishrock. Vortices in the flume were observed by injecting blue dye into the wake of a fishrock.

Rouse (1966) presented an excellent theoretical overview of vortex concepts, and Shen et al. (1969) and Breusers et al. (1977) presented the theory and mechanisms of local scour at bridge piers. Many of the concepts related to pier scour

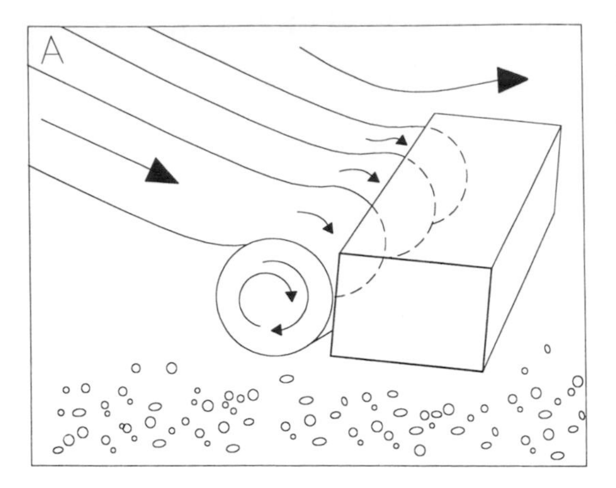

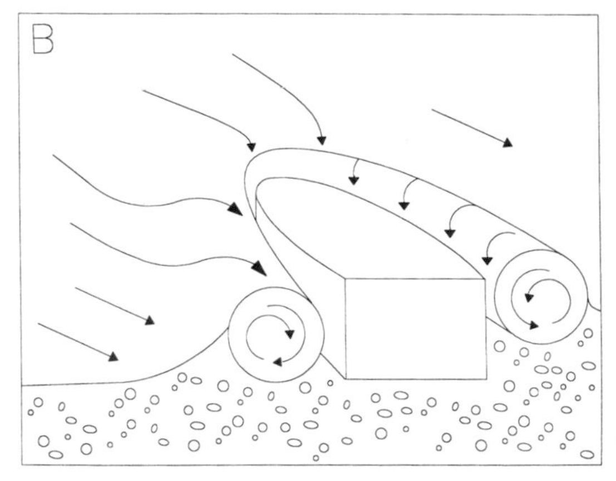

FIGURE 4.—(A) Vortex formation on the upstream face of a large-scale roughness element. Vorticity present in the water is concentrated by the solid face. This horizontal-axis vortex causes high-velocity water to be directed towards the substrate at the base of the element. (B) A horseshoe-shaped vortex forms as the vortex wraps around a roughness element and elongates downstream.

can be applied to local scour at fishrocks. The reader is directed to the works of Jain (1981), Melville (1984), and Raudkivi (1986) for further works on scour at bridge piers.

A horizontal-axis vortex is formed on the upstream face of a fishrock (Figure 4A). As this vortex stretches downstream, the characteristic horseshoe vortex is formed (Figure 4B). In fact, there are three types of vortices associated with fishrocks: Karman vortices, the horseshoe vortex, and trailing vortices (Figure 5). These vortices can cause the sediment particles to lift off the sediment surface and bounce or saltate along the streambed. The active scour region of the upstream part of the horseshoe vortex operates deep

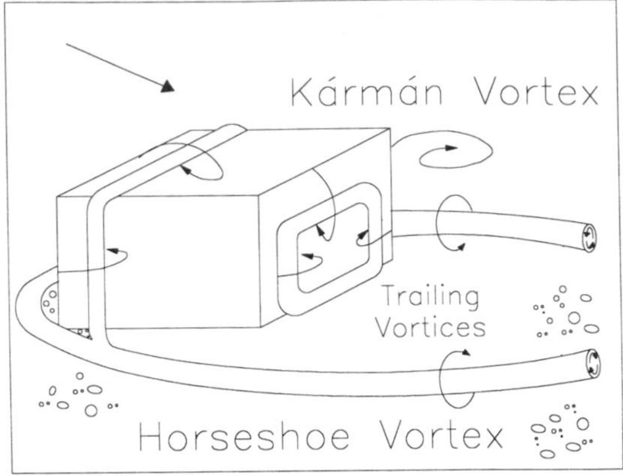

FIGURE 5.—Three types of vortices found at fishrocks. Vertical-axis Karman vortices are shed from the sides of the fishrock. The horseshoe vortex is concentrated on the upstream face of the fishrock and extends downstream. Trailing vortices are shed from the downstream face of the fishrock.

in the scour hole (Figure 6). In surrounding areas, as the angle of repose of the substrate is exceeded, the substrate falls into the active scour region. The vortex in the active scour region picks up individual substrate grains and transports them downstream, where they are deposited to form bars. Figure 7 shows a typical scour and deposition pattern near a fishrock and associated physical parameters.

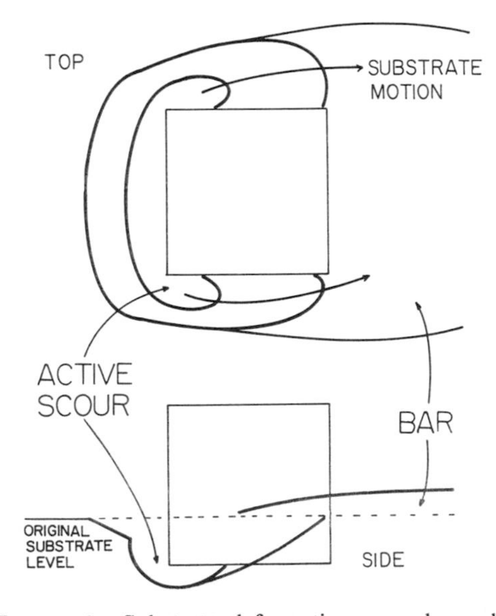

FIGURE 6.—Substrate deformation around a cubic model fishrock. Active scour occurs along the upstream sections of the horseshoe vortex.

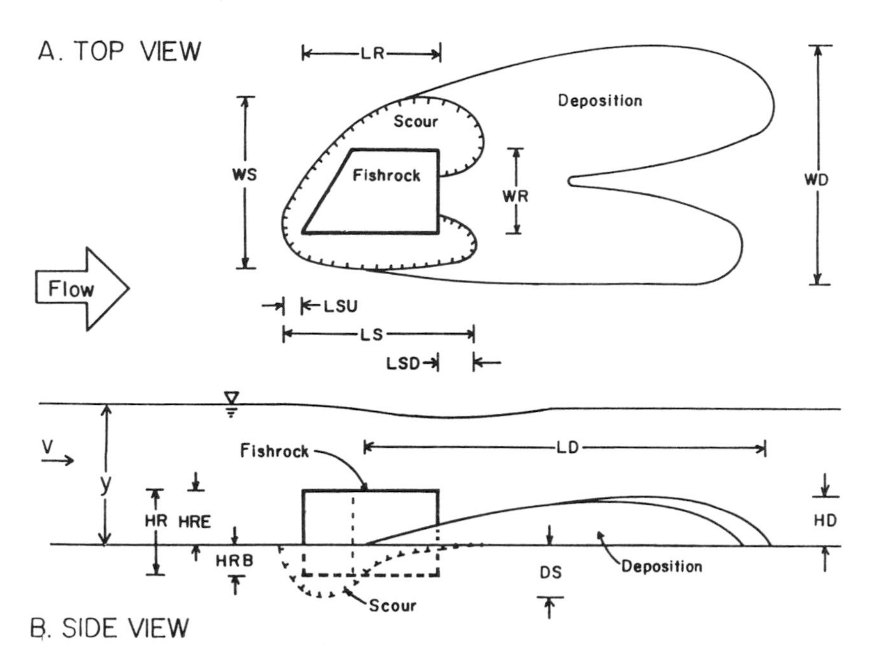

FIGURE 7.—Fishrock, scour pool, and gravel bar definition diagram. HR: height of fishrock; HRE: exposed height of fishrock; HRB: buried height of fishrock; WR: width of fishrock; LR: length of fishrock; LS: length of scour pool; WS: width of scour pool; DS: depth of scour pool; LSU: length of scour pool upstream of the fishrock; LSD: length of scour pool downstream from the fishrock; WD: width of the deposition; LD: length of the deposition; HD: height of the deposition; y: upstream approach flow depth; v: upstream flow velocity.

Conclusions

There are forces acting upon the substrate of a river. The magnitude of these forces is a function of the velocity of the water. Large roughness elements change the velocity of water in the flow field and can direct high-velocity water in the form of vortices towards the substrate. This high-velocity, spiraling mass of water can cause local scour in deformable substrate.

Acknowledgments

I would like to thank the following members of the faculty at Washington State University for their guidance in the preparation of this article: John F. Orsborn, Walter C. Mih, Thomas Bumstead, and Howard D. Copp.

References

Adamek, J. C. 1968. Lift and drag forces on a cube boundary in a finite, three-dimensional flow field with free-surface effects. Doctoral dissertation. Utah State University, Logan.

Bagnold, R. A. 1966. An approach to the sediment transportation problem from general physics. U.S. Geological Survey Professional Paper 422-I.

Breusers, H. N., G. Nicollet, and H. W. Shen. 1977. Local scour around cylindrical piers. Journal of Hydraulic Research 15:211–252.

Daily, J. W., and D. R. F. Harleman. 1966. Fluid dynamics. Addison-Wesley, Reading, Massachusetts.

El-Samni, E. A. 1949. Hydrodynamic forces acting on particles in the surface of a stream bed. Doctoral dissertation. University of California, Berkeley.

Fisher, A. C., and P. C. Klingeman. 1984. Local scour at fish rocks. Pages 286–290 *in* D. J. Schreiber, editor. Water for resource development, proceedings of the conference. American Society of Civil Engineers, Hydraulics Division, Inland Empire Section, Coeur d'Alene, Idaho.

Fontaine, B. L. 1987. An evaluation of the effectiveness of instream structures for steelhead trout rearing habitat in the Steamboat Creek basin. Master's thesis. Oregon State University, Corvallis.

Hall, J. D. 1984. Evaluating fish response to artificial stream structures: problems and progress. Pages 214–221 *in* T. J. Hassler, editor. Proceedings of the Pacific Northwest stream habitat management workshop. Humboldt State University, Arcata, California.

House, R. A., and P. L. Boehne. 1985. Evaluation of instream enhancement structures for salmonid spawning and rearing in a coastal Oregon stream. North American Journal of Fisheries Management 5:283–295.

Jain, S. C. 1981. Maximum clear-water scour around

circular piers. ASCE (American Society of Civil Engineers) Journal of the Hydraulics Division 107: 611–626.

Klingeman, P. C. 1984. Evaluating hydrologic needs for design of stream habitat modification structures. Pages 191–213 *in* T. J. Hassler, editor. Proceedings of the Pacific Northwest stream habitat management workshop. Humboldt State University, Arcata, California.

Klingeman, P. C., S. M. Kehe, and Y. A. Owusu. 1984. Streambank erosion protection and channel scour manipulation using rockfill dikes and gabions. Oregon State University, Water Resources Research Institute Report 98, Corvallis.

Knighton, D. 1984. Fluvial forms and processes. Edward Arnold, Baltimore, Maryland.

Lisle, T. E. 1981. Roughness elements: a key resource to improve anadromous fish habitat. Pages 93–98 *in* T. J. Hassler, editor. Proceedings, symposium on propagation, enhancement, and rehabilitation of anadromous salmonid populations and habitat in the Pacific Northwest. Humboldt State University, California Cooperative Fishery Research Unit, Arcata.

Melville, B. W. 1984. Live-bed scour at bridge piers. ASCE (American Society of Civil Engineers) Journal of the Hydraulics Division 110:1234–1247.

Overton, C. K., W. Brock, J. Moreau, and J. Boberg. 1981. Restoration and enhancement program of anadromous fish habitat and populations on Six Rivers National Forest. Pages 158–168 *in* T. J. Hassler, editor. Proceedings, symposium on propagation, enhancement, and rehabilitation of anadromous salmonid populations and habitat in the Pacific Northwest. Humboldt State University, California Cooperative Fishery Research Unit, Arcata.

Raudkivi, A. J. 1986. Functional trends of scour at bridge piers. ASCE (American Society of Civil Engineers) Journal of the Hydraulics Division 112: 1–13.

Raudkivi, A. J., and R. Ettema. 1977. Effect of sediment gradation on clear water scour. ASCE (American Society of Civil Engineers) Journal of the Hydraulics Division 103:1209–1213.

Reiser, D. W., and T. C. Bjornn. 1979. Habitat requirements of anadromous salmonids. U.S. Forest Service General Technical Report PNW-96.

Roberson, J. A., and C. T. Crowe. 1985. Engineering fluid mechanics, 3rd edition. Houghton Mifflin, Boston.

Rouse, H. 1966. On the role of eddies in fluid motion. American Scientist 51:411–456.

Shen, H. W., V. R. Schnieder, and S. Karaki. 1969. Local scour around bridge piers. ASCE (American Society of Civil Engineers) Journal of the Hydraulics Division 95:1919–1940.

Ward, B. R., and P. A. Slaney. 1979. Evaluation of in-stream enhancement structures for the production of juvenile steelhead trout and coho salmon in the Keough River: progress 1977 and 1978. Pages 8–5 *in* M. E. Seehorn, editor. Proceedings of the trout stream habitat improvement workshop. U.S. Forest Service and Trout Unlimited, Asheville, North Carolina.

Ward, B. R., and P. A. Slaney. 1981. Further evaluations of structures for the improvement of salmonid rearing habitat in coastal streams of British Columbia. Pages 99–108 *in* T. J. Hassler, editor. Proceedings, symposium on propagation, enhancement, and rehabilitation of anadromous salmonid populations and habitat in the Pacific Northwest. Humboldt State University, California Cooperative Fishery Research Unit, Arcata.

American Fisheries Society Symposium 10:219–225, 1991

Increased Densities of Atlantic Salmon Smolts in the River Orkla, Norway, after Regulation for Hydropower Production

N. A. HVIDSTEN AND O. UGEDAL
Norwegian Institute for Nature Research
Tungasletta 2, N-7004 Trondheim, Norway

Abstract.—The River Orkla was regulated for hydropower production in 1983. Estimates of smolt densities of Atlantic salmon *Salmo salar* in the river were conducted from 1983 through 1988. Each year, 3,000–5,000 wild smolts were captured in three sections of the river and marked by fin clipping. Recaptures were made during the smolt run by two traps lowered from a bridge downstream from the area where smolts were tagged. After regulation of the river, the water discharge always exceeded 10 m^3/s, thereby eliminating periods with low discharge (2 m^3/s). Smolt densities appear to have increased gradually from 4.0 (95% confidence interval, 2.7–6.1) per 100 m^2 in 1983 to 7.9 (6.2–10.1) per 100 m^2 in 1987, but dropped to 5.1 in 1988. Because major changes in abiotic conditions are related to stabilized discharges, increase in production area may have been responsible for increased smolt production. High minimum discharge during winter is hypothized to increase winter survival of parr. The estimated increase in smolt densities appears to have resulted in recent higher catches of adult Atlantic salmon by anglers.

The River Orkla was regulated for hydropower production in 1983, when five reservoirs and five hydropower stations were completed. The Svorkmo power plant has its intake downstream of the most important rearing areas at Meldal (Figure 1). To identify seasonal and diel smolt distributions, of Atlantic salmon *Salmo salar* and thereby prevent potential migration of smolts into the power plant, the River Orkla's smolt run has been studied since 1979 (Hesthagen and Garnås 1986). Salmon smolt density was estimated in 1983 just before regulation of the river began that year (Garnås and Hvidsten 1985).

There are few data on the density of Atlantic salmon smolts in rivers (Garnås and Hvidsten 1985), and little is known about the effects of river regulation on smolt production. Before regulation, it was estimated that regulation would lower summer water temperatures in the River Orkla by 2.5°C (Berge et al. 1981). Water temperature is important for salmonid growth (Brett et al. 1969; Elliott 1975a, 1975b). Using Elliott's (1975a) growth model for fish fed maximum rations, Jensen (1987) estimated that a 2.5°C decrease in temperature in the summer would reduce the length of brown trout *Salmo trutta* in the River Orkla by 14%. Hence, smolt age for Atlantic salmon was expected to increase. Because smolt production decreases with smolt age (Symons 1979), we hypothesized that smolt production would decrease after regulation.

Study Area

The River Orkla is in central Norway (Figure 1). The catchment area is 3,092 km^2 and mean annual discharge is 30 m^3/s. About 39% of the river's drainage is controlled for hydropower production. Before regulation for hydropower production, few lakes existed and flood peaks up to 1,000 m^3/s occurred during spring snowmelt runoff. Since regulation, water discharge has been relatively steady throughout the year (Figure 2). Freshets still occur, but they are not as high as formerly. Before regulation, the mean minimum daily discharge during winter was 3.4 m^3/s (range, 1.0–6.6 m^3/s) in 1972–1982. Since regulation, winter discharges have never dropped below 10 m^3/s. Summer discharges have been of the same general magnitude before and after regulation. Atlantic salmon and sea trout (sea-run brown trout) are capable of migrating 92 km up the River Orkla. Because of heavy-metal pollution from mining, smolt production is confined to the upper 70 km of the river. Catches of adult salmon are much higher than brown trout catches and have increased in recent years.

The Atlantic salmon is the dominant species in the river. About 90% of the descending smolts are salmon. The remainder are brown trout.

Methods

To estimate Atlantic salmon smolt abundance, we electrofished in the upper 50 km of the salmon spawning reaches each April before the spring

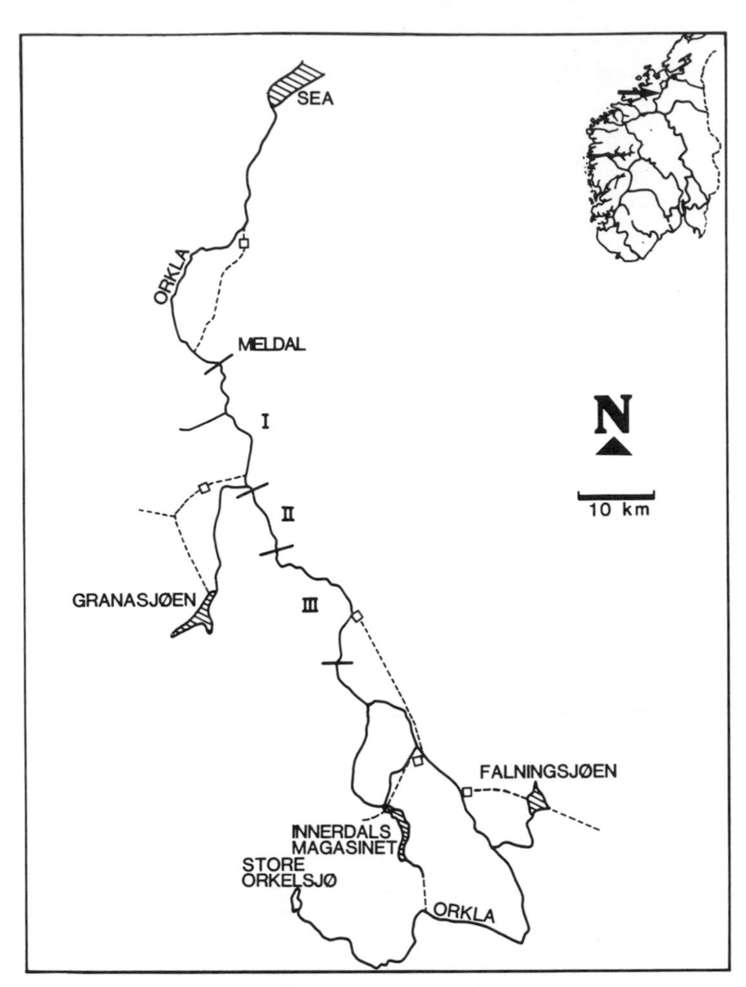

FIGURE 1.—Map showing the River Orkla with the marking sections I, II, and III. Atlantic salmon smolts are caught at Meldal during the smolt run. Adults ascend the river up to the end of section III. The squares represent power stations.

runoff. The river was divided into three sections (Figure 1). Smolts were marked to identify the section of capture by distinctively clipping the adipose and pelvic fins. Smolt densities in the River Orkla have been estimated by the same methods each year following regulation.

The production studies were performed in the period 1983–1988, which included only 1 year of preregulation smolt production. Prior to regulation, the river was covered by ice until the onset of spring flooding, thereby hindering marking by electrofishing.

The smolt run started at the beginning of the spring runoff, which occurred about 1 week before and after May 1, and the run lasted through the first week of June. Migrating smolts were sampled during the run with two traps described by Tyler and Wright (1974) and Hesthagen and Garnås (1986). The traps consisted of a 1-m-square steel frame with a net pouch 5.6 m long (mesh size, 9.0–10.5 mm) and a removable net end. The two traps were operated from racks, which were fastened to Meldal bridge at the downstream end of section I, 43 km from the sea.

Previous investigations revealed that most of the smolts in the River Orkla migrate in the darkest period of the night (Hesthagen and Garnås 1986). Therefore, to avoid clogging with drifting debris during high water, the traps were operated only in the 6-h period between 2100 and 0300 hours. During low discharges, the traps were operated continually. An adjusted Petersen formula—$(M + 1)(C + 1)/(R + 1)$—was used to estimate the size of the salmon smolt population

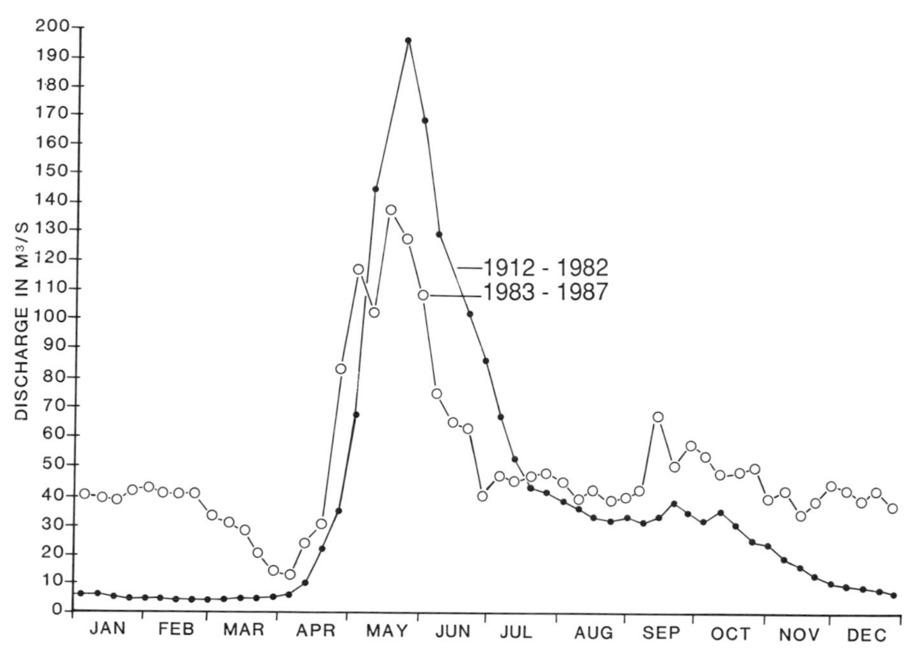

FIGURE 2.—Water discharges in the River Orkla in the periods 1912–1982 and 1983–1987. In the first period discharges were not regulated.

at the time of marking (Ricker 1975): M is the number of smolts marked, C is the number of smolts captured at Meldal bridge, and R is the number of marked smolts recaptured at the bridge. To determine if the marked smolts migrated early, late, or randomly, the smolt population was estimated from the sampling during smolt migration on the basis of the same Petersen method.

The fork lengths of all smolts captured were measured, and smolt age was determined by analyzing otoliths or scales.

Atlantic salmon rearing areas were, for practical reasons, measured from bank to bank with planimeters on maps with a scale of 1:20,000, thereby including areas wetted only at flooding.

Because the smolt production was estimated for only 1 year before the regulation, the catches of adult Atlantic salmon in neighboring rivers as well as in River Orkla were assessed in order to achieve a broader analysis.

Results

Compared with 1983, when hydropower generation started, the estimated densities of smolts produced in the three River Orkla sections were higher ($P = [0.5]^5 = 0.03$) in all 5 years after regulation (Table 1). Densities seem to have increased gradually from 4.0 to 7.9 smolts/100 m^2 during 1983–1987. In 1988 estimated smolt density fell to 5.1 smolts/100 m^2. Because the area of permanently submerged river bottom is less than the area measured from bank to bank, the actual smolt densities were even higher if the rearing areas are considered to have been the minimum submerged areas through the year.

The migration pattern of marked smolts did not differ significantly from that of unmarked smolts. The recapture samples could thus be considered to have been random, and an unbiased estimate of the total smolt density could therefore be obtained by the Petersen method (see Ricker 1975; Figure 3). Mean smolt age was 3.58 years (range, 3.51–3.67 years) in the period 1985–1988. In 1983 and 1984, mean smolt ages were 3.15 and 3.08, respectively. The 1987 catch of adult Atlantic salmon in the River Orkla was twice as high as earlier recorded catches, and 300% higher than the mean catch in the period 1972–1987; in 1985 and 1986 the reported catches were 140 and 142% of the mean catch in the same period (Figure 4). The catch in the River Verdalselv also exceeded previously recorded catches, although the magnitude of the increase was not as great as in the Orkla. In the Rivers Gaula, Stjørdalselv, and Verdalselv,

TABLE 1.—Numbers of Atlantic salmon smolts marked (≥110 mm) and recaptured from different sections and estimated smolt population in the River Orkla between 1983 and 1988. M = number of smolts marked; R = number of recaptured smolts in the sample; C = catch of smolts taken for census; N = size of smolt population at the time of marking; CI = confidence interval.

Year and river section	Area (10^6 m^2)	M	R (% of M)	C	N	Smolt density	
						N/100 m^2	95% CI
			Preregulation of discharge				
1983 I	1.5	1,497	16 (1.1)	1,285	113,319	3.8	2.4–6.3
II	0.5	331	4 (1.2)	1,285	85,390	2.8	1.3–7.1
III	1.0	517	4 (0.8)	1,285	133,230	4.4	2.0–11.1
Total	3.0	2,345	24 (1.0)	1,285	120,687	4.0	2.7–6.1
			Postregulation of discharge				
1984 I	1.5	1,707	17 (1.0)	1,777	168,712	5.6	3.6–9.3
II	0.5	590	6 (1.0)	1,777	150,114	5.0	2.3–10.9
III	1.0	1,094	9 (0.8)	1,777	194,691	6.5	3.6–13.0
Total	3.0	3,391	32 (0.9)	1,777	182,757	6.1	4.4–8.8
1985 I	1.5	2,130	10 (0.5)	779	151,107	5.0	2.9–9.7
II	0.5	660	5 (0.8)	779	85,930	2.9	1.4–6.6
III	1.0	1,420	3 (0.2)	779	277,095	9.2	3.3–18.5
Total	3.0	4,210	18 (0.4)	779	172,872	5.8	3.6–8.8
1986 I	1.5	2,532	10 (0.4)	889	204,943	6.8	3.9–13.2
II	0.5	965	6 (0.9)	889	122,820	4.1	2.0–9.0
III	1.0	1,592	3 (0.2)	889	354,443	11.8	4.8–29.5
Total	3.0	5,089	19 (0.4)	889	226,505	7.6	4.9–12.1
1987 I	1.5	2,435	26 (1.1)	2,848	257,133	8.6	5.9–12.9
II	0.5	1,173	14 (1.2)	2,848	223,060	7.4	4.6–12.7
III	1.0	1,658	22 (1.2)	2,848	205,572	6.9	4.6–10.7
Total	3.0	5,266	62 (1.2)	2,848	238,269	7.9	6.2–10.1
1988 I	1.5	2,082	23 (1.1)	1,778	154,402	5.1	3.5–7.9
II	0.5	1,076	13 (1.2)	1,778	136,856	4.6	2.7–8.1
III	1.0	1,620	19 (1.2)	1,778	144,188	4.8	3.1–7.7
Total	3.0	4,778	55 (1.2)	1,778	151,819	5.1	3.9–6.5

1987 catches were within the limits of previous annual variation.

Discussion

The criteria for estimating smolt densities by mark–recapture experiments (Ricker 1975) were fulfilled in the River Orkla study, as discussed by Garnås and Hvidsten (1985). Sampling areas, procedures, and methods were unchanged throughout the investigation. However, the mesh size of two net pouches was reduced from 10.5 to 9 mm during the sampling in 1987. This change apparantly increased the smolt catch. The fluctuating number of smolts caught in the different years also resulted from difficulties with catch efficiency when flood flows exceeded about 200 m^3, because most smolts migrate at high-water discharges (Hesthagen and Garnås 1986).

After regulation, the salmon smolt population seems to have increased in the River Orkla, in spite of an increase of about 0.5 years in smolt age. A decrease in early growth after regulation was expected based on Jensen's (1987) estimates for brown trout growing in cooler water; thus an increased age of Atlantic salmon smolts also was expected. An increase in smolt density was not anticipated, because smolt age and the number of smolts produced are reported to be inversely related (Symons 1979).

After regulation, the River Orkla's minimum discharge was 10 m^3/s throughout the year. Before regulation, minimum winter discharges were as low as 1.0–6.6 m^3/s (1972–1982). We suggest that this increased winter discharge is the main reason for increase in smolt production.

Young Atlantic salmon seek cover on the bottom during winter (e.g., Lindroth 1955; Gibson 1978; Rimmer et al. 1983), so the availability of suitable cover in winter is important for their survival (Rimmer et al. 1983). Building weirs that raised water levels and created deeper pools has improved the winter survival of salmonids (Saunders and Smith 1962; Näslund 1987).

Gilbert (1978) found increases in the population

American Fisheries Society Symposium 10:227, 1991

FISH PASSAGE

Introduction

As long as there are dams and water withdrawals where fish occur, protection facilities for fish will be needed. Upstream-migrating fish are vulnerable to mortality if their normal passage is blocked or impeded by constructed barriers, and they are liable to injury as they try to pass the barriers. If the fish are on spawning migrations, the population consequences can be severe. Downstream-migrating and rearing fish are killed when they are routed into water intakes and diversion canals or through turbines or control gates that create adverse hydraulic conditions. In some cases, these kinds of passage problems have been the major reason for the demise of fish runs.

Mitigation of passage problems at constructed and natural barriers has gotten increasing priority in recent years. The design and engineering of passage facilities has evolved as successes and failures brought more understanding about the sensitive interactions between biological and engineered systems. Much remains to be learned, but enough project experience has accumulated that there no longer is a need to "reinvent the wheel."

It was in this context that a fish passage section was included in the 1988 Bioengineering Symposium. The papers that follow cover a diversity of large and small projects for the protection and passage of adult and juvenile fish. These projects add to the experience base and should promote the continual evolution toward more biologically successful and economic passage facilities.

William S. Rainey
National Marine Fisheries Service
Portland, Oregon

American Fisheries Society Symposium 10:228–236, 1991

Design, Operation, and Evaluation of an Inverted, Inclined, Outmigrant Fish Screen

JIM A. BOMFORD AND MAURICE G. LIRETTE
British Columbia Ministry of Environment and Parks
Recreational Fisheries Branch
Parliament Buildings, Victoria, British Columbia V8V 1X5, Canada

Abstract.—A unique screening device for juvenile fish has been built on, and is operating in, a hydroelectric canal on Vancouver Island, British Columbia. The canal diverts up to 42.5 m^3/s, and the screen is designed to remove outmigrant smolts of steelhead *Oncorhynchus mykiss* and coho salmon *O. kisutch* from the canal inflow and return them to the Salmon River. The 25-m-long by 6.7-m-wide screen is supported on a removable steel truss suspended in a rectangular section of the canal flowing at a depth of 2.9 m. It incorporates 170 m^2 of slotted woven-wire-mesh screen, and when in service, it inclines downward in the downstream direction, forcing fish into a collector resting on the canal floor at its lower end. The collector diverts the fish laterally out of the canal and into the bypass works. The bypass works consist of a flow-regulating channel, trap, and return pipe to the river. The entire facility is designed to operate passively, in a structurally fail-safe manner, and without power or full-time operator attendance. During two operating seasons (April 15 to June 30, 1987–1989), the facility was 80% efficient, based on mark–recapture studies of small samples of yearling coho salmon smolts. Although the screen provides less than 100% protection, because of compromises required by budgetary constraints, it is a practical, cost-effective alternative to more conventional designs. With appropriate design modifications, this type of screen could function with higher efficiencies at little additional cost, and in streams with fish of all life stages.

The installation of a unique fish screen on the British Columbia Hydro and Power Authority's Salmon River hydroelectric diversion canal is the first attempt in British Columbia to divert downstream-migrating juvenile coho salmon *Oncorhynchus kisutch* and steelhead *O. mykiss* around a major hydroelectric development. Before the development of the Salmon River Canal smolt screen, it was assumed that any form of screened bypass system would involve unprecedented design and operating problems because of the large quantities of water and debris involved (Canada Department of Fisheries and Oceans, unpublished status report).

The Salmon River is on the east coast of Vancouver Island. Its hydrograph is typical of coastal streams in British Columbia; the highest mean monthly flows occur from October to June and low flows from July to September. The mean annual discharge is 61.9 m^3/s, and extremes have ranged from a high of 1,610 m^3/s to a low of 2.6 m^3/s (Water Survey of Canada 1984). Floods result in considerable amounts of floating debris and bedload movement.

The canal intake structure, 50 km from tidal water, was a barrier to both upstream and downstream fish migration. Built in 1957, it consists of a timber crib dam and concrete head works, which incorporate a sluice gate, trimming weir, and radial gate. The radial gate controls the discharge into a concrete-lined trapezoidal canal that diverts up to 42.5 m^3/s of water whenever it is available and maintains a minimum release flow of 1.5 m^3/s back to the river.

Historical notation indicates salmonids were present and migrated upstream of the site of the diversion dam. A 1986 radiotelemetry study confirmed that steelhead successfully migrate up to the diversion (Lirette and Hooton 1988). Upstream and downstream fish passage around the diversion structure were therefore required to facilitate anadromous salmonid production in the upper watershed. The British Columbia Hydro and Power Authority was receptive to the proposal to install the necessary fish passage works around its facility, provided canal inflows were not interfered with and there would be no danger to the existing structures.

Whereas there is a wealth of practical experience with provision of upstream passage for adult fish within the province, we found virtually none with screening downstream-migrating juveniles from large diversions. At smaller industrial, municipal, and agricultural intakes, mechanical and fixed screens are in common use, and there have been studies on the use of louver diverters in hydroelectric intakes in British Columbia (Ruggles and Ryan 1964); however, there are none in operation. Because of the lack of relevant screening experience in the province and restrictions on out-of-province travel to see how similar problems were dealt with elsewhere, the conceptual design had to be developed in relative isolation from other, more-experienced agencies. In this paper, we describe the development of the design in terms of preliminary biological and engineering investigations, conceptual design and model testing, and detailed design, installation, and operation.

Design Development

Biological Investigations

The Salmon River produces both steelhead and salmon (Marshal et al. 1977). The steelhead are the largest found on Vancouver Island and fish of over 8 kg are not uncommon (Hooton et al. 1987). These large fish are highly prized by sport anglers who annually fish 1,500–3,000 d to catch 500–1,500 steelhead in the river (B.C. Steelhead Harvest Analysis, 1970 to 1975).

The potential to increase the production of anadromous salmonids above the Salmon River diversion was estimated in 1980. There are approximately 30 km of available stream length and over 300,000 m^2 of stream area upstream of the diversion (during summer flows). Based on habitat capability estimates of 0.021 steelhead smolts per square meter, the upper Salmon River should produce an average of 6,300 smolts annually (Slaney et al. 1986). The coho salmon carrying capacity was estimated at over 60,000 smolts, based on the equation: coho smolt production = 3.1 (wetted area)$^{0.79}$ (D. E. Marshal, Canada Department of Fisheries and Oceans, unpublished).

A benefit:cost analysis of the proposed screening project was undertaken in 1983 (R. Reid, B.C. Ministry of Environment, unpublished). Assessments indicated that production would be approximately 1,000 adult steelhead and 6,000 adult coho salmon. Production at this level warranted an expenditure of Can$300,000 for the capital cost of the screen.

Preliminary Engineering Investigations

Several variables influence the engineering design of any fish screen, regardless of its type, including canal geometry and hydraulics, fish biology, cost, operating requirements, and fish–screen interactions. Analysis of these variables results in a series of limiting design conditions, which for this project are summarized in Table 1.

Conceptual Design and Model Testing

A literature search provided an overview of the research done on the subject of juvenile-fish screening and the types of screen installations in use, or proposed for use, at similar sites in Canada and the western USA, along with the biological and engineering criteria used to design them (Clay 1961; Ruggles and Ryan 1964; Bell 1973; Skinner 1974; Odenweller and Brown 1982). The major constraints imposed on the design were low capital cost and limited bypass flow. To reduce costs, preliminary investigations were aimed at finding or developing a fish screen that could be installed in the canal without major renovations to the existing structure. It was felt that the cost of canal renovation alone, to allow the incorporation of a proven screen design, would exceed the budget. Restrictions on the allowable bypass flow led to the consideration of a physical barrier rather than a behavioral diverter in an effort to minimize the need for high collector velocities and bypass discharges. It became apparent that existing types of fixed or mechanical screens or louver diverters could be installed for the budgeted capital cost. The project was considered an important part of our fisheries management program, however, and as such warranted continuing the investigations and acceptance of the risks inherent in an unproven screen design.

The chosen design consisted of a rigid screen that would be hinged at its upstream end above the rectangular section of the canal and suspended in the canal, inclining downward in a downstream direction (Figure 1). The inverted, inclined screen would guide the fish into a collector resting on the canal floor at its lower (downstream) end. The collector would divert the fish laterally across the canal floor and through an opening in the canal wall into the bypass works. Based on the maximum allowable sustained swimming speed for the smallest coho salmon smolts, the minimum required screen area would be 152 m^2 (canal discharge divided by fish swimming speed—Table 1). This meant that for a 6.7-m-wide canal, the screen had to be at least 22.7 m long. At the design depth of 2.9 m the inclination would be less than 7°. Theoretically, fish approaching the screen at the maximum approach velocity (V_a) and avoiding the screen at their design swimming speed (V_s) would be required to maintain their position relative to the face of the screen for only 10 s before being

TABLE 1.—Limiting design conditions affecting the Salmon River Canal smolt screen.

Design variable	Value	Comments
	Canal geometry and hydraulics	
Canal cross section		
Trapezoidal		1:1.5 side slopes
Rectangular		6.7 m wide × 2.9 m deep
Canal inflows	0–42.5 m^3/s	Maximum during screen operation
Canal water depth	0–2.9 m	Maximum during screen operation
Canal water velocity	0–2.2 m/s	Maximum during screen operation
Release flow	1.5 m/s	Minimum
	Fish biology	
Design species		
Steelhead		2–4-year-old smolts
Coho salmon		1–2-year-old smolts
Fish fork length (design size)	Maximum, 1,100 mm	Steelhead kelts
	Minimum, 70 mm	Coho salmon smolts
Fish swimming speed (screen approach velocity)	0.28 m/s	Maximum sustained speed of 70-mm coho salmon smolts
Migration period	Apr 1–Jun 30	Maximum canal inflow
	Cost[a]	
Capitol cost	$300,000	1:17 benefit:cost
Operating cost	$5,000 annually	Operation and maintenance
	Operating requirements	
Operation and maintenance	Minimum	See operating cost
Screen cleaning	Manual	No power at site
Debris and bed-load removal	Manual	No control over canal inflows
Bypass flow control	Manual	Self-regulating
Fish counting	Manual	On demand
Operating period	Apr 1–Jun 30	Remove screen when out of service
Structural safety	Fail-safe	Water level controlled
	Fish–screen interactions	
Screen type	Physical or behavioral	≥80% efficiency
Screen area	152 m^2	Canal discharge ÷ swimming speed
Collector	Directional guidance	No still water
Bypass flow regulation	0–0.5 m^3/s	Gated and self-regulating
Trap	0.25 m^3/s	Self-cleaning
Bypass pipe	0.25 m^3/s	Removable

[a]Canadian dollars.

swept into the collector. Further development of this screening concept, however, required that hydraulic model testing be carried out in order to generate the information necessary to produce a workable design.

Model tests were required to generate information on the hydrodynamic loads acting on various types of screening materials and structural geometries subject to differing flow conditions. These tests and a theoretical hydrodynamic analysis provided force coefficients that were used to estimate head loss across the screen and the hydrodynamic lift and drag forces required to structurally design the screen. Additional tests were also undertaken to optimize the geometry of the flow-regulating section of the bypass works (M. C. Quick, University of British Columbia, unpublished). Hydraulic modeling was carried out at the University of British Columbia's Civil Engineering Laboratory. All tests were carried out in a Plexiglas, tilting flume, 16 cm wide, 45 cm deep, and 10 m long, through which water was recirculated.

Tests showed that, for a given percentage open area and inclination to the flow, woven-wire screens had lower force coefficients than perforated-plate screens and passed debris more readily. Tests of different types of woven-wire fabrics showed that force coefficients were roughly proportional to percentage open area. On the basis of these tests, a slotted, triple-shute, woven-wire mesh with 12-mm × 100-mm openings and 2-mm-diameter stainless steel wire was chosen for the screen. The choice of this fabric was a compromise that offered maximum open area (70%), and therefore minimum head loss, and adequate struc-

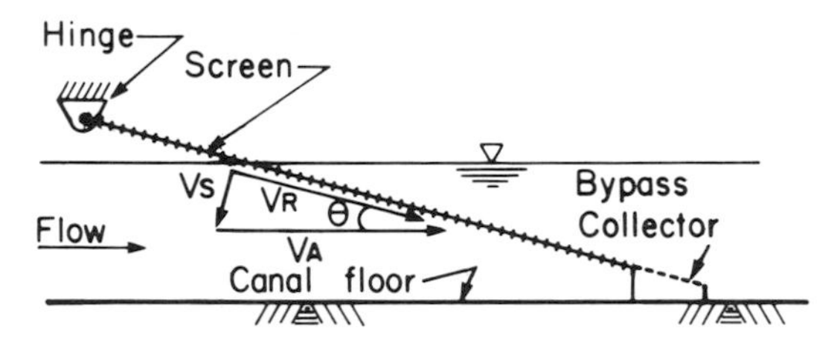

FIGURE 1.—Conceptual design for the Salmon River Canal smolt screen. The screen inclines downward in a downstream direction, guiding the fish into a collector located on the canal floor. The theoretical velocities of the water and of the fish in the canal upstream and under the screen are also shown. (Adapted from Clay 1961.)

tural strength (Figure 2). An equally important consideration in deciding on this material was the trade-off between the needs to pass large quantities of small debris but small numbers of small fish. Although the screen openings were large enough to pass coho salmon smolts, it was felt that the screen would be a behavioral barrier to the fish, given the high approach velocity, the low inclination of the screen, and the orientation of the slots perpendicular to the flow. The screen would be a physical barrier to the larger steelhead smolts. The hydrodynamic loading resulting from flow through both the submerged portions of the screen fabric and its support structure provided the lift and drag forces that determined the structural design of the composite screen and the head loss across it.

In order to provide passive flow regulation through the bypass works to the proposed fish trap, an energy dissipation channel was developed to regulate flow without interfering with fish, debris, or bed-load movement. This channel con-

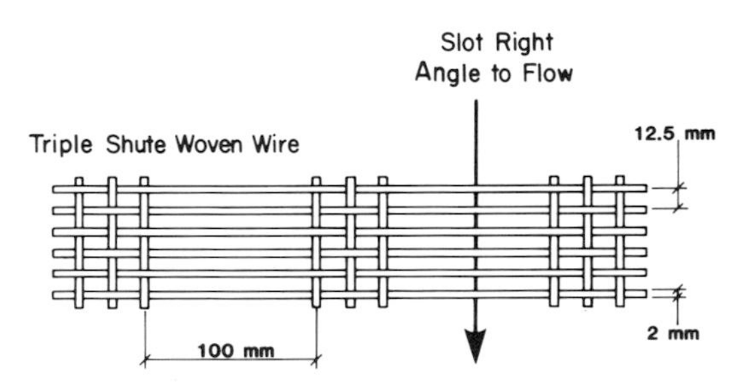

FIGURE 2.—Plan view of the woven-wire screen fabric used on the Salmon River Canal smolt screen.

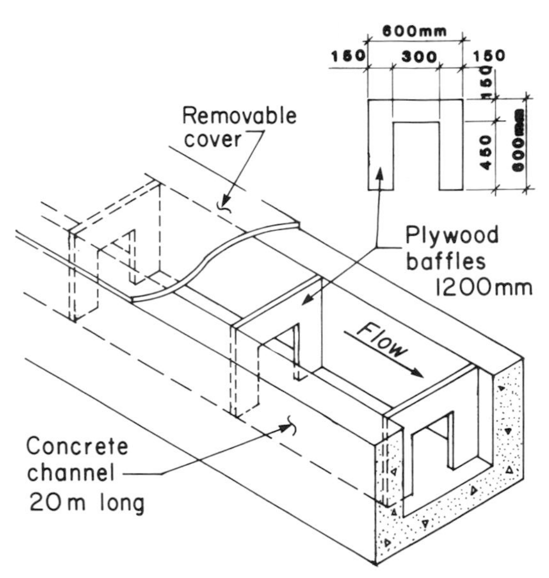

FIGURE 3.—Details of the flow regulator, which controls the bypass discharge from the Salmon River Canal smolt screen.

sisted of a pressurized rectangular conduit, 0.6 m square and 36 m long, containing orifice baffles. Model tests were used to determine optimum baffle and orifice geometries, baffle spacing, and discharge coefficients (Figure 3).

Detailed Design and Operation

The facility can be looked at in terms of a screen assembly and bypass works (Figure 4). The screen assembly is composed of a screen and collector mounted on a removable, hinged carrying truss that is made operational by a counterweighted, water-ballasted, lifting and lowering device. The bypass works consist of an exit well, a flow-regulating channel, a trap, and a bypass pipe.

The Screen Assembly

The 25-m long × 6.7-m-wide woven-wire-mesh screen, as previously described, is continuous across the underside of the downstream portion of a tubular steel frame truss 30 m long × 6.7 m wide. The openings at the edges of the truss adjacent to the canal walls are sealed with 10-mm neoprene rubber flaps. In its out-of-service mode, the screen is suspended above the canal (Figure 5b). The truss is supported by hinges mounted on a catwalk at its upstream end and by a counterweight assembly two-thirds of the way toward its

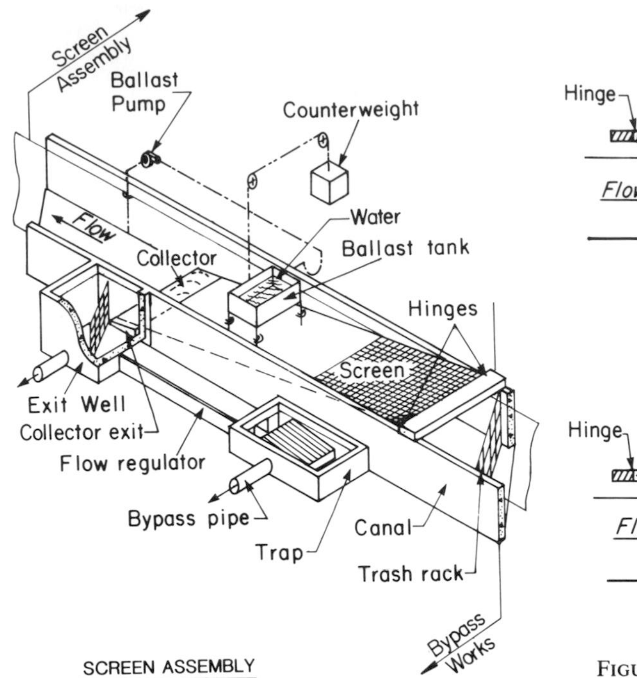

FIGURE 4.—General arrangement of the screen assembly and bypass works for the Salmon River Canal smolt screen.

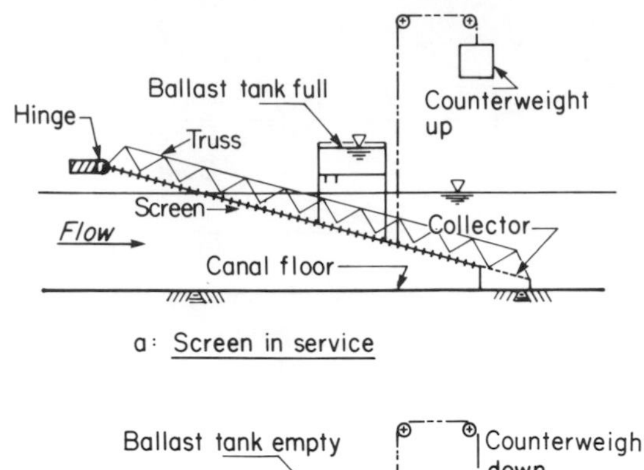

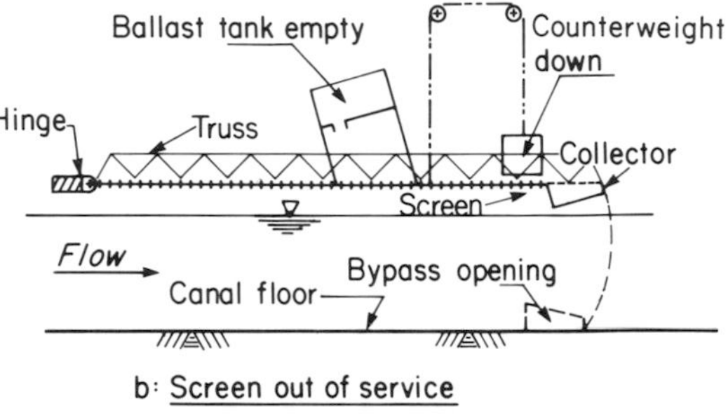

FIGURE 5.—Schematic of Salmon River Canal smolt screen showing **(a)** screen in service, and **(b)** screen out of service.

downstream end. The ballast tank used to lower the assembly is empty and the counterweight used to lift it is down. The counterweight assembly consists of a 4-m^3 (9,600 kg) concrete weight that carries the truss via an overhead gantry and suspension cables.

In its in-service mode, the ballast tank is filled with water, the weight of which counteracts both the surcharge on the counterweight and the hydrodynamic lift on the screen truss and depresses the screen into the canal (Figure 5a). The screen declines at an angle of 6.5° and the collector rests on the canal floor adjacent to the bypass opening in the canal wall. The ballast tank has a capacity of 15 m^3 of water (15,000 kg) and is filled with canal water via a gas-driven pump. It takes roughly 10 min to either fill or drain the tank (lower or lift the screen). The tank can be drained either manually or automatically through valves in its bottom. Manual draining is used to lift the screen for cleaning or maintenance. An automatic drain, actuated by a float valve in a stilling well upstream of the screen, provides emergency lifting of the screen in the event of extreme high water in the canal or backwatering due to debris loading on the screen. The screen will also rise like a huge flap gate, even with a full ballast tank, if excessive amounts of debris accumulate and cause a large head differential across it. Cleaning the screen is achieved by simply lifting it out of the canal. Debris falls off its underside whenever it is raised and is swept away by the flow in the canal. Bed-load accumulation on the canal floor at the downstream end of the screen under the collector is scoured away by the hydraulic action occurring at the constriction of the flow as the screen seats itself.

The collector is located at the extreme downstream end of the screen (Figure 6). It is composed of a 2.1-m-wide panel of 3-mm perforated aluminum plate (3-mm holes, 40% open area) that spans the width of the canal on the plane of the screen. Guide vanes are attached to the underside of the plate. The canal floor acts as a second side to the collector, opposing the perforated plate. The depth of the collector varies due to the inclination of the screen. It is roughly 0.3 m deep upstream and narrows down to 0.1 m at the back (downstream) end, which has a neoprene seal between it and the canal floor. The fish are guided into the collector by the screen, and the collector diverts them laterally across the floor of the canal toward a gated opening cut in the canal wall that

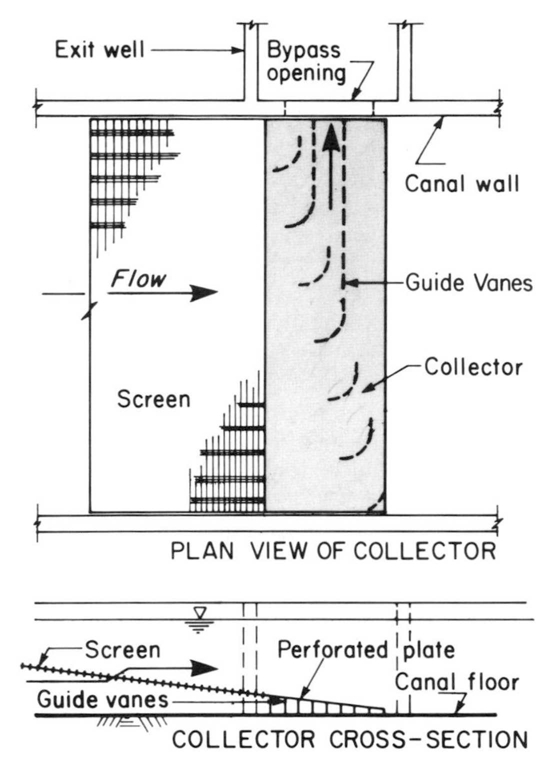

FIGURE 6.—Plan view and cross section of the collector used on the Salmon River Canal smolt screen.

leads to the bypass works. The bypass opening in the canal wall is 1.5 m long and 0.25 m deep at the upstream end, sloping down to 0.1 m at its downstream end. The lateral water velocity in the collector varies with the canal discharge and is roughly equal to the screen approach velocity. The velocity of the water passing through the bypass opening varies from 0.4 to 0.9 m/s, depending on the discharge and depth of water in the canal. At present discharges and velocities through both the collector and the bypass opening are uncontrolled.

The bypass opening in the canal wall discharges into an exit well 2.4 m × 2.4 m × 3.0 m deep. The exit well contains a sluice gate and an outlet to the flow regulator that leads to the trap. A secondary screen, consisting of a vertical perforated-plate screen, diverts a portion of the bypass flow and the fish into the trap via the flow-regulating channel. The water passing through the secondary screen discharges through a sluice gate and back to the river. This secondary screen and sluice gate provide control of supplemental discharges through the collector and bypass opening. The flow regulator, as described in the section on model testing, was developed to regulate the discharge to the trap. This discharge would have been excessive at maximum canal discharge if allowed to pass through the bypass opening uncontrolled. A 35-m length of channel provides from 0.1 to 0.2 m^3/s discharge to the trap for canal water levels from 0.9 to 2.9 m.

The trap contains a 5.75-m^2 inclined, corrugated, perforated-plate screen that discharges the fish into a 0.6-m^3 holding tank for counting. The trap screen is removable, thereby allowing the fish to go directly down a bypass pipe to the river if desired. The trap screen is designed to handle inflows of up to 0.25 m^3/s and is self-cleaning. The aluminum bypass pipe is 20 m long and 0.25 m in diameter and can be removed whenever the screen is not in operation or when the river is in flood.

Discussion and Results

The majority of the screen components were prefabricated in Vancouver and transported to the site by barge and truck. Installation took place in August and September 1986. Initial mechanical and biological evaluations occurred in the spring of 1987 and the screen was operated continuously during the springs of 1988 and 1989 (April 20–June 30).

A program of headwater stocking throughout the watershed above the diversion commenced in September 1986 with the release of young-of-the-year steelhead and coho salmon. The stocking program is expected to continue until adequate adult returns are established. The first outmigrating juveniles were intercepted in the spring of 1988 with the downstream movement of yearling coho salmon and 2-year-old steelhead.

Biological Evaluation

In the spring of 1987, several test releases of smolted, hatchery-reared steelhead were undertaken (Table 2). The fish were released immediately upstream of the screen. Due to a shortage of funds, however, the collector at this time consisted of only a 1.2-m-wide plywood cover fastened to the underside of the screen at its downstream end, and a 0.9-m-wide, 6-mm-mesh woven-wire transition screen upstream of the plywood. General fish behavior under the collector and lower part of the screen was observed via underwater video camera viewing from the exit well through the bypass opening in the canal wall. The fish did tend to orient themselves facing

TABLE 2.—Efficiencies of the Salmon River Canal smolt screen by species and number of marked fish released and recaptured in 1987, 1988, and 1989.

Date	Fish species	Number released	Number recaptured	Efficiency (%)
Apr 1, 1987	Steelhead	561	115	20
Apr 22, 1987	Steelhead	100	44	44
May 5, 1987	Steelhead	51	21	41
May 21, 1988	Coho salmon	27	2	7
Collector modified				
Jun 2, 1988	Coho salmon	9	8	89
Jun 6, 1988	Coho salmon	37	32	86
Jun 13, 1988	Coho salmon	11	7	68
Maximum bypass flow increased from 0.2 to 0.4 m/s				
May 11, 1989	Coho salmon	72	64	89
May 18, 1989	Coho salmon	100	79	79
May 24, 1989	Coho salmon	21	18	86
Jun 2, 1989	Coho salmon	14	10	71

upstream and away from the screen as anticipated (Figure 1). However, most of the fish that reached the collector were holding in a strong upward vortex that passed through the transition screen upstream of the plywood. Some of the fish were impinged against the transition screen momentarily before swimming down and away. What soon became apparent was that the steelhead were not actively or passively finding the bypass opening. There was insufficient transportation flow toward the bypass opening to either attract the fish or sweep them in that direction. The screen was left in the canal for approximately 20 h on three occasions. When the screen was lifted for cleaning, all those fish that had not reached the bypass and were holding under the screen and collector were lost down the canal. These tests resulted in recapture of only 35% of marked fish in 1987. Many of the fish trapped showed signs of eye abrasion and scale loss. It was surmised at the time that the steelhead used in the test had lost the initiative to migrate downstream. The results were not considered realistic due to the stress of transport and release into a hostile environment with high-velocity water and cold temperatures (5–8°C). It was hoped that downstream-migrating juveniles, expected in the spring of 1988, would actively seek out the bypass opening. Steelhead smolts were not considered for use in trap efficiency tests because they can revert to parr after being returned upstream.

Within a few weeks after the screen became operational in 1988, it became apparent that outmigrant coho salmon smolts, like steelhead the previous year, did not seek out the bypass and that there was indeed a serious problem with the collector. The coho salmon smolts, being weaker swimmers than the larger steelhead smolts used in 1987 tests, became impinged on the transition screen. Between April 20 and May 26, 1988, 132 live coho salmon smolts were trapped and 90 were counted dead in the trap or on the transition screen. This indicated a minimum loss of 41% because many dead fish would have been washed off the screen while it was lifted and lost before they were counted. One recapture test of marked coho salmon on May 21, 1988, indicated a loss of over 90%. In May 1988, the plywood cover and woven-wire mesh transition screen were replaced with the perforated plate and guide vane collector (Figure 6). Since that date, the mean recapture rate of marked juveniles released upstream of the screen has been 80% (Table 2). Fish still have a tendency to hold in quiet water within the collector and at the bypass opening, which results in some loss of efficiency because any fish in the collector when the screen is lifted for cleaning are swept down the canal. Fish trapped with the modified perforated-plate collector showed no evidence of scale loss or eye abrasion. The flow regulator also provides quiet water for the fish to hold in, which delays their passage through it. It is too early to say whether this delay (up to a day) is a problem.

These initial evaluations were based on very small samples. Additional, more-comprehensive tests will be required to ascertain the actual causes of fish mortality and locations of fish loss.

Mechanical Evaluation

Mechanically, the facility has functioned well; there were few problems with the lifting or lowering of the screen or in the operation of the flow regulator and trap. The float-controlled automatic drain on the ballast tank was set to drain the tank and lift the screen whenever the canal water level reached 2.75 m (0.15 m below minimum freeboard). This feature has caused the screen to be lifted safely out of the canal on several occasions, without operator attendance, when the screen was allowed to accumulate debris. Screen cleaning was required only once a day on average, indicating that sizing of the screen mesh is adequate for this application. The only types of troublesome debris encountered were alder and maple catkins, most of which passed through the screen.

It was felt that the ability of the screen to pass debris and guide fish may be affected by vibrations set up in the structure as the water passes through it. These audible vibrations increase in intensity with increases in canal discharge, and may induce a secondary behavioral avoidance reaction in the fish. However, further evaluations will have to be done to determine whether or not these vibrations contribute to the screen efficiency. Most debris that did become impinged fell off as planned when the screen was lifted clear of the canal. No other cleaning had to be done, although the canal had to be dewatered once in order to remove large floating debris (branches, logs, etc.) from the coarse trash rack in front of the screen. The initial configuration of the collector led to fish mortality, as discussed already, but the modified collector has rectified most of this problem. Bed load and woody debris were flushed clear of the collector by hydraulic action as the screen was raised and lowered, and jamming or bridging was not a problem. The flow regulator worked well as a passive control on discharges to the trap, passing fish, debris and bed load as designed. Yearly removal of branches and rocks is all that has been required.

Conclusions

The Salmon River Canal smolt screen has met its design objectives and is a practical and cost-effective method of diverting steelhead and coho salmon smolts from a hydroelectric diversion canal; however, it does compromise efficiency in achieving low cost. The screen's main advantages over existing screening technologies include

- low capital, operating, and maintenance costs;
- ease of installation in an existing canal without major structural modifications;
- structurally fail-safe operation during periods of excessive canal inflow or debris accumulation;
- no electric power requirement, an important consideration at remote sites;
- easy cleaning of accumulated debris and bed load, and easy access of screen components for maintenance without interruption of canal inflows; and
- ability to function in high-velocity inflows and variable water depths without operator attendance.

The screen's main disadvantage is that it is less than 100% efficient (80% for fish greater than 70 mm in length). Therefore, the following recommendations are made for improving performance and efficiency in this facility and any similar installations to be built in the future.

- Install finer mesh screen to create a total barrier to fish of all life stages.
- Increase screen area (decrease inclination) to reduce approach velocity.
- Improve the connections between screen and carrying truss to minimize excessive local velocity gradients.
- Provide more effective side seals between the screen, canal walls, and bottom.
- Install screen cleaners to minimize the need to lift the screen, thereby reducing losses of fish holding under the collector and mortalities due to velocity gradients caused by debris blockages.
- Provide a gradation in the percentage of open area of the collector cover to minimize excessive approach velocities within the collector.
- Increase bypass flow and improve guidance within the collector to induce more rapid lateral fish guidance through the collector into the bypass.

Acknowledgments

This project was conceived and coordinated by the British Columbia Ministry of Environment and funded by the Federal/Provincial Salmonid Enhancement Program. We thank Bob Hooton, George Reid, and Art Tautz for their administrative support; Pat Slaney and Ron Ptolemy for their habitat capability studies; Roger Reid for his economic analysis; Michael Quick for the hydraulic model studies and hydrodynamic analysis; Harold Ammundson for the structural design; Harry Reynolds for the mechanical design and engineering coordination; Don Johnson and Walter Brunner for B.C. Hydro and Power Authority's support; Jim Wild, Ken Sunn, and Allan Borham from the Department of Fisheries and Oceans for their contract administration; Bruce Harvey and Lou Carswell for the field operations; and Larry Wells and George Sutcliffe for their assistance in the report preparation.

References

Bell, M. 1973. Fisheries handbook of engineering requirements and biological criteria. U. S. Army Corps of Engineers, Portland, Oregon.

Clay, C. H. 1961. Design of fishways and other related facilities. Canada Department of Fisheries and Oceans, Ottawa.

Hooton, R. S., B. R. Ward, J. A. Lewynsky, M. G. Lirette, and A. R. Facchine. 1987. Age and growth of steelhead in Vancouver Island populations. British Columbia Ministry of Environment, Fisheries Technical Circular 77, Victoria.

Lirette, M. G., and R. S. Hooton. 1988. Telemetric

investigations of winter steelhead in the Salmon River, Vancouver Island. British Columbia Ministry of Environment, Fisheries Technical Circular 32, Victoria.

Marshal, D. E., R. F. Brown, V. D. Chahley, and D. G. Demontier. 1977. Preliminary catalogue of salmon streams and spawning escapement of statistical area 13 (Campbell River). Canada Fisheries and Marine Service Data Report Series AAC/10-77-1.

Odenweller, D. B., and R. L. Brown, editors. 1982. Delta fish facilities program report. California Department of Fish and Game, Technical Report 6, Stockton.

Ruggles, C. P., and P. Ryan. 1964. An investigation of louvers as a method of guiding juvenile pacific salmon. Canadian Fish Culturist 33:7–68.

Skinner, J. E. 1974. A functional evaluation of a large louver screen installation and fish facilities research on California water diversion projects. *In* L. D. Jensen, editor. Proceedings of the second entrainment and intake screening workshop. Johns Hopkins University, Cooling Water Research Project, Baltimore, Maryland.

Slaney, P. A., C. J. Perrin, and B. R. Ward. 1986. Nutrient concentration as a limitation to steelhead smolt production in the Keogh River. Proceedings of the Annual Conference Western Association of Fish and Wildlife Agencies 66:146–155.

Water Survey of Canada. 1984. Historical streamflow summary, British Columbia. Canada Ministry of Supply and Services, En36-418/1984-3, Ottawa.

American Fisheries Society Symposium 10:237–248, 1991

Some Design Considerations for Approach Velocities at Juvenile Salmonid Screening Facilities

ROBERT O. PEARCE AND RANDALL T. LEE

National Marine Fisheries Service, Environmental and Technical Services Division
1002 Northeast Holladay Street, Room 620, Portland, Oregon 97232, USA

Abstract.—The size, and therefore the cost, of screening facilities required at water diversion sites is primarily determined by the allowable approach velocity of water at the screen mesh. General screening criteria established by fisheries agencies specify maximum approach velocities. Biological factors affect the swimming ability of the fish. In addition to the biological factors, proper attention must be given to engineering factors including uniform velocity distribution at the screen facility. Providing basic screen facility hydraulics necessary for effective fish protection requires careful attention to channel configuration and frequently involves use of baffles and training walls to control direction of flow and magnitude of velocity.

Diversions of water from areas that produce Pacific salmon *Oncorhynchus* spp. and steelhead *O. mykiss,* whether the diversions be for power production, irrigation, or municipal, industrial, or other use, present a major threat to juvenile salmonids that are rearing or actively migrating downstream. Numerous types of screens and facility designs have been and continue to be used for protection of these fish, with varying degrees of success. At many of the facilities where significant injury, delay, and predation have been observed, those problems are a result of poor engineering design. The most important general hydraulic engineering factors that affect fish guidance in front of the screens are (1) the allowable approach velocities based on swimming ability of juvenile salmonids, (2) the flow pattern (both magnitude and direction of velocity) immediately upstream of the screen mesh, and (3) the uniform distribution of flow through the total area of screen mesh to optimize facility performance. Biological evaluations carried out at many facilities indicate that inadequate attention is given to these concerns during the design phase of many projects. Some suggestions and examples are presented herein on how to avoid or minimize some common design inadequacies.

Allowable Approach Velocities

Limiting the magnitude of the water velocities approaching juvenile-fish screening facilities is an important concern in the design of these facilities. Water velocities must not be excessive in relation to the swimming ability of the fish. Many factors contribute to the swimming abilities of juvenile salmonids. They include species, size, water temperature, dissolved oxygen level, and other water quality factors.

The National Marine Fisheries Service (NMFS) and other fisheries agencies have developed velocity criteria to be used as guidelines for the design of fish screening facilities (Table 1). These criteria are based upon numerous studies. Screening criteria commonly divide juvenile salmonids into two size-groups: those less than 2.36 in (60 mm) in length are referred to as fry and those 2.36 in (60 mm) or greater in length as fingerlings. These definitions apply here.

Smith and Carpenter (1987) defined sustained swimming speed as the speed that can be maintained for long periods of time without resulting in muscular fatigue. They further defined prolonged swimming speed as higher than sustained speed but sustainable for shorter periods of time. The greatest swimming speed is burst swimming speed, which can be maintained only for a few seconds. Smith and Carpenter studied swimming stamina of five species of salmonid fry in terms of 15-min sustained velocity tests and burst tests. Their results indicate that the temperature of the water, in addition to the size and species of the fish, influences the swimming speed capabilities. Maximum sustained performance of all species increased with increasing temperature between 39 and 54°F, the average increase being 0.13 feet per second (ft/s) for a temperature increment of 9°F. The results of their study were consistent with earlier studies in that absolute swimming speed increased with increasing body length.

Sazaki et al. (1972) concluded that the swimming performance of steelhead and chinook salmon *Oncorhynchus tshawytscha* is directly related to their size. The report also concluded that

TABLE 1.—Agency velocity criteria for screening salmonids. (Sources: EPRI 1986; K. Bates, Washington Department of Fisheries, personal communication.)

Agency	Approach velocity (ft/s)[a]		Transport velocity (ft/s)[d]
	Fry[b]	Fingerlings[c]	
National Marine Fisheries Service	≤0.4	≤0.8	Greater than approach velocity
California Department of Fish and Game	≤0.33 for continuously cleaned screens; ≤0.0825 for intermittently cleaned screens	Same as fry	At least twice the approach velocity
Oregon Department of Fish and Wildlife	≤0.5	≤1.0	Approach velocity or greater
Washington Department of Fisheries	≤0.4	≤0.8	Approach velocity or greater
Alaska Department of Fish and Game	≤0.5	Same as fry	No criterion
Idaho Department of Fish and Game	≤0.5	≤0.5	Sufficient to avoid physical injury to fish
Montana Department of Fish, Wildlife and Parks	≤0.5	≤1.0	No criterion

[a]Velocity component perpendicular to and approximately 3 in in front of the screen face.
[b]Fish less than 2.36 in (60 mm) long.
[c]Fish 2.36 in (60 mm) or longer.
[d]Theoretical velocity vector along and parallel to the screen face; also called sweeping velocity.

the practical velocity limit for fry 1.6 in (40 mm) in length is about 0.5 ft/s and guidance devices should be designed with the swimming capabilities of the fish in mind.

Bates (1988) reviewed the literature on salmonid swimming stamina tests in developing a recommendation for juvenile salmon screening criteria for the Washington Department of Fisheries (WDF). The report proposed conservative criteria in order to protect fish in most situations, with the expectation that more or less stringent criteria may be used depending upon conditions.

The required approach velocity criteria of 0.4 and 0.8 ft/s for fry and fingerlings, respectively (Table 1, NMFS and WDF), are based on expected low water temperature, the sizes of emergent fry and the smallest (2.4-in) fingerlings, and a fish protection efficiency near 100%. The length of time a fish is exposed to a specific velocity at the face of a screen is important to the fish's ability to maintain a position and avoid impingement. The agency criteria in Table 1 do not specify a length of time. However, the common practice in some agencies has been to require that screen bypass systems be designed such that all fish can be expected to reach a bypass within 2 min or less of reaching the screen.

Angle of Approach Flow

Orientation of a juvenile-fish screen can be categorized as perpendicular to the flow or angled to the flow. Screens placed perpendicular to the flow have the disadvantage of requiring the fish to search longer for the bypass entrance. For a fish to stay off the screen and at the same time move laterally in search of the bypass, it must actually swim faster than the approach velocity. This may not create a serious problem at short screens, but fish may tire and be impinged on screens longer than 4–5 ft before finding the bypass. Screens set at an angle to the flow have an advantage of providing a water velocity component V_s parallel to the screen face in addition to a water velocity component V_n normal to the screen (see Figure 1). The parallel component, referred to as the transport or sweeping velocity component, is usually designed to be equal to or greater than the value of the normal velocity component (see Table 1). This requires the angle of the screen face to the direction of the flow not exceed 45°. The sweeping velocity component serves to guide fish along the screen face to the bypass entrance, thus relieving the fish from expending its energy locating the bypass entrance. The fish only has to swim fast enough to overcome the normal component of the velocity to avoid being impinged on the screen mesh.

The hydraulic action associated with the sweeping velocity component also assists in moving the smaller debris that passes through the trashrack downstream along the screen face. Debris is usually allowed into the bypass system and discharged with the fish into an area downstream from the fish screens.

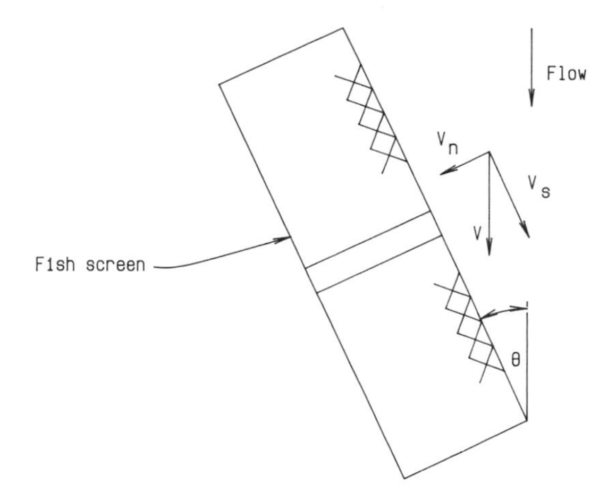

FIGURE 1.—Fish screen oriented at angle θ to the flow. V is the true velocity of flow. V_n is the velocity component perpendicular (normal) to the screen face, $V \sin(\theta)$, the component to which agency criteria for approach velocity apply. V_s is the velocity component parallel to the screen face, $V \cos(\theta)$, the component to which agency criteria for transport velocity apply.

For the above reasons, the flow field in front of the entire screen area should be controlled by design to provide the desired direction of flow to the screen face at all points.

Required Screen Area

The development of the minimum amount of gross screen area required for a facility is a function of the flow rate and the maximum allowable water velocity normal to the screen. For example, to determine the minimum gross screen area required to accommodate fry size fish for a flow of 100 ft³/s with the maximum allowable V_n of 0.5 ft/s,

$$\text{area} = \frac{100 \text{ ft}^3/\text{s}}{0.5 \text{ ft/s}} = 200 \text{ ft}^2.$$

Need for Uniform Flow Distribution

Fisheries resource agencies require that applicable approach velocity criteria be adhered to in the design of the screen facilities and that maximum not be exceeded. The minimum screen area as computed above meets criteria only if there is perfectly uniform distribution of flow over the screen area. Therefore, it is important to maintain uniform flow distribution through the entire mesh area of the screen. This is a simple but significant point. In practice this condition has seldom been achieved. Any nonuniformity of flow will result in some areas of higher-than-acceptable velocities, creating the potential for impingement. Also, juvenile fish can sense changes in velocity, and they avoid moving from lower to higher velocity and vice versa, which affects their passage along the screen face and into the bypass (Bell 1984). For these reasons, particularly close attention must be paid to providing uniform flow conditions that will minimize undesirable delays in fish migration.

Causes of Nonuniform Flow through a Screen

Much study has been made of head losses associated with flow through screen mesh material of various types, including woven wire, perforated plate, and profile wire. As stated by Stefan and Fu (1978), head losses through intake screens are determined by screen geometry, scale effects (Reynolds number), and the orientation of the screen with respect to the direction of flow. For the range of approach velocities and types of mesh used in screening salmonid juveniles, the bulk of the head loss is due to expansion of the jets after they pass through the individual pores of the mesh, and there is little if any dependence on the Reynolds number. For a simplified analysis, therefore, hydraulic theory states that head loss through a screen is approximately proportional to the square of the maximum velocity of fluid going through the pores of the mesh:

$$H_d = \frac{KV_1^2}{2g}; \tag{1}$$

H_d = head loss (ft),
K = the loss coefficient, which is the same for all velocities for a given screen mesh and orientation to flow,
V_1 = velocity of the contracted jet through the pore (ft/s), and
g = the gravitational constant (32.2 ft/s²).

Figure 2 shows a simplified example of flow through woven-wire mesh and profile-wire mesh.

The above equation relates head loss across a screen to the velocity through the mesh and therefore to the approach velocity. A theoretical determination of head loss cannot be made for most screen materials without introducing experimental coefficients. Numerous hydraulic laboratory tests have been performed to empirically determine head losses associated with various types of mesh materials for different porosities and angles of approach flow. Manufacturers of screen mesh material usually have such informa-

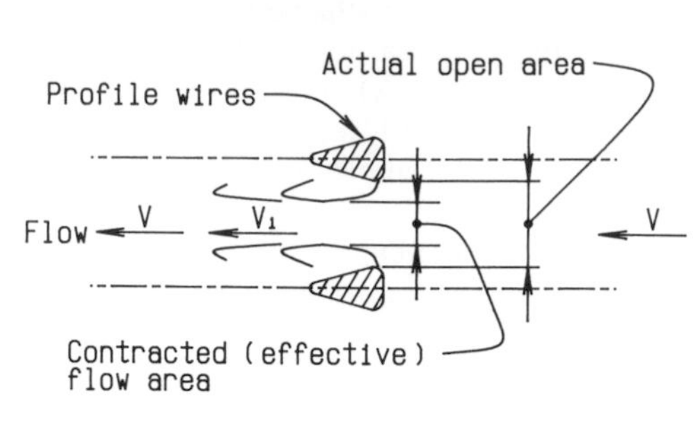

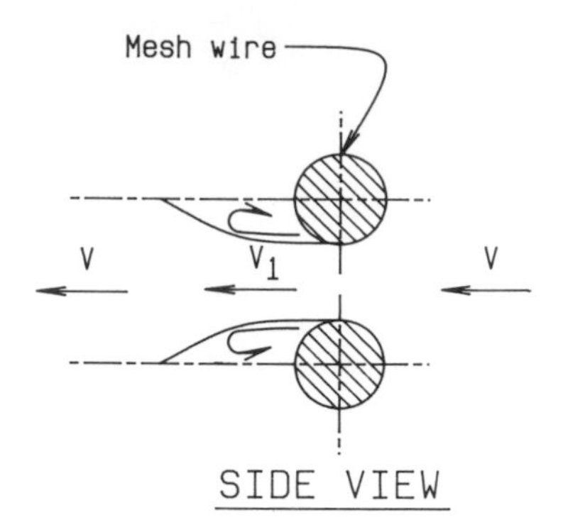

PROFILE WIRE MESH

WOVEN WIRE MESH

FIGURE 2.—Flow contraction through profile-wire and woven-wire meshes of typical fish screens.

tion available. The head loss due to flow through the mesh is relatively small for commonly used meshes with porosities on the order of 40% or greater open area. For an approach velocity component V_n of 1.0 ft/s, the head loss will usually be on the order of 0.1–0.2 ft or less and often not sufficient to create uniform flow through a screen if the approaching flow is nonuniform.

For uniform flow through the screen, the upstream and downstream hydraulic energy profiles along the screen must be parallel to create a constant head differential at all points along the length of the screen. The problem of nonuniform flow through a screen occurs when the head differential across the screen varies with location on the screen face, resulting in greater or lesser pore velocity according to equation (1). With porosity constant, pore velocities (and flow rates) vary as the square root of the variation in H_d.

An example of nonuniform flow occurs at the Leaburg Canal fish screen. This facility is owned by Eugene Water and Electric Board (EWEB) and is located on a hydropower diversion from the McKenzie River east of Eugene, Oregon. Figure 3 shows the screen configuration. The H_d across the screen near the upstream end of the structure was 0.1 ft, whereas the H_d near the downstream end was 0.8 ft as shown. The ratio of velocity components normal to the screen occurring at the two locations was, therefore,

$$V_{n2}/V_{n1} = (0.8/0.1)^{1/2} = 2.83.$$

That is, the velocity component V_{n2} near the downstream end of the screen face was calculated at 2.83 times greater than the velocity component V_{n1} near the upstream end. This screening facility was originally designed on the basis of a maximum allowable approach velocity component V_n of 0.7 ft/s, and the screen area required was determined by dividing the design diversion canal flow (2,200 ft^3/s) by 0.7 to obtain the required area of screen mesh (3,150 ft^2). It can be easily seen that although the average V_n was 0.7 ft/s, the actual V_n values varied greatly along the length of the screen. A range of V_n of 0.4 to 1.1 ft/s, depending on location, was calculated. Actual field velocity measurements indicate this range of V_n was even greater than calculated, varying from less than 0.1 ft/s at the upstream end of the screen to 1.1 ft/s near the downstream end.

The biological evaluation of this facility by Oregon Department of Fish and Wildlife showed that steelhead and chinook salmon larger than fry size passed through the facility with little or no observed injuries or mortalities. However, a major problem with chinook salmon fry, particularly fry less than 2 in long, was mortality due to impingement on the screen at the downstream end near the bypass entrance, where the highest velocities occurred. Some impingement was attributed to inadequate bypass flows, but an important factor in the high mortality was judged to be areas of excessively high approach velocity, which are

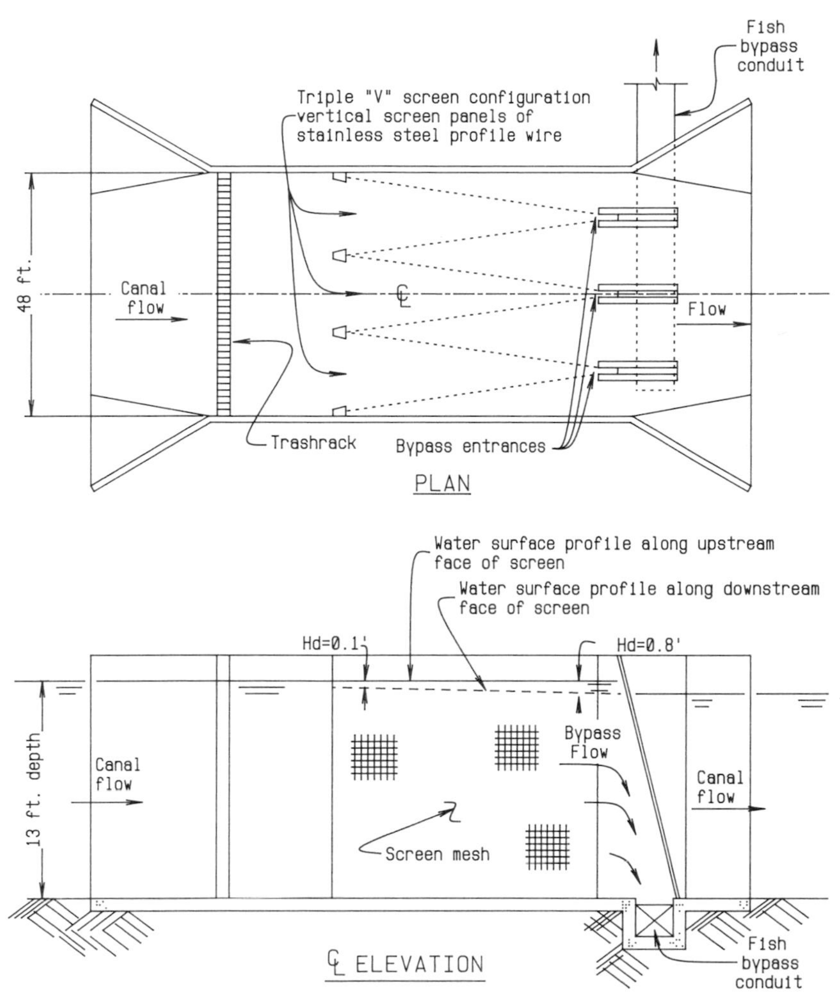

FIGURE 3.—The Leaburg Canal fish screen.

occasionally referred to as "hot spots." Personnel of EWEB have been working to improve fry passage by modifying the bypass system so flows will increase. However, the problem of nonuniform screen velocities had not been addressed at this writing. Further corrective action appears necessary based upon results of tests on the modified bypass during the winter of 1988–1989.

The variations in head differential across a screen that result in nonuniform flow can be caused by several factors. Such variations are especially likely to occur where inadequate attention has been given to the geometry of the approach channel during design. Bends or poorly designed transitions in the channel for some distance upstream of a facility can create such a problem, causing the surface of the approaching flow to build up higher against those sections of the screen where approach velocities are greater. If a negative head differential exists at any point along a screen, reverse flow will occur through the screen, creating eddies in the screen approach channel and seriously compromising fish guidance. A uniform velocity profile across the approach channel is needed. Figure 4 illustrates these points.

The configuration of the escape channel that takes flow downstream away from the screens can also adversely affect flow distribution through the screen by restricting flow behind some areas of the screen more than others. This can create an uneven or excessively sloped hydraulic energy profile along the downstream side of the screens. The Leaburg screening facility (Figure 3), which had an excessive head differential across the downstream end of the screen, has relatively

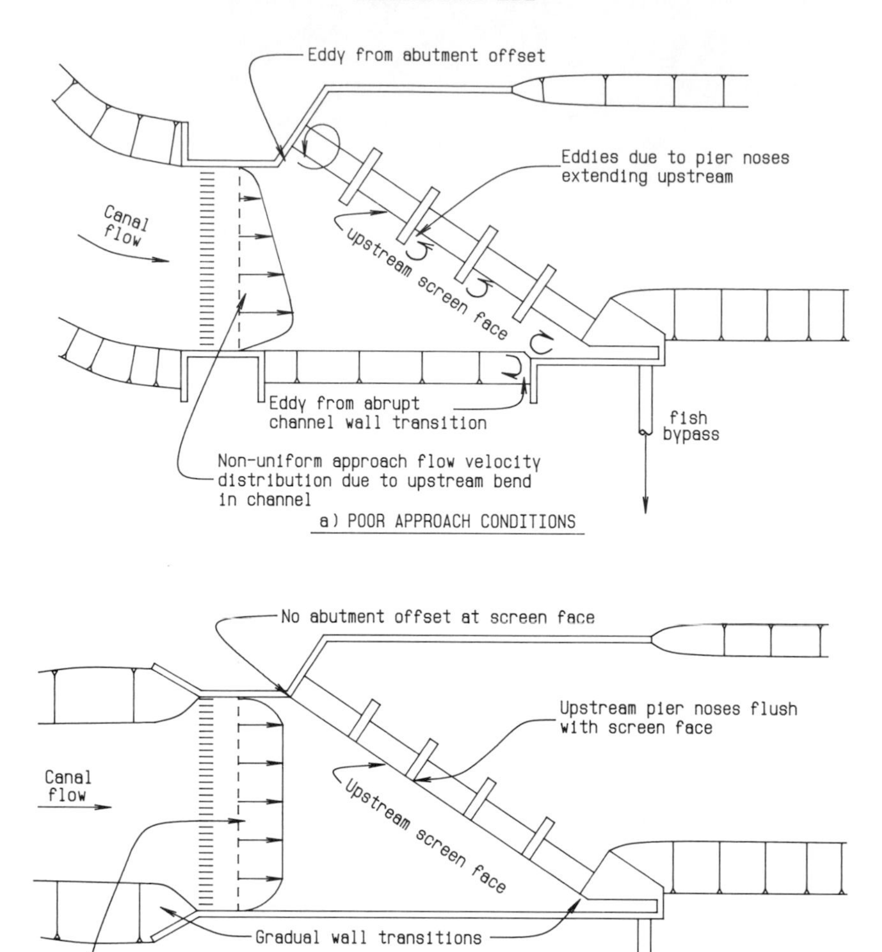

FIGURE 4.—Poor **(a)** and good **(b)** flow conditions for screen approaches.

constricted flow escape channels downstream of the screen. This configuration has resulted in excessive slope of the water surface downstream, a major cause of the nonuniform flow distribution.

Methods to Provide Uniform Flow

Channel Design

To provide the uniform approach flow at the desired angle to the screen face for effective fish guidance, the concerns raised previously about approach and escape channel configuration must be adequately addressed. If a sufficient length of straight and uniform channel cannot be provided upstream of the screening facility, flow vanes or training walls can be designed and installed to correct nonuniform flows. Frequently, a physical hydraulic model study is a worthwhile investment during the design of larger facilities, paying dividends by achieving optimum performance from the screen area provided and avoiding costly structural modifications or operation shutdowns after the facility is put into operation.

Modification of an existing screen facility to correct an unacceptable approach-flow condition was demonstrated at the Water Street Hydroelectric Project in Stayton, Oregon. This project is owned and operated by the Santiam Water Control District. Eighteen 3.5-ft wide by 6-ft high vertical fixed-panel screens in a horizontal align-

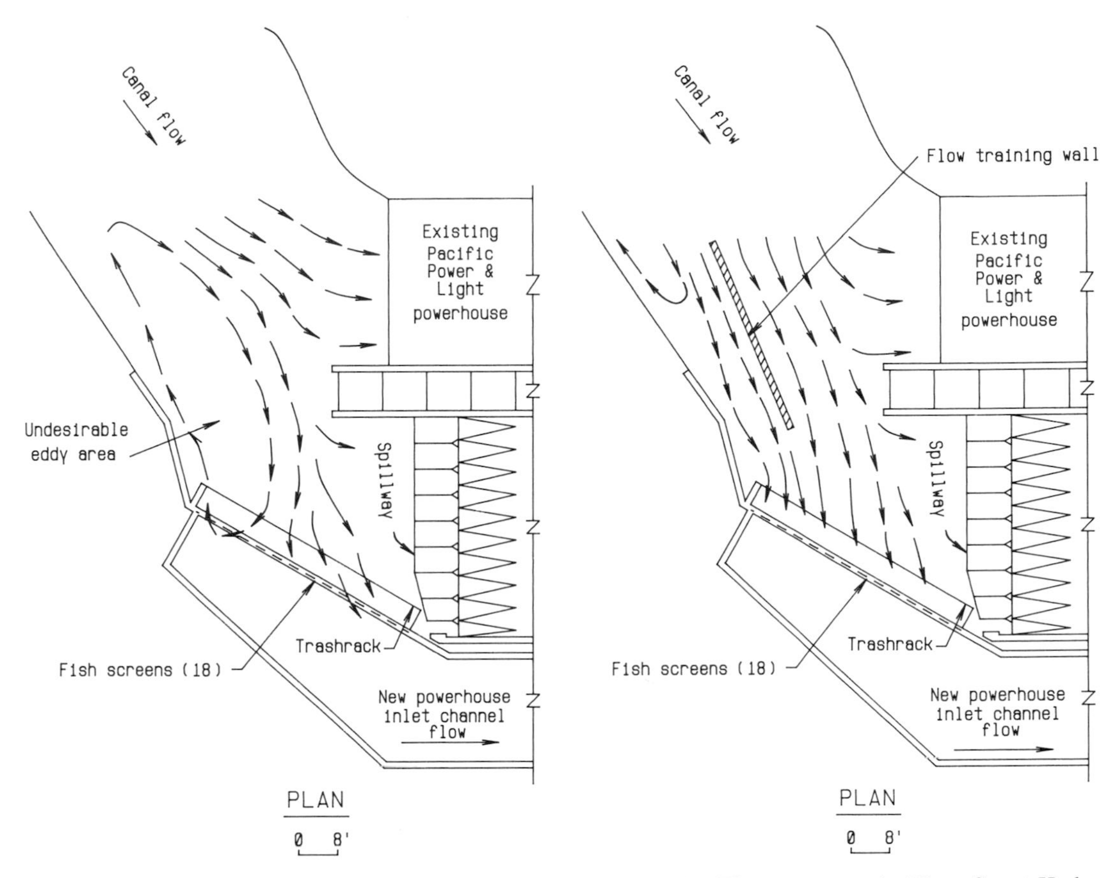

FIGURE 5.—Flow pattern at Santiam Water Control District's Water Street Hydroelectric Project before installation of a training wall.

FIGURE 6.—Flow pattern at the Water Street Hydroelectric Project after installation of a training wall.

ment are used to screen the 185-ft^3/s flow. Forebay flow conditions before corrective actions are shown on Figure 5. In an effort to eliminate a large eddy in front of the four upstream screen panels, a training wall was placed in the forebay, as shown in Figure 6. The bottom section of the wall was made of common concrete highway dividers. A timber section was constructed on the top of the dividers to extend the wall above the water surface. This structure works well; the large eddy along the right bank was reduced and a consistently positive flow toward the screen was created at a favorable angle over the full length of the screen face (Figure 6).

Another example of a modification to improve undesirable approach flows is shown in Figure 7. This was accomplished at the Marmot Dam power canal screens near Sandy, Oregon, owned by Portland General Electric Company. This is an older screen facility that is badly undersized based on current agency screening criteria. The vertical traveling screens are also oriented perpendicular to the canal. Potential for improvement was rather limited. The canal is 38 ft wide with a water depth of 13 ft at the maximum flow of 560 ft^3/s. Total screen area is 350 ft^2. Before modification, a large eddy existed in the center of the canal 10–15 ft upstream of the screens. Large accumulations of juvenile fish were observed in the eddy, where they had little chance of finding the bypass ports located in the walls and piers near the screen face. Installation of flow vanes 12 ft upstream of the vertical traveling screens effectively removed the large upstream eddy and provided a more uniform velocity distribution at the screen face. The vanes also created a slight but desirable angle of flow to the screen face over much of the area to move fish to the left side of the channel where the largest and most effective bypass is located. The flow vanes consist of vertical 3-in by 12-in timbers spaced at 18 in on

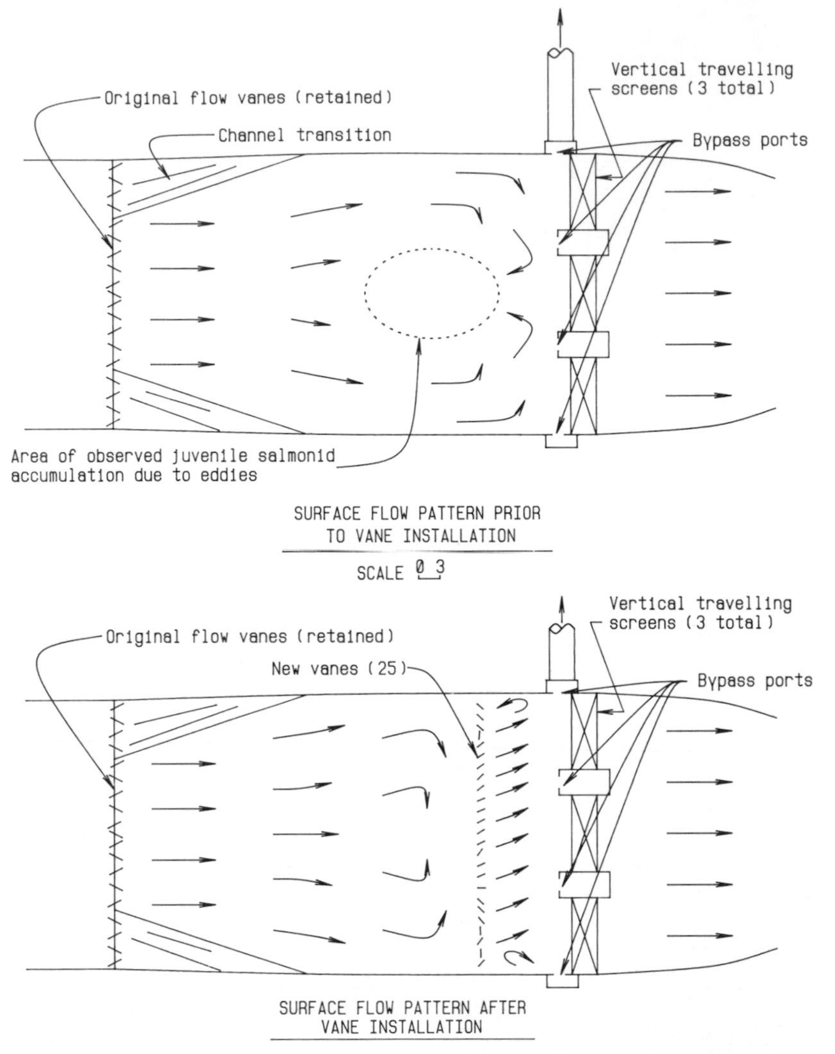

FIGURE 7.—General flow patterns at the Marmot Canal fish screens before and after a second bank of flow vanes was installed.

center across the canal width. They were rotated about their vertical axes and pinned in place when the desired flow conditions were obtained. As a result of this modification, potential for fish delay and predation were greatly reduced even though normal screen criteria are still exceeded. Velocities at 3-ft intervals across the channel at a transect located between the vanes and screens were measured by current meter. As shown in Figure 8, the velocities ranged from 0 to 1.8 ft/s, and 10 of the 12 readings were 0.5 and 1.2 ft/s. The presence of these flow vanes creates a head loss of 0.2 ft at maximum canal flow.

Baffle System Design

In addition to careful design of the channels carrying flow into and away from the screen surface, the screen assembly itself needs to be designed such that a sufficiently uniform head differential through the screen is created over the entire screen area to distribute flow uniformly. As noted above, the head created by commonly used mesh materials is frequently small and not adequate to achieve this objective. Installation of a baffle system behind the screens can ameliorate this problem. In anticipation of potential flow distribution corrections, slots or other accommo-

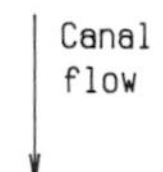

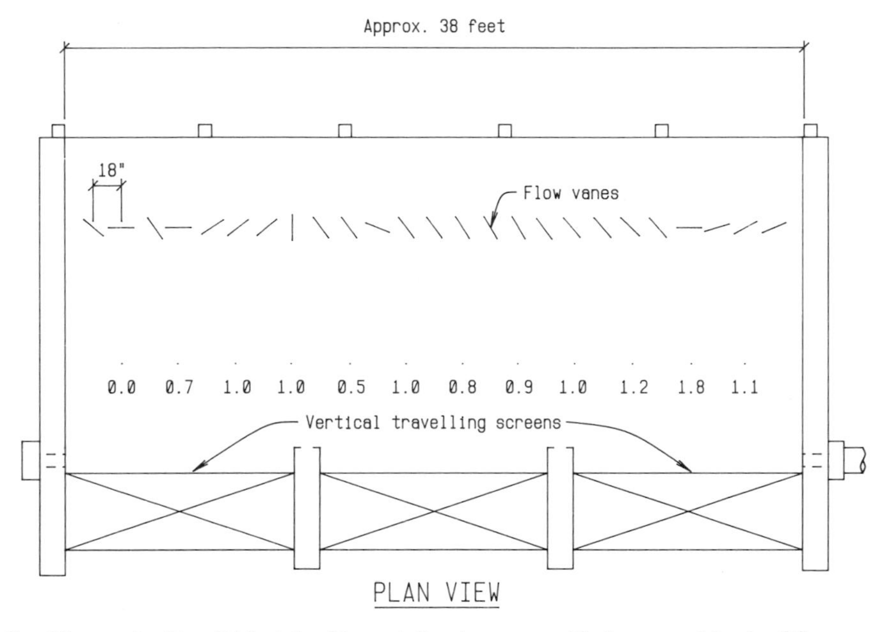

FIGURE 8.—Water velocities (ft/s) at the Marmot Canal screens with the second bank of flow vanes in place. Velocities were measured 5 ft below the water surface and 6 ft in front of the screens. See Figure 7 for flow directions.

dations for baffles of some type should be incorporated into the design of all facilities even if the baffle system is not provided initially.

If the magnitude of the variations in head loss through the screen mesh at different locations is known, an approximate baffle design can be developed that will compensate for or mask those variations, distributing flow more uniformly. Porosity of the baffle system usually will need to be varied over the screen area as necessary to obtain the desired result. The head differential created by a baffle system that uses a single layer of baffles is equal to the energy loss resulting from the sudden expansion of flow jets occurring immediately downstream of the baffle openings. The commonly used equation is

$$H_d = \frac{(V_1 - V)^2}{2g}; \qquad (2)$$

H_d = the head loss (ft) through the baffles,
V_1 = the velocity (ft/s) of the contracted jets through the baffle openings, and
V = the velocity (ft/s) downstream of the baffles where the jets have fully expanded.

The contracted jet velocity V_1 is determined by

$$V_1 = \frac{Q}{A(C_c)};$$

Q = flow rate (ft^3/s) through the baffles,
A = total area (ft^2)of baffle openings, and
C_c = the coefficient of flow contraction at the baffle openings.

Careful consideration is needed in selecting an appropriate coefficient of contraction (C_c) based on the geometry of the openings in the baffle media. Flexibility for adjustment of the baffle porosity should be provided by design to allow fine adjustment of the flow distribution after the facility is placed in operation. Care must be taken so that openings in the baffle media are not too coarse in relation to the distance between it and the screen mesh. This condition can cause undes-

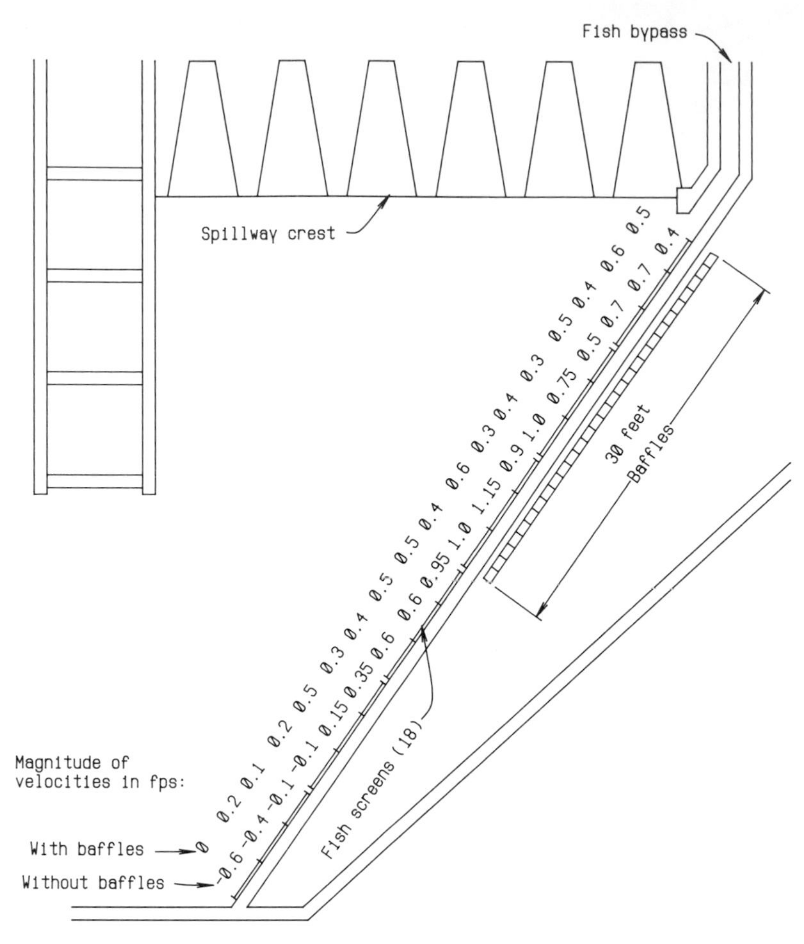

FIGURE 9.—Approach velocities (ft/s) at the Water Street Hydroelectric Project screen before and after flow distribution baffles and a training wall were installed. Velocities were measured 2–3 ft below the water surface and 3 in in front of each screen panel. See Figures 5 and 6 for flow directions and location of the training wall.

ired local areas of higher velocity through the mesh in front of individual baffle openings. The distance between the screen mesh and the baffle media downstream must be sufficiently great that the screen and baffle are hydraulically independent of each other, so total head loss at the two locations is additive, and the presence of baffle media does not cause hot spots at the upstream face of the screen. It is suggested this distance be equal to at least three times the spacing between openings in the baffle system.

The additional head loss created by baffles installed downstream of screens can be a problem if it significantly reduces channel hydraulic capacity or, in the case of hydroelectric projects, the head available for power production. Attention to these hydraulic concerns in the initial project design allows the head loss to be minimized. Retrofitting corrective structural measures frequently results in a less hydraulically efficient facility.

The use of baffles downstream of the screen mesh is preferred to upstream flow vanes for correcting screen velocities. Baffles do not cause formation of eddies or other flow separation zones in an area where fish are present. This minimizes delay and predation potential. It must be recognized, however, that nonuniform flow conditions approaching the screen face as a result of a poor approach channel configuration cannot be fully corrected only by baffling downstream of the screen.

The positive effect of baffles on flow distribution is demonstrated at the Water Street Hydroelectric Project referred to earlier. At this facility, both modification to the forebay flow and instal-

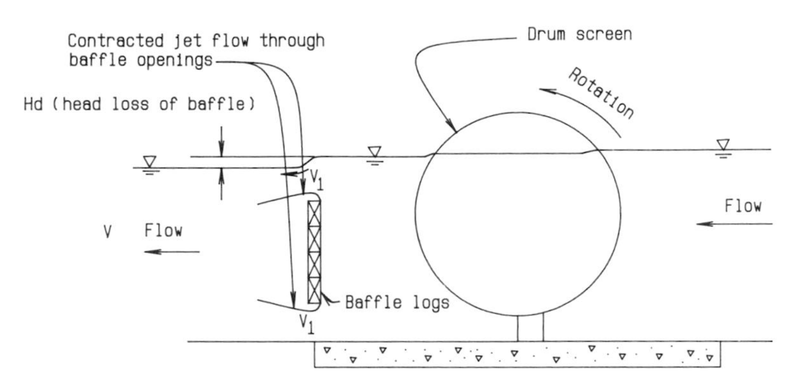

FIGURE 10.—Section through a typical drum screen with baffle.

lation of baffles behind the screens were necessary to correct the poor approach pattern and the nonuniform flow through the screen, respectively. Velocity measurements taken at the screens by the fisheries agencies prior to changes in flow distribution showed velocities ranging from 1.5 to −0.6 ft/s in the directions indicated by Figure 5. The negative velocities indicate reverse flow actually occurred through the screen. This condition occurred at the four screen panels farthest from the bypass entrance. Of the 18 velocities measured initially, 4 met the velocity criteria for fry. Subsequent trial-and-error adjustments in the placement of baffles resulted in a configuration that provided improved velocities at 15 of the 18 screen panels, as shown in Figure 9. Sixteen of the 18 velocities are within the 0.5 ft/s criterion for fry, whereas the two remaining velocities of 0.6 ft/s exceed that criterion only minimally. The trial-and-error adjustment process could have been minimized by the calculation procedure described previously.

In the past 4 years several large drum screens have been installed in the Yakima River basin in Washington State. Several of these facilities are designed for maximum canal flows of 1,300 to 2,200 ft^3/s with an allowable V_n of 0.5 ft/s. Figure 4b is representative of the general layout of these facilities. Hydraulic model studies during the design phase allowed refinement of the proposed facility layouts. Modeling of both the Wapato and Sunnyside canal screens resulted in several modifications to the initially proposed channel configurations, which improved flow distributions. The very short design schedule for the Sunnyside facility, did not allow an adequate model study to be conducted, however. An undesirable hydraulic condition, described below, was found in the completed Sunnyside facility, which could have been avoided with more comprehensive study. It was determined that installation of baffles behind the drums in selected bays would be necessary to obtain the desired flow conditions. Slots were provided in the piers at all the projects at the time of construction to receive baffle logs as needed.

Figure 10 shows a section through a typical drum screen bay with baffle logs in place. It should be noted that flow openings are needed both under and over the baffles to prevent sediment and floating debris from accumulating within the bay. With an 8-in-high bottom opening at the Sunnyside Canal screens, the baffles have provided an added bonus of increasing the bottom velocity to minimize sediment deposits in the area immediately downstream of the bottom drum seal. Upon completion of the Sunnyside screen facility, the baffles were installed, and the V_n and V_s velocity components in front of the screen drums were measured by current meter to verify appropriate baffle installations. Minor adjustments of the baffles were necessary. The required baffling behind 15 of the 17 drums varied from no baffles to baffles that blocked 60% of the water depth. In addition, the two bays farthest upstream have been totally blocked because flow occurred in the reverse (upstream) direction there as a result of a large eddy. The eddy was created by the combined effect of nonuniform approach flow due to a bend in the canal alignment upstream and a poorly designed abutment wall configuration as generally depicted in Figure 4a. The loss of these two bays reduced the total effective screen area of the facility by 12%. This experience indicates how model studies can contribute to cost-effective design if they are sufficiently comprehensive.

At the Wapato Canal screening facility, baffles were initially installed in the configuration recommended in the hydraulic model study report (Engineering Hydraulics, Inc. 1985). In the model, this configuration resulted in very good flow conditions. The V_n component was 0.50 ft/s or less over 72% of the screen area, and 0.60 ft/s or less over 90% of the screen area. Maximum V_n was 1.0 ft/s. The V_s component was twice the V_n component over 96% of the screen area.

The improved flow conditions at both the Sunnyside and Wapato screens resulting from installation of downstream baffles are due to a more uniform flow distribution between bays, not to any localized changes to the flow nets within individual bays as water is redirected towards the baffle openings.

A comprehensive field evaluation of hydraulic conditions at several Yakima Basin fish screens was implemented by the U.S. Bureau of Reclamation at the request of state and federal fishery agencies and Indian tribes during summer 1988 and spring 1989. The hydraulics of the Wapato screen facility were studied in particular detail. Any hydraulic conditions found by this study that adversely affect fish passage will be defined and corrective measures will be developed.

Conclusions

Review of numerous fish screen designs for both proposed and existing facilities shows that a common design flaw is poor flow conditions in the critical area immediately upstream of the screen face. Both the magnitude and direction of water velocities must be controlled within prescribed limits to assure that fish are not impinged upon the screen and are quickly guided by the flow pattern to a bypass. Channel configurations both upstream and downstream of the screen mesh must provide smooth transitions, preventing eddies and other areas of nonuniform velocities. Careful hydraulic analysis, including physical hydraulic modeling when appropriate, can provide desired flow conditions with minimum associated head loss. Providing flow vanes upstream of the screen or baffles immediately downstream of the screen (or both) at facilities where poor flow patterns cause passage problems can frequently rectify damaging situations at minimal cost. Poor approach conditions above the screen facility (such as at the Sunnyside screen facility) cannot be fully corrected by controlling porosity downstream of the screen mesh alone. Corrections must also be made in the channel upstream of the screens.

References

Bates, K. 1988. Screen criteria for juvenile salmon. Washington State Department of Fisheries, Olympia.

Bell, M. 1984. Fisheries handbook of engineering and biological criteria. U.S. Army Corps of Engineers, North Pacific Division, Portland, Oregon.

Engineering Hydraulics, Inc. 1985. Final report: hydraulic model studies of Wapato Irrigation Canal fish screening system. Engineering Hydraulics, Inc., EHI Job 255-004, Redmond, Washington.

EPRI (Electric Power Research Institute). 1986. Assessment of downstream migrant fish protection technologies for hydroelectric application. Electric Power Research Institute, Palo Alto, California.

Sazaki, M., W. Heubach, and J. E. Skinner. 1972. Some preliminary results of the swimming ability and impingement tolerance of young of the year steelhead trout, king salmon and striped bass. Final report (AFS-13) to Anadromous Fisheries Act Project, Sacramento, California.

Smith, L. S., and L. T. Carpenter. 1987. Salmonid fry swimming stamina data for diversion screen criteria. University of Washington, Fisheries Research Institute, Seattle.

Stefan, H., and A. Fu. 1978. Headloss characteristics of six profile wire screen panels. University of Minnesota, St. Anthony Falls Hydraulic Laboratory, St. Anthony Falls.

American Fisheries Society Symposium 10:249–255, 1991

Factors Affecting Performance of the Glenn–Colusa Fish Screen

ROBERT D. CLARK
Glenn–Colusa Irrigation District, Post Office Box 150, Willows, California 95988, USA

JAMES J. STRONG
Aquadyne, Inc., 1730 South Amphlett Boulevard, Suite 208, San Mateo, California 94402, USA

Abstract.—The Glenn–Colusa Irrigation District fish screen consists of 40 screen drums, each 17-ft in diameter and 8 ft wide, arranged in a linear configuration along a dredged side channel of the Sacramento River, and having a total capacity of 3,000 ft^3/s. The original purpose of the Glenn–Colusa fish screen was to protect migrating juvenile chinook salmon *Oncorhynchus tshawytscha* as they moved downstream. The screens have proven troublesome because of changes in water surface elevation and profile brought about by erosion and siltation in the river channel during heavy winter storms. Effective screen area and bypass flows have been substantially reduced. Currently, a single 17-ft prototype drum is being tested. It has been retrofitted with profile-wire screen with 3⁄32-in slots versus wire mesh of 4-by-4 (wires per inch) stainless steel.

The Glenn–Colusa fish screen is in northern California, 3.5 miles north of Hamilton City (Figure 1). A detailed view of the Glenn–Colusa Irrigation District diversion on the Sacramento River (Figure 2) shows the location of the fish screen installed on an oxbow or bypass channel off the main stem of the river. The 3,000-ft^3/s diversion of the district services 175,000 acres of agricultural land and federal wildlife refuges in Glenn and Colusa counties in the Sacramento Valley, where rice is the predominant crop.

The intake channel, shown in Figure 2, is the area where it is necessary to dredge in order to maintain flow to the fish screen. Immediately downstream from the screening structure is a seasonal earthfill dam, emplaced by the district in low-water years, to reduce the pumping head. The bypass outlet conveys excess flows back to the main stem of the Sacramento River.

Diversion History

Water was first diverted from the Sacramento River at the present site in 1905. The addition of pumps over the years has increased the diversion capacity to a maximum of 3,000 ft^3/s. In the 1930s, the California Division of Fish and Game, currently the California Department of Fish and Game, mandated and jointly funded the installation of a fish screen in front of the district's pumping plant.

The district complied in 1935 by installing a screening structure, approved by the California Department of Fish and Game, consisting of 1⁄4-in by 1-in steel bars oriented with the widest dimension parallel to the flow. The bar spacing was set at 3 in. Subsequent flood flows undermined and destroyed the effectiveness of the screen within 3 years. This screen remained in place until the present fish screen was placed in operation in 1972.

Ensuing modifications of the district's pumping capabilities allowed the district to pump approximately 2,600 ft^3/s at the diversion until the old pumping plant was replaced in 1984 with a new pump station. The new pump station has the capacity to pump 3,000 ft^3/s with an 8-ft lift.

Glenn–Colusa Fish Screen Facility

Construction of the Glenn–Colusa fish screen facility was started in 1969 and was designed to protect migrant juvenile chinook salmon *Oncorhynchus tshawytscha* moving downstream. The fish screen was paid for jointly by the California Department of Fish and Game and the federal government, at a cost of US$2.6 million. The operation and maintenance costs are to be shared equally between the district and the California Department of Fish and Game.

Minimal biological testing has been done and is inconclusive in determining cost-to-benefit ratios. This deficiency is significant and has not been addressed. Fish losses have not been adequately documented and screen losses have not been evaluated.

This facility is one of the largest fish screens in the world, consisting of a 450-ft-long concrete abutment housing 40 bays, each 8 ft wide by 27 ft deep (Figure 3). Each bay contains a single removable drum 17 ft in diameter and 8 ft wide (Figure 4). The drums are covered with 4-by-4 (wires per inch) stainless steel mesh cloth with a wire diameter of 0.080-in square openings of 0.17

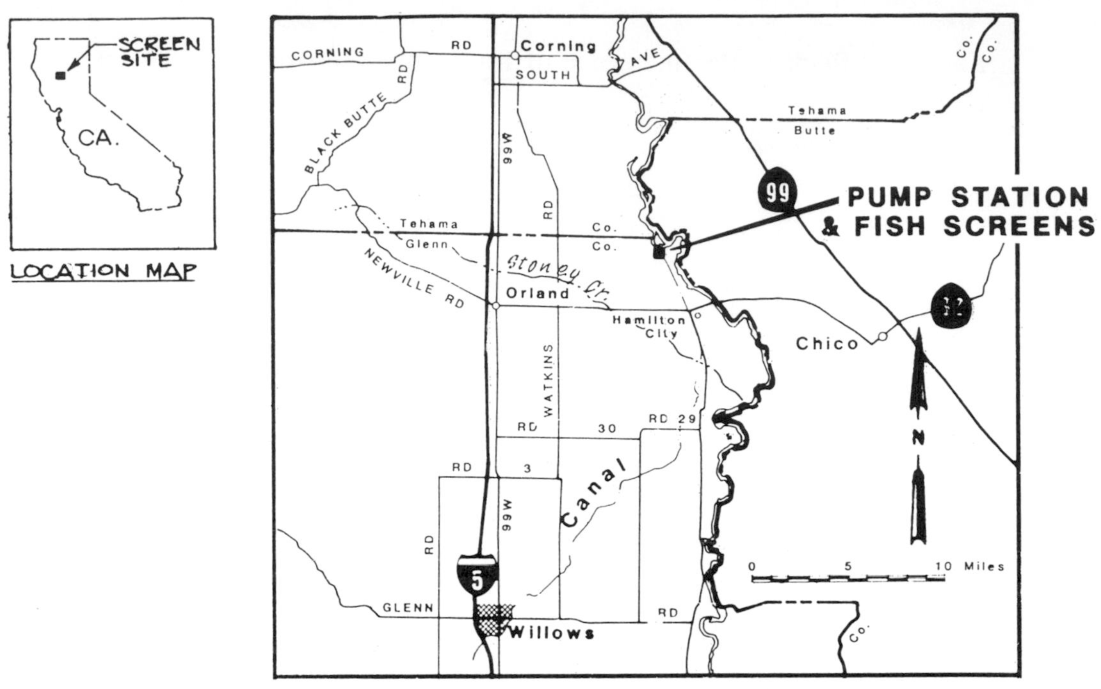

FIGURE 1.—Location of Glenn–Colusa pump station and fish screens.

by 0.17 in, and a diagonal opening of 0.24 in. The effective open area of the mesh is 46 %.

The drums were originally designed to operate at a minimum water surface level of 138.8 ft above mean sea level with submergence of 11.5 ft. A cross section of the screen drum with elevations is shown in Figure 5. Each drum rotates on a horizontal axis as the water flows through the screen drum. This provided for an effective screen surface area of approximately 4,000 ft^2. The maximum design flow was to be 3,000 ft^3/s with a maximum approach velocity of 0.8 ft/s.

To facilitate passage of migrant fish, 10 6-in-wide orifices were incorporated into the screening structure and located between every fourth screen bay. Each orifice empties into a graduated steel pipe buried behind and beneath the screening structure. The steel pipe culminates in a 60-in diameter discharge into the return channel at a point immediately below the seasonal earthen dam. By agreement between between the district and the California Department of Fish and Game, the bypass flow is 90 ft^3/s.

River Degradation

Neutral or reverse flows in the bypass or return channel have been observed on several occasions since the inception of the screening facility. This situation occurred particularly when the intake channel was restricted due to deposition. Beginning in 1981 and continuing sporadically through 1986, such occurrences became more common. Data collected by district personnel show that river elevations have degraded almost 3.5 ft since 1971; degradation since 1981 accounts for almost 1.5 ft.

The net effect upon the operation of the fish screen has been to lower the average working elevation from a design minimum of 139.0 ft above mean sea level to an average working elevation of 135.8 ft during 1985. The net consequence for the fish screen has been an interruption in the bypass flow and increased average velocities at the screen face, as high as 1.37 ft/s. Velocity also varies along the horizontal length of the screen by a factor of three.

The original designers used historical data accumulated at the site in 60 years of operation and designed the screen in the late 1960s to operate at a normal water surface elevation of 140.6 ft above sea level. The Sacramento River, at the point of the intake diversion, is a notoriously meandering river body and has been degrading periodically during floods. At the present time, the average water surface elevation at the fish screen is 136.5 ft above sea level. Water surface levels vary

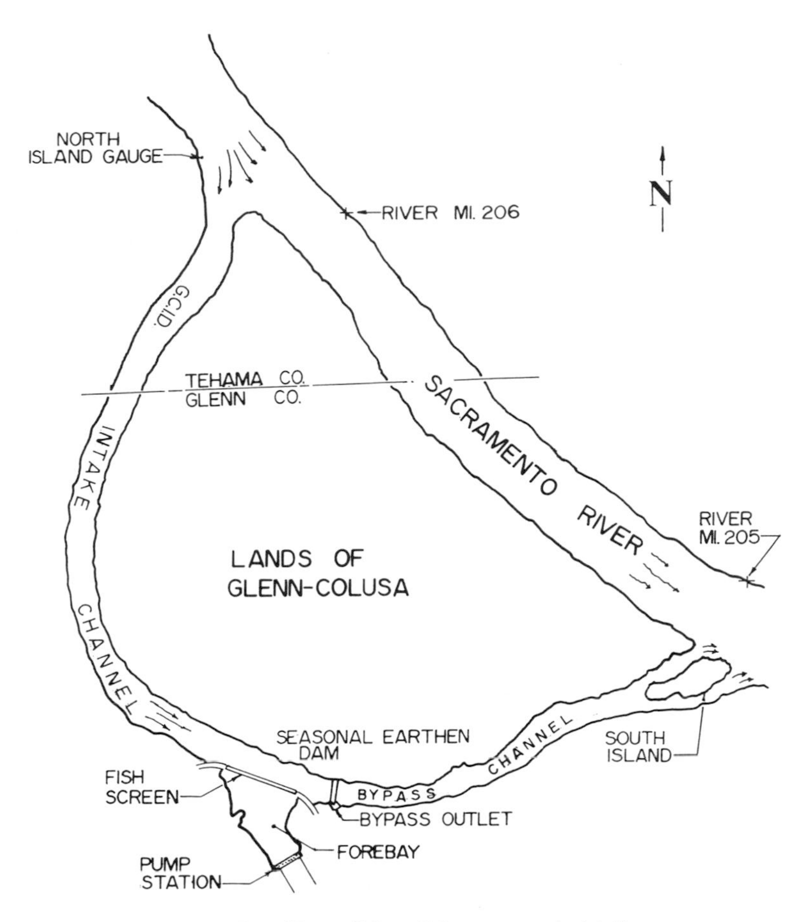

FIGURE 2.—Glenn–Colusa fish screen and vicinity.

during the operating season in relation to river flows.

This drop in water surface has significantly reduced the effective surface area of the screen, causing the average approach velocity to exceed the design value. The intake channel must be dredged regularly to provide the maximum water surface at the screen in order to reduce the approach velocity. In addition, stream degradation compounds the problem associated with fish losses at the screen because bypass flows past the screen are correspondingly reduced.

U.S. Army Corps of Engineers Permit

Early in 1986, the Glenn–Colusa Irrigation District routinely reapplied for a Department of the Army permit, under Section 10 of the Rivers and Harbors Act and Section 404 of the Clean Water Act, to annually place an earthfill dam and perform maintenance dredging for a 10-year period in a lateral channel of the Sacramento River at Hamilton City. The purpose of the dredging is to maintain proper channel depth to provide water supply to the district's main pump station and the California Department of Fish and Game's fish screen.

What ensued was a 2-year struggle before the district was able to secure renewal of its Corps of Engineers permit to dredge the intake channel. Before the dredging permit was granted, a wave of controversy took place involving fishery groups on one side and farmers on the other. Stringent new conditions attached to the permit require the district to cut back on pumping to maintain high bypass flows in the channel to protect fish at the screen during the prime spring migration period, when young smolts are released from an upstream hatchery.

Significance of Channel Bypass Flows

The permit restrictions require the district to reduce pumping so that velocities through the

FIGURE 3.—Fish screen containing 40 drum screens.

screen do not exceed 0.4 ft/s and bypass flows are at least 50 ft^3/s, during a 5-day period in May when the Coleman National Fish Hatchery, upstream on Battle Creek near Redding, releases millions of salmon smolts. The U.S. Bureau of Reclamation provides an additional flushing flow from upstream storage reservoirs. The magnitude of this flushing flow is annually determined relative to the year's water supply.

Adequate channel bypass flows are extremely important to the proper operation of the fish screen. California Department of Fish and Game personnel report that migrating salmon smolts respond to downstream flows, and that neutral or reverse bypass flows create a situation whereby the salmon smolts tend to congregate at the fish screen and are either impinged or become easy prey for predatory fish. Dredging the intake channel at the diversion significantly increases channel bypass flows (Table 1). Data are being gathered on bypass flows and their relationship to river flows and pumping rates.

FIGURE 4.—Individual drum 17 ft in diameter by 8 ft wide.

A September 30, 1986, memorandum report, from U.S. Fish and Wildlife Service field supervisor James McKevitt, Sacramento, to the regional director of the U.S. Bureau of Reclamation, Sacramento, described the effects of the fish flush operation on bypass flows and reported on the observations of U.S. Fish and Wildlife Service scuba-diving operations in the vicinity of the Glenn–Colusa Irrigation District screen. The report stated, "Based on preliminary analysis of marked fish returns, the Service believes that these measures substantially reduced or eliminated salmon losses during [sic] the vicinity of the

TABLE 1.—Effect of dredging on flushing flow through the Glenn–Colusa intake channel during fish release at Coleman National Fish Hatchery (all flows in ft^3/s).

Date, condition	Upstream river flow	Pumping flow	Flow pumped	Bypass flow
1985, before dredging	12,420	1,900	1,420	480
1987, after dredging	13,300	3,183	2,167	1,016

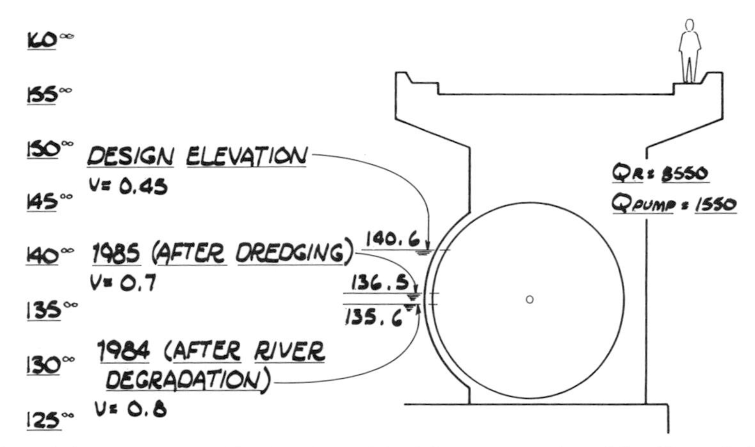

FIGURE 5.—Effect of river changes and intake channel dredging on operation of the Glenn–Colusa fish screen. V is velocity (ft/s), Q_R is river discharge (ft^3/s), and Q_{PUMP} is discharge through the pump (ft^3/s).

Glenn–Colusa Irrigation District fish screens for this critical time period'' (McKevitt 1986).

Although dredging the intake channel is essential to provide adequate bypass flows, these flows can best be achieved by artificially raising the water surface in the river. Presently, it is planned to implement this concept by the construction of submerged barriers or weirs in series across the main river channel. This has been termed ''riffle restoration'' (Figure 6), and could ensure a future water surface elevation near, or at the water level of the original design for the fish screen. It may also be beneficial to modify the return flow channel for better fish passage, improved hydraulic conditions, or both. Future studies will address these issues.

Bypass flows are most sensitive to river elevations and secondarily to rate of diversion through

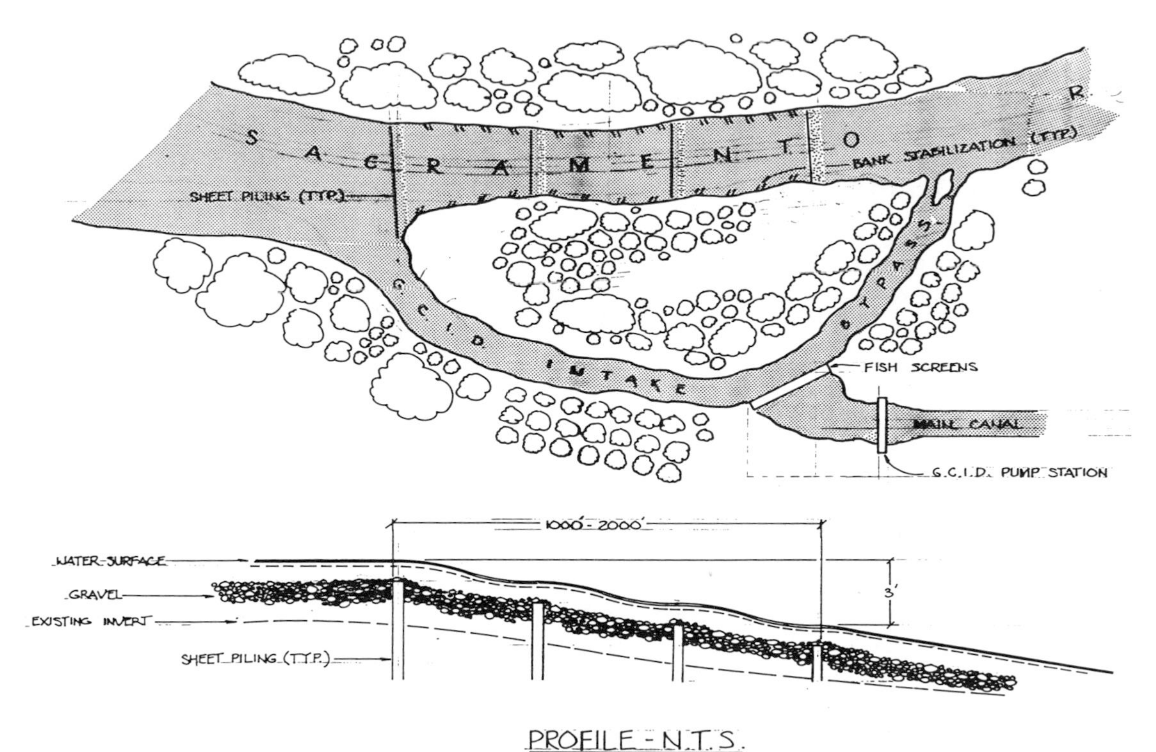

FIGURE 6.—Proposed riffle restoration.

FIGURE 7.—Slot profile-wire screen (3/32 in) presently being retrofitted to the Glenn–Colusa fish screens.

the pump station. River elevations depend primarily on the quantity of water flowing in the river. Artificial modification such as a riffle restoration project could alleviate the dependence on flow to maintain minimum surface levels.

Fish Screen Modification

A relatively simple modification of the fish screen to evaluate the effect of reducing head loss through the screen was undertaken early in 1988. The modification involved replacing the 4-by-4 (wires per inch) stainless steel mesh cloth with profile wire (Figure 7).

The profile wire consists of 0.063-in-wide wire, constructed in a manner to produce a 3/32-in horizontal slot. This profile-wire screen has an open area of 60 % as opposed to the wire mesh, which is 46 % open. This increase in open area will, in itself, reduce the through-screen velocity and is anticipated to greatly aid the operation of the fish screen in reducing fish losses; it will be addressed in future biological studies. In addition, the profile-wire-slot cross section provides a point-to-point contact that reduces blockage due to small gravel, which tends to wedge in woven screen.

A screen study, conducted for the Interagency Ecological Study Program for the Sacramento–San Joaquin Estuary by California Department of Water Resources personnel at the Hood test facility on the Sacramento River, concluded that, "The wedge-wire flat screen clogged at a significantly slower rate than the perforated plate with 3/32 inch slot width and 5/32 inch hole size respectively" (Smith 1982).

Early in the spring of 1988, the profile-wire screen material was installed on one 17-ft-diameter drum before all 40 drums were placed within the structure. The wire mesh material was first removed from the structural frame of the drum. A secondary benefit in the use of the profile-wire material on the screen drum was that the stiffness was greatly improved. Once the drum is installed in the bays, maintenance personnel can stand on the screen and the profile-wire material is able to support their weight much better than the wire mesh, which becomes deformed.

Improvement Study Program

In August 1987, a joint Memorandum of Understanding, summarized below, was entered into by the California Department of Fish and Game and the Glenn–Colusa Irrigation District. This essentially consisted of a study effort aimed at improving overall conditions at the diversion. It contains elements that contain the following:

- implementation of interim measures to reduce fish impact;
- river geomorphic study;
- fishery studies;
- river hydraulics study;
- technical advisory committee; and
- development of the recommended plan.

One of the key elements of this program is a geomorphic study that attempts to predict the meander of the Sacramento River at the point of the district's diversion. This will assist in determining the effects of river degradation on the intake channel and in selection of alternatives in the riffle restoration efforts. Control of the river's minimum level and gradient is the key initial factor in resolution of the problem.

In addition, modification or replacement of the existing screen will be required. The fishery studies will analyze the fisheries trapping data that have been collected at the fish screen. A hydraulic model of the proposed facility will be built and studied to ensure proper performance of a new screening facility.

Conclusion

The difficulties associated with the operation of the Glenn–Colusa Irrigation District fish screen have long been recognized as the most serious screening problem along the Sacramento River. Many questions remain and no simple solution exists to the very complex hydraulic and fishery problems. Resolution of the problems will require the joint efforts of many agencies to restore the river hydraulics and eliminate the fish screen deficiencies. The Glenn–Colusa Irrigation District board of directors has made a wholehearted commitment, jointly with the California Department of Fish and Game, to resolve these issues while continuing the water diversion so economically vital to this area of California.

Acknowledgments

We thank Paul D. Ward, California Department of Fish and Game, and Cynthia F. Davis, Glenn–Colusa Irrigation District, for advice and critical review of the manuscript.

References

McKevitt, J. J. 1986. Memorandum on Central Valley Fish and Wildlife Management Study. Problem A-8: evaluate the need for fish screens on diversion facilities along the Sacramento River. Memorandum report of U.S. Fish and Wildlife Service to U.S. Bureau of Reclamation, Sacramento, California.

Smith, L. W. 1982. Clogging, cleaning, and corrosion study of possible fish screens for the proposed peripheral canal. Report of California Department of Water Resources to Interagency Ecological Study Program for the Sacramento–San Joaquin Estuary, Sacramento, California.

American Fisheries Society Symposium 10:256–263, 1991

Design of Low-Cost Fishways

PETER H. CROOK
Ontario Ministry of Natural Resources
Post Office Box 5463, London, Ontario N6A 4L6, Canada

Abstract.—Fishways have traditionally consisted of concrete flumes built on the side of a dam. This paper presents two alternative constructions that have recently been built in southwestern Ontario for the passage of rainbow trout *Oncorhynchus mykiss* and other salmonid species around mill dam structures. The first structure is a fishlock; after fish enter a lower chamber, water fills a transport pipe, allowing fish to swim above the dam. Concrete sewer pipe was used for the transport and ancillary water supply pipes. The second structure is an earthen bypass channel equipped with concrete baffle blocks. Both bypass methods resulted in considerable cost savings compared with concrete flume construction.

In the early 1960s the Ontario Ministry of Natural Resources (OMNR) started building fishways around dams to enable rainbow trout *Oncorhynchus mykiss* to migrate into coldwater streams. Rainbow trout were introduced to Ontario in the late 1800s but were prevented from migrating up many streams by mill dams. At several sites, conventional concrete flume fishways, such as the weir and vertical slot (Clay 1961), have been constructed around the side of a dam. The high cost of such structures, which often require modifications and repairs to the dam as well as large excavations, reinforced concrete walls, and ancillary works, prompted investigation of other solutions.

Thornbury Fishlock

The Thornbury Dam is located near the mouth of the Beaver River, which flows northward to Georgian Bay of Lake Huron. The old mill structure is 10 m high and is just upstream of Highway 26 (Figure 1). The mill building is on the west side of the dam, which leaves only a small area on the east side to construct a fishpass. Plans were prepared for a weir-type fishway, which required a zigzag configuration because of the tight space. Due to the cost and space limitations, a fishlock scheme (Figure 2) was chosen as a better alternative. Fishlocks had not been constructed in Ontario and there was no local experience to draw on for design and construction. However, drawings were obtained from Scotland where several fishlocks have been constructed. The fishlock was completed in 1978. During the years 1980–1984, over 13,000 rainbow trout were trapped and tagged at the site. The rainbow trout run is expected to grow to about 6,000 annually. Over the past few years a strong run of chinook salmon *Oncorhynchus tshawytscha* has developed and should grow to 7,000–8,000 per annum. Depending on water temperatures, rainbow trout in the area migrate during March, April, and May. There is also a smaller run during part of October and November. The chinook salmon run occurs in August and September.

Design Considerations

The design developed for the fishlock includes three chambers (Figure 3). Chamber A is the lower fish entrance structure. Chamber B is located upstream of chamber A to provide energy dissipation. Chamber C is at the upstream end and provides a fish trap and adjustable weir (gate 2) for flow control (Figure 4). A dividing overflow wall is incorporated in chamber B to maintain a high head during the transport phase. Chamber A has a secondary drainage section that allows flow to be diffused into the river (during the transport phase) and also provides an outlet for the system to be drained down at the end of the transport phase.

The 900-mm-internal-diameter transport pipe was arbitrarily chosen to provide generous swimming area for large fish. All other pipes are sized to carry the required flow. Calculations to consider friction, bend, and junction losses were completed to determine the necessary sizes. Analysis was required to determine the time required to drain the system between transport and attraction cycles. This involved calculating the change in volume on an incremental basis to determine the length of time the drain valve (gate 4) had to remain open and when gate 1 could start its downward opening.

Attraction Phase

During the attraction phase, water flows through the 600-mm pipe from chamber C to

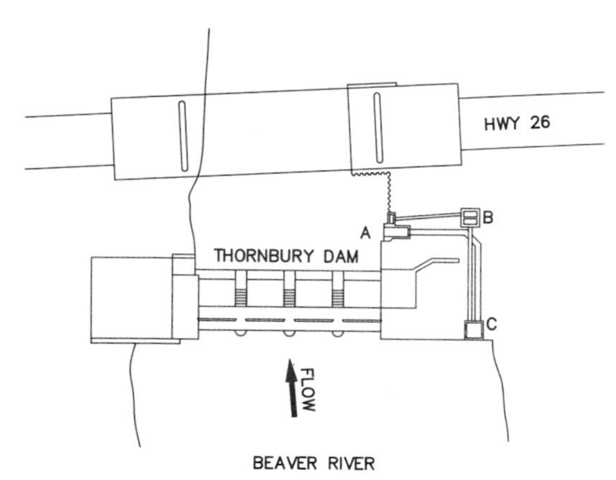

FIGURE 1.—Plan view of Thornbury Dam and fishlock. A, B, and C are fishlock chambers.

chamber B where flow energy is dissipated. Flow continues from chamber B in the second section of 600-mm pipe to chamber A. Flow enters chamber A at a low level under a grating (to prevent fish ingress). Water spills from chamber A over the adjustable weir and fish jump over the weir to enter the fishlock. The weir is maintained at a set level above tailwater for control of sea lampreys *Petromyzon marinus*. A trickle flow is also provided during the attraction phase through the 900-mm transport pipe, for attraction purposes, before the system is filled.

Transport Phase

The weir at the entrance to chamber A also serves as an upward closing gate (gate 1) that is used to seal the entrance. Flow to chamber B through the supply pipe is cut off by closing gate 3. This causes all incoming water to fill the transport pipe and chamber A. Flow in the 600-mm supply pipe reverses. One side of chamber B fills and water spills over the dividing wall into the 450-mm pipe. Water flows from this pipe to the diffusion side of chamber A from where it exits to the river.

Drainage Phase

Upon completion of the transport phase, gate 4 is opened between the two halves of chamber A to drain down the system. After sufficient drawdown, the entrance weir gate is opened and gate 4 is closed to return to the attraction phase.

Construction Features

The fishlock is constructed in the fill at the east side of the dam. All pipes are located underground. The upper surface of chambers A and C are flush with ground level. Chamber B is partially buried in the dam fill. The main part of chamber A is 2.1 m wide, 4.3 m long, and 3.4 m high; chamber B is 2.1 m wide, 2.4 m long, and 9 m high; and chamber C is 3.4 m wide, 3.6 m long, and 4.6 m high. All chambers are made of reinforced concrete. Pipes are class-V reinforced concrete sewer-type pipes with O-ring seals and are supported on class-A concrete bedding. Careful backfill placement and proper compaction were necessary after pipe placement to avoid settling and differential stress on the pipes. Both the 900-mm transport pipe and the 600-mm attraction pipe are 37 m long. All gates are the vertical roller-type operated by electric motors equipped with torque-limiting devices. The cost of the fishlock facility, including control building, fish trap hoist, and fencing was Can$272,000.

Operation and Maintenance Problems

The upward closing of gate 1 during the change from attraction to transport phase caused a high-

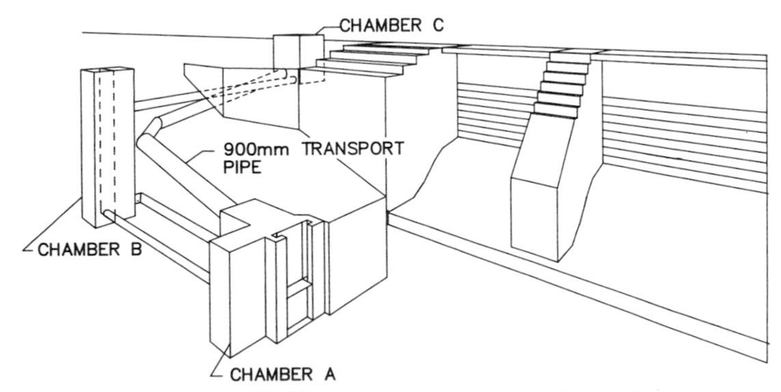

FIGURE 2.—Perspective view of fishlock located on the east side of Thornbury Dam.

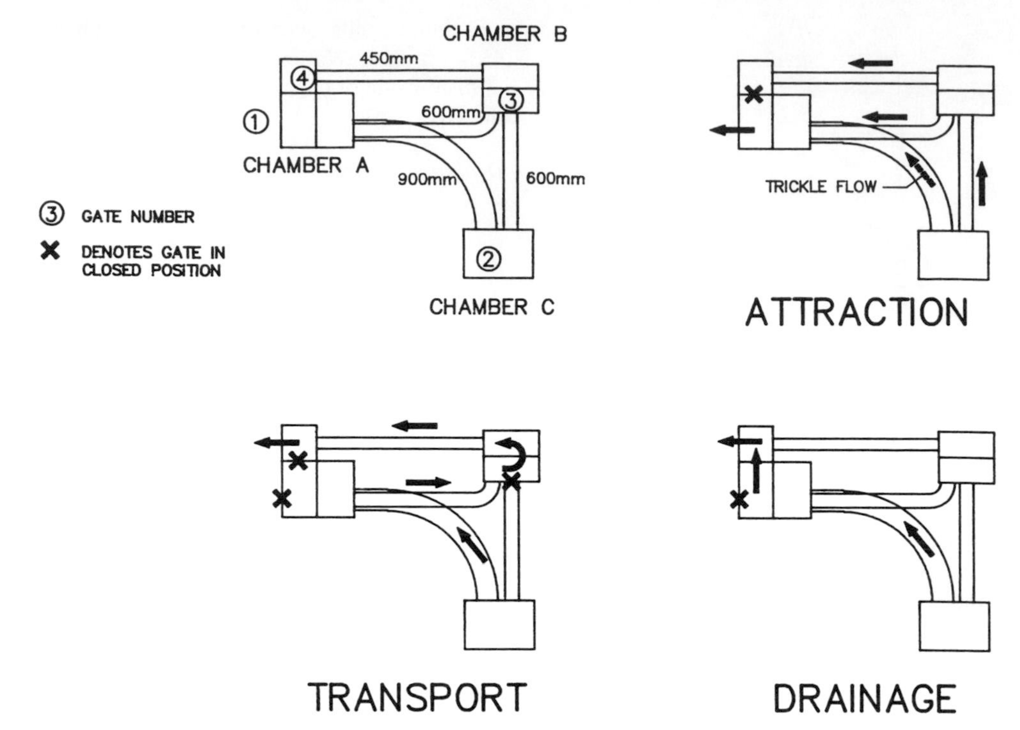

FIGURE 3.—Conceptual arrangement of fishlock gates and chambers in various operating modes. Arrows indicate directions of flow.

velocity sweep-out condition just before closure. Fish that were close to the surface were swept back to the river. To avoid this condition, gate 2 was raised to temporarily cut off the flow. Shutting off the flow also reduced water hammer pressures in the pipes during gate closures.

After several years the pipes were inspected manually and with a video camera. Some movement in the joints and leakage were detected. The movement was probably caused by settling of the fill. A grouting program was completed to seal all joints.

The fishlock had originally been designed for automatic operation; however, during the first few years of operation the system was controlled manually by staff involved in fish survey work. The mechanical timer controller device had corroded during this time and was inoperative. A complete upgrading of the controls was therefore undertaken.

Automatic Controls

The new system consists of a programmable microprocessor controller that operates all gates

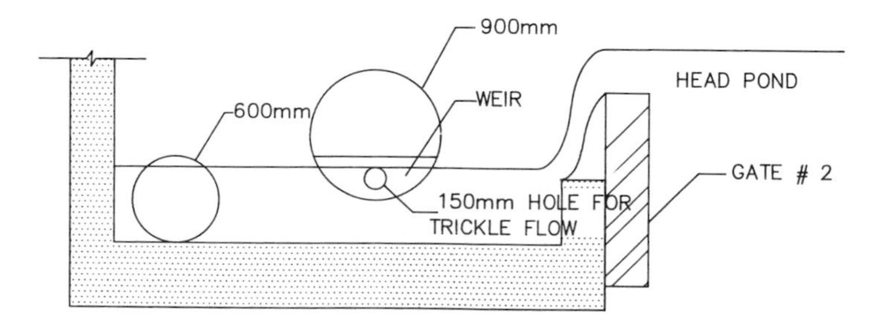

FIGURE 4.—Conceptual arrangement of fishlock flow inlet with inlet weir gate, 600-mm attraction-flow pipe, and 900-mm transport pipe.

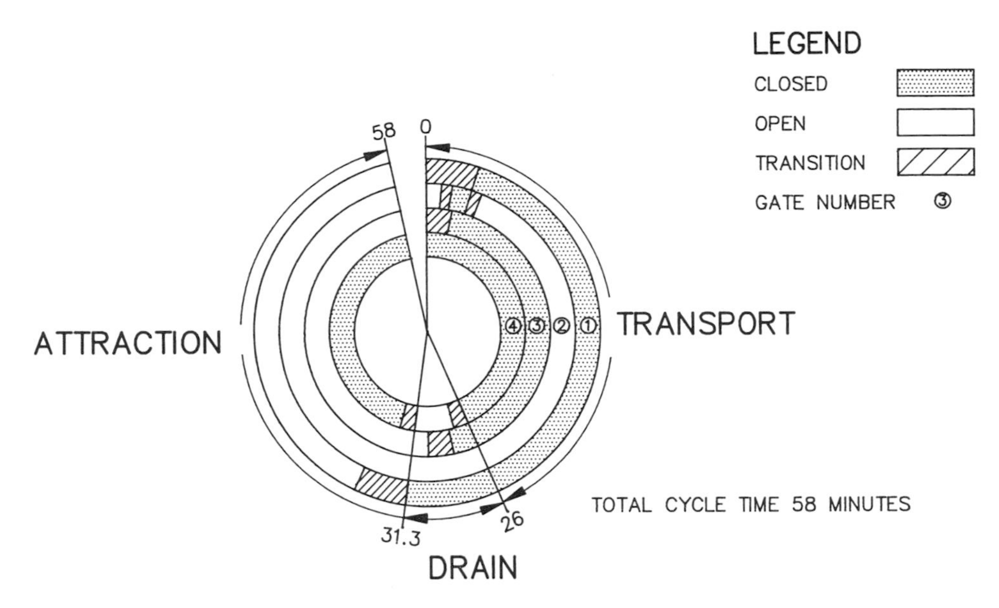

FIGURE 5.—Fishlock operating cycle.

on a time sequence. The weir gates at the upstream and downstream ends are automatically adjusted for water level at regular intervals. Probes control the mechanisms that maintain these gates at the correct settings. The downstream gate is set 385 mm above water level for sea lamprey control and the upstream weir is adjusted to maintain 220 mm of flow into the system. It was found that sufficient attraction flow could be provided with the 220-mm setting rather than a 300-mm setting as originally envisioned. The proposed flow of 0.6 m^3/s was thus reduced to 0.3 m^3/s. To avoid delays and handle reasonable numbers of fish with each lift, an operation cycle of approximately 1 h has been established (Figure 5). The drainage phase takes 5 min and the remaining time is split almost evenly between transport and attraction phases.

If necessary, the control system may be reprogrammed to change the timing sequence or change the gate settings. The control system incorporates procedures for shutdown due to malfunction and start-up.

In spite of the inexperience with this type of facility, the success has been remarkable. Fish are readily attracted to the entrance and are able to sustain themselves in the entrance chamber before climbing during the transport phase. Fish trapped at the outlet have been healthy. When the fish run fully develops, up to 14,000 fish per annum are expected to use the fishlock.

Fish Bypass Channel

In 1982 OMNR introduced a community-involvement fisheries program whereby local angling clubs could qualify for provincial government grants to carry out stream improvements and build fish facilities. This program provided an incentive to design a low-cost fishway for the Walkerton Dam that could be constructed with labor provided by the local angling club. The Walkerton Dam, a low-head mill dam (3 m), blocks migrating rainbow trout on the Saugeen River. The Saugeen River flows westerly through southwestern Ontario and empties into Lake Huron near the town of Southampton.

A fish bypass channel was designed by OMNR and constructed by the South Bruce Game and Fish Association. The club also implemented the necessary legal agreement with the owner of the dam to carry out the works. The design incorporates an earthen channel leading to a 1.6-m culvert pipe and another channel connecting to the headpond area. A similar bypass channel was constructed in 1987 at the 2-m-high Haines Dam, 6 km above the Thornbury Dam on the Beaver River. The Walkerton and Haines Dam bypass channels both incorporate strategically located pairs of 1-m^3 baffle blocks. The blocks, laterally spaced 400 mm apart, induce head loss and thereby maintain the required flow depth in the channel. The designs include slotted headwalls at the up-

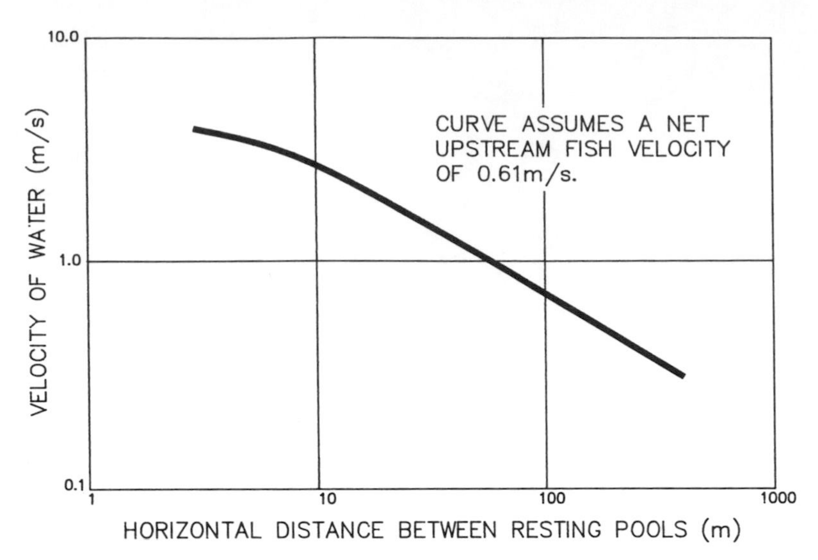

FIGURE 6.—Rainbow trout swimming performance curve.

stream end to control inflow and at the downstream end to provide attraction flow. The downstream headwall also protects the channel from river erosion. The spaces between the baffle blocks and the channel sideslopes are filled with rock riprap to confine flow through the blocks.

The bypass channels are designed to operate during the fish run within the limits of rainbow trout swimming performance (Ziemer 1961), based on sustained swimming speed for various distances (Figure 6). During other periods, particularly in flood times, the bypass channels are closed with stop logs at the upstream entrance. Channel location will vary with site conditions. The usual arrangement is to locate the channel at the downstream toe of the dam embankment and to follow the embankment in an upward direction. When sufficient elevation is obtained, the channel turns through the embankment and exits in the upstream headpond area.

The bypass channel geometry was established with a 2.0-m bottom width and sideslopes with a horizontal:vertical ratio of 2:1. Low-velocity areas require only cobble protection, but heavier riprap is used in the higher velocity areas. The upstream and downstream headwalls also serve as baffles. The upstream slot is limited to 1.0 m in height to act as an orifice and to throttle down flows as headpond levels increase. A minimum flow of 0.6 m^3/s through the downstream wall provides attraction water for the fish. The Walkerton Dam bypass channel (Figure 7) incorporated a culvert section to allow vehicular access to the north side of the dam. Baffle blocks were added to the culvert to reduce velocities and provide fish resting areas. The lengths of the resting pools in the channel may vary; 5.0–7.0 m is suggested as a reasonable length.

The levels upstream and downstream of baffles may be related by

$$H = (k_e + 1)(v^2/2g) + T_w;$$

H is the upstream flow depth (m), v is the average velocity through the baffles (m/s), k_e is the en-

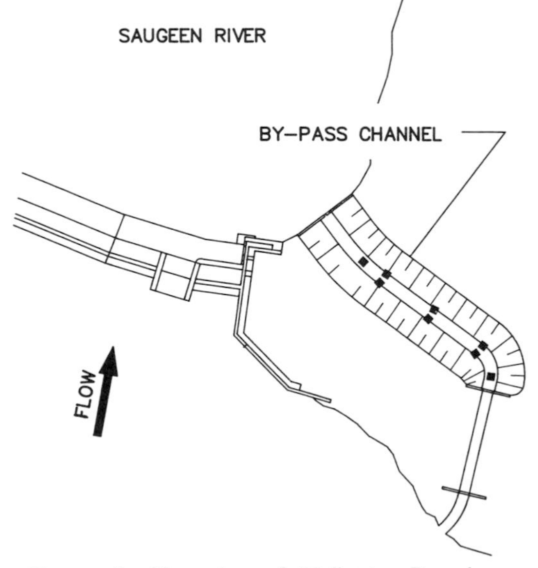

FIGURE 7.—Plan view of Walkerton Dam bypass channel.

trance loss coefficient for the baffles, g is acceleration due to gravity (9.81 m/s^2), and T_w is the tailwater flow depth downstream (m). The entrance loss coefficient is a function of the entrance geometry and flow conditions. For calculation purposes, loss coefficients were based on experience with culvert headwalls: k_e is 0.43 for baffle blocks and 0.10 for headwalls with rounded sides (Portland Cement Association 1964).

This calculation does not take into account the exact profile through baffles. The true profile includes a small dip in the water surface through the transition. However, the calculation allows for a reasonable determination of backwater loss at each set of baffle blocks. Channel friction losses are comparatively small and for practical purposes may be ignored. The upstream flow entrance is controlled by a slot in the headwall 350 mm wide by 915 mm high. The slot acts as an orifice and flow was determined by standard orifice equations. For the upstream slot, which has rounded sides and a square top, a flow coefficient of 0.80 was used (King and Brater 1963).

The calculation procedure is as follows. For a given headpond and river tailwater elevation, assume a trial flow through the bypass channel; carry out a backwater calculation, adding all baffle losses to the river tailwater elevation; calculate the flow through the orifice and compare with the assumed flow; recompute to obtain a reasonable match of flows (successive approximations); check to ensure that tailwater levels below baffles are above critical depth; add additional baffles as required; and recompute.

Critical depth through a rectangular opening is given by

$$y_c = [Q/(3.13b)]^{0.67};$$

y_c is the critical depth through the opening (m), Q is the channel flow (m^3/s), and b is the width of the opening (m). After the channel flow and flow profile have been determined, check the velocities to ensure they are within the limits of fish swimming performance.

When water levels rise in the river and higher flows pass through the fish channel, the baffle blocks will submerge. The majority of flow in such cases will be weir flow with minimal water level difference across baffles. The major head losses will be at the two headwalls, which will not be overtopped. Velocities through the slots in the headwalls will be high, particularly at the upstream entrance orifice. At a maximum flow of 1.1 m^3/s, the velocity will be 3.4 m/s, which is within the limits indicated by the rainbow trout performance curve.

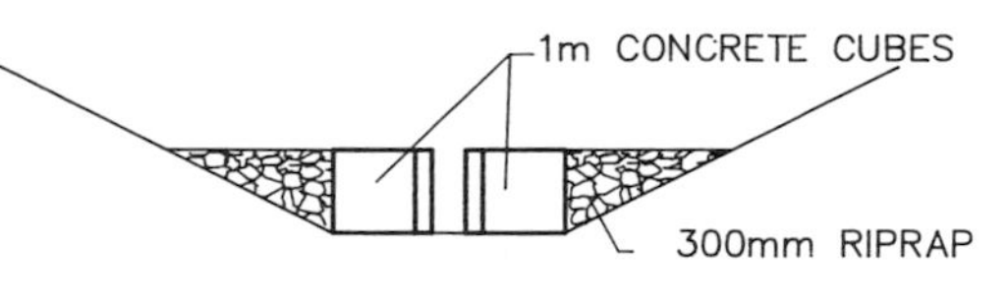

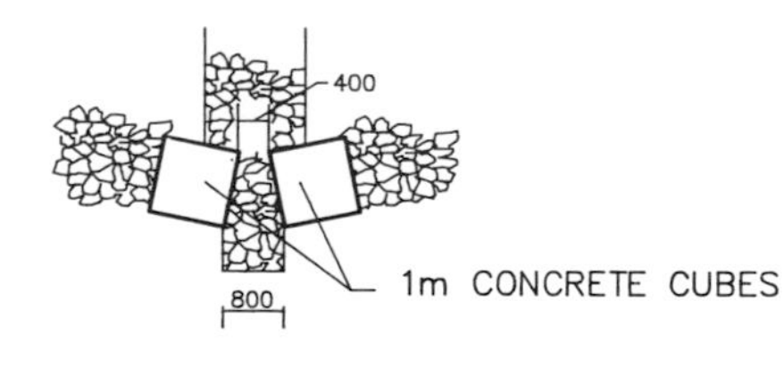

FIGURE 8.—Bypass channel baffle block detail.

Baffle Arrangements

Lateral spacing of baffles (Figure 8) must be large enough for adult fish. A spacing of 400 mm was deemed adequate for this purpose. The blocks are angled so that the gap widens in the upstream direction. This provides a smoother flow transition and makes it easier for the fish.

Riprap is required in the high-velocity areas below baffles. A small pool is also added below the baffles to improve the flow characteristics and assist the fish. Single blocks may also be added below pairs of blocks to provide slight increases in tailwater and to diffuse the downstream flow jets.

The design is quite flexible and adjustments are easily made after construction by adding new blocks or adjusting the spacing of blocks. If flow profiles are in error, because river tailwater or headpond elevations were wrongly assumed, adjustments may be necessary. The loss coefficients may not match the theoretical ones and consequently revisions to block spacing may be desirable.

Construction Features

The Haines Dam bypass channel (Figure 9) comprises 45 m of earthen channel on a 3.7% slope. The Walkerton bypass channel is 42 m long on a 3.0% slope. The channels are lined with cobbles. In high-velocity areas, 300-mm-diameter riprap underlain with filter cloth was used. The Walkerton channel has three pairs of baffle blocks plus single diffusion blocks below the culvert and

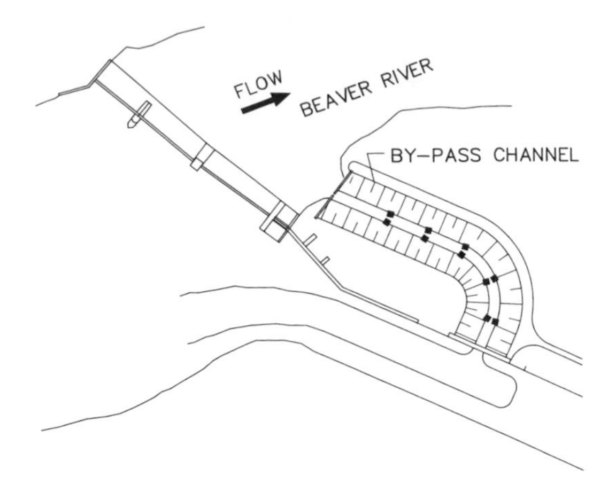

FIGURE 9.—Plan view of Haines Dam bypass channel.

in the most downstream resting pool. Three smaller blocks are located in the culvert. It was originally believed that baffle blocks could be constructed by using surplus concrete from ready-mix batches. However, in the case of both Haines Dam and Walkerton Dam, the contractors found it more convenient to cast the blocks on site.

The Walkerton Dam upstream culvert (Figure 10) is a corrugated steel pipe, 17.0 m long by 1.6 m in diameter, on a 3% slope. Steel baffle plates were bolted to headwalls at the upstream and downstream ends of the culvert. The upstream baffle plate provides upstream orifice flow control, and the downstream baffle provides sufficient backwater to cause the culvert to flow full. Small cubic baffle blocks (0.4 and 0.6 m) are bolted to the culvert floor to reduce velocities in the lower part of the flow area and give the fish resting places. Stop logs can be placed across the upstream headwall to shut off flows outside the fish run period.

The cost of the Haines Dam bypass channel, including ancillary items such as fencing, was Can$ 55,000. The Walkerton Dam bypass channel cost Can$44,000. Because the construction was carried out by a nonprofit group, materials were donated by manufacturers. If all material costs are included as well as volunteer labor, the total project value equals Can$74,000.

Operating Problems

The Haines Dam bypass channel was opened in fall 1987. Rapid removal of the stop logs caused some disturbance to the downstream riprap. Slow and careful removal is required to gradually build

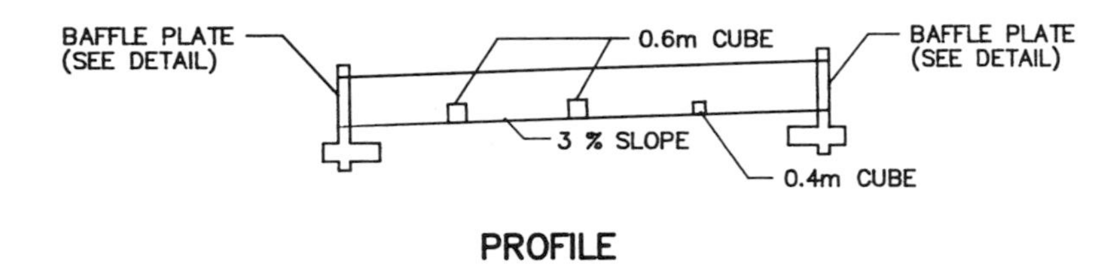

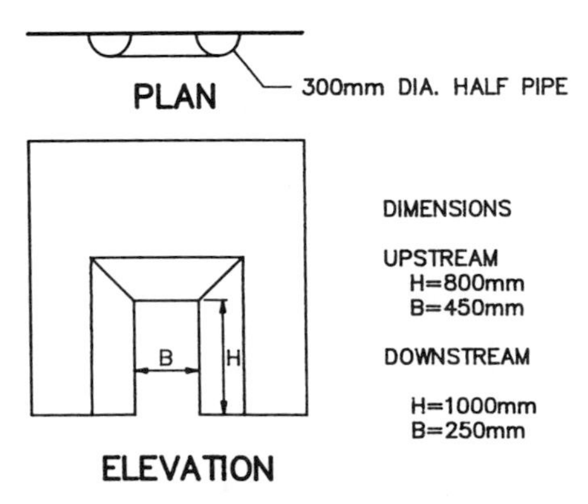

FIGURE 10.—Details of Walkerton Dam bypass channel culvert and baffle plate.

up tailwater as the flow increases. At both the Walkerton Dam site, just below the culvert, and at Haines Dam site, below the entrance orifice, the discharge nappe encouraged fish to jump. Salmonid species appear to have a natural instinct to jump at waterfall areas. Fish jumping over the nappe hit the headwall, but they do not seem to go back downstream and no mortality has been observed. It can therefore be assumed that fish eventually swim through the opening. However, the jumping leads to unnecessary fatigue and injury. Increasing the tailwater with additional baffle blocks to submerge most of the nappe will probably correct this problem.

Flow jets leaving pairs of baffle blocks may be too strong and cause erosion. Such jets are likely at supercritical flow, and rather than maintaining position in the center of the channel, they tend to attach themselves to one bank or the other. The problem can be corrected by increasing the tailwater with additional blocks or adding single downstream diffusion blocks.

Concluding Remarks

The fishlock and bypass channel are viable fish passage facilities. Both can be constructed economically, and in the case of the bypass channel, relatively unskilled labor can be employed.

The fishlock is more likely to be economical for dams higher than 10 m. The bypass channels are more suitable for low dams (less than 5 m). Considerable flexibility to change or add baffle blocks, even after construction, is inherent in the design of the bypass channel. This allows for fine-tuning to obtain the right flow conditions for fish passage.

Electrically operated gates and automatic controls increase the cost of a fishlock, but costs may be reduced if volunteer labor is available to operate the facility. The Ontario Ministry of Natural Resources will be looking into the viability of future fishlocks and their applicability to other species such as walleye *Stizostedion vitreum*. Variations in the pipe and chamber arrangement are possible as well as in the sizes of transport pipe. The fishlock has the advantage of requiring little space and may be constructed in very tight quarters on the side of a dam.

References

Clay, C. H. 1961. Design of fishways and other fish facilities. Department of Fisheries of Canada, Ottawa.

King, H. W., and E. F. Brater. 1963. Handbook of hydraulics. McGraw-Hill, New York.

Portland Cement Association. 1964. Handbook of concrete pipe hydraulics. Portland Cement Association, Skokie, Illinois.

Ziemer, G. L. 1961. Fish transport in waterways. Alaska Department of Fish and Game, Juneau.

American Fisheries Society Symposium 10:264–267, 1991

The Brule River Sea Lamprey Barrier and Fish Ladder, Wisconsin

GALE A. HOLLOWAY

Wisconsin Department of Natural Resources
Post Office Box 7921, Madison, Wisconsin 53532, USA

Abstract.—In the early 1950s, sea lampreys *Petromyzon marinus* devastated fisheries in the Great Lakes. Beginning in the late 1950s, chemical control reduced sea lamprey numbers, but the chemical has inherent drawbacks and its future use may be limited or restricted. In 1985, the Wisconsin Department of Natural Resources designed and installed a combination fish ladder and sea lamprey barrier at a low-head dam on a popular fishing stream in northwestern Wisconsin. This ladder–barrier combination allows free passage of salmonids but eliminates sea lampreys from the upper Brule River, which has been one of the largest producers of sea lampreys among the tributaries to western Lake Superior. Nearly 100% of sea lampreys migrating upstream are trapped and removed from the river before they spawn. However, unless sea lampreys continue to be attracted to the Brule, they may spawn elsewhere; the barrier thus may lose its effectiveness for control of sea lampreys in Lake Superior, and it could contribute to increased sea lamprey populations in other streams.

History

The Bois Brule River, or Brule River as it is commonly known, is one of the most popular trout-fishing streams in northwest Wisconsin. Its stable flow and cool water are ideal habitat not only for salmonids, but also for spawning and larval sea lampreys *Petromyzon marinus*. Since entering Lake Erie through the Welland Canal in the 1800s and spreading throughout the upper Great Lakes in the early 20th century, sea lampreys have dramatically changed the Great Lakes and Brule River fisheries by reducing or eliminating game-fish populations. After spending 3–17 years as ammocoetes in tributary streams, juvenile sea lampreys migrate downstream into the Great Lakes. There, each individual feeds on and destroys approximately 40 pounds of fish during its 12–20 months as a free-swimming parasitic adult (Great Lakes Fishery Commission 1982). In the spring following maturity, adults migrate up a tributary stream to spawn. Sea lampreys have not been found to exhibit a strong homing behavior to their stream of origin (Moore and Schleen 1980).

Early sea lamprey control methods in the Brule River and other Great Lakes tributaries included mechanical and electrical weirs used to remove adults before spawning, but both had drawbacks that limited their use and allowed excessive escapement of adult sea lampreys or mortality of migrating fish, particularly during spring floods (Hunn and Youngs 1980).

The development of lampricides in the late 1950s changed control methods and finally permitted effective control of sea lamprey ammocoetes in selected streams. Although the Great Lakes fishery began to recover after the widespread use of the lampricide 3-trifluoromethyl-4-nitrophenol (TFM) (Smith and Tibbles 1980), concern turned to the chemical's effect on stream fauna, especially in streams with established fisheries. Although it is generally held that there is little significant long-term effect on native fish and other stream inhabitants (Gilderhus and Johnson 1980; Moore and Schleen 1980), local anglers and residents were opposed to continued chemical treatments. Questions about the future cost and availability of TFM, and whether or not it would continue to be registered by the U.S. Environmental Protection Agency (Smith and Tibbles 1980), also were cause for concern.

Barrier Dam Construction

These concerns, intensified by occasional kills of juvenile fish in treated areas, prompted Wisconsin Department of Natural Resources (WDNR) officials to investigate nonchemical methods of control. In 1980, the Great Lakes Fishery Commission reported that an 18-in drop would stop sea lamprey migration because sea lampreys apparently cannot leap clear of the water. Based on this theory, the commission encouraged construction of barrier dams (Hunn and Youngs 1980; Smith and Tibbles 1980), and design was started on a low-head barrier on the Brule at a site about 6 miles from the mouth. The barrier was constructed in 1984 amid controversy about construction of a dam of any type on this primitive, wild river. Soon after construction was completed in the fall of 1984, the local public and anglers (some of whom still resented the construction of the dam) expressed concern that not all fish could negotiate the 4-ft jump that existed during normal river stage. The shallowness of the plunge pool contributed to fish passage difficulties. To address the problem, the WDNR Bureau of Engineering immediately began design of a facility that would incorporate a fish ladder into the existing structure, which was already felt to be an effective sea lamprey barrier. With consideration given to the low available hydraulic head and the operational conditions imposed by a special WDNR task force, the final design consisted of a weir-and-pool ladder that included an adjustable lamprey barrier weir at its midpoint. To accommodate easier fish passage when migrating sea lampreys were not present, and during cold weather (the facility is operated year-round), the entire structure was designed to easily convert to a vertical slot ladder. The facility was built in 1985 (Figure 1) and has been in use continuously since the spring of 1986.

Fishway–Barrier Description

When configured as a weir-and-pool ladder during the sea lamprey spawning run, the adjustable gate midway through the ladder (where the water surface elevation is least affected by river stage) forms a barrier that prevents sea lampreys and other nonjumping fish from ascending farther while allowing salmonids to easily jump the short step. To accommodate fluctuating river stage, flow control into the ladder is provided by two pairs of submerged orifices near the water entrance. During normal operation, ladder flow is stable enough that no adjustments need be made at river flows up to annual flood stage. Above annual flood stage, the gate is raised a few inches to maintain the drop required to block sea lampreys—usually 14–18 in (crest to lower pool). The drop has been effectively kept as low as 12 in when the water temperature is cold enough to repress sea lamprey activity (below 45°F).

After the sea lamprey spawning run, the ladder is completely converted to a standard vertical slot ladder by changing stop logs, and it remains in operation as such throughout the fall and winter. In spring, when the danger of freeze-up is past, conversion is made back to the submerged orifice–weir-and-pool configuration in time for the sea lamprey run. A window in the upper end of the ladder permits observation of passing fish, either by a person stationed there or with a video camera.

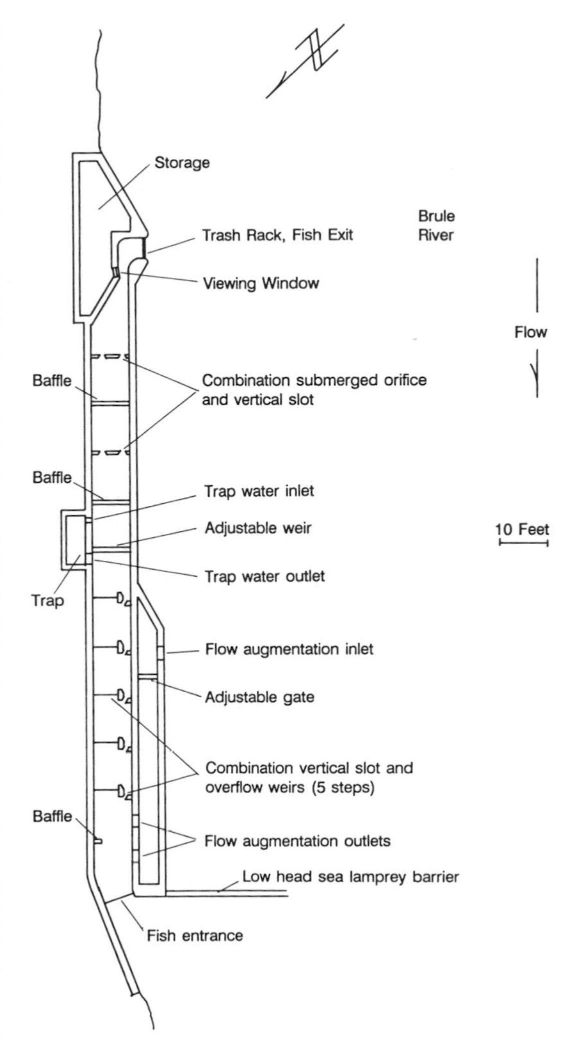

FIGURE 1.—Brule River fish ladder and sea lamprey barrier.

The built-in trap is adjacent to the fishway at the downstream face of the adjustable barrier gate and is readily reached from the access road. Attraction water enters the trap from the upper ladder through a screened opening and exits below the gate through a screen funnel. The size of the opening in the funnel, combined with the height of the drop at the gate, determines what species can negotiate the barrier and continue upstream, and what species enter the trap. When sea lampreys are being trapped, the drop is set at approximately 16 in and a 3-in funnel opening is used. Experience has shown that sea lampreys require very little attraction flow to enter the trap; leakage past the closed water-inlet gate seals usually is adequate. When the trap is used to capture salmonids, however, an attraction flow velocity of 2–3 ft/s is required to draw fish through the screen funnel.

Trapping and Fish Passage Results

Operation of the system as a vertical slot fish ladder began on March 27, 1986, shortly after ice-out, and the first spring-run steelhead *Oncorhynchus mykiss* passed through the ladder the same day. During the next few weeks, many more steelhead passed through the facility with no apparent difficulty. Operation as a weir-and-pool ladder and barrier began on April 9, 1986, and with the installation of the one-way screen funnels in the built-in trap on April 15, the facility was fully functional.

While waiting for the anticipated sea lamprey spawning run, operators and fish managers were kept busy removing hundreds of suckers and small numbers of other miscellaneous species from the trap. Because the importance of these species to the river ecosystem was not well known, all fish except sea lampreys were released back into the river to continue upstream. Most predominant were white suckers *Catostomus commersoni* and longnose suckers *C. catostomus,* which run shortly before sea lampreys. Recording trapped species soon proved to be too time-consuming; consequently, most were simply dipped from the trap and released on the upstream side of the wall. Thousands of brown trout *Salmo trutta*, coho salmon *Oncorhynchus kisutch*, chinook salmon *O. tshawytscha,* and steelhead migrated directly upstream through the ladder during the spring and summer, providing a welcome spectacle by easily negotiating the jump at the barrier gate.

The spring sea lamprey run began in early May and continued into mid-July. Sea lampreys were unable to negotiate the gate and were readily attracted to the trap. During the first few months of operation, nearly 7,000 sea lampreys were captured and removed from the river. The largest 1-d catch was 2,100 on May 26. After a TFM treatment in the fall of 1986, no young-of-the-year ammocoetes were found among the dead sea lampreys above the barrier, indicating that no adult sea lampreys had negotiated the barrier that spring. Because the barrier has thus far proven to be impassable to free-swimming sea lampreys, no ammocoetes are expected in the upper Brule after those that escaped the last TFM treatment mature and migrate to the lake. It is expected, however, that a few sea lampreys will continue to spawn below the barrier.

High Efficiency—An Asset or a Liability?

The efficiency of the built-in trap has been estimated at 95 to 99% by WDNR and the U.S. Fish and Wildlife Service (USFWS), as determined by releasing fin-clipped sea lampreys downstream of the barrier and tabulating recapture rates. The decline in the number of sea lampreys captured between 1986 (6,940) and 1987 (1,897) may be the result of unknown residuals of chemical treatment (Moore and Schleen 1980) in the fall of 1986, but this does not explain the continued decline in 1988 (1,258). Historically, the population has increased the second and third years after treatment (Moore and Schleen 1980). One explanation for the continued drop is the very rapid increase in water temperature and low river stage in the Brule at the beginning of the sea lamprey migration in 1988. However, it could also be due to a decreasing attractiveness of the river caused by a reduction in the number of sea lamprey ammocoetes (Moore and Schleen 1980) above the barrier. In 1989, 3,705 sea lampreys were captured. The threefold increase over 1988 is somewhat encouraging in that it indicates that sea lampreys are still attracted to the Brule (possibly due to the presence of some ammocoetes below the barrier), but discouraging in that it could also indicate an overall increase in the population of sea lampreys in Lake Superior.

Unless sea lampreys continue to be attracted to the Brule River, the new barrier could become a detriment by encouraging spawning sea lampreys to drift to streams where TFM treatment may be prohibitively expensive and physical barriers impractical, such as the St. Louis River in Minnesota (Smith and Tibbles 1980). If this happens, the benefits of structures similar to the one on the Brule could be reduced. The potential for this to occur could increase as river pollution cleanup efforts continue and sea lampreys find decontaminated rivers more desirable as spawning habitats. With recent cutbacks in TFM treatments by the USFWS (due to increasing costs) and the reported increases in sea lamprey scarring on Lake Superior fish (which could indicate a recent increase in the overall sea lamprey population in the lake), this will likely become an increasingly important issue.

Conversely, if attraction to the Brule (or any

other river with a high-efficiency trap) could be improved by the introduction of pheromones (Teeter 1980) or artificial attractants that emulate larval sea lamprey effluvium, the result could be a corresponding reduction in the number of sea lampreys migrating to other area streams. If true, this could reduce dependence on TFM with its related drawbacks (Dahl and McDonald 1980; Smith and Tibbles 1980). Thereafter, traps in strategically located streams might provide a very effective, efficient, and relatively low-cost method of sea lamprey control for the Great Lakes.

References

Dahl, F. H., and R. B. McDonald. 1980. Effects of control of the sea lamprey (*Petromyzon marinus*) on migratory and resident fish populations. Canadian Journal of Fisheries and Aquatic Sciences 37:1886–1894.

Gilderhus, P. A., and B. G. H. Johnson. 1980. Effects of sea lamprey (*Petromyzon marinus*) control in the Great Lakes on aquatic plants, invertebrates, and amphibians. Canadian Journal of Fisheries and Aquatic Sciences 37:1895–1905.

Great Lakes Fishery Commission. 1982. Sea lamprey management program. Great Lakes Fishery Commission, Ann Arbor, Michigan.

Hunn, J. B., and W. D. Youngs. 1980. Role of physical barriers in the control of sea lamprey (*Petromyzon marinus*). Canadian Journal of Fisheries and Aquatic Sciences 37:2118–2122.

Moore, H. H., and L. P. Schleen. 1980. Changes in spawning runs of sea lamprey (*Petromyzon marinus*) in selected streams of Lake Superior after chemical control. Canadian Journal of Fisheries and Aquatic Sciences 37:1851–1860.

Smith, B. R., and J. J. Tibbles. 1980. Sea lamprey (*Petromyzon marinus*) in Lakes Huron, Michigan, and Superior: history of invasion and control, 1936–78. Canadian Journal of Fisheries and Aquatic Sciences 37:1780–1801.

Teeter, J. 1980. Pheromone communication in sea lampreys (*Petromyzon marinus*): implications for population management. Canadian Journal of Fisheries and Aquatic Sciences 37:2123–2132.

American Fisheries Society Symposium 10:268–277, 1991

Pool-and-Chute Fishways

KEN BATES
Washington Department of Fisheries
115 General Administration Building, Olympia, Washington 98504, USA

Abstract.—The pool-and-chute fishway is an economical means of providing fish passage over constructed barriers. Pool-and-chute fishways resemble pool-and-weir fishways at low flows and become baffled chutes at moderate to high flows. The economy of the concept is achieved by exceeding the usual criteria of fishway pool volume based on energy dissipation in each pool. The size and complexity of the structure are thus reduced. Design guidelines covering appropriate application and geometry ensure hydraulic conditions that allow fish passage. Cost comparisons based on actual and estimated construction costs of pool-and-chute and other styles of fishways verify the economic benefit of the concept.

The success of a fishway depends on the range of flows through which it operates successfully, on attraction of fish to the fishway, and on adequate maintenance to keep the fishway operating as intended. A critical element of success of a fishway is its ability to attract and pass fish during periods of high stream flows.

Traditional styles of instream fishways often have limited success at high flows if they lack auxiliary water and flow control systems, which entail substantially greater capital and operating costs. Auxiliary water systems may consist of a water intake, control gate, and diffuser pool to introduce additional water to the entrance pool of the fishway. The additional water enhances the attraction of fish to the entrance. The water supply and diffuser require fine-mesh trash racks. Fishway flow control may consist of orifices with flow depletion or supplementation systems. Mechanical devices such as water surface sensors, automatic flow control gates, tilting or telescoping weirs, and related electronic control systems are often required.

It is assumed during design, sometimes erroneously, that adequate maintenance will be provided. Maintenance demands are often underestimated during design; a fishway owner's commitment to operation and maintenance is obviously influenced by future economic considerations. A design that minimizes operation and maintenance demands is highly desirable.

A hybrid fishway, termed a pool and chute, that includes some advantages of both pool-and-weir fishways and roughened chutes has been designed and constructed. The pool-and-chute fishway is essentially a pool-and-weir fishway with V-shaped weirs that may include ports near the floor. Figure 1 shows plan and elevation views of a pool-and-chute fishway. During low and normal flows, the fishway operates as a pool-and-weir fishway with orifices. At high flows, a high-velocity streaming flow passes down the center of the fishway while a plunging flow is maintained near the sidewalls, providing a zone for fish passage.

Currently Used Fishway Styles and Design Standards

An understanding of currently used fishway styles and their relevant design standards is a basis for design of the pool-and-chute fishway. For the purpose of this discussion, fishways can be divided into three categories: pool fishways with some combination of vertical slots, orifices, and overflow weirs; roughened chutes; and lifts. Lifts include locks, brails, and hoppers. They are rarely used except for fish collection and are not discussed further here.

Fishways are generally designed to operate within design criteria for a specific range of design flows. These design criteria include adequate attraction of fish to the fishway entrance, limited water surface difference between adjacent fishway pools, adequate volume and appropriate geometry to dissipate energy and allow fish to rest in pools, plunging weir flows, minimum water depth, and maximum water velocity within the fishway.

Design Flow

The upper design flow of a fishway is the maximum flow at which the design criteria for fish passage are not exceeded. It is recognized that fish passage during extreme high and low flows is not practical (Bates and Powers 1988). The construction and operating costs of providing passage at all flows is prohibitive, in most cases, due to

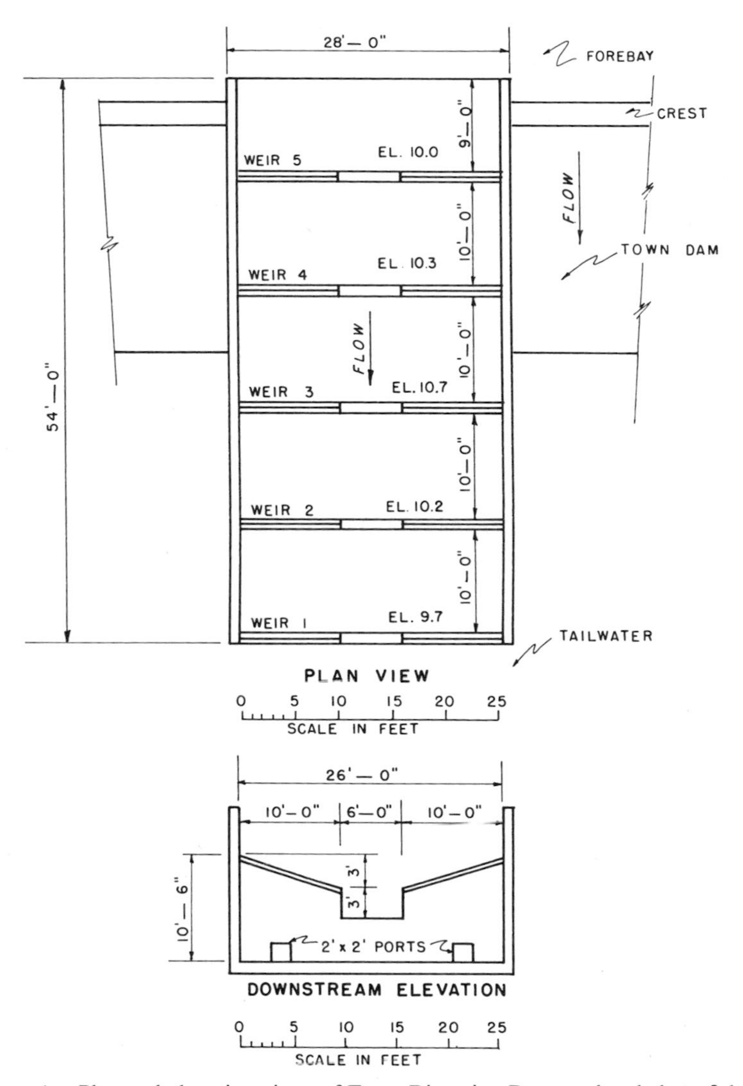

FIGURE 1.—Plan and elevation views of Town Diversion Dam pool-and-chute fishway.

additional requirements for flow control, flood protection, debris clearing, and attraction of fish to the fishway at very high river flows. In designing for fish passage during extreme high flows, passage during midrange flows would often be compromised.

The criteria of fishway design flows vary among agencies. Washington Department of Fisheries generally requires adult passage over artificial barriers 90% of the time during any species' migration season. A benefit of the pool-and-chute fishway is that it can extend the design flow to a higher range than is usually possible at a comparable cost with other styles of fishways.

Pool Fishways

Pool fishways have distinct pools. The flow drops from one pool to the next over a specifically designed weir or through one or more slots or orifices. To move upstream, fish leap or swim over successive weirs or through orifices. Fish can hold or rest within pools. Pool-type fishways are generally limited at their upper design flows by flow instability. Flow instability may include excess turbulence and aeration of the water, oscillating or "galloping" water surface, or shooting flow over the structure. Passage at the upper design flow depends on an adequate effective volume within each pool and a proper pool geometry to dissipate the energy of the flow entering it.

Three criteria that are typically and specifically applied to the design of pool-and-weir fishways and that determine their size, geometry, and flow range are (1) maximum hydraulic drop between pools; (2) minimum length of pool to maintain plunging-flow regime over each weir; and (3) adequate volume in pools to dissipate the energy of the flow entering the pool.

The maximum allowable drop between pools depends on the leaping or swimming ability of the fish intended to be passed and normally ranges from 0.5 to 1.0 ft. The other criteria are discussed below.

Hydraulics of pool-type fishways.—Normal flow circulation in a pool-and-weir fishway is termed plunging regime. Plunging flow is defined as the regime in which the direction of flow on the surface of the pool is upstream. This circulation is set up by the flow from the nappe of the upstream weir plunging to the fishway floor, moving downstream along the floor, and rolling back toward the upstream weir along the surface of the pool. Streaming flow occurs at higher flows than the plunging regime. A surface jet flows over the crests of the weirs and skims over the water surface of the pools; the water accelerates over the weirs without circulating through the pool. Shear forces from the streaming jet cause a circulation in the pool opposite to that in the plunging regime. Rajaratnam et al. (1988) provided a good description of these flow regimes.

Model studies have been performed to determine the characteristics of plunging and streaming flows and the transition between regimes (Rajaratnam et al. 1988; F. Andrew, International Pacific Salmon Fisheries Commission, unpublished). Hydraulic instability occurs in the transition between the upper range of plunging flow and the lower range of streaming flow.

Both the shape of the weir crest and the presence and design of orifices within the weir significantly affect the hydraulics of the downstream pool. They are effective in extending the flow range through which the plunging-flow condition is present and can therefore be used to extend the design flow of the pool-style fishway. Weir shapes similar to an ogee crest are most effective in producing plunging-flow conditions. Studies at the Fisheries-Engineering Research Laboratory at Bonneville Dam (Thompson and Gauley 1965) showed a qualitative improvement in flow conditions in the pool and more rapid fish passage with weirs similar to ogee crests. Model studies for the International Pacific Fisheries Commission (Andrew, no date) identified stable flow ranges as a function of weir-crests shape and orifice configuration.

The flow ranges are based on visual observations of a 1:6 scale model and were recorded for prototype scale. The upper flow limit of plunging-flow conditions in an 8-ft-long pool was increased by 33% (3.9 to 5.2 ft^3/s per foot of weir length) by rounding the weir crest and adding a 6- by 12-in port at the floor. The addition of the ports also eliminated the unstable transition from plunging to streaming flow ranges by lowering the lower limit of streaming flow from 6.1 $ft^3/s \cdot ft$ (square crest, no ports) to 5.2 $ft^3/s \cdot ft$ (round crest, with ports). The upper flow limit of plunging-flow conditions in a 10-ft-long pool was increased by 10% (4.0 to 4.4 $ft^3/s \cdot ft$) by rounding the crest. These findings are close to the dimensionless results presented by Rajaratnam et al. (1988) for normal fishway pool lengths.

Fish passage at high flows is often limited by excess turbulence in pools of the fishway. Excess turbulence eliminates both the steady circulation patterns required to guide fish upstream and the resting or holding areas for fish. Turbulence and aerated water also reduce the thrust a fish can develop to accelerate and move against flowing water. Total energy entering a pool is equal to the product of the head (potential head plus velocity head) between the pool and the next upstream pool, the specific weight of the fluid, and the rate of flow. The efficiency of dissipation of that energy in a pool is a function of the effective volume of the pool. The geometry of the pool determines how much of the actual volume is effective in energy dissipation. A standard used in the Pacific Northwest was described by Bell (1986); the maximum suggested energy dissipation in a fishway pool is 4 foot-pounds per second per cubic foot of pool volume. For water, the volume formula is simplified to

$$V \geq 16 \times Q \times h;$$

V is the effective pool volume in cubic feet, Q is the fishway flow in ft^3/s, and h is the total head of the flow entering the fishway in feet. The energy dissipation criterion was originally intended for application to vertical slot fishways, but it has proven effective in the design of weir-and-pool fishways. Application of this standard limits the flow allowed through a fishway or dictates the volume of fishway pools required. Maximum flow for pool-and-weir fishways ranges from 4 to 25 ft^3/s by this criterion.

The same standard of energy dissipation is applied to pools with orifices, vertical slots, or both. More of the pool volume may be effective for energy dissipation in these cases.

Roughened-Chute Fishways

Roughened chutes have no distinct steps but are channels with enough roughness to reduce the average velocity so fish can pass. Stream channels are examples of natural roughened channels; Denil fishways and Alaska steeppass fishways are examples of roughened-chute fishways.

Rajaratnam and Katopodis (1984) performed prototype hydraulic studies on Denil fishways and presented average and point velocities within the fishway. They also computed the friction coefficient for the assumed shear plane separating the main stream of flow from the recirculating water mass on the sides and bottom enclosed by the fishway baffles. The roughness was based on the imaginary walls of the open channel inside the dimensions of the baffles.

There are several advantages to roughened chutes for fish passage. There is less concern regarding local hydraulics as long as the average velocity is low enough for passage. Roughened chutes theoretically have no limit of maximum depth. They may provide a greater fish passage design range if built with adequately high walls and roughness elements.

Limitations of the roughened-chute fishways often include the need for attraction water similar to pool-and-weir fishways. Floating debris lodged in the fishway can easily alter the flow characteristics enough to make the fishway impassible. High tailwaters submerge the fishway entrance and greatly diminish the attraction of fish.

Pool-and-Chute Fishways

Description

The geometry and nomenclature of a pool-and-chute fishway are shown in Figures 1 and 2. The dimensions shown are for the Town Dam fishway designed by the U.S. Bureau of Reclamation and constructed in 1989 at Town Dam on the Yakima River. The Town Dam fishway was originally intended to be located near the center of the dam with entrances on each side of the lowest pool to provide the best opportunity for fish attraction at low flow. The fishway was constructed at the right bank of the dam at the request of the dam owner.

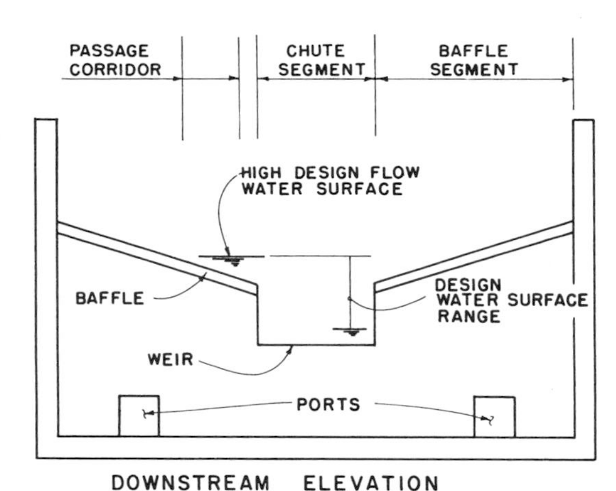

FIGURE 2.—Elevation view of pool-and-chute fishway with nomenclature adopted from weirs.

The geometry of the pool-and-chute fishway is essentially that of a pool-and-weir fishway with V-shaped weirs. During low flow it is a pool-and-weir fishway. The flow plunges over the center horizontal portion of the weir. With increasing flow, the spill over each weir extends over the sloping baffle. At high flows, a high-velocity streaming flow passes down the center of the fishway while a plunging-flow condition persists adjacent to it over the baffles. The width of the plunging-flow regime is termed the passage corridor. Ports through the baffles are included where minimum stream flows are adequate. Flow through the ports helps support the plunging-flow regime as described above.

Hydraulic Analysis of Pool-and-Chute Fishways

The flow through a pool-and-chute fishway is complex and difficult to describe analytically. At high flows, a large portion of the weir acts as a submerged weir, and velocity head is substantial at all except the upstream weirs; flow patterns are complex and include hydraulic jumps and standing waves in the chute segment.

A simplified concept of the fishway was initially used for hydraulic analysis and design. The fishway was divided into three longitudinal segments as shown in Figure 2, one to include the weir and one to include each set of baffles. It was assumed that the segments do not affect each other hydraulically; an imaginary plane separates the chute and baffle segments, preventing an interchange of flow and energy between them. This is an idealization of the actual flow condition only for the purpose of design; there certainly is flow and energy interchange across that plane. The chute segment is analyzed at high stream flow only to calculate the total flow. It is treated as a roughened chute according to the definition of roughness given by the Chezy equation (Chow 1959). In at least one design, the normal energy dissipation volume requirement was reduced in the center segment by 50% (U.S. Army Corps of Engineers 1988).

The baffle segments are not relevant to the low-flow fishway hydraulics because all the flow is concentrated in the center of the fishway. At high flow they are analyzed as a pool-and-weir fishway; thus two design standards are applied. Pool volume in the baffle segments must satisfy the criterion of energy dissipation volume described above, and the distance between pools must be such that plunging flow is maintained.

Observations

Model study.—A physical scale model was constructed to determine the best geometric configuration for fish passage and to test the simplified design concept described above. The model was tested at the McAllister Creek Hatchery near Olympia, Washington. The testing was begun with a model of the Town fishway at a 1:10 scale and a slope of 4.9%. The results described in this section relate to the prototype scale of Town Dam fishway. The model was also tested at slopes of 11.1% (which closely corresponds to a 1:5 model of the Rainbow Creek fishway described in the next section) and at 16.7%.

A range of flows and several weir geometries were tested both with and without ports. Fish passage evaluation was based on visual observation of level of turbulence in the outer-third segments of the fishway and in the uninterrupted circulation patterns intended to guide fish to the next pool. Fish passage ratings were expressed as one of five categories in a range from poor to excellent. Flow circulation patterns were mapped on a horizontal grid at 2 and 4 ft above the floor of the prototype.

The flow at which the circulation in the center segment changed from plunging to streaming was recorded with increasing flow and the reverse with decreasing flow. Water surfaces were recorded in order to determine the Chezy roughness coefficient of the fishway with streaming flow. Velocities of the jet entering the tailwater of the fishway were recorded at 2.0 ft below the water surface at 0, 20, 40, and 60 ft downstream of the fishway.

Streaming flow existed separate from and parallel to the passage corridor. At high flows it spread laterally, only on the surface, over the plunging circulation that persisted within the pools. Passage conditions existed at the passage corridor, where plunging flow was maintained. The passage corridor was consistently about 3 ft wide over each baffle. Flow through the ports was stable and consistent.

Good passage conditions were observed in the 4.9, 11.1, and 16.7% models at flows up to 450, 468, and 136 ft^3/s, respectively, by modifying the cross-sectional shape of the weirs. The weir heights were increased with increasing flow to achieve good passage conditions. The interior weirs (other than two exit weirs) had final heights of 3.3, 5.8, and 6.7 ft, respectively, for the three slopes tested. Additional improvements could likely be made to the 4.9% model to further increase the flow at which good passage conditions exist. The deterioration of passage conditions at the upper limit of "good" passage was due to increased upwelling on the upstream side of the baffles and excessive or unstable standing waves just upstream of the weirs. Standing waves existed in most situations upstream of the weirs at the upper limit of what was considered "good" passage but did not influence the passage corridor.

Specific model study observations are described below with suggested design standards.

Rainbow Creek Fishway.—Rainbow Creek fishway was constructed in 1983 by the U.S. Forest Service. The fishway is 12 ft wide, and pools are 6 ft long with a drop of 0.75 ft/weir for an overall slope of 11.1%. The high design flow of 85 ft^3/s is expected to be exceeded 10% of the time during November through January. At that flow, 76 ft^3/s is intended to pass through the horizontal weir segment and 4.5 ft^3/s through each of the baffle segments. The volume intended for energy dissipation in the outer thirds of each of the pools was 75 ft^3, which is 50% greater than the standard volume criterion described above.

The fishway has been observed through a range of stream flows. On March 26, 1988, a flow of 88 ft^3/s, nearly equal to the design flow of 85 ft^3/s, was measured through the fishway. I observed the fishway at that flow and considered it passable, but at the upper limit of its passage range. The water surface was 0.35 ft higher than the point where the baffles meet the fishway side walls and

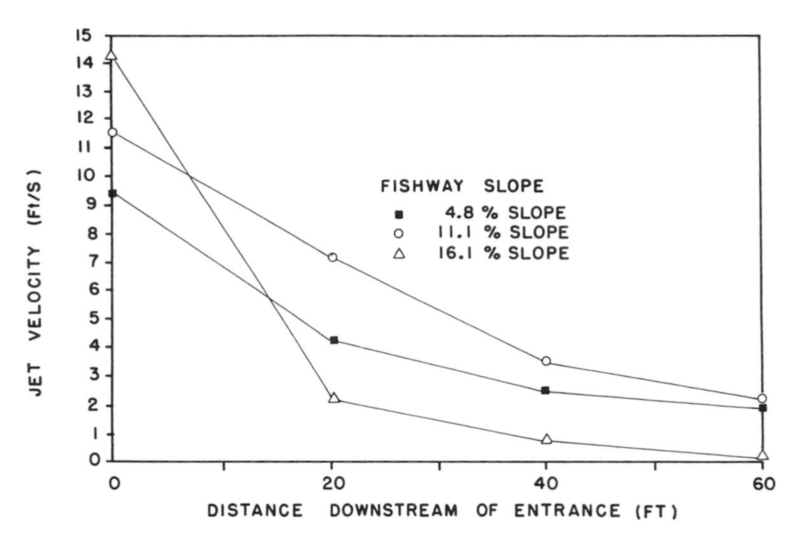

FIGURE 3.—Effect of fishway slope on tailwater jet velocity at prototype scale.

where it was intended to be at that flow. Flow circulation and sediment accumulation within the fishway pools were mapped. Sediment, consisting of sand and small gravel, varied in depth from 1.5 ft near the downstream corners of the pools to 0.2 ft deep at the upstream end of the pools. Accounting for the raised water surface and the sediment accumulation, the volume of the outer segments was decreased from the design volume of 75 ft^3 to 63 ft^3—23% greater than the standard volume criteria would provide.

Applications and Limitations

The intended use of the pool-and-chute fishway is in situations where little or no flow control is available. This is normally the case in small streams such as Rainbow Creek where the entire flow passes through the fishway. Passage through velocity barriers is another logical application. Elevation barriers become velocity barriers when the drop becomes elongated into a steep hydraulic slope.

The limitation on length of passable pool-and-chute fishways is untested. The longest fishway of this style that I know of (Carpenter Creek, Washington) consists of eight 1-ft steps. Peterson and Mohanty (1960) reported the development of roll waves in rough, steep channels where the Froude number (ratio of inertial forces to gravity forces) is greater than 1.6 and the channel is sufficiently long. Roll waves are unstable hydraulic waves that move through the length of the fishway in either the upstream or downstream direction. No roll waves have been observed in existing fishways of this style or in the model. The fishways observed are likely not long enough to produce roll wave instabilities. Froude numbers were limited to 1.18 or less in the model study and to 1.07 in observations at Rainbow Creek.

The pool-and-chute fishway can only be used in a straight configuration unless a very large turning and energy-dissipation pool is provided. The length of turning pool required was demonstrated by mapping the high-velocity plume as far as 60 ft downstream of the prototype Town Dam fishway. The tailrace velocities at a fishway flow of 182 ft^3/s and at three slopes are shown in Figure 3. At the fishway entrance, the velocity of the jet increases with increasing slope, as would be expected. As the slope is increased from 11.1 to 16.7%, the flow exiting the fishway changes from streaming to plunging flow. This is apparently due to the increased vertical component of velocity regardless of the increasing total resultant velocity. The plunging flow dissipates rapidly; at 20 ft below the fishway, the velocity was less than a third that of the jet velocity exiting the 11.1% slope fishway. It dissipates quickly because the vertically plunging flow has a lower downstream velocity component and additional entrained air buoys the jet.

Velocities of the flow leaving a prototype fishway will be less than the prototype velocities determined by the model. The model jet was dissipated only by shear forces in the tailwater pool, which was at least 20 ft deep. The jet from a fishway set in a natural channel will have the additional shear forces of the rough channel boundary.

A second possible limitation is bed-load deposition in the pools of the fishway. Deposition and filling of the pool areas along the walls of the fishway decrease the volume and therefore decrease the high design flow.

Sediment accumulation may be a greater limitation in a pool-and-chute fishway than it would be for other fishway styles. Because of the higher design flow of the pool-and-chute fishway, the flow at which it is scoured clean is likely greater than in a pool-and-weir fishway and is therefore less likely to occur. Large debris may block orifices. Denil fishways are essentially self-cleaning of bedload material. Several cubic yards of gravel and cobble were dumped into a Denil fishway at a 17% slope to evaluate its ability to pass bed material (D. Cagle, Washington Department of Fisheries, unpublished). All but a small portion of material smaller than 4 in passed through the fishway; about half of the 4–6-in rock passed through the fishway. Floating debris, however, can effectively block passage through Denil-style fishways.

Suggested Design Standards

Based on the model studies and observations of existing pool-and-chute fishways, the following design criteria for weirs and baffles are suggested. Dimensions are given in a prototype scale comparable with the Town Dam fishway. These suggested design standards have not been entirely field-verified.

The length, slope, and crest shape of the baffle and the length of the weir determine the fishway flow capacity and high design flow for passage. The width of horizontal weir will determine the allowable quantity of flow passing through the fishway; there is not expected to be a hydraulic or fish passage limitation on the width or flow through the center segment. The width is only limited by cost. The minimum horizontal crest length is based on the flow required at high flow to attract fish for fishways built within a dam crest.

The length of the baffle segments, together with the lateral slope of the baffles, depends on the expected range of forebay water surfaces. The lateral slope of the baffles controls the width of the passage corridor over the weir and establishes the high design flow. The steeper the baffle slope, the narrower the passage corridor but the higher the high design flow. The 1:3 slopes studied provided a passage corridor that was consistently about 3 ft wide; that slope ratio is recommended.

In designing for high flows, the change from static head at the upstream two weirs to velocity head in the interior weirs must be accounted for. The two upstream weirs (exit weirs) should be modified by lowering the upper weir crests to elevations below the grade extended from the downstream weir crests. Unless the exit weirs are modified, additional drop over the upper weirs greatly diminishes the passage rating of the fishway. The number of weirs modified and their modified elevations depend on the slot velocity within the fishway at the high design flow. The greater the difference between the forebay approach velocity and the streaming flow velocity in the fishway, the farther the exit weirs should be lowered. The upstream weir should be lowered an amount equivalent to the gain in velocity head from the forebay to the third weir; the second weir is lowered half as much as a transition.

Figure 1 shows weir elevations of the Town Dam fishway based on model study results. Figure 4 shows water surface profiles at 304 ft^3/s through the model fishway with all weirs 4.2 ft high and with the height of the exit (most upstream) and second weirs modified to 2.5 and 3.3 ft, respectively.

The cross sections of the weir crests can be square in the weir segment to minimize complications of concrete forming. The downstream edge of the crest of the baffle segments should be rounded or truncated to optimize plunging-flow conditions.

Minimum depth is controlled by the elevation of the weirs and should be at least twice the height of ports to prevent the port flow from boiling to the surface. Though not studied, it is expected that a minimum depth must also be provided to maintain plunging flows.

The spacing of weirs was not specifically studied in the model test. The spacing is derived from the desired slope of the fishway and the maximum drop per weir. Close spacing will maintain streaming flow, maximize roughness, and maximize slope of the fishway. At the lower limit of spacing, the design approaches that similar to a Denil fishway. A minimum spacing of 6 ft is suggested to allow adequate resting area within the pools.

Hydraulic Analysis

The depth and velocity, and therefore capacity, of flow within the chute segment during high flows can be determined with the Chezy equation. Fig-

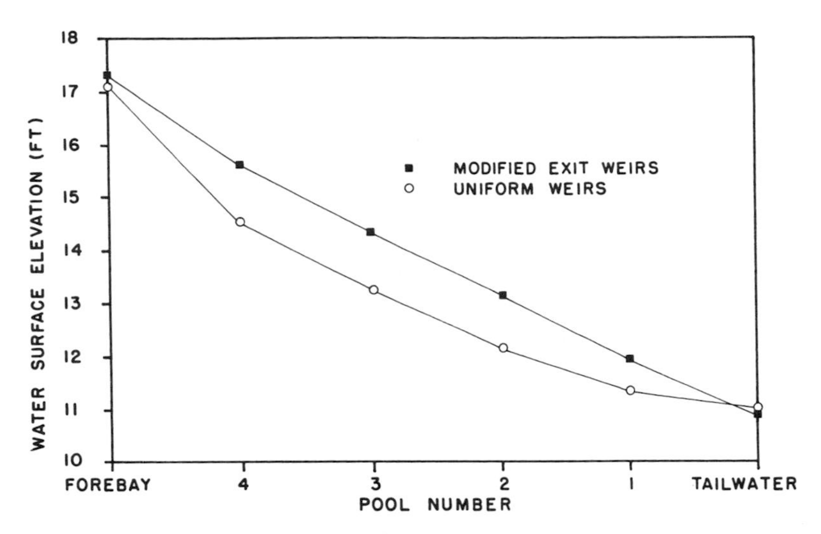

FIGURE 4.—Water surface profiles at flows of 304 ft^3/s through a prototype pool-and-chute fishway with all weirs 4.2 ft high, and with the exit (most upstream) and second weirs modified to 2.5 and 3.3 ft in height, respectively.

ure 5 shows the relationship of the Chezy coefficient to fishway flow with a slope of 11.1% and a horizontal weir height equal to 70% of the weir length.

The feasibility of designing the fishway by segments as described previously was not conclusively demonstrated. With passage conditions considered good or excellent, the pool volume within the weir segment varied from 10 to 40% of what would be required by the standard energy dissipation volume requirement described previously. The volume within the baffle segments with good passage conditions ranged from 30 to 600% of the volume criterion.

Froude numbers were calculated at the weir crests in the model study for the depth and velocity over the crest. No limit of Froude number was determined as a design criterion, although no passage conditions were considered "excellent" with Froude numbers greater than 0.97, and none were better than "good" with Froude numbers greater than 1.03.

Total flow entering the fishway is determined by the geometry and elevation of the upstream weirs. A forebay stage rating curve with the exit (most upstream) and transition weirs 2.5 and 3.3 ft high, respectively, and 6 ft long is shown in Figure 6.

Fishway slope.—Maximum fishway slope depends on desired design flow range, attraction required, and ultimate streaming velocity accept-

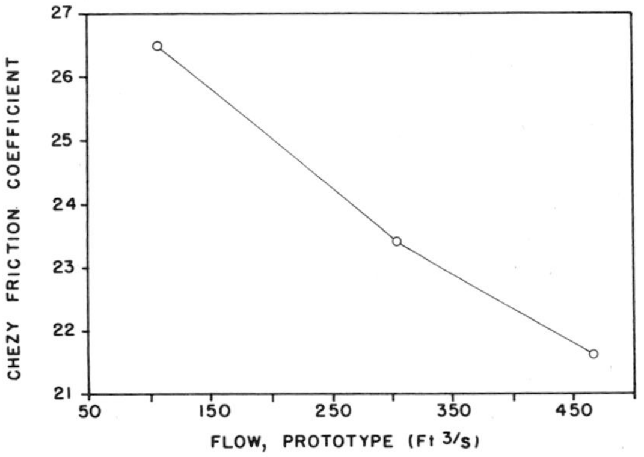

FIGURE 5.—Variation of Chezy coefficient with flow through a prototype pool-and-chute fishway at 11.1% slope and with weir heights 70% of weir length.

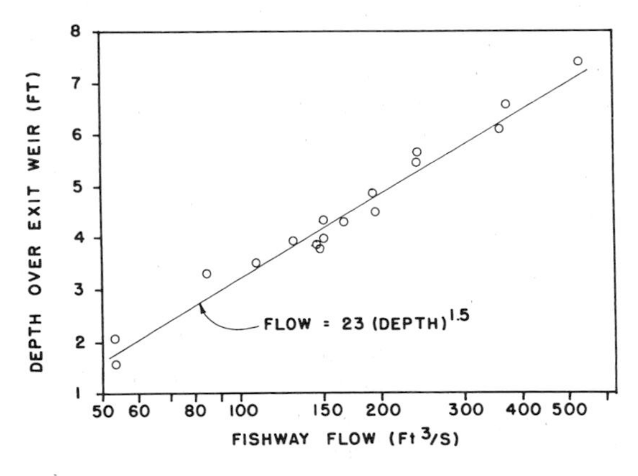

FIGURE 6.—Forebay stage rating curve for pool-and-chute fishway with the exit and transition weirs 2.5 and 3.3 ft high, respectively, and 6 ft long (prototype dimensions).

TABLE 1.—Cost comparison of three fishway styles normalized to total head, fishway entrance flow, and 1988 construction cost.

Location	Fishway rise (ft)	Fishway design flow (ft^3/s)	River design flow (ft^3/s)	Cost (US$) Total construction	Unit construction	Annual maintenance
Vertical slot with auxiliary water						
Town Dam	3.9	104	4,000	337,000	864	5,800
Sunnyside Dam	6.5	104	12,000	322,147	476	7,800
Gibbons Creek	12.0	30	70	222,860	619	
Pool and weir with flow control						
Gibbons Creek	12.0	15	70	166,700	892	
Pool and chute						
Carpenter Creek	8.0	40	40	41,300	129	[a]
Rainbow Creek	8.0	85	85	25,350	36	[a]
Town Dam	3.9	354	4,000	214,000	155	700
Gibbons Creek	12.0	70	70	90,700	108	

[a]No maintenance required in last 5 years.

able within and downstream of the fishway. Slopes exceeding the tested slope of 11.1% should not be attempted without further evaluation.

Design flow.—Maximum design flow is based on width of the weir segment. Allowable flow depth over the baffles should be based on maintaining a plunging-flow regime. The estimate of the upper limit of plunging flow described by Rajaratnam et al. (1988) is a function of weir spacing; that estimate agreed well with the results of this study.

Ports.—Ports do not affect the streaming portion of the fishway. They provide an alternative passage route and should be included any time the low design flow is greater than their combined capacity. Ports also help keep the outer portions of the pool clean from sediment.

Ports should be located as close to the side walls and floor as possible and at least 4 ft laterally from the end of the weir to ensure that turbulence from the weir does not disrupt their operation. Sizing of ports depends on size of fish and amount and size of debris expected at the site; 1.5-ft-square ports or larger are recommended for salmon.

Further Studies

It is suggested that further model and prototype studies be conducted to verify detail design standards and to test effects of sediment deposition. Pool widths and geometric details should be further studied to optimize these details for fish passage.

It is also suggested that a standard be developed to quantify acceptable levels of turbulence in fishway pools of all styles. Pressure fluctuations might be recorded for this purpose. The rapidity and magnitude of pressure fluctuations in a fishway pool may be good indicators of turbulence, and a limit could be determined at which fish passage or guidance is hindered.

Cost Comparisons

Costs of fishway construction currently vary from about US$2,500 per vertical foot for a small, simple pool-and-weir built with force-account labor to as high as $50,000 per vertical foot for a vertical slot ladder with flow control, multiple entrances, auxiliary water supply, and flood and debris protection and constructed under federal contract.

Table 1 shows construction costs for three styles of fishways. An effort was made to normalize the costs for this comparison. Final costs are presented as "unit construction costs," which are total construction cost per foot of rise of the fishway per ft^3/s of design flow at the fishway entrance. Calculating relative costs on the basis of total head and total entrance flow accounts for differences in scale of each fishway and of the river for which each is designed.

Construction costs in Table 1 are presented as 1988 construction costs updated with the Engineering News Record construction cost index. The Town Dam vertical slot fishway cost is from a preliminary design and cost estimate by the U.S. Bureau of Reclamation; the pool-and-chute option was selected for construction.

Flow control at the Gibbons Creek fishway consisted of an orifice control section. Sunnyside,

Carpenter, and Rainbow fishway costs are actual construction costs including federal contracts and force accounts (Carpenter). The Town Dam and Gibbons Creek pool-and-chute fishway costs are from final design cost estimates. All are concrete structures cast in place except for the Gibbons Creek fishway, which is steel sheet pile. Annual maintenance cost for Sunnyside fishway is based on hourly records over several years of operation; maintenance costs of other facilities are expected costs.

Variability of costs among fishways of any specific style is due to site conditions, project scope, contracting considerations, and design and specification details. The large range of unit costs among fishway styles is obvious. Vertical slot fishways require complicated concrete forming, trash racks, and auxiliary water diffusers, flow control and multiple entrance gates, and flood protection. Their higher operation and maintenance costs relate to the same items, especially the manual cleaning required for trash racks and auxiliary water diffusers. Inadequate cleaning maintenance can significantly affect the attraction and passage effectiveness of a fishway that depends on an auxiliary water supply. The maintenance demand of any fishway with trash racks depends on the nature and quantity of debris in the river and the portion of the total river flow passing through the fishway. Vertical slot fishways with auxiliary water systems recently constructed in the Yakima River, including Sunnyside fishway, require at least daily manual cleaning during periods of heavy debris loads.

Pool-and-weir fishways have much simpler concrete details and lower related construction costs. When they are equipped with auxiliary water systems, they have the same construction and operational demands for trash racks and diffusers as do vertical slot fishways.

Pool-and-chute fishways have relatively simple concrete lines requiring less-expensive forming than vertical slots and have the option of precast or sheet-pile construction. They do not need trash racks, gates, and auxiliary water diffusers, which drastically reduces construction and operating costs.

Conclusions

A hybrid design fishway is proposed to exploit the advantages of several standard currently used styles. Preliminary design standards for pool-and-chute fishways have been developed from a hydraulic model study. The streaming flow in the center of the fishway offers excellent attraction for fish and passage for debris. The streaming flow develops due to the depth of flow over the center portion of the weirs and the relatively shallow pools there.

The plunging flow allows fish passage near the walls and adjacent to the high-velocity attraction jet. Plunging flow is reinforced by the shape of the weir crests, depth of flow over the weir, and the ports. Several geometric criteria are used to ensure that plunging flow is maintained along the side walls.

The concept has an economic advantage over other traditional fishway styles. Design guidelines are based on analytic and empirical hydraulic analyses and observations of existing structures.

Design suggestions presented are considered preliminary. Until further work is done, the pool-and-chute fishway should be used only where conditions match those under which it has been tested to date.

Acknowledgments

Pat Powers helped me understand the hydraulics of the pool-and-chute fishway and designed the model study. Bob Gowen assisted with the model study. Tom Stabno provided the figures for this paper.

References

Bates, K. B., and P. Powers. 1988. Design flow criteria for fish passage. Washington Department of Fisheries, Memorandum, Olympia.

Bell, M. C. 1986. Fisheries handbook of engineering requirements and biological criteria. U.S. Army Corps of Engineers, North Pacific Division, Portland, Oregon.

Chow, V. T. 1959. Open channel hydraulics. McGraw-Hill, New York.

Peterson, D. F., and P. K. Mohanty. 1960. Flume studies of flow in steep, rough channels. American Society of Civil Engineers, Journal of the Hydraulics Division 86(HY9):55–76.

Rajaratnam, N., and C. Katopodis. 1984. Hydraulics of Denil fishways. Journal of Hydraulic Engineering 110:1219–1233.

Rajaratnam, N., C. Katopodis, and A. Mainali. 1988. Plunging and streaming flows in pool and weir fishways. Journal of Hydraulic Engineering 114: 939–944.

Thompson, C. S., and J. R. Gauley. 1962. Further studies on fishway slope and its effect on rate of passage of salmonids. U.S. Fish and Wildlife Service Fishery Bulletin 63:45–62.

U.S. Army Corps of Engineers. 1988. Bonneville second powerhouse Steigerwald Lake wildlife mitigation development, feature design memorandum 41. U.S. Army Corps of Engineers, Portland, Oregon.

American Fisheries Society Symposium 10:278–288, 1991

Recent Adult Fish Passage Projects on Tributaries of the Columbia River

WILLIAM S. RAINEY

National Marine Fisheries Service, Environmental and Technical Services Division
1002 Northeast Holladay Street, Room 620, Portland, Oregon 97232, USA

Abstract.—This paper provides an overview of adult fish passage improvements built on tributaries of the Columbia River as a result of the Pacific Northwest Electrical Power Planning and Conservation Act of 1980. The vertical slot design was used at 21 of 27 sites where fishways were constructed between 1984 and 1990. Adult passage facilities are compared, and the suitability of each type is discussed relative to site constraints. Components of each fishway are described, along with discussions of functional requirements and references to design criteria. Design, operation, and maintenance lessons learned are described.

Since the turn of the century, people have profited tremendously from harnessing the Pacific Northwest's greatest natural resource—water. Some of the least expensive electric power in the world is generated at main-stem and tributary projects in the Columbia River basin. Irrigation projects have turned vast arid wastelands into some of the most fertile and productive farmlands in the USA. Flood control dams have greatly reduced losses of life and property.

The cost of these gains has also been great. Dam construction has delayed or entirely blocked the upstream and downstream migrations of Pacific salmon *Oncorhynchus* spp. and steelhead *O. mykiss*. On the upper Columbia and Snake river systems, 1,100 main-stem river miles of habitat are no longer accessible to fish. The once-productive tributaries above Grand Coulee Dam on the Columbia River and Hell's Canyon Dam on the Snake River no longer contain anadromous fish.

As fish returns approached all-time lows in the late-1970s, the importance of this valuable resource began to be realized. The Pacific Northwest Electric Power and Conservation Planning Act of 1980 established the Northwest Power Planning Council, which was directed to develop a program for fish and wildlife restoration in the Columbia basin (Fish and Wildlife Program). Once this momentum had been established, other federal and state fisheries initiatives were also adopted and implemented.

A key component of the Fish and Wildlife Program is the design and construction of passage facilities for adult fish in tributaries of the Columbia River at irrigation and power dams that are barriers to upstream fish passage. It was predicted that by late 1990 27 new fishways for adults were to be completed at a cost of over US$16 million. Most of these barriers had inadequate fishways built over 40 years ago when fish passage requirements were not well defined. Some, including two natural barriers, had no fishway at all. This paper provides an overview of these passage projects on Columbia River tributaries, then comments on significant design, operation, and maintenance issues and lessons learned.

General design criteria used for the referenced fishways are available in literature cited by Clay (1961) and Bell (1984). Some of these criteria are discussed in more detail herein. The three primary sections of a fishway are the entrance, center ladder, and exit. I refer to these components to contrast the various types of tributary fishway designs employed. Water surface profiles will illustrate the similarities and differences in each type of fishway. The purpose of this discussion is to offer aid in choosing the appropriate type of fishway for a given site.

Types of Fishways

Table 1 indicates the type and location of the various fishways used. The vertical slot fishway was the preferred alternative in most cases, except where there were specific reasons for selecting another type.

Pool-and-Weir or Weir–Orifice Fishways

Roza.—One site where the vertical slot fishway was not used was Roza Dam, where canal diversions are for irrigation and power generation (Figure 1). Static head ranges from 27 to 30 ft for 11 months, when the forebay elevation is constant, and is reduced to 15–18 ft during the 1-month maintenance period. The existing ladder was of the notched-weir-type, with a pool-to-pool flow of 5 ft^3/s and a 1:8 slope. Improvements

TABLE 1.—Fish and Wildlife Program fishway types used on tributaries in the Columbia River basin.

Location	Vertical slot	Pool-and-weir or weir-orifice	Sill
Yakima basin	14	3	2
Umatilla basin	4	0	0
Wenatchee basin	3	0	0
West Fork Hood River	0	0	1

consisted of increasing auxiliary water flow, providing a separate low-reservoir exit leg for operation during the maintenance drawdown period, and making other refinements to bring the fishway up to currently accepted standards.

Comparisons with profiles of other fishways will be made later. Because it was believed the existing fishway could be improved to operate satisfactorily and replacement would have been very expensive, the decision was made to modify the existing facility.

Easton.—Easton Dam is another Yakima basin irrigation facility with a static head of 38–41 ft and constant forebay elevation during the 6-month flow diversion period. During the remainder of the year, head is reduced to 25–29 ft. The old fishway needed replacement due to numerous inadequacies, including a 2-ft drop at each weir. A vertical slot fishway was not selected for the new design at Easton due to the substantial seasonal forebay variation. Instead, it was decided the site was most suitable for a modified Ice Harbor-type weir–orifice fishway with 6-ft-wide and 10-ft-long pools and an exit control section that would operate through the full range of forebay elevations (Figure 2). Orifices in the exit section are 1.5 ft wide and 3.0 ft high. A small auxiliary water valve adds flow at lower forebay elevations to maintain a constant center-ladder flow of 20 ft^3/s.

Comparison.—At both Roza and Easton, ladder flow combines with diffused auxiliary flow in the entrance pool. The total flow is then discharged into the tailrace through an entrance gate to attract adult migrants. Rising tailwater creates a backwater effect in the lower fishway, thereby submerging several of the downstream weirs. Some diffused auxiliary water is added above the entrance pool of each of these fishways during high streamflow periods, which allows maintenance of the minimum required attraction velocity of 2.5 ft/s over each submerged weir. This velocity is necessary to attract adult fish away from the entrance pool diffuser and induce upstream movement within the ladder.

The exit sections are also typical of weir-type fishways and are designed especially for large seasonal forebay variations. There are several ways of buffering large forebay changes so that center-ladder flows remain constant. These include orifice control, separate exit legs, or adjustable weirs.

For the orifice control exit (Easton), each orifice is designed for the maximum drop of 1.0 ft at high forebay. As the drop per orifice decreases with lower forebay elevations, more auxiliary water flow is added to the lowest exit pool to maintain a constant center-ladder flow.

Separate exit legs of the fishway (Roza) are frequently used if there are two primary forebay

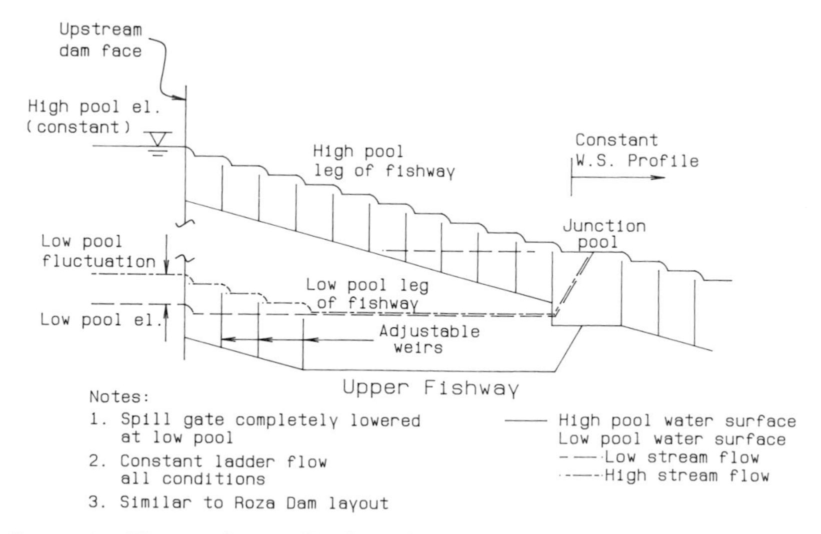

FIGURE 1.—Water surface profile of a typical pool-and-weir fishway with dual exits.

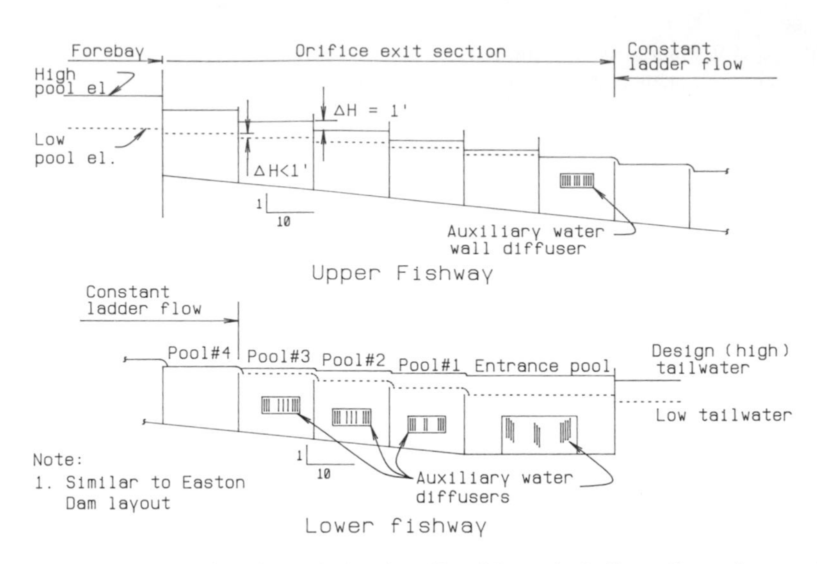

FIGURE 2.—Water surface profile of a typical weir–orifice fishway including orifice exit control section.

ranges during which the fishway must operate. These legs adjoin at a junction pool, below which center-ladder flows are constant.

Adjustable weirs are sometimes employed in the exit sections of larger projects (such as McNary Dam on the Columbia River) to maintain a constant center-ladder flow. This design usually requires sophisticated level control equipment for automated gates. If stop logs are to be installed manually in smaller fishways, constant changes will be required to control center-ladder hydraulic conditions. A slight increase in the forebay elevation may result in excessive flow and turbulence in pools, which retard or block passage. This is the least desirable exit design of those mentioned.

Sill-Type Fishways

For smaller streams, the sill-type fishway should always be considered. Where there is an existing barrier to adult passage, a series of sills can be employed to incrementally backwater the barrier (Figure 3). These sills can be of full- or part-channel width and can be designed to pass fish through a large range of streamflows. The primary benefit of this type of fishway is the reduced operation and maintenance effort necessary to maintain passage, especially with full-width sills. Maintenance is reduced because there are no trash racks to clean or gates to operate. Each weir is designed with a gently sloped V-crest or low-flow notch, which consolidates flow for ease of passage during low-streamflow periods. When flows are high, fish can ascend along the outside extremity of the sills, where there is less turbulence. Sill spacing should generally be 25 ft or more, so there is adequate pool volume to dissipate energy in each pool. Energy dissipation is important and is discussed in more detail later. Fish can usually pass sill-type fishways even when large debris, such as a small tree, partially blocks one of the sills. Sill-type fishways were constructed in 1985 at the West Fork Hood River in Oregon and in 1987 at Salmon Falls on the Little Naches River in Washington. Both of these sites are remote and at natural obstacles. Passage at both locations has opened miles of additional habitat for anadromous fish use.

Vertical Slot Fishway

Vertical slot fishways were constructed at most sites where new fishways were required. Typical barriers are power and irrigation diversion dams with static heads ranging from 4 to 19 ft. The most common design employed 8-ft-wide and 10-ft-long pools with 15-in slot widths and a maximum hydraulic drop of 1 ft per slot. The vertical slot fishway is operationally the most passive with respect to ladder flow. Figure 4 illustrates typical high- and low-flow water surface profiles. Notice the contrast with those shown in Figures 1 and 2. As forebay and tailwater elevations increase with higher streamflows, the hydraulic gradient in the fishway remains approximately the same. Ladder flow increases as pool and slot depths become

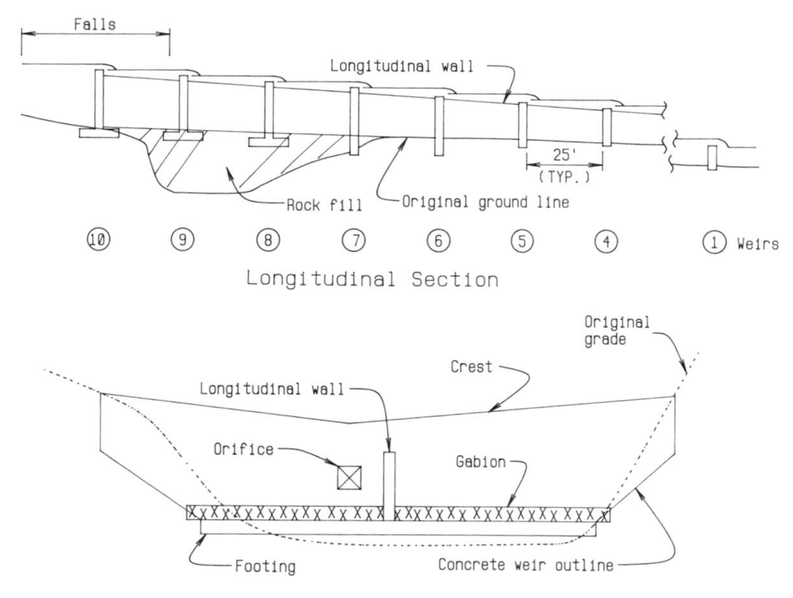

FIGURE 3.—Sill-type fishway.

larger, but pool turbulence remains unchanged because the additional energy is dissipated by the additional pool depth. The result is that the main ladder provides a greater portion of the total entrance attraction flow. This type fishway is not suited to sites where either the forebay or tailwater fluctuates substantially without a corresponding fluctuation of the other water surface.

The formula for the computation of slot flow is

$$Q = W \times D \times C_d \times V;$$

Q = slot flow;
W = slot width;
D = submerged depth (measure from the downstream pool water surface to the slot invert);
C_d = discharge coefficient (0.70–1.10);
V = velocity = $(2g \times \text{slot hydraulic drop})^{1/2}$;
g = gravitational acceleration.

This equation will be discussed in more detail later.

Figures 5 and 6 illustrate typical designs for center and abutment vertical slot fishways.

Auxiliary Water Systems

For each of the larger fishways, auxiliary water systems provide additional flow to improve attraction to entrances. The only exceptions are the sill type and the two smallest vertical slot fishways. The auxiliary systems are essentially the same and include fine trash racks with ⅞-in openings between vertical flat bars of each rack panel.

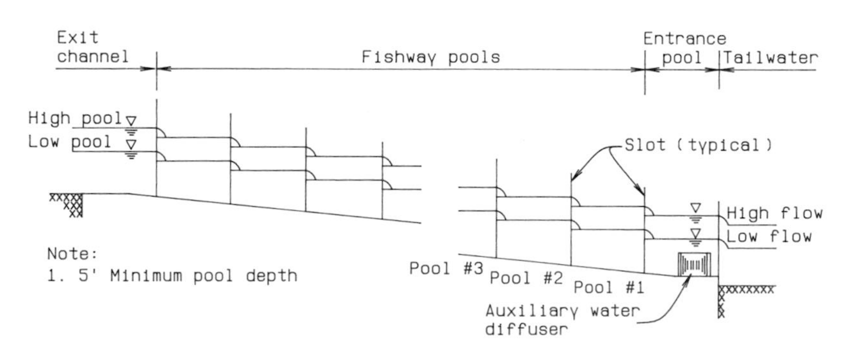

FIGURE 4.—Water surface profile for a vertical slot fishway.

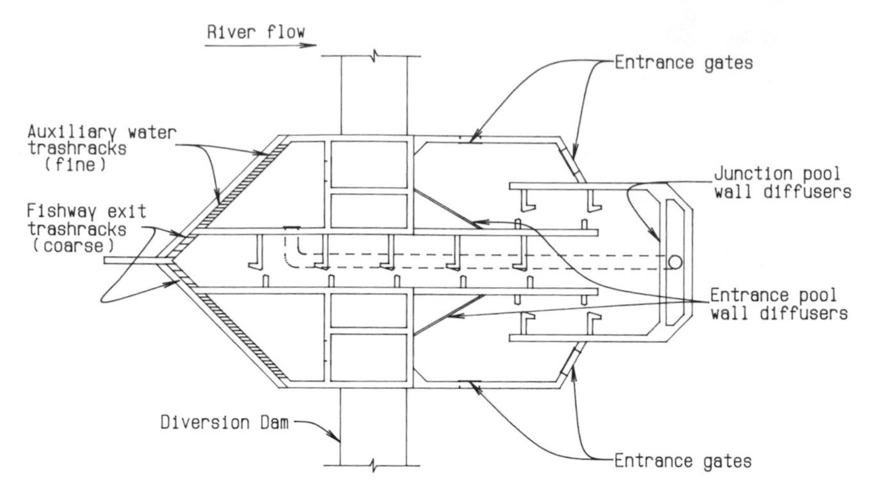

FIGURE 5.—Typical center fishway plan.

These bar spacings are intended to stop smaller debris, thereby reducing additional raking needs lower in each system. Auxiliary water is then routed through a control gate, where it plunges into a stilling chamber. Baffles are frequently employed to further dissipate energy and uniformly distribute flow through the wall diffuser of the main entrance pool. Average wall diffuser velocity should always be as uniform as possible and should not exceed 1.0 ft/s. Excess velocities and surging can induce adult fish to jump, which increases the risk of injuries. Auxiliary and center-ladder transport flow combine to provide fishway entrance attraction flow.

Three existing vertical slot fishways underwent major modifications to increase auxiliary water capacity to a level necessary for satisfactory attraction.

Entrance Requirements

The most important requirement of any fishway is to pass adult migrants safely with a minimum of delay. Most delay occurs in the tailrace of a project, where turbulence associated with spill can obscure relatively low-energy entrance flow. If adults cannot readily find the fishway entrance, excess delay can lead to prespawning mortality (R. Vreeland, National Marine Fisheries Service, unpublished). Impact injuries associated with jumping can cause concussion, lacerations, or abrasions that lead to immediate or delayed injuries and eventual death. The depletion of limited

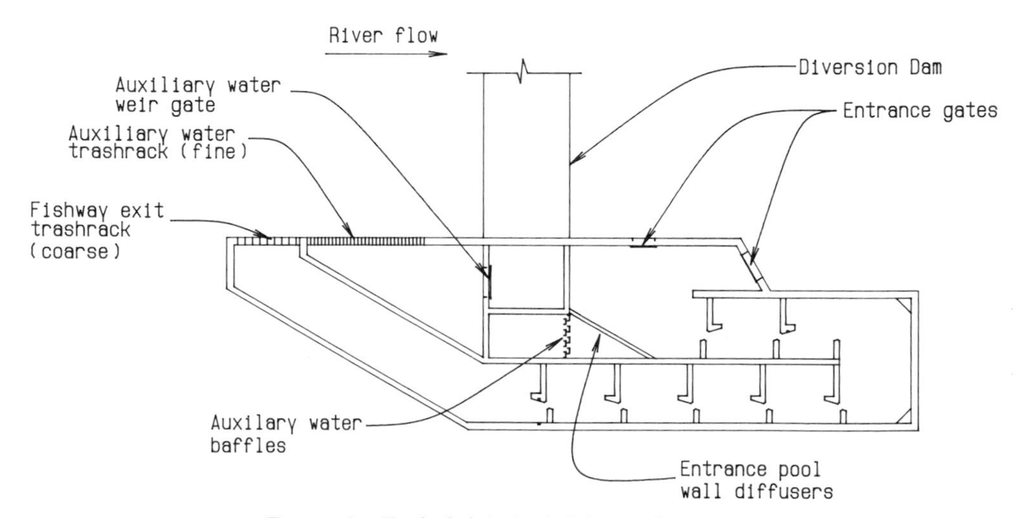

FIGURE 6.—Typical right-bank fishway plan.

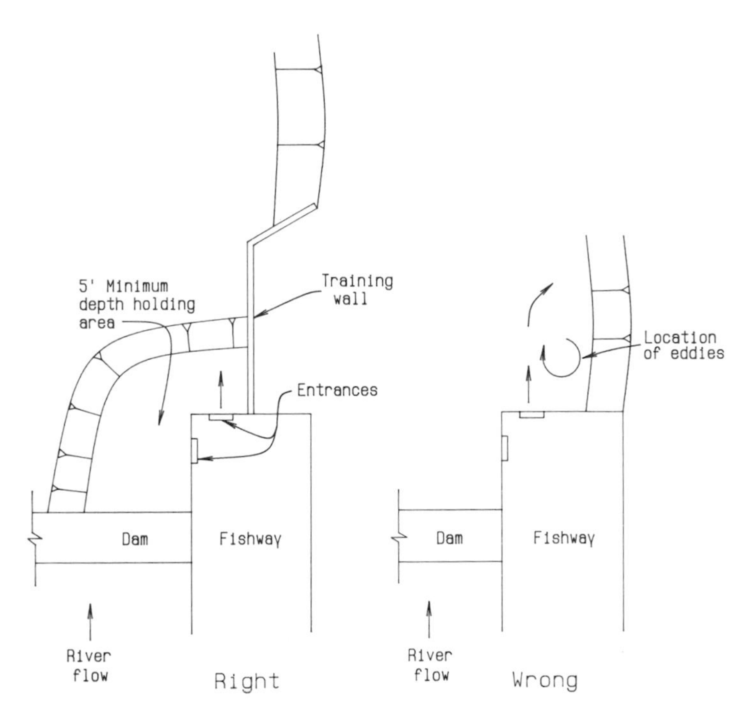

FIGURE 7.—Entrance layouts.

energy reserves associated with numerous failed efforts to pass a barrier may also result in mortality, even if successful passage is ultimately achieved (J. Easterbrooks, Washington Department of Fisheries, personal communication). Excess delay can also prevent fish from reaching cooler water in upper river reaches in a timely manner. If adult fish are delayed until water temperatures in the lower river rise, disease-related mortality will increase.

The most effective way to ensure that entrance flows readily attract adult fish is to jointly consider both behavioral and hydraulic aspects of fish passage. Migration routes and holding patterns below each site should be understood. Field observations by biologists verify that horizontal distribution depends on channel characteristics at lower streamflows, but fish tend to follow the shorelines during high-flow periods. Once migrants reach the dam or barrier, they generally concentrate in less turbulent water at the edge of a highly aerated zone, from where they make periodic ascents toward the barrier. If the fishway can be located near these anticipated holding areas, entrance attraction flows can be routed accordingly for optimum passage results. In some cases, hydraulic conditions must be altered to create desirable holding areas. This can include minor dam crest modifications, tailwater channel excavation, or other measures. The presence of tailwater eddies near the fishway entrance can significantly increase delay. Eddies may cause fish to become confused and disoriented. A downstream retaining wall configuration (Figure 7) has effectively damped eddies near the fishway while providing a guide wall for fish to move along the shoreline and directly into the entrance. The offset of this wall relative to the low-flow entrance wall should be minimized. Excavation next to the entrance is also shown. For these relatively low barriers, a minimum 5-ft pool depth outside the entrance is essential because it provides excellent holding water, aids in reducing spill turbulence, and reduces passage delay. Shallow, high-velocity conditions near the entrance should be avoided.

For the Yakima, Umatilla, and Wenatchee fishways, low-flow entrances are designed to discharge laterally, just downstream from the limit of aerated water below the spillway. High-flow entrances are generally oriented at a 45° angle or parallel to streamflow, depending on the site. Some fishways were designed with midrange entrances, primarily where energy dissipation below the dam extends a considerable distance downstream during higher spill periods. A vital part of the entrance layout for each project was the analysis of tailwater energy dissipation through a range of streamflows. Predesign observations of on-site tailwater conditions at various streamflows were important in this process.

Hydraulic aspects of the entrance attraction flow jet are very important and should also be understood. The entrance flow jet must have an adequate combination of flow and hydraulic drop (e.g., mass and velocity) at the entrance to create momentum adequate to overcome tailwater turbulence. These designs were based on the premise that flow should be more shooting (horizontally oriented) than plunging (vertically oriented) if fish are to perceive the attraction jet at the greatest distance from the entrance. Entrance gate dimensions were sized with slightly greater height than width, and set at an elevation at which the top of the gate blockout is at or slightly below the low tailwater elevation. This contrasts with narrow, deeper entrances that vertically elongate the jet. Tailwater turbulence will dissipate the energy of a narrow jet more readily than the energy of a more compact one.

For these projects, entrances were designed as orifices rather than weirs. Most of the entrances were 3 ft wide and 5 ft high, but some were up to 4 ft wide and 6 ft high. This converts to attraction flows of 90–180 ft^3/s, based on the equation

$$Q = W \times D \times C_d \times V;$$

W = gate width;
D = submerged depth (measured from tailwater to gate invert);
C_d = contraction coefficient;
V = velocity = $(2g \times \text{entrance hydraulic drop})^{1/2}$;
g = gravitational acceleration.

The desired entrance hydraulic drop range is 1.0–1.5 ft. This converts to a velocity of 8.0–9.8 ft/s at the vena contracta, which is the point just downstream from the entrance where velocity is the highest. C_d depends on the entrance pool layout, entrance size, type of fishway, and tailwater elevation. It typically ranges from 0.6 to 1.05, depending on approach velocity upstream from the entrance.

Entrance gates for these fishways are operated either entirely open or closed and do not require frequent adjustments, such as would be necessary for a weir gate-type entrance. The low-streamflow entrance should be closed in favor of the downstream high-flow entrance when river stage increases sufficiently to require it. This greatly reduces the labor commitment and the need for water level control equipment.

The question of how much entrance flow is adequate is frequently asked. Certainly a higher entrance flow, discharged in a shooting manner into a particular tailwater zone, will attract fish more readily than a smaller flow. However, if this zone is not one that is hydraulically satisfactory for fish holding, delay may still occur. Therefore, it is a combination of location of the entrance and the amount and velocity of the attraction flow jet that is important. At the projects referenced in this paper, total attraction flow ranges up to 10% of total streamflow at mid to high river stage. Each larger fishway also has two to four entrances that provide the necessary operational flexibility to ensure that attraction flow is discharged at the optimum tailwater location. Relative to the total cost of each fishway, the addition of one or two additional entrance gates is not a significant expenditure if it ensures optimum attraction.

Lessons Learned

Design

Pool volume requirements for vertical slot fishways.—The basic hydraulic premise in the design of pool-type fishways is that potential energy from each pool is converted to kinetic energy as flow plunges into the next lower pool. Ideally, the receiving pool has adequate submerged water volume to fully dissipate this kinetic energy. If the pool is undersized for the quantity of flow and the hydraulic drop, excessive turbulence will occur, and some velocity energy will carry to the next pool. This causes higher flow and increasing turbulence in lower pools and can reduce important fish-resting volume, thereby delaying or completely blocking passage. The formula that relates these turbulence parameters is

$$Q \times D \times H/\text{Vol} \leq 4.0;$$

Q = slot flow;
D = density of water;
H = slot hydraulic drop;
Vol = submerged pool volume of receiving pool.

The empirical coefficient 4.0 is based on hydraulic and biological observations. Typical pool dimensions of most of the referenced vertical slot fishways are illustrated in Figure 8. Note that 15-in slot widths were employed, rather than the 12-in slots commonly employed in Canada. Designs were for a maximum hydraulic drop per pool

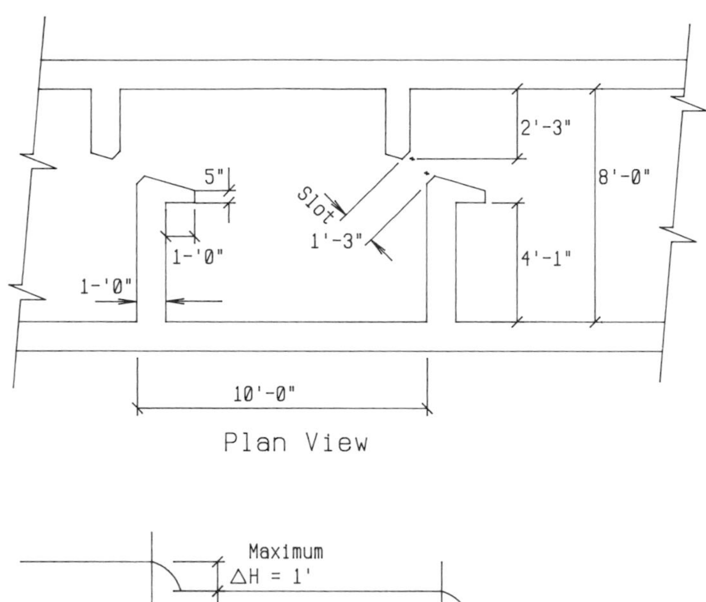

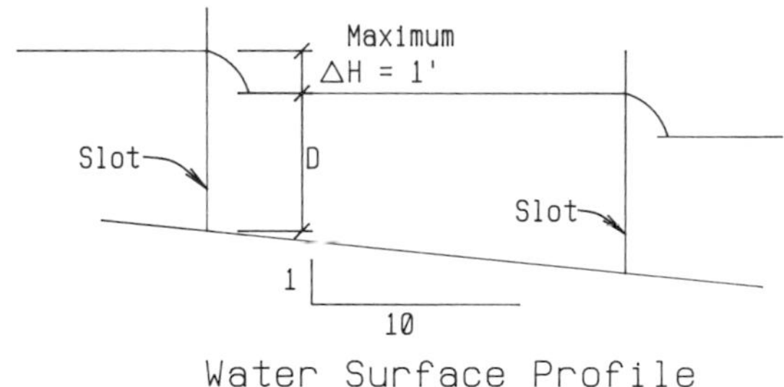

FIGURE 8.—Typical 8-ft-wide and 10-ft-long vertical slot fishway.

of 1.0 ft. For this size fishway, equation (3) gives a coefficient value of 6.3 if the average receiving pool depth is 5.5 ft, the actual pool length is 9.5 ft, and a discharge coefficient of 0.85 is employed in the determination of Q. By comparison, if a 12-in slot is employed and all other parameters are the same, the coefficient drops to 5.1, which still exceeds the empirical 4.0 value.

For the fishways discussed herein, 15-in slots were used to route as much flow through each pool as possible. It was believed that coarse trash racks at the fishway exits would plug less readily than finer trash racks at the auxiliary water intakes, and that higher flow from pool to pool would provide better entrance attraction if auxiliary flow were reduced during high streamflows due to debris problems. It was also believed that the additional slot width would reduce delay in ascent by larger chinook salmon, although this has not been confirmed. There has been no formal assessment of whether 12-in slot widths should have been used, rather than 15-in slots. For most of the fishways, the maximum head of 1 ft occurs only during a narrow range of streamflows, and there is considerable turbulence in each pool. Slot head is typically reduced slightly during the other streamflow periods, which reduces pool turbulence. Counting stations at several of the fishways indicate that fish continue to pass. Delay caused by use of the larger slots has not been apparent. For future designs, it may be prudent to slightly reduce the maximum head per slot or to employ the 12-in slot width. Specific site features and functional requirements will dictate modifications.

Slot discharge coefficient.—The discharge coefficient C_d, from equation (1), is a very important parameter in the design of any type of fishway. If the designer assigns an excessively low value, pool-to-pool flow may be higher than anticipated, and pool turbulence may be excessive. If water is to be conserved, such as at an irrigation or hydropower diversion, fishways will be carefully monitored to ensure excessive flows are not being used. In some instances, choice of an inappropriate C_d may result in the need for a postconstruction modification.

At Sunnyside Diversion Dam on the Yakima

River, the U.S. Bureau of Reclamation measured flow at the right-bank fishway during the peak irrigation demand of the low-streamflow period in 1985. Flashboards were in place on the dam crest, which resulted in the highest average head for each of seven slots at 11 in. Backwater calculations indicated an average slot C_d of 1.07. This was much higher than the range of 0.67–0.84 indicated by Clay (1961), which is commonly employed. Two reasons for this discrepancy were considered. One pertains to a slightly different definition of depth in the application of equation (1). Although the equation is of the exact same form, and Clay (1961) referred to slot depth at a location immediately upstream, rather than downstream, of the slot. The other explanation pertains to the use of a 15-in slot, rather than the previously more common 12-in slot. As indicated by Rajaratnam et al. (1986), the value of C_d is very sensitive to slot geometry. In this case, the greater slot width results in reduced ability of this size pool to dissipate energy, which passes kinetic energy to the next lower pool. The result is a higher slot flow, based on the higher C_d.

Based on a series of field measurements, it was also concluded that for a given vertical slot fishway with a 1:10 gradient, C_d increases if there is a concurrent increase in the average slot depth and hydraulic gradient. It can also be extrapolated that for a fishway pool of a given size, C_d and Q (slot flow) will increase as slot width increases. For fishways of slightly different pool sizes but with the same slot width, the smaller pool can be expected to have a higher C_d when the slot head is constant.

These general conclusions are offered with qualifications. One example of how hydraulic conditions may be altered by conditions in the field pertains to the difficulty in constructing slots of uniform width. Unless templates are employed, one can expect slot width to vary by up to 2 in for normal concrete construction, especially if slot tolerances are not emphasized.

Tumwater fishway slot modification.—Tumwater Dam, on the Wenatchee River in central Washington, was the highest vertical slot fishway constructed, having a maximum head of 19 ft. Due to site constraints, a significant cost savings was realized by increasing the vertical slot fishway gradient to 1:8. If this design modification had not been used, there was some doubt whether funds were available to build the new fishway at all. It was decided that 15-in slots would still be employed, and that volume criteria could be satisfied by increasing the pool width from 8 to 12 ft. All other criteria were the same as for the other 8-ft-wide and 10-ft-long vertical slot fishways being constructed.

A valuable lesson was reinforced at Tumwater concerning the importance of slot jet orientation. When the fishway was operated initially, a pronounced transverse oscillation with a 3-ft amplitude occurred in each pool. The primary problem related to the slot jet centerline orientation. Clay (1961) pointed out the tendency for this jet to turn directly downstream toward the next slot, and that this tendency is the basis for sizing of baffle components. Thus, when the pool width of 12 ft was chosen, the slot jet should have been directed farther away from the downstream slot and more into the primary energy dissipation area behind the downstream baffle. As it was, a range of slot jet orientations occurred—from A to B in Figure 9—during the oscillation cycle. Clay also referred to transverse wave oscillation at the McNary Dam fishway during initial operation, where the wave amplitude was 8 ft. These factors should be considered when accepted pool dimensions or slope must be changed at a proposed fishway site to accommodate site constraints.

The solution at Tumwater was to alter the jet direction and reduce energy to be dissipated in each pool (Figure 9). By extending the stubwall of each slot a distance of 4 in, the width was reduced to 12 in, and the jet centerline direction was changed. A 12-in-high sill was also added, which reduced slot flow depth. Reduction of slot width and depth reduced flow proportionally, thereby reducing kinetic energy imparted to each pool. Oscillations might have been reduced by only changing the direction of the slot jet. However, it was decided that the more prudent approach was to also reduce slot flow, since time and funds were limited. Fortunately, the contractor was still on site and the modification was completed in a matter of weeks. The result was elimination of the oscillation, and provision of satisfactory hydraulic conditions in the pools.

Operations and Maintenance

A point that cannot be overemphasized is the need for competent and committed operations and maintenance staff. The referenced fishways require frequent (sometimes daily) visits to rake trash racks, conduct routine maintenance checks, and ensure that operation of the facility is within criteria. Without this level of oversight, these

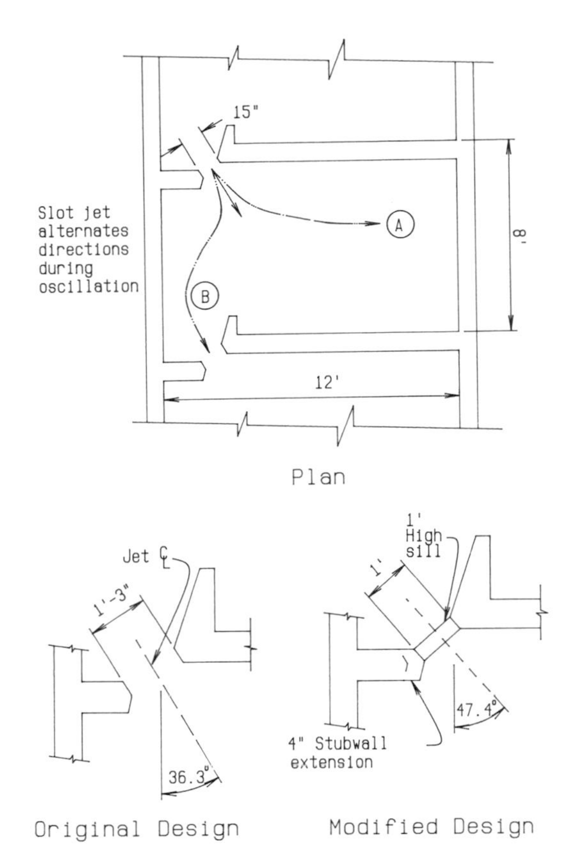

FIGURE 9.—Tumwater fishway, showing the original design (top and lower left) and subsequent modifications (lower right).

facilities will not operate as required to optimize passage.

Debris.—Many people underestimate the operational problems caused by debris accumulations on trash racks. At sites perceived to have greater potential for debris problems, automated debris rakes were employed. Even though this is an expensive alternative, it can be the most economical approach in the long run. The designer should carefully consider historical debris-loading trends in a particular watershed before making a final decision on this issue.

The designer should also consider innovative ways to reduce labor requirements at fishways. The auxiliary water trash racks at these projects consist of flat bar panels, with ⅞-in clear spaces. These are the first to clog with debris and are the most labor intensive to clean.

A design recently employed at Cape Horn Dam on the Eel River, California, greatly reduces labor requirements (Figure 10). It used slightly inclined stainless steel profile-wire screens at a lowered length of dam crest adjacent to the fishway. Cleaning is by periodic manual sweeping or by air-burst releases of compressed air from a manifold under the screens. This option is safer for juvenile fish passage, and experience during winter 1987–1988 indicated the air-burst cleaning system was used infrequently.

Sediment.—Care was taken in the preliminary design phase of each project to locate the auxiliary water intakes and fishway exits where sediment entrainment would be minimized. Some projects had sluice gates adjacent to coarse and fine trash racks in the forebay. Sediment has not caused operational problems to date. None of the fishways have closed because of sediment accumulations.

Operating criteria.—A concise list of operating criteria should be developed prior to the initial operation of each fishway. It should be brief and easily understandable. Staff gages should be located upstream and downstream from trash racks so operators can monitor head differential and rake when necessary. Staff gages should also be located upstream and downstream from entrances to allow necessary auxiliary gate adjustments to ensure the appropriate hydraulic drop. In the Yakima, Umatilla, and Wenatchee basins these gages are the basis for all maintenance and operational activities. Criteria should be reviewed and updated as operational experience is gained.

Organization

The organization of the Yakima Fish Passage Technical Work Group was a vital early step in coordination of the design and construction of these projects. The U.S. Bureau of Reclamation, Bonneville Power Administration, irrigators, federal and state fish agencies, and Indian tribes are represented and have continued to meet every 4–6 weeks since 1983 to discuss biological and technical design-related issues. The Bureau of Reclamation has taken the lead in coordinating meetings, conducting geotechnical and hydrological site investigations necessary for design, and completing preliminary design memorandums based on input from the fish agencies and tribes. Detailed designs were completed by the bureau or by private consultants and reviewed again by work group members. This organization has worked very efficiently in combining available resources to develop successful fish passage designs. All entities are represented, and participants are able to apprise their constituents of pertinent material discussed at each meeting.

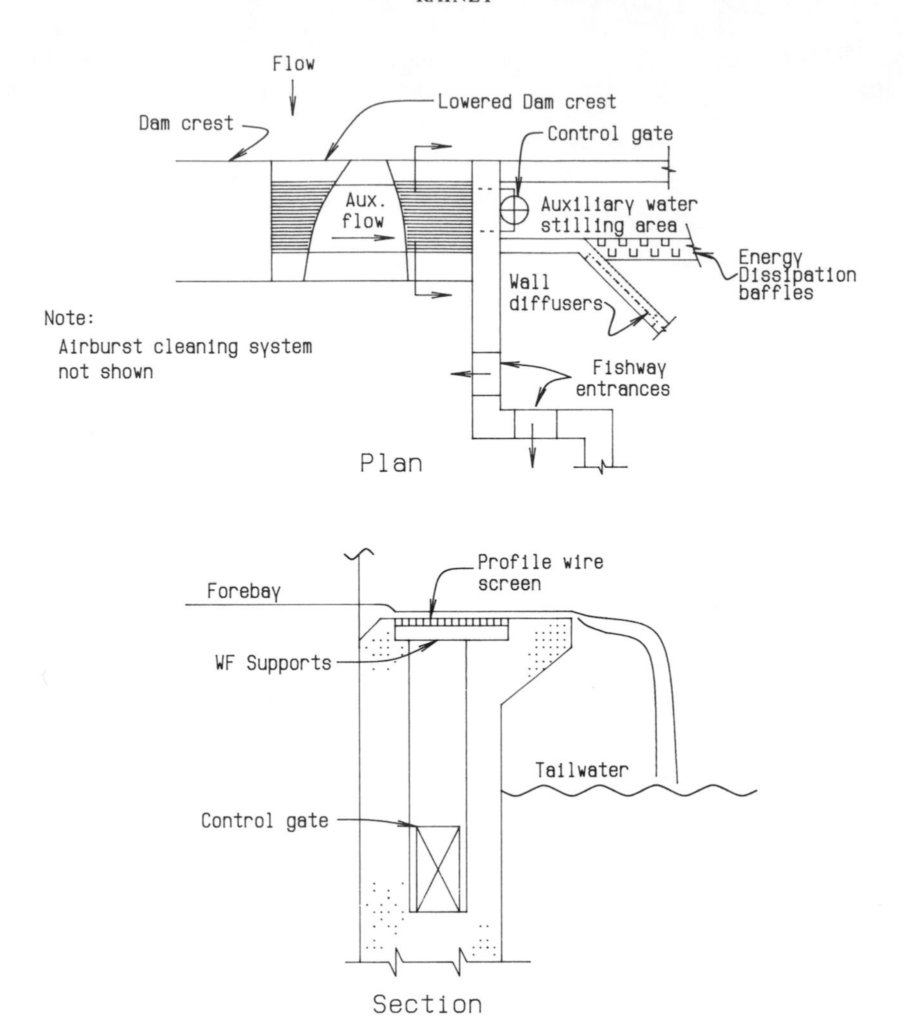

FIGURE 10.—Design for alternative auxiliary water intake.

Summary

The Northwest Power Act of 1980 mandated initiatives to rebuild fish runs in the Columbia Basin. Fish passage improvements on the Yakima, Umatilla, and Wenatchee rivers have been among the most important measures of the Northwest Power Planning Council's Fish and Wildlife Program. This paper provides an overview of adult passage projects on these tributaries of the Columbia River. A wealth of experience has been gained as a result of this ambitious design and construction program. Many of the insights presented herein can be used in planning and designing new fishways at other locations.

References

Bell, M. C. 1964. Fisheries handbook of engineering and biological criteria, 2nd edition. Report to U.S. Army Corps of Engineers, North Pacific Division, Portland, Oregon.

Clay, C. H. 1961. Design of fishways and other fish facilities. Canada Department of Fisheries, Ottawa.

Rajaratnam, N., G. Vander Vinne, and C. Katopodis. 1986. Hydraulics of vertical slot fishways. American Society of Civil Engineers, Journal of the Hydraulics Division 112:909–927.

American Fisheries Society Symposium 10:289–298, 1991

Power and Energy Implications of Passage Structures for Fish

C. E. BEHLKE
University of Alaska Fairbanks
Post Office Box 82230, Fairbanks, Alaska 99708, USA

Abstract.—Fluid mechanic equations are used to show effects of virtual mass force, non-Archimedean buoyant force, and profile drag force on fish in several fish passage structures. Example problems are worked to show computational procedures for calculating net propulsive force, net power, and net energy necessary for fish to swim in a lake, up a steep chute, and through the outlet, barrel, and inlet of a culvert.

Hydraulic Forces Affecting Swimming Fish

Buoyant Force

Fish passage engineers and others responsible for design of fish passage structures have generally assumed that fish surrounded by water are buoyed by a force equal to the weight of the volume of water displaced by the fish, and that the force is directed vertically upward. Thus, weight and buoyant forces appeared to cancel, and both were ignored. Behlke (1987) has shown that this is not the case in fish passage structures where water flows. Fundamental laws of fluid mechanics state that at any point in the fluid, the buoyant force per unit volume of fluid displaced is equal but opposite in direction to the vector gradient of the pressure, ∇p. The buoyant force (B) acting on a fish then is

$$B = (-\nabla p)(\mathrm{Vol}); \qquad (1)$$

∇p is the pressure gradient that formerly occurred in the undisturbed fluid at the instantaneous location of the swimming fish's volumetric centroid and Vol is the volume of the fish's body. In a lake, where hydrostatic pressure conditions exist, $\nabla \mathrm{p} = \gamma$, the specific weight of water. Thus, the buoyant force would be directed vertically upward and would simply be equal to the weight of the volume of water displaced—the classical Archimedean buoyant force. However, Behlke (1987) showed that if a fish swims up through uniform, steady flow in an open channel that slopes at an angle θ with the horizontal, $-\nabla p$ is reduced from the previous value by a factor $1 - \cos(\theta)$, θ being the slope of the channel. Also, the buoyant force is directed normal to the sloping water surface (Figure 1). Thus,

$$B = \gamma(\mathrm{Vol})\cos(\theta). \qquad (2)$$

It should be noted that B is further reduced for fish swimming in aerated water because of a lesser γ.

An interesting example of how B varies in magnitude and direction is that of a free overfall (perched culvert outlet) for subcritical approaching flow. Rouse (1938) showed the pressure distribution in the vicinity of a free overfall for a rectangular channel (Figure 2). If it is assumed that a similar pressure distribution occurs in the longitudinal centerline plane of a culvert at its free overfall, resulting buoyant forces on fish attempting to swim upstream at various locations in the overfall water can be represented as in Figure 3. Clearly, buoyant forces should be considered carefully in design, and great caution must be exercised when buoyant forces acting on fish in passage structures are evaluated.

Profile Drag Force

Traditionally, engineers have been trained to evaluate drag forces on bodies of fixed shape that do not carry their propulsion systems with them. Thus, drag coefficients are determined by laboratory measurements on objects fixed in space so that force measurements are made through the fixation system. Measurements are made at constant fluid velocity. The drag, so measured, is termed profile drag. It is these drag coefficients that adorn engineering fluid mechanics textbooks. However, the development of a system for measurement of drag forces on a body that carries its propulsion system with it is an interesting, challenging, and (to the best of my knowledge) futile exercise.

Biologists have attempted to determine the profile drag on swimming fish by converting measured oxygen inputs to energy inputs (e.g., Brett 1973). However, the assumptions, calculations, and measurements necessary to get from energy input to final net propulsive force, and therefore profile drag force, have not yielded satisfactory results (Webb 1975). Thus, some biologists (and a few mathematicians), well trained in fluid me-

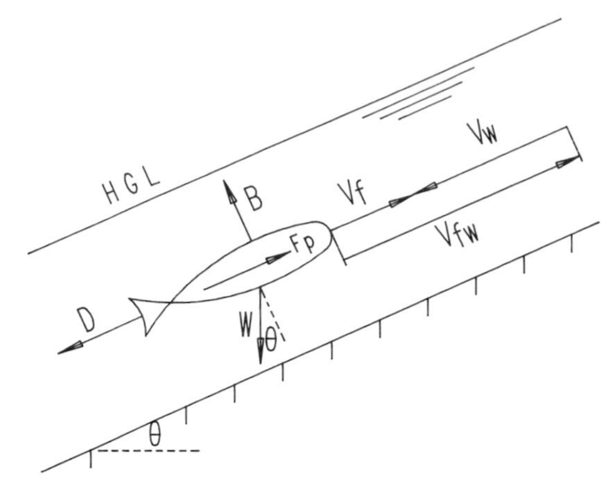

FIGURE 1.—Forces acting on fish swimming in uniform, steady flow in an open channel. B = buoyant force; D = drag; F_p = net propulsive force; HGL = hydraulic grade line; V_f = velocity of the fish; V_w = velocity of the water; V_{fw} = velocity of the fish with respect to the water; W = weight of the fish; θ = angle of the channel with respect to the horizontal.

chanics, have attempted to determine profile drag forces on fish by studying how fish appear to swim and then applying fundamental fluid mechanic concepts to evaluate quantitatively the fluid-generated drag forces on fish. These fluid-dynamicists certainly do not assert that they have the answers, but it is my observation that they, not engineers, are the leaders in this area of research. Thus, I will attempt to use their works to evaluate profile drag and to discuss later the summing of forces that a fish must overcome if it is to move through a fish passage structure.

For an outstanding treatise on swimming hydrodynamics and energetics of fish see Webb (1975). Briefly, however, biologists are presently using as

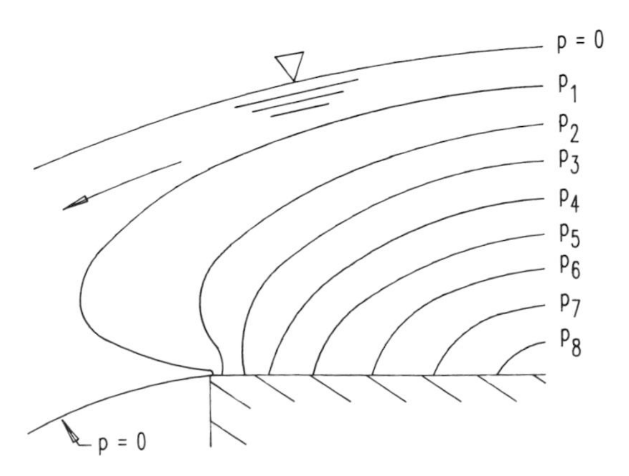

FIGURE 2.—Pressure (p) distribution in vicinity of two-dimensional over-outfall. (After Rouse 1938.)

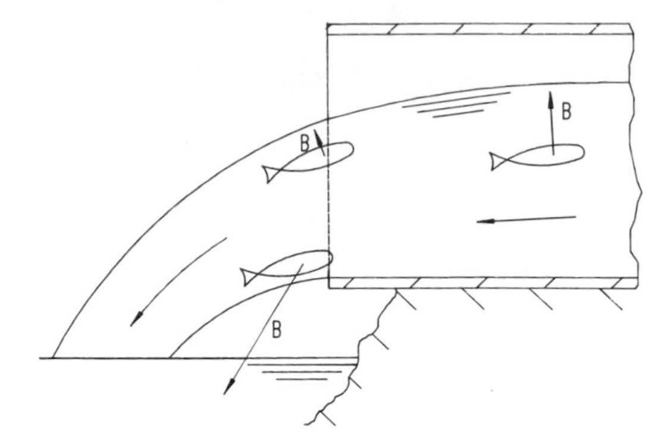

FIGURE 3.—Buoyant force (B) on a hypothetical fish at various locations in the vicinity of a free overfall at the outlet of a culvert.

a reference the drag generated by a turbulent boundary layer on a flat plate, the plate being as long as the fish and of sufficient width to have the same wetted area as that of the body of the fish. This is a severe departure from the traditional engineering use of the projected area (A) of a body in the profile drag equation,

$$D = C_d \rho A V^2/2; \tag{3}$$

C_d is the profile drag coefficient, which depends on Reynolds number (N_R) and the shape of the body for which the drag is to be calculated; ρ is the mass density of the fluid; and V is the velocity of the fluid with respect to the body.

Biologists (Webb 1975) have adopted as their standard (and I recommend it to design engineers) the following:

$$D = C_d \rho S V_{fw}^2/2, \tag{4}$$

for which

$$C_d = k(0.072)N_R^{-0.2}. \tag{5}$$

Here, S is the surface area (not projected area) of the fish, V_{fw} is the velocity of the fish with respect to the water, k is a constant that converts the reference drag coefficient to that of swimming fish, and N_R (the fish's Reynolds number) = $V_{fw}L/\nu$, ν being the dynamic viscosity of the water and L the fish's length. Biologists generally believe that the drag on a swimming fish is 3 to 5 times that of the flat-plate reference, so k varies from 3 to 5 (Webb 1975), depending apparently on the fish. Engineers might prefer 0.074 instead of 0.072 as the constant in equation (5); however, this is a minor point that is obscured by other uncertainties of the question of profile drag on swimming fish.

Equations (4) and (5) assume that the fish swims directly into the oncoming current if the water is moving.

If $S = bL^2$ and equations (4) and (5) are combined, profile drag may be expressed as

$$D = bk(0.072)(\rho)(\nu^{0.2})L^{1.8}V_{fw}^{1.8}/2. \quad (6)$$

Virtual Mass Force

If an object is accelerated in a fluid, if the fluid surrounding an object accelerates, or if both the object and the fluid accelerate, Newton's second law is operative. Because an object accelerating with respect to its surrounding fluid carries some of the fluid with it, an added mass is accelerated with the object. Thus,

$$F_{vm} = (M + M_a)a; \quad (7)$$

F_{vm} is the force necessary to accelerate the object and the fluid that accelerates with it; M is the mass of the object; M_a is the added mass of the fluid accelerating with the object, and a is the acceleration of the object with respect to the surrounding fluid (Daily and Harleman 1965). The term $(M + M_a)$ is called the virtual mass of the object. In relation to fish, M_a is assumed to be approximately 0.2 M (Webb 1975). Thus, the virtual mass force which a fish feels in an acceleration situation is

$$F_{vm} = 1.2\, Ma_{fw}, \quad (8)$$

a_{fw} being the acceleration of a fish with respect to the surrounding water. F_{vm} is in the direction of a_{fw}.

In one-dimensional motion, the acceleration term a_{fw} of equation (8) is

$$a_{fw} = V_{fw}\partial V_{fw}/\partial s + \partial V_{fw}/\partial t; \quad (9)$$

s is distance measured along the streamline that the fish follows in swimming into the current. Usually, fish passage engineers would not be confronted with unsteady flow, so the final term of equation (9) is ignored. It then becomes necessary to evaluate the simplified form of the right side of the equation. I suggest that the equation be put in the following finite difference form:

$$a_{fw} = V_{fw}\Delta V_{fw}/\Delta s; \quad (10)$$

ΔV_{fw} is the difference in V_{fw} between two points Δs distance apart, and V_{fw} may be approximated by the average V_{fw} over the distance Δs.

The virtual mass force usually acts against forward progress, especially at the outlets of culverts if water surface drawdown occurs there (subcritical approaching flow), at the inlet of culverts, in waterfalls where water is free-falling, and through slots and orifices where water accelerates. Also, in the leaping process fish may accelerate sharply, so during the in-water part of that process F_{vm} is an important force opposing motion.

Acceleration situations, which result in the presence of virtual mass forces, also appear to have an additional effect that hinders passage efforts of fish. Unpublished data (P. W. Webb, University of Michigan–Ann Arbor, personal communication) indicate that water or fish accelerations have some interference effects that elevate profile drag and thus may increase the value of k in equation (6) by a factor of 2 or 3. This is startling information that sounds a loud word of warning to design engineers. In most fish passage facilities the design engineer can control water accelerations in locations where fish must swim if they are to negotiate the structure. If at all possible, provisions should be made for fish to avoid zones of downstream-directed water accelerations.

Weight

Fish may change their volumes slightly, thus changing somewhat their specific weight. This process is generally slow, so it probably has little effect on most fish in passage structures. For fish of similar shape and specific weight, weight (W) is proportional to L^3. Specific weight of fish is usually assumed to be that of water.

Because buoyancy often does not completely cancel weight in fluid dynamic situations found in fish passage facilities, fish weight and buoyancy should always be considered jointly in design until it can be shown that they do cancel.

Other Forces

Yawing, centripetal, and turbulent forces and the effects of mucus on drag reduction are not considered here. I have found little information on these topics that can be converted to design principles. However, as the work of bio-fluid-dynamicists progresses, engineers can expect results that may well be incorporated in future design criteria and procedures.

The principal forces acting on fish in flowing water, and their variation with fish size, are summarized in Table 1.

TABLE 1.—Summary of how forces vary with fish length (L), and fluid and dynamic variables for a species of fish.

Force	Determining fish variable	Fluid or dynamic variable
Buoyant	L^3	∇p, γ
Profile drag	$L^{1.8}$	$V_{fw}^{1.8}$, ρ, ν
Weight	L^3	γ_f
Virtual mass	L^3	$a_{fw}\rho_f$

Propulsive Force, Power, and Energy

Propulsive Force

The net propulsive force (F_P) that a fish must generate in order to pass through an element of a fish passage structure is here assumed equal to the sum of the components, in the direction of motion, of the previously discussed forces. However, in some situations the fish must also generate a propulsive component normal to the direction of motion, because of a noncancellation of weight and buoyant force components in that direction (or because of centripetal forces ignored here). That situation is probably most pronounced for fish attempting to enter the mouth of a perched culvert with subcritical flow in the barrel. For a detailed explanation of this and many of the succeeding equations see Behlke (1987) and Behlke et al. (1988).

Some examples follow of net propulsive forces a fish must deliver if it is to move ahead. If it swims upstream in an open channel, culvert barrel, or ramp where uniform, steady flow occurs, profile drag, buoyant, and weight forces impede its forward motion. Behlke (1987) showed that the fish's buoyant force acts normal to the water surface and cancels its weight component normal to the invert but does not cancel the downslope component of weight (Figure 3). Thus,

$$F_P = D + W\sin(\theta), \tag{11}$$

θ being the angle of slope of the channel, water surface, and hydraulic grade line (HGL). If a fish swims through a horizontal, baffled pipe (enclosed flow) or in an open channel inlet or outlet where it swims horizontally but the HGL slopes at angle θ, the buoyant force is greater than the fish's weight and has a downstream component that depends on the slope of the HGL. Here too, B is directed normal to the HGL; thus the propulsive force becomes

$$F_P = D + W\tan(\theta). \tag{12}$$

Where an enclosed (pipe) ladder slopes at an angle ϕ or where a fish follows a streamline at an open channel inlet or outlet that slopes at angle ϕ while the HGL slopes at angle θ,

$$F_P = D + W\{\sin(\phi) + [\cos(\phi)][\tan(\theta - \phi)]\}. \tag{13}$$

A fish attempting to enter a culvert that draws down at its outlet from subcritical flow in the barrel is attempting to move in a zone of accelerating water. If the outlet pool's elevation is sufficient to maintain culvert water depth at or above critical at the outlet lip, approximate hydrostatic conditions exist there, and the HGL slopes approximately at the slope of the water surface. Here the fish is faced also with a virtual mass force in addition to the two forces of equations (12) and (13). Thus, if the streamline (relatively straight) which the fish selects to follow slopes at angle ϕ, then

$$F_P = F_{vm} + D + W\{\sin(\phi) + [\cos(\phi)][\tan(\theta - \phi)]\}. \tag{14}$$

Inspection of this equation and the previous ones that define each of the terms in the right side of the equation (14) reveals that a good deal of information must be measured, calculated, known, or guessed before the equation can be evaluated. W requires a knowledge, usually, of the length of the smallest design fish and its weight–length characteristics. D requires the fish's length, its ratio of surface area to L^2, its velocity with respect to the moving water, and the approximate temperature of the surrounding water. F_{vm} requires the same information as that used to determine W, and it is necessary to know V_{fw} and how it changes through the acceleration zone being studied—for example the vicinity of the outlet or inlet of a culvert.

Power

The instantaneous net propulsive power (Pwr) that a fish delivers to its surroundings is

$$\text{Pwr} = F_P(V_{fw}). \tag{15}$$

Because $V_{fw} = V_w + V_f$, equation (15) can be expressed as

$$\text{Pwr} = F_P(V_w + V_f). \tag{16}$$

Energy

Net energy delivered by a fish in passing through an element of a complex passage structure is

$$E = \int_0^t \text{Pwr}\ dt; \qquad (17)$$

t is the time spent by the fish in traveling through the element. If the fish swims with constant V_f, t equals s/V_f, s being the distance traveled in the element. Thus, if Pwr is described by equation (16) and is not a function of time, equation (17) becomes

$$\begin{aligned} E &= \text{Pwr}\ (V_w + V_f)(s/V_f) \\ &= s\ (\text{Pwr})(1 + V_w/V_f). \qquad (18) \end{aligned}$$

The term V_w/V_f is the price fish pay for swimming. Tilsworth and Travis (1987) reported a 43-min travel time for passage of a single Arctic grayling *Thymallus arcticus* through a culvert of 33.5-m length, so V_f was 0.012 m/s. The water velocity where the fish swam was approximately 0.7 m/s, so $V_w/V_f = 0.7/0.012 = 56.4$. Because it swam and was unable to walk in solid contact with the culvert invert, this fish delivered at least 57.4 times as much energy as it would have if it could have walked through the culvert. (The fish's situation is similar to that of a person who is running to progress slowly upward on an escalator that is moving rapidly downward.)

Equation (18) clearly shows that the faster a fish moves through an element of a structure (fast V_f), the less energy it uses in doing so. On the other hand, it must deliver more power to move quickly through the element. My observations are that fish attempt to get through the most difficult spots in a passage structure as quickly as possible, so they seem to understand equation (18). It is also my observation that if fish cannot see the end in sight, such as in a culvert barrel, they attempt to minimize Pwr. They do so by seeking out the locations where propulsive force (F_P) is minimized, and they reduce (V_f) to some minimum consistent with forward progress, while taking their chances on being able to deliver enough energy to pass through the uncertain element.

Equations (17) and (18) contain the velocity of the fish with respect to a fixed reference (V_f). How fast fish swim in differing situations can only be learned from observations of existing situations. V_f has seldom been recorded when research results have been presented. Inspection of the above two equations clearly reveals how important it is for design engineers to have some knowledge of this parameter if they are to understand why a passage element is good or bad for fish. I suggest that engineers begin to develop catalogs of V_f for different species of fish and different sizes within a species for different difficult fish passage situations. The importance of documenting V_f for different species, sizes, and situations cannot be overemphasized.

Jones et al. (1974), in developing criteria for design of culverts on the MacKenzie Valley highway, assumed that if a fish had the capability of delivering a maximum (for the fish) instantaneous velocity with respect to the water (V_{fw}) of a given value, it actually would deliver that V_{fw} while swimming in a culvert. They then subtracted the estimated water velocity from the instantaneous value of V_{fw} and assumed the difference would be the value of V_f for fish swimming through a design culvert. I question that fish would act as they assumed, though their assumption, if adopted by the fish, would result in a minimal expenditure of energy in passing through long culverts. Fish entering a culvert do not know the culvert length until it is history to them, thus they appear to take power precautions that may or may not bring success in delivering the necessary energy to negotiate the culvert.

Some Biological Implications

The previous equations attempt to present the net propulsive force, power, and energy fish deliver if they are to pass through passage structures. It would appear that each element of the structure should be analyzed and the energy outputs should be summed to determine if the fish is capable of doing the job. However, biological constraints can confuse the efficacy of this summation. For an excellent treatise of the biological (and fluid dynamic) aspects of fish propulsion the reader is referred to Webb (1975). However, a brief, very simplified, overview of the problem is given below.

Modes of Delivery of Energy to Swimming

Fish have two muscle systems for swimming. The red muscle and white muscle systems function quite differently and are capable of delivering vastly different amounts of power and energy, depending on the species and individual fish. Webb (personal communication) described the red and white muscle systems as two different engines in one body.

Red muscle functions aerobically and depends on immediate physiological support systems. The metabolic reactants are generated in small amounts compared with the energy they release and do not accumulate in the tissues. In normal swimming the aerobic, red muscle activity is limited to long-term activity (prolonged and sustained V_{fw}). For many of the fish of interest to engineers, use of the red muscle system results in slow caudal fin movement often of large amplitude. Though this engine that the fish uses for prolonged activity delivers only small amounts of power, it can deliver a great deal of energy over a long period of time. The extent of red muscle in Arctic grayling, as an example, is exceedingly small compared with that of white muscle in the same fish. However, in the Arctic grayling, the red muscle occurs at the outer part of the body next to the skin, where it can deliver a maximum of flutter-bending moment to the caudal fin with a minimum of tension in the contracting muscle.

The white muscle engine in species that engineers are usually interested in accommodating is capable of delivering much more power than is the red muscle. Webb (1975) showed that for cold-water fish the potential power output by white muscle is approximately four times that for red muscle. White muscle functions anaerobically, however, and reactants accumulate in the muscle tissues. This manifests itself by an accumulation of lactic acid in the tissues, a product that diffuses slowly out of the muscles and, through the law of mass action and negative feedback, eventually stops further anaerobic energy production (Eckert et al. 1988). Thus, white muscle activity can only occur for a short time before a long rest is required to eliminate the lactic acid excess in the body. Burst or darting speeds can only be maintained by most species for a few seconds, and if a white muscle energy limit is reached, that muscle cannot soon be used. Negotiating a difficult culvert offers a good example of the importance of red and white muscle activity. If a fish is required to use its limit of white muscle energy to enter a culvert, it might then be able to negotiate the barrel using its red muscle system; but if the culvert inlet presents the need for burst power, the fish probably would not still have enough remaining white muscle capability to negotiate the inlet and would have to fall back downstream.

Example Calculations

Examples of fish energy and power requirements at the two extremes of upstream migration challenges are seen as fish swim through a lake or up a waterfall. Other challenges appear to lie between. Example calculations of power and energy requirements for fish swimming in a lake, a steep chute, and a culvert will be illustrated. For each situation, propulsive force (F_P), net power output (Pwr), and net energy delivered (E) will be calculated.

Lake

Given: $V_w = 0$, $V_f = 1$ m/s at time $t = 0$, $L = 0.5$ m, $b = 0.41$, $a_{fw} = 0.6$ m/s^2, $\nu = 1.55 \times 10^{-6}$ m^2/s, $\rho = 1{,}000$ kg/m^3, $W = 69(\text{N/m}^3) \times L^3 = 8.6$ N, $k = 4$, gravity (g) $= 9.8$ m/s^2.

Determine: Net energy (E) delivered by the fish between times $t = 0$ and $t = 2$ s.

Calculations: Because the HGL does not slope, ∇p is directed vertically upward (equations 1, 2), and B cancels W. Only profile drag (D) and virtual mass (F_m) forces need be considered. Thus,

$$F_P = D + F_{vm} = \text{equation (6)} + \text{equation (7)}$$
$$= bk(0.072)(\rho)(\nu^{0.2})L^{1.8}V_{fw}{}^{1.8}/2 + 1.2Ma_{fw}.$$

Let $C_1 = bk(0.072)(\rho)(\nu^{0.2})(L^{1.8})/2 = 1.17$. Then,

$$F_P = C_1(V_f{}^{1.8}) + 1.2(W/g)(a_{fw}) \quad (19)$$
$$= C_1(1 + 0.6t)^{1.8} + 1.2(8.6/9.8)(0.6)$$
$$= 4.95 + 0.63 = 5.58 \text{ newtons at } t = 2\ s;$$

$$\text{Pwr} = F_P(V_{fw}) \quad [V_w = 0, \text{ so } V_{fw} = V_f = 1 + 0.6t]$$
$$= 1.17(1 + 0.6t)^{2.8} + 0.63(1 + 0.6t)$$
$$= 10.6 + 1.4 = 12 \text{ watts at } t = 2\ s;$$

$$E = \int_0^2 \text{Pwr}\ dt$$
$$= \int_0^2 [1.17(1 + 0.6t)^{2.8} + 0.63(1 + 0.6t)]dt$$
$$= [1.17(1 + 0.6t)^{3.8}/3.8(0.6)] + 0.63(1 + 0.6t)^2/2(0.6)$$
$$= 11.8 \text{ joules net energy delivered from } t = 0 \text{ to } t = 2\ s.$$

Steep Channel or Chute

Reporting on live-fish experiments with chum salmon *Oncorhynchus keta* in good condition, Orsborn and Powers (1985) gave the following information for two chute studies.

Experiment 1: $L = 0.76$ m, chute length (L_c) = 2.3 m, slope of roughened chute (S_o) = 0.27, $V_{fw} = 2.68$ m/s, $V_f = 0.61$ m/s.

Experiment 2: L = 0.76 m, L_c = 2.3 m, S_o = 0.36, V_{fw} = 2.77 m/s, V_f = 0.73 m/s. Weight and water temperature were not given, so I assume $W = 69L^3 = 30.3$ N and a water temperature of 10°C; thus $\nu = 1.31 \times 10^{-6}$ m²/s.

Determine: F_P, Pwr, and E for each of the two experiments of Orsborn and Powers.

Calculations: Assume $a_{fw} = 0$, because no acceleration of water or fish was reported. Equation (11) describes this situation. Previous calculations have shown how to calculate D and W, so only the numerical results are shown here.

For the first experiment (S_o = 0.27),

$$F_P = D + W\sin(\theta) = 14.5 + 7.9 = 22.4 \text{ N};$$

$$\text{Pwr} = F_P V_{fw} = 22.4(2.68) = 60 \text{ W};$$

$$E = \text{Pwr(length of ramp)}/V_f = 60\ (2.3)/0.61 = 226 \text{ J}.$$

For the second experiment (S_o = 0.36), the same sequence of calculations yields

$$F_P = 15.4 + 10.5 = 25.9 \text{ N};$$

$$\text{Pwr} = 25.9\ (2.77) = 71.8 \text{ W};$$

$$E = 71.8\ (2.3)/0.73 = 226 \text{ J}.$$

In the first experiment Orsborn and Powers reported 100% of the chum salmon that attempted the chute successfully negotiated it. In the second experiment, they reported a success rate of only 23%. Because the two values for E above are identical, it is unlikely that the second set of fish were troubled by the net energy that they were capable of delivering. However, the second value for F_P is 16% greater than the first, and the second value for Pwr is 20% greater than the first. Thus, it appears most of the test fish simply could not generate enough power to deliver the propulsive force necessary to climb through the steeper chute. The reported average water velocities down the chute for each of these experiments were virtually identical, 2.07 m/s for the first and 2.04 m/s for the second, so the poor success ratio for the second experiment could not be explained by water velocity and profile drag. The above computations show the $W \sin(\theta)$ term to be 33% greater for the 0.36 slope than for the 0.27 slope, thus illustrating the importance of the fundamental fact that the buoyant force did not cancel the downslope component of the weight force.

Culvert

Hydraulically and as fish passage structures, culverts are very complicated structures, so they command great respect from design engineers. Because fish seek locations of minimum difficulty to swim, average flow velocities in culverts may not be very meaningful except as possible indices to water velocities (V_o) where the fish swim. My experience with Arctic grayling indicates that these fish swim hugging the boundary of culverts, either at the invert or close to the intersection of the water surface with the side of the culvert, whichever is the location of minimum water velocity. In short, in difficult situations that require elevated power outputs (for example, at culvert outlet and inlet), these fish seem to swim with V_f approximately equal to 0.3 m/s almost without regard to V_w so long as their anaerobic limits are not exceeded, though they may swim with much smaller values of V_f in situations of reduced power requirements (e.g., a barrel).

The horizontal angle of skew of water entering a culvert has a profound effect on the horizontal distribution of water velocity in the culvert. My measurements in one such culvert revealed that, along one boundary, water velocities 6 cm from the culvert side and 6 cm beneath the water surface (where fish swam) were only 20% of the average water velocity in the cross section. It appears, from limited observations, that if the angle of skew of water approaching the inlet is 30° to 45°, reduced wall velocity effects may be felt downstream from the culvert inlet a distance of perhaps 8 times the mean water surface width in the culvert barrel. Because this number comes from quite limited data, engineers are encouraged to observe and report their experiences with this extremely important skew effect. Clearly, culvert wall roughness greatly affects the potential for successful passage through the barrel and for success at the outlet and inlet. Multiplate culverts with 5-cm (2-in) corrugations on 15-cm (6-in) wavelengths (Manning n = 0.035), or other artificial or gravel–boulder roughness, generate more favorable boundary conditions than do less-roughened culverts. Waves in the culvert resulting from higher water velocities disorient small fish and frequently bounce them from slower water near the culvert wall to higher-velocity regions where they may be swept downstream.

Given: Culvert length (L_c) = 30.5 m, diameter (D) = 3.05 m, Manning n = 0.036, stream discharge (Q) at fish passage conditions = 2.27 m³/s, culvert slope = 0.005. Design fish is Arctic grayling 240 mm in fork length (L_f), and total length of fish (L) is assumed to be $L_f/0.92$ = 261 mm. Downstream scour pool water surface elevation will match the critical depth of flow in the culvert at the outlet for this Q. Assume velocity of fish with respect to the culvert (V_f) for short distance into culvert (0.6 m) at outlet and inlet is 0.3 m/s.

Determine: Power and energy requirements for fish to enter culvert, pass through barrel, and exit the culvert.

Solution for outlet: Calculate normal depth of flow (y_n) and critical depth of flow (y_c) for this culvert and the given conditions. This requires some reference to appropriate charts or some trial-and-error computations that result in $y_n = 0.9$ m and $y_c = 0.63$ m, so the normal velocity V_n = 1.26 m/s and the critical velocity V_c = 2.08 m/s. Because $y_c < y_n$, this would normally be an outlet control situation. However, because the outlet pool elevation is to match the culvert critical depth at the outlet, the pool elevation forces critical depth to occur at the outlet. If the outlet pool elevation were lower, the critical depth location would move upstream in the culvert, perhaps as much as $4y_c$, so depths farther upstream in the culvert would be somewhat less (and water velocities somewhat greater) than if the critical depth occurred at the outlet. However, the advantages to guaranteeing critical depth at the outlet instead of allowing it to occur further upstream are (1) depths less than y_c and attendant velocities greater than V_c do not occur in the culvert, and (2) the water surface profile drawdown slope and extent at the outlet is controlled, thereby lessening the magnitude of undesirable buoyant effects, and by virtue of smaller water accelerations in the vicinity of the outlet, virtual mass effects are also reduced.

Because of the stated downstream pool elevation and $y_c < y_n$, there exists a hydraulic M-2 water surface profile (Henderson 1966) extending upstream from the outlet. Some quick backwater computations beginning close to the outlet and extending 0.6 m upstream indicate the water surface rise from the outlet to this point is approximately 0.079 m. Thus the slope of the HGL in this zone is approximately 0.079/0.6 = 0.13, so $\theta = \tan^{-1}(0.13) = 7.4°$. The average cross-sectional water velocity at the outlet is the critical velocity of 2.08 m/s, and 0.6 m upstream it is 1.72 m/s. It is assumed water velocities (V_o) where the fish swim average half of the average cross-sectional velocities. Thus, from equation (10) the average value of the water convective acceleration that the fish is subjected to is

$$\begin{aligned} a_{fw} &= [V_f + (V_{o-\text{outlet}} + V_{o-0.6\text{ m}})/2] \\ &\quad \cdot [(V_{o-\text{outlet}} - V_{o-0.6\text{ m}})/0.6] \qquad (20) \\ &= \{0.30 + [0.5(2.08) + 0.5(1.72)]/2\} \\ &\quad \cdot \{[0.5(2.08) - 0.5(1.72)]/0.6\} \\ &= 0.375 \text{ m/s}^2. \end{aligned}$$

So, $F_{vm} = 1.2(W/g)(a_{fw}) = 1.2(69\,L^3/9.8)(0.343)$ = 0.04 N.

Because this short zone at the outlet provides brief, but possibly critical, exposure to the fish, it appears prudent in calculating D to assume the maximum value of V_o (that at the outlet) occurs through the 0.6-m zone. Use equation (6), assume $b = 0.42$, $k = 4$, and water temperature = 4°C, and for V_{fw} substitute $V_o + V_f = 0.5V_c + V_f$ = 0.5(2.08) + 0.3 = 1.34 m/s; then profile drag D = 0.63 N. Equation (14) for $\phi = 0$, becomes

$$\begin{aligned} F_P &= F_{vm} + D + W\tan(\theta) \qquad (21) \\ &= 0.04 + 0.63 + 0.16 = 0.83 \text{ N}. \end{aligned}$$

Then,

$$\begin{aligned} \text{Pwr} &= F_P(V_{fw}) = F_P(V_o + V_f) \\ &= 0.83(1.34) = 1.1 \text{ W}; \end{aligned}$$

$$\begin{aligned} E &= \text{Pwr} \times \text{time to move through the outlet zone} \\ &= 1.1(2) = 2.2 \text{ J}. \end{aligned}$$

Values of F_P = 3.1 N, Pwr = 6.9 W, and E = 13.8 J have been reported by Behlke et al. (1988) for similar Arctic grayling entering a culvert through which they passed successfully, so the calculated values would offer no problem to the given fish entering this culvert.

Solution for barrel: Water accelerations in the main body of the barrel are very small upstream from the outlet zone, so equation (11) can be used for F_P. Quick calculations show that for this very flat slope, $W\sin(\theta)$ is insignificant, so it will be ignored, thus $F_P = D$. Here V_o will be assumed to be 0.4 V_{ave} at any s in the barrel, but any assumption for V_o/V_{ave} would have to rest on the engineer's experience with similar existing culverts on other streams and on the body depth of the fish to be passed through the culvert (the bodies of larger fish extend farther from the wall into areas of higher water velocity). For an assumed depth of y_c at the outlet, the water depths and average cross-sectional velocities (V_{ave}) are as shown in Table 2.

In a previous study of Arctic grayling in a single culvert (L_c = 33.5 m), Behlke et al. (1988) found the average value of V_f through the culvert correlated only with fork length of the fish and did not appear to correlate with any other variable. The relationship was $V_f = 11.7L_f - 0.017$ (m/s), L_f being the fork length of the fish (m); for L_f = 0.24 m, V_f = 0.022 m/s. This may appear to be quite slow, but my experience with these fish is that they often move slowly upstream while swimming through a culvert barrel.

The fish begins its journey through the culvert barrel at s = 0.6 m where it exits the outlet zone, and the inlet zone for the fish will begin 0.5 m downstream from the inlet. Because V_o is 30% greater at s = 0.6 m than at 30 m, average values for Pwr will be calculated for s = 0.6 m to 6 m and for s = 6 m to 30 m, and E will be calculated separately for each of these two reaches of the culvert barrel. Average V_o for s = 0.6 m to 6 m is 0.65 m/s; for s = 6 m to 30 m, average V_o = 0.57 m/s. From equations (6), (11), (16), and (18), and for $V_{fw} = V_{o-ave}$ + 0.022 m/s = 0.672 m/s from s = 0.6 m to 6 m,

$$F_P = D = 0.18 \text{ N};$$

$$\text{Pwr} = D(V_{fw}) = 0.18(0.672) = 0.12 \text{ W};$$

$$E = \Delta s(\text{Pwr})(1 + V_o/V_f)$$
$$= 5.4\ (0.18)(1 + 0.65/0.022) = 29.7 \text{ J}.$$

The second term in the final parenthetical term in the energy equation is the price the fish pays for swimming in a moving fluid, and its value (0.65/0.022 = 29.5) compared with 1 is that price. Similar computations for s = 6 m to 30 m yield F_P = 0.144 N, Pwr = 0.085 W, and E = 54.9 J. The total energy used in the barrel is the sum of the two E values, 84.6 J. Field observations and subsequent computations by Behlke et al. (1988) indicate the values for F_P, Pwr, and E computed here are safe for the design fish.

Solution for inlet: At a sharp-edged culvert entrance, streamlines are contracted and the contracted cross section of high-velocity flow in the center of the culvert leaves low-velocity flow, often with upstream velocities, next to the side walls of the culvert. Thus at the inlet end of the barrel, just before they exit the culvert, fish can usually find a rest area in which they may survey the situation ahead. They need not enter higher velocity flow and the entrance drawdown, which may slope sharply, until they are prepared to do so. I doubt that they remain here long enough to recharge their white muscle engine, but they could.

TABLE 2.—Backwater computations for a culvert. D = 3.05 m, n = 0.036, slope = 0.005, Q = 2.27 m^3/s, and s is measured from the outlet lip.

s (m)	Depth (m)	V_{ave} (m/s)	V_o (m/s)
0.0	0.63	2.08	0.83
0.3	0.71	1.76	0.70
0.6	0.72	1.73	0.69
1.0	0.73	1.69	0.68
1.5	0.74	1.66	0.66
3.0	0.76	1.60	0.64
6.0	0.79	1.52	0.61
10.0	0.81	1.46	0.58
20.0	0.84	1.38	0.55
30.0	0.86	1.33	0.53

Because of the flow contraction due to sharp-edged entrance geometry and because the cross section of flow in the culvert is usually smaller than that of the approaching stream, $V_w^2/2g$ must increase as the water enters the culvert, so the water surface must drop at the inlet by an amount equal to the sum of the increased $V_w^2/2g$ and the entrance loss due to initial acceleration and subsequent deceleration of water entering the culvert. The entrance loss is usually expressed as $K_e(V_w^2/2g)$, K_e being a loss coefficient that depends on geometry and V_w the water velocity for the flow just downstream from entrance contraction–expansion effects. Norman et al. (1985) gave K_e = 0.9 for sharp-edged culverts under outlet control. Thus the water surface at the entrance must drop by an amount $(1 + K_e)\ V_w^2/2g$. I assume this occurs in the first 0.5 m of the inlet end of the culvert. Based on the velocity of flow for s = 30 m from Table 2, the entrance drop in water surface is $1.9\ (1.33^2/2g)$ = 0.17 m. This drop occurs principally in the 0.5-m inlet zone, and the slope of the water surface here is 0.17/0.5 = 0.34. Average water velocities against which the fish swims in the inlet zone are assumed to be the same as those of the culvert downstream from the contraction–expansion zone, i.e., 1.33 m/s, because the fish does not have to brave the fully contracted water velocities. Because this is a zone of short, high-power expenditure for the fish, V_f = 0.3 m/s. If the water acceleration occurs from near-zero approach velocity, $a_{fw} = (V_w + V_f)(\Delta V_w/\Delta s)$ = (1.33 + 0.3)(1.33/0.5) = 4.34 m/s^2. Equation (14), for ϕ = 0, applies to this situation:

$$F_P = 0.37(1.63^{1.8}) + 1.2[69(0.261^3)/9.8] \cdot (4.34) + 69(0.261^3)(0.34) = 0.89 + 0.65 + 0.41 = 1.95\ N;$$

$$\text{Pwr} = F_P(V_{fw}) = 1.95\ (1.63) = 3.17\ \text{W};$$

$$E = L_e(\text{Pwr})(1 + V_w/V_f) = 0.5\ (3.17)(1 + 1.33/0.3) = 8.6\ \text{J};$$

for which L_e is the length of the inlet zone for the fish (0.5 m).

Because the fish moves quickly through this zone, the penalty it pays for swimming is much smaller than for slower movement through the barrel. From a very limited data base (Behlke et al. 1988), these values for F_P, Pwr, and E all appear safe. Here the inlet zone for the fish was assumed shorter than it probably would be in reality in order to illustrate the computational method. As engineers gain experience with this line of thought and the resultant numbers, the assumptions can be further refined.

Conclusions

Virtual mass and non-Archimedean buoyancy forces are shown to be of considerable importance to fish passage under certain hydraulic conditions.

The equations and procedures illustrated give bases, founded on fluid mechanics principles of fish–water interaction, for comparing one fish passage facility option with another. These procedures can be used by experienced engineers, knowledgeable about the swimming characteristics of the fish they are designing for, to design virtually any type of passage device if they also know the hydraulic characteristics of the options.

Because little data exist that can be used except for uniform flow in open channels with negligible slope, a data base of where fish actually swim in moving water masses, how fast they move with respect to the ground, what water and fish velocities and accelerations are, and the simultaneous slope of the hydraulic grade line needs to be developed. A purpose of this paper is to urge engineers knowledgeable about these parameters of fish–water interaction to publish the findings of their experiences.

References

Behlke, C. E. 1987. Hydraulic relationships between swimming fish and water flowing in culverts. Pages 112–132 *in* D. W. Smith and T. Tilsworth, editors. Proceedings of the Cold Regions Environmental Engineering Conference CSCE–ASCE. University of Alberta, Department of Civil Engineering, Edmonton.

Behlke, C. E., D. L. Kane, R. F. McLean, and M. T. Travis. 1988. Spawning migration of Arctic grayling through Poplar Grove Creek culvert, Glennallen, Alaska, 1986. Report FHWA-AK-RD-88-03, of Alaska Department of Transportation and Public Facilities to Federal Highway Administration, Fairbanks.

Brett, J. R. 1973. Energy expenditure of sockeye salmon, *Oncorhynchus nerka*, during sustained performance. Journal of the Fisheries Research Board of Canada 30:1799–1809.

Daily, J. W., and D. R. F. Harleman. 1965. Fluid dynamics. Addison-Wesley, Reading, Massachusetts.

Eckert, R., D. J. Randall, and G. Augustine. 1988. Animal physiology. Freeman, New York.

Henderson, F. M. 1966. Open channel flow. Macmillan, New York.

Jones, D. R., J. W. Kiceniuk, and O. S. Bamford. 1974. Evaluation of the swimming performance of several fish species from the Mackenzie River. Journal of the Fisheries Research Board of Canada 31:1641–1647.

Norman, J. N., R. J. Houghtalen, and W. J. Johnston. 1985. Hydraulic design of highway culverts. U.S. Department of Transportation, Federal Highway Administration, Report FHWA-IP-85-15, Washington, D.C.

Orsborn, J. F., and P. D. Powers. 1985. Analysis of barriers to upstream fish migration. Bonneville Power Administration, Project 82-14, Portland, Oregon.

Rouse, H. 1938. Fluid mechanics for hydraulic engineers. McGraw-Hill, New York.

Tilsworth, T., and M. D. Travis. 1987. Fish passage through Poplar Grove Creek. Alaska Department of Transportation and Public Facilities, Fairbanks.

Webb, P. W. 1975. Hydrodynamics and energetics of fish propulsion. Fisheries Research Board of Canada Bulletin 190.

American Fisheries Society Symposium 10:299–305, 1991

Effects of Spawning-Run Delay on Spawning Migration of Arctic Grayling

Douglas F. Fleming and James B. Reynolds
Alaska Cooperative Fishery Research Unit,[1] *U.S. Fish and Wildlife Service*
University of Alaska Fairbanks, Fairbanks, Alaska 99775, USA

Abstract.—We examined the effects of delay on the spawning run of Arctic grayling *Thymallus arcticus* in Fish Creek, a tributary of the Jack River, near Cantwell, Alaska. The run was sampled at a weir for 25 d during May 1988. Tagged Arctic grayling were delayed in holding pens for 3, 6, or 12 d, then released; control fish were released within 12 h of capture. During delay, a high proportion of females continued to ripen. Males were usually ripe before delay and remained ripe over a longer period than females. Most changes in maturity occurred within the first 3 d of delay. Distribution and migration of delayed and control fish were monitored by recapture in upstream traps. Females released "running-ripe" had higher migratory rates but proportionally fewer reached upstream areas compared with "less ripe" females. Control fish swam farther upstream than fish delayed 3 d or longer. Delay is probably more critical for females than males. Reduction in distances traveled by both sexes as a result of migratory delay may lead to the use of nonpreferred spawning habitats and decreased recruitment. We suggest that spawning delays for Arctic grayling not exceed 3 d.

In central and northern Alaska, Arctic grayling *Thymallus arcticus* migrate from overwintering areas to their spawning grounds during spring, a period of peak streamflow from snowmelt. High water velocities through culverts are normal at this time and can delay upstream spawning migration. Earlier work by MacPhee and Watts (1976), upon which the present Alaska standards for fish passage are largely based, focused upon swimming performance of Arctic grayling and the maximum water velocities in culverts that permit fish passage.

The maximum period of delay that Arctic grayling can experience during spawning migration without significant reduction in spawning success (i.e., critical delay period) is an important consideration in the design and placement of culverts. The effects of delay on spawning migration and success have been recognized as an important aspect of fish passage, but have not received much field work and analysis. For example, declines in abundance of Columbia River chinook salmon *Oncorhynchus tshawytscha* may be attributable to biological effects of spawning-run delay (Haynes and Gray 1980). The goal of our study was to determine effects of delay associated with culvert passage upon the sexual maturation, migratory rate, and spawning distribution of Arctic grayling.

[1]The Unit is jointly sponsored by the U.S. Fish and Wildlife Service, Alaska Department of Fish and Game, and University of Alaska Fairbanks.

Study Area and Methods

The study was conducted on Fish Creek, a second-order tributary of the Jack River in the Alaska Range near Cantwell, Alaska (Figure 1); its watershed area is 110 km^2. Fish Creek passes through a 2.9-m-diameter, 19-m-long culvert under the Denali Highway. The stream hosts an annual spring migration of Arctic grayling upstream from overwintering areas in the Jack and Nenana rivers to spawning and summering habitats in upper Fish Creek. Fish Creek was selected because of its small size and proximity to the Denali Highway. The lower reach includes a series of small, shallow, interconnected ponds, and its headwaters comprise two branches: a warmer, lake-fed branch and a cooler, surface-runoff branch.

Our strategy was to capture adult fish at a weir near the culvert; randomly select individuals and delay them for 3, 6, or 12 d; tag and release them (control fish were released within 12 h of capture); and recapture them in several traps located progressively farther upstream. Multiple upstream recapture sites allowed greater opportunity to evaluate maturation of the delayed and control fish, and to provide checkpoints with which to gauge migratory performance and upstream distribution as a function of delay.

An 18.5-m-long weir was deployed downstream of the culvert on 6 May 1988 to capture the Arctic grayling migrating upstream. Fish were fin-clipped (adipose) and tagged with numbered

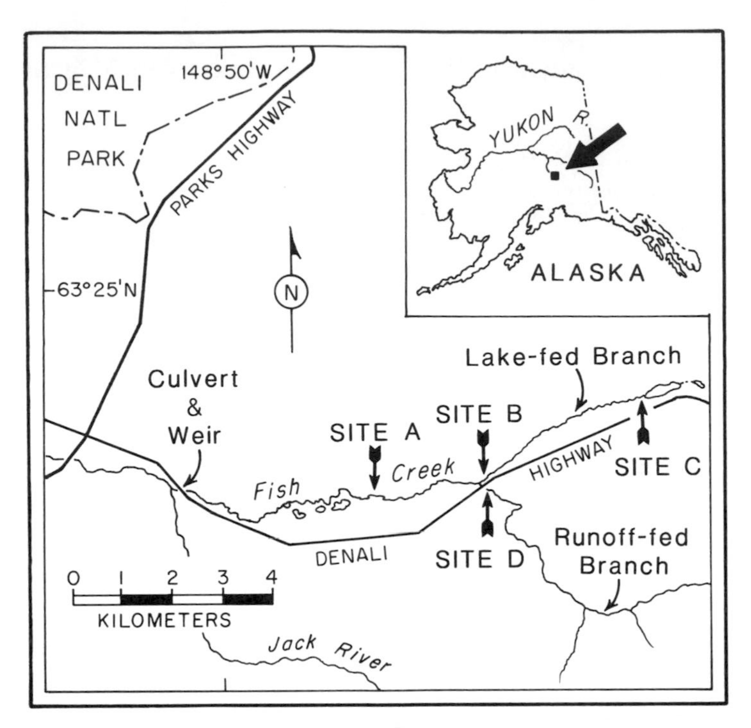

FIGURE 1.—Study area on Fish Creek, along the Denali Highway, south of Fairbanks, Alaska. Recapture sites are shown as A, B, C, and D.

Floy™ FTSL-73 shrimp streamer tags[2] inserted along the anterior edge of the dorsal fin. Streamer tags were tied to increase retention time. Tag numbers, fork lengths, and state of maturity were recorded. Handling was done without anesthesia to avoid introducing additional variables into the assessment of migratory behavior and performance that could affect the results.

To assess maturation status, each fish was examined for protrusion of the urogenital pore and gently squeezed in an attempt to release eggs or milt. Fish were classified in the following italicized categories. Fish of *unknown sex* released no gametes when handled and exhibited no external sexual features. Males were either *partly ripe* (urogenital pore not protruded; little milt released after several squeezes), *ripe* (urogenital pore protruded; much milt released with one or two squeezes) or *spent* (urogenital pore protruded; no milt released; abdomen flacid). Females were *partly ripe* (urogenital pore not protruded; one or few eggs released after several squeezes), *ripe* (urogenital pore often protruded; numerous eggs released after one or two squeezes), *running-ripe* (urogenital pore protruded; numerous eggs released with little or no squeezing), or *spent* (urogenital pore protruded and swollen; no or few eggs released; abdomen flacid). Females that were partly ripe, ripe, or running-ripe were considered to be advancing in maturation status; spent females were declining in status.

Arctic grayling at least 270 mm long were selected for our study; this ensured that all fish in the sample were sexually mature (R. McLean, Alaska Department of Fish and Game, personal communication). Fish were sampled over the entire run (about 25 d) to minimize the effect of unknown differences in early or late-run migrants. Of the Arctic grayling captured at the weir, 159 adults were selected for the delay treatments (mean length, 295 mm; SD, 19 mm), and 164 adults were released as controls (mean length, 288 mm; SD, 16 mm). All fish were released just upstream of the culvert to avoid any additional delay resulting from high discharge through the culvert. Release times and tag numbers of all released fish were recorded. Sample sizes for the delay groups were 56 (3-d delay), 58 (6-d delay), and 45 (12-d delay). Balanced sex ratios within

[2]Use of trade names does not imply product endorsement.

samples were not possible because of a high incidence of individuals of unknown sex captured at the weir.

Delayed fish were held in 1.2 × 3.1-m holding pens (10-mm mesh) staked in water 0.5 m deep and covered with tarpaulins. The three holding pens corresponded to the three delay treatments of 3, 6, and 12 d. Following the completion of the delay period, tagged individuals to be released were removed from the holding pens and reevaluated for maturation status before release.

Four fyke nets were placed at upstream sites, mouth facing downstream and wings bank to bank, and fished continuously until 11 June to monitor the progress of the run (Figure 1). Site A was 9 km upstream from the weir, sites B and D (confluence of the branches) were 12 km upstream, and site C was 18 km upstream. Nets were checked four times daily. Data recorded at the upstream sites were time of recapture, tag number, and maturation status. Time spent by a tagged Arctic grayling in a fyke net never exceeded 8 h. Catches at site D were negligible, confirming that fish did not use the runoff-fed branch for spawning. Data analysis was restricted to catches at sites A, B, and C.

Some tagged fish escaped capture at sites A, B, and C, indicating that the capture efficiency of all fyke nets was less than 100%. Capture efficiencies were calculated as the proportions of tagged fish encountering a fyke net that were captured by that net. The number of tagged fish encountering a net was the sum of tagged fish caught by the net and those that escaped it but were caught farther upstream. Fish escaping the net at site C were captured by dipnetting in the three small spawning tributaries of the lake. Efficiency values with 95% binomial confidence limits were calculated for each delay and maturity group at each fyke net. Actual numbers, with confidence limits, of fish in each group arriving at a site were estimated by dividing the observed catch of a group by the appropriate capture efficiency and binomial confidence intervals.

Results

Maturation Status

At initial capture, 74% of the males were ripe and 52% of the females were ripe or running-ripe. After delays of 3, 6, or 12 d in the holding pens, the maturation status of 68–85% of the females changed from the condition initially observed; only 23–35% of the males showed changes (Figure

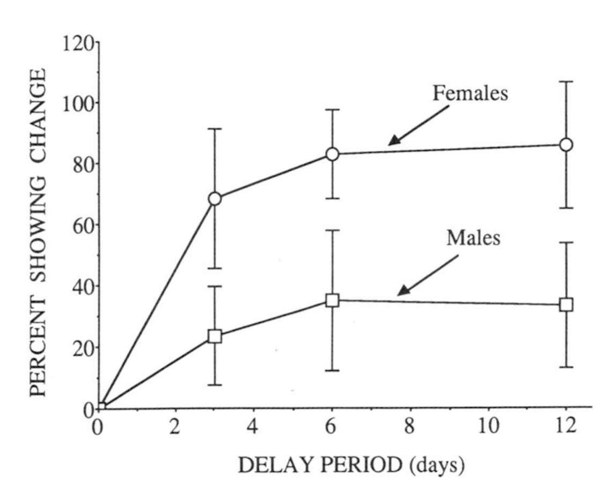

FIGURE 2.—Percentage change in maturation status of Arctic grayling while held in pens for 3, 6, or 12 d. Vertical bars represent 95% confidence limits.

2). Of the control females recaptured upstream, 62–86% had changed maturation status within 8 d after release at the weir; 8–28% of recaptured control males changed in the same period. Among control and delayed fish, more females changed in maturity than did males. Most changes in maturity occurred during 0–3 d after initial capture.

Of the delayed females, 37–64% advanced in maturation status, the percent increase being proportional to the period of delay (Figure 3). The condition of advancing maturity was more prevalent among females delayed 12 d compared with those delayed 3 d (one-tailed t-test, P = 0.10). Among delayed males, 65–75% remained unchanged in maturation status (Figure 4). Proportions among male delay groups were not statistically different ($P > 0.10$). We found some spent fish (declining maturation status) of both sexes in the pens; gamete release, and perhaps spawning, may have occurred during delay but was not observed. Overall, 50% of the females ripened after a mean delay of 6.4 d, and 11% of the males ripened after 6.7 d.

Migratory Performance

Using release and recapture times and stream distances between capture sites, we calculated migratory rates for different segments along the migration route (Table 1). Median rates were used for comparisons because distributions of rates tended toward non-normality. Median migratory rates (sexes and treatments pooled) were 0.22 km/h from the weir to site A, 0.22 km/h from the weir to site B (the rate between sites A and B

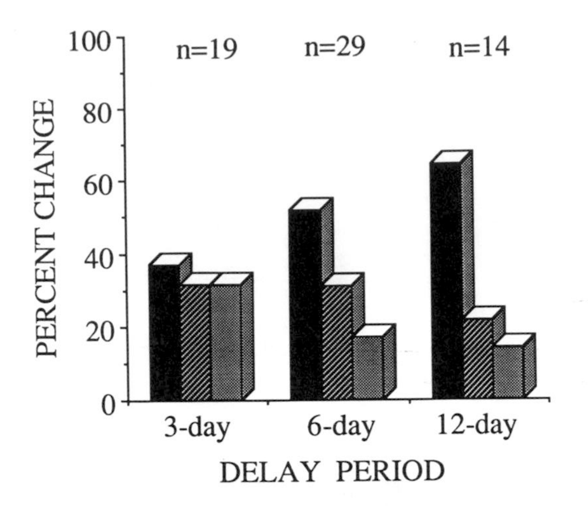

FIGURE 3.—Percentage change in maturation status of delayed female Arctic grayling while held in pens for 3, 6, or 12 d. Advancing status is represented by black bars, declining status by cross-hatched bars, and unchanged status by gray bars.

unknown because the trap at site B was fishing before the trap at site A), and 0.05 km/h from site B to site C. The migratory rate between sites B and C was significantly lower than that in the other two stream segments (Mann–Whitney U-test, $P < 0.0001$).

Migratory rates, analyzed by delay period (sexes and maturity groups combined), were calculated for fish recaptured at site A (Table 1). Median values for the delay groups were 0.37,

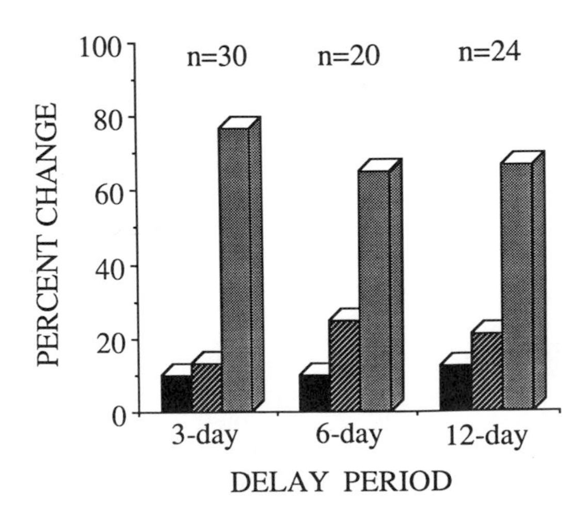

FIGURE 4.—Percentage change in maturation status of delayed male Arctic grayling while held in pens for 3, 6, or 12 d. Advancing status is represented by black bars, declining status by cross-hatched bars, and unchanged status by gray bars.

TABLE 1.—Median migration rates (km/h) for Arctic grayling in various categories.

Category	N	Median
Stream segment		
Weir to site A	62	0.22
Weir to site B	97	0.22
Site B to site C	38	0.05
Delay period		
Control	42	0.37
3 d	24	0.18
6 d	29	0.26
12 d	9	0.30
Sex and maturity		
Running-ripe female	21	0.33
Ripe female	13	0.25
Partly ripe female	3	0.30
Ripe male	35	0.22

0.18, 0.26, and 0.30 km/h, corresponding to the control, 3-, 6-, and 12-d delays. Rates of individual fish ranged between 0.02 and 1.29 km/h. Comparisons of migratory rates for each group showed significant differences between control fish and the 3-d delay fish (Mann–Whitney U-test, $P = 0.02$). Migratory rates, analyzed by sex and maturation status (control and delayed fish combined), were 0.33, 0.25, 0.30, and 0.22 km/h for running-ripe females, ripe females, partly ripe females, and ripe males (Table 1). Running-ripe females exhibited significantly higher migratory rates than ripe females (Mann–Whitney U-test, $P = 0.045$).

Distribution of Spawners

Dropout rates were evaluated according to sex and maturation status upon release at the weir. Dropout rate was the fraction of fish released at the weir that remained resident within a stream section. This was calculated by a regression of the recapture fraction against the recapture site distance and expressed as a percentage per stream kilometer. Dropout rates were 4.5%/km for ripe males and 4.0%/km for ripe females. Running-ripe females dropped out at a mean rate of 5.0%/km ($r^2 = 0.90$, $P = 0.035$), partially ripe females at 3.1%/km ($r^2 = 0.997$, $P = 0.001$). Dropout rate increased with the advancing maturation status of females; the difference between running-ripe and partly ripe females (Figure 5) was significant ($P < 0.05$). A regression based on transformed fractions (arcsine root) to satisfy normality requirements (Zar 1984) gave results identical to those aforementioned.

The estimated numbers and percentages of fish reaching sites A, B, and C were analyzed by delay

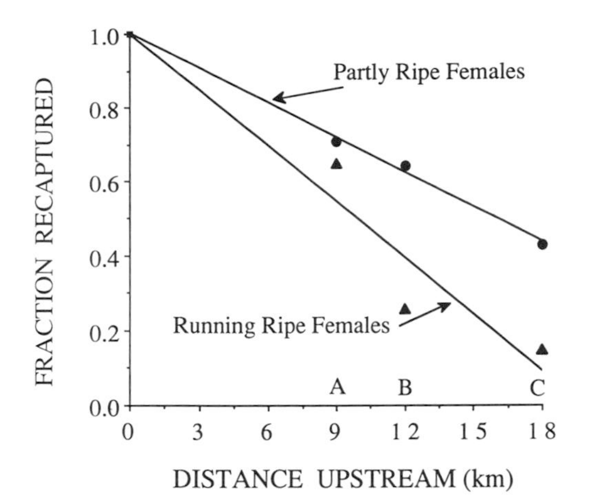

FIGURE 5.—Fraction (Y) of running-ripe (triangles) and partly ripe (circles) female Arctic grayling at upstream recapture sites (A, B, C) as a function of distance upstream (X) of release point (culvert). The fitted regression line for partly ripe females was $Y = 1.0027 - 0.0314X$; $r^2 = 0.99$; $P = 0.001$. The fitted regression line for running-ripe females was $Y = 0.9995 - 0.0503X$; $r^2 = 0.896$; $P = 0.035$.

period and maturation status to further clarify effects on upstream distribution of spawners. Results by delay period (Table 2) indicated that control fish migrated farther upstream (sites B and C) than did fish delayed 3 and 6 d. Fish delayed 12 d were caught upstream in proportions similar to those of control fish, a result not readily understood. Results by maturation status (Table 3) were more straightforward. Upstream distribution of females was directly related to their maturation status upon release. The likelihood of running-ripe females reaching sites B and C was about half that of ripe females and one-third that of partly ripe females. Upstream distribution of ripe males was similar to that of ripe females.

TABLE 2.—Estimated number (upper) and percent (lower) of Arctic grayling reaching upstream recapture sites according to delay period. Confidence intervals ($P = 0.95$) are in parentheses.

Delay period (sample size)	Recapture site		
	A	B	C
Control (122)	79(65–99) 65(53–81)	45(41–51) 37(34–42)	33(29–47) 27(24–38)
3 d (41)	34(27–41) 83(66–100)	13(12–16) 32(29–39)	5(4–7) 12(10–17)
6 d (39)	17(15–21) 43(38–54)	7(6–9) 18(15–23)	6(5–8) 15(13–20)
12 d (31)	15(10–25) 48(32–81)	9(8–9) 29(26–29)	9(8–9) 29(26–29)

TABLE 3.—Estimated number (upper) and percent (lower) of Arctic grayling reaching upstream recapture sites according to sex and maturity at time of release at the downstream weir. Confidence intervals ($P = 0.95$) are in parentheses.

Sex and maturity (sample size)	Recapture site		
	A	B	C
Partly ripe females (14)	10(8–14) 71(57–100)	9(8–14) 64(57–100)	6(5–8) 43(36–57)
Ripe females (41)	27(19–41) 66(46–100)	17(15–22) 41(37–54)	13(12–18) 32(29–44)
Running-ripe females (28)	18(17–22) 64(61–78)	7(6–10) 25(21–36)	4(3–5) 14(11–18)
Ripe males (80)	57(46–74) 71(57–92)	28(26–32) 35(32–40)	19(17–27) 24(21–34)

Ridit analysis (Fleiss 1981) permitted a one-tailed comparison of the spatial distribution of maturity groups relative to each other (Figure 6). The distributions of all males and all females were not significantly different ($P = 1.0$); the same was true for ripe males and ripe females ($P = 0.85$). However, the chances of finding partly ripe females farther downstream than ripe males ($P = 0.20$) or ripe females ($P = 0.36$) were less than even. Partly ripe females were 16% more likely to be found upstream from running-ripe females ($P = 0.08$).

Discussion

Delay Effects

Upon initial capture at the weir, most males and about half the females were ripe. During delay, the proportion of ripe males changed little if any. Up to 64% of the females advanced in maturation status during delay, depending on delay duration; most of these changes occurred during the first 3 d of delay. Delay is probably more critical for females than males.

Delay, per se, probably did not alter the maturation process: control fish swimming upstream matured on a schedule similar to that of delayed fish. Instead, delayed fish responded differently than control fish in their subsequent migratory behavior. These responses were more related to the maturation status of fish upon release at the weir than to the duration of delay. Migratory rates of females increased with advancing maturity: running-ripe females swam faster upstream than less-ripe females. Variances in rates were lowest for running-ripe females, suggesting an increased uniformity in the urge to spawn (Fleming 1989). Ripe males exhibited migratory rates similar to

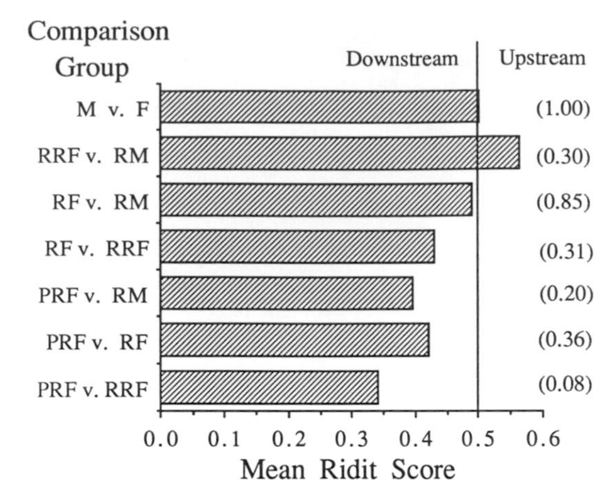

FIGURE 6.—Pairwise comparisons of the upstream–downstream distribution of sex and maturity groups of Arctic grayling relative to each other, based on mean Ridit probability scores. In each pair the reference group appears first and the comparison group second; confidence probabilities are in parentheses. For example, a running-ripe female is 16% (i.e., 0.50 − 0.34, times 100) more likely to be located downstream compared with a partly ripe female (P = 0.08). M = male, F = female, RM = ripe male, RF = ripe female, PRF = partly ripe female, and RRF = running-ripe female.

those of ripe females. However, faster swimming groups were also those with higher dropout rates. As a result, the spatial distribution of spawners, although similar for males and females overall, depended on maturation status when released at the weir. Running-ripe females were likely to stop downstream from less-ripe females. The spawning distribution of ripe males was similar to that of ripe females. In summary, delay did not interfere with maturation, but did result in a handicap during the subsequent upstream migration.

Environmental Effects

The effects of delay were not directly related to the duration of delay. The likely reason is that environmental conditions mediate the delay experience. Water temperature and distance to spawning areas are among the probable factors that complicate the effect of delay duration.

Temperature is well recognized as a mediator in the final stages of maturation before spawning in various species (Billard et al. 1978; Dodson et al. 1985; Morrison and Smith 1986; Beacham and Murray 1988). Hatchery egg-takes in interior Alaska usually require that female Arctic grayling be held in pens as water warms to achieve the running-ripe status necessary for artificial fertilization (Parks et al. 1986). During 1988 in Fish Creek, both ripe and unripe fish were present at site C by May 13, but spawning did not begin until May 17 when water temperature reached 5°C.

Rate of stream warming may be as important as absolute temperature in controlling maturation and affecting the impact of delays. Warming in Fish Creek during the 1988 spawning run was 0.26°C/d (Fleming 1989); during the 1987 spawning run it was 0.20°C/d (Behlke et al. 1988). In Poplar Grove Creek, Alaska, warming rates during spawning runs in 1974 and 1986 were approximately 0.6°C/d (MacPhee and Watts 1976; Behlke et al. 1988), but in 1985 it was 1.25°C/d (Tilsworth and Travis 1986). There was a linear relation between mean daily water temperature and the proportion of ripe females in daily samples (MacPhee and Watts 1976; Behlke et al. 1988). The lower warming rate in Fish Creek may be the result of its larger watershed (110 km^2) compared with that of Poplar Grove Creek (31 km^2). Furthermore, migratory rate of Arctic grayling in Fish Creek (0.20 km/h, pooled average from weir to site C) was five times higher than in Poplar Grove Creek (0.04 km/h; Tilsworth and Travis 1986; Behlke et al. 1988). Thus, in the larger Fish Creek watershed, Arctic grayling appeared to migrate faster, over a longer distance, in a stream that was warming slowly when compared with the smaller watershed of Poplar Grove Creek. However, Poplar Grove Creek has a steeper gradient than Fish Creek. Although water temperature is likely a key factor in determining these differences, the data are insufficient to clarify the exact relationship. Further studies are needed on spawning runs in streams with different thermal regimes and watershed characteristics.

Implications of Delay

The timing of a delay is important because maturation rate may vary within a stream system from year to year. In 1988, 74% of the males and 52% of the females were ripe when captured at the Fish Creek weir. In 1987, the run contained a proportion of ripe males similar to that in 1988, but only 5% of the females were ripe (McLean, personal communication). A delay in 1987 would probably have been less critical for females than one in 1988. The effects of delay timing probably are low when few females are ripe, stream warming rate is low, and distance remaining is short, but serious when conditions are just the opposite.

This study begs the question, "Do eggs of

delayed females, especially those delayed until running ripe, retain their viability?" Bry (1981) reported a significant drop in viability of eggs in rainbow trout *Oncorhynchus mykiss* held 6 d past ovulation. Sakai et al. (1975) held gravid rainbow trout 10 d before artificially fertilizing their eggs and found decreased hatching rate and increased deformity among alevins. Similar effects can be supposed for delayed Arctic grayling, but no such data are available.

Our study indicated that delay caused males and females to stop short of their upstream goal. The displacement of spawners to downstream areas could force them to use spawning habitat of lower quality, resulting in reduced recruitment. Because Arctic grayling spawn in their natal areas (Hop and Gharrett 1989), the effects of delay could be realized not only by displaced spawners, but by their surviving progeny as well.

Although the effects of delay were difficult to measure in terms of delay period, we provisionally recommend that spawning runs of Arctic grayling not be artificially delayed more than 3 d. Our recommendation is based on the observations that most females ripened within the first 3 d of delay; some females declined in maturation status within 3 d of delay; and fish delayed for 3 d or more, especially running-ripe females, failed to migrate to upstream spawning areas.

Acknowledgments

This study was funded by the Alaska Department of Transportation and Public Facilities and by the Federal Highways Administration. We thank Douglas L. Kane and Robert Gieck, Water Resources Center, University of Alaska Fairbanks (UAF); Robert Clark and Robert F. McLean, Alaska Department of Fish and Game; and Michael D. Travis, Alaska Department of Transportation and Public Facilities, for technical guidance and field support; UAF students Alan R. Burkholder, James B. Griswold, J. D. Johnson, Craig Monaco, and Kathleen Roush for field assistance; and Ramona Salonka for procurement of supplies and coordination of personnel.

References

Beacham, T. D., and C. B. Murray. 1988. Influence of photoperiod and temperature on timing of sexual maturity of pink salmon (*Oncorhynchus gorbuscha*). Canadian Journal of Zoology 66:1729–1732.

Behlke, C. E., D. L. Kane, R. F. McLean, J. B. Reynolds, and M. D. Travis. 1988. Spawning migration of Arctic grayling through Poplar Grove Creek culvert, Glennallen, Alaska, 1986. Final report FHWA-AK-RD-88-09 to Alaska Department of Transportation and Public Facilities, Fairbanks.

Billard, R., C. Bry, and C. Gillet. 1981. Stress, environment and reproduction in teleost fish. Pages 185–208 *in* A. D. Pickering, editor. Stress and fish. Academic Press, New York.

Bry, C. 1981. Temporal aspects of macroscopic changes in rainbow trout (*Salmo gairdneri*) oocytes before ovulation and of ova fertility during the post-ovulation period; effect of treatment with 17 alpha-hydroxy-20 beta-dihydroprogesterone. Aquaculture 24:153–160.

Dodson, J. J., Y. Lambert, and L. Bernatchez. 1985. Comparative migratory and reproductive strategies of the sympatric anadromous coregonine species of James Bay. Contributions in Marine Science 27(supplement):296–315.

Fleiss, J. L. 1981. Statistical methods for rates and proportions. Wiley, New York.

Fleming, D. F. 1989. Effects of spawning run delay on spawning migration of Arctic grayling. Master's thesis. University of Alaska, Fairbanks.

Haynes, J. M., and R. H. Gray. 1980. Influence of Little Goose Dam on upstream movements of adult chinook salmon, *Oncorhynchus tshawytscha*. U.S. National Marine Fisheries Service Fishery Bulletin 78:185–190.

Hop, H., and A. J. Gharrett. 1989. Genetic relationships of Arctic grayling in the Koyukuk and Tanana rivers, Alaska. Transactions of the American Fisheries Society 118:290–295.

MacPhee, C., and F. J. Watts. 1976. Swimming performance of Arctic grayling in highway culverts. Final report (Contract 14-16-0001-5207) to U.S. Fish and Wildlife Service, Anchorage, Alaska.

Morrison, J. K., and C. E. Smith. 1986. Altering the spawning cycle of rainbow trout by manipulating water temperature. Progressive Fish-Culturist 48: 52–54.

Parks, D. J., T. Burke, and D. A. Bee. 1986. Arctic grayling culture. Alaska Department of Fish and Game, Division of Fisheries Rehabilitation, Enhancement, and Development, Federal Aid in Sport Fish Restoration, Project F-23-R-1, Final Report, Anchorage.

Sakai, K., M. Nomura, F. Takashima, and H. Oto. 1975. The over-ripening phenomenon of rainbow trout—II. Changes in the percentage of eyed eggs, hatching rate and incidence of abnormal alevins during the process of over-ripening. Bulletin of the Japanese Society of Scientific Fisheries 41:855–860.

Tilsworth, T., and M. D. Travis. 1986. Fish passage through Poplar Grove Creek. Final report of University of Alaska Fairbanks, Water Research Center, to Alaska Department of Transportation and Public Facilities, Juneau.

Zar, J. H. 1984. Biostatistical analysis. Prentice-Hall, Englewood Cliffs, New Jersey.

American Fisheries Society Symposium 10:306–324, 1991

Assessment of Two Denil Fishways for Passage of Freshwater Species

C. Katopodis
Canada Department of Fisheries and Oceans, Freshwater Institute
501 University Crescent, Winnipeg, Manitoba R3T 2N6, Canada

A. J. Derksen
Manitoba Department of Natural Resources, Fisheries Branch
1495 St. James Street, Winnipeg, Manitoba R3H 0W9, Canada

B. L. Christensen
Saskatchewan Parks and Renewable Resources, Fisheries Branch
Post Office Box 3003, Prince Albert, Saskatchewan S6V 6G1, Canada

Abstract.—Fish movements through two Denil fishways were assessed by means of traps at the fish exit (upstream end) of each facility. Located in the Canadian prairies, the Fairford (Manitoba) and Cowan (Saskatchewan) fishways are similar in design and operation. At Fairford, 8,871 fish representing 13 species were caught in the trap, which was operated daily May 6–28 and June 2–12, 1987. White suckers *Catostomus commersoni,* walleyes *Stizostedion vitreum,* and saugers *Stizostedion canadense* made up 93.0% of the run. The Cowan fishway was assessed daily from April 27 to May 11, 1985, and weekly thereafter until June 10, 1985. The four species caught were white suckers, longnose suckers *Catostomus catostomus,* northern pike *Esox lucius,* and walleyes; 11,294 fish were trapped, although it was estimated that over 23,000 fish passed through the fishway. The size range of fish that passed through the fishways was 212–800 mm. The longest Denil fishway section negotiated was 9.5 m at a 12.6% slope. Headwater levels at Fairford were fairly constant, but they decreased at Cowan over the study period. Water depths at the upstream end of the fishways ranged from 0.5 to 1.0 m at Fairford and from 1.2 to 0.8 m at Cowan. Estimated water velocities were low near the bottom of each fishway (0.7–0.9 m/s) and high near the water surface (>1.5 m/s). Although fish movements at both sites were likely obstructed by dams for several decades, all species present ascended the fishways readily. Northern pike, though, appeared to wait 2 to 3 weeks before using the fishway at Cowan. Long residence time by northern pike below this and other dams may be a reflection of behavior in relation to foraging, spawning, or passing through Denil fishways.

Fish movements through the Denil fishways in the Fairford (Manitoba) and Cowan (Saskatchewan) dams were assessed in 1987 and 1985, respectively. The Fairford Dam at the outlet of Lake Manitoba was completed in 1961 and was constructed to maintain lake levels between 247.2 and 247.8 m above sea level (ASL). The Fairford River conveys water from Lake Manitoba to Lake St. Martin, which drains into central Lake Winnipeg. The dam is about 240 km northwest of Winnipeg, Manitoba, and is part of a bridge on Highway 6 (Figure 1). The 73-m-long dam has 11 5.9-m-wide bays, and discharge is regulated by removing or replacing stop logs in one or more of the bays. When the Fairford Dam was constructed, two concrete weirs were incorporated in one of the bays for fish passage. Each weir contained a 610 × 760-mm opening, and stop logs could be placed on the upstream weir. It is unlikely that the weirs were ever effective in passing fish upstream.

As early as 1963, commercial fishermen on Lake Manitoba and the adjoining Lake Winnipegosis expressed concern that walleyes *Stizostedion vitreum* were leaving their lakes via the Fairford River and could not return because of the dam. Plans for a fishway were drawn in 1970 and again in 1974, but no fishway was built. In 1982, plans for a Denil fishway were drafted and the facility was constructed by March 1984. The fishway was incorporated in the third bay from the south bank under the highway bridge. An attraction water flume with a discharge capacity of 3.0 m^3/s was placed alongside of the fishway.

Cowan Lake is a long (50 km), shallow (mean depth, 2.5 m), and narrow (mean width, 645 m) reservoir 120 km northwest of Prince Albert, Saskatchewan (Figure 1). The concrete dam at the

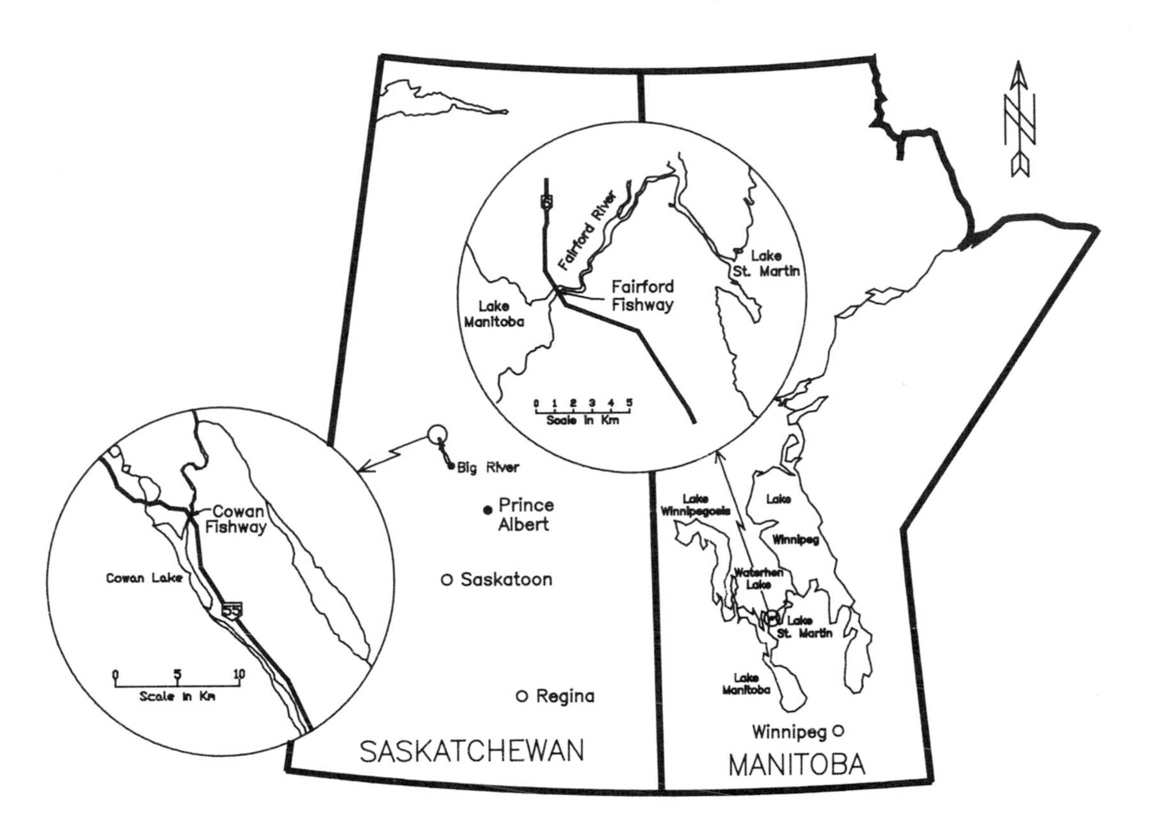

FIGURE 1.—Locations of the Cowan and Fairford fishways.

north end of Cowan Lake was built in 1950 to replace a timber crib dam originally built about 1915 for the purpose of booming logs to sawmills in Big River, Saskatchewan. The dam has four 4.9-m-wide sluiceways, covering a length of 32 m, and the total embankment length is 100 m. Discharge is regulated by a radial control gate in each of the four sluiceways. Cowan Lake has a full supply level of 476.13 m ASL and lake levels are maintained between 475.6 and 476.6 m ASL. The dam was equipped with a fishway that was poorly designed and never successfully passed fish. A Denil fishway was installed during December 1984 and March 1985, while the dam was undergoing repair. The fishway, placed between a sluiceway and the south bank, was built to restore fish access to spawning habitat above the dam and to enhance fish populations within Cowan Lake. The lake is subject to annual summer algae blooms and periodic winterkills of fish. This is due in part to a thick layer of wood debris on the lake bottom from past log booming. Cowan Lake has relatively few fish species (11) compared with other lakes in the area (15–16). Northern pike *Esox lucius* and white suckers *Catostomus commersoni* are the only abundant large species. Walleyes are scarce, probably due to low winter oxygen and a lack of suitable spawning habitat.

The Fairford and Cowan fishways have a similar layout consisting of three flumes equipped with planar baffles, two resting pools, and two vertical lift control gates. Figure 2 shows isometric, plan, and elevation views of the Fairford fishway. The plan view of the Cowan fishway is a mirror image of the Fairford fishway. Table 1 lists the dimensions of each fishway. The net passage width (b) is 400 mm for the Cowan fishway and 300 mm for the Fairford fishway. Fishway slopes are comparable for the upper flumes of each fishway (Fairford: 12.9%; Cowan: 12.6%). For the middle and lower flumes, the slope of the Fairford fishway (12.8 and 12.6%) is higher than the Cowan (10.0 and 10.0%). All three flumes at the Cowan fishway are longer than the corresponding flumes of the Fairford fishway. The resting pools are larger for Cowan than for Fairford. The difference in invert elevations between the water inlet (fish exit) and the water outlet (fish entrance) is slightly greater at

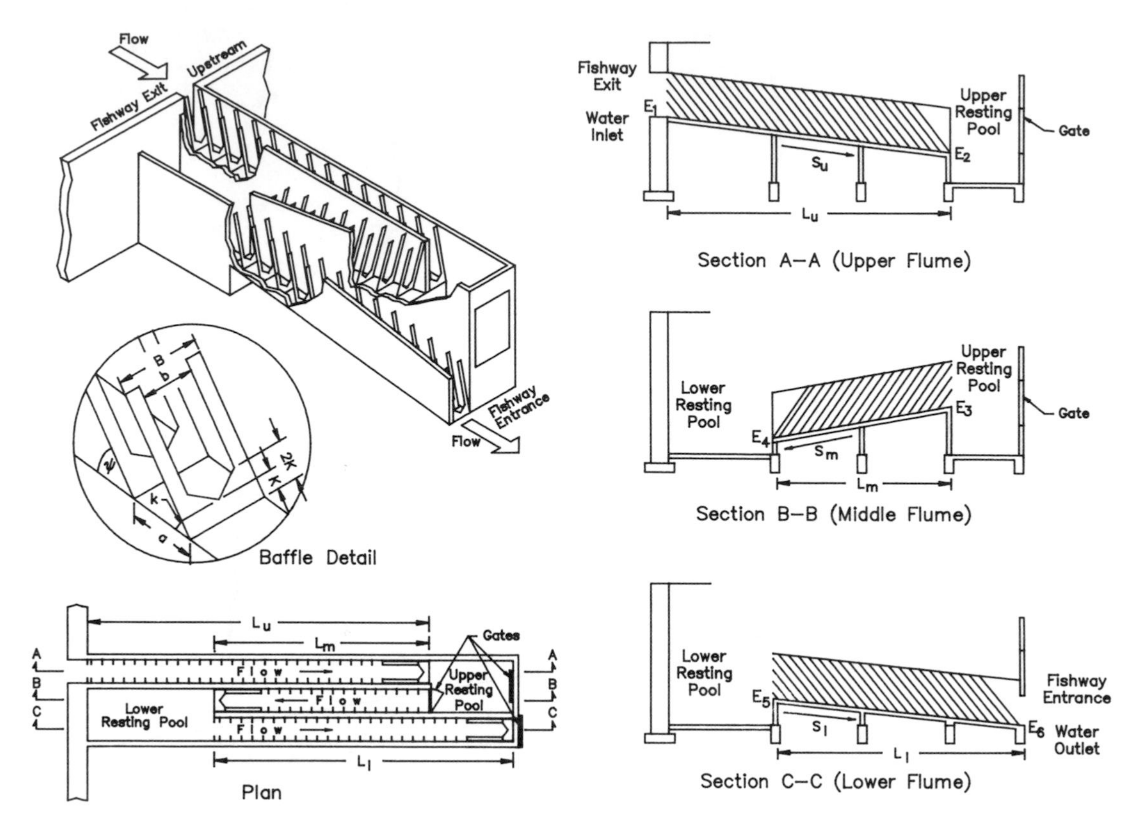

FIGURE 2.—Isometric, plan, and elevation views of the Denil fishways at Fairford and Cowan (mirror images).

Fairford (2.29 m) than at Cowan (2.20 m). The control gates at the outlet of each fishway allow for the operation of either all three fishway flumes when tailwater is low, or only the upper flume when tailwater is high.

Methods

Fairford Fishway

Fish movements through the Fairford fishway were assessed daily during May 6–28 and June 2–12, 1987. Movements were monitored by capturing the fish in a trap fitted to the fishway exit. The trap consisted of a 1.27 × 1.59 × 1.93-m box constructed of two-by-fours and enclosed with a heavy-gauge wire mesh (25 × 50 mm). The top of the trap had a hinged wire-mesh cover to prevent fish from jumping out of the trap. Except on three occasions, the trap was lifted and emptied at least three times per day—in the morning, early afternoon, and evening (0900, 1500, and 1900 hours). Except for May 14–17, all three fishway flumes were operating as a closed system with the top control gate closed and the bottom gate open. On May 14, the lower entrance gate was closed and the top gate was opened so that only the uppermost flume was operating. On May 17, the fishway was converted back to its original setting.

All fish species caught in the fishway trap were counted. The fork lengths (FL) of all walleyes and saugers *Stizostedion canadense* were measured. An attempt was made to tag as many walleyes and saugers as possible with numbered Floy spaghetti tags. Anglers and commercial fishermen were relied upon to voluntarily report tag recaptures. Because of the large numbers of white suckers caught on some days, subsamples of 50 fish were measured. Some lengths were also measured for other incidentally caught species (e.g., cisco *Coregonus artedii,* common carp *Cyprinus carpio,* burbot *Lota lota,* channel catfish *Ictalurus punctatus,* lake whitefish *Coregonus clupeaformis,* and shorthead redhorse *Moxostoma macrolepidotum*).

After each trap lift, water levels were determined at 10 locations within the fishway, at three points in each fishway flume, and at the fishway

TABLE 1.—Dimensions of the Fairford and Cowan fishways. Symbols are defined in Figure 2. Elevations and levels are meters above sea level.

Dimension	Fairford	Cowan
B (mm)	500	634
b (mm)	300	400
a (mm)	300	300
k (mm)	88.4	106.1
K (mm)	125	150
Ψ	45°	45°
Fishway section (upper, middle, lower)		
Length (L, m)	6.3, 5.0, 6.6	9.5, 6.0, 8.5
Slope (S, %)	12.9, 12.8, 12.6	12.6, 10.0, 10.0
Resting pools (length, width, depth)		
Upper (m)	1.45, 1.18, 2.00	235, 1.45, 2.50
Lower (m)	1.39, 1.13, 1.61	3.50, 1.45, 2.50
Control gates (width, length)		
Top (m)	0.536, 1.765	0.9, 1.5
Bottom (m)	0.536, 1.765	0.9, 1.5
Invert elevations (m)		
E_1	246.43	475.00
E_2	245.61	473.80
E_3	245.61	474.25
E_4	244.97	473.65
E_5	244.97	473.65
E_6	244.14	472.80
Total drop ($E_1 - E_6$, m)	2.29	2.20
Full supply level (m)		
	247.76	476.13
Regulated lake levels (m)		
	247.15–247.76	475.6–476.6

exit. The distance from the top of the fishway to the water surface was measured with a metric carpenter's measuring tape. These measurements were subsequently converted into water depths from known fishway dimensions. Upstream and downstream water elevations from staff gauges affixed to the dam were recorded after each trap lift. Water temperatures within the fishway were measured with a maximum–minimum thermometer. Daily discharge records for the Fairford River were obtained from the Water Resources Branch, Manitoba Department of Natural Resources, and the Water Survey of Canada.

Cowan Fishway

Fish movements through the Cowan fishway were assessed daily from April 27 to May 11, 1985, and once a week thereafter until June 10, 1985. The trap fitted to the fishway exit was constructed of square tubular steel, supporting 2.54-mm-square expanded steel mesh walls and floor, and measured 2 m × 2 m × 2 m. Two expanded steel mesh wings in the cage acted as a funnel, allowing fish to enter the cage but preventing them from being swept back down the fishway. This proved effective, although a few suckers were occasionally carried back down the fishway when the trap was crowded. The funnel opening was 2 m tall and 300 mm wide. After an initial period of experimentation, both fishway control gates were left open after April 30, 1985. Due to the high tailwater levels, almost all of the fishway flow passed through the upper flume and very little through the middle and lower flumes. The two main dam discharge gates closest to the fishway were operated in such a manner that fish following along the outside edge of the turbulence created by them would be led to the outlet flow from the fishway entrance. The fish trap was lifted three times daily—morning, afternoon, and evening (at 0900, 1500, and 1900 hours). All fish species were counted. Random samples of each species were weighed and measured (FL) at each lift. Sex and spawning condition (i.e., green, ripe, or spent) were also noted. The length of time the fish trap was out of position was also recorded in order to estimate the total number of fish using the fishway. After each trap lift, water levels were measured at points within the fishway and at the lake and tailrace. Water temperature and pH were also recorded. Some fish were fin-clipped at the fish trap and others were tagged in the tailwater area.

Results and Discussion

Fishway Hydraulics

Fishway depths, discharges, velocities, and water surface profiles are interdependent and relationships between them for Denil fishways have been developed through hydraulic model studies (Katopodis and Rajaratnam 1983, 1984; Rajaratnam and Katopodis 1984; Rajaratnam et al. 1985). Water surface profiles within each fishway were derived from the measurements made along the length of each flume of the Fairford and Cowan fishways. Figures 3 and 4 show examples of such water surface profiles for each fishway. These particular profiles were selected because they represent the highest and lowest water levels recorded in each fishway during the evaluation period. Water depths in the fishway were calculated as the distance from the baffle crest ("V") to the water surface. Water depths near the fishway exit were free of backwater effects and were

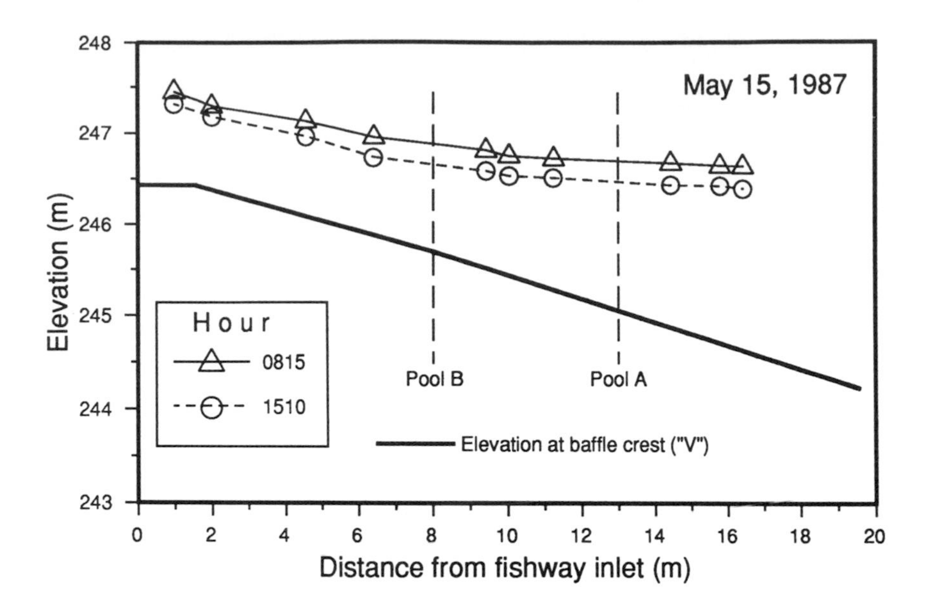

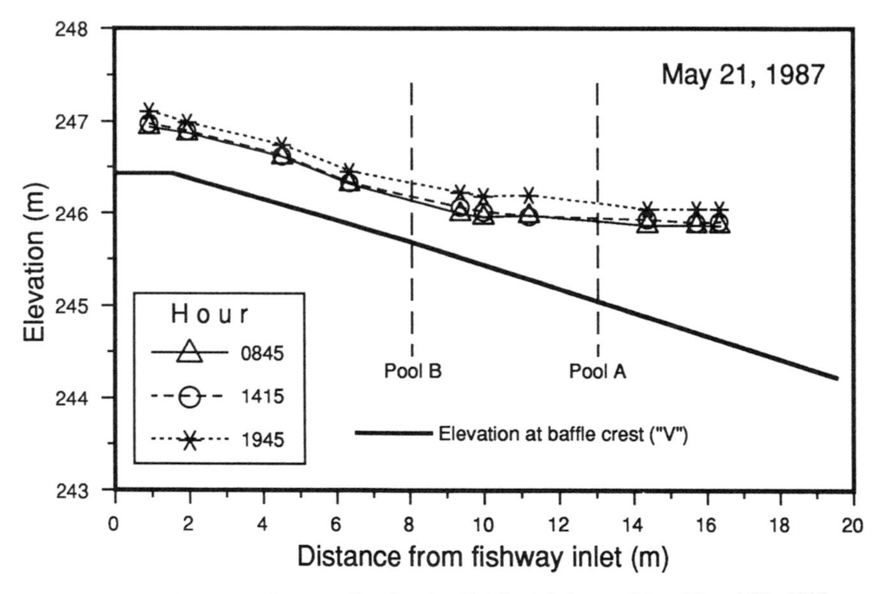

FIGURE 3.—Water surface profiles for the Fairford fishway, May 15 and 21, 1987.

selected to estimate fishway discharge and velocity (Figures 5, 6). At Fairford, the greatest water depth near the fishway exit, 1.01 m, occurred on May 15, 1987; the smallest depth, 0.49 m, occurred on May 21, 1987 (Figure 5). At Cowan, the greatest depth was 1.21 m on April 28 and the least was 0.83 m on May 30, 1985 (Figure 6).

Both the Fairford and Cowan fishways correspond to a standard Denil design for which the following discharge rating curve has been developed through extensive model and prototype testing (Katopodis and Rajaratnam 1983, 1984; Rajaratnam and Katopodis 1984):

$$Q_* = 0.94(d/b)^2; \quad (1)$$

$$Q_* = Q/(gS_o b^5)^{1/2}; \quad (2)$$

Q_* = dimensionless fishway discharge;
Q = fishway discharge (m^3/s);
g = acceleration due to gravity (9.81 m/s^2);

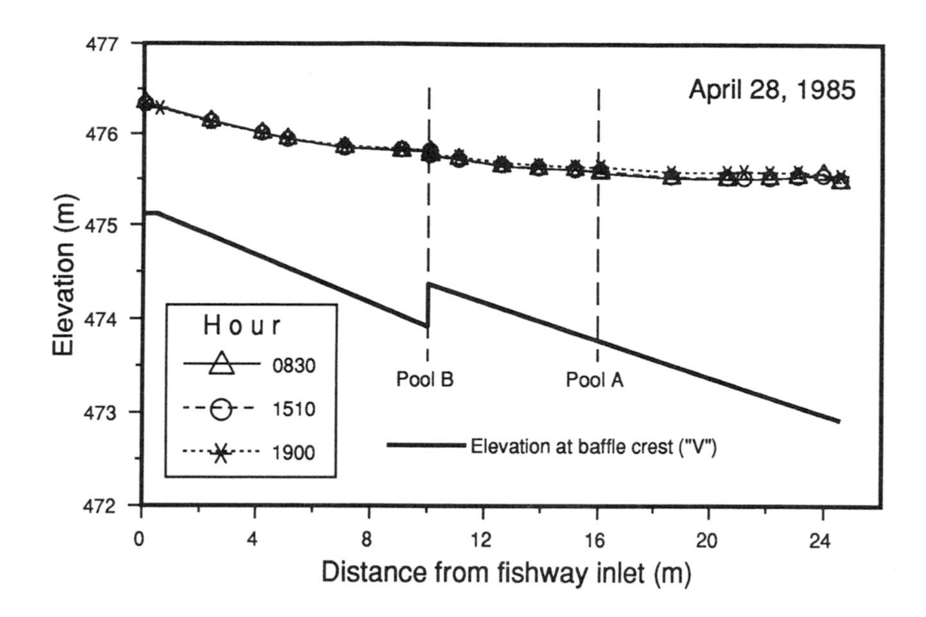

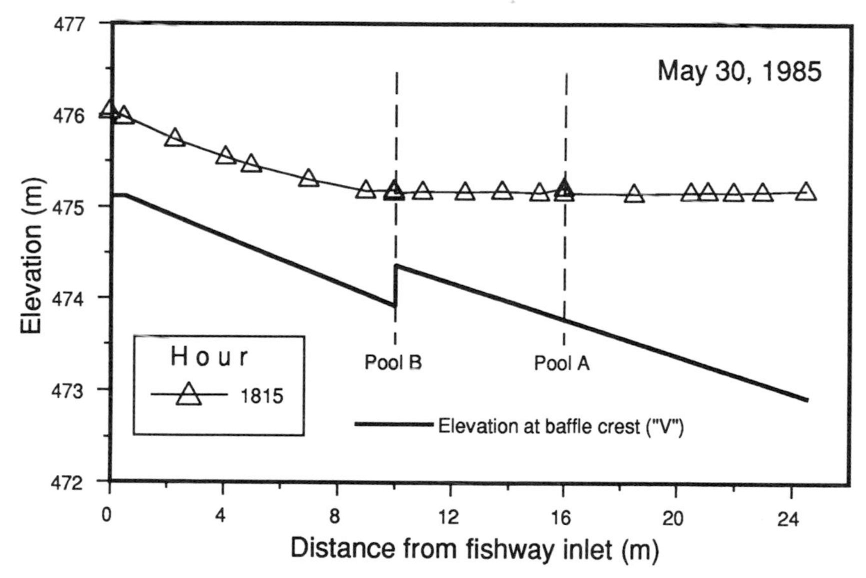

FIGURE 4.—Water surface profiles for the Cowan fishway, April 28 and May 30, 1985.

b = net passage width of the fishway (m);
S_o = fishway slope (m/m);
d = water depth (m; from baffle crest or "V" to water surface).

Given the width (b) of the Fairford and Cowan fishways and the slope of the upper flume in each, equations (1) and (2) reduce to

Fairford fishway $Q = 0.58d^2$; (3)
Cowan fishway $Q = 0.66d^2$. (4)

Equations (3) and (4) were used to estimate the discharge through each fishway from the measured water depths (Figures 5, 6). The Cowan fishway was conveying approximately 14% more discharge than the Fairford fishway for the same depth. Estimated fishway discharge ranged from 0.45 to 0.96 m^3/s at Cowan and from 0.14 to 0.59 m^3/s at Fairford.

The results of the hydraulic model studies referred to earlier were also used to estimate water

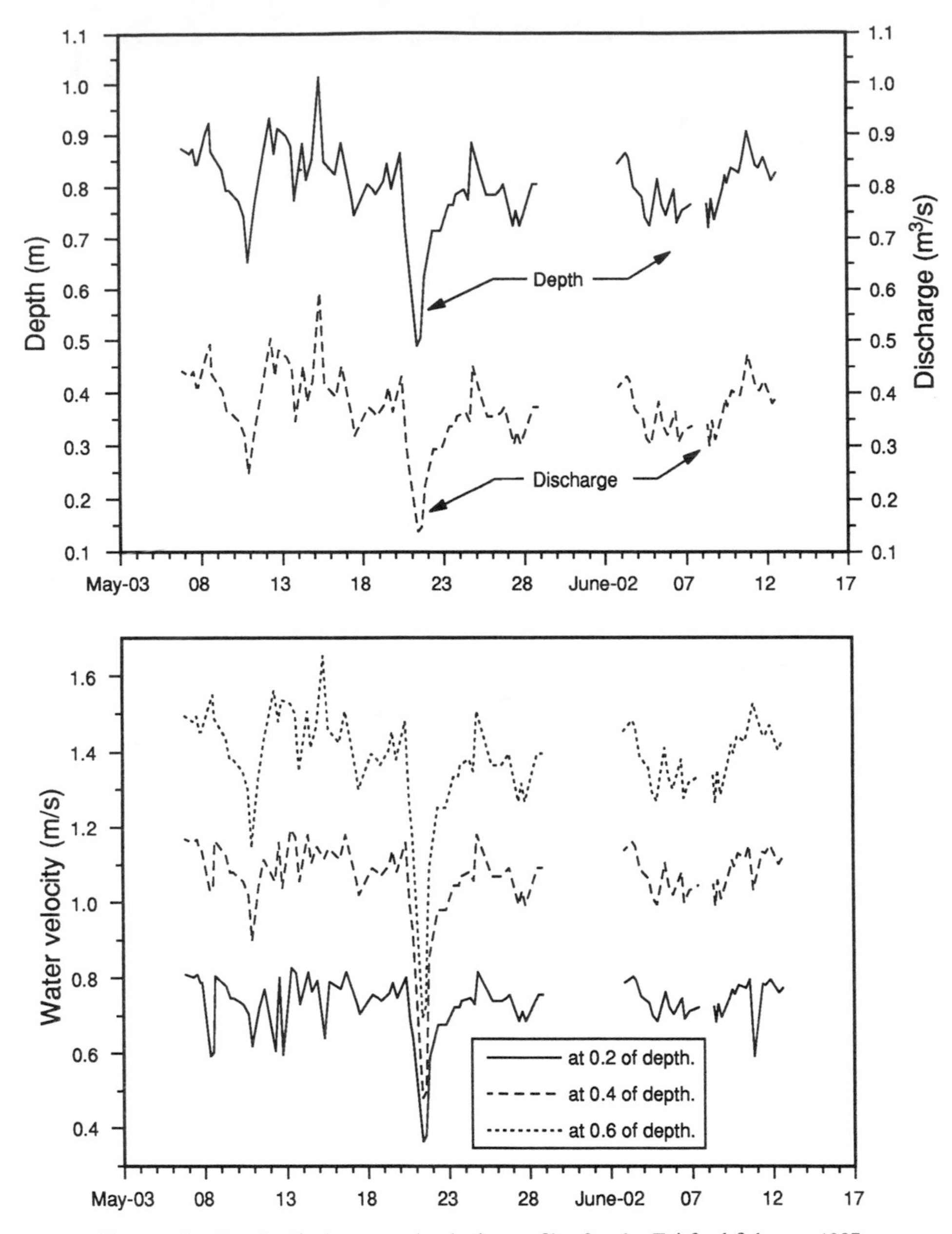

FIGURE 5.—Depth, discharge, and velocity profiles for the Fairford fishway, 1987.

velocities along the centerline of each fishway at a location near the fish exit, where no backwater effect was detected. Water velocities in Denil fishways with the type of baffles used here are low at the bottom of the flume and increase upwards to the water surface. A layer of fast water exists near the water surface. This implies that fish ascending the fishway face varying water velocities dependent on their swimming depth. A representative range of velocities that fish may have negotiated at Cowan and Fairford was estimated by calculating velocities corresponding to $0.2d$, $0.4d$, and $0.6d$, d being the water depth in the fishway (Figures 5, 6).

Fish Passage at the Fairford Fishway

Physical variables from measurements at the Fairford Dam included minimum and maximum

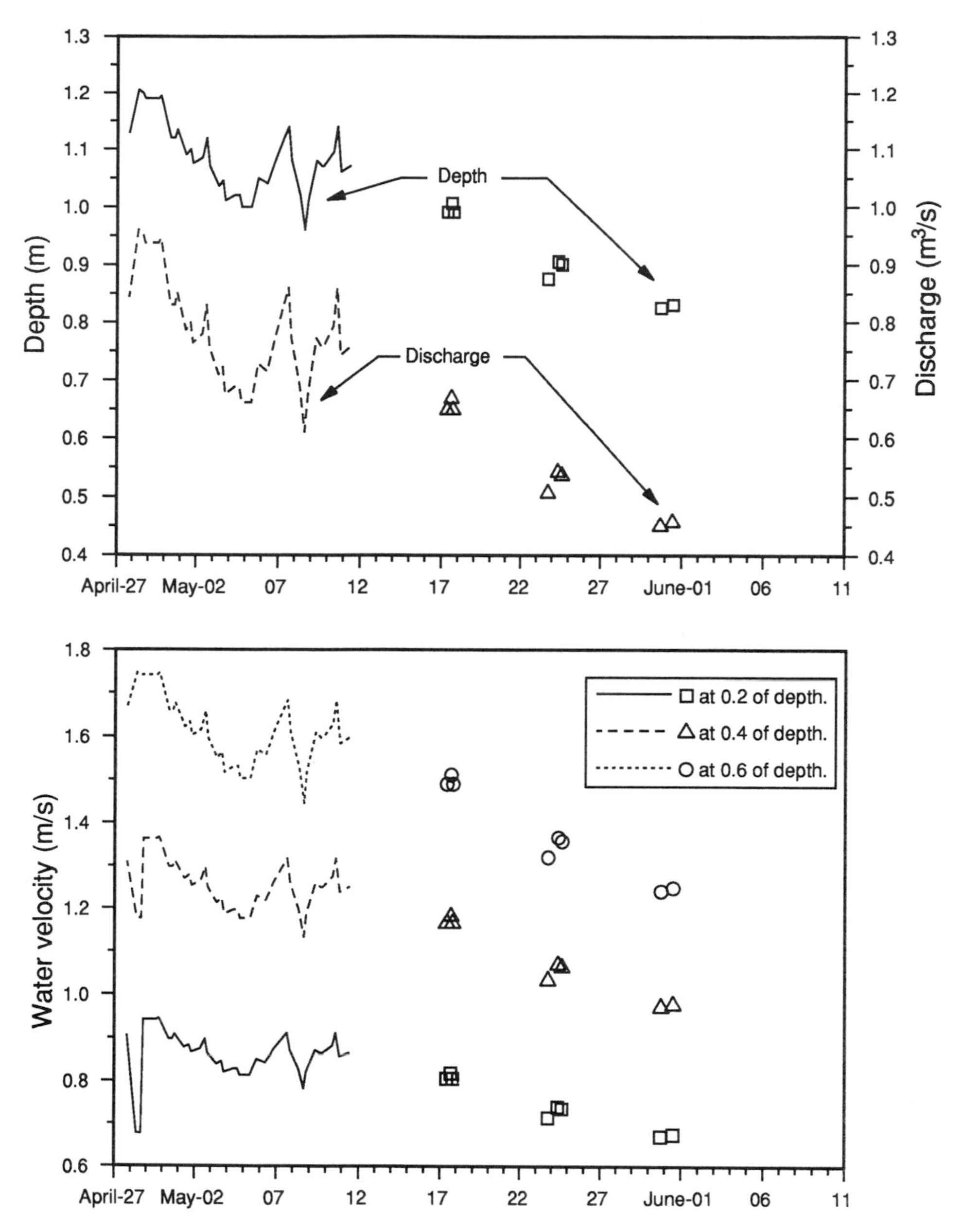

FIGURE 6.—Depth, discharge, and velocity profiles for the Cowan fishway, 1985.

water temperatures in the fishway, water level elevations upstream and downstream of the fishway, and discharge and hydraulic head over the dam (i.e., the difference between upstream and downstream water elevations; Figure 7). Fish passage highlights are presented in Tables 2 and 3. Altogether 8,871 fish, representing 13 species, were captured moving through the Fairford fishway (Table 3). White suckers (57%), walleyes (26%), and saugers (10%) made up 93% of the run. These three species plus cisco, shorthead redhorse, and common carp constituted almost 100% of the catch.

The largest daily catch of white suckers occurred on May 7 (Table 2; Figure 8). This catch is believed to have been part of a spawning run because most individuals were in ripe condition; eggs and milt could be expressed easily with slight

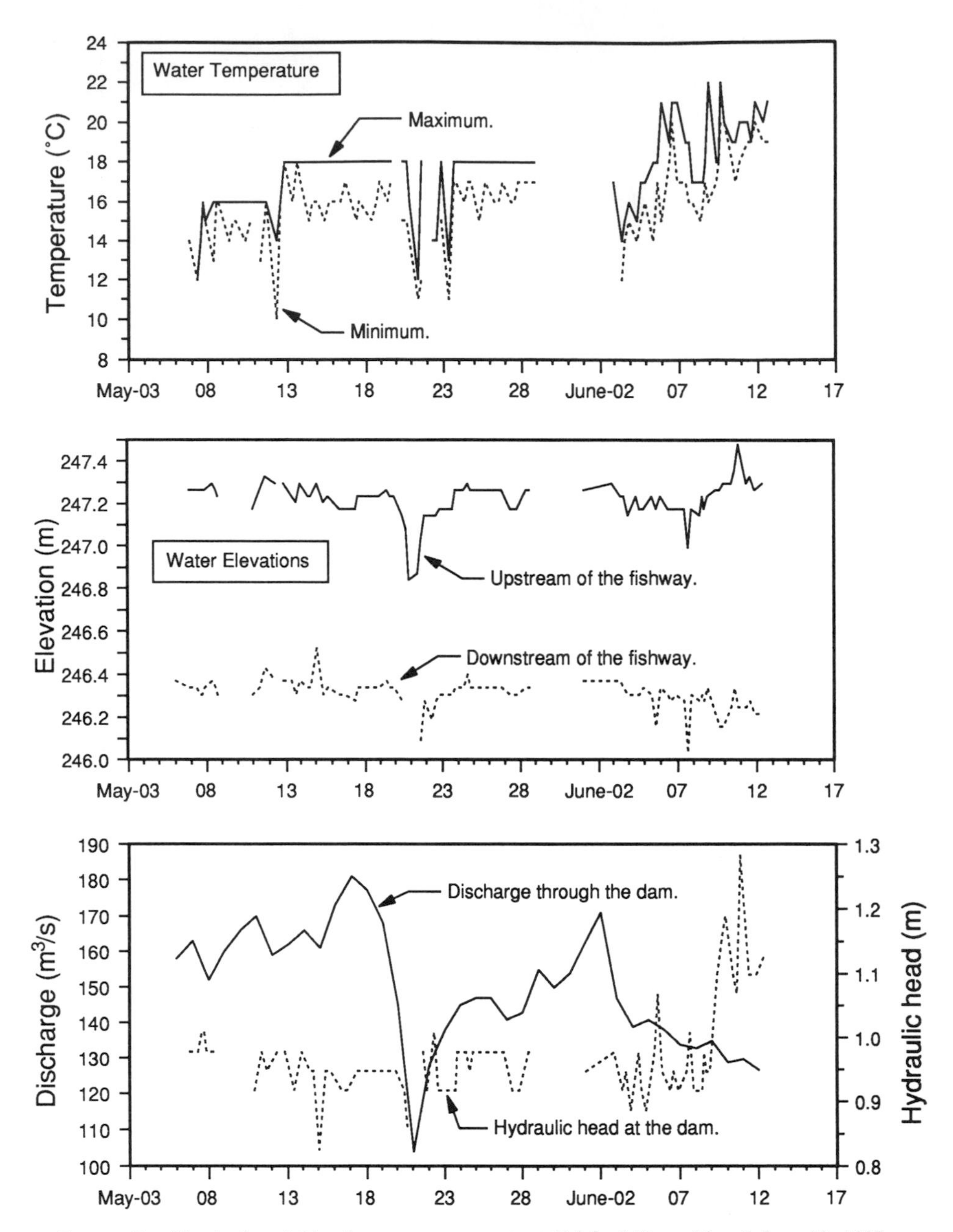

FIGURE 7.—Physical variables from measurements at Fairford Dam, May 6–June 12, 1987.

pressure. After the end of May very few white suckers were caught. The lack of white sucker movements in June may have been due to curtailment of the spawning run by higher water temperatures (Figure 7).

Although some sexually ripe walleyes and saugers were observed, the movement of these species through the fishway was not driven by spawning urges. By the time the fishway assessment began, water temperatures had exceeded the upper limit required for the spawning of these species (Figure 7). The daily catches of both walleyes and saugers were relatively low during the first part of May compared with later catches. Particularly poor catches were made on May 15 and 16. On these dates the gates to the lower and middle fishway flume were closed, the gate to the upper flume was opened, and fish were required to

TABLE 2.—Fish passage highlights for the Fairford Denil fishway from the 1987 field evaluation. Water velocities were estimated from hydraulic model studies based on measured water levels in the fishway. Velocities represent average daily values along the centerline of the fishway, near the exit.

Species	Date	Water temperature (°C)	Water velocities (m/s) in the fishway at 0.2, 0.4, 0.6 of water depth	Fish counted at fish exit		Fork length range (mm)
				Number	%[a]	
White sucker	May 7	16–12	0.8, 1.1, 1.5	1,161	23.1	324–524
	8	16–13	0.7, 1.1, 1.5	588	11.7	308–502
	13	18–16	0.8, 1.1, 1.5	713	14.2	302–520
Walleye	May 18	18–15	0.7, 1.1, 1.4	184	8.0	236–614
	20	18–14	0.7, 1.1, 1.3	123	5.3	250–592
	26	18–16	0.7, 1.1, 1.4	109	4.7	272–590
	28	18–17	0.7, 1.1, 1.4	113	4.9	282–670
	Jun 4	17–14	0.7, 1.1, 1.3	147	6.4	262–546
	6	21–17	0.7, 1.1, 1.3	122	5.3	262–614
	7	19–16	0.7, 1.1, 1.3	180	7.8	254–586
	8	22–15	0.7, 1.1, 1.3	139	6.0	244–458
Sauger	May 10	16–14	0.7, 1.0, 1.3	54	6.0	266–384
	12	18–10	0.7, 1.0, 1.5	49	5.4	248–374
	13	18–16	0.8, 1.1, 1.5	50	5.5	264–356
	18	18–15	0.7, 1.1, 1.4	67	7.4	248–386
	21	18–11	0.4, 0.8, 0.8	72	7.9	230–372
	Jun 6	21–17	0.7, 1.0, 1.3	54	6.0	242–360
	7	19–16	0.7, 1.0, 1.3	103	11.4	230–372
	8	22–15	0.7, 1.0, 1.3	59	6.5	242–386
Cisco	May 11	16–13	0.7, 1.1, 1.4	102	29.0	
	22	18–11	0.7, 1.0, 1.3	90	25.6	220–296
Shorthead redhorse	May 18	18–15	0.7, 1.1, 1.4	21	12.0	
	19	18–16	0.8, 1.1, 1.4	19	10.9	
	26	18–16	0.7, 1.1, 1.4	20	11.4	
	28	18–17	0.7, 1.1, 1.4	32	18.3	
Common carp	Jun 5	21–14	0.7, 1.1, 1.3	22	27.8	

[a]Percentage is based on the number of fish counted for each species.

negotiate only the upper flume (Figure 2). This alteration did not appear to have any significant effect on either the depth of water or velocities within the fishway (Figure 5), and therefore should not have presented a deterrent to fish movement through the fishway. Fish passage may have been delayed for May 15 and 16 because of the change to the conditions at the fishway entrance.

Strong northerly winds on May 20 and 21, accompanied by low temperatures, created a large seiche on Lake Manitoba and significant drops in water levels and discharge at Fairford Dam (Figure 7). A drop in water temperature also occurred at the same time. This event resulted in a substantial drop in water depths, discharges, and water velocities in the fishway on May 21–22 (Figure 5). The reduced water velocities in the fishway, however, did not result in an appreciable increase in fish passage through the fishway. To the contrary, except for saugers on May 21, catches of walleyes and saugers during the low-flow period tended to

TABLE 3.—Fish passage summary for the Fairford Denil fishway from the 1987 field evaluation.

Species	Counted at exit		Measured at exit		Fork lengths (mm)		
	Number	%[a]	Number	%[b]	Range	Interval	%[c]
White sucker	5,032	56.7	496	9.9	214–524	350–500	94.4
Walleye	2,313	26.1	2,193	94.8	236–682	300–400	85.8
Sauger	907	10.2	854	94.2	212–398	250–350	92.3
Cisco	352	4.0	47	13.4	220–296	225–275	83.0
Other[d]	267	3.0					

[a]Percentage is based on total number of fish counted for all species.
[b]Percentage is based on number of fish counted for each species.
[c]Percentage is based on number of fish measured for length.
[d]175 (2.0%) shorthead redhorse; 79 (0.9%) common carp; 4 burbot; 3 lake whitefish; 2 freshwater drum; 1 longnose sucker; 1 silver redhorse; 1 quillback; 1 channel catfish.

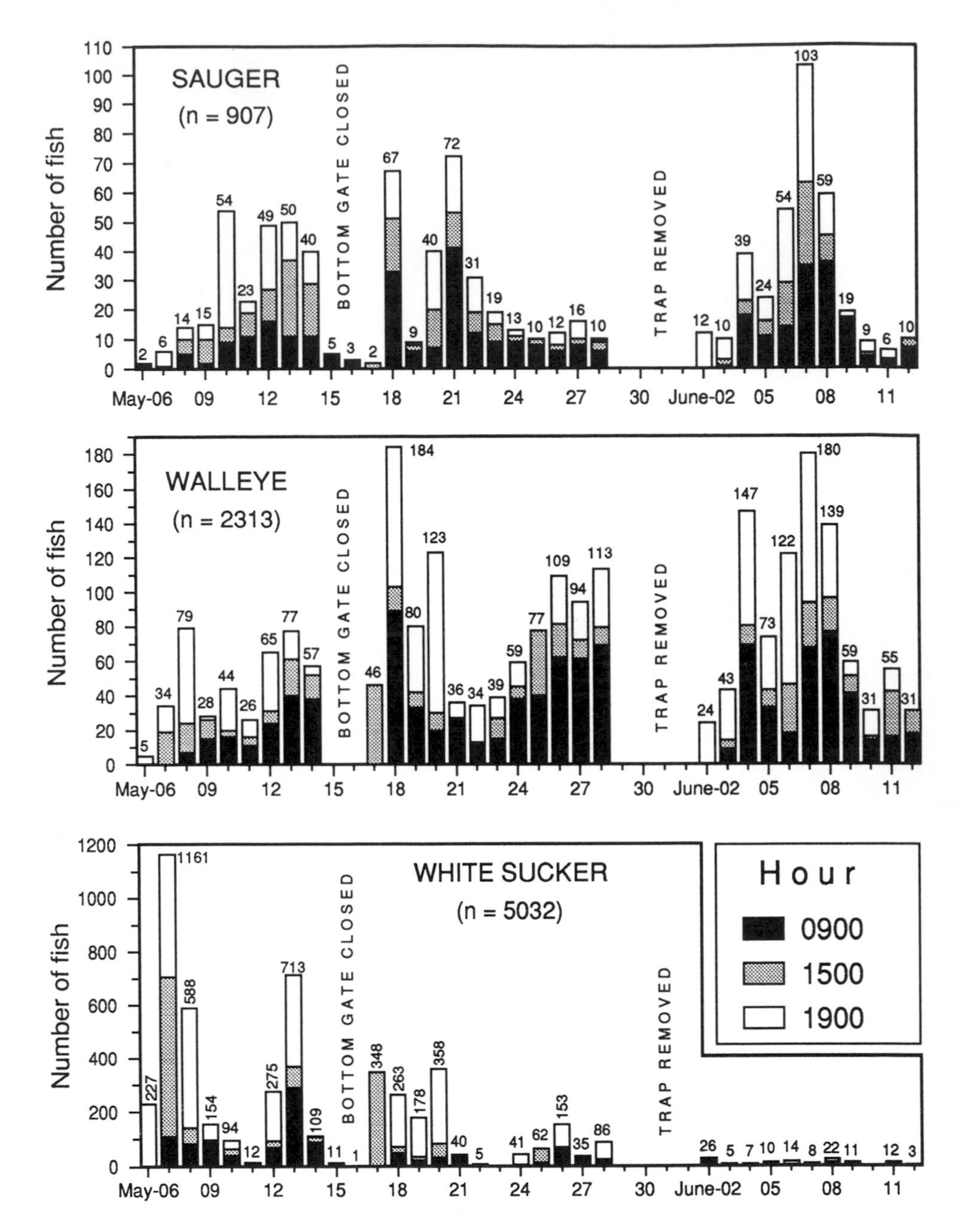

FIGURE 8.—Catch records for the Fairford fishway, 1987.

drop off (Table 2; Figure 8). The reduced movements through the fishway may have been due to the lower temperatures and to the lower flows downstream of the dam (Figure 7), which caused fewer fish to be attracted to the vicinity of the dam and the fishway. Sauger catches continued to decline until the fishway trap was removed on May 28, but walleye catches improved from May 23 to 28.

Good catches of both walleyes and saugers, indicating increased upstream movement, were made for approximately a week after the fishway trap had been reinstalled on June 2 (Table 2; Figure 8). These movements may have been stimulated by a peak in discharge on June 3 (Figure 7). Daily catches of white suckers in June were negligible in comparison with those made during May. Peak movements of ciscoes were observed on May 11 and May 22 (Table 2). These peaks coincided with relatively poor catches of walleyes, saugers, and white suckers. Greater numbers of ciscoes may have passed through the fishway than were observed. Many of the ciscoes caught in the trap were wedged in the wire mesh. This suggests that slightly smaller fish may have been able to squeeze through the wire mesh of the trap.

Shorthead redhorses were caught moving through the fishway primarily during the latter part of May and June (Table 2). Approximately two-thirds of the common carp caught in the Fairford Dam fishway were caught in June. Average water temperature in the Fairford River at this time approached 17°C, which is the optimum spawning temperature for common carp. The catches of all other species in the fishway trap were too small and sporadic to reveal any pattern in movements.

About 86% of the walleyes that used the Fairford fishway ranged in size from 300 to 400 mm FL (Table 3; Figure 9). Relatively small numbers of walleyes less than 250 mm or greater than 500 mm were observed. A similar size distribution was apparent in the walleye population in the Fairford River downstream of the dam. A creel census of anglers fishing below Fairford Dam indicated most walleyes were in the 300–400-mm size range; the ages of these fish were between 4 and 5 years (Derksen 1988). Walleyes greater than 400 mm were 6 years old or older.

Approximately 92% of saugers captured in the fishway trap were between 250 and 350 mm long (Table 3; Figure 9). Few small saugers (<250 mm) were caught. The poor representation of small saugers may be related to the size selection of the mesh used in the fishway trap. Small saugers were frequently found wedged in the wire mesh. Saugers less than 225 mm were probably able to slip through the 25 × 50-mm mesh. The creel census did not reveal whether saugers less than 250 mm were prevalent in the population because anglers did not normally retain saugers or walleyes that were shorter than 300 mm. Only 13% of the ciscoes and 10% of the white suckers caught in the trap were measured (Table 3). Many of the small ciscoes (<225 mm) may have escaped through the wire mesh. The length distribution for white suckers followed a normal distribution with 94% of individual fish falling within the 350–500-mm size range (Table 3; Figure 9). The majority of these fish were mature spawning-run fish.

Except for the period from May 21 to 22, estimated water velocities at the three depths in the fishway remained fairly uniform throughout the assessment period (Figure 5; Table 2). At 20% of the depth (0.20*d*) velocities varied around 0.7 m/s. Although fish could not be observed directly when passing through the fishway, it is suspected that most fish used the deeper part of the fishway flume where the slower velocities existed. This contention is supported by observations when the fishway exit was blocked: fish near the exit were forced toward the surface, could not overcome the faster velocities in the upper layers, and were swept back down the fishway. Common carp were the largest fish to negotiate the fishway; one individual captured in the fishway trap was 780 mm FL and weighed over 10 kg (Derksen 1988). Saugers as small as 212 mm FL were recorded passing through the fishway (Table 3). Smaller fish may have passed through the fishway undetected, escaping through the trap's wire mesh.

Fish Passage at the Cowan Fishway

Physical variables from measurements at the Cowan Dam are presented in Figure 10. Over the duration of the assessment of the Cowan fishway, 1,095 northern pike, 342 walleyes, 5,054 white suckers, and 4,803 longnose suckers *Catostomus catostomus* were counted as they passed through the fishway (Table 4). White and longnose suckers made up 87% of all fish caught. The percentage for northern pike (10%) is believed to be low and is attributed to a bias in the sampling program. Peak passage of northern pike was recorded during weekly counting (May 12–June 10), whereas peak passage for the other species occured during daily counting (April 27–May 11; Figure 11). By interpolating within the periods when the fish trap was out of position for counting and sampling and (toward the end of the study) when the trap was not operated, we estimate that 23,200 fish used the fishway in 1985. These estimates consisted of 6,600 northern pike (28.5%), 600 walleyes (2.6%), 6,000 longnose suckers (25.9%), and 10,000 white suckers (43.1%).

Movement of fish was at its strongest during daylight hours and tended to increase as water temperature rapidly increased in early May. The largest single catch in the trap occurred on May 8 when 1,388 fish ascended the fishway in a 6-h period. May 8 was also the day when the largest

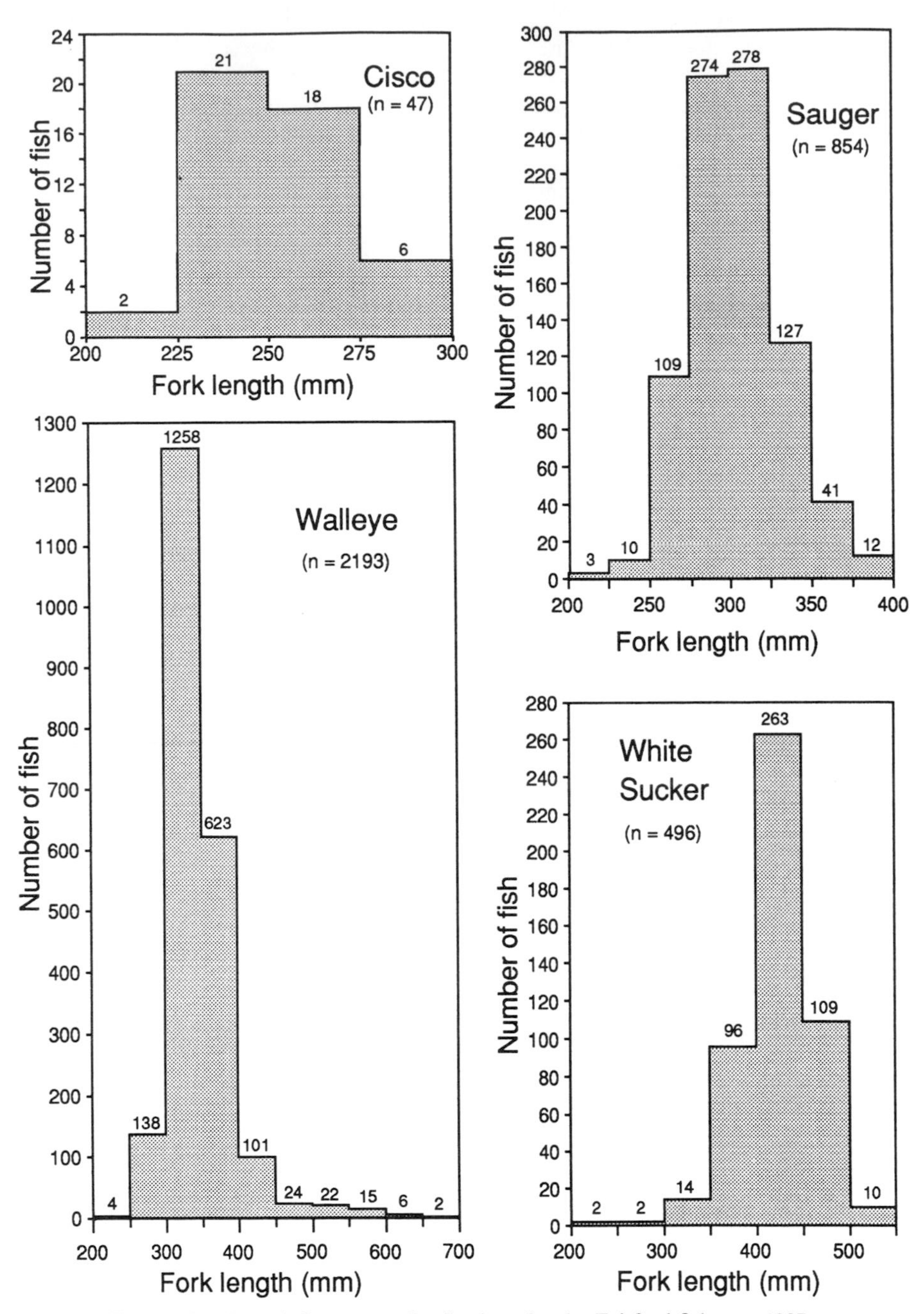

FIGURE 9.—Length-frequency distributions for the Fairford fishway, 1987.

number of fish, 2,604, ascended the fishway. By the time the fishway trap was removed on June 10, longnose suckers had ceased migrating, and only a few walleyes and northern pike were still ascending the fishway. White suckers, based on numbers ascending during 24-h periods (95 on May 23–24, 87 on May 30–31, and 89 on June 10), were still ascending the fishway at a rate of almost 90 per day.

Northern pike moved most strongly through the fishway from May 17 to May 24 (Table 5; Figure 11) at water temperatures between 17 and 18°C (Table 5; Figure 10). All northern pike ascending the fishway at this time were spent. Peak move-

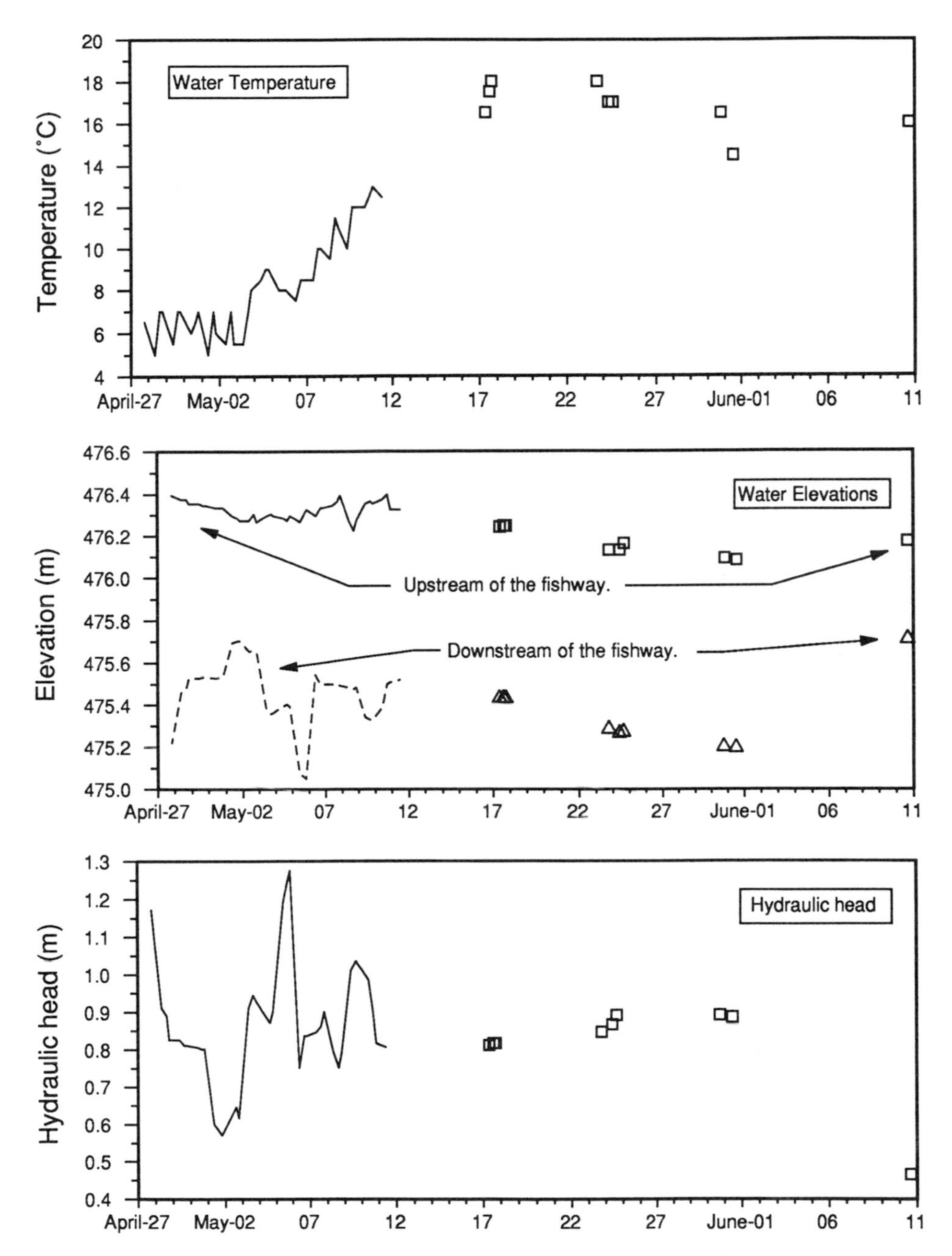

FIGURE 10.—Physical variables from measurements at Cowan Dam, April 27–June 11, 1985.

ment of northern pike occurred when velocities in the fishway were generally decreasing (Figure 6). Walleyes moved through the fishway most strongly on May 4 and May 7, when water temperatures were between 8 and 10°C. Peak movement of longnose and white suckers occurred from May 6 to 10, when water temperatures were between 8.5 and 12.5°C (Table 5; Figure 10). The peak movements of these species appeared to correspond to optimum spawning temperatures. The majority of walleyes, white suckers, and longnose suckers ascending the fishway were in spawning condition but not spent.

The length-frequency distributions of the species counted at Cowan are shown in Figure 12 and summarized in Table 4. It is worth noting that 97%

TABLE 4.—Fish passage summary for the Cowan Denil fishway from the 1985 field evaluation.

Species	Counted at exit		Measured at exit		Fork lengths (mm)		
	Number	%[a]	Number	%[b]	Range	Interval	%[c]
White sucker	5,054	44.8	1,229	24.3	250–498	350–500	96.6
Longnose sucker	4,803	42.5	746	15.5	347–532	350–500	93.7
Northern pike	1,095	9.7	853	77.9	324–800	350–500	96.8
Walleye	342	3.0	341	99.7	265–480	350–450	90.0

[a]Percentage is based on total number of fish counted for all species.
[b]Percentage is based on number of fish counted for each species.
[c]Percentage is based on number of fish measured for length.

of northern pike and white suckers and 94% of longnose suckers were between 350 and 500 mm in length, whereas 90% of walleyes were between 350 and 450 mm. In contrast to the Fairford results, the length frequencies at Cowan reflect all sizes of fish that passed through the fishway, because of the small mesh size used. The smallest fish passing through the fishway was a 250-mm white sucker, whereas the largest fish was an 800-mm northern pike.

Fish Movements

Fish were tagged or fin-clipped at the fishway traps and released upstream of the Fairford or Cowan dams. During the period of study, 2,036 walleyes and 667 saugers were tagged at the fishway trap and released upstream of the Fairford Dam. An average of 16.5 d later (range, 5–41 d), 91 walleyes were recaptured in the trap. Another 78 walleyes were recaptured by anglers below the dam an average of 19.1 d later (range, 0–122 d). Trap recaptures included 14 saugers caught an average 16.5 d after tagging (range, 5–41 d). Anglers caught another 22 saugers below the dam an average 27.5 d after tagging (range, 3–92 d). The recaptures imply that at least 7.6% (205) of all fish tagged—169 walleyes (8.3%) and 36 saugers (5.4%)—had moved downstream over the

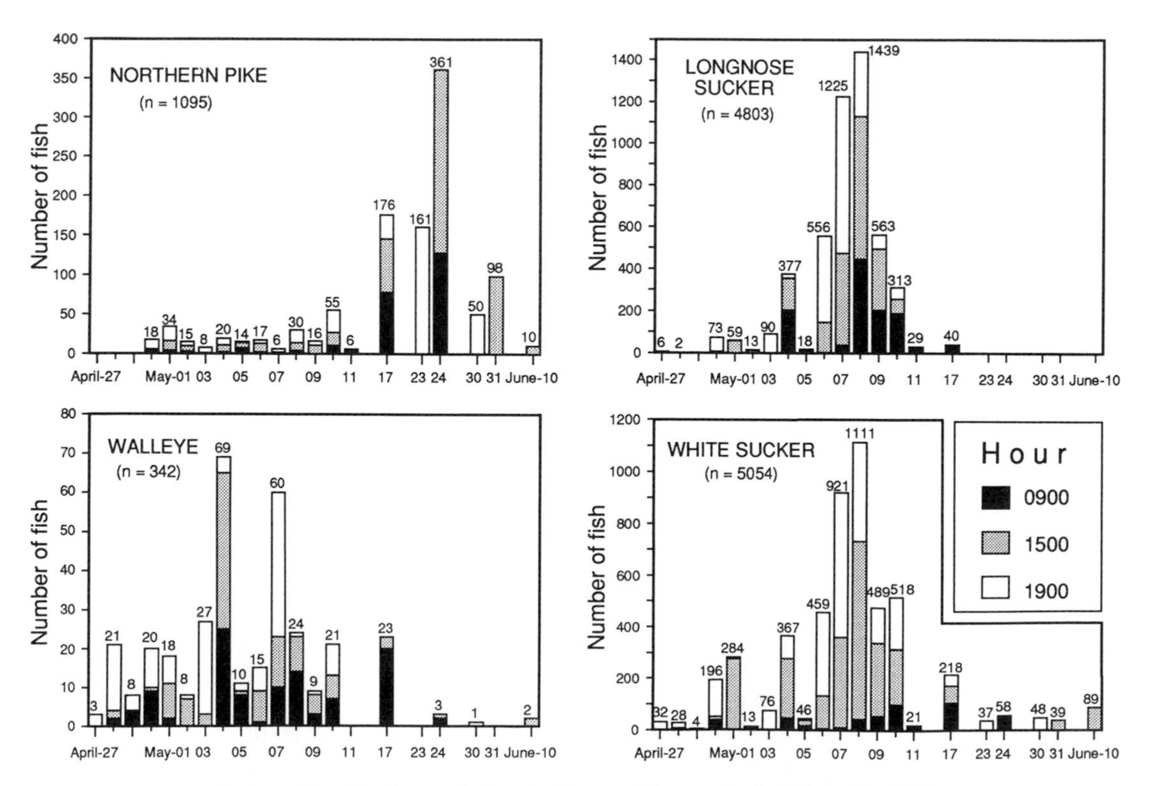

FIGURE 11.—Catch records for the Cowan fishway, April 27–June 10, 1985.

TABLE 5.—Fish passage highlights for the Cowan Denil fishway from the 1985 field evaluation. Water velocities were estimated from hydraulic model studies based on measured water levels in the fishway. Velocities represent average daily values along the centerline of the fishway, near the exit.

Species	Date	Water temperature (°C)	Water velocities (m/s) in the fishway at 0.2, 0.4, 0.6 of water depth	Fish counted at fish exit		Fork length range (mm)
				Number	%[a]	
White sucker	May 4	9.0–8.5	0.8, 1.2, 1.5	367	7.3	272–498
	6	8.5–7.5	0.9, 1.2, 1.6	459	9.1	344–476
	7	10.0–8.5	0.9, 1.3, 1.6	921	18.2	276–454
	8	11.0–9.5	0.8, 1.2, 1.5	1,111	22.0	258–461
	9	12.0–10.0	0.9, 1.3, 1.6	489	9.7	250–465
	10	14.0–12.0	0.9, 1.3, 1.6	518	10.2	288–482
Longnose sucker	May 4	9.0–8.5	0.8, 1.2, 1.5	377	7.8	352–522
	6	8.5–7.5	0.9, 1.2, 1.6	556	11.6	352–506
	7	10.0–8.5	0.9, 1.3, 1.6	1,225	25.5	358–527
	8	11.0–9.5	0.8, 1.2, 1.5	1,439	29.9	347–524
	9	12.0–10.0	0.9, 1.3, 1.6	563	11.7	369–532
	10	14.0–12.0	0.9, 1.3, 1.6	313	6.5	353–526
Northern pike	Apr. 30 to May 16	13.0 to 5.0	0.8, 1.1, 1.4 to to to 0.9, 1.3, 1.7	239	21.8	364–800
	May 17	18.0–16.5	0.8, 1.2, 1.5	176	16.1	360–575
	23	18.0	0.7, 1.0, 1.3	161	14.7	346–514
	24	17.0	0.7, 1.1, 1.4	361	33.0	324–646
	30	16.5	0.7, 1.0, 1.2	50	4.6	334–481
	31	14.5	0.7, 1.0, 1.2	98	8.9	371–568
Walleye	May 3	8.0–5.5	0.8, 1.2, 1.5	27	7.9	268–445
	4	9.0–8.5	0.8, 1.2, 1.5	69	20.2	341–478
	7	10.0–8.5	0.9, 1.3, 1.6	60	17.5	311–468
	8	11.0–9.5	0.8, 1.2, 1.5	24	7.0	340–460
	17	18.0–16.5	0.8, 1.2, 1.5	23	6.7	343–406

[a]Percentage is based on number of fish counted for each species.

dam. Although northern pike were represented in angler creels below the Fairford Dam, no northern pike was observed near or in the fishway. At Cowan, the right pectoral fins of 119 northern pike, 214 walleyes, 366 longnose suckers, and 523 white suckers were fin-clipped between April 27 and May 7, 1985. A few of these fish were sampled a second time, after being carried downstream through the dam gates and ascending the fishway again. Three white suckers, two longnose suckers, one walleye, and no northern pike that had been fin-clipped were recaptured in the trap. The recaptures indicate that of all the fish tagged at Fairford (2,703) and Cowan (1,222), 3.9% (105) and 0.5% (6), respectively, completed a double circuit of moving up the fishway, down over the dam, and up the fishway again.

During the assessment period at the Cowan fishway, thousands of northern pike were observed downstream of the dam. Anglers were so successful that the area was closed to angling. Angling was easier than seining and was used to capture and tag northern pike below the dam. It appeared that there was no suitable spawning habitat or enough food in the tailwater area. It is very likely that the fishway provided fish an opportunity to move into Cowan Lake for the first time in several decades. Although the reasons are not clear, northen pike are known to congregate below weirs and fishways on several other rivers in Alberta (Nelson 1983; Fernet 1984).

Numerous northern pike and a few suckers were observed attempting to ascend the Cowan fishway near the surface, but they were being swept back before they had ascended the first 1–2 m of the upper flume. No fish were observed successfully ascending the fishway by swimming near the surface. From Figure 6 and Table 5 it is evident that water velocities at 60% of the depth (0.6*d*) were almost double those at 20% of the depth (0.2*d*). This implies that if northern pike preferred to swim close to the surface they would have faced much higher velocities than other species who may have preferred to swim at lower levels and could have taken advantage of the lower velocities there. On May 24, 33% of the northern pike moved through the fishway when estimated velocities were 0.7 m/s at 0.2*d*, 1.1 m/s at 0.4*d*, and 1.4 m/s at 0.6*d*. In contrast, it took 17 d, from April 30 to May 16, for 22% of the

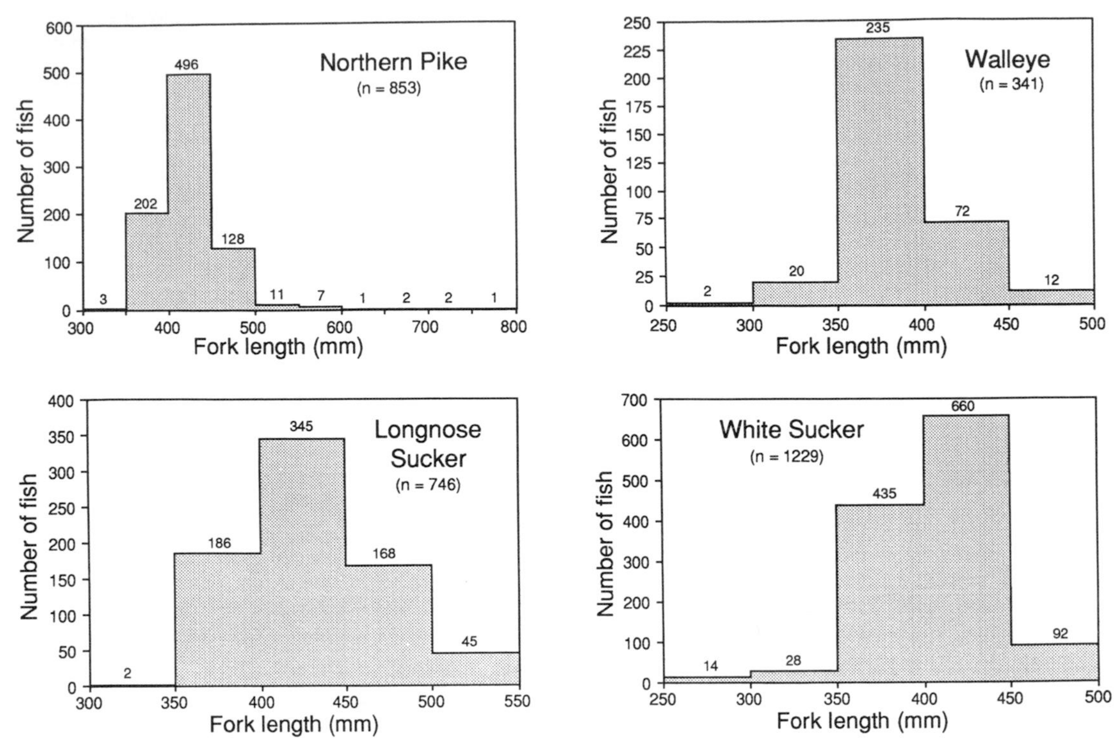

FIGURE 12.—Length-frequency distributions for the Cowan fishway, 1985.

northern pike counted to pass through the fishway; during this period, estimated water velocities were 0.8–0.9 m/s at 0.2*d*, 1.1–1.3 m/s at 0.4*d*, and 1.5–1.7 m/s at 0.6*d*. It is unlikely that the small decrease in water velocities (0.1–0.3 m/s) would have resulted in the significant improvement in northern pike passage through the fishway after May 17. It would appear that hydraulic conditions within the fishway did not hinder the passage of northern pike, unless northern pike were attempting to navigate the fishway close to the water surface at the start of the study period and switched to swimming through the bottom part of the flume later on.

At Cowan, 377 northern pike, 3 walleyes, 26 white suckers, and 17 longnose suckers were tagged below the fishway between April 28 and May 9. Different-colored spaghetti tags inserted on the left or right side of the fish were used to indicate the date of tagging. All fish that ascended the fishway were checked for tags before being released above the dam. There were sufficient tag returns to obtain a rough estimate of the relative residence time of the different species in the tailwater pool before they ascended the fishway. Two of 17 longnose suckers (12%) tagged in the tailwater pool ascended the fishway while the trap was in operation. One ascended 1 d after tagging, and the other ascended 9 d after tagging, for a mean residence time of 5 d. Three of 26 tagged white suckers (12%) ascended the fishway during the counting period; they ascended 6–9 d after tagging, for a mean residence time of 7.7 d. Five of the 377 tagged northern pike (1.3%) were captured in the trap. Residence time in the tailwater pool ranged from 12 to 24 d after tagging for an average residency of 16.2 d.

Fernet (1984) investigated the effectiveness of a standard Denil fishway (b = 360 mm; d = 0.60–0.83 m) at the outlet of Fawcett Lake in Alberta. Upstream and downstream fish movements were monitored at a counting fence installed downstream of the Fawcett Lake weir. Fish ascending the fishway were monitored at a trap installed at the fishway exit. A substantial number of fish were present between the counting fence and the weir at the time the fishway was installed (May 27, 1983), and an unknown number of fish passed through holes in the counting fence before June 17.

Between May 28 and June 29, 1983, 293 northern pike ascended the Fawcett River and entered the upstream trap at the counting fence, while 94 were caught in the fishway trap. The total number of northern pike tagged in the study area was 384, whereas the number of tagged northern pike that ascended the fishway was 40, indicating that 10.4% of the marked population used the Denil fishway. Of the tagged northern pike, 11 or 2.9% of the marked population moved downstream through the counting fence, whereas no northern pike were found to move upstream through the fishway and return downstream over the weir. Fernet (1984) observed that the northern pike that could be seen from the edge of the river were not heavily motivated to migrate upstream, but appeared to hold and feed downstream of the weir. Residence time, or the time between release of tagged northern pike from the upstream trap at the counting fence and recapture in the fishway trap, was an average of 9.9 d (range, 1–26.5 d).

Fernet (1984) also reported that during the study period (1983), 1,432 white suckers moved upstream through the counting fence trap while 1,936 passed through the fishway. Of 1,545 white suckers marked, 903 or 58.5% ascended the fishway. Only one white sucker was found to have returned downstream. Mean residence time for tagged white suckers was 2.2 d (range, 0.5–14.5 d). Forty-five longnose suckers, six walleyes, and two mountain whitefish *Prosopium williamsoni* used the fishway as well.

The results of the Fawcett Lake study proved that for all species present, the Denil fishway performance was superior to that of an experimental Denil installed in 1982, and far superior to the pool-and-weir fishway previously operated at this location. A Denil fishway on the Beaverlodge River near Beaverlodge, Alberta, similar to the one at Fawcett Lake, is used by substantial numbers of Arctic grayling *Thymallus arcticus,* longnose suckers, white suckers, and lesser numbers of northern pike (Alberta Fish and Wildlife, unpublished data). Fernet (1984) concluded that white suckers used the fishway well, and he attributed the low passage rate for northern pike to lack of motivation. He recommended that future studies should attempt to confirm the existence of suspected northern pike spawning areas downstream of the weir.

Schwalme et al. (1985) studied (May 12–June 25, 1984) the effectiveness of a fish passage facility built into a weir on the Lesser Slave River in Alberta. Three fishways—a vertical slot and two standard Denil fishways side by side in a single enclosure constituted the fish passage facility. Of the two Denil fishways (b = 360 mm, d < 1.2 m), one is set at a 10% slope and is 8 m long, and the other is set at a 20% slope and consists of two 2-m-long sections with a resting pool in between. Schwalme et al. (1985) reported that fish using the fishways included thousands of spottail shiners *Notropis hudsonius,* substantial numbers (>100) of northern pike, longnose suckers, white suckers, and immature yellow perch *Perca flavescens,* and lesser numbers of burbot, adult yellow perch, lake whitefish, and trout-perch *Percopsis omiscomaycus.*

High water levels throughout the study period allowed most fish to surmount the weir without using the fishways (Schwalme et al. 1985). Of the fish that chose to use the fish passage facility, northern pike strongly preferred to ascend the Denil fishways and the two sucker species preferred to ascend the vertical slot. From plasma glucose and lactate measurements, Schwalme et al. (1985) concluded that ascending the Denil fishways was at most only moderately stressful for northern pike. Seining immediately below the entrance to the fishways and from a location 100 m downstream of the weir indicated that as a proportion of all fish caught, northern pike were about equally represented at both locations, whereas the two sucker species, burbot, and lake whitefish were much more common near the fishway entrance. Also, northern pike were not underrepresented among the fish caught in the traps at the fishway exits compared with their abundance in the seine hauls. Stomach pumping of seine-caught northern pike indicated that this species was feeding heavily. Schwalme et al. concluded that northern pike were as motivated as other fish in the river to travel upstream past the weir.

Summary and Conclusions

Observations of fish movements through the fishways at Fairford (1987) and Cowan (1985) show that a wide variety and size range of freshwater species successfully use Denil fishways over a range of water levels. Species varied between the two sites, and 14 species in all were trapped at the fishway exits. Similar results with the same and other freshwater species have been reported by other investigators. At Fairford, where the net fish passage width was 300 mm, the size range of fish that negotiated the fishway was 212 mm (sauger) to 780 mm (common carp). Similarly, at Cowan, where the net passage width was 400 mm, fish size ranged from 250 mm (white sucker) to 800 mm (northern pike). The Denil fishways operated over about a 0.5-m range of headwater levels. At Cowan, fish negotiated a 9.5-m-long section at a 12.6% slope; at Fairford, fish passed through a 6.3-m section at a 12.9% slope. The studies at Fairford and Cowan allowed only a partial assessment of the effectiveness of Denil fishways, because most data collected related only to fish that passed through the fishways. The limited tagging data indicate that some fish passed through the fishways twice (3.9% of tagged fish at Fairford and 0.5% at Cowan), and that 7.6% of the fish returned downstream over the Fairford Dam after successfully negotiating the fishway once. Thousands of northern pike were observed downstream of the Cowan Dam. It was estimated from tag returns that northern pike waited 2–3 weeks before using the fishway at Cowan. Long residence times below dams and apparent low passage rates for northern pike have also been observed by other investigators. Although a full explanation has not emerged, there are indications that the factors involved may vary from site to site and appear to involve northern pike behavior in relation to foraging, spawning, or swimming through Denil fishways. More field studies are needed to clarify this and are presently under consideration. These studies should provide observations on fishway entrance conditions on spawning and foraging behavior of northern pike, and they should include comprehensive estimates of fish populations downstream of dams, residence times in tailwaters, numbers of fish using the fishways, and fish swimming depths in Denil fishways.

Acknowledgments

We thank R. H. Gervais for data, text processing, and computer plotting, and S. A. Smith, J. N. Stein, and G. McKinnon for reviewing the manuscript.

References

Derksen, A. J. 1988. An evaluation of the Fairford Dam fishway May–June, 1987 with observations on fish movements and sport fishing in the Fairford River. Manitoba Department of Natural Resources, Fisheries Branch, Manuscript Report 88-6, Winnipeg.

Fernet, D. A. 1984. An evaluation of the performance of the Denil 2 fishway at Fawcett Lake during the spring of 1983. Environmental Management Associates, Calgary, Alberta.

Katopodis, C., and N. Rajaratnam. 1983. A review and laboratory study of the hydraulics of Denil fishways. Canadian Technical Report of Fisheries and Aquatic Sciences 1145.

Katopodis, C., and N. Rajaratnam. 1984. Similarity of scale models of Denil fishways. Pages 2.8-1 to 2.8-6 *in* H. Kobus, editor. International Association for Hydraulic Research symposium on scale effects in modelling hydraulic structures. Technische Akademie, Esslingen, West Germany.

Nelson, R. L. 1983. Northern pike (*Esox lucius*) and white sucker (*Catostomus commersoni*) swimming performance and passage through a step and pool fish ladder. Master's thesis. University of Alberta, Edmonton.

Rajaratnam, N., and C. Katopodis. 1984. Hydraulics of Denil fishways. Journal of Hydraulic Engineering 110:1219–1233.

Rajaratnam, N., C. Katopodis, and G. Van der Vinne. 1985. M1-type backwater curves in Denil fishways. Proceedings of the seventh Canadian hydrotechnical conference, volume IB:141–156, Canadian Society for Civil Engineering.

Schwalme, K., W. C. Mackay, and D. Lindner. 1985. Suitability of vertical slot and Denil fishways for passing north-temperate, nonsalmonid fish. Canadian Journal of Fisheries and Aquatic Sciences 42:1815–1822.

American Fisheries Society Symposium 10:325–334, 1991

Evaluation of Rotating Drum Screen Facilities in the Yakima River Basin, South-Central Washington State

DUANE A. NEITZEL, C. SCOTT ABERNETHY, AND E. WILLIAM LUSTY

Pacific Northwest Laboratory[1]
Post Office Box 999, Richland, Washington 99352, USA

Abstract.—Field-test data from four rotating drum screen facilities indicate that juvenile salmonids are safely returned from the fish screen facility to the river from which the fish were diverted. This conclusion is based on five observations: (1) release–recapture tests with branded salmonids indicated fish that passed through the screening facility were not killed or injured at different rates from control groups; (2) predators were not concentrated within the screen facility; (3) test groups of fish were not delayed within the screen facility; (4) screens with properly maintained seals prevented fish from passing through the screen structure; and (5) altered operating flow conditions did not adversely affect test conclusions. Tests were conducted with smolts of steelhead *Oncorhynchus mykiss* and spring chinook salmon *Oncorhynchus tshawytscha*, and with fall chinook salmon fry. More than 11,000 fish were released during the tests. Conclusions are based on the condition of nearly 8,000 test fish captured. Tests were conducted at the Sunnyside, Richland, Toppenish/Satus, and Wapato Canal Fish Screening Facilities in south-central Washington State.

The Yakima River basin once supported large runs of salmonids. During the late 1800s, between 500,000 and 600,000 adult salmon *Oncorhynchus* spp. returned to the Yakima River and its tributaries (Bureau of Reclamation 1984). Runs of salmon included several species and races: spring, summer, and fall chinook salmon *O. tshawytscha*, coho salmon *O. kisutch*, sockeye salmon *O. nerka*, and steelhead *O. mykiss*.

Reduced salmonid runs to the Yakima River basin result from many factors. Some of the runs are now extinct or near extinction. There is no sockeye salmon run in the Yakima River basin today, and only 37 coho salmon passed the Prosser Diversion Dam in 1983 (Hollowed 1984). Spawning escapement averaged about 2,000 salmonids in the early 1980s (U.S. Bureau of Reclamation 1984). Recent improvements in efforts to manage and enhance salmonid runs in the Yakima River increased the total spawning escapement to 8,000 adults in 1986 (Fast et al. 1986).

The Pacific Northwest Electric Power Planning and Conservation Act (Public Law 96-501) was passed to enable preparation and implementation of a regional Conservation and Electric Power Plan. The Northwest Power Planning Council (NPPC) administers the plan and is charged with developing a program to protect and enhance fish and wildlife populations and to mitigate adverse effects from development, operation, and management of hydroelectric facilities (NPPC 1984).

The Yakima River basin was selected as one site for enhancement of salmon and steelhead runs. Under the Plan, the Bonneville Power Administration (BPA) and the U.S. Bureau of Reclamation (BR) are funding the construction of fish passage and protection facilities at 20 existing irrigation and hydroelectric diversions in the Yakima River basin (Figure 1). The BPA is also providing funds to the Yakima Indian Nation to increase production of spring chinook salmon in the Yakima River basin.

The Sunnyside, Richland, Toppenish/Satus, and Wapato Canal fish screening facilities (Sunnyside, Richland, Toppenish/Satus, and Wapato screens) are part of the passage and protection facilities being constructed by BPA and BR. We evaluated the effectiveness of these facilities for safely passing juvenile salmonids downstream.

Description of the Fish Screening Facilities

The fish screening facilities include a canal, trash rack, drum screens, fish bypass, and fish return pipe (Figures 2, 3). The complexity of the facilities varies from site to site. Fish enter the canal through headgates at the diversion. A trash rack in the canal, perpendicular to the canal flow, removes large debris upstream of the drum screen facility. The concrete screening facility houses cylindrical screens with axes parallel to the length of the structure. Water depth at the screens varies with canal flow. However, the average depth

[1]The Pacific Northwest Laboratory is operated for the U.S. Department of Energy by Battelle Memorial Institute.

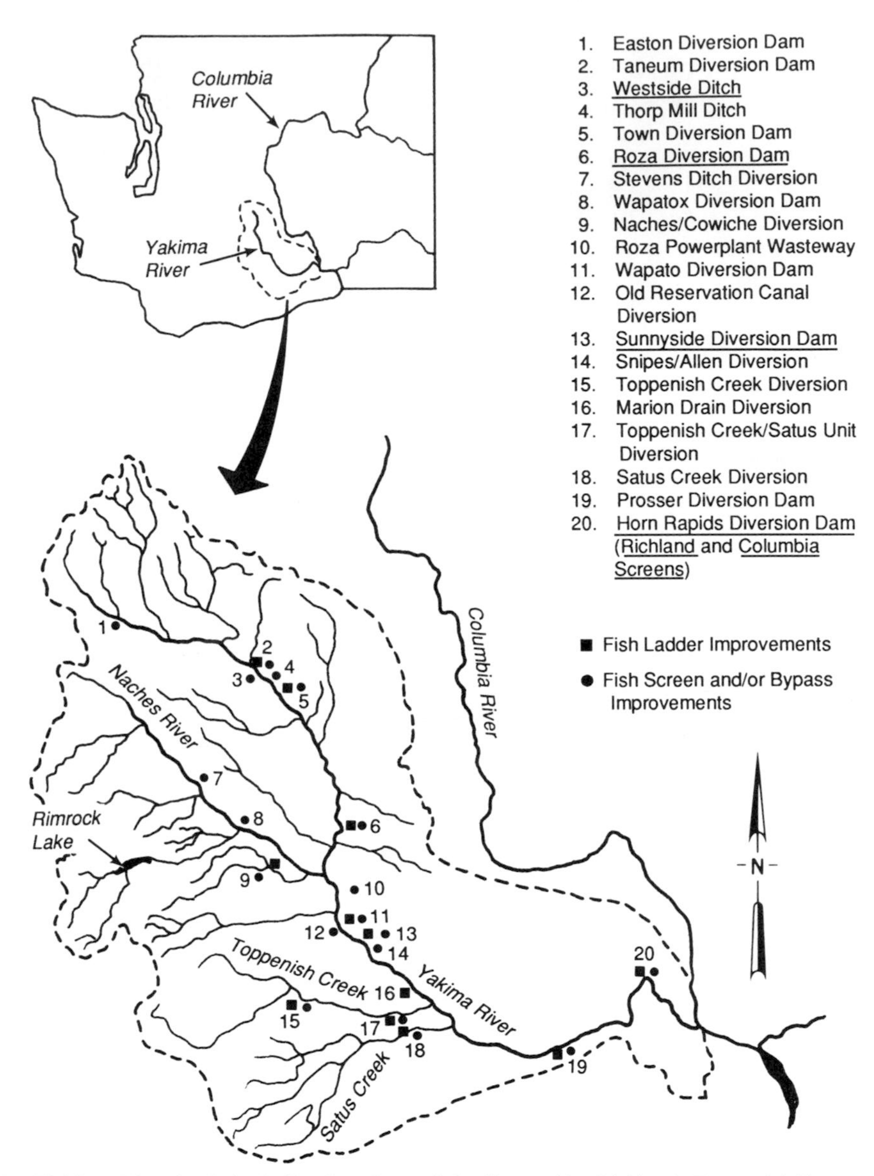

FIGURE 1.—Yakima River basin including locations of the Sunnyside, Richland, Toppenish/Satus, and Wapato fish screening facilities and other fish protection and passage facilities.

across the face of the screens is about 75–80% of the screen diameter. Screen mesh openings are 3.18 mm.

The screen facilities are installed at an angle of 26° to canal flow. This orientation is designed to provide a ratio of sweeping velocity to approach velocity equal to or exceeding 2:1 (Easterbrooks 1984). The maximum allowable approach velocity is 0.15 m/s. Screen orientation and flow velocity differential help direct fish to the fish bypass.

The fish bypass facility provides passage through the screen facility and prevents fish from returning to the area upstream of the drum screens. At the terminus of the fish passage facility, a pipe returns fish to the river from which they were diverted. Some facilities include a separation chamber and a pumpback system that allows most of the diversion water to be used for irrigation after fish have passed through the fish screening facility.

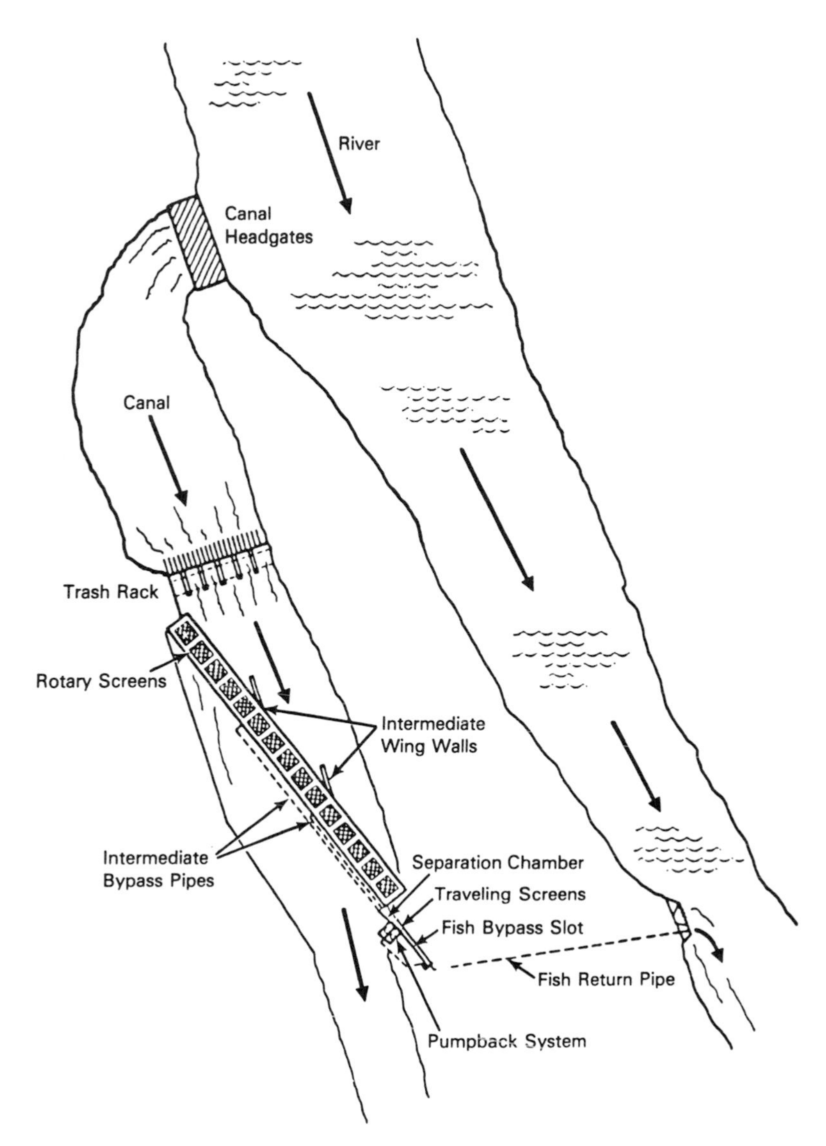

FIGURE 2.—Fish screening facility design used in the Yakima River basin.

Sunnyside Screens

The Sunnyside Diversion Dam and Canal are on the Yakima River at kilometer 167 (km 167) from the Yakima's junction with the Columbia River. Water is diverted into the Sunnyside Canal from a reservoir behind the dam. Canal flow varies from 17 to 37 m^3/s during the irrigation season. Canal flow begins each year in late March or early April with the opening of the canal headgates. Canal flows are lowest in the spring and usually peak in July. Flows remain near maximum until irrigation demand is reduced in late summer. The canal is emptied in October. The Sunnyside screens are located about 360 m downstream of the Sunnyside Canal headgates. The facility includes 17 screens, each about 3.5 m wide and 7.5 m in diameter. A separation chamber is located at the terminus of the bypass pipes and the beginning of the fish return pipe. Two bypass water return pumps, each with a capacity of 14 m^3/s, are located behind traveling screens. The traveling screens are equipped with screen washers to prevent fish and debris from being entrained in the pumpback system.

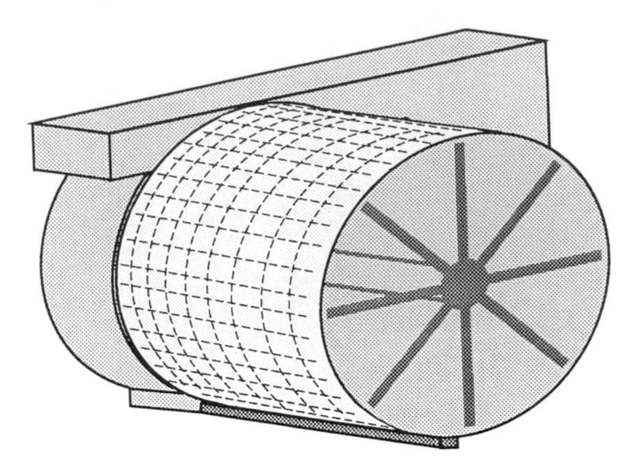

FIGURE 3.—Rotary drum screen used in the Yakima River basin fish screening facilities.

Richland Screens

The headgate of the Richland Canal is located at the Horn Rapids Diversion Dam on the Yakima River at km 29. The capacity of the Richland Canal is about 2.5 m^3/s. Canal flow behind the screens is maintained at 0.8–1.4 m^3/s during the irrigation season (April–October) and at about 0.6 m^3/s during the remainder of the year. Canal flow is regulated at the canal headgates about 1 km upstream of the Richland screens. The screening facility houses four rotary drum screens, each about 3 m wide and 1.7 m in diameter. The fish bypass is located in the flow control structure at the downstream end of the screening facility. Water and fish diverted past the front of the screens pass through the fish bypass slot and out the fish return pipe. Flow through the fish return is adjusted to about 0.7 m^3/s.

Toppenish/Satus Screens

The Toppenish/Satus Unit Diversion is located at km 6 on Toppenish Creek, just downstream of the confluence of Toppenish Creek and Marion Drain. The diversion directs water from Marion Drain and Toppenish Creek to the Satus Canal. Canal operation begins in late March or early April and continues throughout the irrigation season, usually to mid-October. Canal capacity is about 18 m^3/s. The Toppenish/Satus screens are located about 200 m downstream of the headgates of Satus Canal. The facility diverts fish entering the canal and directs them back to Toppenish Creek. The facility houses eight rotary drum screens, each about 5.5 m wide and 3.7 m in diameter. The fish bypass slot is located at the downstream end of the screening facility. Approximately 0.6 m^3/s of water pass through the fish bypass slot and out the fish return pipe.

Wapato Screens

The Wapato Diversion is located at km 172 on the Yakima River. The diversion directs water from the Yakima River into the Main Canal. Canal operation begins in early March and continues throughout the irrigation season, usually until mid-October. Canal capacity is about 57 m^3/s. The Wapato screens are located about 1 km downstream of the headgates of the Main Canal. The screening facility houses 15 rotary drum screens, each about 7.3 m wide and 4.6 m in diameter.

A flow control structure and the separation chamber are located at the downstream end of the screen facility. Two fish bypass pipes and the terminal bypass, each with a flow of about 1.4 m^3/s, feed into the separation chamber. During normal operation, about 4.2 m^3/s of water enter the separation chamber. About 0.9 m^3/s of water and all fish that are diverted in front of the screens pass through the flow control structure and out the fish return pipe. Two bypass water return pumps, each with a pumping capacity of 1.4 m^3/s, are located behind traveling screens near the terminus of the separation chamber. The traveling screens are equipped with screen washers to prevent fish and debris from being entrained in the pumpback system. The pumpback system is not used during normal operation. Adequate flows are maintained in the fish bypass by discharging 3.4 m^3/s of water back to the Yakima River over adjustable weirs in the pump basin. When the pumps are operating, flow over the weirs is reduced.

Methods

Two types of testing or observing were conducted to evaluate the effectiveness of the screens: mark–recapture and monitoring. Mark–recapture tests were used to determine (1) if fish passing through the fish bypass structure were descaled or killed, (2) if fish were trapped or delayed within the screen facility, and (3) if fish could pass through the screens. Monitoring was used to qualitatively assess the presence of predators within the screening facility.

Test Fish

The species of fish selected for tests were recommended by fisheries biologists from the

Washington State Department of Fisheries, U.S. Fish and Wildlife Service, and the Yakima Indian Nation. The species were selected on the basis of occurrence in the Yakima River and the likelihood of the fish encountering the screen facilities.

Fall chinook salmon fry (<60 mm fork length, FL) were used for screen integrity tests. Steelhead, 0.015–0.022 fish/g and 15–23-cm FL, and spring chinook salmon, 0.025–0.033 fish/g and 12–16-cm FL were used for descaling tests. Fish were acclimated to temperatures at each test site before release.

Sampling Equipment

Fish were captured within the screening facility, at the terminus of the primary fish return pipe, and in the canal behind the screens. Inclined planes were custom-built to fit each structure. Fyke nets were used to collect fish behind the screens. Electrofishing equipment was used in some canals and at the terminus of some pipes.

Inclined plane.—The inclined plane was placed in the fish return between the last rotary drum screen and the beginning of the fish return pipe. The inclined plane, consisting of an aluminum frame covered with a perforated aluminum sheet (0.32-cm-diameter holes, staggered centers, 40% open), varied in length and width according to the structure. Adjustable wings or siding were fastened to the sides of the plane to compensate for irregularities in the walls of the concrete bypass structure and a live-box was attached at the end. Flow was directed over the plane surface by inserting dam boards in the upstream stop log slot in the fish bypass. Height of dam boards relative to water depth determined water volume through the fish bypass.

Fyke nets.—Fyke nets were used to capture fish in screen integrity tests. A fyke net was set in Richland Canal about 75 m downstream of the screening facility. At the Wapato screens, the nets were fished immediately downstream of selected screens during each test.

Descaling Evaluation

The evaluation system developed by the U.S. Army Corps of Engineers (Basham et al. 1982; Neitzel et al. 1985) was used to evaluate the condition of fish. Evaluation criteria included modifications established in 1985. Descaling was evaluated in each of 10 surface areas of the fish, 5 on each side. When 40% or more scale loss was observed in any 2 areas on one side of a fish, the fish was classified as descaled.

Test Procedure

Fish were released downstream of the diversion headgates and upstream of the screen structure and were recaptured at the terminus of the fish bypass, at the terminus of the pipe, or in nets downstream of the screens. Condition of test fish was determined before test release. Test fish were cold-branded to identify specific test groups; some test fish were held for posttest observation.

Screen integrity tests were conducted at the screens by releasing branded groups of fall chinook salmon in front of and behind the rotary screens. Fish were collected as they appeared either on the inclined plane in the fish return or in fyke nets placed in the canal behind the screens.

Statistical Analysis

Estimates of the percentage of fish descaled or killed were based on the number of test fish caught. Descaled fish were included in the total dead count in evaluations of results. Confidence intervals for the mortality estimates were calculated from Mainland's tables (Mainland et al. 1956). Data for replicate tests were combined to obtain a mean estimate. The estimate assumes each fish behaved independently (i.e., fish within a test did not behave more similarly than fish between tests, and no interactions occurred among fish within a test). Although some interaction is expected among fish, independent behavior is an assumption necessary for the analytical methods used. All tests were conducted in the same manner to reduce the effect of nonindependent behavior of fish.

To estimate the percentage of fish that might pass through the screens, the inclined plane efficiency, net capture efficiency, and the net retention efficiency were calculated. Each efficiency is based on the percentage of fish captured from the test fish that were released for a given test. The test fish were released upstream and downstream of the screens and into the fyke nets. Confidence intervals were computed with a standard normal approximation method (Mood et al. 1974).

Results

Fish passing through the fish bypass facilities were not descaled or killed. Predators did not concentrate within the facility as a result of facility operations. Fish were not flushed from the screen forebays, but appeared to move on their own volition. The angled rotary drum screen design effectively prevented fish from entering the

TABLE 1.—Descaling and mortality data from release and recapture tests with steelhead and chinook salmon smolts at fish screening facilities in the Yakima River basin. Numbers inside parentheses are 95% confidence intervals.

Screen facility	Number of fish		Percent descaled or killed	
	Released	Recaptured	Before test	After test
Steelhead test results				
Sunnyside	1,647	507	0 (0–3.9)	0 (0–0.3)
Richland	600	363	0.3 (0–1.8)	1.1 (0.3–2.8)
Toppenish/ Satus	1,560	1,388	26.4 (22.0–32.9)	18.9 (17.4–21.6)
Wapato	1,715	1,158	0.3 (0–1.4)	1.4 (0.8–2.2)
Chinook salmon test results				
Sunnyside	4,492	3,625	0 (0–2.3)	1.2 (1.4–2.1)
Richland	990	712	0 (0–1.2)	1.3 (0.7–1.8)
Wapato	1,814	1,421	0 (0–0.9)	2.4 (1.7–3.3)

canal behind the screens, so long as the seals on the sides and bottom of the screens were properly installed and maintained. Designed changes in operation of the screens did not appear to affect the efficiency of the screens.

Fish Condition after Passage through the Screen Bypass System

Release–recapture tests were conducted with steelhead and spring chinook salmon. The condition of the fish that passed through the screen bypass system and were captured by our inclined plane was similar to the condition of the fish that were checked before the beginning of the tests (Table 1).

Sunnyside screens.—Thirty groups of fish were released at five locations within the screen structure. Of 4,492 chinook salmon released, 3,625 were recaptured. Of 1,647 steelhead released, 507 were recaptured. Less than 2% of chinook salmon were descaled or killed. These losses do not differ from the 95% confidence interval for the condition of the controls. None of the captured steelhead were descaled or killed.

Richland screens.—Of the three groups each of steelhead and spring chinook salmon released downstream of the trash rack (200 fish/group; 1,200 total), 363 (61%) steelhead and 560 (93%) spring chinook salmon were recaptured. About 1% of the steelhead and 0.7% of the spring chinook salmon were descaled or dead. Of the steelhead and spring chinook salmon held for 96-h observation, two steelhead and no spring chinook salmon died. Combined losses from descaling and delayed mortality were within the 95% confidence interval for the condition controls.

Toppenish/Satus screens.—Because of the construction schedule, the tests at the Toppenish/Satus screens were started 2 months after the canal was filled and well after the native steelhead smolts had migrated past the Toppenish/Satus Diversion. The canal water was about 17°C during the tests; consequently, the test fish had to be acclimated at this temperature. Fish scales dislodged easily, and many fish lost scales during acclimation and transport.

Of the three groups of steelhead smolts released behind the trash rack (520 fish/group; 1,560 fish total), 1,388 (89%) were recaptured during the 96-h sampling period. The descaling rate for recaptured steelhead was 18.9%, based on all steelhead that were captured, and 26.4% for the control group. There was no increase in the percentage of descaling for test fish as compared with control fish.

Wapato screens.—Tests at the Wapato screens were conducted during low canal flow (14 m^3/s) and during full canal flow (57 m^3/s). Altogether, 1,775 marked fish were released in the low-flow tests conducted early in the irrigation season at flows typical of those during canal start-up. A further 1,754 marked fish were released in later tests during full canal flow to evaluate fish passage conditions during peak salmonid migration in the Yakima River.

Of the 835 steelhead planted during low-flow tests, 361 (43.2%) were recaptured on the inclined plane in the fish return during the next 96 h. Based on the number of descaled fish that were captured, we estimated that 0.8% of the steelhead were descaled. No mortalities were observed among 55 steelhead held for 96 h of observation. Of 440 steelhead released in the morning during full-flow tests, 398 (90.5%) were caught in the following 36 h. Based on the number of captured fish that were descaled, we estimated that about

TABLE 2.—Estimated time (h) to catch 50% and 95% of test fish captured during release–recapture tests.

Test site and species	Time (h) to catch		Number of fish		Percent recaught
	50%	95%	Released	Recaught	
Richland					
Steelhead	23.0	52.0	600	365	61
Spring chinook salmon	0.8	5.0	600	559	93
Fall chinook salmon	8.3	32.5	3,300	2,129	65
Toppenish/Satus					
Steelhead	11.5	43.3	1,560	1,389	89
Spring chinook salmon	0.5	1.5	1,030	999	97
Fall chinook salmon	0.5	0.5	2,460	1,760	72
Wapato (day release, low canal flow)					
Steelhead	17.5	85.0	835	361	43
Spring chinook salmon	10.5	86.0	940	579	62
Wapato (day release, high canal flow)					
Steelhead	11.5	12.5	440	403	92
Spring chinook salmon	2.0	11.0	470	456	97
Fall chinook salmon	~0.5	~2.0	7,214	6,587	91
Wapato (night release, high canal flow)					
Steelhead	0.5	4.0	440	399	91
Spring chinook salmon	<0.5	0.5	404	404	100

1.8% were descaled or dead. Of the 440 steelhead released just before dark in the full-flow tests, 399 were captured during the following 24 h, and we estimated 1.5% were descaled or dead. Overall, the mean loss from descaling was 1.4% (Table 1), which is within the range of the 95% confidence interval for the condition controls (Table 1).

Marked spring chinook salmon were also released during low canal flow, and in the morning and just before dark during high canal flow. Of 940 fish released during low flow, 579 (61.6%) were recaptured on the inclined plane in the following 96 h, and 1.4% were descaled or dead. No mortalities were observed among 88 salmon held for 96 h of observation. Of 470 salmon released in the morning during full-flow tests, 456 were recaptured in the following 36 h, and 0.4% were descaled. Of 404 chinook salmon released just before dark during full flow, 386 were recaptured during the next 24 h, and 6.2% were descaled or dead. Overall, the loss resulting from descaling was 2.4% (Table 1). The descaling that occurred resulted from the simultaneous catch of large numbers of hatchery-released fish. By not counting the injured fish caught during the hatchery release, the descaling estimate is within the range of the 95% confidence interval for the controls.

Concentration of Predators

Few predacious fish (largemouth bass *Micropterus salmoides*, smallmouth bass *M. dolomieui*, northern squawfish *Ptychocheilus oregonensis*) were caught in the fish return during our tests. Based on the stomach content analysis, limited predacious feeding activity was apparent in the canal during our tests. The guts of one smallmouth bass and one squawfish contained two of our branded fall chinook salmon fry. Gulls *Larus* spp. were not common at the sites except at the Richland screens. Feeding gulls at Richland did not appear to be related to screen operation or design.

Downstream Movement Delay

The downstream movement of steelhead and spring chinook salmon released for descaling evaluations was monitored each half hour as the fish appeared on our sampling plane in the fish return. The rate and percentage of recovery for steelhead and spring chinook salmon indicated that salmonid smolts are not flushed from the screen forebays (Table 2); rather, they move through the screen forebay of their own volition. Movement rate varied in response to factors such as canal flow, smolting condition, and species-dependent behavior. Movement rates were slower during low canal flow than during high canal flow (Wapato screens; see Table 2). Spring chinook salmon vacated the screen forebay more rapidly than steelhead, and thus had a slightly higher recapture rate in our tests.

TABLE 3.—Recapture data for fall chinook salmon fry released at Richland Canal fish screening facility, spring 1987.

Number released	Number recaptured in			Percent recaptured in	
	Plane	Fyke	Shocker	Bypass	Canal
Three groups of test fish released upstream of the screens					
1,008	490	0	0	48.6	0
1,004	462	0	0	46.0	0
1,009	444	0	0	44.0	0
Total or average					
3,021	1,396	0	0	46.2	0
Three groups of test fish released downstream of the screens					
1,001	0	584	17	0	60.0
1,010	0	550	39	0	58.3
1,010	0	609	45	0	64.8
Total or average					
3,021	0	1,743	101	0	61.0

TABLE 4.—Recapture data for fall chinook salmon fry released during screen integrity tests at Wapato fish screening facility, spring 1987.

Number released	Number recaptured in			Percent recaptured in	
	Plane	Fyke	Other	Bypass	"Canal"
723	695	2	0	96.1	0.3
724	700	1	0	96.7	0.1
723	631	26[a]	0	87.3	3.6
1,470	1,278	59	39	86.9	4.0
1,472	1,311	9	0	89.1	0.6
1,502	1,396	2	0	92.9	0.1
Total or average					
6,614	6,011	99[b]	39	90.9	2.1

[a]Eleven test fish from test 1 were caught in the net during test 2.

[b]The total increases to 110 fish if the 11 test fish released in test 1 and caught in test 2 are included.

Screen Integrity Tests

The integrity of the screen facilities for preventing fish from passing through the screens was related to the design and maintenance of the seals on the sides and bottom of the screens. Rollover and impingement of juvenile salmonids were infrequent.

Richland screens.—At the Richland facility, 3,021 fall chinook salmon fry were released in front of the screens and 3,021 behind the screens. During 41 h after release, 1,396 fish (46.2%) of the fish planted in front of the screens were recaptured in the fish return structure. During the 94-h period after the release, none of the fish released in front of the screens (0%) and 1,845 (61.1%) of the fish released behind the screens were recaptured by fyke net (1,743 fish) or electrofishing (101 fish) in the canal behind the screens (Table 3). No fish released behind the screens were captured on the inclined plane in the fish return. Fall chinook salmon fry (52.1 mm FL) were not flushed from the forebay. Most fish were captured on the inclined plane either immediately after their release or after sunset on the first night.

Wapato screens.—At Wapato, 9,314 fall chinook salmon fry were released in screen integrity tests (Table 4). Fish were released in front of the screens, in the intermediate and terminal fish bypasses, and in the mouth and cod end of fyke nets positioned behind the screens.

Of 600 fish planted in the intermediate and terminal bypasses, 571 were recaptured in the fish return, indicating a catch efficiency for the incline plane trap of about 95%, assuming no losses to predation or to passage through the traveling screens in the separation chamber. Catch efficiency of the fyke nets varied from 33 to 93%. The net retention efficiency ranged from 55 to 97%.

Of 6,614 fish planted in front of the screens, 6,011 (about 91%) were caught in the fish return, and 110 (1.7%) were caught in the fyke nets behind the screens. Given the catch efficiency estimates for the plane and the fyke nets, we can account for almost all of the fry released in front of the screens.

Designed Change in Operation

Descaling evaluations were conducted at the Wapato screens at canal surface elevations of 283.8 and 284.9 m, corresponding to canal flows of 29.5 and 48.1 m^3/s, respectively. Canal level did not affect descaling rate among our test fish; however, movement of fish from the forebay was much slower during low canal flows.

Discussion

Our data indicate there is no significant descaling or delayed mortality as juvenile salmonids are diverted by the screen facilities. The angle of the screen facility to the flow of water in the canal sweeps fish along the face of the screens and into the bypass system. The angle of the facility also provides guidance toward the bypass openings.

During our studies, we did not observe increased predation on juvenile salmonids in or near screen facilities that could be attributed to the screen design or operation. At some screens, gulls fed on the test fish as the fish were released. This feeding appears to be a result of the timing and location of release. The design and operation of the screens in the Yakima River basin do not appear to enhance predation on juvenile fish as they pass through the screen facilities.

Fish are not involuntarily delayed at or within the screen facilities when bypass flows are set according to the operating criteria. Salmonids that have not completed smolt transformation may reside in screen facility forebays when canal flows are low or designed flows through the fish bypass are not achievable. Efforts should be made to minimize abnormal flow at each screening facility by incorporating fish bypass flow into canal start-up operations.

Screen integrity tests completed at the Richland and Wapato screens indicated that screen effectiveness can vary in preventing fish from entering the irrigation canal. The Richland screens were very effective at preventing fish from entering the canal, primarily because of low approach velocities in the screen forebay. However, at the Wapato screens, poor seals were responsible for some fish passing by the screens. Annual inspection and replacement of faulty seals should alleviate the problem, but a new screen seal design may be necessary.[2] Screen seals at the Sunnyside screens are similar to those at the Wapato screens and might also require improvement. Screen integrity tests with zero-age chinook salmon should be conducted at the Sunnyside screens.

Some chinook salmon fry passed over the rotary screens at the Wapato facility. Water flows did not appear to be uniform through all screens, thus resulting in a higher approach velocity at some screens. Passage over the screens appeared to be related to the presence of driftwood or other floating matter at the water surface in front of screens with high water flow. Stop log adjustments behind the screens to achieve uniform flow might eliminate the problem; however, modifications in front of the screen, such as the addition of a skimmer or spray system, might also be necessary.[3]

The operating criteria for each screening facility should be written to cover the entire range of potential flow conditions of each canal. The criteria should be written to correspond with measurement facilities at the screens. Additionally, the operating criteria should address full canal flow conditions and specifically address operations during canal start-up or during low canal flow.

Conclusions

The angled rotary drum fish screens designed and used at the Sunnyside, Richland, Toppenish/Satus, and Wapato screens physically exclude juvenile salmonids from the water diversion. The screening facility safely returns fish to the parent water body. This is accomplished by screen design, which provides an approach velocity at the screens that is slow enough to preclude juvenile fish impingment. Additionally, the angled structure guides fish toward the fish bypass system so fish are not trapped in the bypass facility. Problems that we have encountered at the screens, such as ineffective side and bottom seals, can be ameliorated. At the Wapato and Sunnyside screens, seals were redesigned and a preventive maintenance program was initiated. As the result of both programs, the screens are effective barriers to prevent fish from entering the water diversion downstream of the screens.

Acknowledgments

This work was funded by the Bonneville Power Administration (BPA), U.S. Department of Energy. The involvement and cooperation of many people helped this project succeed. Thomas J. Clune of the BPA was the Project Manager. Robert T. Tuck and David E. Fast of the Yakima Indian Nation, Gary Malm of the U.S. Fish and Wildlife Service (USFWS), and John Easterbrooks of the Washington State Department of Fisheries (WDF) reviewed test procedures. William E. James (WDF), Ralph Malson of the Leavenworth National Hatchery (USFWS), James L. Cummins of the Washington State Department of Wildlife, and Richard Nelson of the Chelan County Public Utility District helped with procurement and rearing of test fish. The manuscript was reviewed by Susan A. Kreml. Dale Becker, Dennis D. Dauble, R. William Hanf, and Theodore M. Poston assisted with field tests.

References

Basham, L. R., M. R. Delarm, J. B. Athern, and S. W. Pettit. 1982. Fish transportation oversight team annual report, fiscal year 1981: transport operations on the Snake and Columbia rivers. U.S. National Marine Fisheries Service, Technical Services Division, Northwest Regional Office, Portland, Oregon.

[2]Screen seals were redesigned and changed at the Wapato screens in 1988. Preliminary analysis of test data collected during 1988 indicates the new seals prevent fish from passing by the screens.

[3]Stop log adjustments were completed during 1988. Flow measurements were conducted in 1988 to assess the effectiveness of the stop logs.

Easterbrooks, J. A. 1984. Juvenile fish screen design criteria: a review of the objectives and scientific data base. Washington Department of Fisheries, Habitat Management Division, Yakima.

Fast, D., J. Hubble, and B. Watson. 1986. Yakima River spring chinook enhancement study: fisheries resources management, Yakima Indian Nation. Report to the Bonneville Power Administration, Portland, Oregon.

Hollowed, J. J. 1984. 1983 Yakima River fall fish counts at Prosser Dam. Yakima Indian Nation, Fisheries Resource Management Technical Report 84-11, Toppenish, Washington.

Mainland, D., L. Herrera, and M. I. Sutcliffe. 1956. Tables for use with binomial samples. Mainland, Herrera, and Sutcliffe, New York.

Mood, A. M., F. A. Graybill, and D. C. Boes. 1974. Introduction to the theory of statistics. McGraw-Hill, New York.

Neitzel, D. A., C. S. Abernethy, E. W. Lusty, and L. A. Prohammer. 1985. A fisheries evaluation of the Sunnyside Canal fish screening facility, spring 1985. Report to Bonneville Power Administration, Division of Fish and Wildlife, Portland, Oregon.

NPPC (Northwest Power Planning Council). 1984. Fish and wildlife program (as amended). Northwest Power Planning Council, Portland, Oregon.

U.S. Bureau of Reclamation. 1984. Finding of no significant impact: fish passage and protective facilities, Yakima River basin, Washington. U.S. Bureau of Reclamation, Pacific Northwest Region, Boise, Idaho.

American Fisheries Society Symposium 10:335–346, 1991

Developments in Transportation from Dams to Improve Juvenile Salmonid Survival

JAMES B. ATHEARN
U.S. Army Corps of Engineers, North Pacific Division
Post Office Box 2870, Portland, Oregon 97208, USA

Abstract.—Transportation of migrating smolts was first considered in the late 1960s to protect declining stocks of Snake and Columbia river salmon *Oncorhynchus* spp. and steelhead *O. mykiss* that were suffering lower survival rates during a period of increased hydroelectric development and degraded inriver passage conditions. Research conducted by the National Marine Fisheries Service showed higher adult return rates from transported groups of smolts than from smolts that migrated in the river. Mass transportation, including the use of barges, to protect large numbers of outmigrants began as a response to the 1977 drought. Because of predicted low river flows and resultant adverse outmigration conditions, all smolts that could be collected at Lower Granite, Little Goose, and McNary dams were transported to release sites below Bonneville Dam. Subsequent adult returns from that outmigration were almost entirely from transported fish. Success of that operation in terms of both fish survival and demonstration of mass transportation as a viable alternative to poor inriver passage conditions, supported by continued positive research results, led to conversion of the research program to a fully operational mode in 1981. Since that time, the U.S. Army Corps of Engineers' Juvenile Fish Transportation Program has grown in response to increased fish numbers and technological advances. Major improvements to facilities and operational methods are discussed.

Decline of the once abundant salmon *Oncorhynchus* spp. and steelhead *O. mykiss* of the Columbia River basin has been attributed to a combination of many factors including irrigation, logging, mining, grazing, pollution, overharvest in the Columbia River, rise of the ocean fisheries, and dam construction (Gunsolus 1977; Raymond 1988). Problems resulting from construction of hydroelectric dams include blocked passage routes, inundation of spawning areas, delays in migration, passage through turbines, predation in reservoirs, and gas bubble disease (Ebel et al. 1973; Collins et al. 1975; Bentley and Raymond 1976; Collins 1976; Ebel 1977; Raymond 1979; Park 1980; Weitkamp and Katz 1980; McCabe et al. 1983). In the past 15–20 years, reduced survival of downstream migrants has probably been the most critical problem hindering rebuilding of depleted anadromous fish runs in the Columbia and Snake rivers.

In the late 1960s and early 1970s, attention focused on fish mortality caused by elevated dissolved gas levels created by spill. Raymond (1979) and Weitkamp and Katz (1980) reviewed problems resulting from gas bubble disease. Ebel (1979) described the use of spillway deflectors to reduce gas supersaturation at U.S. Army Corps of Engineers dams and concluded that they eliminated the problem except during periods of very heavy spill.

Dam-related losses increased as more turbines were installed in Snake and Columbia river dams (Collins et al. 1975; Park 1980; Raymond 1988). Between 1968 and 1985, generation capacity at the dams was increased to the point that all the river flow in a typical water year could be passed through the powerhouses. This paper will not address the evolution of powerhouse collection and bypass systems except to say that successful development of these facilities resulted in more fish that could be collected for mass transport, if desired, or for safe bypass back to the river.

National Marine Fisheries Service researchers believed smolt losses due to other dam-related problems could be reduced by collecting smolts at the uppermost dam, transporting them around as many as eight dams, and releasing them into the Columbia River below Bonneville Dam (Ebel et al. 1971; Collins et al. 1975; Park 1985; Figure 1). Transport research was first conducted on the Snake River at Ice Harbor Dam between 1968 and 1970, then continued at Little Goose Dam during 1971–1973 and 1976–1978, at Lower Granite Dam during 1975–1980, and at McNary Dam during 1978–1980 (Park 1985). The studies compared adult return rates for steelhead and chinook salmon *Oncorhynchus tshawytscha* that had been transported as juveniles by truck and barge with return rates for nontransported fish. Return ratios of transported to nontransported chinook salmon

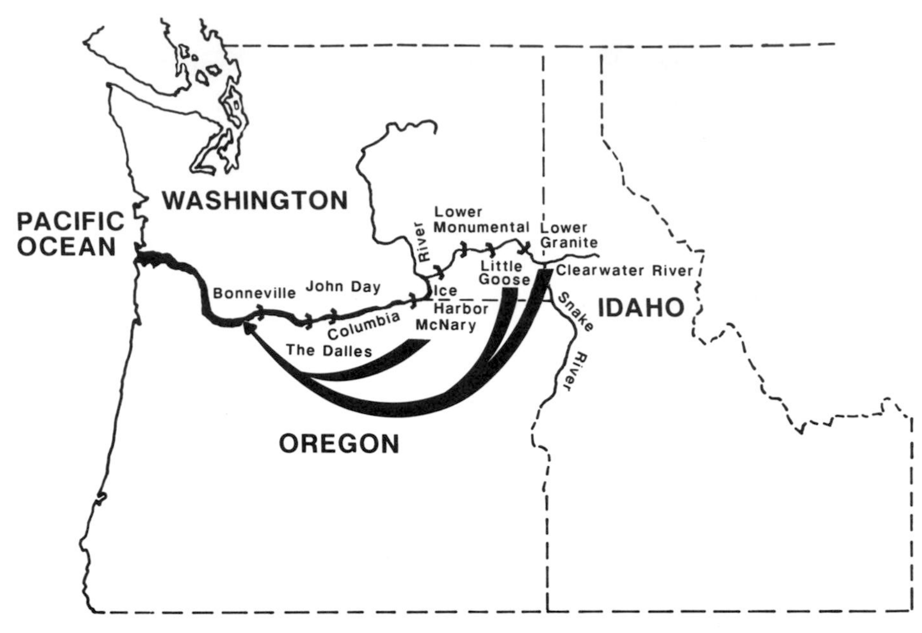

FIGURE 1.—Locations of fish collection facilities at Lower Granite, Little Goose, and McNary dams, transportation direction, and release site downstream of Bonneville Dam.

(transport benefit ratios) ranged from 0.7:1.0 to over 18.1:1.0 and estimated adult return rates ranged from 0.1 to 9.0%. In similar tests for steelhead, transport benefit ratios ranged from 1.3:1.0 to 17.5:1.0 and estimated adult return rates were 0.4 to over 4.0% (Park 1985). Park (1985) reported transport benefit ratios at McNary of 4.0:1.0 for fall chinook salmon, 2.0:1.0 for coho salmon *O. kisutch*, and 2.5:1.0 for steelhead. All transported groups of all species returned in significantly higher numbers than nontransported groups except trucked steelhead in 1980, which returned at a measurably but not significantly greater rate than nontransported fish. Fall chinook salmon transport tests continued through 1983 at McNary Dam and a transport benefit ratio of more than 2.0:1.0 was realized for each year (Matthews et al. 1988). The lower transport benefit ratios for the latter tests were attributed to better inriver survival of nontransported control fish.

In studies on adult chinook salmon and steelhead that were captured during their seaward migration as juveniles and then transported downstream, homing ability was not seriously diminished (Ebel et al. 1973; Ebel 1974, 1980; Slatick et al. 1975). McCabe et al. (1983) barged coho salmon directly from Willard National Fish Hatchery (Little White Salmon River) to a release site downstream of Bonneville Dam. Subsequent adult returns to the Little White Salmon River were lower than expected, leading McCabe et al. to conclude that homing ability may have been impaired.

Bjornn and Ringe (1984) compared adult returns for juvenile hatchery spring and fall chinook salmon and steelhead allowed to migrate to the ocean with juveniles only allowed to migrate a short distance (up to 4 km) before being collected and transported to the lower Columbia River. More of the juvenile fish that migrated only a short distance and then were transported were recaptured as they passed through the estuary than those that migrated downstream normally. Conversely, adult returns to hatcheries or Snake River dams were higher for normal-migration groups than for short-migration transport groups. The authors concluded that some of the chinook salmon and steelhead smolts allowed to migrate only short distances voluntarily before being transported did not acquire sufficient cues for satisfactory homing back to hatcheries or release sites. Other researchers have also concluded that juveniles transported directly from hatcheries to downstream release sites had some homing impairment (Harmon and Slatick 1987; Slatick et al. 1988). Even when homing is impaired, Slatick et al. (1988) noted that increased survival of trans-

TABLE 1.—Numbers (in thousands) of juvenile salmonids transported from Lower Granite (1975–1988), Little Goose (1971–1988), and McNary (1978–1988) dams to release sites downstream of Bonneville Dam, 1971–1988.

Year	Steelhead	Chinook salmon Yearling	Chinook salmon Subyearling	Coho salmon	Sockeye salmon	Total
1971	154	109				263
1972	227	360				587
1973	176	247				423
1974	0	0				0
1975	549	414				963
1976	435	751				1,186
1977	906	1,373				2,279
1978	1,376	1,655	40	22	7	3,100
1979	1,880	2,458	449	83	209	5,079
1980	3,070	4,049	652	33	57	7,861
1981	3,104	2,692	2,101	102	309	8,308
1982	2,624	1,371	1,819	73	202	6,089
1983	1,994	1,040	4,482	2	44	7,562
1984	3,098	1,606	4,165	40	121	9,030
1985	4,302	3,538	6,481	64	403	14,788
1986	4,751	2,556	5,902	33	254	13,496
1987	4,608	5,144	6,665	183	437	17,037
1988	6,433	6,445	6,693	212	248	20,031

ported steelhead can allow more fish to return to hatcheries and fisheries than would return from nontransported hatchery fish.

Carlson et al. (1987) are studying the effectiveness of transporting chinook salmon and sockeye salmon *Oncorhynchus nerka* collected in gatewells at Priest Rapids Dam to areas below Bonneville Dam. Their preliminary data suggest a negative benefit ratio for spring chinook salmon and a generally positive benefit ratio for sockeye salmon; however, they believe some of the positive benefits may be offset by homing loss or unknown mortality factors during the adult migration.

Low survival of naturally migrating smolts in 1972–1973 prompted a decision by northwestern fisheries agencies to mass-transport steelhead from Little Goose Dam in 1975 (Park and Ebel 1975). Mass transportation was expanded in 1976 to include chinook salmon and to transport from Lower Granite Dam (Park et al. 1977). During the 1977 spring outmigration, severe drought produced record low flows. An emergency mass transportation program was implemented and barging was included for the first time as an integral part of the overall transportation plan (Park et al. 1978). Survival of nontransported smolts was low and transport operations allowed a limited return of at least 7,000 chinook salmon and 12,000 steelhead, whereas virtually no adults would have returned without transportation in 1977 (Park et al. 1980).

Mass transportation, including barging, and transport research continued from 1978 through 1988 (Park et al. 1979, 1980, 1981; Smith et al. 1980, 1981). Numbers of juvenile salmonids transported have increased dramatically from about 250,000 in 1971 to over 20,000,000 in 1988 (Table 1). Beginning in 1981, U.S. Army Corps of Engineers biologists supervised the juvenile fish collection and transportation program as it became fully operational. Also in 1981, the Fish Transportation Oversight Team (FTOT) took over from National Marine Fisheries Service researchers the role of monitoring mass transportation and recommending improvements in facilities and operational procedures. The FTOT is an interagency coordination body representing federal and state fish agencies, Indian tribes, and the U.S. Army Corps of Engineers. Remarkable improvements in collection and transport operations have been a major contribution to the success of transportation in recent years (Park 1985). The purpose of this paper is to describe the recent improvements and developments in collection and transportation of juvenile salmonids and to briefly discuss the future role of mass transportation.

Collection Facility Improvements

The major components of all current collection and transportation facilities include a powerhouse collection channel and transport pipe, a separator, a counting and distribution system, raceways, a fish-handling facility, transport vehicles, and release sites (Figure 2).

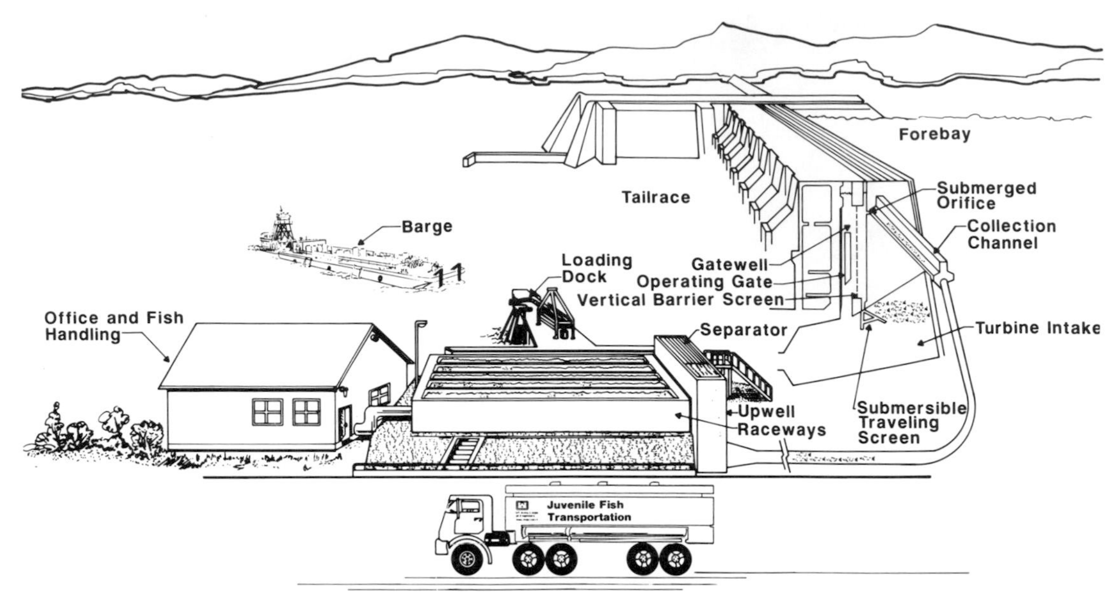

FIGURE 2.—Juvenile salmonid collection and transportation system.

Powerhouse Collection Channel and Transport Pipe

In 1968 the U.S. Army Corps of Engineers installed the first submerged orifices in the turbine intake gatewells of McNary Dam. These orifices successfully passed juvenile salmon and steelhead from the gatewells into an ice and trash sluiceway and then into the tailrace of the dam (Bentley and Raymond 1969). A similar system installed in Ice Harbor Dam in 1970 was used to capture large numbers of juvenile salmon and steelhead. Further, 20-cm-diameter pipes were placed over the orifices to direct water and fish into a narrow wooden flume (50 cm high, 85 cm wide, 80 m long) at the end of which fish could be collected with an inclined screen trap and marked for transportation studies (Park and Farr 1972).

Smith and Farr (1975) described the juvenile bypass system designed for Little Goose Dam. It had an enclosed conduit to pass fish around the turbines, unlike the open ice and trash sluiceway flumes at Ice Harbor Dam. The transport pipe for bypassing fish to the tailrace was subsequently coupled to a fish-holding facility on the south shore below the dam, where National Marine Fisheries Service researchers collected juvenile salmonids for transport evaluation studies.

In 1978 and 1979, the U.S. Army Corps of Engineers reconstructed the Little Goose bypass system to reduce injuries to juvenile fish and reduce fish-holding time in gatewells (Park et al. 1980). The gatewell orifices were enlarged from 15 to 30 cm, and the powerhouse collection channel was changed to an open flume that emptied into an enclosed, pressurized transfer pipe to the fish-holding facility. Evaluation of this system led to improvements that were incorporated into the design of a similar facility at Lower Granite Dam, described by Matthews et al. (1977).

The only serious problems with the Lower Granite powerhouse collection channel have been infrequent failures of auxiliary water supply valves and debris plugging of the gatewell orifices. Debris problems at Lower Granite Dam are compounded by the small-diameter (20-cm) gatewell orifices, which are more prone to plugging than the 30-cm orifices at Little Goose and McNary dams. During periods of high river-borne debris loads, the gatewell orifices require almost constant cleaning. Floating debris has covered over 6.7 hectares in the Lower Granite forebay (Basham et al. 1983). Forebay debris is now diverted away from in front of the powerhouse by a log boom installed in 1984. Diligent cleaning of turbine intake trash racks and gatewells has reduced debris-related problems.

In 1978, a wooden flume was suspended on the wall of the McNary Dam ice and trash sluiceway to divert juvenile salmonids exiting the gatewell orifices through a separator to holding raceways.

Problems later noted with the flume design caused debris to accumulate and subyearling chinook salmon (less than 50 mm long) to impinge on the water elimination screens, and caused adult American shad *Alosa sapidissima* to accumulate at the north (downstream) end of the flume near the entrance to the transfer pipe (Delarm et al. 1984). The American shad buildup was thought to delay subyearling chinook salmon, substantially increasing their mortality rate in the system. A stationary perforated-plate screen was installed in the downstream end of the flume to enlarge the total screen area and constrict flow to increase water velocity and prevent fish buildup or delay. Park et al. (1984) determined that chinook salmon passage after the modification was relatively rapid.

Excess hydraulic energy exists in pressurized fish-transport pipes, where head losses may equal 21 m. This energy may injure and stress fish and cause water surges at the pipe discharge. Matthews et al. (1985) found that spring chinook salmon smolts were significantly stressed in the bypass pipe at Lower Granite Dam. Restrictor rings and pinch valves, installed to dissipate excess hydraulic energy, also caused injury as fish hit them or the debris that lodged against them (Smith et al. 1980; Basham et al. 1983; Delarm et al. 1984). Currently, the Little Goose collection facility is being modified to correct the excess hydraulic energy and several other serious problems. Three open-flumes (one of corrugated metal and two types with baffles) to replace the pressurized pipe were studied by Congleton and Ringe (1985) and Congleton et al. (1988). Based on their results, the U.S. Army Corps of Engineers selected the corrugated metal flume to transfer juvenile fish from the powerhouse collection channel to the holding facilities.

Separator

A separator, used to segregate juvenile chinook salmon from the larger steelhead juveniles for transport studies, was first installed in the collection facility at Little Goose Dam. As described by Smith and Farr (1975), the separator consisted of aluminum tubes 3.2 cm in diameter progressively spaced from narrow to wide openings. Spacing could be varied and graded fish fell through the bars into one of three water-filled hoppers below. A similar but improved system was later installed in the collection facility at Lower Granite Dam. Improvements included increased length of separator bars from 3 to 6 m to provide a finer degree of size grading, a trash-sweeping device to prevent accumulation of water-borne debris, and a water-spraying assembly to provide a constant film of water on the grader bars and allow easier and faster fish movement across them (Matthews et al. 1977). Regardless of these improvements, these "dry" separators stressed juvenile fish by forcing them out of the water and across a set of bars, frequently tailfirst, until they reached a wide enough space to fall through.

To alleviate this stress, a "wet" separator was developed to grade the fish under water (Gessel et al. 1985). The design included submerged bars and water flow directed up from the hoppers. Because of a smolt's normal reaction to water current and a tendency to sound as an avoidance reaction, the wet separator proved effective, separating 90% of the smaller chinook salmon and 88% of the steelhead in model studies at Lower Granite Dam, and 73% of the chinook salmon in prototype studies at Little Goose Dam (Gessel et al. 1985). Even with the substantial improvement from a dry to wet separator, subsequent stress studies identified this part of the collection and transport process as potentially stressful for chinook salmon at Lower Granite Dam (Matthews et al. 1986). Congleton et al. (1984) also noted higher stress levels in fish leaving the separator, which they believed may have been a delayed response to passage from the gatewells to the separator.

Distribution and Counting Systems

Distribution systems to deliver fish from the separator to holding raceways evolved from straight 10-cm-diameter polyvinyl chloride pipes to a pressurized system used at Lower Granite Dam to open flumes. The pressurized system with air-operated gates was designed to allow quick operation by a single worker, increase sampling frequency (to reduce counting error), reduce fish jumping at the surface disturbance caused by water falling into raceways from the counting tunnels, and reduce delay of juveniles in the distribution system (Basham et al. 1982). When counting errors were still too high, the system was converted to an open flume with automatic gates to divert sample fish at a predetermined interval (Basham et al. 1983).

Other important modifications to the distribution system included replacing all 15-cm-diameter piping used to move fish around the facility with 25-cm-diameter pipe and replacing all sharp pipe angles with sweeping 90° turns (1.8-m radius). The larger pipes enabled workers to reduce the time required to transfer all fish from a fully loaded raceway into a truck or barge from 60 min per raceway to about 10 min (Basham et al. 1983). Also, the fish no longer had to be forcefully crowded.

Until 1982, all fish exiting the separator hoppers passed through 10-cm-diameter electronic-counting tunnels for enumeration. Increased numbers, including large (commonly up to 40 cm) steelhead juveniles, prompted development of an automatic sampling system that only counted 5 to 10% of the hourly collection. The rest of the fish passed via open flumes directly to raceways. This method was chosen to reduce stress levels and simplify fish sampling (Basham et al. 1983).

Raceways

Smith and Farr (1975) described the five fish-holding raceways constructed at Little Goose Dam (1.2 m wide, 2.1 m deep, and 24.4 m long). Similar raceways were constructed at Lower Granite Dam (Matthews et al. 1977) and an additional five raceways were added in 1983 to increase holding capacity.

Juvenile fish-holding raceways at McNary Dam have a unique design developed through value engineering. They are approximately 13 m long, 1.2 m wide, 2.4 m deep at the upstream end, and 1.2 m deep at the downstream end. The bottom of the upstream half of each raceway is horizontal, whereas the downstream half slopes toward the water surface at an 18% grade. The barge-loading line is at the upstream end and trucks are loaded from the downstream end. Basham et al. (1982) described truck-loading problems, particularly with fall chinook salmon. The fish had to be forcefully crowded into the shallow end and out through a 15-cm port. They also tended to hold in the pipes after leaving a raceway. Installation of an overflow into an open flume eliminated the loading problems. The seven raceways at McNary Dam had a combined capacity of about 9,500 kg of fish, which was not adequate for the large numbers of fish being collected for mass transportation. Installation of two temporary plastic-lined aluminum raceways (10.7 m long, 2.4 m wide, and 1.0 m deep) increased the total facility capacity to just over 11,300 kg, still not adequate for peak daily collection days but there is no room near the facility to add more temporary ponds.

Air-operated control valves were installed on all water-supply and fish-release pipes to replace hand-operated valves. This allows workers more flexibility to operate remote valves quickly and easily. They have also been added to gatewell orifice valves at McNary and Little Goose dams and installation at Lower Granite Dam was projected for 1989.

Handling Facilities

A small fish sample is inspected daily at the collection facilities to determine species composition, average weight, fork length, and general fish condition (descaling and physical injury), and to verify electronic counts. Incidental recoveries of branded fish are also noted. The sample also provides large numbers of fish for handling and marking for transport evaluation and smolt monitoring. Stress research on chinook salmon smolts has shown potential adverse effects resulting from handling and marking (Congleton et al. 1985; Matthews et al. 1986; Park et al. 1986). Schreck et al. (1985) reported that anesthetization, handling, and marking of fall chinook salmon smolts at McNary Dam apparently did not cause stress in addition to that experienced by fish going directly into the raceways. They recommended, however, a 24–48-h recovery period for the fish before release.

A system that has been called "preanesthesia" was installed at Lower Granite Dam in 1986 (Koski et al. 1987) and at McNary Dam in 1987 (Koski et al. 1988). The system provides for anesthetization of smolts before they are dipnetted and it has improved the handling and marking operation for juvenile fish used in transport evaluation studies. Research has shown that stress (Schreck et al. 1985; Matthews et al. 1986) and delayed mortality (Matthews et al. 1988) of chinook salmon and steelhead were reduced with the preanesthetic systems.

Fish Transport Equipment

Trucks

Smith and Ebel (1973) described the first two tanker trucks used for fish transport from the Snake River. They were modified surplus aircraft-refueling tanks with a capacity of about 18,900 L. Life-support systems included a water-recirculating pump and spray bars, sand filter, liquid and compressed oxygen, and refrigeration unit. Juvenile fish were successfully hauled 560 km from Little Goose to Bonneville Dam at a loading rate of 120 g/L.

As mass transportation evolved, five new tankers were acquired. They had similar life-support systems and a capacity of 13,250 L (1,590 kg of fish). The loading rate established for the transport operation in effect today is 60 g/L. Under special circumstances, steelhead may be hauled at the original 120-g/L rate.

Barges

The two barges first used for emergency mass transportation in 1977 were temporarily modified as described by McCabe et al. (1979). Each had a steel tank 33.2 m long, 8.5 m wide, and approximately 1.5 m deep. Water was pumped from sea chests through a spray bar at the forward end of the tank. At a maximum pumping rate of 330 L/s, one complete turnover was provided every 20 min. Supplemental oxygen was added through air stones in the main water supply line. Barge capacity was estimated by McCabe et al. (1979) to be approximately 20,400 kg of chinook salmon or 26,850 kg of steelhead.

Success of the 1977 operation prompted the U.S. Army Corps of Engineers to acquire and modify two surplus barges (41.1 m long by 10.7 m wide) in 1978. They have three fish compartments each with a capacity of 321,700 L and a single-pass pumped water supply of 328 L/s. At a loading rate of 6.9 g/(L·s), the fish-hauling capacity is 11,800 kg. These barges do not have recirculation capability.

As the numbers of downstream migrant juvenile salmonids continued to increase, more transport capacity was needed. Two more barges, specifically designed and constructed for mass transportation, were acquired, one in 1981 and another in 1982. These new barges (45.7 m long by 10.7 m wide) have four compartments each with a combined capacity of 378,500 L and can haul up to 22,680 kg of fish. Water is pumped through aeration–degassing columns at a rate of about 630 L/s. These barges have recirculation capability in the event they must pass through an area of poor water quality.

A round-trip from Lower Granite Dam to Little Goose and McNary dams, and down to the release site 8 km below Bonneville Dam takes 4 d. Therefore, by 1982, there were enough barges to have one leave Lower Granite Dam every day during the peak juvenile fish outmigration period.

Release Sites

Use of inadequate release sites could negate all efforts to safely collect and transport juvenile fish. All barge releases are made in midchannel about 8 km downstream from Bonneville Dam between midnight and 0300 hours. Towboat operators are instructed to vary the exact location every night to reduce predation.

Several sites have been used for release of trucked fish, including Bradford Island between Bonneville First Powerhouse and the spillway, into the juvenile bypass system downwell at the Bonneville Dam Second Powerhouse, into the adult fish ladder by the Washington shore, and at public boat ramps below Bonneville Dam at Dalton Point and Hamilton Island. During an early transport evaluation study, fish were released 5 km downstream of John Day Dam (Slatick et al. 1975). Adult returns from these fish indicated it was generally a poor release site, presumably because juvenile fish still had to pass The Dalles and Bonneville dams.

Characteristics of a good release site include accessibility, good water quality, adequate depth so fish do not hit the bottom when released, and sufficient water velocity to preclude predator populations from congregating and to allow the juvenile fish to quickly orient to continue their downstream migration. Koski et al. (1987) described seasonal problems at Dalton Point during low summer flows and summer predation by northern squawfish *Ptychocheilus oregonensis* and gulls *Larus* spp. at Bradford Island. Also, the volume of the submerged portion of the 30.5-cm discharge pipe at Bradford Island equals the volume in the transport tanker. To assure good water quality and encourage displacement of fish from the discharge pipe, water is pumped into the line (about 17 L/s) for 4 h after a fish release.

Operational Improvements

A major key to successful mass transport is to keep juvenile salmonids moving through the system as safely and quickly as possible. As previously mentioned, additional barges were acquired to cope with increased fish collection in recent years and the barging season was extended to include more of the spring and fall chinook salmon outmigrations. Summer barging of fall chinook salmon from McNary Dam began in 1983 and about 2.2 million were barged that year.

Studies to determine a difference in survival between trucked and barged fish have provided variable results. In studies at Lower Granite Dam in 1978, significantly fewer adult returns were noted from juvenile spring chinook salmon that were trucked than from those that were barged (Park 1985). However, Park (1985) noted that similar comparisons for steelhead and fall chinook

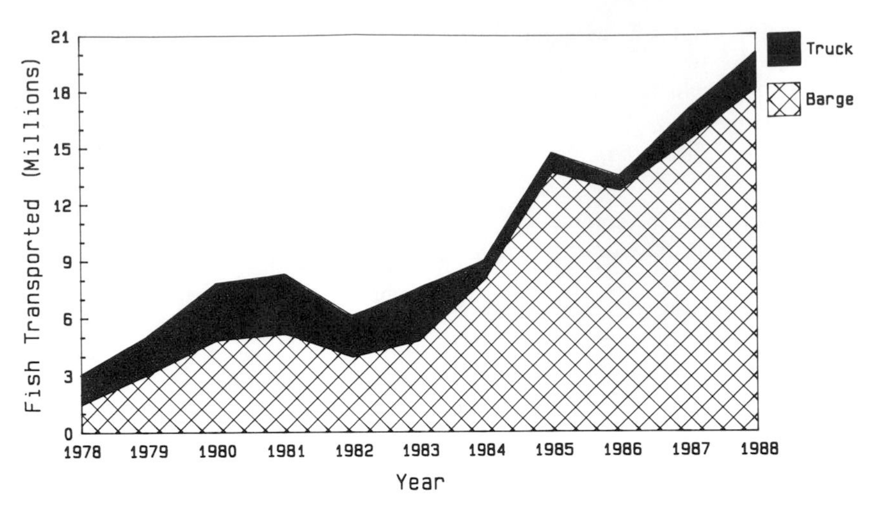

FIGURE 3.—Summary of juvenile fish transported by truck or barge from Lower Granite, Little Goose, and McNary dams, 1978–1988.

salmon have shown no significant difference. In tests at McNary Dam during 1978 to 1980, Park (1985) found no significant difference in adult returns in truck–barge comparisons. In 1978, only about half of the fish were transported by barge, the remainder by truck. Since 1985, the portion transported by barge has exceeded 90% (Figure 3).

In a further effort to reduce stress and increase the speed of fish movement through the system, a line was installed at the Lower Granite Dam fish collection facilities to load fish directly from the separator to a waiting barge rather than via raceways. Between 1983 and 1987, 31–74% of the fish barged from Lower Granite Dam have been loaded directly (Koski et al. 1985, 1988).

Generally, transport operations begin during the last week of March at all three collector projects. Transport ceases in mid to late July at Lower Granite and Little Goose dams and continues through September at McNary Dam. Late-season sampling at Lower Granite Dam was requested by state and federal fish agencies and tribes to determine if significant numbers of juvenile salmonids migrated later in the season. Gatewell sampling between 29 August and 28 September 1984 confirmed minimal juvenile fish movement (Koski et al. 1985). A similar request was made in 1985 because of unusual runoff during the spring migration period and warm summer reservoir temperatures (Koski et al. 1986). Gatewells were sampled between 30 September and 31 October and, again, few migrants were collected. This follows the findings of Raymond (1979) who noted significant fall migrations of chinook salmon in tributaries but not in the main Snake River.

Collection and transport from McNary Dam was extended a month (October) in 1987 at the request of state and federal fish agencies. Because of low river flows and higher-than-normal water temperatures that summer, agency personnel believed a large number of juvenile fall chinook that might migrate late remained upstream. As was observed previously in the Snake River, a late-season outmigration failed to materialize. During October, the total daily collection averaged 615 juvenile salmonids, well under the 1,000-fish trigger established by the Fish Transportation Oversight Team as the minimum for continuing transport.

A mid-to-late July temperature-related problem, apparently aggravated by powerhouse operation, has been reported at McNary Dam for several years (Koski et al. 1986, 1987, 1988). Elevated mortality rates for fall chinook salmon juveniles coincided with powerhouse peaking operations. Apparently, when additional turbine units are brought on line, warm forebay water is introduced into a section of the powerhouse collection flume where cooler water enters from gatewell orifices of unoperating units. Temperature differences of 4.2°C were measured coincident with mortality rates as high as 10% (up from a normal rate of 1 to 2%). Sequential operation of the northern turbine units (7–14) and minimizing load changes were initiated in an effort to alleviate this condition. This effort reduced but did not completely solve the problem.

Future of Mass Transportation

Positive research results and strong regional fish agency and tribal support for transporting steelhead and fall chinook salmon give the juvenile fish transportation program a future. Transportation was included in the Northwest Power Planning Council's 1987 amendments to its Fish and Wildlife Program. However, much work is needed to improve facilities and operations to further reduce stress and injury to juvenile fish. Facility modifications have been recommended by Athearn (1985). Construction of permanent collection and holding facilities at Little Goose Dam was to be completed for the 1990 outmigration season. A new facility at McNary Dam is several years behind schedule, although the existing facilities are essentially obsolete. Two larger (34,000-kg fish capacity) transport barges were to be constructed for the 1989 outmigration. Their increased capacity will not, however, reduce overcrowding problems at Little Goose and McNary dams, which presently lack enough raceway capacity to hold a single peak day's collection of fish.

Correction of problems causing low spring chinook salmon survival is essential for the perpetuation of this race. Benefits from transportation have not been realized for this race as they have for fall chinook salmon and steelhead. Bacterial kidney disease is the likely culprit particularly in hatchery fish (Fryer and Sanders 1981; Banner et al. 1983, 1986; Matthews et al. 1985, 1987, 1988; Park et al. 1986). The role transportation may play in transmission or expression of this disease must be thoroughly researched and understood.

Interactions between transported juvenile chinook salmon and steelhead have been studied. Results reported by Park et al. (1984) and Matthews et al. (1986) show a higher stress level in chinook salmon held with steelhead than those held with conspecifics. In similar studies, Congleton et al. (1984) reported mixed results. In chinook salmon smolts confined with steelhead, plasma cortisol concentrations were significantly elevated in 4 of 11 tests conducted at Lower Granite Dam. The barges and trucks are compartmentalized to allow segregation of chinook salmon and steelhead if desired during transport. This may be given more serious consideration after further research is conducted.

A temporal method of segregating juvenile chinook salmon from steelhead was successfully implemented through the Idaho hatchery release programs in 1983 (Delarm et al. 1984). Steelhead hatcheries delayed releases of fish so that the daily collection peaks at Lower Granite Dam for chinook salmon and steelhead were about 1 month apart. Separating the daily collection peaks for the two species in this way also reduces the total number of fish that arrive on the peak days. Delayed releases of steelhead were continued until the two drought years of 1987 and 1988. Steelhead were released early in those years to benefit from whatever spring runoff occurred. There appears to be sufficient evidence of benefits from segregating chinook salmon and steelhead for Idaho to continue this practice.

With the overwhelming emphasis on restoring salmon and steelhead runs in the Columbia River basin and the obvious success of the juvenile fish transportation program, there should be a continued effort to improve and expand transport capabilities. The Corps of Engineers must continue to work closely with state and federal fish agencies and Indian tribes to develop a program compatible with other basinwide restoration efforts.

References

Athearn, J. B. 1985. A review of juvenile salmonid transportation operations from 1981 to 1984. Pages 3-01 to 3-108 *in* Comprehensive report of juvenile salmonid transportation. U.S. Army Corps of Engineers, Portland, Oregon.

Banner, C. R., J. J. Long, J. L. Fryer, and J. S. Rohovich. 1986. Occurrence of salmonid fish infected with *Renibacterium salmoninarum* in the Pacific Ocean. Journal of Fish Diseases 9:273–275.

Banner, C. R., J. S. Rohovich, and J. L. Fryer. 1983. *Renibacterium salmoninarum* as a cause of mortality among chinook salmon in salt water. Journal of the World Mariculture Society 14:236–239.

Basham, L. R., M. R. Delarm, J. B. Athearn, and S. W. Pettit. 1982. Fish Transportation Oversight Team annual report—FY 1981 transport operations on the Snake and Columbia rivers. NOAA (National Oceanic and Atmospheric Administration) Technical Memorandum NMFS (National Marine Fisheries Service) F/NWR-2, Portland, Oregon.

Basham, L. R., M. R. Delarm, S. W. Pettit, J. B. Athearn, and J. V. Barker. 1983. Fish Transportation Oversight Team annual report—FY 1982 transport operations on the Snake and Columbia rivers. NOAA (National Oceanic and Atmospheric Administration) Technical Memorandum NMFS (National Marine Fisheries Service) F/NWR-5, Portland, Oregon.

Bentley, W. W., and H. L. Raymond. 1969. Passage of juvenile fish through orifices in gatewells of turbine intakes at McNary Dam. Transactions of the American Fisheries Society 98:723–727.

Bentley, W. W., and H. L. Raymond. 1976. Delayed migrations of yearling chinook salmon since completion of Lower Monumental and Little Goose dams on the Snake River. Transactions of the American Fisheries Society 105:422–424.

Bjornn, T. C., and R. R. Ringe. 1984. Homing of hatchery salmon and steelhead allowed a short-distance voluntary migration before transport to the lower Columbia River. Idaho Cooperative Fishery Research Unit, Technical Report 84-1, Moscow.

Carlson, C. D., and six coauthors. 1987. Fish transportation studies—Priest Rapids Dam 1986. Public Utility District Number 2 of Grant County, Ephrata, Washington.

Collins, G. B. 1976. Effects of dams on Pacific salmon and steelhead trout. U.S. National Marine Fisheries Service Marine Fisheries Review 38(11):39–46.

Collins, G. B., W. J. Ebel, G. E. Monan, H. L. Raymond, and G. K. Tanonaka. 1975. The Snake River salmon and steelhead crisis—its relation to dams and the national energy crisis. U.S. National Marine Fisheries Service, Northwest Fisheries Center, Seattle.

Congleton, J. L., T. C. Bjornn, B. H. Burton, B. D. Watson, J. I. Irving, and R. R. Ringe. 1985. Effects of handling and crowding on the stress response and viability of chinook salmon parr and smolts. Idaho Cooperative Fishery Research Unit Completion Report (contract DE-AC79-83BP11196) to Bonneville Power Administration, Portland, Oregon.

Congleton, J. L., T. C. Bjornn, C. A. Robertson, J. L. Irving, and R. R. Ringe. 1984. Evaluating the effects of stress on the viability of chinook salmon smolts transported from the Snake River to the Columbia River estuary. Idaho Cooperative Fishery Research Unit, Technical Report 84-4, Moscow.

Congleton, J. L., and R. R. Ringe. 1985. Response of chinook salmon and steelhead trout smolts to three flumes tested at Lower Granite Dam, 1985. Idaho Cooperative Fish and Wildlife Research Unit, Technical Report 86-3, Moscow.

Congleton, J. L., E. J. Wagner, and R. R. Ringe. 1988. Evaluation of fishway designs for downstream passage of spring chinook salmon and steelhead trout smolts, 1987. Final Report (contract DE-AI79-86BP64234) to Bonneville Power Administration, Portland, Oregon.

Delarm, M. R., L. R. Basham, S. W. Pettit, J. B. Athearn, and J. V. Barker. 1984. Fish Transportation Oversight Team annual report—FY 1983 transport operations on the Snake and Columbia rivers. NOAA (National Oceanic and Atmoshperic Administration) Technical Memorandum NMFS (National Marine Fisheries Service) F/NWR-7, Portland, Oregon.

Ebel, W. J. 1974. Snake River runs of salmon and steelhead trout: collection and transportation experiments at Little Goose Dam, 1971–74. U.S. National Marine Fisheries Service, Northwest Fisheries Center, Seattle.

Ebel, W. J. 1977. Major passage problems. American Fisheries Society Special Publication 10:33–39.

Ebel, W. J. 1979. Effects of atmospheric gas supersaturation on survival of fish and evaluation of proposed solutions. U.S. National Marine Fisheries Service, Northwest and Alaska Fisheries Center, Seattle.

Ebel, W. J. 1980. Transportation of chinook salmon, *Oncorhynchus tshawytscha*, and steelhead, *Salmo gairdneri*, smolts in the Columbia River and effects on adult returns. U.S. National Marine Fisheries Service Fishery Bulletin 78:491–505.

Ebel, W. J., D. L. Park, and R. C. Johnson. 1973. Effects of transportation on survival and homing of Snake River chinook salmon and steelhead trout. U.S. National Marine Fisheries Service Fishery Bulletin 71:549–563.

Ebel, W. J., H. L. Raymond, C. W. Long, W. M. Marquette, R. Krcma, and D. Park. 1971. Progress report on fish-protective facilities at Little Goose Dam and summaries of other studies relating to the various measures taken by the Corps of Engineers to reduce losses of salmon and steelhead in the Columbia and Snake rivers. U.S. National Marine Fisheries Service, North Pacific Fisheries Research Center, Seattle.

Fryer, J. L., and J. E. Sanders. 1981. Bacterial kidney disease of salmonid fish. Annual Review of Microbiology 35:273–298.

Gessel, M. H., W. E. Farr, and C. W. Long. 1985. Underwater separation of juvenile salmonids by size. U.S. National Marine Fisheries Service Marine Fisheries Review 47(3):38–42.

Gunsolus, R. T. 1977. Status of the salmon and steelhead runs entering the Columbia River. American Fisheries Society Special Publication 10:21–22.

Harmon, J. R., and E. Slatick. 1987. Use of a fish transportation barge for increasing returns of steelhead imprinted for homing. U.S. National Marine Fisheries Service, Northwest and Alaska Fisheries Center, Seattle.

Koski, C. H., S. W. Pettit, J. B. Athearn, and A. L. Heindl. 1985. Fish Transportation Oversight Team annual report—FY 1984 transport operations on the Snake and Columbia rivers. NOAA (National Oceanic and Atmospheric Administration) Technical Memorandum NMFS (National Marine Fisheries Service) F/NWR-11, Portland, Oregon.

Koski, C. H., S. W. Pettit, J. B. Athearn, and A. L. Heindl. 1986. Fish Transportation Oversight Team annual report—FY 1985 transport operations on the Snake and Columbia rivers. NOAA (National Oceanic Atmospheric Administration) Technical Memorandum NMFS (National Marine Fisheries Service) F/NWR-14, Portland, Oregon.

Koski, C. H., S. W. Pettit, J. B. Athearn, and A. L. Heindl. 1987. Fish Transportation Oversight Team annual report—FY 1986 transport operations on the Snake and Columbia rivers. NOAA (National Oceanic and Atmospheric Administration) Technical

Memorandum NMFS (National Marine Fisheries Service) F/NWR-18, Portland, Oregon.

Koski, C. H., S. W. Pettit, J. B. Athearn, and A. L. Heindl. 1988. Fish Transportation Oversight Team annual report—FY 1987 transport operations on the Snake and Columbia rivers. NOAA (National Oceanic and Atmospheric Administration) Technical Memorandum NMFS (National Marine Fisheries Service) F/NWR-22, Portland, Oregon.

Matthews, G. M., D. L. Park, S. Achord, and T. E. Ruehle. 1986. Static seawater challenge test to measure relative stress levels in spring chinook salmon smolts. Transactions of the American Fisheries Society 115:236–244.

Matthews, G. M., D. L. Park, J. R. Harmon, and T. E. Ruehle. 1988. Evaluation of transportation of juvenile salmonids and related research on the Columbia and Snake rivers, 1987. National Marine Fisheries Service, Northwest and Alaska Fisheries Center, Seattle.

Matthews, G. M., D. L. Park, J. R. Harmon, C. S. McCutcheon, and A. J. Novotny. 1987. Evaluation of transportation of juvenile salmonids and related research on the Columbia and Snake rivers, 1986. National Marine Fisheries Service, Northwest and Alaska Fisheries Center, Seattle.

Matthews, G. M., D. L. Park, T. E. Ruehle, and J. R. Harmon. 1985. Evaluation of transportation of juvenile salmonids and related research on the Columbia and Snake rivers, 1984. National Marine Fisheries Service, Northwest and Alaska Fisheries Center, Seattle.

Matthews, G. M., G. A. Swan, and J. R. Smith. 1977. Improved bypass and collection system for protection of juvenile salmon and steelhead trout at Lower Granite Dam. U.S. National Marine Fisheries Service Marine Fisheries Review 39(7):10–14.

McCabe, G. T., Jr., C. W. Long, and S. L. Leek. 1983. Survival and homing of juvenile coho salmon, *Oncorhynchus kisutch*, transported by barge. U.S. National Marine Fisheries Service Fishery Bulletin 81:412–415.

McCabe, G. T., Jr., C. W. Long, and D. L. Park. 1979. Barge transportation of juvenile salmonids on the Columbia and Snake rivers, 1977. U.S. National Marine Fisheries Service Marine Fisheries Review 41(7):28–34.

Park, D. L. 1980. Transportation of chinook salmon and steelhead smolts 1968–80 and its impact on adult returns to the Snake River. National Marine Fisheries Service, Northwest and Alaska Fisheries Center, Seattle.

Park, D. L. 1985. A review of smolt transportation to bypass dams on the Snake and Columbia rivers. Pages 2-1 to 2-66 *in* Comprehensive report of juvenile salmonid transportation. U.S. Army Corps of Engineers, Portland, Oregon.

Park, D. L., and W. J. Ebel. 1975. Snake River runs of salmon and steelhead trout: collection and transportation experiments at Little Goose Dam, 1971–75. U.S. National Marine Fisheries Service, Northwest Fisheries Center, Seattle.

Park, D. L., and eight coauthors. 1979. Transportation activities and related research at Lower Granite, Little Goose, and McNary dams 1978. U.S. National Marine Fisheries Service, Northwest and Alaska Fisheries Center, Seattle.

Park, D. L., and W. E. Farr. 1972. Collection of juvenile salmon and steelhead trout passing through orifices in gatewells of turbine intakes at Ice Harbor Dam. Transactions of the American Fisheries Society 101:381–384.

Park, D. L., G. M. Matthews, T. E. Ruehle, J. R. Harmon, E. Slatick, and F. Ossiander. 1986. Evaluation of transportation of juvenile salmonids and related research on the Columbia and Snake rivers, 1985. U.S. National Marine Fisheries Service, Northwest and Alaska Fisheries Center, Seattle.

Park, D. L., G. M. Matthews, J. R. Smith, T. E. Ruehle, J. R. Harmon, and S. Achord. 1984. Evaluation of transportation of juvenile salmonids and related research on the Columbia and Snake rivers, 1983. U.S. National Marine Fisheries Service, Northwest and Alaska Fisheries Center, Seattle.

Park, D. L., T. R. Ruehle, J. R. Harmon, and B. H. Monk. 1980. Transportation research on the Columbia and Snake rivers 1979. U.S. National Marine Fisheries Service, Northwest and Alaska Fisheries Center, Seattle.

Park, D. L., and six coauthors. 1981. Transportation research on the Columbia and Snake rivers, 1980. U.S. National Marine Fisheries Service, Northwest and Alaska Fisheries Center, Seattle.

Park, D. L., J. R. Smith, E. Slatick, G. M. Matthews, L. R. Basham, and G. A. Swan. 1978. Evaluation of fish protective facilities at Little Goose and Lower Granite dams and review of mass transportation activities 1977. U.S. National Marine Fisheries Service, Northwest and Alaska Fisheries Center, Seattle.

Park, D. L., J. R. Smith, E. Slatick, G. A. Swan, E. M. Dawley, and G. M. Matthews. 1977. Evaluation of fish protective facilities at Little Goose and Lower Granite dams and review of nitrogen studies relating to protection of juvenile salmonids in the Columbia and Snake rivers, 1976. U.S. National Marine Fisheries Service, Northwest and Alaska Fisheries Center, Seattle.

Raymond, H. L. 1979. Effects of dams and impoundments on migrations of juvenile chinook salmon and steelhead from the Snake River, 1966 to 1975. Transactions of the American Fisheries Society 108:505–529.

Raymond, H. L. 1988. Effects of hydroelectric development and fisheries enhancement on spring and summer chinook salmon and steelhead in the Columbia River basin. North American Journal of Fisheries Management 8:1–24.

Schreck, C. B., and six coauthors. 1985. Columbia River salmonid outmigration: McNary Dam passage and enhanced smolt quality. Oregon Cooperative Fishery Research Unit Completion Report (contract DACW68-84-C-0063) to U.S. Army Corps of Engineers, Corvallis, Oregon.

Slatick, E., D. L. Park, and W. J. Ebel. 1975. Further studies regarding effects of transportation on survival and homing of Snake River chinook salmon and steelhead trout. U.S. National Marine Fisheries Service Fishery Bulletin 73:925–931.

Slatick, E., and seven coauthors. 1988. Imprinting hatchery reared salmon and steelhead trout for homing, 1978–1983. U.S. National Marine Fisheries Service, Northwest and Alaska Fisheries Center, Seattle.

Smith, J. R., and W. J. Ebel. 1973. Aircraft-refueling trailer modified to haul salmon and trout. U.S. National Marine Fisheries Service Marine Fisheries Review 35(8):37–40.

Smith, J. R., and W. E. Farr. 1975. Bypass and collection system for protection of juvenile salmon and trout at Little Goose Dam. U.S. National Marine Fisheries Service Marine Fisheries Review 37(2): 31–35.

Smith, J. R., G. M. Matthews, L. R. Basham, S. Achord, and G. T. McCabe. 1980. Transportation operations on the Snake and Columbia rivers 1979. U.S. National Marine Fisheries Service, Northwest and Alaska Fisheries Center, Seattle.

Smith, J. R., G. M. Matthews, L. R. Basham, B. H. Monk, and S. Achord. 1981. Transportation operations on the Snake and Columbia rivers, 1980. U.S. National Marine Fisheries Service, Northwest and Alaska Fisheries Center, Seattle.

Weitkamp, D. E., and M. Katz. 1980. A review of dissolved gas supersaturation literature. Transactions of the American Fisheries Society 109:659–702.

American Fisheries Society Symposium 10:347–351, 1991

Approach to Facility and Design Modifications: Stress Monitoring of Migratory Salmonids[1]

CARL B. SCHRECK

U.S. Fish and Wildlife Service, Oregon Cooperative Fishery Research Unit[2]
Oregon State University, Corvallis, Oregon 97331, USA

Abstract.—Knowledge of the physiological status of migratory salmon and trout can be used to establish the effect on fish of specific features of a passage, collection, transportation, or rearing facility. Assessments conducted at the McNary Dam fish collection and transportation system on the Columbia River and at several hatcheries suggested the appropriateness of physiological and performance measures of stress, development (smoltification), and general health for use in evaluating such facilities and their design, and in formulating rearing strategies. The ultimate performance test for an anadromous species is the return rate of adults, which often cannot be evaluated. Tag-return tests have an unacceptably long lag between the time a facility should be evaluated and the availability of the results, and many years of replication are required to sort out the effects of environmental variation. Measures of various physiological factors associated with the generalized stress response, metabolic fluxes, and osmotic balance, in concert with short-term performance tests, can provide strong insights into how facilities or management systems affect the well-being of fish. Such tests can be highly sensitive, have short information turnaround times, are comparatively inexpensive, and require relatively few fish. I discuss the use and values of physiological assessment based on site-specific examples, consider the inferences that can be drawn from such data, and provide cautionary notes about the potential misuse of physiological information.

Knowledge of the physiological status of migratory salmon and trout can be used to establish how specific features of a facility for passing, collecting, transporting, or rearing fish might affect the animals. Physiological measurements can indicate how certain structural features of a system or management protocol might affect the well-being or general quality of fish. My objectives here are to discuss the use of physiological assessment of fish performance capacity based on examples from research at the McNary Dam fish bypass, collection, and transportation system on the Columbia River and at several anadromous salmonid hatcheries; to consider what inferences one might draw from such data; and to provide cautionary notes about the potential for misuse of physiological information.

The ultimate performance test for an anadromous species is the return rate or realized fecundity, but it is usually not feasible to evaluate return rates. Mark–release studies require many years to complete, and the lag between the time a facility should be evaluated and the availability of results is often unacceptable. Many years of replication are needed to sort out the effects of environmental variation. Measures of specific physiological factors associated with the generalized stress response, metabolic fluxes, and osmotic balance—in concert with short-term performance tests—can be powerful tools enabling insights into how facilities or management systems affect the quality of fish (Schreck 1981; Wedemeyer et al. 1990). Further, such knowledge can aid in designing effective facilities or in managing them—allowing the enhancement of the quality of fish handled or reared.

Background

Several aspects of physiological systems must be understood when stress measures are applied to fishery issues. One needs to carefully select the physiological tests for a particular assessment based on the specific temporal dynamics (i.e., response times and durations) for each physiological system (Wedemeyer et al. 1990). For example, if someone were interested in evaluating the effects of an acute stress seconds or minutes after the event, stress hormones (catecholamines and corticosteroids) would be appropriate factors to measure (Mazeaud et al. 1977; Schreck 1981). For a more chronic stressor, or one extending over hours to days, one could examine corticosteroids,

[1]Oregon Agricultural Experiment Station Technical Paper 8695.

[2]Supported cooperatively by the U.S. Fish and Wildlife Service, Oregon State University, and the Oregon Department of Fish and Wildlife.

metabolic factors such as sugar levels, or performance tests such as osmoregulatory capacity, swimming capacity, or disease resistance. If one were interested in responses that occurred in durations of days to weeks, one might evaluate growth (perhaps in seawater) and disease resistance (Maule et al. 1988, 1989). The most powerful approach would be to run a select suite of tests that, in concert, would provide a fairly precise and accurate reflection of the physiological status of the fish—much akin to a battery of stress tests that physicians use (Selye 1976).

Several factors render such physiological tests particularly powerful in assessing physical, facility-associated stresses. First, cumulative stressors appear to cause cumulative stress responses (Barton et al. 1986; Sigismondi and Weber 1988). That is, as fish encounter a series of different stressful events, each discrete stressor imparts its own effect on top of any previous responses by that physiological system. This accumulation apparently holds up to the point where fish are no longer capable of responding and thus are under severe stress that is lethal or nearly lethal.

Another important consideration is that the magnitude of the stress response appears to change as the fish undergo parr–smolt transformation (Barton et al. 1985). Either the same stressor is perceived as being more severe, or the capacity of the fish to respond to the same stressor is amplified as they go through smolting. For example, the same handling stress applied to coho salmon *Oncorhynchus kisutch* at various times during their smolting season resulted in sequentially higher concentrations of plasma cortisol, one of the primary teleostean stress hormones (Barton et al. 1985).

It is also important to bear in mind that unhealthy individuals may be unable to generate typical stress responses. Salmon that were sick, for instance, were incapable of generating a cumulative physiological stress response when subjected to a series of exposures to handling (Barton et al. 1985). Environmental factors such as temperature can affect the stress response of salmonids (Strange et al. 1977; Barton and Schreck 1987). Such environmentally induced modifications of the physiological response to stress can be important in assessing facilities; for example, fall chinook salmon *O. tshawytscha* at McNary Dam appeared to be most stressed when collected late in their run in August, when the temperature of the Columbia River was highest (Schreck et al. 1985b; Maule et al. 1988). Other factors that can alter the response to stress are exposure to toxicants (Schreck and Lorz 1978), habituation to relatively mild stress by repeated exposure (Barton et al. 1987), and feeding (Barton et al. 1988). There may also be differences between species or between stocks in the response to a given stressful situation. For example, Maule et al. (1988) suggested that spring chinook salmon may be more susceptible than fall chinook salmon to the stresses associated with collection or transportation at McNary Dam.

Facility Design Evaluation

My colleagues at Oregon State University and I used clinical and physiological performance factors to identify particularly stressful elements and practices in the collection and transportation system at dams, to provide insights into fish-marking procedures as practiced at McNary Dam and certain hatcheries, and to evaluate the effects of fish-loading densities at Columbia River hatcheries. Our objectives were to use the evaluations to suggest methodologies or alterations that might moderate stress experienced by the fish. The following synopsis of the results of these investigations is provided to illustrate how such physiological information can be used.

McNary Dam Collection and Bypass Facility

The juvenile bypass facility at McNary Dam, designed to collect or bypass downstream-migrating salmon and trout, consists of a screened gatewell system that collects the fish at the upstream side of the dam and funnels them through a pipe to the downstream side of the dam. The fish emerge from an upwelling box, cross a dewatering pan to a bar sorter, and then fall through into a flume that leads them to a raceway. The fish may spend more than a day in a raceway before they are crowded into a flume that loads them onto a transport truck or a pipe that leads to a barge for later transport for release below Bonneville Dam (Maule et al. 1988). Specific techniques and results for evaluation of this system were presented by Schreck et al. (1985b) and Maule et al. (1988). Basically, the runs of fall chinook salmon were assessed in three years and at three times (early, middle, and late) during each run. Less-detailed evaluations were made for spring chinook salmon and sockeye salmon *O. nerka*. Tests consisted of assaying clinical factors such as plasma concentrations of cortisol, glucose, and lactic acid; hepatic glycogen; interrenal cell nuclear dimensions; and white and red blood cell quantities. Performance tests used included seawater challenge, physiological response to a second stressor, and swimming performance.

Each element of the dam's collection system imparted its own degree of stress, in a cumulative sequence. The use of a black plastic cover over the collection system at the downstream side of the dam showed that fish in the dark recovered from the stresses associated with collection much more rapidly than did others collected in full daylight. It was also evident that the fish collected in the normal system remained stressed for up to a day after they were collected. It thus appears reasonable to suggest that the fish be allowed to recover for a day before being subjected to a subsequent stress such as transportation.

Further, these tests demonstrated in the first year of the study (1982) that fall chinook salmon in the latter part of their run were already apparently stressed during passage through the gatewell. This stress was attributed to the presence of large numbers of American shad *Alosa sapidissima* that had packed the gatewell on their postspawning migration and through which the salmon would have to pass. In subsequent years when the U.S. Army Corps of Engineers increased the flow of water through the gatewell system to flush out the shad, the salmon smolts did not exhibit this stress. Indeed, the increased flow appeared to moderate the stress of collection at all times during the outmigration of the fall chinook salmon.

Fish Transportation Systems

The smolts collected at McNary Dam are loaded onto trucks or barges and transported for 3–4 h or about 18 h, respectively, for release below Bonneville Dam. A similar assessment for these procedures was conducted by using the tests mentioned above. In addition, a disease challenge test was conducted to determine the capacity of the fish to resist pathogens (Schreck et al. 1985b; Maule et al. 1988, 1989). From these observations and earlier work with transported coho salmon (Specker and Schreck 1980), the following conclusions became evident.

The loading procedure (i.e., getting the fish onto the truck or barge) appears to be the main cause of stress associated with transportation. Fish were highly stressed immediately after entering the transport vehicle; consequently, this practice should be carefully examined to moderate the stress effects of transportation.

As long as water quality remained good, no further stress associated with transport was detected en route, and the fish often demonstrated some degree of recovery from loading as they were being transported. It was surmised from observations of fish that had been transported that the unloading procedures (i.e., liberating the fish) could be as stressful as loading. Fish were stressed after transportation and their capacity to perform such activities as resisting diseases, osmoregulating, and swimming were impaired. It also appeared to take about a day for the fish to recover from these stresses.

Fish-Marking Systems

Many of the fish collected at McNary Dam are marked with coded wire tags. These smolts are passed from the flume into a holding tank where they may spend up to a day before being crowded, dipnetted, transferred to an anesthetic bath, anesthetized, marked, and then channeled through a pipe to a raceway. The stress-associated effects of this practice, based on plasma concentrations of cortisol, were examined (Schreck et al. 1985b). In another study, clinical examinations were made of chinook salmon undergoing coded-wire tagging or fin clipping at Warm Springs National Fish Hatchery and Round Butte Hatchery, Oregon Department of Fish and Wildlife, and coupled with a physiological assessment of immunocompetence (Maule et al. 1989).

The marking procedure at McNary Dam was stressful to the fish; again, about a day was needed for recovery. The same trend was observed at the hatcheries, supported by data on the immune system suggesting that the fish could not withstand pathogens as well soon after marking but that their immune systems actually appeared enhanced the day after the stress. Conclusions that can be drawn from these observations are that recovery times of a day or longer are appropriate before the fish are further stressed by liberation, transportation, or a similar activity.

Physiological data from some of our earlier laboratory studies (Strange and Schreck 1978) revealed that preanesthetization could mitigate the stresses associated with handling practices. That is, fish anesthetized before they experienced crowding, netting, handling, etc., would not undergo a major stress response, and their performance capacities would be unimpaired. The possibility of applying this concept to the marking system at McNary Dam was evaluated (Schreck et al. 1985b). Fish that passed directly from the flume into an anesthetic bath recovered significantly faster from the marking stress than did those experiencing the normal crowding protocol in a holding tank.

Hatchery Operations

Hatchery practices, in addition to those involving handling, affect fish physiologically, and clinical measures are consequently of value in assessing such rearing strategies. For example, in studies conducted with coho salmon at Eagle Creek (Schreck et al. 1985a) and Willard (Patiño et al. 1986) national fish hatcheries, physiological measures of stress were used to evaluate the effects of rearing density and metabolic load (related to water flow rate). Because both stress and rearing conditions can affect the developmental trajectories of salmon, this investigation was coupled with an examination of physiological factors commonly associated with the parr–smolt transformation; these included gill ATPase and blood thyroid hormones. The various rearing conditions greatly affected the developmental physiology of the fish, and the data supported speculation that smolt quality was affected as well. The interpretation of the clinical and physiological performance data was correct for both hatcheries; development was affected by both density and water flow, and this ultimately affected return rates.

Inferences from Physiological Data and Cautionary Notes

On the basis of both my experiences with physiological assessment of quality in salmonids and information in the literature such as that published by Pickering (1981) and Wedemeyer et al. (1990), I believe that clinical and physiological performance data can clearly indicate the presence of stress in fish. In other words, fish that exhibit certain physiological characteristics are stressed. Further, this information is of value in making fishery management and facility design decisions. However, not all stressors result in a typical stress response, and physiological data without prior validation should not be used to infer that fish are not stressed; negative data can be misleading. It is thus essential that the stress response be fully characterized before data are applied in management. Such characterization necessitates having available some basis against which to judge physiological patterns; background information is essential. Further, the data should be interpreted by those trained in this field so that conclusions will be drawn that realistically reflect the situation (e.g., developmental stage of the fish, prior history of stress, environmental conditions).

It is possible to use physiological data to judge whether a particular situation is better or worse than another, but not by how much. For example, it was possible to anticipate from physiological information the relative return rates of transported and nontransported coho salmon and to predict that fish allowed to recover from the trucking before release would have a higher return rate (Schreck et al., in press). Physiological assessments have not evolved to the point where the severity of a stressful situation can be determined without considerable background work; stress cannot as yet be quantified. Nonetheless, physiological evaluation can be an extremely powerful source of information that can withstand rigorous statistical assessment. Data are rapidly available and can be obtained rather inexpensively; they can be used as the basis for immediate decisions. Such data can also be of considerable value for optimizing release–recapture studies that are much more costly and do not lend themselves to examination of many treatment groups; after a physiological examination one would have a basis for determining the treatment groups into which money should be invested for a mark-and-release trial.

Physiological data can also be useful in judging the developmental processes of salmonids. Resultant data can indicate the quality of rearing facilities and practices. Although it is difficult to define precisely what a smolt is, physiological data can be sound measures of the effects of hatchery conditions on the fish and can provide both standards against which to compare different practices or facilities and predictors of fish quality.

Acknowledgments

I gratefully acknowledge the contribution of numerous colleagues, students, and technicians who helped with the work described here and the formulation of ideas. A. Maule, as project leader, B. Barton, S. Bradford, and G. R. Bouck have my thanks for their efforts associated with the McNary Dam study. Funding for the studies mentioned in this paper were provided by the Bonneville Power Administration (Project 82-16), the U.S. Army Corps of Engineers (contract DACW68-84-C-0063), and Sea Grant, U.S. Department of Commerce (grant NA85AA-D-SG-

095). R. Patiño made major contributions to projects involving rearing densities; that work was funded by the National Marine Fisheries Service (contract 41 USC 252(C)(3)) and by Sea Grant (grant 81AA-D-00086).

References

Barton, B. A., and C. B. Schreck. 1987. Influence of acclimation temperature on interrenal and carbohydrate stress responses in juvenile chinook salmon (*Oncorhynchus tshawytscha*). Aquaculture 62:299–310.

Barton, B. A., C. B. Schreck, and L. D. Barton. 1987. Effects of chronic cortisol administration and daily acute stress on growth, physiological conditions, and stress response in juvenile rainbow trout. Diseases of Aquatic Organisms 2:173–185.

Barton, B. A., C. B. Schreck, R. D. Ewing, A. R. Hemmingsen, and R. Patiño. 1985. Changes in plasma cortisol during stress and smoltification in coho salmon, *Oncorhynchus kisutch*. General and Comparative Endocrinology 59:468–471.

Barton, B. A., C. B. Schreck, and L. G. Fowler. 1988. Fasting and diet content affect stress-induced changes in plasma glucose and cortisol in juvenile chinook salmon. Progressive Fish-Culturist 50:16–22.

Barton, B. A., C. B. Schreck, and L. A. Sigismondi. 1986. Multiple acute disturbances evoke cumulative physiological stress responses in juvenile chinook salmon. Transactions of the American Fisheries Society 115:245–251.

Maule, A. G., C. B. Schreck, C. S. Bradford, and B. A. Barton. 1988. The physiological effects of collecting and transporting emigrating juvenile chinook salmon past dams on the Columbia River. Transactions of the American Fisheries Society 117:245–261.

Maule, A. G., R. A. Tripp, S. L. Kaattari, and C. B. Schreck. 1989. Stress-induced changes in glucocorticosteroids alter immune function and disease resistance in chinook salmon (*Oncorhynchus tshawytscha*). Journal of Endocrinology 120:135–142.

Mazeaud, M. F., F. Mazeaud, and E. M. Donaldson. 1977. Primary and secondary effects of stress in fish: some new data with a general review. Transactions of the American Fisheries Society 106:201–212.

Patiño, R., C. B. Schreck, J. L. Banks, and W. S. Zaugg. 1986. Effects of rearing conditions on the developmental physiology of smolting coho salmon. Transactions of the American Fisheries Society 115:828–837.

Pickering, A. D., editor. 1981. Stress and fish. Academic Press, London.

Schreck, C. B. 1981. Stress and compensation in teleostean fishes: response to social and physical factors. Pages 295–321 *in* A. D. Pickering, editor. Stress and fish. Academic Press, London.

Schreck, C. B., and H. W. Lorz. 1978. Stress response of coho salmon (*Oncorhynchus kisutch*) elicited by cadmium and copper and potential use of cortisol as an indicator of stress. Journal of the Fisheries Research Board of Canada 35:1124–1129.

Schreck, C. B., R. Patiño, C. K. Pring, J. R. Winton, and J. E. Holway. 1985a. Effects of rearing density on indices of smoltification and performance of coho salmon, *Oncorhynchus kisutch*. Aquaculture 45:345–358.

Schreck, C. B., and six coauthors. 1985b. Columbia River salmonid outmigration: McNary Dam passage and enhanced smolt quality. Completion Report (contract DACW68-84-C-0063) to U.S. Army Corps of Engineers, Portland, Oregon.

Schreck, C. B., M. F. Solazzi, S. L. Johnson, and T. E. Nickelson. In press. Transportation affects performance of coho salmon. Aquaculture.

Selye, H. 1976. Stress in health and disease. Butterworth, Boston.

Sigismondi, L. A., and L. Weber. 1988. Changes in avoidance response time of juvenile chinook salmon exposed to multiple acute handling stresses. Transactions of the American Fisheries Society 117:196–201.

Specker, J. L., and C. B. Schreck. 1980. Stress response to transportation and fitness for marine survival in coho salmon (*Oncorhynchus kisutch*) smolts. Canadian Journal of Fisheries and Aquatic Sciences 37:765–769.

Strange, R. J., and C. B. Schreck. 1978. Anesthetic and handling stress on survival and the cortisol concentration in yearling chinook salmon (*Oncorhynchus tshawytscha*). Journal of the Fisheries Research Board of Canada 35:345–349.

Strange, R. J., C. B. Schreck, and J. T. Golden. 1977. Corticosteroid response to handling and temperature in salmonids. Transactions of the American Fisheries Society 106:213–218.

Wedemeyer, G. A., B. A. Barton, and D. J. McLeay. 1990. Stress and acclimation. Pages 451–490 *in* C. B. Schreck and P. B. Moyle, editors. Methods for fish biology. American Fisheries Society, Bethesda, Maryland.

American Fisheries Society Symposium 10:353, 1991

FISH HATCHERIES

Introduction

During the 9 years between the 1979 Bioengineering Symposium in Traverse City, Michigan, and the 1988 conference in Portland, Oregon, the bioengineering of aquatic production systems has seen many advancements. Many of these important advancements and improvements were presented at the 1988 Fisheries Bioengineering Symposium and are contained in this section.

Economic considerations and limited water resources have caused a strong trend toward the intensification of hatchery operations, in terms of both numbers and weight of fish cultured per unit volume of water. Recent research on the effects of critical variables such as gas supersaturation, dissolved oxygen, carbon dioxide, pH, and ammonia has produced more appropriate water quality criteria for both the design and operation of aquatic production facilities. The use of supplemental oxygen to increase carrying capacity has been the subject of intense research and experimentation. Several papers were presented on different types of absorber systems and configurations.

Increased hatchery intensity has also increased the need for on-line monitoring and control of critical hatchery functions. Computer-aided production modeling, scheduling, and reporting are fast becoming integral parts of hatchery management. To combat disease problems, many hatcheries require disinfection of water before and after it passes through fish chambers and raceways.

Tighter environmental regulations in the USA and Europe have resulted in the formulation of feeds that cause less pollution and the development of improved systems to remove solid wastes. Restrictions on the discharge of therapeutic compounds approved by the U.S. Food and Drug Administration are becoming a serious problem and may require major modifications of incubation and rearing systems to reduce discharges.

Improved siting criteria and evaluation of large hatcheries constructed during the past 20 years have shown that large centralized hatcheries may be inappropriate in some areas. Smaller, less-costly hatcheries are more economic in many cases. This type of hatchery may be operated by a local authority, sportsmen's organization, or fisheries agency. Alternative rearing system and release strategies show promise in increasing the returns of anadromous salmonids.

JOHN COLT

James M. Montgomery, Consulting Engineers
Bellevue, Washington

American Fisheries Society Symposium 10:354–364, 1991

On Choosing Well: Bioengineering Reconnaissance of New Hatchery Sites

BRUCE G. SHEPHERD[1]
Department of Fisheries and Oceans, Resource Enhancement Branch
555 West Hastings Street, Vancouver, British Columbia V6B 5G3, Canada

Abstract.—Site selection can significantly affect biological performance and economic costs and is critical in determining the ultimate success of an enhancement facility. Bioengineering reconnaissance should start with systematic and extensive surveys of potential hatchery sites. Sites showing the most promise should then be subjected to more intensive field investigations. The validity of all key factors and assumptions used in the decision-making process should be regularly reviewed, because these factors actually control the site-selection process. In addition, the differing approaches and knowledge bases used by biological and engineering specialists can result in serious communication problems.

British Columbia's Salmonid Enhancement Program (SEP) was approved in 1977 (Lill 1981). Since 1977, over two dozen pilot and production-scale enhancement facilities, having a combined capacity in excess of 250 million eggs per year, have been built by the Engineering Division and operated by the Enhancement Operations Division (EOD) of SEP. Overall, the EOD hatcheries have been meeting their targets quite well (Figure 1). Such a smooth and rapid approach to capacity is due in part to the bioengineering reconnaissance process that was used.

There are many aspects to achieving sound design, but site selection is critical to the ultimate success of an enhancement facility, both biologically and economically. The SEP staff has had to make many site-selection choices in the last decade. Most of those decisions were made correctly, but some were not; others probably would have been made differently with the benefit of hindsight. No one likes to make mistakes, but they can be valuable learning tools in that they force one to review criteria and assumptions and to find ways to improve the decision-making process.

This paper outlines an approach to site selection developed during the first phase of SEP, and it presents some of the lessons learned along the way. It is a personal perspective; others within SEP may have quite different views from those expressed here. However, a review of the U.S. experience draws many similar conclusions (Colt 1988).

[1]Present address: British Columbia Environment, 3547 Skaha Lake Road, Penticton, British Columbia V2A 7K2, Canada.

Organization of the Salmonid Enhancement Program

The Salmonid Enhancement Program is a part of the Canadian Department of Fisheries and Oceans (DFO); its organizational history was reviewed in more detail by Shepherd (1984). The program is struggling to meet several objectives and to please diverse user groups, some of which are inherently in conflict. Multiple-objective programs can be a setup for failure; rarely is everybody pleased (Peters and Waterman 1982). For SEP, the problem is complicated by communication gaps that inevitably develop when any organization expands beyond a certain size.

For the last decade, SEP subscribed to the specialist approach, including formation of a New Projects Unit to act as a bioengineering communications interface in the development of major hatcheries; Burrows (1981) felt that such an interface is a necessity. The specialist mode worked well during the formative growth years of SEP but has been less satisfactory in the last few years. Due to a decrease in hatchery construction, the New Projects Unit was disbanded in 1988 and its staff was reassigned to assessment and operations support duties within EOD. Three operations units of EOD now provide any biological input needed for new hatchery projects.

How future needs for bioengineering reconnaissance and evaluation services will be met remains uncertain within SEP. Possible options include a formal team that exists outside the three SEP divisions, a formal group attached to the Engineering Division or to the Special Projects Division, or informal teams within the most appropriate division. The optimum organizational structure depends largely on objectives. When

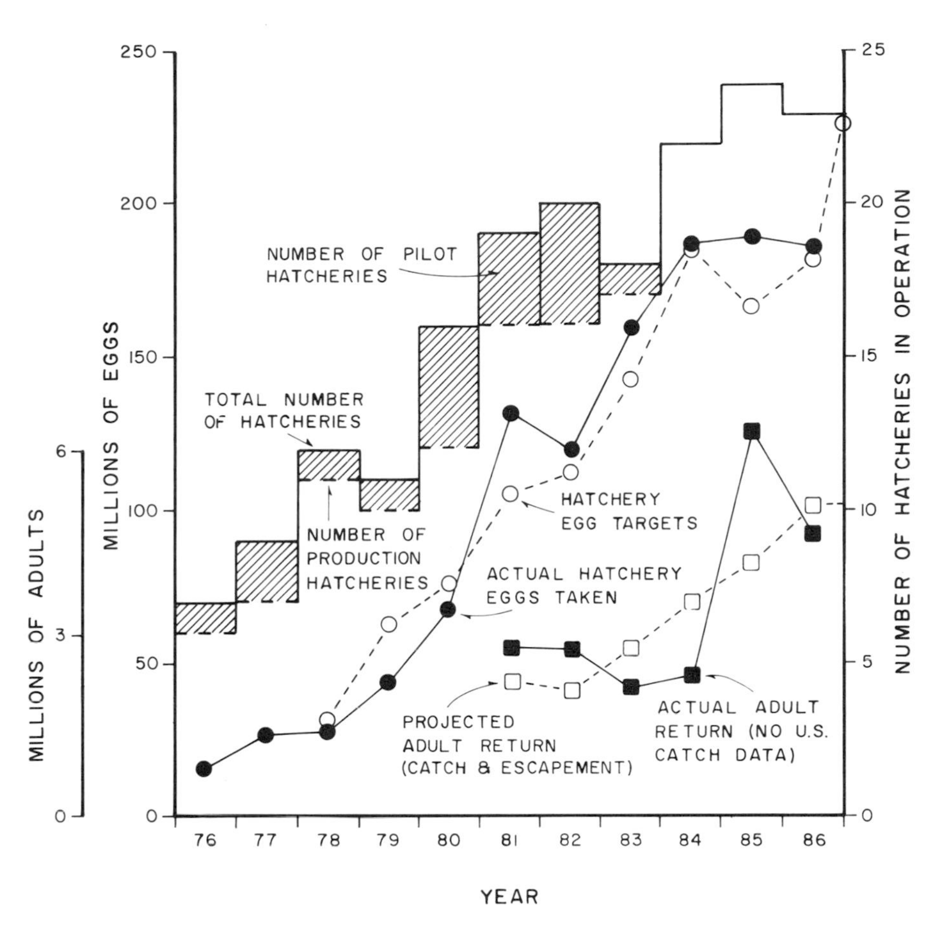

FIGURE 1.—Performance of Enhancement Operations Division's facilities over time. Adult returns include production from spawning channels.

technologies have to be actively developed, the specialist approach has inherent advantages; where local consensus and commitment to projects are more important, the geographic grouping is likely to work better.

Bioengineering Reconnaissance Process

Bioengineering reconnaissance can be split into two levels: overview reconnaissance to array and rank enhancement opportunities; and follow-up feasibility studies of highly ranked opportunities.

Overview Reconnaissance

Most of the major salmon-bearing streams of British Columbia were surveyed by bioengineering teams, which normally consisted of the New Projects coordinator and a senior engineer. Whenever possible, fishery officers were asked to participate in these surveys and impart their additional local knowledge of the systems. Almost all of these surveys were done by helicopter, which offers the speed and range to cover large areas quickly, yet allows close-up examination of any key features noted during the surveys. The surveys were meant to provide an overview so the most promising sites could be flagged for more intensive groundwork.

Ideally, the reconnaissance team should survey the area in each of the four seasons (see Table 1). More recently, the bioengineering team often made only one joint survey, thereafter breaking into their specialist groups to do the remaining surveys. In some ways this approach may be more efficient, but team participation in all four surveys gives a far better understanding of the area as well as of the other group's concerns. Depending on circumstances, surveys have had combined purposes. The dangers in reducing the

TABLE 1.—Timing and rationale for bioengineering reconnaissance surveys at the overview level 2 (modified from Shepherd 1984).

Survey number[a]	Survey type	Timing	Objectives
1	Orientation	Preferably during good weather and clear water, normally midsummer, unless systems are glacial (which are better done in early spring)	Basic familiarization with area; during this and all subsequent surveys, keep notes on the factors outlined in Table 2; eliminate totally unsuitable systems from further surveys
2	Spawning	During peak of spawning for the key stocks in the area, normally late summer to late fall	Examine numbers and distribution of spawners in relation to the physical potential of the habitat; determine vulnerability of adults to capture; identify holding areas and potential fence sites
3	Midwinter	During period of extreme cold, normally between December and February	Identify and measure groundwater outflows; evaluate potential problems associated with low winter flows (i.e., dewatering of redds, potential for icing of intakes)
4	Flood	During peak runoff, normally during spring melt or fall freshet	Check for scouring and erosion problems; evaluate extent of flooding on potential sites; gauge turbidity of potential water supplies; depending on circumstances, this survey may be optional

[a]The order in which surveys 2–4 are done depends upon when the first orientation survey is made.

number of survey trips are that important reconnaissance items (Table 2) can be forgotten or looked at too superficially, and that seasonal variation in those factors can be missed.

TABLE 2.—Important factors to check during bioengineering reconnaissance surveys (from Shepherd 1984).

Size of watershed (reflects water storage capacity)

Drainage pattern of watershed (e.g., dendritic versus radial or parallel networks can indicate stability of subsurface geology)

Number and size of lakes (buffering of sediment load, temperature, and magnitude of discharges)

Type and extent of vegetation (buffering effects similar to lakes)

Actual stream flow versus flood channel size and amount of meandering (indicators of stream stability)

Water color, turbidity, and temperature (water quality indicators)

Type, location, and height of any obstructions or high-gradient sections (useful in defining limits to salmon distribution as well as potential for gravity supply[a] of surface water to a facility for fish or power)

Potential for groundwater (e.g., springs and delta areas indicate potential; rock outcrops do not)

Competing resource activities in watershed (agriculture, logging, mining, industrial or urban development)

Location and type of human settlement (potential for labor and logistical support)

Type and proximity of access to potential sites

Type (single-phase or three-phase) and proximity of power

[a]A net head of 0.5–5.0 m within 0.5 km of the facility site is needed to economically develop an adequate gravity supply to a facility.

Upon completion of these surveys, a report was prepared summarizing the observations made and identifying those enhancement opportunities worthy of further feasibility studies. Such formal reporting is important, because it forces the team to draw their observations together, treat them consistently, and document the rationale behind system and site-selection decisions.

Feasibility Studies

Those systems and sites identified as having the best potential in overview reconnaissance surveys were subjected to more intensive investigations by the Engineering Division and the New Projects Unit (Table 3). In this paper, emphasis is placed on components that should normally involve the biological sector. In terms of the various feasibility studies identified in Table 3, the biological baseline and water quality monitoring studies require the greatest amount of time to complete, and they are seasonally inflexible. When projects must be fast-tracked for financial or other reasons, the resulting substandard data base imparts costs in higher risk and overly conservative designs. Similar problems will occur when the time and funds for basic planning are not provided in advance of funds for construction.

Ground inspection.—Ground inspection of sites was undertaken to obtain additional information on site topography, available hydraulic head, feasibility of access, type of vegetative cover and amount of merchantable timber, soil types and

TABLE 3.—Feasibility studies required for development of a Salmonid Enhancement Program facility, in approximate chronological order (from Shepherd 1984).

Type of study	Minimum duration	Responsibility: New Projects Unit	Responsibility: Engineering Division	Responsibility: Other
Ground inspections	1 week	×	×	
Biological baseline studies	1–2 years	×		
Aerial photography			×	
Authority to enter land			×	
Ground control survey	3 months		×	
Topography mapping			×	
Economic overview			×	×
Surface water monitoring	1 year	×	×	
Groundwater potential	1 month		×	
Establish access	variable		×	
Test well(s)	2 months		×	
Groundwater monitoring	1 year	×	×	
Alternate-site analysis	6 months		×	
Acquisition, zoning of land	6 months		×	×

rock outcrops, groundwater seepages, use by wildlife, etc.

Biological baseline studies.—Biological baseline studies involve two major activities: collation of existing data and generation of new data through fieldwork. Initially, the New Projects Unit attempted to collate all existing data of biological value for bioengineering reconnaissance and facility planning into in-house backgrounder reports. A "Primary Sources Manual" was compiled listing key references and contacts by agency and type of information. This manual was kept in loose-leaf form to facilitate updating.

Only a few backgrounders were completed, but they can improve planning and avoid duplication of fieldwork. Time and staff constraints resulted in only cursory and informal reviews of key data items, which were incorporated into contract specifications for proposed fieldwork.

Also due to staff constraints, the majority of biobaseline studies were done through contracts with consultants. There are several drawbacks, as well as some advantages, to the contracting out of fieldwork (Table 4). Depending on the situation, the field studies incorporated general biophysical reconnaissance for adult and juvenile phases, as well as site-specific feasibility work (Table 5). At least 2 years' adult and juvenile data should be collected in order to allow some evaluation of variation between years.

Initially, much effort went into attempting to estimate juvenile and adult populations accurately, and to collect and rear fry for coded-wire tagging. These program components were expensive and often conflicted with other program objectives, such as definition of the distribution and duration of rearing. For the purposes of facility design, the start–peak–end dates of the wild fry migration are crucial, but accurate enumeration of wild fry populations is needed only where facility fry may be outplanted to augment natural rearing. Similarly, adult migration timing is critical, but highly accurate numbers of spawners are only useful to determine whether earlier abundance estimates can be used to project the average availability of brood stock (independent fishery officer estimates often were half of study estimates). Coded-wire tagging is of no direct use for facility design but was included to provide information on stock contributions to fisheries. The first tag returns from wild chinook salmon stocks that had been pen-reared to taggable size (under primitive fish culture conditions) were very poor, and our management biologists requested that such tagging be terminated. Elimination of these components resulted in significant cost savings and allowed better coverage of other items and systems.

Logistical and cost savings can be made both by consolidating neighboring systems into a regional study package and by coordinating with other groups when possible. There have been embarrassing moments when two and even three different DFO study teams, each with their own chartered helicopter, surprised each other on the same day at the same remote site. Maximizing savings through joint studies requires good communications during the planning of fieldwork. In large and diverse organizations such as DFO, the production of a fieldwork bulletin to regularly update the reconnaissance plans and activities of the various interest groups could help.

TABLE 4.—Advantages and disadvantages to contracting out of fieldwork, as perceived by both government staff and contractors.

Advantages	Disadvantages
Government perspective	
(1) The job gets done despite staff constraints	(1) The selection process is lengthy and tedious (small jobs take longer to set up than to do)
(2) The number of government employees is controlled	(2) There is a greater need for inspectors
(3) Private-sector expertise is established	(3) In-house local knowledge is lost
	(4) In-house training grounds lost for technicians and contract managers
	(5) Expertise is often lacking initially
(4) Inefficient use of capital equipment is avoided	(6) Equipment pool is lost (1-year rental fee often equals purchase price)
(5) No dilution of objectives or diversion of staff	(7) Study flexibility is lost (modifications are made at arm's length)
	(8) Staff continuity is lost due to high turnover rates
(6) Specialized expertise is available on short-term commitment	(9) Many tasks can be illegal yet are necessary to get the job done (which also is required)
(7) Costs of procedures are more obvious, providing more incentive to streamline sampling	(10) Government trades biological expertise for administrative expertise
(8) There is flexibility in termination of projects	(11) There can be field crew safety and morale problems
	(12) Government's image gets linked to contractors' nonpayment of suppliers
(9) Reporting is more timely	(13) There is risk of bankruptcy and litigation
(10) There are no in-house office politics	(14) Contractors' results are viewed with suspicion
	(15) Alienation or lack of interest by government staff—not "our" project
Contractor perspective	
(1) The government is a significant source of revenue	(1) Low profit and high risk forces concentration on cash flow (i.e., bids for new work take priority)
	(2) The bidding process is expensive
(2) The government can transfer its problems (simpler administration in the private sector)	(3) The government seems to prefer to raise rather than relieve paperwork and payment is slow
(3) Contractor experience and level of performance are increased	(4) Contract performance measures or incentives are lacking
	(5) A lack of communication, together with a desire for a high degree of control by government staff, can cause supervisory confusion for the contractor
(4) Expertise is developed and lodged in the private sector	(6) There is government "raiding" of trained staff
	(7) Contracts are generally too short-term (1–2 years) to provide enough incentive to develop highly specialized staff and equipment capabilities

Water quality monitoring.—Water quality monitoring of both surface and groundwater sources should be done for at least one full year in order to gauge seasonal variations. Where possible, sites were geographically grouped, and sampling circuits were undertaken four times over the year. During each visit to a site, a "hatchery series" water quality sampling was done (see Table 6 for variables sampled). In addition, a thermograph was installed on the first trip and serviced on succeeding trips. The dissolved gas content (total gas pressure and oxygen) of the water source was measured at least once. On every visit, water temperature and pH was measured. High accuracy in measuring pH and oxygen is not required except when readings fall near the recommended limits.

Where wells were developed specifically for enhancement facilities, we followed a slightly different water quality sampling sequence. In general, each well was pumped at a minimum 1,200 L/min over 96 h and sampled every 24 h on the initial visit. Given the expensive and one-time nature of this testing, the on-site sampling equipment was upgraded (with backups) to allow more accurate measurement of temperature, oxygen, pH, and conductivity. At the same time that water quality was tested, aquifer yield usually was evaluated by a hydrogeologist. A thermograph was then installed in the well. Where possible, quarterly water quality samples were taken with a submersible domestic pump (after it was run for a few hours to clear the well).

The interpretation of water quality data re-

TABLE 5.—Checklist of biological baseline study items that may be required to confirm the potential of a site for major enhancement.

Type of information	Value of information
A. Reviews of available information	
(1) Watershed description (physical features, land use, access, etc.)	(1) Assists in defining logistics and competing uses for area
(2) Summary of climatic and discharge records	(2) Defines normal weather and flow patterns
(3) Review of available data on salmon populations	(3) Allows studies to focus on data gaps
B. Weather and water	
(1) Weather during studies (mainly precipitation and air temperature; wind at net pen sites)	(1) Gauge normality of weather during studies
(2) Daily or monthly water levels and flows (current and ice patterns at net pen sites)	(2) Used in calibration of fish migrations, judging normality of flow, flooding of site, and flow limitations as a water supply
(3) Stream water quality (sampled monthly and during high or turbid flows)	(3) Determine potential limitations to fish, and water treatment needed for design
(4) Daily average and extreme stream and lake temperatures (summed on a monthly basis) at probable intake sites	(4) Per B.3 above; also allows timing of fish culture events (e.g., hatching, ponding) to be defined for design
(5) Weekly temperature and salinity profiles at flood and ebb tides in estuarine and marine areas	(5) Delineate extent of estuarine influence for migrating fish, and judge potential for sea pen rearing
(6) Seasonal temperature, oxygen, and water quality profiles in lakes	(6) Gauge suitability as water supply and need for aeration
C. Juvenile fish	
(1) Survey of physical habitat features (streams and estuaries)	(1) Together with C.4 and C.7, defines natural and artificial rearing potentials
(2) Downstream migration index trapping	(2) Establish start–peak–end timing for fry or smolt hatchery releases
(3) Size, condition, and growth patterns of migrant and rearing fish	(3) Set minimum standards for juvenile releases from hatchery
(4) Proportions or numbers of migrant and rearing fish	(4) Actual numbers are important only where outplanting of hatchery fry into natural populations is proposed (define unused capacity)
(5) Incidence of disease or abnormalities (all species caught)	(5) Can have implications for feasibility and design of hatchery
(6) Relative abundance of predators and competitors	(6) Gauge degree of predation and competition
(7) Instream and estuarine duration and distribution of rearing fish	(7) Per C.4; helps to define strategies for location, time, and sizes of releases
(8) Trapping and tagging of wild juveniles	(8) Provide information on migration patterns and contributions to fisheries
D. Adult fish	
(1) Biophysical survey of holding and spawning areas	(1) Estimate natural spawning capacity
(2) Survey of all possible obstacles to upstream passage of adults	(2) Gauge availability of spawning habitat
(3) Migrant and spawner numbers (repetitive surveys, mark–recapture where necessary)	(3) Establish start–peak–end timing for hatchery egg takes; judge accuracy of earlier records and actual availability of brood stock for hatchery
(4) Distribution of holding and spawning fish	(4) Establish logistics for brood-stock capture
(5) Sex and age compositions; length, weight, and fecundity relationships	(5) Useful in forecasting production from hatchery as well as biomass and holding flows for design
(6) Egg-retention rates; incidence of diseases and abnormalities in all species caught	(6) Point up problems in natural spawning and can have implications for hatchery feasibility and design
(7) Average diameter of water-hardened eggs	(7) Sizing of incubator screens and egg-picking equipment
(8) Observations of predators and competitors	(8) Assist in defining logistics of brood-stock capture and holding
E. Miscellaneous	
(1) Morphometry of estuaries and lakes	(1) For estuaries, defines suitable areas for rearing and holding net pens; for lakes, allows estimation of water storage
(2) Plankton and algae sampling (vertical hauls at midpoint of lake and near possible intake sites)	(2) Gauge lake productivity for outplanting of fry; presence of nuisance organisms in potential hatchery water supply
(3) Zooplankton watch in estuarine areas	(3) Assists in defining time of release for fish going directly to salt water

TABLE 6.—Checklist of water quality variables that should be sampled at proposed freshwater salmonid hatchery sites.

Variable	How often has variable been found to be a problem?			Recommended sampling frequency		
	Often	Seldom	Never	Every quarter	At least once	Only under unusual conditions
Alkalinity (total)	×			×		
Ammonia (un-ionized)	×				×	
Carbon dioxide[a]		×				×
Chloride		×			×	
Conductivity	×			×		
Cyanide		×				×
Hardness	×			×		
Hydrogen sulfide[a]	×					×[b]
Nitrate	×			×		
Nitrite	×			×		
Oxygen[a]	×			×		
pH[a]	×			×		
Phosphate	×			×		
Residues—filtrable, nonfiltrable, total	×			×		
Silica		×			×	
Sulfate			×			×
Total gas pressure[a]	×			×		
Turbidity	×[c]					×[c]
Metals						
(1) Al, Ca, Cd, Cu, Fe, Hg, Pb, Zn	×			×		
(2) Na		×		×		
(3) As, Cr, K, Mg, Mn, Si		×			×	
(4) Ba, Co, Mo, Ni P, Sb, Se, Sn, Sr, Ti, V,			×			×

[a]Must be measured on site.
[b]Odor is obvious in enclosed space at very low concentrations.
[c]Indirect measure of nonfiltrable residue; not necessary to do both.

quires good judgment. If the water supply is deemed acceptable but is not, a facility can be built that will suffer poor production and costly water treatment retrofits. On the other hand, if sites that are actually suitable are rejected through an overly conservative approach, additional unnecessary reconnaissance or construction costs can result. Because of their importance, our water quality criteria were updated regularly. Where water quality seems marginal, pilot operations can be useful if undertaken carefully.

Pilots and Pitfalls

For the purposes of this paper, a pilot facility is defined as a test facility designed to answer a specific question and reduce uncertainty in the design of the production-scale facility. In the face of uncertainty, a pilot often seems like a good idea. However, a pilot facility too often ends up delaying a site decision, at great expense, without really reducing the level of uncertainty. This can happen for a variety of reasons, but can be generalized as a lack of commitment to the objectives for the pilot. Problems that have confounded the results of SEP pilots on more than one occasion are discussed below.

Poor Experimental Design

In setting up the experimental design, all possible outcomes have to be considered. Our pilot at Devereux Creek was set up to determine whether the local stock of chinook salmon *Oncorhynchus tshawytscha*, which coexists with stocks of kokanee and sockeye salmon *Oncorhynchus nerka*, was vulnerable to the infectious hematopoietic necrosis (IHN) virus. It was demonstrated that the juvenile chinook salmon could be infected quite easily when the IHN virus was deliberately introduced into the water supply. Chinook salmon reared on the water supply containing sockeye and kokanee salmon did not come down with IHN. The lack of an IHN outbreak during the pilot did not constitute a "no" answer; it only meant that no outbreak occurred during the operation of pilot. Because there was no way of detecting the number of IHN virus particles in the water supply, it was impossible to determine the degree of risk involved.

Staff Turnovers

Promotion or transfer of biologists associated with a pilot project can result in a lack of continuity and insufficient scientific input to the experiment.

Facility Faults

The use of makeshift equipment for a limited-budget pilot increases the risk of failure or (worse) of coming to the wrong conclusion. For instance, the aeration tower of a pilot on Mathers Creek was not shielded from sunlight and shed globs of periphyton and *Saprolegnia* fungus into the gravel incubators (Fedorenko and Shepherd 1985). Mortality of chum salmon *Oncorhynchus keta* was high prior to emergence, but it was impossible to determine if this was due to the elevated ammonia levels (the purpose of the pilot) or to suffocation due to the plugging of the incubator.

A second example was at the Chehalis pilot, which was intended to test the effect of soft well water on the culture of salmonids. Rainbow trout *Oncorhynchus mykiss* were chosen as the test animals, because they were considered the most sensitive species. However, local chinook salmon eggs were brought in prior to trout eggs, for lack of a better incubation site. The pumps failed one snowy night, subjecting the chinook salmon eggs to stagnant water and freezing temperatures. Not surprisingly, mortalities were high after this incident and continued until shortly after ponding. The rainbow trout did not do well initially either, but the eggs were from low-quality brood stock (after all, what manager wants to give good eggs to a project when all the fish will be destroyed at the end of it?). The survivors fed and grew well, despite receiving some rough handling by trainee fish culturists. Thus it was concluded that, given the other stresses faced by the fish, there was no detrimental effect attributable to the soft-water problem, and construction proceeded. Nevertheless, chinook salmon—and no other species—have suffered heavy mortalities during incubation ever since start-up of this hatchery (the hatchery staff refer to the problem as The Plague). In hindsight, this problem was probably present during the pilot operation, but the reduced survival was attributed to equipment failure.

Changing Strategies

Major changes in the proposed strategy for the production facility can cause havoc with the pilot operation. Such uncertainty can cause staff to focus attention on the most recent strategy proposed, to mirror that strategy in the pilot operation, and not to answer more fundamental questions.

Production Given Priority

It is preferable to involve regional hatchery staff heavily in a pilot operation, because such collaboration can lead to better fish culture and an improved in-house local knowledge base useful in start-up of the production facility. However, there are also risks. Without close scientific supervision of the experiment embodied in the pilot, hatchery staff tend to focus on production. The Mathers pilot was designed to determine the effects of marginally high ammonia levels in the groundwater. As actually operated, the well water was replaced with a creek–well water mix. This approach minimized fry mortality, but little was learned about the effects of the ammonia and the pilot was turned into an uneconomic miniproduction facility.

Such a production attitude can be aggravated by the piggybacking of other groups of fish onto pilot operations. Quite often there will be excess capacity available at a pilot and it is tempting to make use of it by adding another species or a tag group to estimate fishery contributions, etc. Such additions to the program can have benefits, not just the planned ones but also some unanticipated ones, if the personnel are observant (for example, the Chehalis pilot). With the temptation there comes a price; the objectives of the pilot become more diffuse, and staff attention can be diverted from ensuring experimental protocol is followed to maintaining the extra groups of fish.

Budget Constraints

In periods of budget or staff reductions, permanent production facilities often were given funding priority and pilots tended to be viewed as expendable. This can have detrimental effects on staff interest and thus on the project as a whole. Cutbacks in operating staff and budgets can result in the elimination of monitoring essential for the proper evaluation of the pilot. Also, there is often a temptation to attempt too many experiments at one time, which can confound experiments and dilute limited resources at the expense of adequate evaluation.

The real value generated from a pilot hatchery is in the knowledge gained. All changes of a pilot and its operations should be approached cautiously and with a conscious bias towards satisfying experimental rather than production demands. Failure to do so may trade short-term production gains for the long-term viability of the production hatchery.

Dynamics of the Bioengineering Reconnaissance Process

The site-selection process often can be strongly influenced by several external factors. It is imperative that there be a regular review of the validity of the key factors and assumptions used in decision making. Some of the ways in which process dynamics can alter the situation are discussed below.

Changes in Production Objectives

Between the end of bioengineering reconnaissance and the start of actual production, it is not uncommon for production objectives to radically change. Budget constraints may seriously limit the flexibility of the existing design to meet such changes. For example, many of the SEP hatchery sites were chosen primarily because they offered both groundwater and gravity-feed surface water supplies. At the detailed design stage for most of these hatcheries, the stream intake and pipeline became a major cost and the surface water supply usually was deferred to stay within budget. At some sites this deferral became something more permanent in fairly short order. For instance, the Quesnel River hatchery site near Likely, British Columbia, was chosen as the first production-scale hatchery for chinook salmon in the upper Fraser River watershed. Besides having excellent groundwater and gravity-feed surface water potential, the site was adjacent to the spawning grounds of a major chinook salmon stock recommended for enhancement. These advantages were believed to outweigh the disadvantages of access and logistics that this remote high-snowfall canyon area had, and site development began. At the same time, fisheries management staff requested that the production objectives of the hatchery change from rearing one local stock to a central station for six geographically separated chinook salmon stocks; only 20% of the production would come from the Quesnel River stock. The potential for a gravity-feed water supply was eliminated forever by building the hatchery too high on the site; but possible disease problems for nonlocal stocks made a stream supply unattractive at this point anyway. In retrospect, what was required was a centrally located site offering established infrastructure and a well-water supply. Elimination of the surface water source criteria could have allowed siting of the hatchery in the city of Quesnel rather than at the isolated site that it now occupies.

Effective Communication

The object of communication is the reduction of uncertainty (Kelly 1969). Thus efficient communication is essential to optimal facility design. Without rapid and accurate feedback on several levels, outdated or inappropriate criteria will continue to be used for new facilities. Complaints regarding inadequate lines of communication seem to be common in large organizations, whether government or private. It appears that there is a critical size of organization beyond which information and communication systems must be developed and standardized. In addition to being large in size, both DFO and SEP are complex, having several geographic and specialist groups. As specialists, biologists and engineers have very basic differences in approach, which seem to stem largely from their professional training rather than from individual personalities (Shepherd 1984; Colt 1988). Recognition of these differences are critical in any effort to improve communications.

There is no pat approach that will make communication effective in all situations but the following are important in the development of enhancement facilities (Shepherd 1984).

- Use a task-force management approach for implementing large and complex projects requiring interdisciplinary collaboration.
- Appoint a project manager with clearly defined authority and responsibilities; the latter must include developing effective documentation and communication procedures for both staff and clients.
- Ensure the task force has both regional and specialist expertise, and that all members are

American Fisheries Society Symposium 10:365–367, 1991

Application of the Production Capacity Assessment Bioassay

JAMES W. MEADE

U.S. Fish and Wildlife Service, Tunison Laboratory of Fish Nutrition
3075 Gracie Road, Cortland, New York 13045, USA

Abstract.—Production capacity assessment (PCA) is a bioassay used to quantify limits for the safe use of a water supply. The PCA is compared with the common carrying-capacity determination methods. Examples are developed for use of PCA for comparison of water supplies and for production system design. The method is dynamic and thus useful for estimating changes in production associated with changes in water quality, such as might result from high fish loading achieved with the addition of levels of aeration or oxygen injection.

The first limit to the carrying capacity of an intensive fish production system is usually dissolved oxygen (DO) availability, given a clean water source of appropriate temperature. The second limiting factor is often thought to be ammonia (NH_3), followed by nitrite (in ponds and recirculation systems) and solids. Typical carrying-capacity calculations go something like those that follow.

Carrying Capacity Based on O_2

How many fish can be produced under a given oxygen constraint? The answer depends on the oxygen consumption rate of the fish, which in turn depends (at a given temperature) on fish size and feeding rate. Westers (1981) eliminated the need to consider fish size by assuming that a unit quantity of dispensed food results in a relatively predictable oxygen demand. It does not matter whether a kilogram of food is fed to 15-cm-long fish or 50-cm fish; the oxygen demand associated with a type of food is predictable. For salmonid foods, it is about 200 g O_2/kg of food.

Given an available DO level, say 3.9 mg/L (assume the minimum incoming DO to be 10.4 and the minimum safe outflow DO to be 6.5 mg/L), the daily oxygen budget is the available concentration multiplied by the flow, in liters per hour, times 24 h per day, or

$$\text{daily } O_2 \text{ budget} = (3.9 \text{ mg/L}) \times (60 \text{ L/h}) \times 24 \text{ h}$$
$$= 5{,}616 \text{ mg}$$
$$= 5.6 \text{ g } O_2.$$

The actual feeding schedule may not cover a 24-h period. If one assumes feeding is over 12 h then

$$\text{daily } O_2 \text{ budget} = (3.9 \text{ mg/L}) \times (60 \text{ L/h}) \times 12 \text{ h}$$
$$= 2.8 \text{ g } O_2.$$

The duration of the food-associated oxygen demand may be greater than the feeding period. If one assumes 70% of the food demand occurs during the feeding period, the O_2 required per kilogram of food fed during this time is (200 g/kg) $\times$ 0.7 = 140 g/kg. The "capacity" then in terms of food is

$$\text{food/day} = 2.8 \text{ g } O_2/(140 \text{ g } O_2/\text{kg food})$$
$$= 0.02 \text{ kg food.}$$

This calculation may be adjusted to reflect a known duration and level of oxygen demand. For instance, it can be shown under some conditions that there is a peak demand for a short period of time. With relatively continuous feeding the oxygen demand will be relatively even.

In terms of fish weight, the capacity is the food/day (0.02 kg) divided by the body weight (BW) to be fed, as

% body weight fed/d	Weight of fish (kg)
1.0	2.0
1.5	1.3
2.0	1.0

Under an oxygen constraint of 3.9 mg/L, one can produce up to 0.02 kg/% BW of fish for each liter per minute (Lpm) of water flow, or 1.0 kg of fish per Lpm at 2% BW.

Carrying Capacity Based on Ammonia

But what about the ammonia constraint? First, one sets some limit, such as 0.02 mg/L of un-ionized ammonia, measured as nitrogen (NH_3-N/L), and then determines the amount of fish at a given feeding rate, or the amount of food offered, that will produce that amount of ammonia requried to reach the concentration limit.

To find the allowable total ammonia, one needs to know the NH_3 percentage, or the un-ionized ammonia fraction, at the given pH and temperature. The ammonia ionization fraction can be calculated (Emerson et al. 1975) or located in one of many available tables. At a pH of 7, and at 11°C, the un-ionized ammonia fraction is 0.201; thus the total ammonia limit will be roughly 0.1 mg/L (0.02 ÷ 0.201). Next, one needs to know how much ammonia the fish generate. One also needs to know when they generate the ammonia. That is, is ammonia excreted at a constant rate over 24 h (probably not), or is there a spike, or is there some other pattern? If excretion rate is not constant, is it the mean or the maximum concentration that is important, or should both concentrations be considered in relation to duration of exposure? Let us assume the fish generate 30 g of total ammonia nitrogen per kilogram of food, and that the ammonia is excreted in a pattern similar to that of oxygen consumption.

Now one can estimate the amount of food that will produce 0.1 mg ammonia/L. For a 12-h feeding day, during which an assumed 70% of the total ammonia production will be generated, the 30 g of NH_3-N/kg food distributed will be diluted into 1,140 L/h × 12 h = 13,680 L. Because 0.1 mg/L is allowed, 1,370 mg of ammonia may be generated, and 1.37 g/(30 g/kg) = 0.047 kg of food. Dividing food amount by the feeding rate in percent body weight, one can rear 2.35 kg of fish per Lpm at 2% BW.

To calculate these estimates, one must know the safe ammonia concentration limit, the ammonia ionization relation, and the ammonia production rate. One should also know the rate of change of ammonia production with time after feeding, or the peak production rate. There are many assumptions involved in this methodology, and a reliable universal limit for un-ionized ammonia has not been established (Meade 1985). Furthermore, reported ammonia production rates vary from 20 to 78 g/kg of diet (Fyock 1977; Westers 1981).

There are numerous ways to define and calculate carrying capacity. Piper (1970) and Piper et al. (1982) outlined conventional guidelines and methods. H. Westers (Michigan Department of Natural Resources, personal communication) suggested a general loading equation. The user chooses appropriate values, based on data from the actual facility or from appropriate literature, for the following:

- allowable un-ionized ammonia (UA_{max}) in mg/L;
- oxygen required per kilogram of food (O_{food}) in g/kg food;
- total ammonia nitrogen produced per kilogram of food (TAN_{food}) in g/kg food;
- available oxygen (O_{avail}) in mg/L.

Wester's equation is

$$\text{kg/Lpm} = \frac{(UA_{max}) \times (O_{food}) \times 144}{(O_{avail}) \times (TAN_{food}) \times \%\ UA}.$$

There has been sufficient research to show that the effects of ammonia vary greatly with DO concentration, with water quality (especially the concentration of sodium, calcium, or the ionic strength and the alkalinity), with the environment (culture system) or level of stress, and with fish species (Meade 1985). And the influences of other metabolites, of nitrites, of solids, and perhaps of other factors can be synergistic and should be considered.

A site-specific bioassay can be used to avoid assumptions and unknowns associated with a single-limiting-factor approach such as that for ammonia. Only the aggregate effects (of the water quality, fish excreta, waste food, and other constituents in the system) on growth need to be measured.

Production Capacity Assessment (PCA) Bioassay

The PCA bioassay (Meade 1988) is a short-duration growth study based on a series of rearing units. The underlying idea is to mimic the actual, or proposed, production situation by using, in metabolic load increments, the water and fish that will be used in production. No effort is made to measure or estimate ammonia. The DO effects and constraints are eliminated by aeration before each serial use of the water.

Generally, as water is reused without treatment other than aeration, fish growth is reduced (as one effect of the increasing total or aggregate metabolites, or of water quality change). The manager must choose an acceptable or tolerable growth reduction, for instance no more than 50%. The mean growth rates of fish in each water use of a series of units are regressed against the cumulative levels of DO removed from the water flow, and that relation is used to predict the limit of water use in terms of mg DO/L. For instance, an effective cumultive oxygen consumption at which growth might be reduced to 50% of the maximum (ECOC50) is used as the capacity limit for the facility. Practical applications could include

- evaluating sites or water sources,
- comparing sites or water sources,
- assigning a quantitative value, which can translate to monetary value, to a water source, and
- assigning realistic production objectives and limits.

Examples

Determine the capacity of systems, or potential systems, based on a PCA that resulted in an ECOC50 = 12.5 mg DO/L.

(1) A series of aerated jars is used for initial feeding studies. The food and fish will remove only 2 mg DO/L per jar. How many jars can be placed in a series?

$$\frac{\text{ECOC}}{\text{DO required}} = \frac{12.5}{2} = 6.25 = 6 \text{ jars.}$$

(2) An indoor, total water reuse system is planned for catfish raceway production. The catfish will tolerate a 0.06-mg/L concentration of un-ionized ammonia. Catfish food demands 180 g O_2/kg. Each raceway will be stocked and fed so as to remove at least 4 but no more than 6 mg DO/L from the flow. How many raceways can be installed per series?

$$\frac{12.5}{6} = 2.$$

(3) A manager wants to add oxygenation to a series of three raceways. The raceways are presently stocked at 1 lb/ft^3 and the fish remove about 1.9 mg DO/L in each. What will the allowable fish density be in the raceways after oxygen is installed?

$$\frac{12.5 \text{ mg DO/L}}{3 \text{ uses}} = 4.17 \text{ mg/L per use.}$$

Then,

$$\frac{4.17}{1.9} = 2.19, \text{ and } 2.19 \times 1 \text{ lb/ft}^3 = 2.19 \text{ lb/ft}^3.$$

When all time and materials are considered, the total cost to determine a PCA could range from several hundred dollars in some existing facilities to several thousand dollars for undeveloped water supplies. However, the PCA provides information specific to the system that can be used quickly, reliably, and repeatedly for design, retrofit, and management.

References

Emerson, K. R., R. C. Russo, R. E. Lund, and R. V. Thurston. 1975. Aqueous ammonia equilibrium calculations: effect of pH and temperature. Journal of the Fisheries Research Board of Canada 32:2379–2383.

Fyock, L. O. 1977. Nitrification requirements of water re-use systems for rainbow trout. Colorado State University, Special Report 41, Fort Collins.

Meade, J. W. 1985. Allowable ammonia for fish culture. Progressive Fish Culturist 47:135–145.

Meade, J. W. 1988. A bioassay for production capacity assessment. Aquacultural Engineering 7:139–146.

Piper, R. G. 1970. Know the proper carrying capacity of your farm. American Fishes and U.S. Trout News (May–June):4–7.

Piper, R. G., I. B. McElwain, L. E. Orme, J. P. McCraren, L. G. Fowler, and J. R. Leonard. 1982. Fish hatchery management. U.S. Fish and Wildlife Service, Washington, D.C.

Westers, H. 1981. Fish culture manual for the State of Michigan. Michigan Department of Natural Resources, Lansing.

American Fisheries Society Symposium 10:368–371, 1991

Hatchery Management of Lake Trout Exposed to Chronic Dissolved Gas Supersaturation

William F. Krise

U.S. Fish and Wildlife Service, National Fishery Research and Development Laboratory Rural Delivery 4, Box 63, Wellsboro, Pennsylvania 16901, USA

Abstract.—Lake trout *Salvelinus namaycush* can be successfully reared in hatcheries with incompletely degassed water if they are not exposed to supersaturation during their most susceptible ages. Tolerances of lake trout eggs and sac fry for gas supersaturation are high; eggs withstand excess gas pressures of 80 mm Hg (ΔP, or mm Hg pressure above that of saturation), and sac fry tolerate a ΔP of 148 mm Hg for 40 d, provided the gas pressure in the water supply does not fluctuate rapidly. Among lake trout reared from eggs at chronic ΔP levels of 10–80 mm Hg, growth decreased and mortality increased 60–80 d after the swim-up stage. For alevins first exposed to high gas levels after 30 d of feeding, growth was slowed and mortality increased within 60 d. Mortality varied with the age of fish, level of supersaturation, and rate and magnitude of change in gas supersaturation.

Lake trout *Salvelinus namaycush* are reared in hatcheries primarily to supplement populations in the Great Lakes. Restoration requires that survival after stocking be high. Dissolved gas supersaturation has been implicated as a stressor (Wedemeyer 1981) that reduces the health and ability of fish to survive once they are stocked. Some considerations of fish fitness for stocking are survival, growth rate, condition, duration of exposure to supersaturation, and resistance to disease. In hatcheries with low levels of gas supersaturation, it is beneficial to manage stocks during the juvenile rearing period in a way that minimizes the effect of gas supersaturation.

Lake trout fit a general salmonid model (Jensen et al. 1986) that describes the susceptibility of juveniles to supersaturation. Data combined from acute and chronic exposures to supersaturation are useful in making decisions about the most susceptible ages of lake trout reared for stocking as yearlings. Jensen et al. (1986) estimated 40-d ET50s (exposure time to 50% mortality) for salmonids in water 15 cm deep. The tolerable ΔP (pressure above equilibrium saturation gas pressure) was 120 mm Hg for 4.5-cm fish and only 60 mm Hg for 20-cm fish; thus, smaller fish were more tolerant of supersaturation than larger ones. Data from chronic exposures are useful for determining minimum or threshold tolerance levels of the species.

Lake trout are often reared in hatcheries where the fish are exposed to sublethal gas supersaturation during their captive life span. Effects of supersaturation may reduce growth or either predispose the fish to disease or to poststocking mortality. Rearing of lake trout in hatchery waters in the absence of damaging gas supersaturation would result in more effective hatchery rearing for population restoration programs.

The purpose of this paper is to review the supersaturation tolerance of lake trout from the egg stage through the juvenile stage, and to collate these results into a management plan describing maximum allowable levels of supersaturation for fish culture.

Eggs

Eggs tolerate high levels of dissolved gases. Krise and Meade (1988) showed no increase in egg mortality at ΔP levels up to 81 mm Hg. Eggs of steelhead *Oncorhynchus mykiss* and chinook salmon *O. tshawytscha* tolerated ΔP levels to 167 mm Hg (Meekin and Turner 1974; Nebeker et al. 1978); even a ΔP of 213 mm Hg has been tolerated by chinook salmon eggs (Rucker and Kangas 1974). Alderdice and Jensen (1985) hypothesized that internal hydrostatic pressures of 50–90 mm Hg within the egg chorion raised the threshold of tolerance of salmonid eggs, offsetting the effect of gas supersaturation until levels become higher than the internal pressure in eggs. Lake trout eggs, like those of other salmonids, appear to be highly resistant to supersaturation.

Sac Fry

Lake trout sac fry exposed within 2 d after hatch to a ΔP of 148 mm Hg (Krise and Herman 1989) and held for 40 d exhibited 99% survival at initial feeding. However, most of these fish showed external gas bubbles or signs of gas bubble trauma and were moribund. At a ΔP of 42 mm Hg, 40% of the fish had bubbles around the rim of the eye, indicating that low excess gas pressures are also stressful. Nebeker et al. (1978) found no mortality among steelhead sac fry held at a ΔP of 203 mm Hg, suggesting that salmonid sac fry may tolerate supersaturation better than do older life stages.

In chronic exposures, lake trout hatched and held through the swim-up or alevin stage at excess pressures of 10–80 mm Hg showed no difference in mortality or incidence of blue-sac disease (Krise and Meade 1988). Although sac fry and small lake trout suffer little mortality at elevated gas pressures, tissue damage occurs at sublethal gas levels. Among fry exposed to excess pressures for 15 d, Krise and Herman (1989) found subcutaneous emphysema at ΔP = 8 mm Hg, bubbles on the eye and head at ΔPs of 42–148 mm Hg, and rectal bubbles at ΔPs of 119–148 mm Hg. The most significant damages were hemorrhaging in the eye and in the orbit or cranium, which occurred at ΔPs of 8 mm Hg and higher.

Juveniles

Lake trout fingerlings have a relatively high resistance to supersaturation. Thorn et al. (1978) found lower mortality of lake trout exposed 7 h to ΔPs up to 350 mm Hg than of similarly exposed brook trout *Salvelinus fontinalis*, brown trout *Salmo trutta*, and rainbow trout *Oncorhynchus mykiss*. In 29-d exposures, lake trout fingerlings grew and converted food better at a ΔP of 7 mm Hg than at a ΔP of 100 mm Hg (Thorn et al. 1978). In a 30-d exposure of 7.7-cm lake trout under three feeding schedules, mortality ranged from 2.6 to 10.6% at ΔP = 108 and from 44 to 55% at ΔP = 159 (Krise et al. 1990). Specific growth rates of fish on full rations were significantly higher at ΔPs of 11 and 46 mm Hg than at ΔPs of 78–159 mm Hg.

Eye damage resulting from bubble formation around the rim of the orbit is usually the first sign of gas bubble trauma in lake trout fingerlings (Thorn et al. 1978; Krise et al. 1990). In the absence of external bubble formation, small hemorrhages become visible in the eye and increase in size as gas pressure increases (Krise et al. 1990).

Fluctuations in gas pressures, as well as continual exposure to supersaturation, may affect the health and survival of lake trout. Fish that are subjected to rapid changes in dissolved gas pressure and are not allowed exposure to equilibrated gas pressures, may be adversely affected by further changes in gas pressures. The effects of rapid changes in pressure have been simulated in bioassays. Thorn et al. (1978) noted that lake trout exposed to ΔPs of 45–350 mm Hg for 7 h and returned to a ΔP of 7–8 mm Hg developed eye hemorrhages and corneal cataracts. Lake trout held at ΔPs of 7–15 mm Hg developed eye hemorrhages when transferred to ΔPs of 42–160 mm Hg (Krise et al. 1990), but fingerlings from the same egg hatch did not develop hemorrhages when kept at excess gas pressures up to 81 mm Hg from hatch (Krise and Meade 1988). Machado et al. (1987) showed that rainbow trout subjected to high levels of supersaturation (ΔP = 76 mm Hg) for 46 d rapidly developed signs of gas bubble trauma, and their mortality increased as oxygen levels were increased. These fish were not placed into an equilibrated water supply. Apparently gas bubble formation and some potential for lingering physical damage occur with rapid changes in gas pressures when the fish are denied access to gas-equilibrated water after exposure.

Intermittent exposures, however, can result in increased resistance to supersaturation if recovery time in equilibrated water is sufficient. Meekin and Turner (1974), who exposed chinook salmon and steelhead to high and low gas pressures for alternating periods of 4 h and 16 h, showed that juveniles could tolerate a ΔP of 167 mm Hg if they were held in equilibrated water for 8 h/d. Blahm et al. (1976) and Weitkamp (1976) also reported that intermittent exposure to supersaturation increased resistance to supersaturation because of the time the fish were exposed to equilibrated water. Damage depended on ΔP and length of exposure; recovery seemed to be fastest after short exposures to high gas pressure. Weitkamp and Katz (1980) suggested that it takes a fish longer to equilibrate to decreasing gas pressure than to increasing pressure.

Two important considerations arise when the effects of intermittent exposure on resistance of salmonids to gas supersaturation are analyzed. The first is determination of recovery gas pressure, because recovery may not be complete unless the water supply's gas pressure equilibrates. The second is the use of mortality to measure recovery from exposure to supersaturation (Meekin and Turner 1974; Blahm et al. 1976; Weitkamp 1976). In previously cited work (Thorn et al. 1978; Machado et al. 1987; Krise and Meade 1988) mortality was not used as the measure of tolerance; instead, tolerance was measured by sublethal incidence of gas bubble trauma and subsequent visible physical damage (or lack of damage) to fish. Bubble formation indicates the potential suitability and general health of the cultured fish for survival in the wild. Rapid changes in gas pressure, even at sublethal levels, may predispose fish to secondary effects of supersaturation, thereby reducing long-term survival.

Studies in which lake trout were exposed to low levels of supersaturation for long periods indicate that mortality is unaffected, but growth may be affected. Exposure of swim-up lake trout to chronic modest levels of supersaturation (ΔPs of 13–81 mm Hg) resulted in statistically similar mortality (33% at ΔP = 13 and 37% at ΔP = 81 mm Hg) (Krise and Meade 1988). In another study (my unpublished data), 1-month-old lake trout alevins grew less during the next 56 d at ΔP exposures of 17–75 mm Hg than at ΔP = 4 mm Hg. Length and weight of fish were statistically greater (3 and 11% higher, respectively) at ΔP = 4 than at ΔP = 17 mm Hg, and growth reduction was even greater for fish held at ΔPs of 33–75 mm Hg. Fish held at ΔPs of 4 and 17 mm Hg were significantly larger than fish at ΔP = 33 mm Hg or more after 112-d exposures. These studies indicate that age of exposure may affect growth of salmonids exposed to sublethal supersaturation.

Suggested Total Gas Pressure Limits

Juvenile lake trout stocks exposed to chronic gas supersaturation are managed by preventing potentially damaging dissolved gas pressures and limiting rearing to water with supersaturation below that affecting the fish. Analysis of the biological tolerance of the species enables the development of management guidelines. Variables for consideration include survival, time of exposure, life stage, growth rate, and physiological changes that occur within the fish. Prolonged exposure of fingerlings to high levels of sublethal gas pressures should be avoided.

Many hatcheries rear fishes at sublethal levels of supersaturation, and therefore the usefulness of acute bioassays for long-range planning is limited to partitioning age- or size-groups into "resistant" and "less resistant" groups and to describing resistance of different ages of a species reared under the same conditions (Thorn et al. 1978; Gray et al. 1985). This partitioning enables the manager to estimate ages of least resistance to gas supersaturation and to avoid exposing fish at times of low tolerance. Because sublethal gas supersaturation is typically chronic in hatcheries, actual data from long-term exposures are most applicable to rearing conditions and procedures.

TABLE 1.—Recommended maximum gas supersaturation levels for culture of lake trout in terms of ΔP, the dissolved gas pressure above that at saturation.

Stage or size	Maximum ΔP (mm Hg)	Reason
Egg	45	Internal hydrostatic pressure in eggs
Sac fry	35	Bubbles around the orbit at ΔP = 42
Juvenile <5.0 cm	10	Reduced growth rate above ΔP = 4
Juvenile 9.6 cm	10	Reduced growth rate, low tolerance of acute ΔP
Juvenile 15.3 cm	<30	Increased tolerance of acute ΔP

Guidelines for maximum levels of chronic supersaturation are shown in Table 1. Recommendations for eggs (ΔP = 45 mm Hg) and sac fry (ΔP = 35 mm Hg) are based on the high tolerance of these life stages. Treatment of water supplies so that ΔP remains below 10 mm Hg for growth of fingerlings should eliminate losses in growth. Management of lake trout stocks reared in hatcheries for restoration of wild stocks requires reduction in dissolved gases to meet minimum tolerance levels. Without reduction in dissolved gas levels, lake trout are susceptible to losses of growth and fitness. Total elimination of gas supersaturation is the preferred method of managing water quality; however, in facilities where this has not been accomplished, a reduction of excess total gas pressure to less than 10 mm Hg will provide safe rearing for all fish during the first 2 years of life.

References

Alderdice, D. F., and J. O. T. Jensen. 1985. Assessment of the influence of gas supersaturation on salmonids in the Nechako River in relation to Kemano completion. Canadian Technical Report of Fisheries and Aquatic Sciences 1386.

Blahm, T. H., R. J. McConnell, and G. R. Snyder. 1976. Gas supersaturation research, National Marine Fisheries Service Prescot Facility. ERDA (Energy Research and Development Administration) Conf-741033:11–19.

Gray, R. H., M. G. Saroglia, and G. Scarano. 1985. Comparative tolerance to gas supersaturated water of two marine fishes, *Dicentrarchus labrax* and *Mugil cephalus*. Aquaculture 48:83–89.

Jensen, J. O. T., J. Schnute, and D. F. Alderdice. 1986. Assessing juvenile salmonid response to gas supersaturation using a multivariate dose–response model. Canadian Journal of Fisheries and Aquatic Sciences 43:1694–1709.

Krise, W. F., and R. L. Herman. 1989. Tolerance of lake trout sac fry to dissolved gas supersaturation. Journal of Fish Diseases 12:269–273.

Krise, W. F., and J. W. Meade. 1988. Effects of low level gas supersaturation on lake trout (*Salvelinus namaycush*). Canadian Journal of Fisheries and Aquatic Sciences 45:666–674.

Krise, W. F., J. W. Meade, and R. A. Smith. 1990. Effect of feeding rate and gas supersaturation on survival and growth of lake trout. Progressive Fish-Culturist 52:45–50.

Machado, J. P., D. L. Garling, Jr., N. R. Kevern, A. L. Trapp, and T. G. Bell. 1987. Histopathology and the pathogenesis of embolism (gas bubble disease) in rainbow trout (*Salmo gairdneri*). Canadian Journal of Fisheries and Aquatic Sciences 44:1985–1994.

Meekin, T. K., and B. K. Turner. 1974. Nitrogen supersaturation investigations in the mid-Columbia River. Washington Department of Fisheries Technical Report 12.

Nebeker, A. V., J. D. Andros, J. K. McCrady, and D. G. Stevens. 1978. Survival of steelhead trout (*Salmo gairdneri*) eggs, embryos, and fry in air-supersaturated water. Journal of the Fisheries Research Board of Canada 35:261–264.

Rucker, R. R., and P. H. Kangas. 1974. Effect of nitrogen supersaturated water on coho and chinook salmon. Progressive Fish-Culturist 36:152–156.

Thorn, W., C. Lessman, and R. Glazer. 1978. Some effects of controlled levels of dissolved gas supersaturation on selected salmonids and other fishes. Minnesota Department of Natural Resources Section of Fisheries Investigational Report 347.

Wedemeyer, G. A. 1981. The physiological response of fishes to the stress of intensive aquaculture. Pages 3–18 *in* Proceedings of the world symposium on aquaculture in heated effluents and recirculation systems. European Inland Fisheries Advisory Commission, Stavanger, Norway.

Weitkamp, D. E. 1976. Dissolved gas supersaturation: live cage bioassays at Rock Island Dam, Washington. ERDA (Energy Research and Development Administration) Conf-741033:24–36.

Weitkamp, D. E., and M. Katz. 1980. A review of dissolved gas supersaturation literature. Transactions of the American Fisheries Society 109:659–702.

American Fisheries Society Symposium 10:372–385, 1991

Water Quality Considerations and Criteria for High-Density Fish Culture with Supplemental Oxygen

JOHN COLT

James M. Montgomery, Consulting Engineers
2375 130th Avenue NE, Suite 200, Bellevue, Washington 98005, USA

KRIS ORWICZ

Aquatic Biotics
14150 NE 20th Street, Suite 365, Bellevue, Washington 98007, USA

GERALD BOUCK

Division of Fish and Wildlife, Bonneville Power Administration
Post Office Box 3621, Portland, Oregon 97208, USA

Abstract.—The use of pure oxygen can significantly increase the carrying capacity of a culture system both by supporting increased fish density and by allowing reuse of the water. The minimum allowable dissolved oxygen level is well defined, but considerably less information is available on the maximum allowable dissolved oxygen level. The maximum allowable dissolved oxygen level appears to be influenced primarily by (1) oxygen toxicity, (2) physiological and developmental problems caused by high or rapidly fluctuating dissolved oxygen levels, (3) gas bubble trauma, and (4) biological effects of increased loading and density. Consideration must be given to the effects of oxygen supplementation on other water quality variables such as total gas pressure and carbon dioxide levels. Dissolved gas criteria are suggested for both coldwater and warmwater conditions.

The use of supplemental oxygen can increase the carrying capacity of a fish culture system if dissolved oxygen is the most limiting factor. This can allow increased fish density, water reuse, or a reduced water requirement, all of which can have a significant impact on production economics (Gowan 1987; Severson et al. 1987).

Oxygen supplementation commonly increases carrying capacity but this is not always the case and the conditions under which this does not occur are not fully understood by fish culturists (Colt and Watten 1988). The use of supplemental oxygen can increase the total gas pressure and, if it is accompanied by greater fish loadings, result in increased carbon dioxide levels. Thus some water quality criteria for oxygen-supplemented systems may be substantially different from those for traditional hatcheries as well as for the environmental protection of rivers and lakes. Developmental and physiological problems may result from the improper operation or failure of pure oxygen systems.

The purpose of this paper is to describe some biological effects resulting from the use of supplemental oxygen and to suggest critical areas where additional research is needed. Specific water quality criteria are presented for coldwater and warmwater conditions.

Measurement of Intensity in Fish Culture

The use of supplemental oxygen may support more fish per individual rearing unit or allow serial reuse of the water. With increased rearing intensity, the density, loading, and exchange rate may be important and are described as follows:

$$D = M/V; \tag{1}$$

$$\text{DI} = D/\text{TL}; \tag{2}$$

$$L = M/60{,}000Q; \tag{3}$$

$$\text{EXC} = 3{,}600Q/V; \tag{4}$$

$$\text{OC} = (\text{DO}_{\text{in}} - \text{DO}_{\text{out}}); \tag{5}$$

D = density of fish (kg/m^3 of rearing unit);
M = mass of fish in rearing unit (kg);
V = volume of rearing unit (m^3);
DI = density index [kg/m^3 · (fish length, cm)];
TL = total length of fish (cm);
L = loading (kilograms of fish divided by flow to rearing unit in liters per minute);
Q = flow to rearing unit (m^3/s);
EXC = water exchange rate (rearing unit volumes per hour);
OC = oxygen consumption in a rearing unit (mg/L);

TABLE 1.—Concentrations of dissolved oxygen in mg/L corresponding to various partial pressures (Colt 1984). The C^* value corresponds to the equilibrium concentration of oxygen at 1 atmosphere pressure of air.

Temperature (°C)	Partial pressure of oxygen (mm Hg)							
	C^*	200	300	400	500	600	700	800
5	12.8	16.2	24.3	32.4	40.4	48.5	56.6	64.7
10	11.3	14.3	21.5	28.7	35.9	43.0	50.2	57.4
15	10.1	12.9	19.3	25.7	32.2	38.6	45.1	51.5
20	9.1	11.7	17.5	23.4	29.2	35.0	40.9	46.7
25	8.2	10.7	16.0	21.4	26.7	32.1	37.4	42.8
30	7.5	9.9	14.8	19.8	24.7	29.7	34.6	39.6

DO_{out} = effluent dissolved oxygen from a rearing unit (mg/L);
DO_{in} = influent dissolved oxygen to a rearing unit (mg/L).

In a serial reuse system, the cumulative loading, density, density index, or oxygen consumption can be computed by summing the values for each individual unit. The production of ammonia or consumption of oxygen is proportional to total ration (Westers 1981), so either total feed consumption or cumulative oxygen consumption (COC) can be used to describe the degree of reuse (Meade 1988). The maximum COC may be based on observed growth reduction or specific water quality criteria such as ammonia (COC_{amm}), carbon dioxide (COC_{car}), or suspended solids (COC_{ss}).

Loading, water exchange rate, and density (Westers 1981) are related by

$$L = 0.06D/\text{EXC}. \quad (6)$$

Therefore, increasing the density of fish in a rearing unit also increases loading if the water exchange rate does not increase proportionally. Much of the published information on the effects of density is questionable because both density and loading were statistically confounded.

Reporting of Dissolved Gas Levels

Individual Gases

Individual dissolved gas can be expressed in terms of concentration (mg/L, mL/L, or μM/L), percent of saturation, and partial pressure. Biologists and engineers most commonly use mg/L or percent of saturation. Oxygen toxicity depends on the partial pressure of oxygen, so it is necessary to be able to convert between the different units:

$$\text{percent of saturation (\%)} = \frac{C}{C^*}100; \quad (7)$$

$$\text{partial pressure (mm Hg)} = \frac{C}{\beta}A; \quad (8)$$

C = concentration of dissolved gas (mg/L);
C^* = equilibrium concentration (mg/L);
A = constant depending only on the specific gas;
β = Bunsen coefficient [L/(L · atmosphere)].

The value of C^* depends on temperature and gas composition whereas β depends only on temperature. The value of A for oxygen is 0.5318. The concentrations of dissolved oxygen corresponding to various oxygen partial pressures are presented in Table 1 as a function of temperature.

Gas Supersaturation

Gas supersaturation should be reported in terms of ΔP, the difference between the sum of the partial pressures of all the dissolved gases and the local barometric pressure (Colt 1984). This parameter can be measured directly by the membrane-diffusion method (APHA et al. 1989).

Biological Effects of Oxygen

Low DO Concentrations

Minimum allowable DO levels are based on critical dissolved oxygen concentrations or pressures, which may be defined differently for different life stages. For eggs, embryos, and alevins, the critical DO is generally defined as the dissolved oxygen level at which oxygen uptake becomes dependent on the oxygen concentration. The critical DO for the incubation and hatching of salmonid eggs is highest just before hatching. For eggs of steelhead *Oncorhynchus mykiss*, the critical DO varies from 7.6 mg/L at 6.0°C to 10.2 mg/L at 15.1°C (Rombough 1986). If the influent DO is 90% of saturation at sea level, the DO available for metabolic use ranges from 3.6 mg/L at 6.0°C to −1.1 mg/L at 15.1°C. The available DO is significantly reduced at higher elevations. The

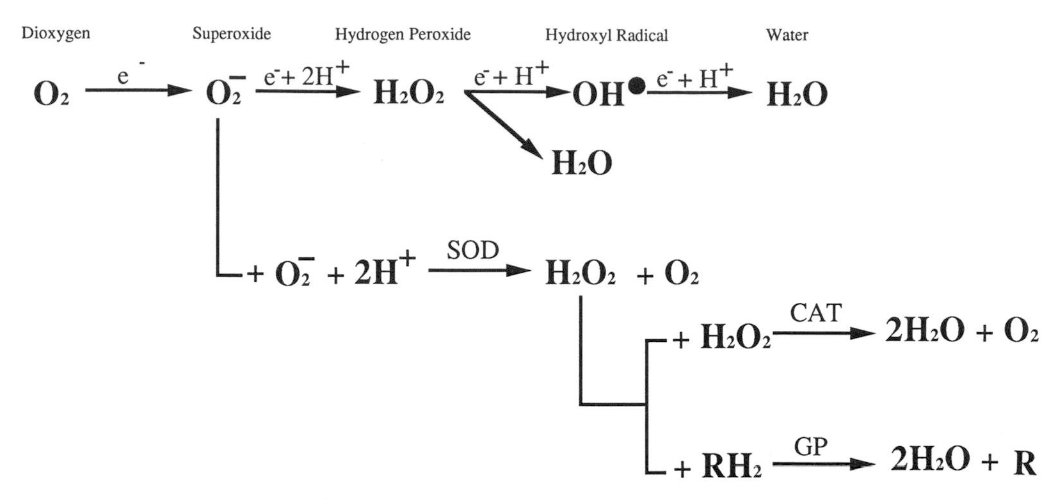

FIGURE 1.—Univalent pathways of oxygen reduction (adapted from Fridovich 1978). SOD = superoxide dismutase, CAT = catalase, GP = glutathione peroxidase, R = reductants, and e = electron.

critical DO for alevins (Rombough 1986) is in the range of 2.5 mg/L (6.0°C) to 4.5 mg/L (15.1°C).

For juvenile fish, the critical dissolved oxygen concentration is based on growth reduction. This critical concentration is influenced by temperature and feeding level and, under laboratory conditions, ranges from 5 to 6 mg/L for salmon and trout and from 3 to 4 mg/L for warmwater fish such as channel catfish *Ictalurus punctatus* (Andrews et al. 1973; Brett 1979).

Improved growth of Atlantic salmon *Salmo salar* has been achieved at DO concentrations in the range of 6.0–7.0 mg/L (Poston and Williams 1988; P. Christensen, A. S. Birger Christensen, personal communication). The use of oxygen supplementation can also increase survival and improve fish health and quality (Severson et al. 1987; Westers et al. 1987). Some of the beneficial effects of oxygen supplementation may be due to the stripping of other gases that may be chronically supersaturated (Severson et al. 1987; Westers et al. 1987).

High DO Concentrations

The maximum allowable DO level depends on several factors, including oxygen toxicity, physiological dysfunctions, and developmental problems. Much of the literature has been generated from studies directed toward basic research in physiology and much applied research remains to be done in this area.

It is important to distinguish between normbaric hyperoxia (elevated oxygen levels at atmospheric pressure) and hyperbaric hyperoxia (elevated oxygen levels at total pressures greater than atmospheric pressure). Because of the high pressures, short exposure times, and problems with gas bubble trauma, the results of most hyperbaric hyperoxia experiments are not relevant to intensive culture and so are not covered in detail in this review.

Oxygen Toxicity

Although oxygen is required for the survival of aerobic organisms such as fish, some of the byproducts of oxygen metabolism are highly toxic. Aerobic organisms have developed defenses against these toxic compounds, but high levels of dissolved oxygen can overwhelm these defenses.

The electronic structure of dioxygen (O_2) in its ground state and principles of quantum mechanics lead to a spin restriction that favors the univalent pathway of oxygen reduction (Fridovich 1981). This involves the reduction of dioxygen (O_2) to the superoxide radical (O_2^-), in turn to hydrogen peroxide (H_2O_2), then to the hydroxyl radical ($OH^\bullet$), and finally to water (Figure 1). Some of these compounds damage cells; they peroxidize unsaturated lipids in biological membranes, depolymerize polysaccharides, cleave DNA, inactivate enzymes, and lyse erythrocytes (Fridovich 1981; Halliwell and Gutteridge 1984). The hydroxyl radical is thought to cause the greatest damage because of its extremely high reactivity with almost every type of biological molecule.

The primary defenses against oxygen toxicity at the molecular level are the enzymes superoxide dismutase (SOD), catalase (CAT), and glutathione peroxidase (GP) (Fridovich 1978). The enzyme SOD eliminates O_2^- and produces H_2O_2, which in turn can be removed by CAT or GP. The production of the hydroxyl radical ($OH^\bullet$) can be controlled by keeping the concentration of superoxide radical and hydrogen peroxide low by the action of SOD, CAT, and GP.

Most of the oxygen consumed by an aerobic organism is used by cytochrome oxidase, which reduces oxygen to water without releasing either the superoxide radical or hydrogen peroxide. Several important univalent reactions are known (Figure 1) and the proportion of oxygen reduced by this pathway in bacteria may be as high as several percent (Fridovich 1981).

The toxicity of oxygen depends on species, life stage, and environmental condition, as well as on physiological and nutritional history, but a hyperoxic toxic threshold exists for all organisms. Common clinical signs of oxygen toxicity in fish include reduced excitability and sudden loss of equilibrium, rigidity of the fins, active swimming in place, spasms, pelting with caudal fins, and flotation problems (D'Aoust 1969; Sebert et al. 1984).

For mammals, it is commonly assumed that the toxicity of oxygen depends on the oxygen consumption rate (Barthelemy et al. 1981). Sebert et al. (1984) found no significant effect of temperature on oxygen toxicity when working with temperature-acclimated fish. In work with amphibian tadpoles, oxygen toxicity did not depend on the oxygen consumption rate, but was inversely correlated with the concentration of SOD and CAT (Barja de Quiroga and Gutierrez 1984; Gil et al. 1987). Acclimation to hyperoxia significantly increased both SOD and CAT activities in amphibian tadpoles. The swim bladder and retina of fish are commonly exposed to very high oxygen partial pressures (under hyperbaric hyperoxia conditions) and are very resistant to oxygen toxicity (D'Aoust 1969; Baeyens et al. 1973).

Several environmental and nutritional factors can modify the effects of oxygen under hyperbaric hyperoxic conditions. The sensitivity of rats to oxygen toxicity is inversely related to ambient light illumination (Bitterman et al. 1986). Fasting increases the tolerance of rats and mice to oxygen toxicity (Haugaard 1968). Other nutritional factors such as vitamins, amino acids, metals, and chelating agents can have a significant effect on oxygen toxicity (Haugaard 1968; Cowey 1986). In goldfish *Carassius auratus* infected with *Vibrio anquillarum*, hyperbaric hyperoxia in combination with sodium sulfisoxazole increased survival (Keck et al. 1980). The significance of these factors to fish under normoxic hyperoxia conditions remains to be demonstrated.

Information on "safe" dissolved oxygen concentrations is presented in Table 2 for several fasted aquatic species. Commercial salmon facilities in the Pacific Northwest routinely use pure oxygen to increase the influent oxygen gas pressures to 230–310 mm Hg (Gowan 1987; Severson et al. 1987). Based on these data, oxygen pressures up to 400 mm Hg appear safe for most salmonid species. The concentrations corresponding to an oxygen partial pressure of 400 mm Hg would range from 32.4 mg/L at 5°C to 19.8 mg/L at 30°C (Table 1).

Highly active and excitable fish such as striped bass *Morone saxatilis* may be more susceptible to oxygen toxicity than less active fish. Striped bass are typically found in warmer water than salmonids. High water temperature results in a higher oxygen partial pressure corresponding to a given DO concentration. Water is better mixed in circular tanks than in raceways, so oxygen toxicity is much less of a problem in the tanks (Colt and Watten 1988).

Physiological Problems

Acclimation to hyperoxia or a return from hyperoxia to normoxia results in significant changes in blood acid–base conditions. These changes have been studied widely in aquatic animals by respiratory physiologists. For example, at an oxygen pressure of 500–600 mm Hg in rainbow trout *Oncorhynchus mykiss*, hyperoxia induced a marked drop in arterial pH (respiratory acidosis) due to a threefold elevation in arterial CO_2 pressure. Over a 72-h period of hyperoxia, the arterial pH drop was compensated for by a rise in plasma bicarbonate (Hobe et al. 1984). Transfer from hyperoxia to normoxia results in rapid transfer of blood CO_2 to the water and an increase in blood pH; the resulting metabolic alkalosis is largely compensated for within 24 h. Crayfish appear unable to adjust blood pH to compensate for increases in arterial CO_2 (Dejours and Beekenkamp 1977).

The ability of fish to tolerate rapid changes in oxygen partial pressure may be an important consideration in designing facilities and in fish-release strategies. Potential mortality could occur after failure of oxygen supplementation systems or upon transfer of fish from a facility using pure oxygen supplementation. It may be prudent to

TABLE 2.—Effects of normbaric hyperoxia conditions (elevated oxygen concentrations at atmospheric pressure) on aquatic animals. Dissolved oxygen levels are expressed both as mg/L (DO) and partial pressures (Po_2, equation 8).

Species	DO (mg/L)	Po_2 (mm Hg)	Exposure (d)	Reference
		No effects		
Anguilla anguilla	27	420	21	Bornancin et al. (1977)
Astacus leptodactylus	47	700	100	Dejours and Beekenkamp (1977)
Discoglossus pictus (tadpoles)	41	710	15	Gil et al. (1987)
Morone saxatilis	18–19	370–390	8	Chiba (1988)
Oncorhynchus mykiss	30–40	500–600	3	Hobe et al. (1984)
Salmo trutta	20	340	35	Swift (1963)
Scyliorhinus stellaris	25	500	6	Heisler et al. (1988)
Various freshwater fish	20–30	300–450	10–50	Wiebe and McGavock (1932)
Various freshwater fish (egg-hatching)	<35	<500	6–9	Gulidov (1969, 1974); Gulidov and Popova (1977)
		Increased mortality		
Oncorhynchus mykiss	>(42–46)	>700	7	Sebert et al. (1984)
Rana ridibunda (tadpoles)	33	570	15	Gil et al. (1987)
Various freshwater fish (egg-hatching)	35–45	500–700	6–9	Gulidov (1969, 1975)
		Reduced growth		
Mercenaria mercenaria	13.7	280	1	Huntington and Miller (1989)

limit oxygen supplementation to normoxic or low-level hyperoxic conditions. Further, it seems prudent to acclimate fish to normoxic conditions well in advance of their release.

Developmental Problems

High dissolved oxygen concentrations can have significant developmental effects on eggs and larvae (Gulidov 1969, 1974, 1975; Gulidov and Popova 1977). Interpretation of experiments with high DO is difficult due to the lack of replicates and high variability between experiments. In general, oxygen levels up to approximately 500 mm Hg had little effect on hatching and larvae development, but there was increased mortality of some species between 500 and 700 mm Hg. More interestingly, at DO concentrations above 550–700 mm Hg, the number of erythrocytes was reduced, and above 725 mm Hg, no erythrocytes were reported. These changes are probably detrimental if the larvae are to be transferred to normoxia conditions. Therefore, the use of extreme hyperoxia during incubation seems imprudent until more data are available.

Biological Effects of Carbon Dioxide

Assuming that the respiratory quotient of fish is equal to 1.0 (Kutty 1968), the mass of dissolved carbon dioxide (DC) added to the water in a rearing unit is

$$\text{DC added (mg/L)} = 1.375\,(DO_{in} - DO_{out}). \quad (9)$$

Because of the high solubility of carbon dioxide, little DC will be removed by reaeration across the water surface or by artificial aeration (Watten et al. 1991, this volume), and the cumulative carbon dioxide added in a system can be computed from equation (9) by using the overall ($DO_{in} - DO_{out}$) or cumulative oxygen consumption. The actual carbon dioxide gas in solution depends on temperature, total carbonate concentration, carbonate alkalinity, and pH (Stumm and Morgan 1981).

High concentrations of carbon dioxide (hypercapnia) may reduce the capacity of the blood to transport oxygen and cause formation of calcareous deposits in the kidneys. Relatively high levels of carbon dioxide (20–40 mg/L) are required to significantly reduce oxygen transport (Basu 1959). The prevalence and severity of nephrocalcinosis depends primarily on the carbon dioxide concentration and, to a lesser degree, the mineral composition of the diet (Smart et al. 1979). The kidney deposits are composed mainly of calcium phosphate, apatite, and brushite.

Rainbow trout exposed to carbon dioxide in the range of 12–60 mg/L for a minimum of 275 d developed nephrocalcinosis, the prevalence and severity of which increased with carbon dioxide concentration (Smart et al. 1979). The growth and feed conversion ratios of rainbow trout reared at 12 and 24 mg/L of carbon dioxide were similar over a 275-d exposure. Fish reared at 55 mg/L of carbon dioxide showed reduced growth and higher feed conversion ratios, but it was only after

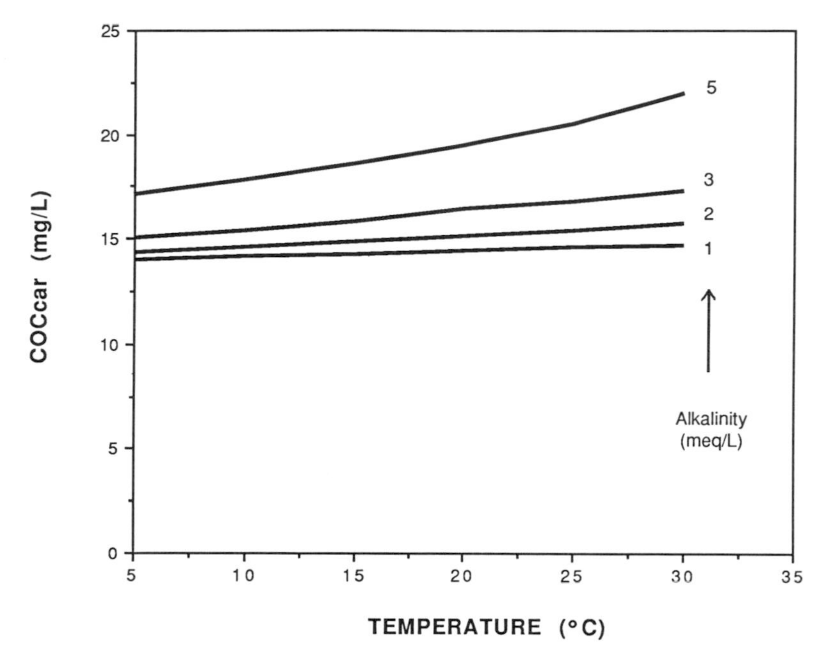

FIGURE 2.—Cumulative oxygen consumption based on carbon dioxide limitations (COC_{car}) as a function of carbonate alkalinity and temperature. Curves are based on an influent carbon dioxide concentration equal to equilibrium, a 20-mg/L concentration of carbon dioxide in solution, no transfer of carbon dioxide from the water, and a respiratory quotient of 1.0.

about 330 d at this exposure that growth was seriously impaired (Smart et al. 1979).

Hypercapnia resulting from hyperoxia produces physiological changes similar to those produced by elevated ambient carbon dioxide concentrations (Thomas et al. 1983; Wheatly et al. 1984). In addition, potential problems with nephrocalcinosis may occur at lower ambient carbon dioxide concentrations under hyperoxic conditions (Schlotfeldt 1980, 1981). This is because the increase in blood carbon dioxide is metabolic in origin, and carbon dioxide levels may be higher inside the fish than in the ambient water and strongly influenced by the level of hyperoxia.

Carbon dioxide concentrations should be maintained below 20 mg/L in culture systems for salmon and trout (Sigma Environmental Consultants 1983), although specific criteria for oxygen supplementation systems are lacking. Allowable carbon dioxide levels in the culture of warmwater species are largely unknown, but carbon dioxide concentrations in intensive channel catfish culture routinely exceed 40–50 mg/L in pure oxygen systems (A. Schuur, Agrifuture Inc., personal communication). Because carbon dioxide affects oxygen transport (Basu 1959), it may be desirable to increase the DO criteria by 3–4 mg/L when carbon dioxide concentrations are high.

The cumulative dissolved oxygen consumption when carbon dioxide becomes limiting is termed the COC_{car}. The COC_{car} only varies from 14 to 22 mg/L as a function of temperature (5–30°C) and alkalinity (1.0–5.0 meq/L; Figure 2). The COC_{car} increases as the alkalinity increases, especially above 3 meq/L. For conditions typical of trout and salmon hatcheries (alkalinity, 1 meq/L), COC_{car} plots corresponding to 20, 40, and 60 mg/L carbon dioxide criteria are presented in Figure 3; at 20°C, the COC_{car} is approximately 72% of the respective carbon dioxide criterion. The accumulation of metabolic carbon dioxide can significantly reduce the ambient pH (Figure 4).

Because of diel fluctuations in the carbon dioxide production rate resulting from feeding, the average value of achievable COC_{car} depends strongly on feeding procedures. For salmonids fed once or twice a day, Westers (1981) suggested a peaking factor (maximum oxygen consumption/average oxygen consumption) of 1.44. Therefore, the average achievable COC_{car} would be reduced to approximately 9–10 mg/L to maintain the car-

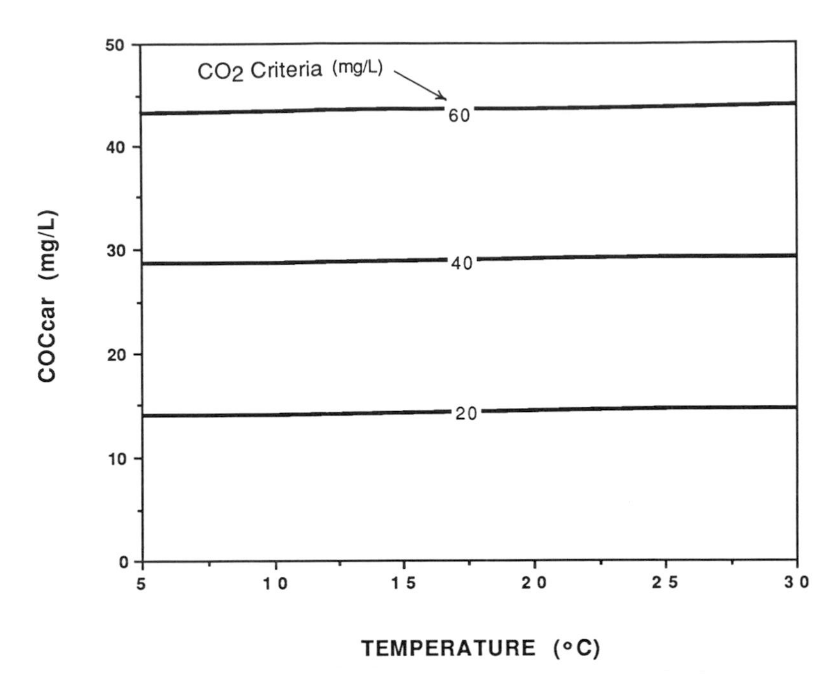

FIGURE 3.—Cumulative oxygen consumption based on carbon dioxide limitations (COC_{car}) as a function of carbon dioxide criterion and temperature. Curves are based on an influent carbon dioxide concentration equal to equilibrium, no transfer of carbon dioxide from the water, a carbonate alkalinity of 1 meq/L, and a respiratory quotient of 1.0.

bon dioxide criterion during the period of maximum metabolic activity.

Biological Effects of Gas Supersaturation

The biological response to gas supersaturation at a given ΔP depends on species, life stage, water quality, and the animal's depth in the water column (pressure). The following discussion is based primarily on chronic effects (>30 d); recent information on acute effects has been presented by Jensen et al. (1986), and Weitkamp and Katz (1980) summarized the previous literature.

Species

Little comparative information is available for chronic exposure of species to gas supersaturation. Extrapolation from short-term lethal experiments is difficult, if not invalid. Based on general hatchery experience, it appears that lake trout *Salvelinus namaycush* and Atlantic salmon are typically more sensitive than other salmonids (J. Meade, U.S. Fish and Wildlife Service, personal communication). Marine or estuarine species such as striped bass (Cornacchia and Colt 1984) and striped mullet *Mugil cephalus* (Kraul 1983) appear to be very sensitive during their larval stages. Fish with smaller larval forms are commonly less tolerant of overinflation of the swim bladder.

Age or Life Stage

Based on work with lake trout, eggs and sac fry are less sensitive to gas supersaturation than feeding fry (Krise 1991, this volume). Small juveniles are more sensitive than larger juveniles. Krise (1991) suggests the following criteria for lake trout:

Life stage	ΔP
Eggs	45
Sac fry	35
Juveniles 9.6 cm	10
Juveniles 15.3 cm	<30

Therefore, a ΔP less than 10 mm Hg may be generally safe for most fish during their first 2 years of life.

Other species have shown increased mortality of fry and fingerlings under hatchery conditions due to gas bubble trauma at ΔPs in the range of 20–40 mm Hg (Cornacchia and Colt 1984; Wright

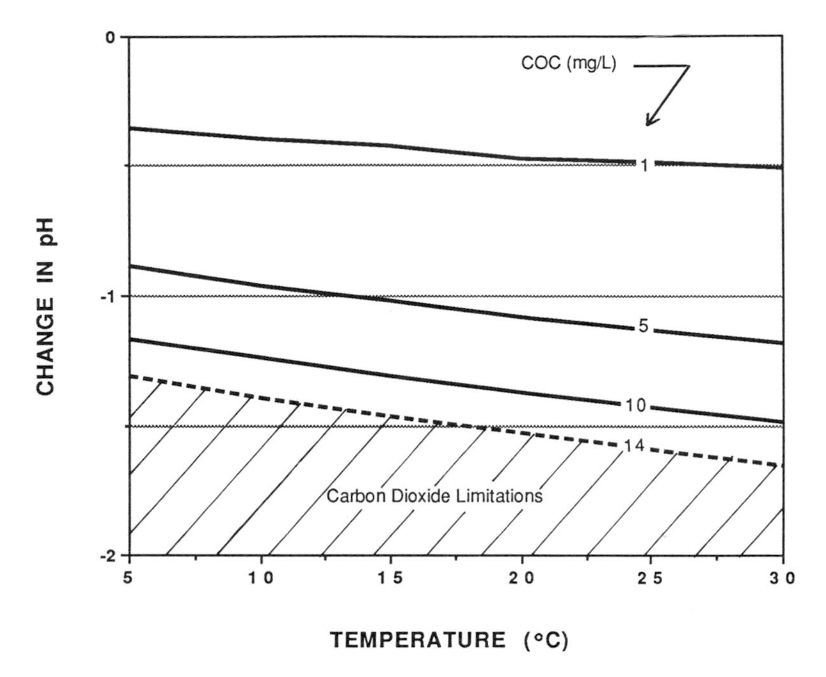

FIGURE 4.—Reduction in pH due to metabolic carbon dioxide production as a function of cumulative oxygen consumption (COC) and temperature. Curves are based on an influent carbon dioxide concentration equal to equilibrium, no transfer of carbon dioxide from the water, a 20-mg/L concentrate of carbon dioxide in solution, a carbonate alkalinity of 1 meq/L, and a respiratory quotient of 1.0.

and McLean 1985), although there were no significant effects on growth. Krise and Meade (1988) thought that this could be the result of deviations from a normal size distribution due to the mortality of smaller fish. Juvenile channel catfish are less sensitive to gas supersaturation, only showing an increased incidence of gas bubble trauma and mortality at a ΔP = of 76 mm Hg (Colt et al. 1985).

Water Quality

At a given ΔP, the biological effects of gas supersaturation are reduced as the nitrogen:oxygen ratio is decreased (Nebeker et al. 1976). Most of the research on chronic effects of gas supersaturation has been conducted with nitrogen:oxygen ratios (based on partial pressures) near the atmospheric value (3.77). Although increasing the oxygen level at a constant ΔP increases the tolerance of acute gas bubble trauma (Nebeker et al. 1976), little is known about its effect on chronic gas bubble trauma.

Tolerance of gas supersaturation may be influenced by other water quality variables, but as yet the issue remains largely unresolved. Average time to 50% mortality at a ΔP of 190 mm Hg was not significantly different between seawater and fresh water for steelhead, but average survival time was lower in seawater (Bouck and King 1983). Similarly, tolerance of supersaturation was not altered by the addition of low levels of carbon dioxide (Nebeker et al. 1976). The effects of un-ionized ammonia and temperature on tolerance of gas supersaturation have not been determined but require consideration.

Depth

The animal's position in the water column has a significant effect on its tolerance of gas supersaturation. The actual ΔP that the animal experiences is reduced by 73.42 mm Hg/m at 20°C (Colt 1984). Fish do not appear to detect and avoid acutely lethal gas levels (Weitkamp and Katz, 1980; Lund and Heggberget 1985), but they do sound to restore buoyancy when exposed to a ΔP of less than approximately 84 mm Hg (Chamberlain et al. 1980; Alderdice and Jensen 1985). Regardless of whether or not fish can detect and avoid gas supersaturation, deeper rearing units offer passive protection due to the greater average depth of the fish in the water column (Lund and

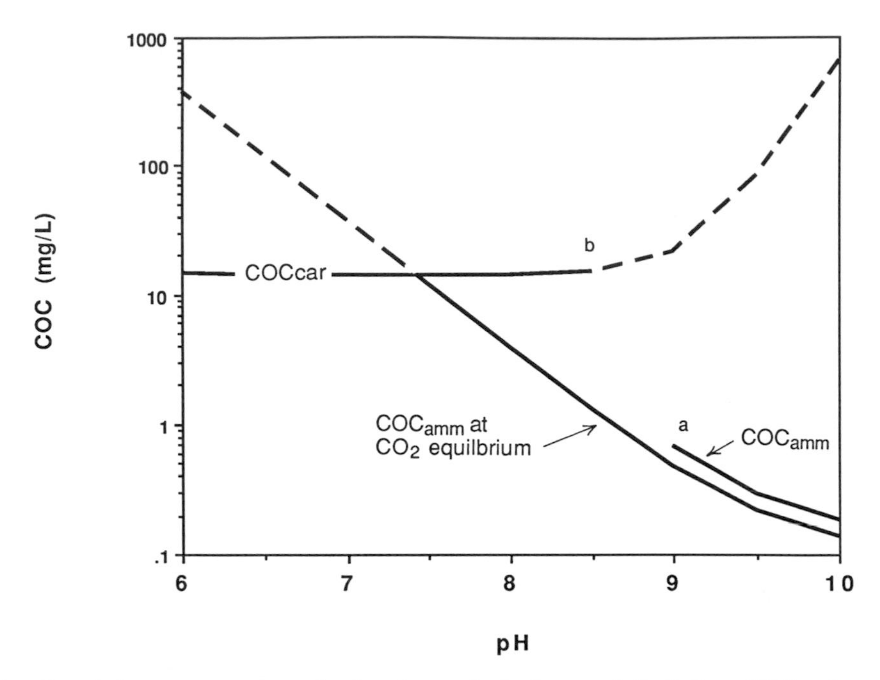

FIGURE 5.—Cumulative oxygen consumption based on un-ionized ammonia limitations (COC_{amm}) as a function of pH. Separate curves are presented for COC_{amm} for equilibrium carbon dioxide concentrations and maximum COC_{amm}. Cumulative oxygen consumption based on carbon dioxide (COC_{car} = 20 mg/L) is also included for comparison. Below a pH of 9.00–8.50, the maximum COC jumps from the COC_{amm} curve at point (a) to the COC_{car} curve at point (b). Curves are based on an influent ammonia concentration equal to zero, an un-ionized ammonia criterion of 16 μg/L, an ammonia production of 30 g/kg feed, an oxygen consumption of 250 g/kg feed, a respiratory quotient of 1.0, and a temperature of 15°C. The contribution of excreted ammonia to alkalinity and the effects of ionic strength on activity were not considered.

Heggberget 1985). The estimation of potential protection by hydrostatic pressure will depend on species behavior and available water depth.

Biological Effects of Increased Rearing Intensity

Metabolic Wastes

The use of pure oxygen to achieve increased rearing density or in the reuse of water will result in an increase in the concentration of metabolic wastes such as ammonia, carbon dioxide, and suspended solids (fecal solids and uneaten feed). Therefore, dissolved gas criteria for high-intensity culture systems that incorporate oxygen supplementation may need to allow for the joint effect of all these water quality variables. Because high levels of one variable may reduce or increase the toxicity of the others, water quality criteria for systems using oxygen supplementation may be quite different from those for typical rearing systems.

Ammonia

Under some conditions, the reuse of water will be limited by un-ionized ammonia. Cumulative oxygen consumption based on un-ionized ammonia limitations (COC_{amm}) is presented in Figure 5 as a function of pH. Separate COC_{amm} curves are presented for zero cumulative oxygen consumption–equilibrium carbon dioxide conditions and cumulative oxygen consumption corresponding to 20 mg/L of carbon dioxide in solution.

At pH values typical for trout and salmon, $COC_{amm} > (C^* - DO_{out})$ and supplemental oxygen can be used to increase the carrying capacity. Under warmwater conditions (especially with geothermal waters) or marine conditions, ammonia may be the most limiting variable and supplemental oxygen may be of little value.

The computation of COC_{amm} in a pure oxygen system must account for the effects of carbon dioxide on pH (Figure 5). Two separate COC_{amm} curves are presented. The curve for COC_{amm} at

CO_2 equilibrium represents the maximum oxygen consumption if the carbon dioxide is maintained at equilibrium with the atmosphere. At high pH, the accumulation of metabolic carbon dioxide reduces the pH and increases the COC_{amm} (the COC_{amm} curve ends at point a). Between pH 9.00 and 8.50, the buffer intensity (Stumm and Morgan 1981) sharply decreases and metabolic carbon dioxide drops the pH enough that the ambient un-ionized ammonia concentration does not significantly increase. Therefore, below a pH of 9.00–8.50, the maximum COC jumps from the COC_{amm} curve at point a (ammonia-limited) to the COC_{car} curve at point b (carbon-dioxide-limited). Un-ionized ammonia limits oxygen consumption at high pH, whereas carbon dioxide limits oxygen consumption at low pH.

The COC_{amm} value depends strongly on the un-ionized ammonia criterion. Based on recent laboratory research, the commonly used un-ionized ammonia criterion in the range of 10–15 μg NH_3-N/L is excessively conservative (Meade 1985a). Hypercapnia and hyperoxia have a significant effect on nitrogen excretion in some species and may depend strongly on pH or ionic composition of the water (Heisler 1984; Randall and Wright 1987). Some of the observed differences in ammonia toxicity between laboratory dosing studies and production conditions may be due to differences in carbon dioxide levels.

Low DO concentrations increase the toxicity of ammonia (Thurston et al. 1981), but it is not known if DO levels above saturation reduce ammonia toxicity. Reassessment of un-ionized ammonia criteria for DOs and COCs in the range of 10–30 mg/L is needed.

Suspended Solids

Little information is available on the effects of uneaten feed and fecal matter on the growth and mortality of fish. The bacterial count of commercial fish feeds ranges from 10^3 to 10^7 bacteria per gram (Trust 1971) and includes some potentially pathogenic species. The use of oxygen supplementation systems and water reuse may require the use of more efficient control methods for solids such as described by Boersen and Westers (1986).

Density

In raceways, the density is commonly computed (Piper 1975) from a density index of 3.2 $kg/m^3 \cdot cm$ (0.5 $lb/ft^3 \cdot in$). Therefore, the allowable density will vary from 24 kg/m^3 with fish 7.6 cm long (3 in) to 80 kg/m^3 with fish 25.4 cm long (10 in). Interpretation of much of the data on effects of density is difficult because of the interrelationship between density and loading (equation 6). If loading rates are maintained low (high exchange rate), high densities can be achieved, but the allowable density depends on the ability of the particular species to tolerate crowding and on other factors such as diseases. Densities as high as 540 kg/m^3 have been achieved in experimental salmonid culture (Buss et al. 1970).

Recent work has shown that densities up to 245 kg/m^3 (density index = 12.6 $kg/m^3 \cdot cm$) has no effect on the fin condition of lake trout in cages (Soderberg and Krise 1987). Densities up to 234–253 kg/m^3 had no effect on growth and mortality (Soderberg and Krise 1986; Soderberg et al. 1987). The allowable fish densities under low loading conditions may not be related to fish length as predicted by equation (2), and small fish may be able to do as well at higher density indexes as larger fish (Soderberg et al. 1987). Densities of 24 kg/m^3 and 48 kg/m^3 are routinely attained in oxygen-supplemented systems with smolts of coho salmon *Oncorhynchus kisutch* and juvenile channel catfish (Collins et al. 1984; Severson et al. 1987). The evaluation of densities for stockable or saleable fish is straightforward, but the subsequent return of adults is the only valid measure for anadromous fish and evaluation may require at least 7–10 years. Other concerns with the use of pure oxygen systems are that supplementation might facilitate both infectious and noninfectious diseases, and that high oxygen levels might impair modification, postrelease survival, and migrational behavior of salmon and steelhead smolts. Thus, the optimum density for a hatchery is more an operational research problem than a pure biological problem. An increased total number of released smolts may be able to compensate for a reduced percent return of the adults (Banks 1987).

Specific rearing-density criteria for oxygen-supplemented systems are unavailable for most species and much more research is needed. This is particularly true for anadromous salmonid smolts. The as yet undetermined potential influence of social stress, altered physiological functions, or subclinical diseases may temporarily debilitate the fish at liberation time. Given the opportunity to reacclimate to more typical conditions, fish reared at high density may become indistinguishable from their counterparts in traditional hatcheries.

TABLE 3.—Preliminary water quality criteria for intensive culture systems. Dissolved oxygen and carbon dioxide are expressed as concentrations (mg/L) or partial pressures (equation 8). Gas supersaturation levels are expressed in terms of ΔP, the difference between the total dissolved gas pressure and local barometric pressure.

Parameter	Criteria for chronic exposure: Cold water (12°C)	Warm water (25°C)
Dissolved oxygen (low)	6–7 mg/L	4–5 mg/L
Dissolved oxygen (high)	300 mm Hg 21 mg/L	300 mm Hg 16 mg/L
Dissolved carbon dioxide	20 mg/L	20 mg/L
ΔP (all life stages)	10 mm Hg	20 mm Hg
ΔP (specific life stages)		
Eggs	45 mm Hg	
Sac fry	35 mm Hg	20 mm Hg
Early juveniles	10 mm Hg	50 mm Hg
Advanced juveniles	<30 mm Hg	

Loading

Given water quality criteria for ammonia, dissolved oxygen, and carbon dioxide, the loading rate in kilograms of animal per liter per minute (Lpm) (or effective cumulative oxygen consumption) can be directly computed. The loading parameter is then considered a measure of performance rather than a design criterion.

Repeated reaeration and reuse of water can result in adverse physiological changes such as reduced growth or tissue damage even though both dissolved oxygen and un-ionized ammonia criteria have not been exceeded (Meade 1985b). Loading in excess of 6 kg/Lpm reduces growth and causes tissue damage. This may be due to the synergistic effects of un-ionized ammonia, fecal solids, and other metabolites. Meade (1988) suggested that production capacity should be directly determined at a given site by a 1–2-month chronic bioassay. More information is needed on loading criteria, but several production hatcheries are being operated with COCs in the range of 5–15 mg/L (Gowan 1987; Severson et al. 1987).

Criteria

Preliminary water quality criteria for intensive culture conditions are presented in Table 3. When un-ionized ammonia and carbon dioxide concentrations approach their recommended maxima, considerations should be given to increasing the minimum dissolved oxygen criteria by 3–4 mg/L. The criterion for high dissolved oxygen levels are conservative and most species should be able to tolerate higher oxygen pressures. Rapid changes in oxygen pressures should be avoided, especially from hyperoxic to normoxic conditions. The criteria for gas supersaturation are based on lake trout for the coldwater conditions and on striped bass and channel catfish for warmwater conditions. When information about particular species and life stages is not available, criteria of 10 and 20 mm Hg should be used for coldwater and warmwater conditions, respectively.

Acknowledgments

We thank Larry Fidler, Joseph Cech, and William Krise for critical reviews of earlier versions of this paper, and Dan Chang for assistance with the sections on carbonate chemistry.

References

Alderdice, D. F., and J. O. T. Jensen. 1985. Assessment of the influence of gas supersaturation on salmonids in the Nechako River in relation to Kemano completion. Canadian Technical Report of Fisheries and Aquatic Sciences 1386.

Andrews, J. W., T. Murai, and G. Gibbons. 1973. The influence of dissolved oxygen on the growth of channel catfish. Transactions of the American Fisheries Society 102:835–838.

Baeyens, D. A., J. R. Hoffert, and P. O. Fromm. 1973. A comparative study of oxygen toxicity in the retina, brain and liver of the teleost, amphibian and mammal. Comparative Biochemistry and Physiology 45A:925–932.

Banks, J. 1987. Effects of raceway inflow and rearing density interactions on adult returns of coho and spring chinook salmon. Pages 40–49 *in* Papers on the use of supplemental oxygen to increase hatchery rearing capacity in the Pacific Northwest. Bonneville Power Administration, Portland, Oregon.

Barja de Quiroga, G., and P. Gutierrez. 1984. Superoxide dismutase during the development of two amphibian species and its role in hyperoxia tolerance. Molecular Physiology 6:221–232.

Barthelemy, L., A. Belaud, and C. Chastel. 1981. A

comparative study of oxygen toxicity in vertebrates. Respiration Physiology 44:261–268.

Basu, S. P. 1959. Active respiration of fish in relation to ambient concentrations of oxygen and carbon dioxide. Journal of the Fisheries Research Board of Canada 16:175–212.

Bitterman, N., Y. Melamed, and I. Perlman. 1986. CNS oxygen toxicity in the rat: role of ambient illumination. Undersea Biomedical Research 13:19–25.

Boersen, G., and H. Westers. 1986. Waste solids control in hatchery raceways. Progressive Fish-Culturist 48:151–154.

Bornancin, M., G. De Renzis, and J. Maetz. 1977. Branchial Cl transport, anion-stimulated ATPase and acid-base balance in *Anguilla anguilla* adapted to freshwater: effects of hyperoxia. Journal of Comparative Physiology 117:313–322.

Bouck, G. R., and R. E. King. 1983. Tolerance to gas supersaturation in fresh water and sea water by steelhead trout, *Salmo gairdneri* Richardson. Journal of Fish Biology 23:293–300.

Brett, J. R. 1979. Environmental factors and growth. Pages 599–675 *in* W. S. Hoar, D. J. Randall, and J. R. Brett, editors. Fish physiology, volume 8. Academic Press, New York.

Buss, K., D. R. Graff, and E. R. Miller. 1970. Trout culture in vertical units. Progressive Fish-Culturist 32:187–191.

Chamberlain, G. W., W. H. Neill, P. A. Romanowsky, and K. Strawn. 1980. Vertical responses of Atlantic croaker to gas supersaturation and temperature change. Transactions of the American Fisheries Society 109:737–750.

Chiba, K. 1988. The effects of dissolved oxygen on the growth of young striped bass. Bulletin of the Japanese Society of Scientific Fisheries 54:599–606.

APHA (American Public Health Association), American Water Works Association, and Water Pollution Control Federation. 1989. Standard methods for the examination of water and wastewater, 17th edition. APHA, Washington, D.C.

Collins, C. M., G. L. Burton, and R. L. Schweinforth. 1984. Evaluation of liquid oxygen to increase channel catfish production in heated water raceways. Tennessee Valley Authority, TVA/ONRED/AWR-84/8, Gallatin, Tennessee.

Colt, J. 1984. Computation of dissolved gas concentrations in water as functions of temperature, salinity, and pressure. American Fisheries Society Special Publication 14.

Colt, J., K. Orwicz, and D. Brooks. 1985. The effect of gas supersaturation on the growth of juvenile channel catfish, *Ictalurus punctatus*. Aquaculture 50: 153–160.

Colt, J., and B. Watten. 1988. Application of pure oxygen in fish culture. Aquacultural Engineering 7:397–441.

Cornacchia, J. W., and J. E. Colt. 1984. The effects of dissolved gas supersaturation on larval striped bass, *Morone saxatilis* (Walbaum). Journal of Fish Diseases 7:15–27.

Cowey, C. B. 1986. The role of nutritional factors in the prevention of peroxidative damage to tissues. Fish Physiology and Biochemistry 2:171–178.

D'Aoust, G. G. 1969. Hyperbaric oxygen: toxicity to fish at pressures present in their swim bladders. Science (Washington, D.C.) 163:576–578.

Dejours, P., and H. Beekenkamp. 1977. Crayfish respiration as a function of water oxygenation. Respiration Physiology 30:241–251.

Fridovich, I. 1978. The biology of oxygen radical. Science (Washington, D.C.) 201:875–880.

Fridovich, I. 1981. Superoxide radical and superoxide dismutases. Pages 250–272 *in* D. L. Gilbert, editor. Oxygen and living processes—an interdisciplinary approach. Springer-Verlag, New York.

Gil, P., M. Alonso-Bedate, and G. Barja de Quiroga. 1987. Different levels of hyperoxia reversibly induce catalase activity in amphibian tadpoles. Free Radical Biology and Medicine 3:137–146.

Gowan, R. 1987. Use of supplemental oxygen to rear chinook in seawater. Pages 35–39 *in* Papers on the use of supplemental oxygen to increase hatchery rearing capacity in the Pacific Northwest. Bonneville Power Administration, Portland, Oregon.

Gulidov, M. V. 1969. Embryonic development of the pike [*Esox lucius* L.] when incubated under different oxygen conditions. Problems of Ichthyology 9:841–851.

Gulidov, M. V. 1974. The effects of different oxygen conditions during incubation on the survival and some of the development characteristics of the "Verkhovka" [*Leucaspius delineatus*] in the embryonic period. Journal of Ichthyology 14:393–397.

Gulidov, M. V. 1975. Responses of carp (*Cyprinus carpio* L.) eggs of different viability to the action of chronic hyperoxia. Doklady Biological Sciences 219:496–498.

Gulidov, M. V., and K. S. Popova. 1977. The influence of increased O_2 on the survival and hatching of the embryos of the bream, *Abramis brama*. Journal of Ichthyology 17:174–177.

Halliwell, B., and J. M. C. Gutteridge. 1984. Oxygen toxicity, oxygen radicals, transition metals and disease. Biochemical Journal 219:1–14.

Haugaard, N. 1968. Cellular mechanisms of oxygen toxicity. Physiological Reviews 48:311–373.

Heisler, N. 1984. Acid–base regulation in fishes. Pages 315–401 *in* W. S. Hoar and D. J. Randall, editors. Fish physiology, volume 10, part A. Academic Press, New York.

Heisler, N., D. P. Toews, and G. F. Holeton. 1988. Regulation of ventilation and acid–base status in the elasmobranch *Scyliorhinus stellaris* during hyperoxia-induced hypercapnia. Respiration Physiology 71:227–246.

Hobe, H., C. M. Wood, and M. G. Wheatly. 1984. The mechanisms of acid–base and ion regulation in the freshwater rainbow trout during environmental hyperoxia and subsequent normoxia. I. Extra- and intracellular acid–base status. Respiration Physiology 55:139–154.

Huntington, K. M., and D. C. Miller. 1989. Effects of suspended sediment, hypoxia, and hyperoxia on larval *Mercenaria mercenaria* (Linnaeus, 1758). Journal of Shellfish Research 8:37–42.

Jensen, J. O. T., J. Schnute, and D. F. Alderdice. 1986. Assessing juvenile salmonid response to gas supersaturation using a general multivariate dose–response model. Canadian Journal of Fisheries and Aquatic Sciences 43:1694–1709.

Keck, P. E., S. F. Gottlief, and J. Conely. 1980. Interaction of increased pressures of oxygen and sulfonamides on the *in vitro* and *in vivo* growth of pathogen bacteria. Undersea Biomedical Research 7:95–106.

Kraul, S. 1983. Results and hypotheses for the propagation of the grey mullet *Mugil cephalus* L. Aquaculture 30:273–284.

Krise, W. F. 1991. Hatchery management of lake trout exposed to chronic dissolved gas supersaturation. American Fisheries Society Symposium 10:368–371.

Krise, W. F., and J. W. Meade. 1988. Effects of low-level gas supersaturation on lake trout (*Salvelinus namaycush*). Canadian Journal of Fisheries and Aquatic Sciences 45:666–674.

Kutty, M. N. 1968. Respiratory quotients in goldfish and rainbow trout. Journal of the Fisheries Research Board of Canada 25:1689–1728.

Lund, M., and T. G. Heggberget. 1985. Avoidance response of two-year-old rainbow trout, *Salmo gairdneri* Richardson, to air-supersaturated water: hydrostatic compensation. Journal of Fish Biology 26:193–200.

Meade, J. W. 1985a. Allowable ammonia for fish culture. Progressive Fish-Culturist 47:135–145.

Meade, J. W. 1985b. Effects of loading levels on lake trout (*Salvelinus namaycush*) in intensive culture. Doctoral dissertation. Auburn University, Auburn, Alabama.

Meade, J. W. 1988. A bioassay for production assessment. Aquacultural Engineering 7:139–146.

Nebeker, A. V., G. R. Bouck, and D. G. Stevens. 1976. Oxygen and carbon dioxide and oxygen–nitrogen ratios as factors affecting fish in air-supersaturated water. Transactions of the American Fisheries Society 105:425–429.

Piper, R. G. 1975. A review of carrying capacity calculations for fish hatchery rearing units. U.S. Fish and Wildlife Service, Bozeman Informational Leaflet 1, Bozeman, Montana.

Poston, H. A., and R. C. Williams. 1988. Interrelations of oxygen concentration, fish density, and performance of Atlantic salmon in an ozonated water reuse system. Progressive Fish-Culturist 50:69–76.

Randall, D. J., and P. A. Wright. 1987. Ammonia distribution and excretion in fish. Fish Physiology and Biochemistry 3:107–120.

Rombough, P. J. 1986. Mathematical models for predicting the dissolved oxygen requirements of steelhead (*Salmo gairdneri*) embryos and alevins in hatchery incubators. Aquaculture 59:119–137.

Schlotfeldt, H. J. 1980. The increase of nephrocalcinosis (NC) in rainbow trout in intensive aquaculture. Pages 198–205 *in* W. Ahne, editor. Fish diseases—third COPRAQ session. Springer-Verlag, Berlin.

Schlotfeldt, H. J. 1981. Some clinical findings of a several years survey of intensive aquaculture systems in Northern Germany, with special emphasis on gill pathology and nephrocalcinosis. Pages 109–119 *in* K. Tiews, editor. Aquaculture in heated effluents and recirculation systems, volume II. Heenemann Verlagsgesellschaft, Berlin.

Sebert, P., L. Barthelemy, and C. Peyraud. 1984. Oxygen toxicity in trout at two seasons. Comparative Biochemistry and Physiology 78A:719–722.

Severson, R. F., J. L. Stark, and L. M. Poole. 1987. Use of oxygen to commercially rear coho salmon. Pages 25–34 *in* Papers on the use of supplemental oxygen to increase hatchery rearing capacity in the Pacific Northwest. Bonneville Power Administration, Portland, Oregon.

Sigma Environmental Consultants. 1983. Summary of water quality for salmonid hatcheries. Report to Canada Department of Fisheries and Oceans, Vancouver.

Smart, G. R., D. Knox, J. G. Harrison, J. A. Ralph, R. H. Richards, and C. B. Cowey. 1979. Nephrocalcinosis in rainbow trout *Salmo gairdneri* Richardson; the effect of exposure to elevated CO_2 concentrations. Journal of Fish Diseases 2:279–289.

Soderberg, R. W., D. S. Baxter, and W. F. Krise. 1987. Growth and survival of fingerling lake trout reared at four densities. Progressive Fish-Culturist 49:284–285.

Soderberg, R. W., and W. F. Krise. 1986. Effects of rearing density on growth and survival of lake trout. Progressive Fish-Culturist 48:30–32.

Soderberg, R. W., and W. F. Krise. 1987. Fin condition of lake trout, *Salvelinus namaycush* Walbaum, reared at different densities. Journal of Fish Diseases 10:233–235.

Stumm, W., and J. J. Morgan. 1981. Aquatic chemistry, 2nd edition. Wiley, New York.

Swift, D. R. 1963. Influence of oxygen concentration on the growth of brown trout, *Salmo trutta* L. Transactions of the American Fisheries Society 92:300–301.

Thomas, S., B. Fievet, L. Barthelemy, and C. Peyraud. 1983. Comparison of the effects of exogenous and endogenous hypercapnia on ventilation and oxygen uptake in the rainbow trout (*Salmo gairdneri* R.). Journal of Comparative Physiology 151:185–190.

Thurston, R. V., G. R. Phillips, R. C. Russo, and S. M. Hinkins. 1981. Increased toxicity of ammonia to rainbow trout (*Salmo gairdneri*) resulting from reduced concentrations of dissolved oxygen. Canadian Journal of Fisheries and Aquatic Sciences 38:983–988.

Trust, T. J. 1971. Bacterial counts of commercial fish diets. Journal of the Fisheries Research Board of Canada 28:1185–1189.

Watten, B., J. Colt, and C. E. Boyd. 1991. Modeling the effect of dissolved nitrogen and carbon dioxide on the performance of pure oxygen absorption systems. American Fisheries Society Symposium 10: 474–481.

Westers, H. 1981. Fish culture manual for the state of Michigan. Michigan Department of Natural Resources, Lansing.

Westers, H., V. Bennett, and J. Copeland. 1987. Michigan's experience with supplemental oxygen in salmonid rearing. Pages 12–16 *in* Papers on the use of supplemental oxygen to increase hatchery rearing capacity in the Pacific Northwest. Bonneville Power Administration, Portland, Oregon.

Weitkamp, D. E., and M. Katz. 1980. A review of dissolved gas supersaturation literature. Transactions of the American Fisheries Society 109:659–702.

Wheatly, M. G., H. Hobe, and C. M. Wood. 1984. The mechanisms of acid–base and ion regulation in the freshwater rainbow trout during environmental hyperoxia and subsequent normoxia. II. The role of the kidney. Respiration Physiology 55:155–173.

Wiebe, A. H., and A. M. McGavock. 1932. The ability of several species of fish to survive on prolonged exposure to abnormally high concentrations of dissolved oxygen. Transactions of the American Fisheries Society 62:267–274.

Wright, P. B., and W. E. McLean. 1985. The effects of aeration on the rearing of summer chinook fry (*Oncorhynchus tshawytscha*) at the Puntledge Hatchery. Canadian Technical Report of Fisheries and Aquatic Sciences 1390.

American Fisheries Society Symposium 10:386–392, 1991

Biological Rock Filter Design for Closed Aquaculture Systems

DON P. MANTHE
Manthe and Associates
7725 East Luke Lane, Scottsdale, Arizona 85253, USA

Abstract.—The efficiency of a biological filter can be measured by determining the amount of oxygen consumed by bacteria biomass in the filter. The supply of dissolved oxygen is a principal factor limiting filter performance in high-density aquaculture systems. A mass balance technique is used in this paper to quantify relationships defining the filter volume and flow-through rates required for biological filter design. Filter designs employing flow recirculation with supplemental aeration are the most successful. Filter acclimation by chemical addition and filter performance techniques are also discussed.

Closed systems with little or no discharge are gaining wide acceptance for the intensive culture of various aquatic species. Recirculation systems are characterized by the reuse of water and typically contain filtration components that process animal wastes to a relatively inert and harmless state. Although the filtration components vary, most systems employ biological filtration for the critical reduction of biological oxygen demand (BOD) and nitrification.

Biological filter design for aquaculture can be considered an empirical science. The difficulties associated with biological filtration can invariably be explained by examining design criteria and maintenance routines. Thus, the need for design methodologies and performance-monitoring techniques is widely recognized (Wheaton 1977). This paper presents a methodology for biological filter design based on oxygen consumption in the filter. In addition, various management and performance techniques are examined.

Filter Function

Biological filters are designed to optimize natural waste degradation processes whether aerobic or anaerobic. Although anaerobic filters have been used, most biological filter schemes are based upon enhancing aerobic processes. This is not surprising when one recognizes that the most critical degradation step, nitrification, depends upon bacteria species that are strictly aerobic. These bacteria are responsible for transforming the toxic nitrogen forms of ammonia and nitrite to the relatively harmless form of nitrate. Furthermore, the many types of bacteria usually present in biological filters can break down carbonaceous organic compounds through aerobic pathways (Metcalf and Eddy, Inc. 1979).

The ability of water to hold dissolved oxygen is limited, so oxygen can quickly limit filter performance even when flow rates are moderate. This is particularly true for the submerged rock filters that have been widely applied in aquaculture. Figure 1 shows oxygen traces through experimental closed systems in states of equilibrium and system failure (undesirable water quality). All traces reveal that the highest oxygen demand is incurred by the biological filter. In the state of system failure, dissolved oxygen concentrations decrease to as low as 1 mg/L after the biological filter and ammonia and nitrite concentrations increase to toxic levels.

If oxygen is not present in the required stoichiometric amount, the nitrification rate decreases. This can create a kinetic bottleneck and result in elevated concentrations of toxic nitrogen forms. The evolution of more sophisticated filtration forms, including pure oxygen injection, trickling filters, and the rotating biological disk, can be explained largely by the need to increase oxygen availability for the degradation processes. However, this increase in oxygen availability is usually achieved at a subsequent energy cost.

Filter Design

Oxygen Demand Relationships

The amount of oxygen consumed by a biological filter is measured by the biological activity, which in turn is related to the waste loading imparted to a recirculating system. The relationship of oxygen consumption to waste loading, coupled with the recognition that oxygen frequently limits filter performance, strongly indicates that oxygen can be used as a design parameter. The analysis of filter oxygen consumption is critical, because an evaluation by nitrogen utilization alone might not take into consideration oxygen limitations unknowingly placed on a biologi-

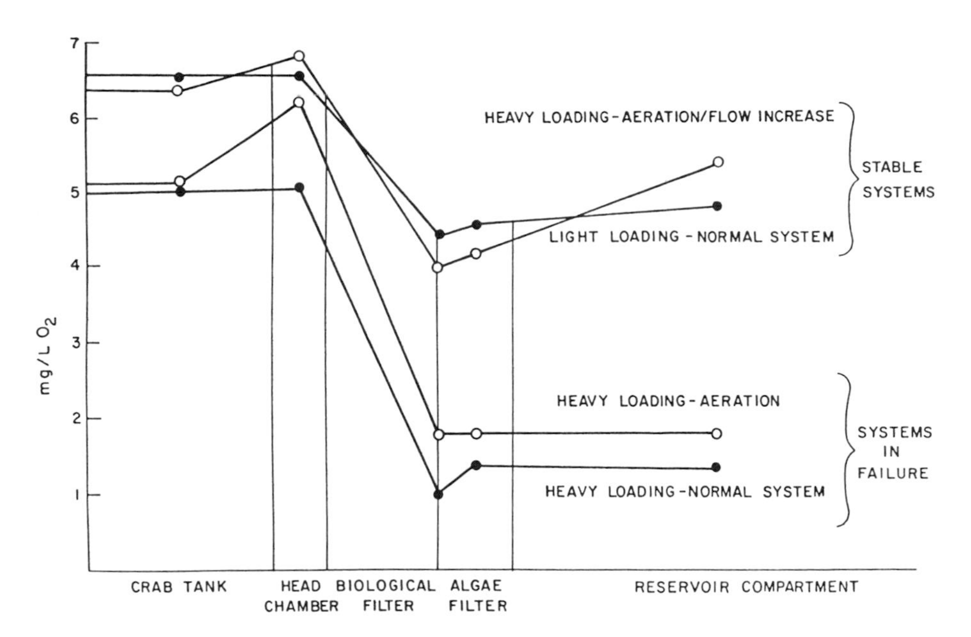

FIGURE 1.—Oxygen traces showing closed systems in states of equilibrium and failure with differing system modifications.

cal filter. Temperature and reaeration are among the factors that affect dissolved oxygen concentration and thus affect the filter's ability for nitrogen utilization. Because nonnitrogenous oxygen demand can be more than half the filter oxygen demand, oxygen seems to be a more reasonable evaluation parameter than nitrogen. In my experience, nonnitrogeneous demands can account for up to 80% of the filter oxygen consumption.

Using oxygen as a design parameter was first proposed by Hirayama (1965, 1974). Recognition of oxygen as the factor controlling the capacity of closed recirculation systems (Manthe et al. 1984) has led to further development of this concept, and the reader is directed to Manthe et al. (1988) for a more detailed delineation of the following discussion of design equations.

The oxygen consumed during filtration for a biological filter can be defined as

$$\text{OCF} = Q(C_i - C_o); \quad (1)$$

OCF = oxygen consumed in filtration (mg O_2/d); Q = flow through the filter (L/d); C_i = filter influent oxygen concentration (mg O_2/L); C_o = filter effluent oxygen concentration (mg O_2/L); and OCF reflects the amount of oxygen consumed in the filter as a result of instantaneous exertion of biochemical oxygen demand and respiration of the bacteria. The OCF of a filter can be expected to vary hourly because of fluctuations in animal loading, feeding, and changes in environmental factors. Furthermore, the bacterial population must be in balance with the population of animals for OCF determinations to be accurate. An accurate determination of mean OCF values for design purposes requires many empirical measurements from acclimated filtration systems.

If a minimum dissolved oxygen concentration of 2 mg/L is required to prevent inhibition of nitrifying bacteria (Gaudy and Gaudy 1978), the oxygen carrying capacity of a filter (OCC: mg O_2/d) is

$$\text{OCC} = Q(C_i - 2.0). \quad (2)$$

The OCC for a given flow and influent oxygen concentration is the maximum amount of oxygen that will be supplied to the filter in a day. Figure 2 illustrates schematics of different filter configurations with governing expressions defining the oxygen carrying capacities. For successful operation, the amount of oxygen made available to the filter (OCC) must be greater than the oxygen demand on the filter (OCF).

The mean oxygen demand of the filter per unit biomass can be computed empirically by the relationship

$$\overline{\text{OLR}} = \sum_{j=1}^{n} (\text{OCF}_j/W_j); \quad (3)$$

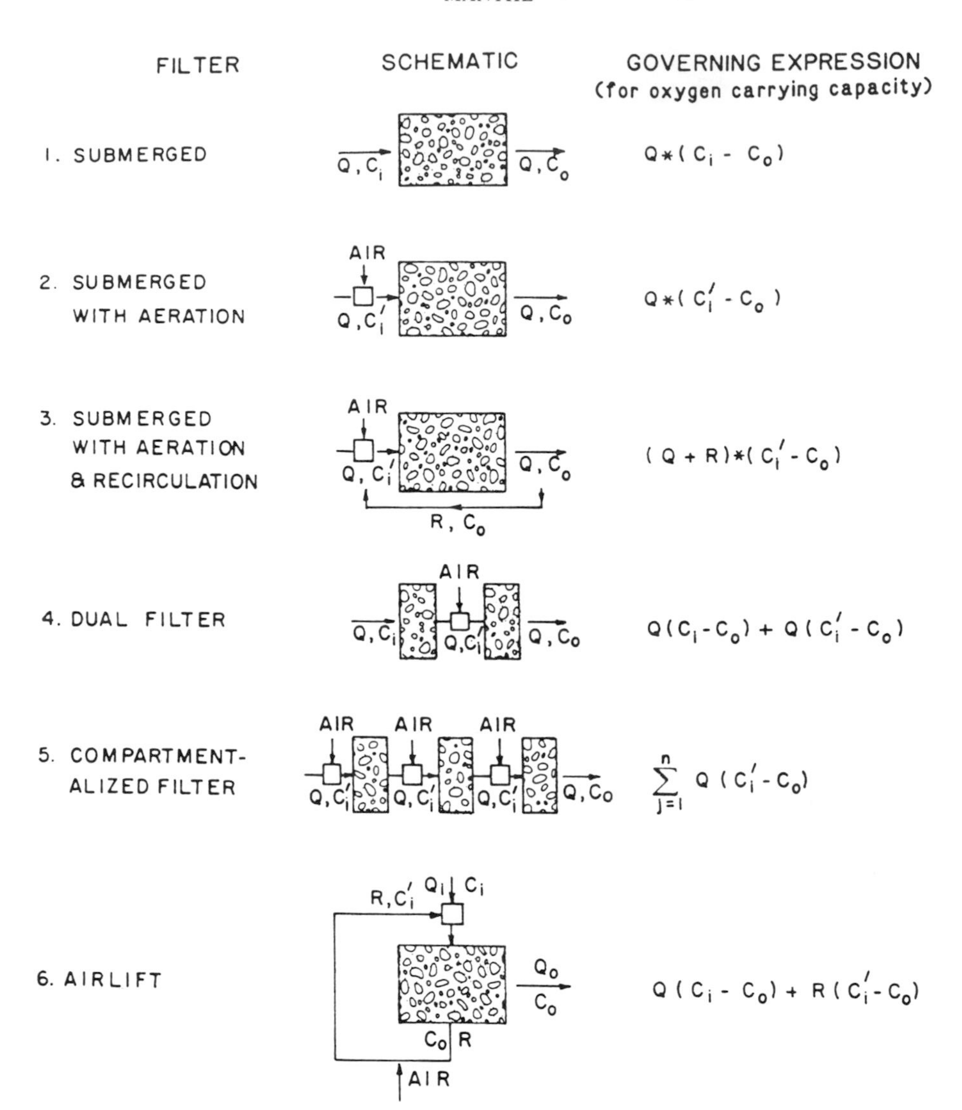

FIGURE 2.—Schematics of different biological filter configurations with governing expressions defining the oxygen carrying capacities (OCC). Q = flow to and through the filter; Q_i and Q_o are flows into and out of the filter when some of the flow enters a recirculation system directly from the filter; R = recirculation flow; C_i and C_o are oxygen concentrations in the inflow and outflow water; C_i' = is the oxygen concentration after aeration; $j = 1, 2, \ldots, n$ = number of filters; asterisks denote multiplication.

$\overline{\text{OLR}}$ = mean oxygen loading rate for organism (mg O_2/kg · d); and W_j = weight of organisms (kg) in the system for the jth observation.

To determine the theoretical organism carrying capacity of any given filter in terms of oxygen demand, the following equation may be used:

$$\text{FCC} = \text{OCC}/(\overline{\text{OLR}} + z\sigma); \quad (4)$$

FCC = filter carrying capacity (kg of organisms); z = standard normal deviate corresponding to the selected degree of confidence; and σ = standard deviation of OLR (mg O_2/kg · d).

Inclusion of the appropriate z-value adds to the robustness of the filter because it includes the probable occurrence of variation in the $\overline{\text{OLR}}$. For example, selecting a z-value of 2 gives a 95% confidence interval that the $\overline{\text{OLR}}$ will be in the accepted range of values with only a 2.5% probability that the upper limit will be exceeded on any given day. As the filters become limited by some other physical factor, such as clogging, the relationship between oxygen and filter evaluation will fail. However, in the range of oxygen limitation, the z-value is an important consideration.

Selection of an appropriate z-value is important to the stability of the system as a whole. Factors that can increase stability include system volume and removal of suspended solids. Factors that can lead to instability include shock loading and the presence of decaying matter. The magnitude of the selected z-value will increase or decrease in response to the above conditions. In commercial aquaculture systems where shock loading and decrease of the organism-to-volume ratio is the norm, I think that a z-value of at least 1 is appropriate.

By the rearrangement of equations (2), (3), and (4), the required flow rate through the biological filter can be determined by

$$Q = W(\overline{\mathrm{OLR}} + z\sigma)/(C_i - 2.0). \quad (5)$$

The volume of media required to support oxygen utilization in a rock filter can be expressed as

$$V = W(\overline{\mathrm{OLR}})/\mathrm{MOCF}; \quad (6)$$

V = volume of media needed to support oxygen utilization (m^3); and MOCF = maximum oxygen consumed in filtration (kg $O_2/m^3 \cdot d$). The MOCF is a naturally robust parameter because the filter requires several days of continuous higher animal loadings before any observed increase of microbial biomass occurs. Thus, the standard deviation associated with $\overline{\mathrm{OLR}}$, σ, is not expressed for filter volume considerations.

As an example of the analysis of oxygen utilization under the conditions imposed in this paper, it was observed experimentally that 1 m^3 of 2-cm limestone pieces could support an MOCF of 1.2 kg O_2/d without limitations from other parameters (Manthe et al. 1988). However, for the longevity and reliability required in the commercial sector, I believe that no more than 60–80% of the MOCF value should be used in the design equation.

Equations (5) and (6) provide a rational basis for determining both the volume of a filter and the required recirculation rate. This approach for analyzing biological filters in terms of oxygen should be applicable to a wide variety of aquaculture organisms, if oxygen consumed during filtration (OCF) under the specific culture conditions is determined. By using a small pilot system or aquarium and small numbers of animals, this could be easily done. The methodology described provides a quantitative analysis of filter oxygen supply needs for the particular aquaculture venture undertaken. A filter then can be designed and incorporated that will meet oxygen needs in the most efficient and economical manner.

Filter Media

Filter efficiency, once bacteria populations are established, is controlled by the surface area of the medium the flow contacts as the water passes through the filter or by the amount of oxygen available to the bacteria. Surface area is important because it affects the quantity of bacteria that can colonize the medium for BOD reduction and nitrification. The surface area available is principally determined by the size and shape of the medium. The following discussion centers around the more traditional media, such as rock, but recognizes the many other types of artificial media.

As filter medium size decreases, the MOCF increases until physical clogging becomes the limiting factor. Clogging can be due to entrapment of gross solids or to excessive bacterial growth. In what I term biological blinding, the growth of bacteria can be so great that the bacteria themselves can clog the interstitial pores and affect mass transfer by biofilm diffusion. This will also cause water traveling through the filter to be confined to rapid narrow pathways, thereby greatly reducing the surface area of exposure.

As the size of the medium increases, clogging is not a factor, but the MOCF decreases as surface area limitations occur at larger medium sizes. The decrease in surface area and, consequently, in the bacteria populations, will result in a decline of water quality both in the transitory and peak periods of loading. Rock media with diameters of 2.0–3.0 cm prevent biological blinding while providing suitable surface areas for bacterial colonization (Manthe et al. 1988).

The shape of the medium can directly affect the flow patterns in the filter. Filter media that maintain uniform interstitial spaces between elements are advantageous because flow across the filter is normalized. For example, elliptical media tend to channelize flow and reduce filter performance, whereas spherical media tend to maintain spacing between elements and therefore maximize surface area and flow equalization.

In closed systems in which the water is reused, the pH of the system water will decline because of the bacterial processes in the filter. In many cases, the filter medium used is calcareous (lime-

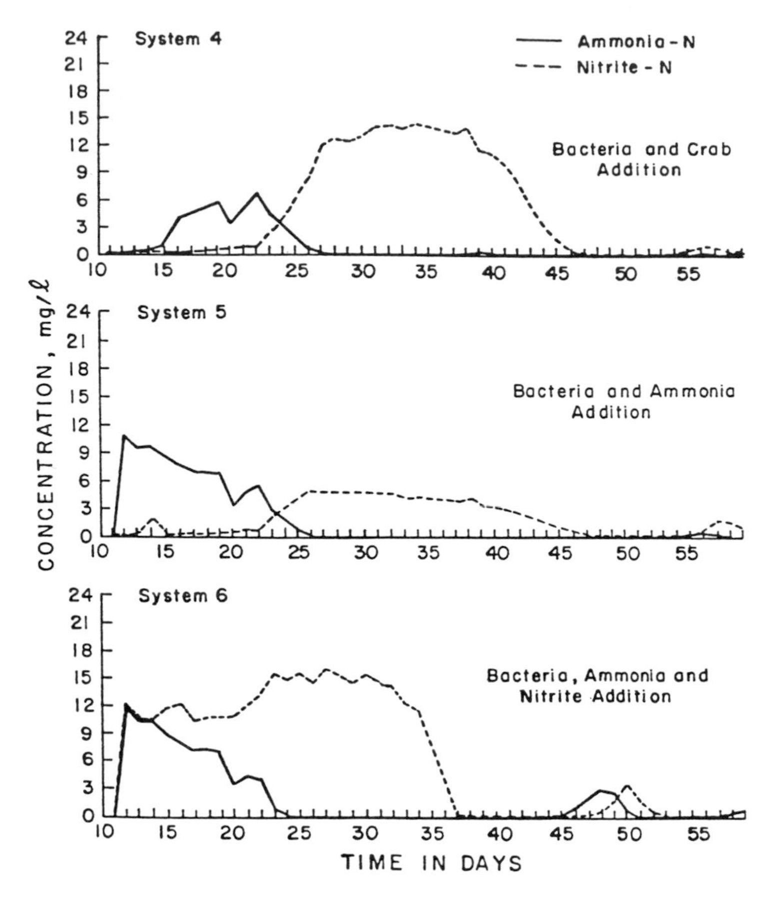

FIGURE 3.—Filter acclimation curves of systems including a control and ammonia and nitrite additions.

stone, shell) to provide a prophylactic buffer for the overall system pH. The pH level (approximately 7.0) maintained by dissolution of the medium can be significantly below the optimum for nitrifying bacteria during periods of heavy system loading. There is some disagreement on pH values for optimum nitrification rates, but it appears that *Nitrosomonas* sp. has a high constant oxidation rate between pH 7.0 and 9.0, and *Nitrobacter* sp. conversion rates are satisfactory between 6.5 and 8.5 (Wheaton 1977). However, nitrification in closed systems can be increased in times of heavy loading by raising the pH or by driving off the excess carbon dioxide produced by the oxidation process. If a noncalcareous substrate is used as the filter medium in a closed system, pH must be controlled by other means.

Filter Management

Filter Acclimation

Heavy or increased loadings of commercial aquaculture systems are a major problem if the biological filters are not properly acclimated. According to conventional theory, as the animals excrete ammonia (their primary nitrogen metabolite), ammonia concentrations increase. Populations of *Nitrosomonas* sp. increase in the filter and consume ammonia. As concentrations of ammonia decrease and nitrite increases, *Nitrobacter* sp. becomes established and begins converting nitrite to the much less toxic nitrate. Traditionally, biological filter acclimation for aquaculture systems takes from 30 to over 100 d (Hirayama 1974; Mevel and Chamroux 1981; Manthe et al. 1984) because of differences in temperature, salinities, etc. If the aquaculture system is not properly acclimated to accept heavy animal loadings, system failure and animal mortalities can result from declining water quality. Thus, methods to decrease biological filter acclimation time during this crucial period should increase productivity as well as revenues and provide safe operation of the system during heavy animal loadings.

Manthe and Malone (1987) demonstrated that

the addition of ammonia and nitrite could stimulate animal loading and acclimate a biological filter. As shown in Figure 3, chemical acclimation with ammonia addition did not reduce start-up time, but it did acclimate the filter without using live animals, thus avoiding high mortalities associated with that practice. Animal additions after the chemical acclimation demonstrated the ability of the filter to adapt to loadings under commercial conditions. Chemical acclimation with a combination of ammonia and nitrite decreased biological filter acclimation time by 10 d (or 28%). In the system containing chemical ammonia and nitrite, the bacteria initially had a complete substrate to feed on. *Nitrobacter* sp. did not have to wait approximately 10 d to have enough nitrite to increase its population. Further research in the area of artificial start-up by chemical addition is needed for other aquaculture applications, but the preliminary results are promising. In addition, it is thought that chemical addition could be used to maintain biological filters at peak efficiency during times of little or no animal loading.

Filter Operation

The design analysis above reveals the superiority of a filter design that incorporates a high flow rate and aeration. These processes substantially increase the amount of oxygen available for biological filter functions. Figure 4 illustrates two methods for increasing the oxygen supply to rock filters. Airlift aeration uses a standpipe and airlift pump to recirculate and aerate water in the filter. Recycle aeration employs an auxiliary recirculation pump (or excess flow from the main pump) to spray water through an aeration head on top of the filter, thus increasing the flow rate. From a theoretical perspective, either of these approaches can be used to enhance performance of an existing biological filter whose carrying capacity (FCC) has been exceeded, and either approach minimizes the total system energy requirements by recirculating only on the filter proper.

Because the carrying capacity of a filter is inversely proportional to the oxygen loading rate on it, any reduction in the $\overline{OLR}$ leads to an increase in the estimated carrying capacity. A well-maintained aquaculture system (no decaying organic materials present) will have a lower $\overline{OLR}$ and therefore a higher theoretical filter carrying capacity. Removal of gross particulates and organics before water reaches the filter will decrease the $\overline{OLR}$. This can be accomplished by using techniques such as sedimentation, mechanical filtration, or foam fractionation. In my experience, the FCC can be increased more than 30% by mechanical prefiltration (Manthe et al. 1988). Prefiltration will increase the FCC by removal of suspended particles that the filter would ultimately have to process until limiting factors (pH, clogging) preclude any filter carrying capacity gain.

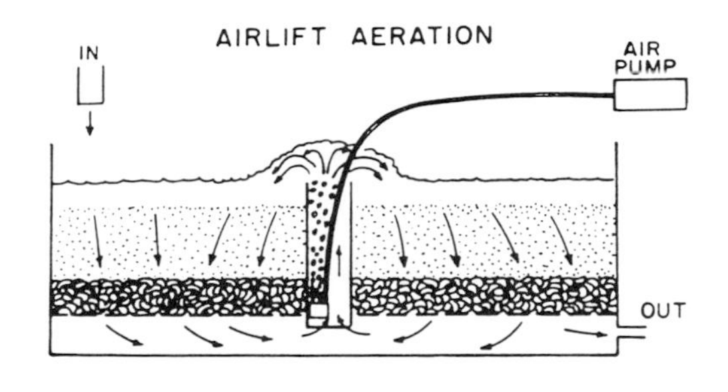

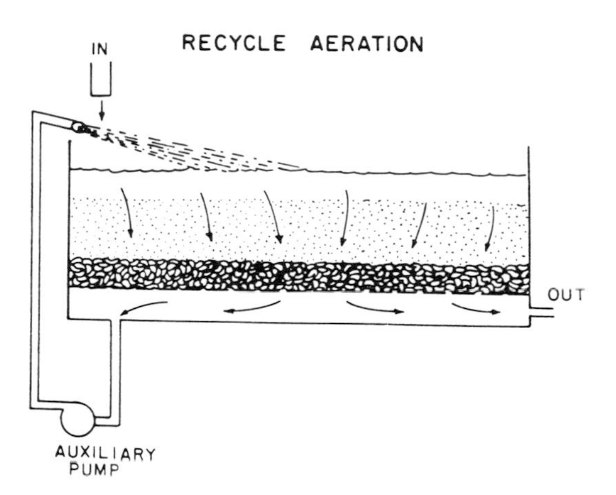

FIGURE 4.—Two methods for increasing oxygen supply to rock filters.

Clogging due to excessive bacterial growth can appear on the surface of a filter bed at higher loading densities, which will effectively reduce the FCC. The higher $\overline{OLR}$ due to the higher animal loadings is thought to increase the heterotropic bacteria populations, which perhaps compete with nitrifiers for available surface area and clog the bed. In closed systems, pH limitations can appear in response to increased acid production by the breakdown of wastes from higher animal loadings. In some closed systems the pH is regulated by the dissolution of the calcium carbonate filter bed, and pH control becomes inhibited by the limited kinetic dissolution rate and the increase of biofilm covering the carbonate substrate.

Continuous monitoring of dissolved oxygen should enhance the ability to automate intensive commercial production facilities (Malone et al. 1986). Oxygen consumption in a biological filter is a very dynamic parameter. Increases in system loading caused by stock addition or excess feeding can be detected by response of the filter, generally within a short period of time. Oxygen monitoring in biological filter components will provide advanced warning of impending waste buildup due to system overloading or failure of biological filtration units. Coupled with robust design criteria, filter monitoring could virtually eliminate loss of stock due to filtration upsets.

References

Gaudy, M. S., and E. T. Gaudy. 1978. Microbiology for environmental engineers. Wiley, New York.

Hirayama, K. 1965. Studies on water control by filtration through sand bed in a marine aquarium with closed circulating system—I. Oxygen consumption during filtration as an index in evaluating the degree of purification of breeding water. Bulletin of the Japanese Society of Scientific Fisheries 31:877–892.

Hirayama, K. 1974. Waste control by filtration in closed systems. Aquaculture 4:369–385.

Malone, R. F., D. G. Burden, and D. P. Manthe. 1986. A strategy for the automatic determination of filter efficiency in recirculating systems. *In* J. G. Balchen and A. Tysso, editors. Symposium of the International Federation of Automatic Control—Aquaculture '86. Trondheim, Norway.

Manthe, D. P., and R. F. Malone. 1987. Chemical addition for accelerated biological acclimation in closed blue crab shedding systems. Aquacultural Engineering 6:227–236.

Manthe, D. P., R. F. Malone, and S. Kumar. 1984. Limiting factors asociated with nitrification in closed blue crab shedding systems. Aquacultural Engineering 3:119–139.

Manthe, D. P., R. F. Malone, and S. Kumar. 1988. Submerged rock filter evaluation using an oxygen consumption criterion for closed recirculating systems. Aquacultural Engineering 7:97–111.

Metcalf and Eddy, Inc. 1979. Wastewater engineering: treatment, disposal, and reuse, 2nd edition. McGraw-Hill, New York.

Mevel, G., and S. Chamroux. 1981. A study on nitrification in the presence of prawns in marine closed systems. Aquaculture 23:29–43.

Wheaton, F. W. 1977. Aquacultural engineering. Wiley, New York.

American Fisheries Society Symposium 10:393–401, 1991

A Bioenergetic Model for Application to Intensive Fish Culture System Management

ANTHONIE M. SCHUUR
2408 18th Street, Bakersfield, California 93301, USA

Abstract.—Intensive fish culture operations, especially those incorporating temperature control systems to enhance growth rates and pure oxygen injection devices to increase culture densities, require continual and accurate allocation of inputs in order to operate efficiently. Management of culture space, feed, oxygen, water, and thermal resources is complex, dynamic, interdependent, and of direct economic importance. The field of physiological energetics provides a highly useful framework of mathematical relationships between metabolism, growth, and feeding that can be used to predict the outcome of various management strategies. The use of energetic concepts to manage a commercial intensive culture system resulted in the application of a mathematical management model, which is presented in this paper.

The application of bioenergetic modeling processes to commercial aquaculture production has lagged behind the wealth of physiological studies available because of a perceived lack of precision by physiologists, who often are reticent to extend their findings beyond a particular group of species and experimental conditions (Brett and Groves 1979). Additional resistance arises from the traditional agricultural research approach, which is heavily weighted toward empirical observation, and from the general tendency of the commercial aquaculture industry to avoid more complex, often computer-based management methods. A sobering indicator of the level of the energetic sophistication prevalent in the aquaculture industry is the typical feed bag instruction to "feed 3% of of live weight once or twice per day." That type of feeding regimen may be satisfactory for extensive pond management but is clearly unsatisfactory for intensive system management where space and other input resources are at a premium.

The proliferation of intensive production systems is the inevitable consequence of diminishing land and water resources as aquaculture becomes a larger and established industry. These systems may be characterized by high culture densities, mechanical aeration or oxygen injection systems, water conservation by regenerative devices or other means to extend use to maximum metabolite levels, and programmed feeding schedules. Because output per unit of space or unit of flow is a primary measure of economic efficiency, many intensive production facilities share the following management conditions.

- Oxygen is a critical production variable. Biomass consumption peaks must be balanced with oxygen inputs.
- Fish generally exhibit postprandial (after feeding) oxygen consumption peaks.
- The peak production of metabolites (e.g., ammonia, carbon dioxide) lags somewhat behind but also follows feeding.
- Fish are stocked at relatively high densities to use available space but must be periodically thinned to avoid overloading.
- Fish growth, especially among fingerlings and smaller stockers, can be very dynamic.
- Excessive sampling of fish in production is stressful and interrupts feeding cycles and growth.
- Even the most rudimentary calculation of ration requires knowledge of biomass to be fed. The biomass can either be predicted by a model or sampled. Often a model is used with less frequent sampling.
- Weight is constantly changing in a system because of growth, harvesting, and transfer. Not only is weight changing in the absolute sense, it is also changing in relation to specific average animal size, which has implications for feeding, metabolism, and water quality maintenance.
- When energy budgets in calories are correlated with operation budgets in dollars, feed consumption is consistently the highest single cost of aquaculture production. Furthermore, other costs such as those for pumping energy, thermal inputs, and pure oxygen are closely correlated with feed consumption.
- To completely confound managers of inten-

sive aquaculture systems, all of the above are related to one another.

Energy Budget Approach

The model presented here has its basis in energy budgets developed by several authors including Winberg (1956), Paloheimo and Dickie (1965), Warren and Davis (1967), and Brett and Groves (1979). These energy budget models were devised for purposes of ecological studies or for physiological research, although their potential application to aquaculture production is evident. Allen et al. (1984) proposed a feed ration model based on energy budgets, which was used in an earlier computer simulation of lobster growth (Botsford et al. 1975). The present model is similar to that of Allen et al. in content and application, although some components have been redefined.

The format used by Brett and Groves (1979) best describes the essential logic of the energy budget and ration relationship.

$$I = E + M + G;$$

I = the total energy ingested,
E = the fecal and nonfecal energy excreted,
M = the energy expended in the chemical transformation of feed into growth, activity, and heat, and
G = the energy captured in the fish tissue;

$$M = M_s + aM_{r-s} + bM_{f-s} + cM_{a-s} + H;$$

M_s = standard metabolism (resting, unfed),
M_r = routine metabolism (mobile, unfed),
M_f = feeding metabolism (maintenance ration),
M_a = active metabolism (maximum activity),
H = heat increment (also called specific dynamic action or SDA), and
a, b, c = proportions of time.

A balanced energy equation (per 100 calories of ingested ration, I, 95% confidence levels included) for an average carnivorous fish was developed from composite data for several species as follows:

$$100I = (44 \pm 7)M + (29 \pm 6)G + (27 \pm 3)E.$$

Brett and Groves (1979) also provided an excellent conventional energy budget schematic that explains energy partition in typical carnivorous fishes.

The Model

The energy budget models discussed above provide the concept and considerable experimental documentation for modeling fish energetics and are very useful for understanding the fate of energy in ecological and physiological systems. However, they lack the mathematical means to track the energetics of groups of fish in aquaculture systems that are constantly changing in average size and metabolic rate over the culture period. There is a need, therefore, for a dynamic means of computing continuous energetic equations for fish of various average sizes, growth rates, and feeding rates under different environmental conditions. With such an algorithm it would bc possible to simulate, with a personal computer spreadsheet program, an almost infinite variety of culture patterns and to use the values computed for operations planning and daily regulation of culture inputs.

The model consists of the following relation between the ration (df/dt), growth (dW/dt), and metabolism (dO_2/dt):

$$df/dt = \frac{C(dW/d) + H \cdot O \cdot T \cdot (dO_2/dt)}{R};$$

$$dW/dt = KW^a,$$
$$dO_2dt = MW^m,$$
$$T = Q_{10}^{(t_1 - t_0)/10}.$$

Model variables for the baseline condition are

W = variable: instantaneous wet weight (g),
M = 0.02: feeding metabolic rate (g O_2/d),
m = 0.80: allometric exponent for metabolism,
a = 0.70: allometric exponent for growth,
R = 2.30: kcal metabolizable energy per gram of ration,
O = 3.24: oxycalorific equivalent (kcal/g O_2),
H = 1.35: heat increment and activity loss ratio,
C = 1.25: kcal per gram wet weight fish,
K = 0.06: baseline growth constant, and
T = 1.0: temperature adjustment constant at 25°C, with Q_{10} = 2.3, t_1 = 25°C, and t_0 = 25°C, and
Q_{10} = 2.3: factor by which metabolic rates increase for a 10°C rise in temperature.

The principal model expression above provides, in a single equation, a quantitative relationship between ration, growth, and metabolism.

The dividend of the expression is in units of energy for instantaneous values of growth and metabolism as a function of a specific weight. Weight, as discussed in the next section, can be computed as a function of time by integrating the dW/dt expression over a time period. Thus for a particular growth curve designated by the constant K, the value of the energetic components can be calculated over time in any time increment useful for management purposes. By manipulating the components of the expression, the following information about fish culture performance on a projected growth curve may be calculated:

- instantaneous growth rate;
- weight as a function of time;
- ration as a function of weight or time;
- specific feeding rate as a function of weight or time;
- feeding metabolism as function of weight or time and temperature;
- total oxygen consumption as a function of weight or time and temperature;
- instantaneous feed conversion ratio; and
- instantaneous metabolism: growth ratio.

Table 1 shows calculations of various model values over selected growth cycles from 5 to 900 g. Table 2 shows comparative oxygen consumption rates at 25 and 30°C. A composite of the values for the model baseline case computed at 15-d intervals over a 360-d production cycle is shown in Table 3. Composite values may be used to compute values for total feed consumption, feed conversion, oxygen consumption, and other ratios of management interest.

The Growth Component

The differential growth equation, $dW/dt = KW^a$, was derived by both Parker and Larkin (1959) and Winberg (1971). It is a convenient and tractable expression to express the curvilinear growth of young, rapidly growing fish that typifies aquaculture growth cycles. The value of 0.7 assigned to exponent a is typical of many growth curves of a wide variety of fish and other classes of animals in an "unlimited" growth phase. Parker and Larkin discussed the concept of growth stanzas that differentiate growth rates of various life stages in a fish. Young, sexually immature fish tend to accurately track a growth curve at a given ration, but then assume a new growth constant when energy is diverted to sexual development. The stanza concept is important to the present modeling exercise, because it may be more accurate to model a production cycle with two or more periods with correspondingly different growth constants, particularly when fish are subject to developmental changes. The growth differential can be integrated over a time interval in the following manner:

$$W = [K \text{ (time) } (1 - a) + w^{(1 - a)}]^{[1/(1 - a)]};$$

time is the growth period, in days, between the initial weight, w, and the instantaneous weight, W.

Weight can then be determined as a function of time for any growth constant K; conversely, K can be calculated if the initial weight, final weight, and time interval are known. In growth studies the growth constant K is a very useful index for the comparison of curvilinear growth performance; K is directly proportional both to instantaneous growth at the same weight and to the time to attain a target weight from the same size.

For example, as is often the case with comparative growth studies in aquaculture, one may conduct a feeding trial in which two groups of fish, initially of similar size, are fed different diets at the same rate for some time, which results in a different final average weight for the two groups. Then one may want to extrapolate the findings of the feeding trial to other groups of fish with different initial weights. The following numerical example might be typical of such an exercise for two groups of fish fed two feeds for a 2-week trial.

	Group 1	Group 2
Feed % protein	36	28
Number of individuals	30	30
Initial average weight (g)	103.6	105.7
Final average weight (g)	124.6	114.6
Average weight gain (g)	21.0	8.9
Calculated value of K	0.0545	0.0237

The calculated values of K can then be applied to an estimate of the grow-out days required to take the fish from a typical stocking size of 100 g to a harvest size of 850 g. The required culture times are 219 and 505 d, respectively, for the growth rates determined or groups 1 and 2. It is important to note that K might also be used as a comparative index of growth for fish of a very dissimilar size, if it can be assumed that the exponent a remains constant over the subject size range. If, for example, another group of fish were tested at an initial size of 211.4 g, or twice that of group 2, and their weight gain was exactly double that of group 2, would it mean that the growth performance was equal? No—actually it would be

TABLE 1.—Instantaneous weights, growth rates, feeding rates, and food conversion ratios for culture of fish modeled with the baseline model values and various values of the growth constant, K.

Time (d) or fish weight (g)	Growth constant (K)				
	0.020	0.040	0.060	0.080	0.100
			Instantaneous mean fish weight (g)		
0 d	5.0	5.0	5.0	5.0	5.0
30 d	7.1	9.8	13.0	17.0	21.8
60 d	9.8	17.0	27.4	41.6	60.3
90 d	13.0	27.4	50.4	84.2	131.4
120 d	17.0	41.6	84.2	150.5	246.8
150 d	21.8	60.3	131.4	246.8	419.4
180 d	27.4	84.2	194.5	379.7	662.6
210 d	34.0	114.0	276.4	556.0	990.5
240 d	41.6	150.5	379.7	782.8	1,418.0
270 d	50.4	194.5	507.5	1,067.6	
300 d	60.3	246.8	662.6		
330 d	71.5	308.3	848.3		
360 d	84.2	379.7	1,067.6		
			Daily growth rate (g/fish)		
10 g	0.10	0.20	0.30	0.40	0.50
50 g	0.31	0.62	0.93	1.24	1.55
100 g	0.50	1.00	1.51	2.01	2.51
200 g	0.82	1.63	2.45	3.26	4.08
300 g	1.08	2.17	3.25	4.34	5.42
500 g	1.55	3.10	4.65	6.20	7.75
700 g	1.96	3.92	5.88	7.85	9.81
900 g	2.34	4.68	7.02	9.36	11.69
			Daily ration (g feed/fish)		
10 g	0.25	0.30	0.36	0.41	0.46
50 g	0.86	1.03	1.20	1.37	1.54
100 g	1.48	1.76	2.03	2.30	2.58
200 g	2.55	3.00	3.44	3.88	4.33
300 g	3.51	4.10	4.68	5.27	5.86
500 g	5.23	6.07	6.92	7.76	8.60
700 g	6.81	7.88	8.94	10.01	11.08
900 g	8.30	9.57	10.84	12.11	13.38
			Daily ration (% of mean fish weight)		
10 g	2.46	3.01	3.55	4.10	4.64
50 g	1.73	2.06	2.40	2.74	3.07
100 g	1.48	1.76	2.03	2.30	2.58
200 g	1.28	1.50	1.72	1.94	2.16
300 g	1.17	1.37	1.56	1.76	1.95
500 g	1.05	1.21	1.38	1.55	1.72
700 g	0.97	1.13	1.28	1.43	1.58
900 g	0.92	1.06	1.20	1.35	1.49
			Food conversion ratio[a]		
10 d	2.46	1.50	1.18	1.02	0.93
50 d	2.79	1.67	1.29	1.11	0.99
100 d	2.95	1.75	1.35	1.15	1.03
200 d	3.13	1.84	1.40	1.19	1.06
300 d	3.23	1.89	1.44	1.22	1.08
500 d	3.38	1.96	1.49	1.25	1.11
700 d	3.47	2.01	1.52	1.28	1.13
900 d	3.55	2.05	1.54	1.29	1.14

[a]Grams of feed fed/grams of weight gained by fish.

considerably better, corresponding to a growth constant of 0.0291 and a culture time of only 410 d for the projected grow-out period.

The Metabolism Component

The metabolism component of the model is a standard expression for metabolism as a function of weight. The exponent m is determined by the slope of the line of the log–log regression of oxygen consumption as a function of weight. Although the exponent varies experimentally from approximately 0.6 to 0.9, the value is close to 0.8 for the majority of species and for a wide variety of experimenters. The size effect of me-

TABLE 2.—Comparison of model oxygen consumption at 25°C (baseline case) and 30°C expressed in calories, and the resultant effect on the instantaneous value of the growth constant K.

Mean fish weight (g)	Daily oxygen consumption (kcal/fish)		Growth constant K	
	25°C	30°C	25°C	30°C
10	0.55	0.84	0.060	0.014
50	2.00	3.03	0.060	0.007
100	3.48	5.28	0.060	0.003
200	6.06	9.20	0.060	−0.001
300	8.39	12.72	0.060	−0.004
500	12.62	19.14	0.060	−0.007
700	16.52	25.05	0.060	−0.010
900	20.20	30.63	0.060	−0.011

tabolism is very significant for energetic budgets in that the metabolism of a 1-g fish per unit weight may be 3–5 times greater than that of a 1-kg fish.

The value of M used in calculating the model results here is an estimate for metabolism of striped bass *Morone saxatilis* derived from values determined by Klyashtorin and Yarzhombek (1975). The value is adjusted to correspond to the active feeding metabolism level at which the fish displays normal feeding activity and feeds at a rate that maintains its body weight.

Total oxygen consumption for a fish is estimated to be about 15–24% greater than the presumed feeding metabolic rate due to the heat increment of chemical reactions to deaminate amino acids in the synthesis of fish tissue protein and to other exothermic reactions that result in heat loss unaccounted for in the metabolic rate. The activity of fish in typical high-density culture situations may realistically account for another 10–15% of oxygen consumption. These losses correspond to a model value of H estimated at 1.35. This value is presumed to be fairly accurate by deduction from actual production experience, but is by far the least well documented of the model's numerical values.

Temperature Effect

The temperature effect on metabolism is represented in the constant T, which is Q_{10} raised to the power t_1 minus t_0 divided by 10; Q_{10} is the increase in rate for an increase of 10°C , t_0 is some baseline temperature within the Q_{10} range, and t_1

TABLE 3.—Composite model calculations for baseline values with the growth constant K equal to 0.06 over a culture period of 360 d.

Time (d)	Mean Weight (g)	Growth rate (g/d)	Metabolic rate (g O_2/d)	Ration (g/d)	Metabolism-to-growth ratio (caloric ratio)	Food conversion ratio
0	5.0	0.19	0.10	0.21	1.37	1.14
15	8.4	0.27	0.15	0.31	1.44	1.17
30	13.0	0.36	0.21	0.43	1.51	1.20
45	19.3	0.48	0.29	0.58	1.57	1.23
60	27.4	0.61	0.38	0.76	1.62	1.25
75	37.7	0.76	0.49	0.97	1.68	1.27
90	50.4	0.93	0.62	1.21	1.73	1.29
105	65.8	1.12	0.77	1.48	1.77	1.31
120	84.2	1.34	0.94	1.78	1.82	1.33
135	105.9	1.57	1.13	2.12	1.86	1.35
150	131.4	1.82	1.34	2.50	1.90	1.37
165	160.8	2.10	1.57	2.91	1.94	1.39
180	194.5	2.40	1.83	3.37	1.98	1.40
195	232.9	2.72	2.11	3.86	2.01	1.42
210	276.4	3.07	2.42	4.40	2.05	1.43
225	325.2	3.44	2.76	4.98	2.08	1.45
240	379.7	3.83	3.13	5.61	2.11	1.46
255	440.3	4.25	3.52	6.28	2.14	1.48
270	507.5	4.70	3.94	7.00	2.17	1.49
285	581.4	5.17	4.39	7.76	2.20	1.50
300	662.6	5.66	4.88	8.58	2.23	1.51
315	751.4	6.18	5.40	9.44	2.26	1.53
330	848.3	6.73	5.95	10.36	2.29	1.54
345	953.5	7.31	6.53	11.33	2.32	1.55
360	1067.6	7.91	7.15	12.35	2.34	1.56

Composite ratios

Food conversion (weight fed/weight gain)	1.49
Oxygen to feed (weight oxygen/weight fed)	0.56

is a culture temperature (also within the Q_{10} range) at which the effect is to be calculated. The Q_{10} is experimentally derived from the measurement of metabolism at various temperatures in the appropriate range.

Brett (1964) and Beamish (1964) determined that a Q_{10} of approximately 2.3 was useful in determining the temperature effect on standard metabolism for several species of temperate-water fishes. The Q_{10} concept has a long history but it is suspect among physiologists, who have been unable to accurately apply it with any uniformity to wide temperature ranges and different species. The value of Q_{10} is quite sensitive to species, the actual temperature range examined for a particular species, and other factors such as size and activity rate.

Although the temperature effect constant is applied to the metabolism component of the model, higher-than-optimal temperatures can have a profound effect on growth. If the assumption is made that there is a temperature at which a species of fish grows at a maximum rate, given an adequate ration to sustain growth, then a further elevation of temperature is likely to cause a decline in growth. The rationale for this general assumption is that the metabolic rate is directly related to the rate of chemical reactions within the fish, which are mediated by temperature. Energy for growth, on the other hand, is the residual energy left after metabolic needs have been met and is physically limited by the total feed consumption capacity of the fish. Thus if optimal growth rates occur when feeding approaches the maximum that the fish can consume because of gut size or other physical constraints, and the maximum growth ration is held constant, an increasing proportion of energy is diverted from growth to metabolism when the temperature rises above the optimal growth temperature. Table 2 shows a comparison of metabolic energy requirements for fish of various weights at 25°C, which is assumed (for this example) to be the maximum growth rate temperature, and at 30°C, which approaches lethal limits for culture. The instantaneous growth coefficient, K, remains constant for the fish at 25°C. At 30°C, the model predicts a severe reduction in K at the lower fish weights and even negative growth rates for heavier fish. The reduction in growth rate in relation to size is consistent with the changing ratio of metabolism to growth shown as a function of size in Table 3.

Actual culture system performance at our farm in Paso Robles, California, during a summer heat wave when we were unable to adequately regulate the temperature of our geothermal water source, closely paralleled the result predicted by the model. A decline in growth was experienced by different size-classes of fish and was also realized in distinct degrees related to size. The smaller young of the year demonstrated only minor growth declines. Larger, market-size yearlings barely displayed maintenance levels of growth, and some lost more than 5% of their body weight per month over the period of peak summer heat when culture temperatures often exceeded 30°C. Attempts to provide more calories and offset the metabolic increase by increasing rations were unsuccessful because the fish accepted little more than optimal growth level amounts of feed.

This high-temperature episode provided an excellent example of the methodology for refining the model through iterative evaluation of production results. In an earlier form of the model, which was used to generate actual production feeding tables, a Q_{10} of 2.0 was used. This value was accepted as an average applicable to a wide range of species from diverse environments. Production results much more severe than anticipated by the model prompted further review of the literature, which resulted in the 2.3 value now used. That refinement resulted in dismally accurate prediction of poor growth performance. Unfortunately, no further research was able to change the weather or provide cooler water. Given the actual results and model confirmation, high temperatures were a severe constraint to our culture operations.

Ration Energy

The constant for ration energy, R, is an estimate of the metabolizable energy (ME) in a unit of feed. The use of ME in place of gross energy (GE) as the energy unit in energy budgets accomplishes two things within the model. Firstly, it somewhat simplifies the model by eliminating the need to calculate the loss of energy to excretion. By definition, ME is the gross energy less the energy lost to excretion. Secondly, it provides a more precise estimate of the actual energy available to fish for growth and metabolism if the excretion losses have been experimentally determined. Even if ME is not known for a particular feed component or species, examination of existing ME values does provide a useful insight into available metabolizable energy that may be used to deduce or estimate ration ME in a particular culture situation. "Nutrient Requirements of

Coldwater Fishes" (National Research Council 1981) and its companion publication for warmwater fishes (National Research Council 1983) provide a listing of known GE, digestible energy, and ME values for rainbow trout *Oncorhynchus mykiss* and channel catfish *Ictalurus punctatus*. Salmonid ME values are by far the best known and must be used with caution in estimation of values for other fish families, especially for feeds with high percentages of carbohydrates that salmonids metabolize very poorly.

An estimate of the metabolizable energy available in a typical production fish feed is as follows

Component	Percentage of ration	Estimated ME (kcal/g)	Estimated ME (kcal/g ration)
Protein	35	3.10	1.09
Fat	7	8.00	0.56
Carbohydrate	32	2.10	0.67
Ash	10	0.00	0.00
Fiber	8	0.00	0.00
Moisture	8	0.00	0.00
Total	100		2.32

For purposes of the present model, a value of 2.3 kcal/g ration presented is assumed.

Feeding Schedules and Food Conversion Ratio

The most immediate use of the model by production managers is to create feeding schedules that provide the quantity of rations for routine daily allocation of feed. The feeding schedule used at Aquafuture Farms of Paso Robles was generated in tabular form by a personal computer spreadsheet program. The table used was similar to the abbreviated version shown in Table 1. For normal growth rates, we found that weekly recalculation of feeding schedules for each of 12 culture tanks provided an accurate and gradual increment of feed to sustain growth in accordance with projected growth rates. In the case of small fingerlings less than 20 g, biweekly or even daily increases in feed may be used to systematically keep feeding in equilibrium with the very high specific growth rates typical of small fish.

The feeding schedule was used in association with a food conversion ratio (FCR) table (Table 1) and a weekly growth calculation table that was completed at the first of each week. The growth calculation table was used to calculate the weekly weight gain on the basis of the ration actually fed, not the projected growth rate. Transactions among tanks were also calculated into the new week's individual tank inventory data, which consisted of total tank weight, number, and average size of fish.

During actual operations, projected feeding schedules may be disrupted by harvests, transfers, power outages, and temperature control problems. Under all of these conditions it is usually appropriate to reduce or withhold rations because of unfavorable water quality conditions. On the basis of the feed actually fed, the feeding table was used in a reverse fashion to estimate the actual model growth rate corresponding to the average daily ration fed for the week. The new average weight for the tank was then calculated and used as the starting weight for the programmed feeding rate in the coming week. Projected growth rates for the coming week were assigned to each tank as appropriate for production needs and were calculated with a selected value of the constant *K*. In practice, the value of *K* was selected from growth curves A–F, which designate *K* values of 0.10 to 0.00 in decrements of 0.02. The A-curve corresponds to very rapid growth, whereas the F-curve is, by definition, a maintenance or no-growth condition. Typical production growth rates under routine conditions fell into the intermediate B, C, and D categories.

A significant feature of feed rate and FCR tables and the weekly growth computation is that, although the mechanics seem somewhat complex, all the values are generated by a single model and are therefore internally consistent. A key advantage of the model-based method used at Paso Robles is that it was extremely accurate in predicting growth rates and harvest weights over long periods without sampling. The average difference between actual weights when a tank batch was harvested or otherwise weighed out completely ranged between 1 and 5% for culture periods of 3 to 6 months. Because minor errors in the weighing of small fish aliquots used to calculate the number of fish stocked in a tank were magnified in time with increasing weights, occasional anomalies did occur, but backward calculation almost invariably proved that the flaw was an error in the counting of fish stocked or the initial average weight sample.

A major reason for the accuracy of prediction is that great care was taken to accurately calculate the FCR over relatively small increments of time. The model was used to generate a refined estimate of the instantaneous FCR, which in turn resulted in the most accurate estimate of the new weight for which a new feed rate was calculated. More simplistic models might use an overall FCR for the entire culture period, thereby inducing an error that will, ironically, increase the FCR. The model makes it possible to calculate small incremental changes in feeding rate in direct relation to a corresponding growth rate.

Oxygen Consumption Rates

Model oxygen consumption rates for the baseline case and 30°C are shown in Table 2. The accurate prediction of oxygen consumption has important implications for all aquaculture systems but is of particular interest for those that employ pure oxygen systems. Bioengineering criteria for system design, the operation of oxygen distribution systems, and the operational cost of using pure oxygen as a system input are all related to the estimation of oxygen consumption in a culture system.

The oxygen:feed ratio shown in Table 3 might be considered a hypothetical goal for oxygen consumption in culture applications. In practice, because of system losses, excess fish activity, and temperature variation, design parameters may be two to three times as high as the calculated ratio. Yet the calculation provides a fixed point by which to judge the efficiency of oxygen injection systems. The 0.56 ratio calculated in Table 3 compares quite favorably with the fundamental biochemical basis for energy transformation (Hoar 1966) and with other studies of fish energetics. Hogendoorn's (1983) experimental results for the African catfish *Clarias lazera* compute to a ratio of 0.40, and measurements of sockeye salmon *Oncorhynchus nerka* (Brett and Zala 1975) show a ratio of 0.61. The similarity of these experimental results with the model value is not unexpected because of the equilibrium of oxidative phosphorylation, which is the underlying biochemical mechanism of energy transformation:

$$C_6H_{12}O_6 + 6O_2 \rightarrow 6CO_2 + 6H_2O + 38ATP.$$

For this reaction, the ratio of the molecular weights of oxygen (192) to the foodstuff glucose (180) is 1.07. With glucose as the energy source, 1 mol of ATP (adenosine triphosphate) yields 18.7 kcal of metabolic energy or 3.94 kcal/g of glucose. If one assumes that about 20% of the feed energy is incorporated in the tissues and that typical feed has 2.3 kcal/g, the theoretical weight ratio of oxygen to the feed is

$$(2.3/3.94) \times 0.8 \times 1.07 = 0.500.$$

Conclusion

An aquaculture management model based on principles of physiological energetics provides a means of quantitatively relating feeding, growth, and metabolism in a dynamic production situation. The following applications of the model presented are suggested.

- The model can be used in simulation of probable growth and temperature regimes to quantitatively define weight in process and the inputs required over a production cycle.
- The model can be used in real-time production situations to predict the outcome of a management strategy. A typical management question would be something like, "How many days will it take fish of a given size to reach a harvest size of 800 g on the present growth curve and what will be the feed rate on the harvest day?"
- The model can be used to calculate hypothetical ideals of efficiency. It provides a methodical framework for evaluating the cost-effectiveness of improving a culture system by spending on capital facilities or changes in operational procedure.
- Used essentially the same as for simulation (though somewhat differently), tabular forms of the model can provide routine, daily aids to accurately allocate culture system inputs.

This model, like any other, is subject to error through the application of values or relationships that are not true. The model can be a very useful means of comparing and adjusting production performance with a quantitative standard. Over time values can be refined, which might be thought of as calibrating the model with the fish. This iterative procedure should lead to more precision in management practices and a better understanding of the physiological processes upon which fish culture depends.

References

Allen, P. G., L. W. Botsford, A. M. Schuur, and W. E. Johnston. 1984. Bioeconomics of aquaculture. Elsevier, Amsterdam.

Beamish, F. W. H. 1964. Respiration of fishes with special emphasis on standard oxygen consumption. II. Influence of weight and temperature on several species. Canadian Journal of Zoology 42:177–188.

Botsford, L. W., H. E. Rauch, A. M. Schuur, and R. A.Shleser. 1975. An economically optimum aquaculture facility. Proceedings of the World Mariculture Society 6:407–420.

Brett, J. R. 1964. The respiratory metabolism and swimming performance of young sockeye salmon. Journal of the Fisheries Research Board of Canada 21:1183–1226.

Brett, J. R., and T. D. D. Groves. 1979. Physiological energetics. Pages 280–352 *in* W. S. Hoar, D. J. Randall, and J. R. Brett, editors. Fish physiology, volume 8. Academic Press, New York.

Brett, J. R., and C. A. Zala. 1975. Daily pattern of nitrogen excretion and oxygen consumption of sockeye salmon (*Oncorhynchus nerka*) under controlled conditions. Journal of the Fisheries Research Board of Canada 32:2479–2486.

Hoar, W. S. 1966. General and comparative physiology. Prentice-Hall, New Jersey.

Hogendoorn, H. 1983. Growth and production of the African catfish, *Clarias lazera* (C. & V.). III. Bioenergetic relations of body weight and feeding level. Aquaculture 35:1–17.

Klyashtorin, L. B., and A. A. Yarzhombek. 1975. Some aspects of the physiology of the striped bass, *Morone saxatilis*. Journal of Ichthyology 15:985–989.

National Research Council. 1981. Nutrient requirements of coldwater fishes. National Academy Press, Washington, D.C.

National Research Council. 1983. Nutrient requirements of warmwater fishes and shellfishes. National Academy Press, Washington, D.C.

Paloheimo, J. E., and L. M. Dickie. 1965. Food and growth of fishes. I. A growth curve derived from experimental data. Journal of the Fisheries Research Board of Canada 22:521–542.

Parker, R. R., and P. A. Larkin. 1959. A concept of growth in fishes. Journal of the Fisheries Research Board of Canada 16:721–745.

Warren, C. E., and G. E. Davis. 1967. Laboratory studies on feeding, bioenergetics, and growth of fish. Pages 175–214 *in* S. D. Gerking, editor. The biological basis of freshwater fish production. Blackwell Scientific Publications, Oxford, UK.

Winberg, G. G. 1956. Rate of metabolism and food requirements of fishes. Nauchnye Trudy Belorusskogo Gosudarstvennogo Universiteta Minsk. Translated from Russian: Fisheries Research Board of Canada Translation Series 194, 1960, Ottowa.

Winberg, G. G. 1971. Methods for the estimation of the production of aquatic animals. Academic Press, New York. (Translated from the 1968 Russian edition by A. Duncan.)

American Fisheries Society Symposium 10:402–409, 1991

Effect of Diet on Growth and Survival of Coho Salmon and on Phosphorus Discharges from a Fish Hatchery

H. George Ketola

U.S. Fish and Wildlife Service, Tunison Laboratory of Fish Nutrition
3075 Gracie Road, Cortland, New York 13045, USA

Harry Westers and Walter Houghton

Michigan Department of Natural Resource, Fisheries Division
Post Office Box 30028, Lansing, Michigan 48909, USA

Charles Pecor

Michigan Department of Natural Resources, Platte River Fish Hatchery
15200 Honor Highway, Beulah, Michigan 49617, USA

Abstract.—In two hatchery experiments, coho salmon *Oncorhynchus kisutch* were fed a low-phosphorus diet or Oregon Moist Pellets (OMP) for 45–46 weeks from starting weights of 4–5 g to their release as smolts to Lake Michigan. In the first experiment, fish fed the low-phosphorus diet grew 85% as well as and returned from Lake Michigan at 79–106% the rate of fish fed OMP, and their hatchery effluent contained 33–80% less phosphorus. In the second study, the low-phosphorus diet was modified to contain 10% fish meal in an attempt to increase growth rate. Fish fed the modified diet (T2M) grew as well as those fed OMP and discharges of phosphorus were significantly reduced. The amounts of phosphorus discharged per 1,000 kg of fish production were 4.8 and 12.2 kg for the T2M and OMP diets, respectively. The amounts of phosphorus discharged per 1,000 kg of feed fed were 4.2 and 8.2 kg for the T2M and OMP diets, respectively. Overall, when coho salmon were fed the T2M diet, the amount of phosphorus discharged into effluents was reduced by 49–61% compared with discharges from the OMP treatment.

Interest in the effect of fish hatchery diets on the discharge of phosphorus into effluent waters began when the Fisheries Division of the Michigan Department of Natural Resources pointed out the need to reduce discharges of phosphorus in the effluents from a hatchery that raised coho salmon *Oncorhynchus kisutch*. Grant (1979) of the State Water Quality Division in Michigan reported that effluent discharges from the Platte River Salmon Hatchery significantly increased the total amount of phosphorus going into Platte Lake, thereby hastening eutrophication of this mesotrophic lake (area, 1,018 hectares; average depth, 9.1 m). The average concentration of phosphorus in the river was 12 μg/L above the hatchery and 33 μg/L below. Thus the hatchery discharge increased phosphorus by 21 μg/L in the river as it flowed through the hatchery. The total annual discharge of phosphorus was estimated to be 1,500 kg. A later report (Kanaga and Evans 1982) estimated the annual discharge to be 760 to 1,000 kg. Most, if not all, of this increase in phosphorus apparently came from the fish feed. This report recommended that the hatchery be permitted to discharge no more than about 760 kg of phosphorus per year. The Fisheries Division believed that treatment of the effluent itself probably could not meet this recommendation. We therefore sought ways to reduce the loss of dietary phosphorus into the hatchery effluent. Estimates made from published data on Atlantic salmon *Salmo salar* (Ketola 1975) indicated that the retention of dietary phosphorus in salmon carcasses (0.4% P on a fresh basis) sometimes ranged from 14 to 22%. Therefore, about 78–86% of the dietary phosphorus was discharged (in dissolved and solid forms) into hatchery water. This suggested the potential to markedly increase retention of phosphorus and thereby reduce phosphorus discharges caused by fish feeds.

A preliminary report of laboratory studies showed that the diet and source of phosphorus significantly influenced the amount of feed phosphorus that was discharged into the effluent water or appeared in the sludge (Ketola 1982). Defluorinated rock phosphate (18% P) was promising as a source of phosphorus because of its low solubility in water and its nutritional value to poultry. Subsequent laboratory studies (Ketola 1985) demonstrated that a new economical plant-protein diet (diet 7) containing defluorinated rock phosphate resulted in growth of rainbow trout *Oncorhynchus mykiss* equal to 86% of control fish growth and reductions of phosphorus discharge by 46%. This diet contained 3.5% defluorinated rock phosphate and a total level of approximately 1.15% total phosphorus. When it was fed to coho salmon *Oncorhynchus kisutch* in an 8-week hatchery experiment, growth was 81% of that of coho salmon fed the control diet, Oregon Moist Pellet (OMP), and phosphorus discharges in effluents were decreased by 50% or more (Ketola 1985).

The present paper describes a long-term test of a slight modification of diet 7 (Ketola 1985) fed to coho salmon in a hatchery. The modified diet is designated diet T2. A second long-term hatchery test was conducted in which this diet was further modified (diet T2M) to contain 10% fish meal in order to improve growth rate. In both studies, the influence of these diets on the levels of phosphorus discharged in the effluents was determined and compared with those for the commercial salmon feed, OMP.

TABLE 1.—Composition of experimental diets fed to coho salmon.

Ingredient	Diet: T2 %	Diet: T2M %
Blood flour	9.7	9.7
Corn gluten meal, 60%	29.5	29.5
Soybean meal, 48% (finely ground)	41.0	31.0
Herring meal (minimum protein 65%, minimum fat 10%)	0.0	10.0
Menhaden oil[a]	11.0	11.0
L-lysine · HCl, feed grade (98% as lysine · HCl)	0.4	0.4
Vitamin premix 30[b]	0.5	0.5
Choline, 60%	0.4	0.4
L-ascorbic acid	0.1	0.1
Ethoxyquin[c]	+	+
KCl, feed grade	0.5	0.5
NaCl, feed grade	0.5	0.5
Mineral mixture[d]	+	+
Defluorinated rock phosphate (finely ground)[e]	3.5	2.2[f]
Wheat middlings	2.9	to 100%

[a]Six percent oil added "inside" diet and 5% as "top dress" after diet was pelletized.

[b]Except for choline and ascorbic acid, U.S. Fish and Wildlife Service premix 30 provides vitamins at levels in Table 1 of Lemm (1983).

[c]Antioxidant at 120 mg/kg of diet.

[d]Mineral mixture provided the following (in mg/kg diet): Mn, 100; Zn, 100; Cu, 10; Fe (ferrous), 100; I, 5; Se (selenite), 0.1; Mg, 990.

[e]Defluorinated rock phosphate containing 18% P.

[f]Pending the phosphorus content of herring meal determined by analysis, the level of defluorinated rock phosphate was calculated to provide, in combination with herring meal, 0.66% phosphorus (i.e., the level of dietary P provided by defluorinated rock phosphate plus herring meal was 0.66%).

Methods

Two large-scale experiments were conducted with 6-month-old fingerling coho salmon at the Platte River Fish Hatchery, Beulah, Michigan, to determine the long-term effects of the T2 diet and a modification of it on phosphorus discharges in the hatchery effluent, growth rates of fingerlings up to smolt size, and on the survival and return of fish after release from the hatchery as marked smolts.

Experiment 1.—In experiment 1, two diets—T2 and OMP—were assigned in a completely randomized design and fed to fingerling coho salmon for 46 weeks. The commercially manufactured OMP diet (OP4 and OP2 formulas) served as a practical control. The composition of diet T2 is shown in Table 1. It contained by calculation 0.8% available phosphorus. Each diet was initially assigned at random to triplicate lots of 140,000 fingerlings (initial body weight, 5 g). Each lot of 140,000 fish consisted of two groups of 70,000 fish held in adjacent raceways and fed by a common automatic feeder. Because of hatchery constraints, the three replicate lots in each diet treatment were regrouped into two lots of 220,000 fingerlings after 11 (OMP) and 13 weeks (T2). Most regrouping was done by transferring fish from the initial third replicate groups and from an extra lot of fish fed the same diet. Total weights of fish and average weights were determined by weighing repeated samples of counted fish from each lot before and after regrouping of fish, to minimize the introduction of variation in the total average weights during regrouping.

After the feeding phase of the study was completed, 40,000 smolts fed the T-2 diet and 60,000 fed OMP were distinctively marked by removing the adipose fin and one of the ventral fins. Because of marking limitations, fish were marked according to diet (not replicates) and then released into the Platte River in the spring to migrate to Lake Michigan. Released fish that returned to the river to spawn during the following and subsequent fall seasons were captured in traps and enumerated according to their marks.

Experiment 2.—In experiment 2, the T2 diet was modified (T2M) to contain 10% fish meal (Table 1). The T2M diet contained 0.8% available phosphorus, by calculation. Cost of the T2M feed, commercially manufactured and delivered, ranged from US$0.40 to $0.68/kg for pellets and crumbles, whereas cost for the OMP ranged from $1.04 to $1.11/kg. Experiment 2 was conducted in two periods totaling about 45 weeks. Period 1 began on June 1 and lasted, on the average, about 17 weeks. Period 2 lasted about 28 weeks. There were four replicates for each diet during period 1 and three during period 2. Initially, fingerling coho salmon were stocked in each of eight pairs of raceways and all fish were fed the OMP diet for a few weeks to precondition the fish to their new environment. Each diet was assigned at random to four replicate lots of approximately 199,000 fish each. Fish having a mean weight of 4 g were reared in paired or adjacent raceways fed by a shared automatic feeder in a randomized complete block design (paired variates). Therefore, the mean of all the fish fed in two adjacent raceways fed by the same feeder was considered as one independent experimental replication or observation. At the end of period 1, fish were weighed according to blocks and regrouped into three blocks by transferring (according to diet) fish from the fourth block to the other three blocks. After the transfers, 231,000 fish remained in each replicate. These three blocks were retained until the end of the study. The average weight of coho salmon fed T2M was 14% more than that of fish fed OMP, both before and after regrouping.

In period 1, total phosphorus balance was measured in this study to estimate the total amount of phosphorus lost in the soluble and suspended fractions in the effluents. The total amount for each raceway was determined by computing the total amount of phosphorus fed (as determined from analysis of phosphorus in feed) and then subtracting the total phosphorus deposited in the fish carcasses (both dead and alive) and the total phosphorus lost in the settleable solid wastes.

Phosphorus was analyzed in the feed and in whole fasted coho salmon at the beginning and end of periods 1 and 2. Fish were fasted for 72 h to ensure that intestinal tracts were devoid of food phosphorus. Also, all settleable solid wastes in the sludge were collected and sampled from each raceway for 52 d during period 1 (from July 7 to August 28). The wastes were collected with a portable suction water pump that pumped settleable solid wastes in a slurry (approximately 1,100–3,500 L) from each raceway onto a truck with a large holding tank calibrated to determine the total volumes of slurries removed from each raceway. Total collections of settleable solids were obtained by pumping each raceway. On July 7 (day 1), we began the process by pumping each raceway to remove all the settleable solids. Afterward, pumping and sampling of the settleable solid wastes was done five times up to day 52. Samples were analyzed for phosphorus content, and total phosphorus removed from each raceway was computed. The weights of fish that died in each raceway were estimated to compute the loss of phosphorus by mortality.

Measurement of growth.—Mean weights were determined at the beginning of each experiment and periodically thereafter. The fish were not weighed in the summer because water temperatures were high—up to 19°C. Weights were determined by the conventional hatchery inventory method. This method involved weighing and counting three random samples of at least 5 kg of fish per raceway—from the influent end, middle, and effluent end. At the beginning and end of the experiment, we determined the total weight of all fish and computed the total number of fish from inventory data. Otherwise, the total weight was computed from the sample weights and counts, and the number of fish known to be in the raceway from the initial determination and mortality records.

Feeding levels.—Feed was fed to fish in pairs of raceways by common automatic feeders (model 2L-100-6-HS, Garon Company, Vancouver, Washington) that dispensed feed daily at 30-min intervals throughout the daylight hours in each experiment. In experiment 1, the daily amounts of OMP feed were determined by the hatchery constant (7.1) method of Buterbaugh and Willoughby (1967). The T2 diet was fed at the same level of dry matter. In experiment 2, OMP was fed to satiation as determined by the presence of wasted feed on the bottom of the raceways. The T2M diet was fed at the same level as OMP. During the winter months when water temperatures decreased markedly, feed was offered at 0.4% of body weight at a reduced frequency—21, 8, 5, 10, and 20 d per month during November, December, January, February, and March, respectively.

Chemical analyses.—Analyses of phosphorus in the diets (Table 2) and fish carcasses were made at the beginning and end of the feeding study. Analyses of phosphorus also were performed on periodic samples of sludge representative of the total sludge generated in each raceway of experimental fish. This was done in an attempt to estimate total phosphorus balance in the hatchery.

Whole carcasses of fish (fasted for 72 h), sludge,

TABLE 2.—Chemical analyses of diets; OMP is Oregon Moist Pellet (formula designation is in parentheses).

Diet	Granule (gn) or pellet		Protein (% dry matter)	Fat (% dry matter)	Ash (% dry matter)	Moisture (% as fed)	Phosphorus (% as fed)
	Designation	mm					
Experiment 1							
T2	1/16	1.6	44	13	8	9	1.14
	3/32	2.4	44	13	8	8	1.22
	1/8	3.2	45	12	8	9	1.18
	1/8	3.2	47	12	8	10	1.18
	1/8	3.2	47	12	8	10	1.22
OMP (OP-2)	1/16[a]	1.6	32	15	8	27	1.32
	3/32	2.4	35	13	8	30	1.22
	1/8	3.2	35	14	8	27	1.27
	1/8	3.2	36	14	8	29	1.23
Experiment 2							
T2M	gn 3	1.6	49	12	7	8	1.14
	gn 4	2.4	46	14	8	11	1.11
	3/32	2.4	47	13	7	10	1.08
	1/8	3.2	49	13	7	10	1.10
OMP (OP-2)	1/16[a]	1.6	40	14	8	30	1.33
	3/32	2.4	38	14	8	27	1.37
	1/8	3.2	36	15	8	29	1.46

[a]OMP 1/16 designation, OP-4 formulation.

and feeds were frozen and stored until analyzed. Samples were dried under vacuum with heat before chemical analyses were made. Phosphorus in diets, carcasses, sludge, and water was analyzed as follows. Each sample of diets, carcasses, and sludge was prepared by wet ashing with nitric acid over a hot plate (400°C), over a flame, and in a muffle furnace at 600°C for 16 h. Then the ash was cooled, wetted with 50% nitric acid, heated to 204°C, and ignited over flame until dry and burned to a gray-white ash. The ash was cooled and dissolved in hot 25% nitric acid (93°C) and quantitatively transferred to a volumetric flask. The dissolved ash was diluted to volume with deionized distilled water. An aliquot of each sample was then analyzed for phosphorus by the method of Kitson and Mellon (1944) modified to entail mixing an aliquot of sample with 1 mL of 2.5 N HNO_3, 1 mL of 0.25% NH_4VO_3 in 2% HNO_3, 1 mL of 5% $(NH_4)_6 Mo_7O_{24} \cdot 4H_2O$, and enough deionized water to make a total volume of 10 mL. After half an hour of color development, samples were read with a spectrophotometer at a wavelength of 490 μm against standard solutions made with K_2HPO_4.

Proximate analyses of the whole carcasses were conducted for crude protein (N × 6.25), fat, ash, and moisture. Moisture in carcass samples that were frozen, minced, and lyophilized was determined under reduced pressure in a refrigerated Virtis freeze dryer, model 10-100 (Gardiner, New York). Lyophilized samples were ground, dried again in a vacuum oven, and analyzed. Total ash content was determined by combustion of dried samples at 600°C for 2 h according to method 7,010 of the Association of Official Analytical Chemists (AOAC 1975). In experiment 1, crude protein was determined by the Kjeldahl method 2.049 (AOAC 1975), and fat was determined by extraction with diethyl ether in a Soxhlet apparatus for 16 h. In experiment 2, automated procedures for crude protein and fat were used after verification for accuracy by use of known calibration samples. Crude protein was analyzed by combusting the samples in pure oxygen and determinating gaseous nitrogen by thermal conductivity in an FP-228 Nitrogen Determinator, model 601-700 (Leco Corporation, St. Joseph, Michigan). Crude fat was extracted from samples placed in filter thimbles by boiling for 15 min and by rinsing for 45 min in hot diethyl ether in a Soxtec System HT6 (Tecator Inc., Herndon, Virginia).

Samples of influent and effluent water were collected periodically to determine the influence of diet on phosphorus levels in effluent water. Water samples were allowed to settle for 30 min in large vessels; the upper layer of water was then carefully decanted into acidified sample bottles and analyzed by a commercial laboratory for total phosphorus by the ascorbic acid method 4500-P,E (APHA et al. 1989).

TABLE 3.—Influence of low-phosphorus (T2) and Oregon Moist Pellet (OMP) diets on growth and carcass composition of coho salmon (experiment 1). Values are means ± SE of three replicates (140,000 fish per replicate) during weeks 1–10 and of two replicates (240,000 fish per replicate) thereafter. Asterisks denote significant differences between diets (*t*-tests, $P < 0.05$*, or $P < 0.005$**).

Time and variable	Diet	
	T2	OMP
Start (June 5)		
Body weight (g)	4.8	4.7
Week 10		
Body weight (g)	12 ± 0.09	15 ± 0.26*
Feed/gain (g/g)	1.53 ± 0.03	1.39 ± 0.07
Mortality (%)	4.0 ± 0.06	1.4 ± 0.06**
Week 46		
Body weight (g)	22 ± 1.25	26 ± 0.30*
Mortality (%)	20 ± 1.0	14 ± 4.1
Packed blood cell volume (%)	37 ± 0.5	38 ± 1.2
Carcass composition		
Protein (% of dry weight)	65 ± 0.7	62 ± 0.5
Fat (% of dry weight)	22 ± 0.04	26 ± 2.0*
Ash (% of dry weight)	10 ± 0.4	10 ± 0.1
Moisture (% of fresh weight)	76 ± 0.3	74 ± 0.9
Bone		
Ash (% fat-free dry)	46 ± 0.01	47 ± 0.5*
P (% fat-free dry)	9.6 ± 0.01	9.8 ± 0.07

Percentage of packed red blood cell volumes (hematocrits).—We determined hematocrits for 24 fish randomly sampled per raceway by microcentrifugation of blood for 10 min at 14,000 × gravity in heparinized microcapillary tubes (75 mm long, 1.0–1.2 mm inside diameter).

Statistical analyses.—The data were analyzed by the *t*-test in experiment 1 and by the *t*-test for paired variates in experiment 2. We performed *F*-tests for homogeneity of variances, and when variances were heterogeneous, the *t*-values (and degrees of freedom) were determined by the method for unequal variances of Snedecor and Cochran (1980).

Results and Discussion

Experiment 1

At 23 weeks, the weight of coho salmon fed the T2 diet was 15% less ($P < 0.005$) than that for fish fed the T2 diet and 20% lower ($P < 0.05$) at 46 weeks in experiment 1 (Table 3). Mortality of coho salmon fed the T2 diet was low (4%), but it was significantly greater than the 1.4% mortality at 10 weeks for fish fed the OMP diet. At 46 weeks, mortality did not differ significantly between treatments ($P > 0.05$). Salmon fed OMP had significantly more carcass fat than those fed the T2 diet. However, diet had no significant influence on packed blood cell volume, on carcass protein content or ash, or on phosphorus contents of bone or carcass. Ketola (1975) showed that deficiency of phosphorus in Atlantic salmon reduced growth and bone ash content. Andrews et al. (1973), studying channel catfish *Ictalurus punctatus*, demonstrated the same signs of phosphorus deficiency and a significant reduction in hematocrits (packed blood cell volumes) in one of two experiments. Therefore, the results presented in this report indicate that the level of available dietary phosphorus (0.8%) was adequate in the T2 diet. Although the minimum requirement of coho salmon for phosphorus is not known, the level of available phosphorus (0.8%) in the T2 diet meets the requirement of rainbow trout determined by Ogino and Takeda (1978), which further supports the belief that the T2 diet contained an adequate level of phosphorus.

TABLE 4.—Influence of low-phosphorus (T2 or T2M) and Oregon Moist Pellet (OMP) diets on concentrations of phosphorus in hatchery effluent. Asterisks denote significant differences between diets ($P < 0.05$*, $P < 0.01$**, $P < 0.005$***).

Week (month)	Inlet concentration (μg/L)	Effluent concentration (μg/L)			SE of difference
		T2 diet	OMP diet	T2M diet	
Experiment 1[a]					
4 (Jul)	10	30 ± 3.6	46 ± 12.1		
13 (Sep)	12	32 ± 6.5	113 ± 9.9***		
23 (Nov)	<2	16 ± 1.0	24 ± 1.1***		
30 (Dec)	<2	7 ± 1.0	7 ± 1.6		
Experiment 2					
8 (Jul)	26		41**	26	2.2
26 (Dec)	9		25*	13	1.5
38 (Feb)	12		19	8	5.1

[a]Effluent concentrations in experiment 1 are means ± SEs.

Phosphorus discharges into effluent waters from coho salmon fed the T2 diet were markedly and significantly lower than those for fish on the OMP diet at 13 and 23 weeks, but not at 4 and 30 weeks (Table 4). The probable reason for the lack of significance at 4 weeks was the high variability in the measurements for the OMP diet. At 30 weeks, phosphorus levels and feeding rates were both low because of the low water temperatures and reduced feeding rates during winter. Overall, the T2 diet markedly reduced phosphorus discharges into the effluents—probably by 33–80% compared with the OMP diet. These phosphorus discharge values and growth results agree closely with previous results of Ketola (1985) from feeding diet T2 (then called diet 7) and OMP to coho salmon during an 8-week pilot study.

During the first fall after 100,000 marked smolts were released to the wild, 94 marked jacks returned from Lake Michigan to the river in a T2:OMP treatment ratio of 0.71, approximately 6% higher than the ratio of 0.67 expected if the diet had had no effect on the rate of survival in the lake and returns to the river. During the second fall, 1,861 marked spawning fish returned in a ratio of 0.53. These ratios indicated that the rate of survival and return of coho salmon fed the T2 diet was between 79 and 106% of that for fish fed the OMP diet. The average weight of fish fed the T2 was 2.1 kg and that of fish fed the OMP diet was 2.2 kg. Coho salmon thus can apparently be fed the T2 diet that significantly reduces phosphorus discharges in hatchery effluents, and yet supports growth and survival in the hatchery and after release at rates close to that for conventional hatchery feed.

Experiment 2

When coho salmon were fed the T2 diet supplemented with 10% fish meal (diet T2M) for 45 weeks, their mean final body weight was 27 g—not significantly different from that of fish fed the OMP diet, 26 g (Table 5). Furthermore, diet had no significant influence on 45-week mortality, feed conversion, packed blood cell volume (hematocrit), or the protein, fat, ash, moisture, or phosphorus contents of the carcass. In contrast, ash, moisture, and phosphorus contents of carcasses were significantly increased (by 3–9%) in fish fed the OMP diet during the first part of the study (period 1). The reason for these small increases is not clear.

During period 1 of the study, the growth of coho salmon fed the T2M diet was approximately 20% greater ($P < 0.01$) than that for fish fed the OMP diet, and feed conversion (feed/gain) was numerically lower for fish fed the T2M diet (Table 4). However, when calculations were made to correct for differences in dietary moisture content, the resulting feed conversions (weight of dry feed/gain in weight of fish) were nearly identical—1.05 for diet T2M and 1.07 for OMP. Furthermore, the average dry weights of feed fed per replicate were 2,539 kg for the OMP diet and 3,211 kg for the T2M diet. Thus, the coho salmon fed the T2M diet received approximately 26% more feed (dry weight) than did those fed OMP. This explains why fish fed T2M gained more weight even though corrected feed/gain values were

TABLE 5.—Influence of low-phosphorus (T2M) and Oregon Moist Pellet (OMP) diets on growth and carcass composition of coho salmon (experiment 2). Values are means of four replicates (199,000 fish per replicate) at week 17 and of three replicates (231,000 fish per replicate) thereafter. Asterisks denote significant differences between diets (paired t-test, $P < 0.05$*, $P < 0.01$**).

Time and variable	Diet: T2M	Diet: OMP	SE of difference
Start (June 1)			
Body weight (g)	4	4	
Week 17 (end period 1)			
Body weight (g)	18	15**	0.41
Feed/gain (g/g)	1.15	1.48**	0.03
Mortality (%)	14	16*	0.59
Carcass composition			
Protein (% of dry weight)	52	51	0.60
Fat (% of dry weight)	35	35	0.50
Ash (% of dry weight)	7.7	8.4**	0.09
Moisture (% of fresh weight)	70	72*	0.31
P (% of fresh weight)	0.50	0.52**	0.01
Week 45 (end period 2)			
Body weight (g)	27	26	2.38
Feed/gain[a] (g/g)	4.6	4.9	0.69
Mortality (%)	21	23	0.35
Packed blood cell volume (%)	32	32	0.70
Carcass composition			
Protein (% of dry weight)	64	62	0.98
Fat (% of dry weight)	20	22	1.50
Ash (% of dry weight)	10	11	0.56
Moisture (% of fresh weight)	75	76	0.63
P (% of fresh weight)	0.53	0.53	0.03

[a]Feed/gain values apparently reflect wasted feed.

nearly identical. Therefore, the two feeds, T2M and OMP, appeared to have nearly the same nutritional value even though the T2M contained only 10% fish meal.

Concentrations of phosphorus in effluent waters from coho salmon fed the T2M diet were markedly and significantly lower than those for fish fed the OMP diet at 8 and 26 weeks, but not at 38 weeks (Table 4). The phosphorus concentrations in inlet water were as high as or higher than those in effluents for the T2M diet at 8 and 38 weeks, thereby reflecting some sampling or analytical variability. This variability may explain the reason for the lack of a significant difference at 38 weeks. Overall, the T2M diet significantly reduced phosphorus concentrations in the effluent water when compared with the OMP diet.

Measurements of total phosphorus balance (Table 6) showed that nearly 40% of the phosphorus consumed by coho salmon fed T2M was retained in the body tissues elaborated by surviving fish during the experiment—significantly ($P < 0.001$) more than that for fish fed OMP (26%). Only about 2.4–2.7% of the phosphorus consumed was lost in the carcasses of fish that died during the study. If the amounts of phosphorus retained in the fish that died during the study were combined with those that survived, approximately 42 and 28% of the dietary phosphorus was retained (or deposited) in the carcasses of fish fed the T2M and OMP diets, respectively. The combined retention value for the T2M diet (42%) is considerably lower than values (68–81%) calculated from data reported by Ogino and Takeda (1978) for rainbow trout fed a purified diet containing only available phosphorus at or near the minimum required level. For a better comparison, retention of phosphorus by coho salmon fed the T2M was computed on the basis of available phosphorus consumed (0.8% of diet) instead of total phosphorus (1.1% of diet). On this basis, coho salmon retained 58% of phosphorus consumed in the T2M diet, which approaches but is still less than the values calculated from data for rainbow trout as reported by Ogino and Takedo (1978). This difference may be related to differences between species of fish, diet, or growth rates. The rainbow trout appeared to grow only at 41–48% of their potential based on their calculated temperature unit growth rates at 15 and 18°C (0.0027–0.0032 cm/degree-day) when compared with the higher growth rate (0.0066 cm/degree-day) for rainbow trout reared at 15°C as reported by Westers (1987).

Settleable solid waste accounted for 19.4% of the phosphorus consumed by coho salmon fed T2M—significantly ($P < 0.02$) more than that for the OMP treatment (12%). By difference, it was computed that about 38% of the phosphorus in the T2M diet was discharged in the effluents as soluble and suspended phosphorus, which was significantly ($P < 0.001$) less than that in the OMP diet (about 60%). When the differences in dietary phosphorus contents and the total phosphorus intakes are considered, the differences in the amounts of phosphorus discharged in the effluents become even greater. Overall, the amounts of phosphorus discharged, in dissolved and suspended form in the effluent, per 1,000 kg of feed were about 4.2 kg for the T2M diet and 8.2 kg for the OMP diet. When expressed as amount per 1,000 kg of fish production, these values were 4.8 kg for the T2M diet and 12.2 kg for the OMP diet. Therefore, feeding the T2M diet to coho salmon in the hatchery significantly ($P < 0.01$) reduced the amounts of phosphorus discharged in the effluents—by approximately 49 or 61%, depending on the method of comparison with the OMP diet. These discharge values compare favorably with values obtained in a laboratory study by Ketola

TABLE 6.—Influence of feeding low-phosphorus (T2M) and Oregon Moist Pellet (OMP) diets to coho salmon on discharge of phosphorus in hatchery effluents and on phosphorus balance (experiment 2). Asterisks denote significant differences between diets ($P < 0.02$*, $P < 0.01$**, $P < 0.001$***).

Variable (period 1)	Diet		SE of difference
	T2M	OMP	
Effluent			
Settleable solids (sludge), dry matter			
Percent of feed weight	12.6	11.6	1.68
P content (%)	1.59	1.57	0.12
Phosphorus discharge, kg/1,000 kg production			
"Dissolved"[a] in water	4.8	12.2**	0.64
Settleable solids	2.5	2.4	0.24
Phosphorus discharge, kg/1000 kg of feed			
"Dissolved"[a] in water	4.2	8.2***	0.26
Settleable solids	2.1	1.7	0.19
Total phosphorus balance			
P in feed consumed			
kg P/replicate	38.7	51.3	
% of feed P	100	100	
P deposited in new tissues of surviving fish			
kg P/replicate	15.3	13.3*	0.43
% of feed P	39.7	26.0***	0.42
P deposited in new tissues of fish that died			
kg P/replicate	1.03	1.22**	0.026
% of feed P	2.7	2.4	0.12
P in settleable solids (sludge)			
kg P/replicate[b]	7.51	6.16	0.75
% of feed P[c]	19.4	12.0*	1.47
P in effluent[d]			
kg P/replicate	14.8	30.6**	1.78
% of feed P	38.2	59.6***	1.72
P in waste (effluent + settleable)			
kg P/replicate	22.2	36.7**	1.53
% of feed P	57.5	71.5***	0.35

[a]Dissolved and suspended in water.

[b]Computed (extrapolated) from the amounts of P in sludge during July 8 to August 28 expressed as a percentage of feed P.

[c]Measured for the period July 8 to August 28.

[d]Determined by difference: P dissolved and suspended in effluent = feed P − (P gained by survivors + P gained by dead fish + settleable P).

(1982) with rainbow trout fed similar diets. In the laboratory study, 5.7 kg of phosphorus were discharged into the effluents per 1,000 kg of production when rainbow trout were fed an experimental diet as compared with 15 kg when they were fed a commercial salmon feed. When laboratory values were expressed per 1,000 kg of feed fed, 4.5 kg of phosphorus were discharged in the effluent when rainbow trout were fed the experimental diet, as compared with 12.4 kg when they were fed the commercial feed (Ketola 1982).

Settleable solid wastes amounted to about 12–

13% of the weight of feed fed; the phosphorus content of this waste was about 1.6% of the dry matter and was not significantly influenced by diet (Table 6).

At the end of the experiment, fish were marked and released; however, data on survival after release and recapture are not yet available.

In summary, two production-scale hatchery experiments with coho salmon fingerlings showed that use of two economical diets (T2 and T2M) containing little excess phosphorus markedly reduced phosphorus discharges in the effluent water. The T2 diet contained no fish meal and supported growth equal to about 85% of that for a conventional hatchery feed (OMP). Coho salmon fed the T2 diet survived in the hatchery and in the wild after release to Lake Michigan at rates comparable to those of fish fed OMP. Supplementing the T2 diet with 10% fish meal (T2M diet) improved growth and survival of coho salmon reared in the hatchery to the levels supported by the OMP diet and reduced phosphorus discharges by 49–61%.

References

Andrews, J. W., T. Murai, and C. Campbell. 1973. Effects of dietary calcium and phosphorus on growth, food conversion, bone ash and hematocrit levels in catfish. Journal of Nutrition 103:766–771.

AOAC (Association of Official Analytical Chemists). 1975. Official methods of analysis, 12th edition. AOAC, Washington, D.C.

APHA (American Public Health Association), American Water Works Association, and Water Pollution Control Federation. 1989. Standard methods for examination of water and wastewater, 17th edition. APHA, Washington D.C.

Buterbaugh, G. L., and H. Willoughby. 1967. A feeding guide for brook, brown, and rainbow trout. Progressive Fish-Culturist 29:210–215.

Grant, J. 1979. Water quality and phosphorus loading analysis of Platte Lake 1970–1978. Michigan Department of Natural Resources, Water Quality Division, Publication 4833-9792, Lansing.

Kanaga, D., and E. D. Evans. 1982. The effect of the Platte River Anadromous Fish Hatchery on the fish, benthic macroinvertebrates and nutrients in the Platte Lake. Michigan Department of Natural Resources, Water Quality Division, Lansing.

Ketola, H. G. 1975. Requirement of Atlantic salmon for dietary phosphorus. Transactions of the American Fisheries Society 104:548–551.

Ketola, H. G. 1982. Effect of phosphorus in trout diets on water pollution. Salmonid 6(2):12–15.

Ketola, H. G. 1985. Mineral nutrition: effects of phosphorus in trout and salmon feeds on water pollution. Pages 465–473 *in* C. B. Cowey, A. M. Mackie, and J. G. Bell, editors. Nutrition and feeding in fish. Academic Press, London.

Kitson, R. E., and M. G. Mellon. 1944. Colorimetric determination of phosphorus and molybdivanadophosphoric acid. Industrial and Engineering Chemistry Analytical Edition 16:379–383.

Lemm, C. A. 1983. Growth and survival of Atlantic salmon fed semimoist or dry starter diets. Progressive Fish-Culturist 45:72–75.

Ogino, C., and H. Takeda. 1978. Requirements of rainbow trout for dietary calcium and phosphorus. Bulletin of the Japanese Society of Scientific Fisheries 44:1019–1022.

Snedecor, G. W., and W. G. Cochran. 1980. Statistical methods, 7th edition. Iowa State University Press, Ames.

Westers, H. 1987. Feeding levels for fish fed formulated diets. Progressive Fish-Culturist 49:87–92.

American Fisheries Society Symposium 10:410–416, 1991

Ration Optimalization to Reduce Potential Pollutants—Preliminary Results

TORBJØRN ÅSGÅRD[1] AND ROY M. LANGÅKER
Institute of Aquaculture Research (AKVAFORSK)
The Agricultural Research Council of Norway
N 6600 Sunndalsøra, Norway

KARL D. SHEARER
National Marine Fisheries Service, Northwest Fisheries Center
2725 Montlake Boulevard East, Seattle, Washington 98112, USA

ERLAND AUSTRENG AND ARNE KITTELSEN
Institute of Aquaculture Research (AKVAFORSK)
The Agricultural Research Council of Norway
N 6600 Sunndalsøra, Norway

Abstract.—The aim of the study was to use the nutrient and energy budgets from a practical farming situation for rainbow trout *Oncorhynchus mykiss* and Atlantic salmon *Salmo salar* to show how ration size and feed utilization determined the amount of nutrients in the effluent water. Separate budgets for nitrogen, phosphorus, and energy were determined showing the fractions of the effluent nutrients that came from feed loss, feces, and other excreta. The interrelationships between these fractions were compared between a group of fish whose ration size was well adapted to its biomass and whose feed loss was low and a group of fish fed in excess whose feed loss was high. This comparison indicated the uncertainty in composition of the effluent water in situations where the amount of feed loss is not determined. A Triangle filter was installed for filtering the effluent water, and the filter retained sludge containing about 0.1% dry matter, which represented 19% of the effluent nitrogen, 23% of the phosphorus, and 31% of the energy. This was less than the amount of these nutrients in the feed loss. Overall, the flow diagrams pinpointed how optimalization of ration level and feed composition contribute to reduce pollution.

Weston (1986) summarized studies in which estimates of total effluent loadings from salmonid culture were quantified on the basis of the weight of fish, annual fish production, and amount of feed provided. In other studies, the composition of the effluent was measured without evaluation of whether pollutants originated from feed loss or excreta, or whether the excreta were feces or other excreta (Bergheim et al. 1982, 1984). Such differentiation in amount of effluent under different production conditions has been done theoretically (Maroni 1985; Åsgård 1986; Stigebrant 1986).

In this study we partitioned the nutrient budget under hatchery conditions. We wanted to measure, indirectly, the amount of phosphorus, nitrogen, and energy (closely related to potential oxygen consumption or chemical oxygen demand, COD) in the effluent water and show the amounts contributed from feed loss, feces, and other excreta. It is necessary to distinguish between these contributors to nutrient effluent when measures for effluent reduction are to be developed or evaluated. It is also important to consider the ratio between bound and dissolved nutrients. In this study we did not look further into the effluent's pollution potential, which will depend on recipient characteristics such as volume, water exchange, surface area, and topography (Stigebrant 1986).

Methods

The experiment was carried out at AKVAFORSK, Sunndalsøra, Norway, in 30 tanks (1.5 m × 1.5 m) in a barn supplied with artificial light 24 h/d. The fish were production lots of Atlantic salmon *Salmo salar* L. and rainbow trout *Oncorhynchus mykiss*, which were grouped according to size into six groups (Table 1). The water depth in each tank was 25 cm. The tanks received a constant, total water flow of 1,000 L/min; the amount of water flowing into each tank was adjusted according to the biomass to maintain the amount of oxygen in the outlet from each tank

[1]This paper was prepared while the author was on leave to the National Marine Fisheries Service, Northwest Fisheries Center in Seattle, Washington, USA.

TABLE 1.—Growth data for rainbow trout and Atlantic salmon in six experimental groups.

Fish group	Number of tanks	Initial mean weight (g)	Total weight (kg)	
			Initial	Final
		Rainbow trout		
1	2	906	58	69
2	12	85	284	459
3	3	35	71	125
		Atlantic salmon		
4	5	168	151	188
5	2	52	42	56
6	6	14	124	178
Total	30		730	1,074

over 7 mg/L. The mean temperature was 8.7°C. The growth experiment lasted for 28 d and was succeeded by a feed digestibility trial that lasted for 7 d. Twice a week the tanks were cleaned. A Triangle filter (TF-24RS from AB Hydrotec, Sweden) with 63-μm mesh size was used for collecting particles (sludge) in the effluent water.

The number and size of Atlantic salmon and rainbow trout in the six groups are shown in Table 1. Fish were weighed at the start and at the termination of the experiment. Automatic disc feeders fed the fish continuously 24 h/d. Feeding amounts for the individual groups of fish were based on the preceding 2 weeks' growth and on published growth tables (Austreng et al. 1987). The amount of feed fed to each tank was adjusted daily and the feed used was commercial TESS Elite feed from T. Skretting A/S (Table 2).

Initial and final samples for proximate composition of fish were collected. The feed was identical in the growth study and in the digestibility study, and the feed samples were collected along with fecal samples from the digestibility study.

Chemical analyses of feed, feces, fish, and sludge were carried out according to standard procedures. Nitrogen was determined as Kjeldahl-N, fat by HCl–ether extraction, and phosphorus by spectrophotometry. Langåker (1988) presented some additional analysis from this study. Calculated energy contents of samples were confirmed by bomb calorimetry done on the same samples. For effluent sludge material caught by the Triangle filter, the energy content was determined by bomb calorimetry.

The following standard equations and calculations were used in this study. Feed conversion was calculated as (kilograms feed offered)/(kilograms fish produced). Feed efficiency was calculated as (kilograms fish produced)/(kilograms feed offered). Feed loss was (feed offered) − (feed eaten). Digested nutrient was (nutrient eaten) × (digestibility coefficient for the nutrient). Gross energy was calculated with the factors presented by Phillips and Brockway (1959): 39.5 kJ/g fat, 23.6 kJ/g protein, and 17.2 kJ/g carbohydrate (fish, feed, and feces), or it was, measured by bomb calorimetry (sludge). Metabolizable energy (ME) consumed was [(gross energy eaten) × (digestibility coefficient for the energy)] − (energy loss from deamination and ammonia excretion). This loss was, accordingly to Elliot and Davidson (1975), 3.97 kJ/g protein, which was digested but not deposited in the fish. The amount of feed eaten for producing 1 kg of fish was considered to be constant for each of the two species if the chemical composition did not change during the study and was equal for the different groups of fish.

TABLE 2.—Chemical composition of feed and fish, and apparent feed digestibility coefficients (FDC), for Atlantic salmon and rainbow trout.

Component	Atlantic salmon	Rainbow trout	Feed	FDC[a] Atlantic salmon	FDC[a] Rainbow trout
Dry matter (%)	28.2	27.8	91.8	76	74
% of dry matter					
Ash	7.9	7.6	7.5	29	32
Crude protein	57.7	51.2	50.7	85	88
Crude fat	33.5	39.2	15.5	93	90
Carbohydrates			26.3[b]	40[c]	46[c]
Total phosphorus	1.46	1.34	1.23	48	48
Total energy (kJ/g)	25.8	26.4	22.5	75	78

[a]FDC = 100 − 100 (% nutrient in feces/% nutrient in feed) × (% Cr in feed/% Cr in feces).
[b]Calculated as 100 − (% protein + % fat + % ash).
[c]Calculated from the equation (% carbohydrates × FDC) = (% dry matter × FDC) − [(% protein × FDC) + (% fat × FDC) + (% ash × FDC)].

TABLE 3.—Feed offered, estimated feed eaten, feed loss, fish produced, feed conversion, and feed efficiency for six experimental groups of rainbow trout and Atlantic salmon.

Feed and production statistic	Total	Rainbow trout group			Atlantic salmon group		
		1	2	3	4	5	6
Feed offered (kg)	407	40.5	204.1	51.0	48.5	14.1	48.9
Feed eaten (kg)	308	10.2	159.8	49.3	31.2	11.6	45.6
Feed loss (%)	24	75	22	3	36	18	7
Fish produced (kg)	344	11.1	174.3	53.8	36.8	13.6	54.1
Feed conversion[a]	1.18	3.65	1.17	0.95	1.32	1.04	0.9
Feed efficiency[b]	0.85	0.27	0.85	1.06	0.76	0.97	1.11

[a]Feed conversion = kg feed offered/kg fish produced.
[b]Feed efficiency = kg fish produced/kg feed offered.

Calculations of the amount of feed eaten when 1 kg of Atlantic salmon or rainbow trout was produced were based on the feed consumption and the fish production in the tanks where the highest feed efficiencies were achieved for both species. Feed loss was assumed to be zero in these tanks. The digestibility coefficients were calculated according to the indirect Cr_2O_3 method described by Austreng (1978).

The amounts of nitrogen and phosphorus in nonfecal excrement were calculated as the difference between the amount digested and the amount deposited in the fish. Budgets for energy, phosphorus, and nitrogen were calculated according to the equations above and the chemical composition and energy content.

Results and Discussion

The mean initial fish density was 57 kg/m^3. This increased to 85 kg/m^3 by the termination of the experiment. Chemical composition of the fish and feed as well as feed digestibility coefficients are shown in Table 2. The amount of feed fed and fish produced were 407 kg and 344 kg, respectively, which gives a feed conversion of 1.18 and feed efficiency of 0.85 (Table 3). In the tanks where the feed loss was considered to be zero, the amount of ME required for producing 1 kg of rainbow trout was 13.8 MJ, which is close to the 13.5 MJ reported by Austreng (1976b). We found that 12.5 MJ were required for producing 1 kg of Atlantic salmon. According to the amount of fish produced and the amount of ME required for producing 1 kg of fish, we estimated that about three-fourths of the feed offered was eaten and one-fourth was lost. This shows that there is a considerable feed loss with the feed conversion achieved in this experiment.

There were big differences in performance among the groups of fish. For most of the groups there was a close relationship between predicted growth and the growth achieved. Group 3, which performed well, had a feed conversion of 0.94, a feed efficiency of 1.06, and only 4% feed loss. On the other hand, group 1 showed poor growth. This is assumed to be the result of stress that the fish underwent immediately before the experiment started. The authors were first informed about the stress situation after the experiment was ended. This was not accounted for when growth was predicted and the amount of feed was calculated, and nearly no feed loss was observed because water flows were high and the tanks were self-cleaning. Feed conversion was 3.68, feed efficiency was low (0.26), and as much as 76% of the feed was lost. This situation is parallel to that of practical farming conditions in which the biomass estimate has to be reestablished after unforeseen stress occurs to prevent excess feeding for a longer period of time.

In fresh water and seawater, phosphorus and nitrogen, respectively, are usually the limiting nutrients for algal blooms under Norwegian conditions. Nutrient discharge from the tanks was 9.8 g of phosphorus and 66.2 g of nitrogen per kilogram of fish produced. Figure 1 shows the difference between group 1 and group 3 in phosphorus loss and retention. Corresponding levels for nitrogen are shown in Figure 2. The amounts show similar patterns for the two nutrients and confirm how important it is not to overfeed. It has been shown that overfeeding does not result in increased growth (Storebakken and Austreng 1987a, 1987b). To avoid excess feeding but still achieve maximum growth, guidelines for expected growth of Atlantic salmon and rainbow trout (Austreng et al. 1987) are useful.

The compositions of the effluent from group 1 and group 3 were remarkably different. For group 1, feed loss dominated, whereas feces and other

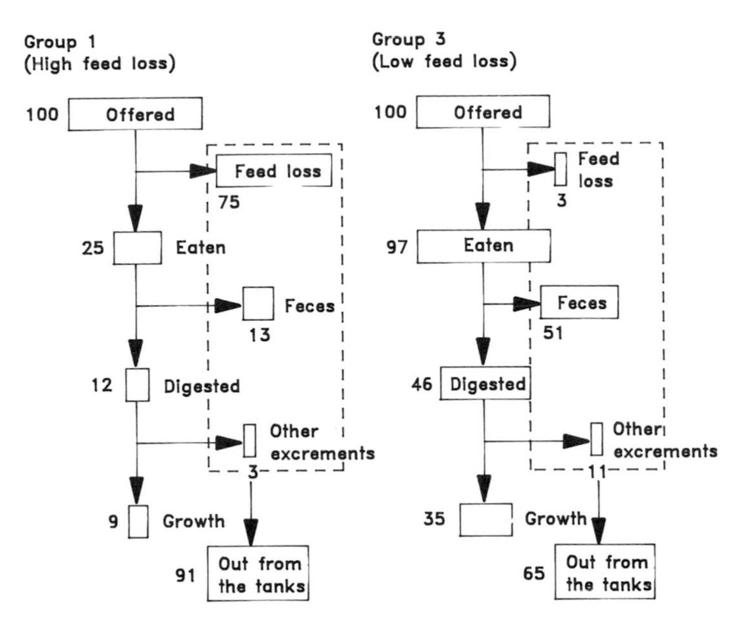

FIGURE 1.—Phosphorus retention and loss by rainbow trout, expressed as percentages of the amount offered: comparison between group 1 (high feed loss) and group 3 (low feed loss).

excreta dominated losses from group 3. Phosphorus and nitrogen in other excreta (e.g., from gills or in urine) are easily dissolved in the water, whereas the phosphorus and nitrogen in feces and feed loss dissolve to a lesser degree. Figures 1 and 2 show differences between the two nutrients in the way they are excreted. Whether the nutrient is dissolved or particle-bound is important for the efficient operation of a cleaning device. The efficiency of the Triangle filter used in this study depended on the level of particle-bound nutrients and the size and the strength of the particles. The filtered sludge contained 19% of the effluent nitrogen, 23% of the phosphorus, and 31% of the

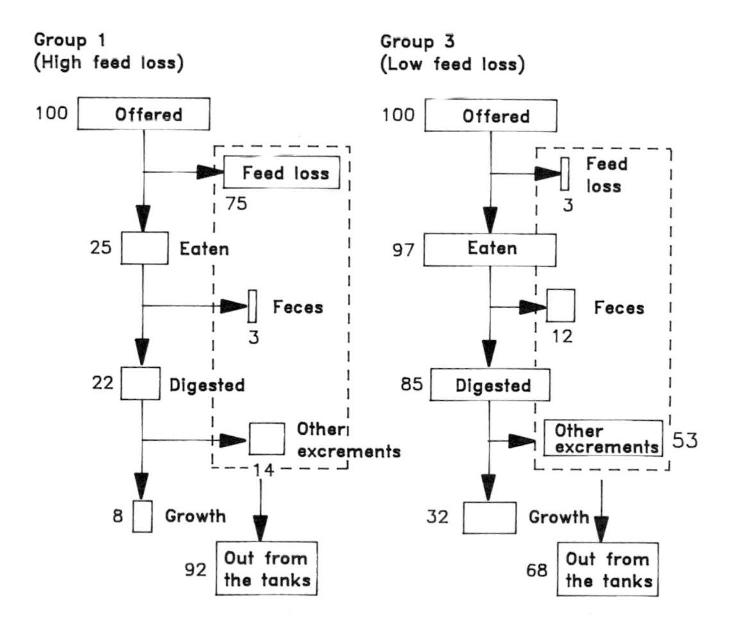

FIGURE 2.—Nitrogen retention and loss by rainbow trout, expressed as percentages of the amount offered: comparison between group 1 (high feed loss) and group 3 (low feed loss).

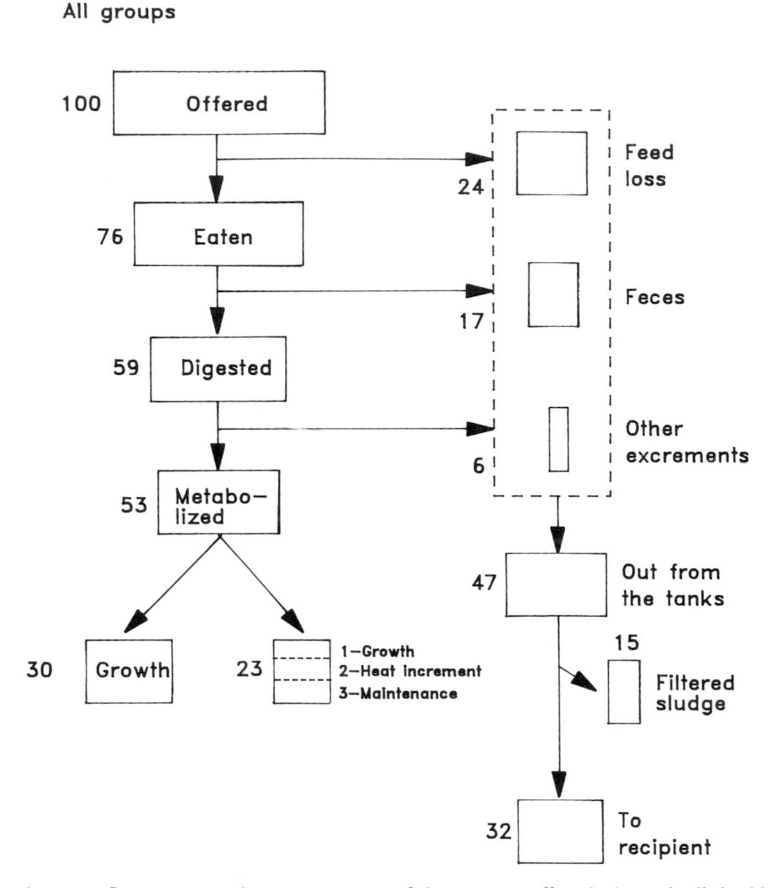

FIGURE 3.—Total energy flow expressed as percentages of the amount offered, through all the 30 tanks containing the six groups of rainbow trout and Atlantic salmon.

energy. The filter should be efficient in a situation where there is a high feed loss and much less efficient when feed loss is low. Improved retention of lost food on the filter will depend on factors such as the consistency of feed, time from feed release until recovery and type of current in the pipe. When feces are the major effluent source, the efficiency of the filter can be improved by feed manipulation that improves feces stability. In general, emphasis should be placed on preventing the nutrients from entering the water rather than trying to extract them afterwards.

Figure 3 shows that we collected less energy in the sludge than the amount present in the feed loss. For group 1, 92% of the energy in the effluent represented feed loss, whereas only 12% of the effluent group-3 energy was from feed loss (Figure 4). If it is assumed that the oxygen consumption during degradation of the organic matter in the effluent was 1.1, 2.8, and 1.4 g oxygen per gram of carbohydrate, fat, and protein, respectively (Åsgård 1986), the potential oxygen consumption during degradation of the effluent was 620 g oxygen per kilogram fish growth. More than 60% of this was due to feed loss even though the feed conversion was 1.18. Feed has a high energy content and 1 kg of dry feed demands 45% more energy for degradation than 1 kg of dry feces. This is due to the high digestibility of the energy-rich protein and fat and poorer digestibility of carbohydrates and ash. If we want to know the potential oxygen consumption from the effluent, we will have to know the relative contributions of the feed and the excreta.

In an optimal diet all the nutrients are absorbed and few are excreted, but we must deal with available feedstuffs at acceptable prices and with the conversion losses that exist. For salmonids, the optimum protein level is between 35 and 50% of dry matter, depending on energy concentration

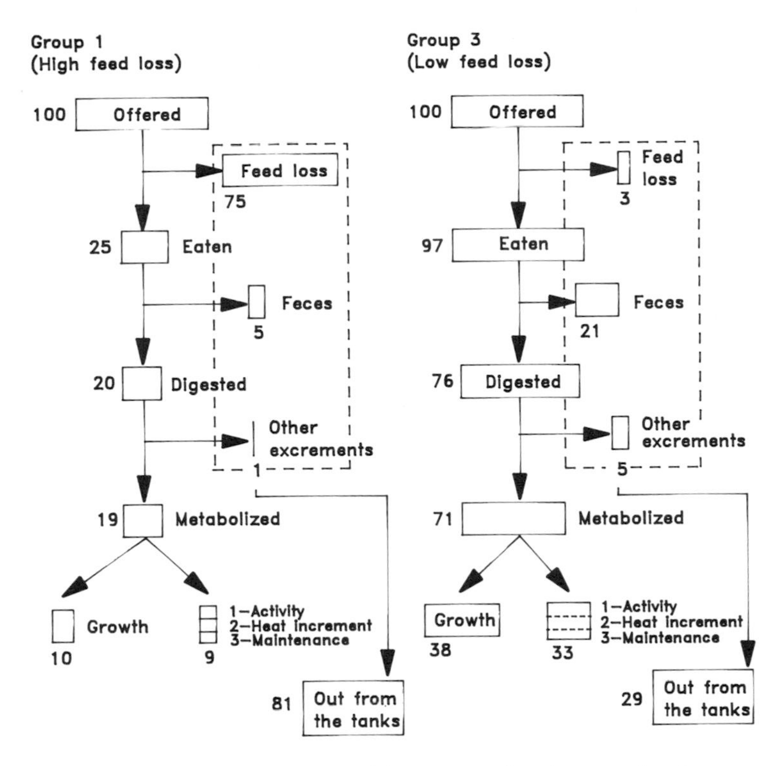

FIGURE 4.—Energy flow, expressed as percentages of the amount offered: comparison between group 1 (high feed loss) and group 3 (low feed loss) of rainbow trout.

in the diet and life stage of the fish (Austreng et al. 1988). In our study, group 3 retained 32% of the nitrogen, whereas retentions as high as 40–45% have been reported for practical diets (Austreng 1979; Storebakken and Austreng 1987b). It seems difficult to avoid losses of ammonia, but optimal feeding (Storebakken and Austreng 1987a, 1987b) and use of high-fat energy and low-protein feeds (Austreng 1976a, 1976b, 1979; Andorsdottir 1986) will be important in reducing this loss. Carbohydrates contributed considerably to the fecal loss of energy, because their digestibility was below 50% in this experiment. To reduce the amount of effluent, only highly digestible carbohydrates should be used and then only in low amounts. Our dry feed contained 12.3 g phosphorus per kilogram dry matter, which means that we spent 13.6 g phosphorus per kilogram fish growth while only 3.8 g was recovered in the fish even though the digestibility was 48%. According to Wiesmann et al. (1988), 4 g of phosphorus per kilogram of dry matter are sufficient at a feed conversion of 1 kg dry matter per kilogram of fish growth. The apparent retention has to be close to 100% if this amount of phosphorus in the diet is to be sufficient for normal growth. As long as there is sufficient phosphorus in the diet, deposition will be determined by digestibility and the requirements for maintenance and growth. This explains the low retention we observed.

When the fish are fed more than is required for growth, feed will be lost; digestibility of the nutrients is then of less importance than when feeding is optimal. A diet resulting in low amounts of excreta is important for reducing pollution. As long as practical feed conversions are between 1.5 and 2.0, however, it is more important to keep feed loss to a minimum and adjust the ration according to what the fish requires.

Acknowledgments

This study was part of a project carried out in cooperation with two additional Norwegian institutions, the Norwegian Institute for Water Research and the Institute for Georesources and Pollution Research. Langåker (1988) has presented some results from this study in his University thesis (in Norwegian).

References

Andorsdottir, G. 1986. Protein quality, methionine supplementation and fat levels in starter diets for salmon fry. Thesis. University of Oslo, Oslo.

Åsgård, T. 1986. Forureining frå smoltanlegg—fôrspill eller gjødsel—eksempel. Norsk Fiskeoppdrett 11(7–8):50–51.

Austreng, E. 1976a. Fat and protein in diets for salmonid fishes. I. Fat content in dry diets for salmon parr (*Salmo salar*, L.). Meldinger fra Norges landbrukshøgskole 55(5):1–16.

Austreng, E. 1976b. Fat and protein in diets for salmonid fishes. III. Different types of fat in dry diets for rainbow trout (*Salmo gairdneri*, Richardson). Meldinger fra Norges landbrukshøgskole 55(7):1–17.

Austreng, E. 1978. Digestibility determination in fish using chromic oxide marking and analysis of contents from different segments of the gastrointestinal tract. Aquaculture 13:265–272.

Austreng, E. 1979. Fett og protein i fôr til laksefisk. VI. Fordøyelighet og fôrutnyttelse hos regnbueaure (*Salmo gairdneri*, Richardson) ved ulikt fettinnhold i fôret. Meldinger fra Norges Landbrukshøgskole 58(6):1–12.

Austreng, E., B. Grisdale-Helland, S. J. Helland, and T. Storebakken. 1988. Feed evaluation and nutritional requirements. 6. Farmed Atlantic salmon and rainbow trout. Livestock Production Science 19: 369–374.

Austreng, E., T. Storebakken, and T. Åsgård. 1987. Growth rate estimates for cultured Atlantic salmon and rainbow trout. Aquaculture 60:157–160.

Bergheim, A., H. Hustveit, A. Kittelsen, and R. Selmer-Olsen. 1984. Estimated pollution loadings from Norwegian fish farms. II. Investigations 1980–1981. Aquaculture 36:157–168.

Bergheim, A., A. Sivertsen, and A. R. Selmer-Olsen. 1982. Estimated pollution loadings from Norwegian fish farms. I. Investigations 1978–1979. Aquaculture 28:347–361.

Elliot, J. M., and W. Davidson. 1975. Energy equivalents of oxygen consumption in animal energetics. Oecologia (Berlin) 19:195–201.

Langåker, R. M. 1988. Forureining fra smolt- og setjefiskanlegg. Thesis. Agricultural University of Norway, Ås.

Maroni, K. 1985. Forurensning fra fiskeoppdrett i relasjon til fôrtilførsel. Norwegian Institute for Water Research, Information Report 0-85 266, Oslo.

Phillips, A. M., and D. R. Brockway. 1959. Dietary calories and the production of trout in hatcheries. Progressive Fish-Culturist 21:3–16.

Stigebrant, A. 1986. Modellberäkningar av en fiskodlings miljøbelastning. Norwegian Institute for Water Research, Information Report 1823, Oslo.

Storebakken, T., and E. Austreng. 1987a. Ration level for salmonids. I. Growth, survival, body composition, and feed conversion in Atlantic salmon fry and fingerlings. Aquaculture 60:189–206.

Storebakken, T., and E. Austreng. 1987b. Ration level for salmonids. II. Growth, feed intake, protein digestibility, body composition, and feed conversion in rainbow trout weighing 0.5–1.0 kg. Aquaculture 60:207–221.

Weston, D. P. 1986. The environmental effects of floating mariculture in Puget Sound. University of Washington, Seattle.

Wiesmann, D., H. Scheid, and E. Pfeffer. 1988. Water pollution with phosphorus of dietary origin by intensively fed rainbow trout (*Salmo gairdneri*, Richardson). Aquaculture 69:263–270.

American Fisheries Society Symposium 10:417–420, 1991

Ozone for Disinfecting Hatchery Rearing Water

DAVID E. OWSLEY

*U.S. Fish and Wildlife Service, Dworshak National Fish Hatchery
Post Office Box 18, Ahsahka, Idaho 83520, USA*

Abstract.—Ozone has been widely used in Europe for many years to disinfect water. Ozone is a viable means of disinfecting water for aquaculture. It has been proven effective against fish and human pathogens at very low dosage rates. Ozone is not as restricted as chlorination or ultraviolet radiation in the quality of water that can be treated. On-site production of oxygen will double the capacity of ozone generators and reduce costs accordingly as compared with compressed air systems. It will also eliminate the buildup of nitrous oxide and reduce annual maintenance. The aquatector could improve efficiency and eliminate the need for a contact chamber. Packed columns and stripping towers remove residual ozone and eliminate the cost of detention tanks.

Ozone is a three-atom allotrope of oxygen. It is a colorless gas and can be readily detected by its odor at very low concentrations. It is an unstable gas and the strongest oxidizing agent commercially available. Ozone is produced by passing air or oxygen through a high-frequency electric field. Ozone has to be generated at the point of application.

Commercial ozone generators are available in many designs. Most generators use the high-voltage corona discharge system. This system consists of two surfaces separated by a space. A high voltage is impressed across this space. Air or oxygen is passed between the surfaces, where the oxygen molecules are excited sufficiently to form ozone. Ozone can be produced by ultraviolet radiation. Ultraviolet radiation does not produce the quantity of ozone that an ozone generator can achieve. Ozone is also produced by lightning or processes such as welding that bring air and high voltages together.

Ozone is directly toxic to aquatic organisms and to humans. Ozone is highly effective as a disinfectant, having about twice the oxidizing capability of chlorine. Ozone is toxic on contact, killing both bacteria and viruses with equal effectiveness and speed. Ozone reacts very quickly compared with compounds such as chlorine. Ozone effectiveness is much less affected by pH and temperature than chlorine. Use of ozone as a disinfectant is not limited to low-turbidity water, as is use of ultraviolet radiation. Ozone does not appear to leave harmful residues in water, such as the chloramines produced by chlorine.

Both organic and inorganic materials exhibit a demand for ozone, which is used to remove color, odor, and turbidity. Water containing organic matter must be treated with a higher ozone level than similar water without organic matter to achieve the same disinfection level. Inorganic materials such as iron and manganese can be oxidized to insoluble oxide forms by ozone.

Ozone decomposes back to oxygen. The decomposition rate is temperature-dependent, rapidly increasing with increased temperature. An ozone destruction unit is basically a heater. Ozone has been widely used in Europe for many years for water disinfection. Europe does not have the luxury of abundant clean water that exists in the USA and Canada. The use of chlorine and ultraviolet radiation is more economical for clean water, but they are ineffective against certain diseases. One example for humans is giardiasis, which is caused by the protozoan *Giardia lamblia* commonly found in high-mountain streams and lakes. Ozone kills *G. lamblia* and its cysts, whereas chlorine is ineffective. An example in aquaculture is the infectious hematopoietic necrosis (IHN) virus. This epizootic virus causes high mortality, and fish that survive become carriers. Adult fish shed the virus during spawning, and natural transmission occurs through the water. External symptoms of fish with IHN include hemorrhaging under the skin, exophthalmia (protruding eyes), swollen abdomens, lethargy, darkening of skin color, and hemorrhaging at the base of the fins. Internally, the liver, spleen, and kidneys are usually pale. The stomach and intestine may be filled with fluid.

In recent years, the IHN virus has plagued the Pacific Northwest and killed millions of juvenile salmonids, including many at Dworshak National Fish Hatchery. This hatchery, at the confluence of the North Fork of the Clearwater River and the main-stem Clearwater River in north central Idaho, is the world's largest hatchery for steelhead *Oncorhynchus mykiss*; annual production is approximately 2.9 million steelhead smolts weighing over 190,000 kg. The virus was identified in the hatchery in 1982 when 48% of the juvenile fish were lost in the nursery building. In 1983, the hatchery suffered a 98% loss of juveniles and one group of fish being reared in water treated with ultraviolet radiation suffered 100% loss. The fish in the treated water originated from parent fish tested to be IHN-negative. Ovarian fluid, sperm, and spleen samples were assayed for the presence of IHN virus by a plaquing procedure. The cells were incubated for 8 d to get a positive or negative confirmation of the virus.

Production goals at Dworshak were met by transferring eyed eggs to Kooskia National Fish Hatchery, 48.3 km upstream on the main-stem Clearwater River. The rearing-water source for the steelhead eggs and fry was well water. No IHN-related mortality was documented and the juvenile fish were transported back to Dworshak for final rearing and release.

In 1984, Dworshak hatchery suffered a 70% loss of steelhead fry in the nursery building. This loss came in spite of a full-scale brood-stock culling program, in which single females were used with single males and only after each fish tested negatively for the virus. The recurring disease problem at Dworshak hatchery provoked a joint study by the Idaho Department of Fish and Game, U.S. Army Corps of Engineers, and the U.S. Fish and Wildlife Service to seek a safe, reliable, water-sterilization system. Ozone was selected based upon the work conducted at the U.S. Fish and Wildlife Service laboratory in Seattle, which indicated that ozone concentrations as low as 0.01 mg/L destroyed the IHN virus in 10 min or less (Wedemeyer et al. 1978).

Review of Studies and Results

During the 1985 production year at Dworshak hatchery, ozone was applied to a 2,271-L/min raw water supply. The water was treated with a 4.5-kg/d generator manufactured by Emery Industries. A liquid oxygen system was used to supply oxygen to the ozone generators. A 10-min contact time with an ozone residual of 0.20 mg/L was used for the study. An untreated raw water supply was used for a control. Both water supplies were 12°C. The ozonated water was degassed in packed columns (Owsley 1981) to reduce the residual ozone to a safe level for rearing fish. Fish in 11 of the 14 control tanks had to be destroyed because of the IHN virus. Although IHN was found in fish reared in ozonized water, only slight mortality occurred, and it was assumed that the IHN came from the fish rather than the water (Owsley 1984). The ozone study has been repeated every year since 1985 with similar results. Beginning in 1986, compressors have been used to supply oxygen for the ozone generators, because liquid oxygen was too expensive to use after the first year.

Other aquaculture studies have shown positive results when ozone was used to treat water supplies. *Ceratomyxa shasta* was controlled at the Cowlitz Hatchery in Washington (Tipping and Kral 1985). Coleman National Fish Hatchery in California has had some success in controlling *Myxosoma cerebralis* (whirling disease) with ozone (Baker 1986). Ozone was superior to chlorine for inactivating the fish pathogens *Aeromonas salmonicida* (furunculosis), *Yersinia ruckeri* (enteric redmouth), and infectious pancreatic necrosis (IPN) virus (Wedemeyer et al. 1977).

Future plans for ozone at Dworshak include a water-reuse study. The combination of ozone and reuse is not a new concept. Morrison (1977) achieved a 70% decrease in ammonia by using ozone on a pilot reuse system at Dworshak. Plate counts showed that ozone consistently provided better disinfection of make-up water than the existing ultraviolet system. A similar study by Oakes et al. (1978) at Dworshak showed that nitrite could be virtually eliminated in reuse water. Williams et al. (1982) demonstrated at the U.S. Fish and Wildlife Service's Tunison Laboratory in New York that ozone and reuse were an excellent combination requiring only 1% make-up water.

Rosenthal and Kruner (1985) determined ozone oxidation rates for ammonium, nitrite, and biological oxygen demand (BOD) separately and in combination by using an improved contact chamber design. The upflow design with an improved foam removal system produced good results. However, foam can be a problem in an ozone system. The foam is made up of protein, suspended and dissolved solids, and organic and inorganic compounds. It is mainly found in the contact chamber, and most systems are designed with some type of removal device, such as a protein skimmer. Foam can hinder fish culture operations during incubation and especially during nursery rearing. In nursery tanks, the foam prevents the starter feed from getting to the fish. A foam remover such as Dow Corning Anti-Foam FG-10 can be used to clear the water.

Sampling and Monitoring

Ozone is unstable and samples should be fixed at the sampling site. There are several procedures available to sample ozone.

The DPD (*N,N*-diethyl-*P*-phenylenediamine) method (Paulin 1967) is easy to use but has limitations. The test must be completed between 1 and 6 min after the sample is fixed. It has numerous interferences and should only be used as an indicator in water that is similar to distilled water.

A modified DPD method (Wedemeyer et al. 1978) showed good results at lower ozone levels. This method used a longer sample cell for increased accuracy.

The indigo blue method (Bader and Hoigne 1982) is the standard procedure for measuring ozone in water. The preparation is stable up to 4 h and has the fewest interferences of any method available. Accuracy is good and color development is excellent.

Feasibility

The feasibility of treating lesser-quality water compared with developing good-quality supplies was studied at Coleman National Fish Hatchery in northern California (Sverdrup and Parcel Engineering 1986). Four water supply systems were evaluated: wells, chlorination–dechlorination, ultraviolet sterilization, and ozonation. The findings of that study, with ultraviolet sterilization used as the base are presented below. Based on cost-effectiveness, ozonation was the logical choice to pursue.

Water supply	Construction costs (%)	Annual operation and maintenance (%)
Wells	366	45
Chlorination–dechlorination	70	61
Ultraviolet sterilization	100	100
Ozonation	94	36

As new developments arrive, ozonation will become more cost-effective. One new development, the aquatector (Schutte 1986), could greatly improve or eliminate the need for ozone contact basins. The aquatector uses microbubble technology, which could improve the efficiency of getting ozone into the water. This would in turn reduce the size of equipment needed and reduce costs.

On-site oxygen-generating equipment can double the production of ozone and reduce costs accordingly. Liquid oxygen is also an alternative that can be considered.

New technology from the ozone industry is appearing. Better, more-reliable equipment that will be smaller, more energy efficient, and produce more ozone is being developed.

Safety

Ozone is a very toxic compound. The toxicity of ozone is a function of time and concentration. The maximum allowable concentration for an 8-h day is 0.1 mg/L. Ozone can normally be detected in the air by the human nose in the range of 0.05 mg/L. It is very important that the work area be free of ozone. Ozone can be converted back to oxygen with a heat system before being discharged into the atmosphere.

Ozone is also very toxic to fish and must be removed from the water supply. Wedemeyer et al. (1979) determined that the permissible safe exposure level was 0.002 mg/L. Several other studies have verified this level, whereas others have reported much higher exposure levels. This discrepancy from Wedemeyer's work is probably due to sampling accuracy and differences between production and laboratory conditions. The conventional method to remove the ozone is with detention chambers. Ozone has a short half-life and can be removed by holding the water in a chamber for a period of time.

A faster, more economical way to remove the ozone is to strip it out of the water by using packed columns (Owsley 1981). Efficiency of packed columns varies from 70 to 95% removal. Complete removal can be accomplished by using stripping towers (Montgomery Engineers 1987). A stripping tower is a packed column with a counter-current air flow. Stripping towers raise the energy costs because of the additional height and blower requirements.

Carbon filters also remove ozone very effectively. These filters accommodate small systems and are not as economical as packed columns or stripping towers.

Conclusions and Recommendations

Ozone is a viable means of disinfecting water for aquaculture. It has been proven effective against fish and human pathogens at very low dosage rates.

Ozone is not restricted, as is chlorination or ultraviolet radiation, in the quality of water that can be treated. It has been used successfully in

Europe for many years. Pilot studies in U.S. aquaculture have shown a lot of promise.

On-site production of oxygen will double the capacity of ozone generators and reduce costs accordingly, compared with compressed air systems. It will also eliminate the buildup of nitrous oxide and reduce annual maintenance.

The aquatector could improve efficiency and eliminate the need for a contact chamber. This, in turn, would reduce the size of the oxygen-generating equipment.

Packed columns and stripping towers remove residual ozone and eliminate the need and costs of detention tanks.

By incorporating the above features in the design of an ozone system, ozone can be feasible as well as effective.

References

Bader, H., and J. Hoigne. 1982. Determination of ozone in water by the indigo method: a submitted standard method. Ozone; Science and Engineering 4:169–176.

Baker, B. 1986. Ozonation of hatchery water supply at Coleman National Fish Hatchery. Proceedings of the Northwest Fish Culture Conference, Eugene, Oregon.

Montgomery Engineers. 1987. Ozone removal methods and review of the design concept. U.S. Army Corps of Engineers, Information Report, Walla Walla, Washington.

Morrison, T. J. 1977. A pilot plant trial for ozone sterilization of fish hatchery water. Master's thesis. University of Idaho, Moscow.

Oakes, D., P. Cooley, and L. Edwards. 1978. Ozone disinfection of make-up and recycle water at Dworshak National Fish Hatchery. University of Idaho, Chemical Engineering Deparment, Moscow.

Owsley, D. E. 1981. Nitrogen gas removal using packed columns. Pages 71–82 *in* L. J. Allen and E. C. Kinney, editors. Proceedings of the bioengineering symposium for fish culture. American Fisheries Society, Fish Culture Section, Bethesda, Maryland.

Owsley, D. E. 1984. Operation of the ozone pilot system at Dworshak National Fish Hatchery. Proceedings of the Northwest Fish Culture Conference, Tacoma, Washington.

Paulin, A. T. 1967. Procedures for measuring ozone in water. Journal of the Institution of Water Engineers 21:537.

Rosenthal, H., and G. Kruner. 1985. Treatment efficiency of an improved ozonation unit applied to fish culture situations. Ozone; Science and Engineering 7:179–189.

Schutte, A. R. 1986. Evaluation of the aquatector, an aeration system for intensive fish culture. U.S. Fish and Wildlife Service, Information Leaflet 42, Bozeman, Montana.

Sverdrup and Parcel Engineering. 1986. An evaluation of alternative water supply systems. U.S. Fish and Wildlife Service, Information Report, Portland, Oregon.

Tipping, J. M., and K. B. Kral. 1985. Evaluation of a pilot ozone system to control *Ceratomyxa shasta* at the Cowlitz Trout Hatchery. Washington State Game Department, Bulletin 85-18, Olympia.

Wedemeyer, G. A., N. C. Nelson, and C. A. Smith. 1977. Survival of two bacterial pathogens *Aeromonas salmonicida* and the enteric redmouth bacterium in ozonated, chlorinated, and untreated waters. Journal of the Fisheries Research Board of Canada 34:429–432.

Wedemeyer, G. A., N. C. Nelson, and C. A. Smith. 1978. Survival of the salmonid viruses infectious hematopoietic necrosis (IHNV) and infectious pancreatic necrosis (IPNV) in ozonated, chlorinated and untreated waters. Journal of the Fisheries Research Board of Canada 35:875–879.

Wedemeyer, G. A., N. C. Nelson, and W. T. Yasutake. 1979. Potentials and limits for the use of ozone as a fish disease control agent. Ozone; Science and Engineering 1:295–318.

Williams, R. C., S. G. Hughes, and G. L. Rumsey. 1982. Use of ozone in a water reuse system for salmonids. Progressive Fish-Culturist 44:102–105.

American Fisheries Society Symposium 10:421–426, 1991

Specifying and Monitoring Ultraviolet Systems for Effective Disinfection of Water

JAMES D. CAUFIELD

James D. Caufield, P.E.
2075 S.W. First Avenue, Suite 2N, Portland, Oregon 97201, USA

Abstract.—The operational characteristics to be identified in a specification for an ultraviolet disinfection unit include water flow rate, temperature, radiation absorption coefficient, and minimum radiation dosage level. Material, equipment, and monitoring requirements also need to be stated. Minimum dosage required to treat water for fish culture is usually taken as 33,000 $\mu W \cdot s/cm^2$, unless there is reason to guard against a more resistant pathogen. Absorption of ultraviolet radiation by water is a seasonal variable because of changes in color and turbidity. A year of testing the water to be treated is recommended to determine the maximum absorption coefficient. In response to these specifications, suppliers can submit technical data on specific disinfection units whose effectiveness can then be tested by mathematical analysis. Relatively complex formulas have been reduced to a family of curves that can be used to accomplish this analysis by simple arithmetic summation.

Disinfection of surface water supplies for fish hatcheries is frequently necessary because of the presence of known or suspected pathogens in the water. Groundwater supplies are rarely subject to similar concerns. Occasionally disinfection of water within recirculation systems is appropriate to guard against the introduction of pathogens from outside sources.

The disinfection agent of choice for potable water supplies is chlorine because it can provide a residual that assures safety. Ozone is used to a limited extent, but is usually followed by chlorination to obtain the chlorine residual. Such residuals are not toxic to humans but are toxic to fish. Chlorination followed by dechlorination has been used in some aquaculture facilities, and limited use of ozonation has been reported. Yet, ultraviolet radiation remains the preferred means of disinfection for fish hatchery water supplies because it adds nothing to the water and provides no toxic residual.

Effectiveness of Ultraviolet Radiation

Ultraviolet light is a component of sunlight that occupies the portion of the spectrum between 13.6 and 400 nm. It is most effective as a disinfectant at a wavelength of 253.7 nm, which is commercially produced by low-pressure mercury lamps. In a previous study, T. Fadaak (UMA Engineers, Inc., unpublished) stated that the mutagenic action of these rays involves upsetting the ultrastructures of living microorganisms by disrupting unsaturated bonds in the primary structure of nucleic acids. It forms thymine dimers that halt nucleic acid replication and consequently impair the genetic activity of the organism. These rays have slight penetrating powers, so their action is only effective on surfaces or thin films. The more protected an organism's genetic material is, the higher the required dosage will be to destroy that organism. Protection can be afforded by the organism's shell or structural configuration. The organism can also be shielded from the rays by a screen, such as the quartz tubes that house the ultraviolet lamps, by particles, turbidity, and color in the water around the organism, and by the water itself. Thus, to effectively use ultraviolet disinfection one must know the pathogenic organism of concern and the environment in which it will be found.

Ultraviolet radiation dosage is measured as a function of sensed light intensity and time per unit area in units of microwatt-seconds per square centimeter ($\mu W \cdot s/cm^2$). Because the output of ultraviolet lamps is rated in microwatts per square centimeter at a distance of 1 m in air at 100 h of lamp life, two other dimensions are introduced: distance between lamp and object, and lamp life. The intensity of ultraviolet lamps declines with age, and 7,500 operating hours is the accepted life of a lamp. Though the intensity reduction varies with lamp types, and among lamps of the same type, an estimated 40% reduction at end of life is recommended as a design assumption. Minimum dosage is therefore calculated on the basis of 60% of the 100-h rating.

Dosages required to destroy fish pathogens have been reported by Hoffman (1974), Kimura et al. (1976), Bullock and Stuckey (1977), and oth-

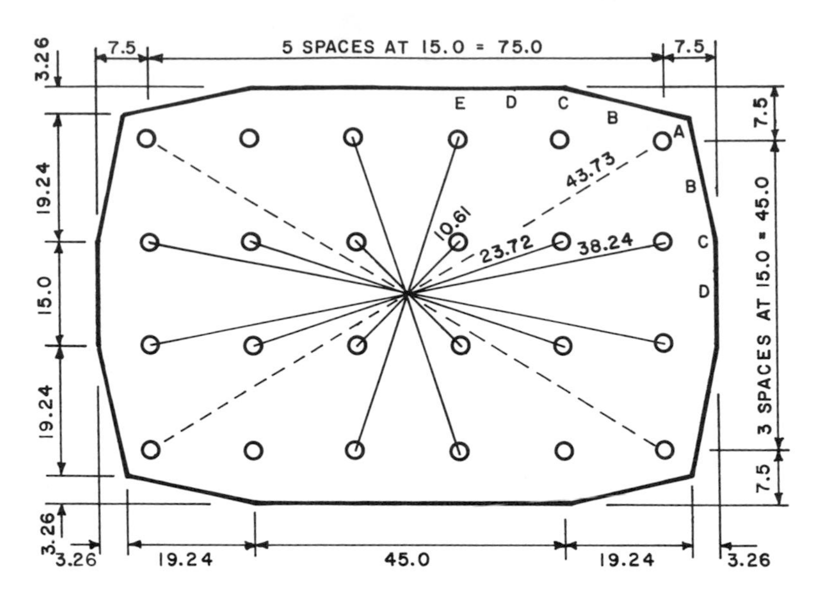

FIGURE 1.—Cross section of a gravity ultraviolet disinfection unit. Dimensions are in centimeters.

ers. Many common fish pathogens can be destroyed at dosages of less than 30,000 $\mu W \cdot s/cm^2$, but recognition must be given to those requiring higher doses. Inactivation dosages for *Ichthyophthirius* tomites and *Costia necatrix* are in excess of 300,000 $\mu W \cdot s/cm^2$. The 700-μm^2 size of the *Ichthyophthirius* tomites suggests removal by filtration before ultraviolet disinfection, whereas the 70-μm^2 size of *C. necatrix* may suggest other, additional means of disinfection. The generally accepted design dosage for fish culture is 33,000 $\mu W \cdot s/cm^2$ at end of lamp life, or 55,000 $\mu W \cdot s/cm^2$ at 100 h. This, or some other pathogen-specific design dosage establishes the first criterion of a specification for an ultraviolet disinfection unit. Other primary criteria are the geometry of the unit and the transmissivity of the water.

Ultraviolet Disinfection Units

Disinfection units may be pressurized or nonpressurized (gravity) assemblies. The former usually consist of a cylinder housing a radial array of lamps enclosed in quartz sleeves parallel with the length of the cylinder. The latter usually have a rectangular boxlike housing but otherwise are similar. Some form of baffles is provided to eliminate laminar flow along the housing walls, which are necessarily remote from the majority of the lamps. Some units provide reflectors in an attempt to reinforce the dosage. The flow of water is usually parallel with the length of the lamps, although one manufacturer provides flow perpendicular to the lamps. Units are in service that incorporate Teflon® tubes to carry the water, and lamps lacking quartz sleeves are located in air outside the Teflon tubes.

Based on the usual configuration, the net volume of the housing divided by the flow rate gives the retention time during which the water is subject to disinfection. The cross-sectional area of the housing is reduced by the area of the quartz sleeves. The length of the housing is taken as the arc length of the lamp, unless the housing length does not permit uniform water flow over the full arc length, in which case a shorter length should be used. Figure 1 is a dimensioned cross section of a gravity disinfection unit similar to that recently installed in a hatchery. The water entry and exit were parallel with the lamps, centrally located at baffled ends of the housing.

Ultraviolet Dosage Analysis

The disinfection unit depicted in Figure 1 was supplied with lamps designated as G64TL5 (Voltarc Tubes Inc.). The ultraviolet output (ϕ) of each of these lamps was 26.7 W with an incident radiant flux density at 1 m of 190 $\mu W/cm^2$, an arc length of 147.3 cm, a lamp diameter of 15 mm, and a rated life of 7,500 h. The surface area (A) of each of these lamps was $1.5 \cdot \pi \cdot 147.3 = 694.14$ cm^2.

Thus, the emitted radiant flux density of each lamp was

$$\phi/A = \frac{(26.7 \times 10^6)}{694.14} = 38{,}465\ \mu W/cm^2.$$

Each lamp was protected by a quartz sleeve with an outside diameter of 25 mm. Quartz has a coefficient of refraction of 0.95, which reduced the effective emitted radiant flux density (I_o) to

$$\frac{38{,}465 \times 0.95 \times 1.5}{2.5} = 21{,}925\ \mu W/cm^2.$$

The incident radiant flux density (I), the irradiance upon a remote particle that provides the disinfection, derives from the foregoing in indirect proportion to the distance between the particle and the source. Air has little absorptive quality for ultraviolet radiation and the irradiance is, effectively, geometric. On the other hand, water absorbs ultraviolet radiation in varying degrees as described above. This characteristic is defined by the absorption coefficient (α) in the following form: $I = I_o\, e^{-\alpha r}$; r is the distance from the point source to the particle.

Figure 2 shows the irradiance relative to distance from the source for ultraviolet radiation in air and in water with various coefficients of absorption. As a basis for interpretation of these

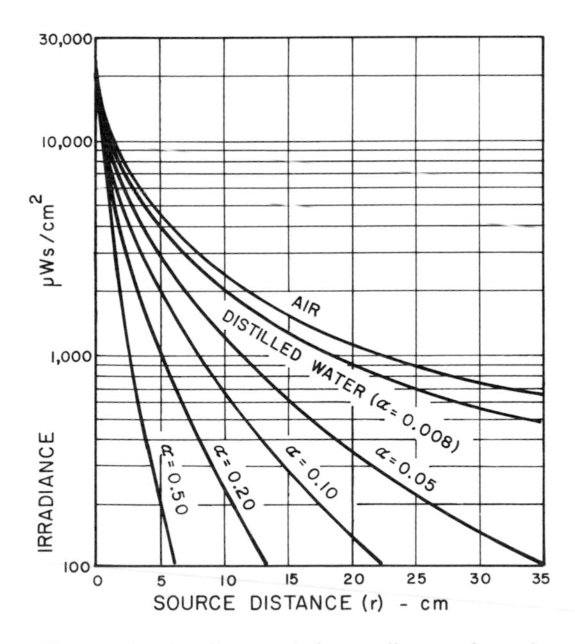

FIGURE 2.—Irradiance relative to distance from the source in air and in water with various coefficients of absorption (α).

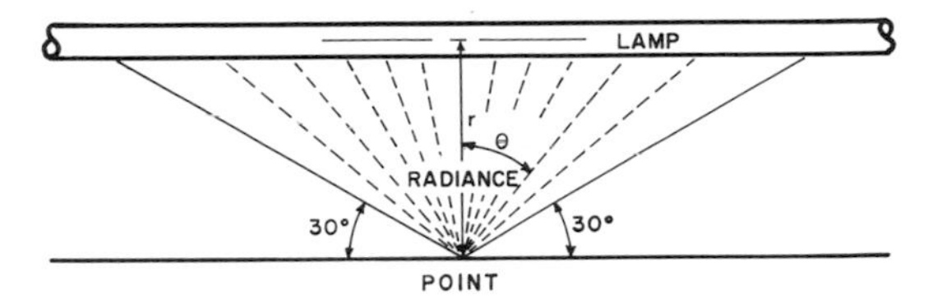

FIGURE 3.—Radiance impinging upon a particle of water from a linear source.

coefficients, Luckiesh (1944) reported that the city of Portland water supply, an unfiltered surface supply, has a coefficient of 0.083. Coefficients for most public water supplies are below 0.20, higher values relating to wastewaters. The coefficient for the water supply for a recent hatchery was found to be 0.067. That value is used in the analytical example below.

Not reproduced herein are curves showing percentages of absorption, for these are believed to be misleading. If such curves are used, it should be remembered that the reduction is a percentage of the irradiance in air at the given distance, rather than a percentage of the emitted radiant flux density.

In commercial ultraviolet disinfection units, the distance from the lamp to the most remote particle of water is small relative to the length of the lamp. The emitted radiant flux density does not come from a point source but from a series of points along a cylinder of radiant flux. Accordingly, a particle of water receives irradiance from some length of the tube as indicated in Figure 3. As a realistically conservative limit, the maximum angle of incidence (Θ) is taken as 60°. Thus, the total irradiance received by a particle of water is that caused by the line of effective emitted radiant flux density integrated over the 120° central angle:

$$\sum I = I_o \int_{-\pi/3}^{\pi/3} \cos\Theta\ e^{-\alpha r/\cos\Theta}\, d\Theta.$$

Numerical integration of the foregoing resulted in Figure 4. Note that for either $r = 0$ or $\alpha = 0$, the value of the integral is the square root of three. Cruver (1981), whose earlier paper proposed the foregoing, extended the integration another step: "To account for the presence of turbulence in the two-dimensional model, the area-weighted average scalar irradiance was determined." The effect of this is to multiply the value by two, which is believed to give too much credit to the admittedly beneficial effects of such an indeterminate as turbulence.

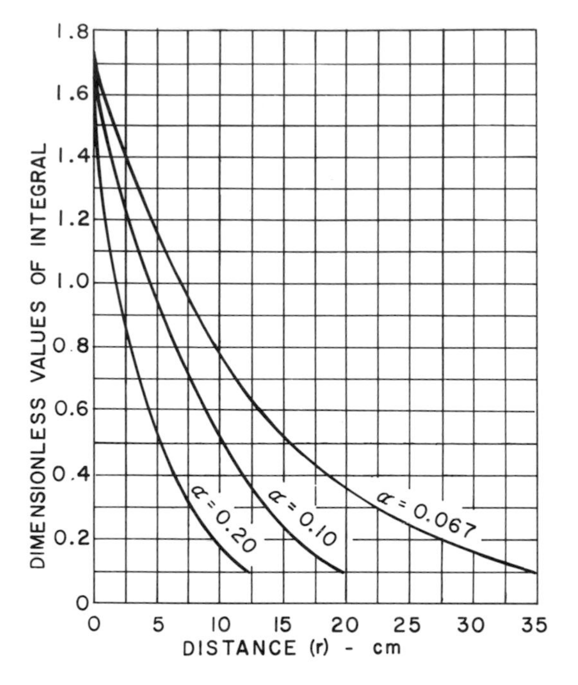

FIGURE 4.—Variation of $\int_{-\pi/3}^{\pi/3} \cos \Theta \ e^{-\alpha r/\cos \Theta} \ d\Theta$ with distance r and absorption coefficient α; Θ is the angle of incidence.

Continuing the mathematical analysis of the unit depicted in Figure 1, note that critical intensity points are indicated by A, B, C, D, and E. The net cross-sectional area of this unit is 0.4900 m^2. Twenty-four lamps are indicated. At the ends of the lamps the central angle reduces to the 60° angle of incidence, and the arc length needs to be reduced by cot 30° times 7.50 cm, giving an effective length of 134.3 cm. The flow through the unit was rated at 0.070 m^3/s for a theoretical detention of 9.40 s.

Dimensions are indicated on Figure 1 between each lamp that contributes to the dosage and the center of the unit where the dosage is a maximum. These dimensions are values of r. For each value of r, the value of the corresponding integral is taken from the $\alpha = 0.067$ curve of Figure 4. The incident radiant flux density at the center of the unit (I) caused by each lamp is the product of the emitted radiant flux density (I_o) at the surface of the quartz sleeve times the value of the integral, times the detention, times the ratio of the radius of the quartz tube, divided by r:

$$\frac{21{,}925 \times 1.25 \times 9.4}{r} \int = I.$$

These are summed as follows:

r	$\int$	I	Number of lamps	Total dosage
10.61	0.76	18,453	4	73,812
23.72	0.27	2,932	8	23,456
38.24	0.06	404	4	1,616
43.73	→ 0		4	
Total				98,884

Similarly, dosages for points A, B, C, D, and E have been calculated at 57,448, 50,552, 50,838, 50,250, and 51,836 $\mu W \cdot s/cm^2$, respectively. Though these range as low as 91% of the desired dosage, acceptance would be recommended, for it is realized that the velocity of the water along the wall of the unit will be less than that at the center of the unit, and the average detention of 9.4 s would be exceeded at these critical points.

Specifying the Unit

To accomplish the foregoing calculations, the specifications must require that a dimensioned drawing of the proposed unit be provided with the bid. Then it may be analyzed for effectiveness. The balance of the specification needs to address operational characteristics as well as material and equipment criteria.

Operational characteristics include water flow rate, water temperature variations, site conditions that will affect the temperature of the unit, and the most critical absorption coefficient for the water to be used. The importance of the operating temperature at the ultraviolet lamp wall may be seen by referring to Figure 5 provided by Nagy (1955). The supplier should be required to provide heat loss calculations to ensure operations at the optimum 41°C lamp wall temperature, and should also be required to provide heating or cooling, if necessary, to achieve that optimum temperature under the anticipated environmental conditions in which the unit will operate.

Absorption coefficients can vary with the seasons for surface waters. High values may occur in the spring when runoff is high in turbidity, or in the fall when deciduous leaves may add tannic acid to the water. A year of sampling the water and measuring absorption coefficients is recommended before specifying a disinfection unit. Considering the lead time for design and construction of most projects, the year of testing can usually be accommodated.

Material and equipment requirements to be specified include the lamps, the quartz sleeves, the seals, the housing material, flow control (if

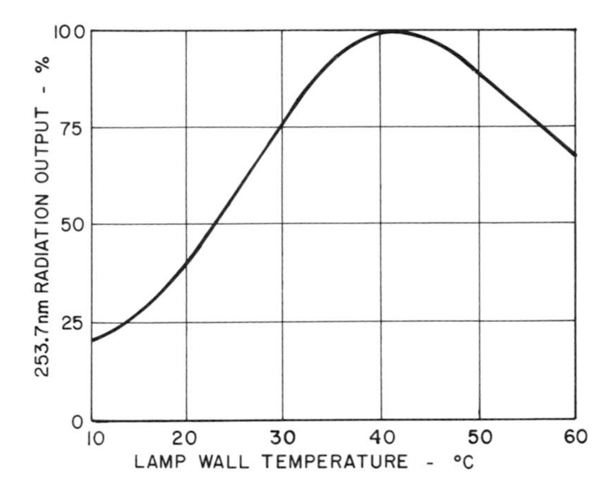

FIGURE 5.—Relative output of the G36T6 Sterilamp at various temperatures. (After Nagy 1955.)

any), lamp failure control, an ultraviolet intensity meter, spare parts, the recommended cleaning procedure, and operation and maintenance manuals.

The heart of the ultraviolet disinfection unit is, of course, the ultraviolet lamps. These should be specified as low-pressure mercury vapor lamps constructed of type 7910 or 7912 glass to avoid production of ozone or hydrogen peroxide. The supplier should provide the following specific information regarding the lamps:

- number of lamps,
- manufacturer's designation,
- length and diameter,
- bulb lamp watts,
- base,
- operating amps and volts,
- starting current,
- rated average life in hours,
- total watts of UV output at 100 h,
- microwatts per square centimeter at 1 m from lamp,
- ballast type and designation, and
- current list price of replacement lamps.

Required for the quartz sleeves are dimensions, confirmation of the coefficient of refraction (0.95), and current list price of replacement sleeves.

Seals should be capable of withstanding at least 1.0 MPa in pressure units, but 0.1 MPa is usually adequate in gravity units.

Stainless steel is the preferred housing material.

An intensity-sensitive, automatically activated solenoid shutoff valve is usually provided in units for potable water supplies. For fish culture this is generally not recommended.

A meter or other device should be provided to indicate that all lamps are burning, and should one or more lamps go out, a rotary switch or other means should be provided to identify which lamps are out.

Each section of the purification chamber should be provided with at least one accurately calibrated ultraviolet-intensity meter that is not responsive to visible or infrared radiation. It should provide a visible display of dosage (intensity times detention at design rate of flow) and it should be provided with an alarm to indicate a dosage below 33,000 $\mu W \cdot s/cm^2$. The appropriateness of the location of the meter can be confirmed by the foregoing calculations.

Cross and Peterson (1987) as well as Tobin et al. (1983) have separately reported problems with intensity meter accuracy. It is suggested that the problem is simply one of calibration, which calculations can correct. Properly calibrated, the intensity meter will automatically monitor the absorbance of the water as well as the intensity of the lamps.

The supplier is usually requested to provide an additional 10% of the operating number of lamps, quartz sleeves, seals, and ballasts as spare parts.

The procedure for cleaning the quartz sleeves should be satisfactory to the purchaser. Cleaning should be possible without disassembly of the unit, but opinions differ on the precise method to be used. Many units are provided with internal wipers. The author's experience with eyeglasses and windshields suggests that wipers may be ineffective without some form of detergent. Preferred is the capability of citric acid bathing. This leads to an arrangement of parallel units so that one may be taken out of operation for cleaning.

As a minimum, operation and maintenance manuals should provide complete wiring diagrams and detailed procedures for calibrating all meters and detectors. A complete list of all parts should be provided with instructions for removal and replacement of the parts. Sources of parts, if other than the original supplier, should be listed. An emergency telephone number for repair advice is desirable.

Within the low-bid, general contract system faced by all public and most private agencies, there is a way to deal with something that needs to be evaluated on the basis of performance rather than price. The disinfection units can be separated from the general contract by providing a fixed allowance for the units with the purchaser either being credited or debited for the difference between the allowance and the actual cost of the acquired units, or by providing a separate contract for acquisition of the units and limiting the general contract to the installation of such units.

In this way the acquisition can ensure that the units will meet the technical requirements, will provide those operational features preferred by the purchaser, and will provide the least lifetime cost. Annual lamp replacement and power costs will likely exceed 15% of the purchase price of an ultraviolet disinfection unit, so lifetime cost rather than first cost should be the criterion.

Downstream Piping

At the beginning of this paper, I addressed the practical limitations of ultraviolet disinfection. It is disinfection, not sterilization. Some pathogens will pass through the best system. Accordingly, downstream recontamination and bacterial buildup should be carefully considered in the overall piping design. Cross and Peterson (1987) recommended avoiding dead-water areas in the piping. McGarvey (1988) recommended a minimum piping velocity of 1.5 m/s to prevent bacterial growth. Pittner and Bertier (1988) argued that bacterials cannot be completely scoured from piping through hydrodynamic design. On the other hand, high velocities may make it difficult for the bacteria to adhere to the piping walls initially. Stable velocities are also important because a sudden increase in velocity will dislodge some bacterial buildup and pass it on to the rearing units. Fortunately, flows to rearing units are relatively constant for weeks at a time. When an additional rearing unit is put on line, its water supply should be run to waste for several hours before fish are stocked in the unit.

Periodically, it may be found necessary to disinfect piping with strong oxidants. A piping layout that permits isolation of segments for this purpose should be incorporated in the initial design.

Bioassays

Determinations of lethal dosages of ultraviolet radiation are based on carefully controlled bioassays. It is the premise of this paper that once such a dosage level has been confidently established for a particular pathogen of concern, repetition of a bioassay becomes unnecessary. The calculation method of assessing the effectiveness of a proposed ultraviolet disinfection unit is preferred. Aside from the tedious numerical integration needed to develop curves as shown in Figure 4, the calculations are simple, and they provide a clear understanding of what goes on inside the unit. They are not subject to transient variables or experimental error. They are not subject to extrapolation from bench scale to prototype, or from one pathogen to another. They are less expensive and can simplify the bidding procedure.

References

Bullock, G. L., and H. M. Stuckey. 1977. Ultraviolet treatment of water for destruction of five gram-negative bacteria pathogenic to fish. Journal of the Fisheries Research Board of Canada 34:1244–1249.

Cross, V. K., and L. Peterson. 1987. Efficacy of ultraviolet water treatment at the Green Lake, Maine, National Fish Hatchery. Progressive Fish-Culturist 49:233–235.

Cruver, J. E. 1981. Ultraviolet disinfection of wastewater. *In* proceedings of the 25th Great Plains Wastewater Design Conference, Omaha, Nebraska.

Hoffman, G. L. 1974. Disinfection of contaminated water by ultraviolet irradiation, with emphasis on whirling disease (*Myxosoma cerebralis*) and its effects on fish. Transactions of the American Fisheries Society 103:541–550.

Kimura, T., M. Yoshimizo, K. Tajima, Y. Exura, and M. Sakai. 1976. Disinfection of hatchery water supply by ultraviolet (U.V.) irradiation: I. Susceptibility of some fish pathogenic bacterium and microorganisms inhabiting pond water. Bulletin of the Japanese Society of Scientific Fisheries 42:207–211.

Luckiesh, M. 1944. The transmission of germicidal energy by water from various cities of the United States. General Electric Review 47:26. (General Electric Company, Cleveland, Ohio.)

McGarvey, F. X. 1988. Maintenance of high purity ion exchange beds. Ultrapure Water 5(4):51–55.

Nagy, R. 1955. Water sterilization by ultraviolet radiation. Westinghouse Electric Corporation, Lamp Division Research Report BL-R-6-1059-3023-1, Bloomfield, New Jersey.

Pittner, G. A., and G. Bertier. 1988. Point-of-use contamination control of high purity water through continuous ozonation. Ultrapure Water 5(4):16–22.

Tobin, R. S., D. K. Smith, A. Horton, and V. C. Armstrong. 1983. Methods of testing the efficacy of ultraviolet light disinfection devices for drinking water. Journal of the American Water Works Association 75:481–484.

American Fisheries Society Symposium 10:427, 1991

Development of Carbon Filtration Systems for Removal of Malachite Green[1]

LEIF L. MARKING

U.S. Fish and Wildlife Service, National Fisheries Research Center
Post Office Box 818, La Crosse, Wisconsin 54602, USA

Abstract.—The U.S. Fish and Wildlife Service was granted an Investigational New Animal Drug permit (INAD number 2573) by the U.S. Food and Drug Administration to allow the use of malachite green at selected state and federal fish hatcheries. However, the INAD permit requires that the fungicide be removed from all treated water after March 1989. A study was designed to (1) determine the type of filter and kind of carbon that was most efficient and (2) demonstrate that carbon filters can be used to remove malachite green from water used for egg incubation or to hold adult salmon before spawning. Minicolumn simulation studies showed that 8 × 30-mesh granular carbon was effective for continuously removing malachite green from water for 230 d at a flow rate of 500 gal/min and for only 62 d at a flow rate of 1,000 gal/min. The removal capacity at the slower flow rate was 69 mg of malachite green per gram of carbon. A filter system that contained 20,000 lb of activated carbon in each of two chambers was effective for removal of malachite green from treated water in adult salmon holding ponds at flows of 500 gal/min (6.4 gal/min per ft^2) and greater. The removal efficiency was 99.8% after 105 h of operation, and the adsorption capacity of the system was projected to be 20 or more years of routine hatchery operation. A filter system that contained 2,000 lb of activated carbon in each of two chambers was effective for removal of malachite green from treated water in salmon egg incubation units at the designated flow rate of 50 gal/min (4.0 gal/min per ft^2) and also at faster flow rates. Removal efficiency decreased only slightly for faster flows in both filter systems, and the efficiency improved when treated water was passed through two filter chambers in series.

[1]The full text of this paper has been published elsewhere. See: L. F. Marking, D. Leith, and J. Davis. 1990. Development of a carbon filter system for removing malachite green from hatchery effluents. Progressive Fish-Culturist 52:92–99.

American Fisheries Society Symposium 10:428–436, 1991

Evaluation of High- and Low-Pressure Oxygen Injection Techniques[1]

WILLIAM P. DWYER, GREG A. KINDSCHI, AND CHARLIE E. SMITH
U.S. Fish and Wildlife Service, Fish Technology Center
4050 Bridger Canyon Road, Bozeman, Montana 59715, USA

Abstract.—Tests were conducted to evaluate high- and low-pressure techniques for adding pure oxygen to water. A high-pressure oxygen injection unit, the Aquatector, increased the dissolved oxygen in water flowing at a rate of 8,706 L/min by 3.3 mg/L with an oxygen absorption efficiency of 84%. Nitrogen gas saturation was reduced from 102.1 to 99.2%, while total gas saturation was increased from 98.9 to 104.1%. Oxygen absorption efficiencies and the dissolved oxygen of the incoming water were inversely related. Similar efficiencies were obtained with an Aquatector and with a pure oxygen line fed into a pressurized reuse water line. A series of low-pressure tests were conducted with sealed columns to evaluate absorption efficiencies at various water and oxygen flow rates. The tests were conducted with a sealed 10.2-cm column at water flows of 30, 40, and 50 L/min versus oxygen flows of 50, 150, 250, and 350 mL/min at each water flow. Dissolved oxygen in the output was monitored and absorption efficiencies were calculated. Absorption efficiencies ranged from 43 to 78% over the range of flows tested with the 10.2-cm column. Tests were also conducted with a 15.2-cm column; however, it was determined later that data collected with the 10.2-cm sealed column cannot be directly extrapolated to predict performance of large, 61-cm, production-size columns.

Typically, production hatcheries are pushed to the limit by the number of fish they are requested to produce. Conditions are most limiting when loading is highest just prior to stocking. Increasing the oxygen content of the water improves environmental conditions, which can lead to higher quality fish as well as increased production. Several systems have been designed and tested that use either air or oxygen to increase the oxygen content in fish-rearing water (Speece et al. 1971; Colt and Tchobanoglous 1981; Owsley 1981; Speece 1981; Hackney and Colt 1982). The use of atmospheric air to increase dissolved oxygen (DO) may also increase nitrogen and total gas supersaturation, which can be detrimental to the culture of fish (Weitkamp and Katz 1980; Colt and Westers 1982). Westers (1981) demonstrated that pure oxygen can be used to increase fish production economically while maintaining acceptable total gas pressures.

Dissolved oxygen levels can be increased by two basic processes, which can be categorized as high-pressure and low-pressure techniques.

The Aquatector can be classified as a high-pressure technique for adding oxygen to water. It was designed in West Germany by France-Josef Damann for use in the champagne industry. Since its invention, several other applications have been made, including additions of oxygen to water in fish culture.

Pumps supply water to the top of the Aquatector column, where the water is mixed with pure oxygen under pressure and flows back ("sidestreams") into the untreated water. The oxygen-rich water with small bubbles (about 0.05 mm in diameter) flows from the bottom of the column through a brass needle valve that regulates pressure and flow. A bypass valve on the pumps can also control water flow and pressure. The internal pressure in the Aquatector is indicated by a pressure gauge on the column itself.

The control unit regulates oxygen flow into the column through a solenoid valve. If the water level in the column fluctuates, the oxygen is switched on or off automatically to maintain the desired level. An oxygen regulator is used to maintain pressure needed in excess of the internal column pressure.

Low-pressure oxygen injection techniques have been pioneered by biologists from Michigan. Pure oxygen has been used to reduce nitrogen and increase DO concentrations in hatchery source water since the early 1980s (H. Westers, Michigan Department of Natural Resources, personal communication). Sealed columns are used with water flow rates of 1.0 L/min per cm^2 of cross-sectional area. These columns are similar to the packed

[1]The Fish and Wildlife Service makes neither recommendation nor endorsement of any commerical product and has made no determination with respect to patent rights involved in the devices mentioned herein.

columns described by Owsley (1981) except that the tops are sealed and the lower end of the column is reduced in diameter and extends below the water surface.

Packed columns are adequate in most situations for degassing (Owsley 1981), but they are limited in others because they neither increase the DO content of the water above 100% nor reduce nitrogen pressure to less than 100% saturation. Effluents from packed columns typically have a positive ΔP, the difference between the actual partial pressure of a gas in water (measured in millimeters of Hg) and the partial pressure of that gas in the surrounding atmosphere. When water is pumped through a packed column, the air–water interface is increased by the packing; as air is drawn through the column the exchange of gases results in an approach to equilibrium.

The sealed column, which does not draw air in from the top, creates a partial vacuum at water flow rates near 1.0 L/min per cm^2 of cross-sectional area. The effect of the vacuum increases the ΔP, and both oxygen and nitrogen are removed from the water. However, when oxygen is injected into the column, the ΔP for nitrogen increases because there is little nitrogen in the atmosphere within the column. This causes a release of large amounts of nitrogen from the water—much more than would occur if atmospheric gases filled the column. Oxygen injection causes the ΔP for oxygen to become negative because there is less oxygen gas pressure in the water than in the atmosphere within the column. This in effect causes more oxygen to be dissolved in the water.

Thus, the nitrogen content of the water decreases markedly, often to levels below 100% saturation, and oxygen may increase to levels above 100% saturation. Oxygen absorption efficiency for this type of column is commonly about 50–60%. However, when water flow rate is decreased, oxygen absorption efficiency can be increased considerably.

The objective of this study was to evaluate high- and low-pressure oxygen injection systems to determine oxygen absorption efficiencies. In laboratory tests, Schutte (1988) demonstrated that a small Aquatector increased DO at high oxygen absorption efficiencies. In this test a larger unit (model 50) was used in fish production situations (Figure 1). The unit was evaluated for ease of operation, for oxygen absorption efficiencies, for performance at various oxygen flow rates and pressure levels, and for determining effects on nitrogen and total gas saturation levels. A sealed column was used to evaluate oxygen absorption efficiencies in a low-pressure system. Sealed columns were tested at a range of water and oxygen flow rates.

Tests were conducted at the U.S. Fish and Wildlife Service Fish Technology Center and at the Jackson (Wyoming) National Fish Hatchery (NFH).

Methods

High-pressure techniques.—The Aquatector model 50/225 consists of three components: a 150-cm column with pressure gauge, a brass needle valve, and an oxygen flow control unit (Figure 1). Other components are two 2-horsepower pumps (Jacuzzi model 2HJM-S/B), an oxygen flowmeter (Puritan series C), an oxygen regulator (Victor Equipment Co., model SR450D), an oxygen hose, a water bypass valve, and miscellaneous plumbing supplies. Cost for the complete system was about US$5,600 (in 1986). Liquid oxygen could be purchased in 131.1-m^3 bottles that cost $95 to $175, depending on locality.

Oxygen absorption efficiencies (OE, %) were determined with the following formula:

$$OE = 100 \times \frac{\text{water flow (L/min)}}{\text{1,428 mg DO/L}} \times \frac{\text{DO increase (mg/L)}}{\text{oxygen added (L/min)}}.$$

The purity of liquid oxygen used was greater than 99.5%.

At the Jackson NFH, once-used water flowing at 8,706 L/min was available for testing oxygen absorption efficiencies and the effects of adding pure oxygen on total gas and nitrogen gas levels. Reuse water from the raceways was pumped through a pipe, 30 cm in diameter, from a sump area at the lower end of the raceways to a headbox at the upper end of the raceways, 50 m away. The Aquatector was tested at the sump and headbox. Water in the sump was under a pressure of 74.9 cm of Hg because two 15-horsepower reuse pumps were used to return the water to the headbox.

At the upper end of the raceways, the Aquatector was placed at the reuse headbox (47.8 m long) and tested by adding the flow from the unit to 8,706 L/min of pumped reuse water. This allowed for several measurements of DO along the reuse headbox. The Aquatector was tested at water

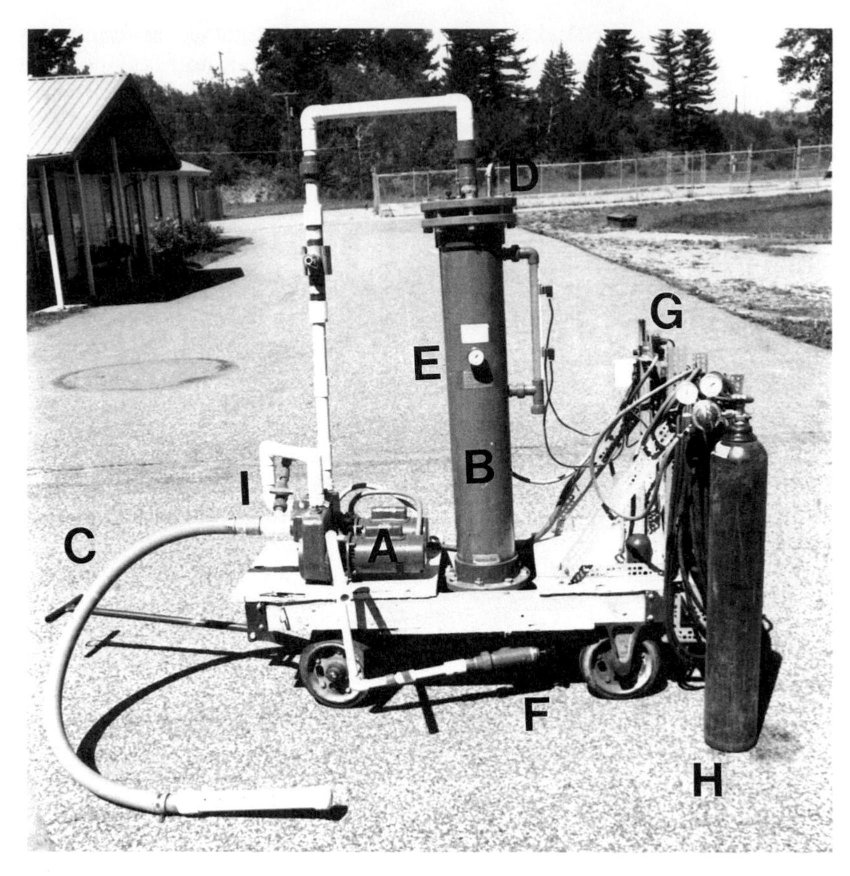

FIGURE 1.—Components of the Aquatector model 50/225: pumps (A), column (B), water supply line (C), oxygen supply line (D), column pressure gauge (E), brass needle valve (F), oxygen control unit (G), oxygen source (H), and water bypass valve (I).

flows of 190 and 250 L/min versus column pressures of 179.9–269.9 cm of Hg at 15-cm intervals. Three oxygen injection rates (8.7, 16.9, and 24.0 L/min) were tested at each flow and pressure combination. These flow rates are equal to gas: liquid ratios [(L/min oxygen : L/min water)100] of 4.6, 8.9, and 12.6 at the 190-L/min water flow and 3.48, 6.76, and 9.6 at the 250-L/min water flow. The effects of these variables on DO, oxygen absorption efficiency, DO retention time, and nitrogen and total gas levels were monitored.

The pressure and gas flow rates evaluated were based on manufacturer's recommendations and also on the failure of the Aquatector to operate properly at higher or lower settings. Delta-P and three DO measurements were recorded at each setting. Dissolved oxygen was measured every 10–15 min and gas pressure every 30 min (42 observations) 14.5 m down-channel from the Aquatector over a 4-d period. The modified Winkler titration method (APHA et al. 1985) was used to test water in which DO exceeded 20 mg/L. A DO meter (Yellow Springs Instrument Co., Inc., model 50), calibrated by the standard Winkler titration method, was also used. Nitrogen and total gas levels were monitored with Weiss saturometers (models ES-2 and ES-3).

In analyzing the data, the Statistical Analysis System (SAS Institute 1982) was used. The general linear models procedure was used to test for main effects (column pressures, water flow rate, oxygen addition rate) and interactions.

Low-pressure techniques.—Tests were conducted with a packed column 10.2 cm in diameter. Water flows were 30, 40, and 50 L/min, and at each water flow rate oxygen flows of 50, 150, 250, and 350 mL/min were tested. These represent gas:liquid ratios ranging from 0.16 to 1.17 at 30 L/min, from 0.125 to 0.875 at 40 L/min, and from 0.10 to 0.70 at 50 L/min. Three samples were taken at each setting. Oxygen measurements were made with and without oxygen injection, and absorption efficiencies were calculated. Oxygen content of the water used in these tests was at 90–95% of saturation. Temperature ranged from 7.2 to 7.9°C and DO of incoming water was 9.6–9.9 mg/L. Concentrations were calculated based on temperature and barometric pressure, described by Colt (1984). The column (height, 1.25 m) was packed with 1 m of 2.54-cm Koch rings. This system was set up in the laboratory to collect baseline data on oxygen absorption efficiency. The data collected in this phase of the study were compared with data collected with the 15.2-cm column. The 15.2-cm column was 1.82 m long with 1 m of 2.54-cm packing.

Tests were also conducted to determine the effect of the use of packing and of the oxygen injection site on absorption efficiency. In addition to tests with a full 1 m of packing, we also experimented with reductions in column height by restricting the outflow, thus backing the water up into the column. This procedure in effect decreased the column height and the amount of air–water interface.

To analyze the data, we used the NCSS Statistical Program (Hintze 1987). Regression analysis was performed to describe the relation between flow rates and oxygen added to the water, and Students t-test was used to compare means.

TABLE 1.—Dissolved oxygen increase, oxygen absorption efficiency, and nitrogen and total gas pressure of 8,706-L/min reuse water injected with pure oxygen from the Aquatector at Jackson National Fish Hatchery. Untreated water contained 7.0–8.3-mg/L concentrations of dissolved oxygen with 102.1% nitrogen gas and 98.9% total gas pressure.

Oxygen added, L/min (gas:liquid ratio)	Dissolved oxygen increase (mg/L)	Oxygen absorption efficiency (%)	% gas saturation	
			Nitrogen	Total
Oxygen injected upstream from reuse pump				
8.3(0.09)	0.94	69.4	102.0	100.5
17.2(0.19)	1.97	69.0	100.9	102.0
25.5(0.29)	2.46	58.8	100.7	103.0
Oxygen injected downstream from reuse pump				
8.3(0.09)	1.06	78.3	102.0	100.5
17.2(0.19)	2.05	72.7	100.1	100.8
25.5(0.29)	2.47	59.0	99.7	101.3
Oxygen injected at Aquatector, downstream from reuse pump				
7.5(0.11)	0.95	77.0	101.9	101.5
15.4(0.18)	1.81	71.6	102.0	103.3
23.5(0.27)	2.71	70.3	100.7	104.8

Results and Discussion

High-Pressure Systems

A column pressure of 254.9 cm of Hg, a water flow of 250 L/min, and an oxygen flow of 24 L/min provided the best oxygen absorption efficiency and DO increase with the Aquatector. Such an optimum did not occur at Jackson NFH when the Aquatector effluent was pumped into a pipeline already pressurized at 75 cm of Hg. There were no differences ($P > 0.05$) in oxygen absorption efficiencies between the Aquatector water flows, column pressures, or oxygen flows tested (Table 1). The various Aquatector settings yielded oxygen absorption efficiencies from 70.3 to 77.0%. Without the Aquatector, oxygen absorption efficiencies ranged from 58.8 to 78.3%. Although the differences were not significant ($P > 0.05$), oxygen absorption efficiencies were higher and total gas pressures less when oxygen was injected downstream from one of the pumps than when it was injected upstream from the pump (Table 1). In addition, use of the Aquatector did not significantly ($P > 0.05$) improve oxygen absorption efficiencies over direct injection in this situation. The absence of significantly higher increases in DO, coupled with the high installation and operational costs of the Aquatector, makes its use uneconomical when oxygen can be injected into pressurized water supplies.

With the Aquatector at the reuse headbox channel at Jackson NFH, oxygen-rich water from the effluent of this unit reached 58 mg/L DO. Sidestreaming this oxygen-rich water into 8,706-L/min water flow increased the DO from 7.8 to 11.1 mg/L without any reduction in DO over a distance of 47.8 m in the headbox.

Analysis of the data recorded from the various Aquatector settings and the effects on 8,706-L/min water flow indicated that optimal increases in DO were 3.3 mg/L when 24 L/min of pure oxygen were added (Figure 2); the Aquatector water flow was 250 L/min and the column pressure was 254.9 cm of Hg. Increasing the water flow through the Aquatector significantly increased the DO ($P < 0.01$). This most likely happened because optimal conditions were maintained in the Aquatector for best bubble formation. Schutte (1988) also noted greater DO concentrations and improved oxygen absorption efficiencies at higher water flows and a direct relationship of DO to increased oxygen flows.

There was a positive effect ($P < 0.01$) on DO as Aquatector pressures and water flows were increased (Figure 2). This effect was greater as oxygen flows increased at each water flow. For

FIGURE 2.—Average dissolved oxygen increase in 8,706-L/min flow at a series of Aquatector pressures for water and oxygen flow rates. Aquatector water flows were 189 L/min (A) and 250 L/min (B).

example, at 8.7-L/min oxygen flow there was no correlation (r = 0.00) between increasing column pressures and DO with 250-L/min column water flow. However, there was a positive correlation between increasing internal pressures and DO (r = 0.71) at an oxygen flow of 24.0 L/min and a water flow of 250 L/min. At the lower water flow of 189 L/min, the Aquatector would not maintain a suitable water level when the higher two levels of oxygen (16.9 and 24.0 L/min) were added. Consequently there was no increase in DO at these settings.

Oxygen absorption efficiencies followed the same trend as mentioned above for increases in DO (Table 2). As column pressures and flows of pure oxygen increased, oxygen absorption efficiencies increased at both water flows ($P < 0.01$).

The optimum oxygen absorption efficiency for each water flow was obtained at column pressures of 254.9–269.9 cm of Hg ($P < 0.01$). Column pressures higher than 269.9 cm of Hg decreased oxygen absorption efficiencies because an adequate ratio of water to oxygen level could not be maintained in the Aquatector column. Nitrogen gas concentration was reduced and total gas concentration increased when the DO was increased (Figure 3). Decreased nitrogen gas concentration with increases in DO in a sealed column was also noted by P. Boersen (Michigan Department of Natural Resources, personal communication). He was also able to maintain low total gas pressures by using a column with a low vacuum. With the Aquatector, lethal levels of total gas may develop if the DO is increased too much. This can be prevented by controlling oxygen and water flows.

The Aquatector model 50/225 efficiently increased DO and decreased nitrogen gas concentration in fish-rearing facilities. Although not

TABLE 2.—Oxygen absorption efficiency when Aquatector effluent water at flows of 190 or 250 L/min were mixed with 8,706 L/min of reuse water, at each of three oxygen flows.

Oxygen flow (L/min)	Column pressure (cm Hg)						
	180	195	210	225	240	255	270
			Effluent flow: 190 L/min				
8.7	74.1	46.7	62.9	60.8	47.8	75.5	70.8
16.9	72.2		75.4	67.4	63.3	75.3	76.1
24.0				64.5	66.0	72.2	76.1
			Effluent flow: 250 L/min				
8.7	56.2	57.6	43.8	65.5	52.3	83.2	60.6
16.9	50.6	49.8	47.0	74.3	68.7	79.7	72.9
24.0	51.3	74.1	52.9	75.3	72.3	84.0	80.5

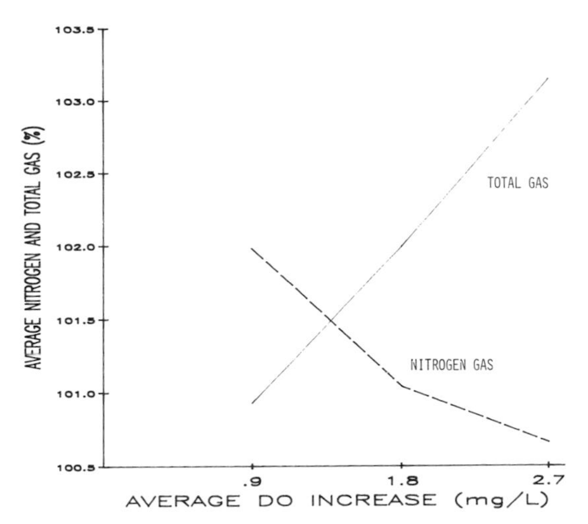

FIGURE 3.—Mean percent nitrogen and total gas pressures versus dissolved oxygen increase when oxygen-rich water from the Aquatector was added to 8,706-L/min reuse water.

suited for use in the pressurized water distribution system at Jackson NFH, it significantly increased DO in open-water systems. As with most mechanical systems, this unit needs to be monitored and maintained frequently to make sure it is operating properly. It should be emphasized that without standby oxygen sources and electrical systems, units such as this should not be relied on too heavily for fish life support systems. Users of the Aquatector should be aware that supersaturating water with oxygen can cause gas bubble disease if total gas pressure exceeds 100% saturation.

Low-Pressure Systems

In studies with the 10.2-cm column, as much as 9.1 mg/L of DO were added to the outflowing water. The lowest water flow produced the highest DO in the outflowing water—up to 20 mg/L at an oxygen flow of 350 mL/min (Figure 4). The DO declined considerably when water flow was increased to 50 L/min. Because these tests showed that there was little difference between water flow rates of 30 and 40 L/min, lower water flows were not evaluated. When the rate was increased to 50 L/min, a slight negative pressure developed in the column, which accounted for the observed decrease in DO of the outflowing water. Optimum water flow based on 1 L/min per cm^2 of cross-sectional area is 81 L/min for a 10.2-cm column. This recommended rate is similar to that presented by Owsley (1981) for nitrogen removal and degassing of water.

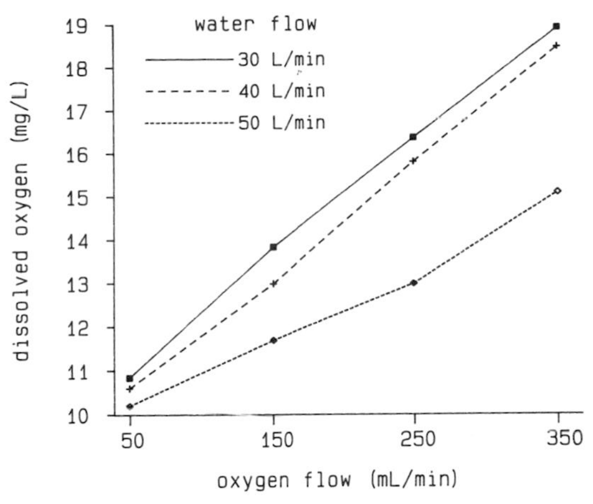

FIGURE 4.—Oxygen flow versus oxygen content in the effluent water at 30-, 40-, and 50-L/min water flow. These tests were made with the 10.16-cm column.

The objective of these tests was to evaluate oxygen absorption efficiency—not nitrogen removal. Absorption efficiency was highest at the 40-L/min water flow and the 350-mL/min oxygen flow (Figure 5). Absorption efficiency was up to 78%, which is high, especially when one considers that this efficiency can be achieved by using a sealed column without high-pressure pumps. Absorption efficiency was also good at the water flow of 30 L/min. Efficiency increased from 44% at the 50-mL/min oxygen flow to about 60% at the 150-, 250-, and 350-mL/min flows. Oxygen absorption efficiency at the 50-L/min water flow rate remained between 54 and 59% at all of the oxygen flows tested. Very small changes in DO measure-

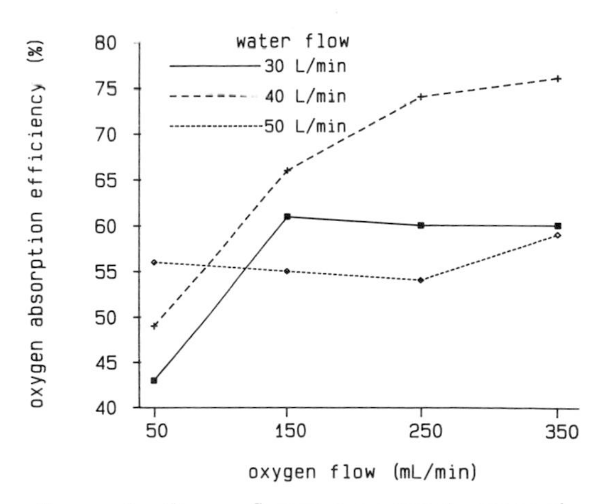

FIGURE 5.—Oxygen flow versus percent oxygen absorption efficiency at 30-, 40-, and 50-L/min water flows. A 10.16-cm column was used for these tests.

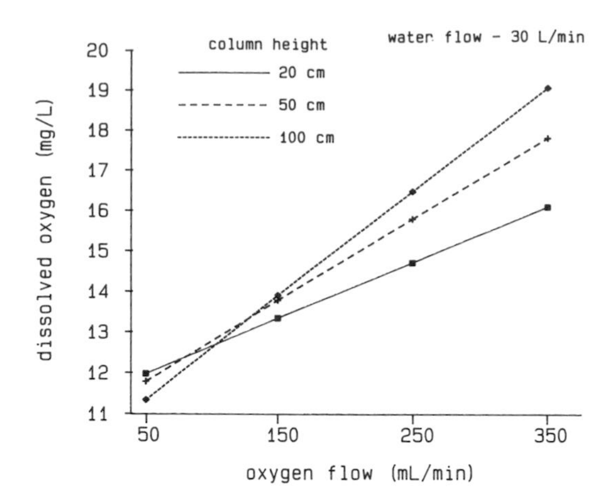

FIGURE 6.—Regression lines showing oxygen flow versus oxygen content in the effluent water at three column heights. A 10.16-cm column was used and water flow was 30 L/min.

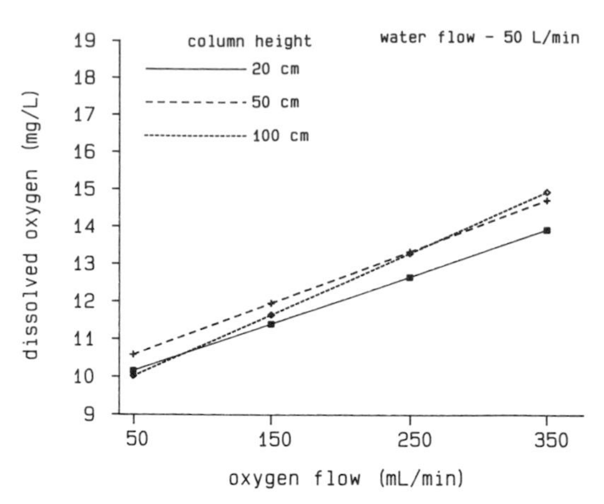

FIGURE 8.—Regression lines showing oxygen flow versus oxygen content in the effluent water at three column heights. A 10.16-cm column was used and water flow was 50 L/min.

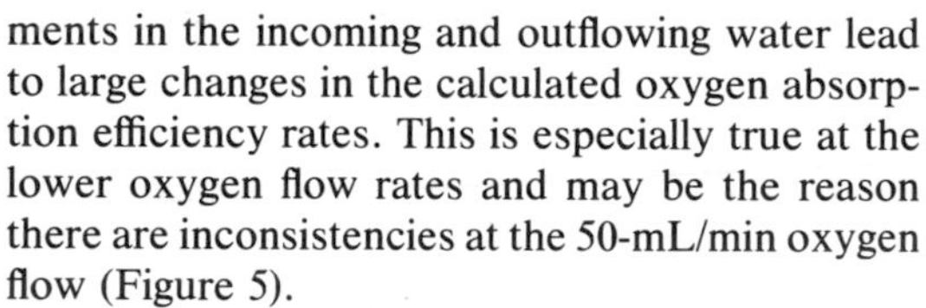

ments in the incoming and outflowing water lead to large changes in the calculated oxygen absorption efficiency rates. This is especially true at the lower oxygen flow rates and may be the reason there are inconsistencies at the 50-mL/min oxygen flow (Figure 5).

Figures 6–8 show the oxygen content of the water versus oxygen flow in mL/min for each of the three levels of head or column sizes. The measurements made at the 30- and 40-L/min water flows showed the same trends. At low-oxygen flows, there were no differences in the DO of the outflow water at any column length. However, there were differences in the water DO levels as the oxygen flow and column length increased (Figures 6, 7).

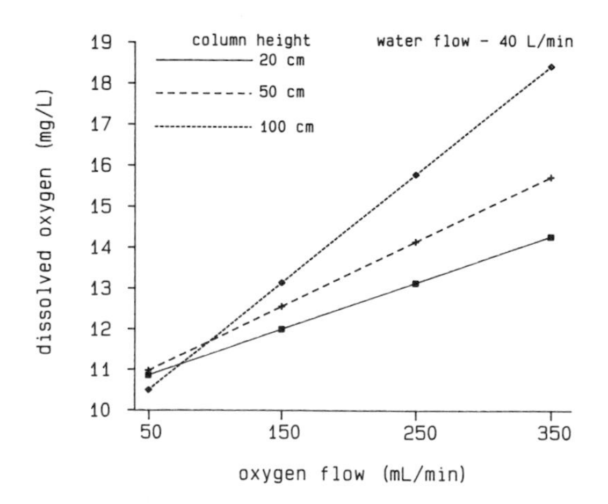

FIGURE 7.—Regression lines showing oxygen flow versus oxygen content in the effluent water at three column heights. A 10.16-cm column was used and water flow was 40 L/min.

Results of the tests at the 50-L/min water flow rate were not affected by column length (Figure 8). Although DO increased in the outflow water with each successive increase in oxygen flow rate, column length had no obvious effect. These data show that at a low-oxygen flow there is no difference in oxygen absorption efficiency based on column length, but there is an increasing effect as the oxygen injection rate increases. The higher water flow probably drew the oxygen through the column at a higher rate with less being dissolved.

A series of tests was conducted to compare oxygen absorption efficiency when the column was operated with or without media. Flow rates were set at 40 L/min for water and 350 mL/min for oxygen; when oxygen was injected at the top of the column, the oxygen content of the effluent water was 18.5 mg/L with media and 12.7 mg/L without media. Results were similar when oxygen was injected into the bottom of the column: 17.9 mg/L oxygen was measured in the outflow when operated with media and 12.9 mg/L without media. When the oxygen flow was reduced to 150 mL/min, the trends continued to be the same. DO levels were 13.0 mg/L with and 11.0 mg/L without media. These tests show that the use of media was very important in improving oxygen absorption efficiency. Media provided a means to break up the water and created a larger amount of oxygen–water interface, thus allowing for greater transfer

TABLE 3.—Mean dissolved oxygen content in effluent water when oxygen was injected at the top or bottom of a sealed column for various oxygen and water flow rates. Asterisks denote significant differences between injection sites (*t*-test, $P < 0.05$).

Flow rate			Effluent dissolved oxygen (mg/L)	
Water (L/min)	Oxygen (mL/min)	Gas:liquid ratio	Top injection	Bottom injection
30	50	0.17	10.8	10.9
30	150	0.50	14.4	14.4
30	250	0.83	16.4	16.5
30	350	1.16	18.9	19.2
40	50	0.125	10.6	10.6
40	150	0.375	13.0	13.0
40	250	0.625	15.8	14.7
40	350	0.875	18.5	17.9*
50	50	0.10	10.1	10.1
50	150	0.30	11.1	10.9*
50	250	0.50	12.4	12.9*
50	350	0.70	14.5	13.5*

of gas. This effect was most pronounced at the higher oxygen flows.

Under the conditions of this study there was little difference between the two oxygen injection points (that is, into the top or bottom of the column). Oxygen content of the water remained essentially the same when corresponding tests were compared at low-water and -oxygen flows (Table 3). For example, at the 30-L/min water flow and 50-mL/min oxygen flow, mean DO was 10.8 mg/L for top injection and 10.9 mg/L for bottom injection. An overall mean for all oxygen flows at the 30-L/min water flow was 15.1 mg/L for top injection and 15.25 mg/L for bottom injection. At higher water flows, however, efficiency was better when oxygen was introduced into the top of the column. At the 40- and 50-L/min water flows and the 350-mL/min oxygen flow, DO of the effluent water was up to 1 mg/L greater when injected at the top of the column than when injected at the bottom.

Tests were conducted with columns of larger diameter to ensure that data collected with the smaller 10.2-cm column could be extrapolated to larger sizes and flows. A 15.2-cm-diameter packed column was tested. Water flows were 67, 90, and 112 L/min versus oxygen flows of 112, 337, 562, and 786 mL/min. These are proportional to the 10.2-cm column flow rates, on the basis of cross-sectional area.

Data collected at the 90-L/min water flow produced oxygen levels of 10, 12, 15.8, and 17.3 mg/L when the 15.2-cm column was used, compared with 10.6, 13.0, 15.8, and 18.5 mg/L at the proportional 10.2-cm column water flow of 40 L/min.

Tests performed since this work was done indicate that data from a 10.2-cm column may not be extrapolated to large production-size units. The reason for this may be the relationship of the cross-sectional surface area to circumference. Small columns have a lower ratio of cross-sectional area to circumference than larger columns. For example, the 10.2-cm column has a ratio of 2.54, and a large production-size column 61 cm in diameter has a ratio of 15.2. Perhaps a larger proportion of the water flows down the column walls in smaller columns than in larger sizes.

The sealed packed columns have much potential in aquaculture, because they are simple, effective devices that can economically add a considerable amount of oxygen to water. Increasing the oxygen content of the water will allow aquaculturists to increase their fish loads, improve the rearing environment, and increase profits. Sealed packed columns can be installed in an existing facility with a minimum of head and used only as needed through the production year. If a gravity flow system that uses liquid oxygen is installed, there is little to go wrong; however, it is advisable to incorporate an alarm system. Oxygen injection systems are becoming more common and will be considered an essential part of fish production in the near future.

Acknowledgments

We thank Ziegler Brothers, Inc., for supplying the Aquatector model 50/225, Bill Mebane for his help with operation, Frederick Barrows for help with data analysis, and particularly Randy Elliott for his assistance during the initial tests. We are also grateful to the staffs of the Bozeman Fish Technology Center and the Jackson NFH for their technical assistance, and to Engineering Products for supplying the Model SL 6 15.2-cm column used in these tests.

References

APHA (American Public Health Association), American Water Works Association, and Water Pollution Control Federation. 1985. Standard methods for the examination of water and wastewater, 16th edition. APHA, Washington, D.C.

Colt, J. 1984. Computations of dissolved gas concentrations in water as functions of temperature, salinity, and pressure. American Fisheries Society Special Publication 14.

Colt, J. E., and G. Tchobanoglous. 1981. Design of aeration systems for aquaculture. Pages 138–148 *in*

L. J. Allen and E. C. Kinney, editors. Proceedings of the bio-engineering symposium for fish culture. American Fisheries Society, Fish Culture Section, Bethesda, Maryland.

Colt, J. E., and H. Westers. 1982. Production of gas supersaturation by aeration. Transactions of the American Fisheries Society 111:342–360.

Hackney, G. E., and J. E. Colt. 1982. The performance and design of packed column aeration systems for aquaculture. Aquacultural Engineering 1:275–295.

Hintze, J. L. 1987. Number cruncher statistical system, version 5.1. Hintze, Kaysville, Utah.

Owsley, D. E. 1981. Nitrogen gas removal using packed columns. Pages 71–82 *in* L. J. Allen and E. C. Kinney, editors. Proceedings of the bio-engineering symposium for fish culture. American Fisheries Society, Fish Culture Section, Bethesda, Maryland.

SAS Institute. 1982. SAS user's guide: statistics. SAS Institute, Cary, North Carolina.

Schutte, A. R. 1988. Evaluation of the Aquatector: an oxygenation system for intensive fish culture. Progressive Fish-Culturist 50:243–245.

Speece, R. E. 1981. Management of dissolved oxygen and nitrogen in fish hatchery water. Pages 53–62 *in* L. J. Allen and E. C. Kinney, editors. Proceedings of the bio-engineering symposium for fish culture. American Fisheries Society, Fish Culture Section, Bethesda, Maryland.

Speece, R. E., M. Madrid, and K. Needham. 1971. Downflow bubble contact aeration. American Society of Civil Engineers, Journal of the Sanitary Engineering Division 97:433–441.

Weitkamp, D. E., and M. Katz. 1980. A review of dissolved gas supersaturation literature. Transactions of the American Fisheries Society 109:659–702.

Westers, H. 1981. Fish culture manual for the State of Michigan—principles of intensive fish culture. Michigan Department of Natural Resources, Lansing, Michigan.

American Fisheries Society Symposium 10:437–444, 1991

Evaluation of Two Methods for Oxygenating Hatchery Water Supplies

B. LUDWIG AND G. GALE
British Columbia Ministry of Environment, Recreational Fisheries Branch
Victoria, British Columbia V8V 1X6, Canada

Abstract.—Two methods for oxygenating hatchery water supplies were examined. A preliminary test of an oxygen absorption column was conducted to determine outflow concentrations of dissolved gases over a range of water (100–445 L/min) and oxygen gas (0.25–5.0 L/min) flows. The oxygen absorption column was 1.9 m long and 0.45 m in diameter. Water and oxygen gas were introduced in a countercurrent arrangement. As oxygen gas flow increased, dissolved oxygen in the outflow water increased from approximately 100 to 290%, while dissolved nitrogen declined from 102 to 60%; total gas pressure in outflow water varied from 101 to 108%. Oxygen transfer efficiency varied from 65 to 93%, declining at the higher water flows while the efficiency of nitrogen stripping increased. Oxygen was also introduced directly to a hatchery water supply line. At a constant water flow of 660 L/min and oxygen gas flows between 5 and 8 L/min, dissolved oxygen was increased from 68.6 to 224%, while dissolved nitrogen decreased from 108.4 to 85.0%; total gas pressure increased from 100.0 to 114.2%. Oxygen transfer efficiencies exceeded 95%. Both methods of oxygenation provided efficient means to increase dissolved oxygen and decrease dissolved nitrogen in hatchery water supplies. Long-term studies are required to examine the influence of hyperoxygenated water on blood physiology.

Recently, there has been much interest among fish culturists concerning the use of hyperoxygenated water—that is, water supersaturated with dissolved oxygen (DO)—for rearing salmonids. With more oxygen available for the fish, the fish culturist theoretically can rear more fish with a given water flow. The addition of pure oxygen has the added benefit of reducing dissolved nitrogen (DN) levels to less than 100% saturation.

Because the Fish Culture Section, Recreational Fisheries Branch, is responsible for producing wild trout for approximately 1,000 lakes in British Columbia, we were interested in, but skeptical about, claims of higher loading rates with oxygenation. Our approach to oxygenation, however, was not necessarily to increase production but to improve the quality of the product at facilities constrained by water flow or quality. We saw an immediate use for the technology to improve oxygen levels in reuse water without having to pump water to the top of an aeration tower. As well, there was a potential for in-line aeration. Our initial objective was to familiarize ourselves with the oxygenation technology by installing an oxygen absorption column and testing its performance under a variety of conditions. Arising from this was a desire to introduce oxygen to a hatchery water supply line as a more simplistic and less costly approach to oxygenating water.

This report summarizes the results of two studies: a preliminary test of an Ewox 450 oxygenator and the injection of oxygen into a hatchery water supply line.

Ewox 450 Oxygenator

Methods

An Ewox 450 oxygenator was purchased from Ewos AB, Sodertalje, Sweden, and installed at the Fraser Valley Trout Hatchery, Abbotsford, British Columbia (B.C.). The oxygenator was a vented packed column filled with media resembling 3.8-cm-diameter pall rings (Figure 1). Water was introduced to the top of the unit and oxygen gas to the bottom in a countercurrent arrangement. Bottled oxygen gas was metered into the oxygenator through a pressure-compensated Chemetron flow gauge (Field's Welding and Industrial Supplies Ltd., New Westminster, B.C.). The flow gauge was calibrated at an air temperature of 21°C. The purity of the oxygen gas exceeded 99.5% (B.C. Welding Supplies Ltd., Richmond, B.C.).

Dissolved gases in the inlet and outlet water were measured at water flows between 100 and 440 L/min and oxygen gas flows between 0.25 and 5.0 L/min. Loading rates varied from 0.06 to 0.28 L/(min · cm^2). Total gas pressure was measured with a Novatech tensionometer (Novatech Designs, Victoria, B.C.). Dissolved oxygen was measured with a Leeds Northrup dissolved-oxygen meter and by the Winkler method with the azide reagent modified to measure DO levels up to

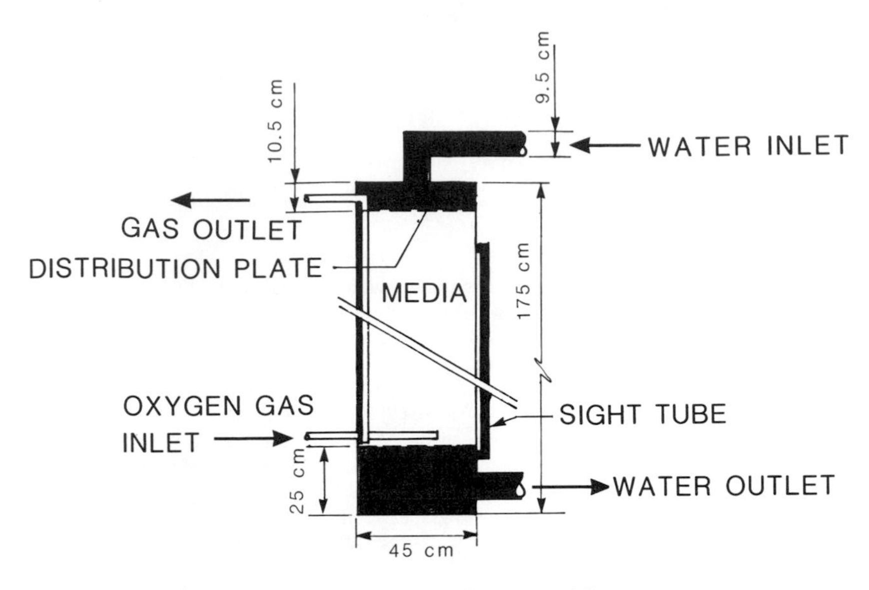

FIGURE 1.—Schematic drawing of an Ewox 450 oxygenator.

40 mg/L (APHA et al. 1975). Total gas pressure and nitrogen levels were calculated with the total gas pressure model developed by the Canada Department of Fisheries and Oceans (A. Kling, unpublished data). The tensionometer and a Gischard pocket altimeter were used to measure barometric pressure. Before we measured the dissolved gases, the oxygenator was allowed to stabilize for at least 4 h. Water flows were measured by timing the filling of a rearing trough of

TABLE 1.—Summary of water quality measurements for an Ewox 450 oxygenator at Fraser Valley Trout Hatchery.

Flow (L/min)		Temperature (°C)		Barometric pressure (mm Hg)	Dissolved oxygen			
					Inflow		Outflow	
Water	Oxygen	Water	Air		mg/L	%	mg/L	%
113	0.25	10.1	7.0	747	11.2	101.5	13.7	124.1
102	0.50	10.2	8.8	747	10.7	97.2	16.3	148.0
101	0.75	9.9	10.6	756	11.3	100.7	20.5	182.6
97	1.0	10.2	8.4	745	11.1	101.1	23.0	209.4
164	1.0	7.4		773	12.1	99.2	20.2	165.6
159	2.0	7.8		769	11.8	98.2	24.7	205.6
159	3.0	7.8		769	11.8	98.2	29.3	243.9
162	4.0	7.6		769	11.8	97.7	34.9	289.0
197	0.75	10.7	11.2	748	10.6	97.2	15.4	141.3
197	1.0	10.5	10.2	743	11.1	102.0	17.6	161.8
199	1.5	10.2	6.2	742	11.2	102.4	20.4	186.5
207	2.0	11.0	16.1	759	11.0	100.2	22.2	202.2
307	2.0	9.6	10.7	750	11.2	99.8	19.2	171.2
302	3.0	10.1	12.0	754	11.4	102.3	23.2	205.7
309	4.0	9.3	8.5	751	11.4	100.8	26.0	229.8
311	5.0	9.5	8.5	738	11.4	103.0	29.4	265.7
415	1.0	10.5	14.3	744	11.1	101.5	13.6	124.4
398	2.0	10.0	5.9	747	11.2	101.6	16.5	149.7
392	3.0	10.0	5.6	746	11.3	102.2	19.2	173.7
392	4.0	10.0	8.6	746	11.0	99.5	21.0	190.0
440	2.0	9.3	7.2	767	11.7	101.3	16.4	142.0
445	3.0	9.3	7.4	767	11.7	101.3	18.9	163.6
436	4.0	9.2	7.0	765	11.7	101.3	21.4	185.3
432	5.0	9.4	6.8	762	11.5	100.4	24.4	213.1

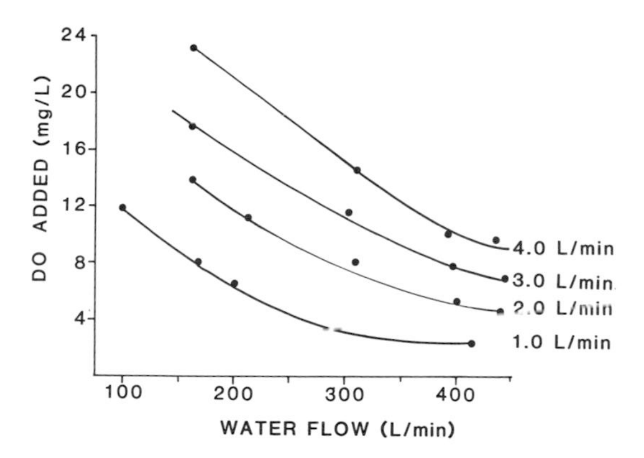

FIGURE 2.—Influence of water flow on outflow dissolved oxygen (DO) from an Ewox 450 oxygenator supplied with various flows of oxygen gas.

known volume. Water and air temperatures were measured with a hand-held thermometer.

Results and Discussion

As water flow increased, the amount of DO added and DN stripped decreased (Table 1; Figures 2, 3). Water flows between 100 and 300 L/min showed linear inverse relationships with dissolved gas levels, but the slope of the lines decreased, especially for DO, at flows exceeding 300 L/min. Relationships between oxygen gas flow and dissolved gas levels also were linear but over the entire range of gas flows tested (Figures 4, 5).

The relationship between DO added and DN stripped (Figure 6) was described by the expression DO added = 2.63(DN stripped) + 1.07. For every 3 mg DO/L added, roughly 1.0 mg DN/L was stripped. The ratio of DO to DN varied between 2.5 and 4.2 (mean 3.1 ± 0.55, 1 SD) (Table 2). The higher ratios occurred when water flows of 300 L/min or less were combined with oxygen gas flows of 1.0 L/min or less. As well, for the water flows of 100, 160, 200, and 300 L/min, the ratio of DO to DN decreased as the oxygen gas flow increased.

For water flows less than 100 L/min, small additions of oxygen gas resulted in very high levels of DO (Table 1). The original literature supplied with the Ewox 450 indicated that the unit was designed to handle a maximum flow of 450 L/min. The manufacturer subsequently advised

TABLE 1.—Extended.

Dissolved nitrogen				Total gas pressure (%)		Tensionometer reading (mm Hg)		Height of water in Ewox (cm)	Oxygen transfer efficiency (%)
Inflow		Outflow							
mg/L	%	mg/L	%	Inflow	Outflow	Inflow	Outflow		
18.6	101.0	18.0	97.8	101.1	103.2	8	24	62	79.1
18.8	102.2	17.4	94.7	101.1	105.8	8	43	41	80.7
19.0	101.4	16.3	87.1	101.2	107.1	9	53	29	88.4
18.6	101.5	14.7	79.8	101.4	106.9	10	51	27	81.4
20.5	101.1	18.5	91.4	100.6	106.9	5	53	27	93.2
20.3	101.7	16.2	81.0	100.9	107.1	7	54	27	72.1
20.3	101.7	14.2	70.9	100.9	107.1	7	54	27	65.2
20.4	101.5	12.1	60.2	100.7	108.2	5	62	27	65.7
		17.4	95.4		105.0		37	27	90.3
18.4	101.1	16.5	90.7	101.2	105.6	9	41	27	91.2
18.5	101.0	15.4	84.2	101.2	105.6	9	41	27	85.1
		14.6	79.2		104.9		37	27	84.4
19.0	101.3	16.7	89.2	100.9	106.3	7	47	27	87.7
		14.9	80.8		107.0		52	27	84.1
19.1	101.0	14.1	74.8	100.9	107.3	7	54	27	79.6
18.6	100.4	12.1	65.7	101.0	107.5	7	55	27	79.0
18.3	100.0	17.3	94.6	100.3	100.8	2	6	150	73.4
18.5	100.6	16.4	89.1	100.8	101.8	6	13	140	75.5
18.6	100.8	15.5	83.9	101.1	102.7	8	20	130	71.7
18.8	101.5	14.8	80.1	101.1	103.1	8	23	123	71.0
19.4	100.5	17.7	91.6	100.7	102.1	5	16	141	72.5
19.4	100.5	16.7	86.5	100.7	102.6	5	20	127	74.9
19.4	100.7	15.8	81.8	100.8	103.4	6	26	121	74.0
19.1	100.1	14.4	75.4	100.1	104.2	1	32	109	78.0

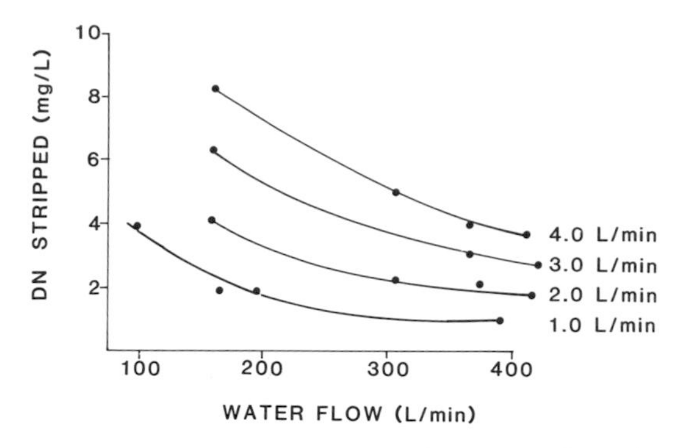

FIGURE 3.—Influence of water flow on outflow dissolved nitrogen (DN) from an Ewox 450 oxygenator supplied with various flows of oxygen gas.

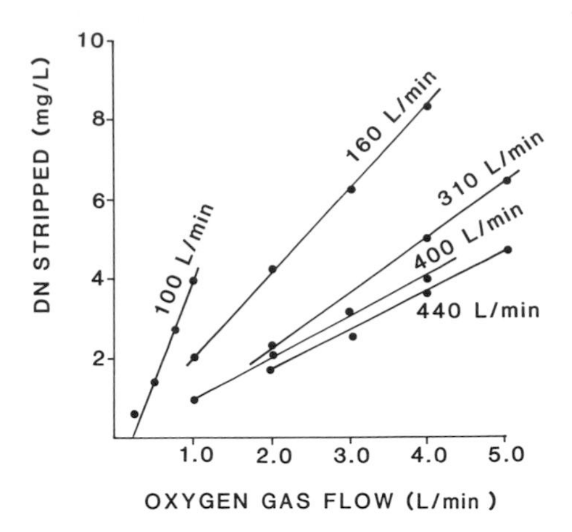

FIGURE 5.—Influence of oxygen gas flow on outflow dissolved nitrogen (DN) from an Ewox 450 oxygenator at various water flows.

that the unit could handle a flow of 600 L/min (K. Carlsson, Ewos AB, personal communication). Limitations in the available water flow prevented us from testing at the suggested flow, but if small flows of water are to be treated, the unit must be sized accordingly to avoid excessive levels of DO and high total gas pressure.

To compare the performance of the Ewox oxygenator with other types of oxygenators and to identify the optimum operating conditions for the Ewox 450, the oxygen transfer efficiency of the unit was calculated. Efficiency refers to the weight of DO added to a given flow of water, expressed as a percentage of the weight of oxygen gas introduced to the unit. The gas flow, which was measured in liters per minute, was converted to weight (g/min). According to the manufacturer (Carlsson, personal communication), the flow meter had to be adjusted for calibration pressure and temperature.

$$C_P = [(1 + P_r)/(1 + P_e)]^{1/2};$$

$$C_t = [(273 + t_e)/(273 + t)]^{1/2};$$

P_r is actual pressure in atmospheres, P_e is calibration pressure in atmospheres, t_e is calibration temperature in °C, and t is air temperature in °C. The flow of oxygen gas (V) was then corrected by $V = (C_P)(C_t)(F)$, F being oxygen gas flow in liters per minute as measured by a flowmeter. The conversion of the gas flow to weight was based on a derivation of the ideal gas law (Keenan and Wood 1971): $m = (P)(V)(M)/(0.082T)$; m is flow of

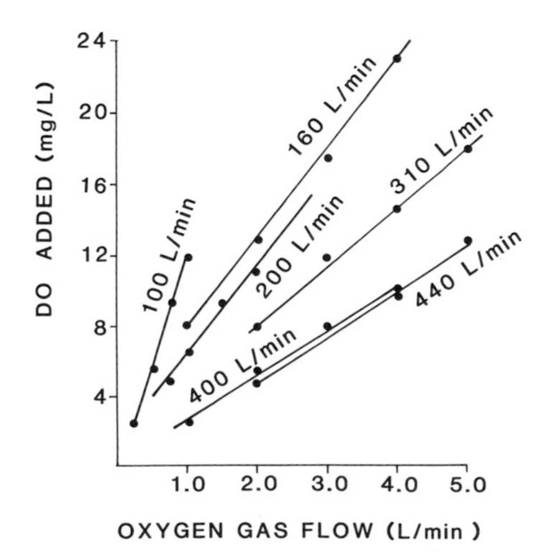

FIGURE 4.—Influence of oxygen gas flow on outflow dissolved oxygen (DO) from an Ewox 450 oxygenator at various water flows.

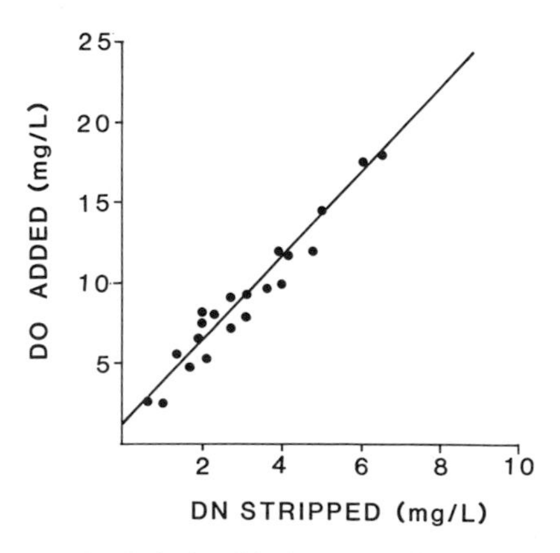

FIGURE 6.—Relationship between dissolved oxygen added (DO) and dissolved nitrogen stripped (DN) in outflow water from an Ewox 450 oxygenator.

TABLE 2.—Ratio of dissolved oxygen (DO) added to dissolved nitrogen (DN) stripped in outflow water from an Ewox 450 oxygenator.

Water flow (L/min)	Oxygen gas flow (L/min)	DO added (mg/L)	DN stripped (mg/L)	Ratio DO:DN
113	0.25	2.5	0.6	4.2
102	0.50	5.6	1.4	4.1
101	0.75	9.2	2.7	3.4
97	1.0	11.9	4.0	3.0
164	1.0	8.1	2.0	4.1
159	2.0	12.9	4.1	3.1
159	3.0	17.5	6.2	2.8
162	4.0	23.1	8.3	2.8
197	1.0	6.5	1.9	3.5
199	1.5	9.2	3.1	3.0
307	2.0	8.0	2.3	3.5
309	4.0	14.6	5.0	2.9
311	5.0	18.0	6.4	2.8
415	1.0	2.5	1.0	2.5
398	2.0	5.3	2.1	2.5
392	3.0	7.9	3.1	2.5
392	4.0	10.0	4.0	2.5
440	2.0	4.7	1.7	2.7
445	3.0	7.2	2.7	2.7
436	4.0	9.7	3.6	2.7
432	5.0	12.9	4.7	2.8

gas in g/min, T is air temperature in °K, P is pressure inside the unit in atmospheres, V is corrected gas flow in L/min, and M is molecular weight of oxygen. Efficiency was thus described by $E = 100(f)(\mathrm{DO})/m$; DO is dissolved oxygen added in g/L and f is water flow in L/min. Pressure inside the Ewox was not measured in this study, but according to the manufacturer (K. Carlsson, personal communication), a typical pressure was 0.5 atmospheres.

There were several potential problems with this calculation including (1) the pressure inside the unit varied, depending on the operating height of water, and (2) the assumption that the temperature of compressed gas was the same as the air temperature may not have been valid (J. Colt, J. M. Montgomery Consulting Engineers, personal communication). In an attempt to resolve the question of pressure inside the unit, a flowmeter was calibrated by expelling oxygen gas into an inverted volumetric flask filled with water and measuring the time taken to replace the water with gas. At a calibrated flow setting, the change in weight over time for a bottle of compressed oxygen was measured. This was done with the oxygen line connected to the Ewox and then with it disconnected. At a flow setting of 1.0 L/min, the change in weight for the bottle was 1.39 g/min when the line was connected and 1.34 g/min when the line was disconnected. At a known air temperature and with use of the ideal gas law without the calibration pressure adjustment, the gas flow was 1.36 g/min. The close similarity of these three figures indicated that one need not account for calibration pressure or pressure inside the Ewox. A simpler and possibly more accurate method to convert the flow measurements to weight and to calculate efficiency is to weigh the oxygen bottles and compute a weight change over time. This method was not tested until after the Ewox had been dismantled and thus could not be applied to the Ewox data. The conversion of gas flow from L/min to g/min was described by

$$m = (V)(M)/0.082T,$$

V being corrected for air temperature. A sample calculation of oxygen transfer efficiency for a water flow of 307 L/min and an oxygen gas flow of 2.0 L/min (Table 1) is as follows:

$$\mathrm{DO} = 8.0 \text{ mg/L or } 0.008 \text{ g/L};$$
$$\text{air temperature} = 10.7°\mathrm{C};$$

$$V = (C_t)f = [(273 + 21)/(273 + 10.7)]^{1/2}(2.0) = 2.036;$$

$$m = (V)(M)/(0.082)T = 2.036(32)/0.082(273 + 10.7) = 2.8006;$$

$$E = 100(f)(\mathrm{DO})/m = 100(307)(0.008)/2.8006) = 87.7\%.$$

Efficiencies determined by the above method varied from 65.2 to 93.2% (Table 1). Efficiencies for DO addition tended to decline as water flow increased from 100 to 400 L/min, although between 400 and 440 L/min the average efficiency increased slightly from 72.9 to 74.8%. This slight increase in performance for DO addition is also reflected in the decrease in slope for the relationship between DO added and water flow (Figure 2).

The efficiency for the Ewox was considerably higher than the 50% experienced by G. Boersen (Michigan Department of Natural Resources, unpublished data) for a modified, sealed packed column, but was lower than the 90% quoted by Speece (1981) for an enclosed packed column.

What makes the Ewox oxygenator unique is that it is vented, yet only certain operating conditions allow the vent to function. Most oxygenators of a similar design are completely sealed.

The addition of a sight tube allowed the height of water inside the unit to be monitored. Under normal operating conditions, the water height inside the unit was established by the level of the gas outlet tube. The bottom of the gas outlet tube was 27 cm from the bottom of the unit. Water dropped through the packing and was exposed to an oxygen-rich atmosphere, where gas exchange occurred. The water then pooled in the bottom of the unit before it was expelled. As the gas pressure inside the unit increased, the water height gradually decreased until it was below the level of the gas outlet tube. A mixture of gas and water was then expelled from the tube. The reduction in internal pressure allowed the level of water to increase abovc the bottom of the tube and the process was repeated. Thus the levels of water cycled approximately 1–2 cm in height as the vent tube expelled the gas–water mixtures.

At very low gas flows or high water flows, the water height rose well above the level of the outlet tube (Table 1). Under some conditions, the unit was very nearly or completely filled with water. It often took several hours for the operating water height to stabilize when changes to water or gas flows were made. At these flows, the gas outlet tube expelled water only. When the unit was filled with water, the mode of operation changed. Instead of gas exchange occurring between an oxygen-rich atmosphere and a thin film of water spread over the media, gases were exchanged at the bubble–water interface as the bubbles rose through the column and were broken up by the media. Poorer performance might be expected from the oxygenator because the nitrogen and excess oxygen gas could not escape from the gas outlet tube under these conditions and, in fact, the oxygen transfer efficiency was lower for the two higher flows of 400 and 440 L/min. However, the DO:DN ratio was also lower, as was the total gas pressure (TGP) (Tables 1, 2). Total gas pressure generally increased as dissolved oxygen was added but the amount of increase was greater for higher flows. The decline in oxygen transfer efficiency was not of sufficient magnitude to account for the changes in the DO:DN ratio and TGP, implying that an increase in nitrogen-stripping efficiency was also occurring. When the oxygenator was operated at high flow, the level of water inside the unit was above the level of the gas outlet tube (Table 1). Although the pressure inside the unit was not measured in this study, under some flow conditions, and depending on the elevation of the outlet pipe compared with the operating height of the water, the unit may actually operate under a slight vacuum; thus oxygen transfer efficiency would be expected to decline and nitrogen stripping would be promoted.

By expressing DO as "mg/L added" and DN as "mg/L stripped," we had hoped to use the relationship between dissolved gases and flows to predict the performance of the oxygenator at other locations with different levels of incoming gases and different targets for outflow gases. However, when the oxygenator was subsequently installed at a new location, we were unable to obtain the same performance (B.L., unpublished data). A comparison of gas levels in the outflow water indicated poorer performance at the new location, in particular for DN and TGP. Because of space constraints at the new location, the inflow and outflow plumbing was less direct and water had to pass through several elbows and valves. Also, there was less pressure on the water supplied to the top of the column. At the previous location, the difference in elevation between the top of the unit and the top of the head tank supplying the water was about 4.0 m, compared with 2.95 m at the new location. The manufacturer recommended that the Ewox be installed with as few valves and elbows in the inlet and outlet piping as possible and that the water be delivered under low pressure (Carlsson, personal communication). The ideal setup, according to the manufacturer, is to have the water drop through the unit and directly into the ponds, but for most applications this may be difficult. The apparent sensitivity of the unit to plumbing configuration of the inlet and outlet makes it more difficult to predict performance of the units at new locations, especially for DN stripping.

In summary, Ewox oxygenators are a relatively efficient way to oxygenate either large flows or sidestream flows. They are easy to install and operate, and they appear to be well constructed. Their performance will depend on the method of installation, because the units seem to be sensitive to plumbing configuration and possibly to water pressure. Direct plumbing connections with a minimum number of valves and bends, and low inlet water pressures are recommended. Depending on the desired dissolved oxygen levels in outflow water, the Ewox may be operated as a highly efficient oxygenator or as an oxygenator–vacuum degasser simply by adjusting incoming water and oxygen gas flows and the configuration of inlet and outlet plumbing.

TABLE 3.—Introduction of oxygen gas into the upper spring water supply line at Summerland Trout Hatchery at a water flow of 660 L/min and a water temperature of 10.5°C.

Oxygen gas flow (L/min)	Oxygen gas flow (g/min)	Barometric pressure (mm Hg)	Tensionometer reading (mm Hg)	Dissolved oxygen mg/L	Dissolved oxygen %	Dissolved nitrogen mg/L	Dissolved nitrogen %	Total gas pressure (%)	Oxygen transfer efficiency (%)
0	0	755	0	7.6	68.6	20.8	108.4	100	
5	8.67	755	101	20.8	187.7	17.4	93.9	113.6	100.5
6	9.33	755	102	21.9	197.7	17.0	91.5	113.7	101.0
7	10.67	755	105	23.2	209.4	16.5	88.9	114.1	96.5
8	12.0	755	106	24.9	224.8	15.8	85.0	114.2	95.2

Injection of Oxygen into a Water Supply Line

Methods

The Summerland Trout Hatchery is adjacent to Okanagan Lake in Summerland, B.C. The water is taken from two small springs and flows to the hatchery through gravity-feed supply lines. The two springs differ in water quality. The lower spring has the best water quality and is the primary water supply. The upper spring is only used to supplement the supply from the lower spring.

The upper spring line consists of a 20-cm-diameter line that is reduced to a 7.6-cm line. There was insufficient water velocity in the 20-cm line to carry the oxygen down the line, so oxygen was introduced through a coupler at the point where the line was reduced. A 46-cm-long piece of perforated plastic tubing (0.6 cm, inside diameter) was connected to an oxygen bottle and threaded through the coupler. The difference in elevation between the point of gas introduction and the hatchery was 28 m. The distance between the same two points was approximately 229 m. The contact time was approximately 1.6 min at a water flow of 660 L/min. The water temperature was 10.4°C. Oxygen was supplied from a pressurized cylinder (1.4 m^3). Oxygen gas flow was measured with a pressure-compensated flow gauge (0–15 L/min). The flow gauge was calibrated by the method discussed earlier. Dissolved-gas measurements were made for oxygen gas flows between 5 and 8 L/min, based on methods similar to those used at Fraser Valley Trout Hatchery. To determine the weight of oxygen added to the spring line, the oxygen bottle was weighed before and after each set of measurements, and the elapsed time was recorded. Oxygen transfer efficiency was calculated by expressing the weight of DO added to the water as a percentage of the weight of oxygen added to the line.

Results and Discussion

As the oxygen gas flow increased from 5 to 8 L/min, there were significant increases in DO and decreases in DN (Table 3). Total gas pressure increased from 100% in unoxygenated upper spring water to 114.2% at the highest oxygen gas flow. Oxygen transfer efficiencies exceeded 95%.

The average ratio (±SD) of DO added to DN stripped was 4.4 ± 0.4. This was considerably higher than the ratio for the Ewox, which indicated that the system was not as efficient as the Ewox at stripping nitrogen. The reason for this is unknown but may be related to water pressure.

The advantages of introducing oxygen gas directly to a supply line are the low installation cost, the ease with which the system can be installed and operated at many hatchery sites, and the high oxygen transfer efficiency. All that is required is a convenient access point to introduce the oxygen, plus sufficient contact time. One potential problem is the risk of a partial blockage of flow in the line due to coalescing of gas. Unless the line is under extremely low pressure, the potential of a total flow blockage is thought to be minimal. No flow reductions were noted during our tests.

At Summerland Trout Hatchery, the high total gas pressure does not present a problem because the oxygenated water will be mixed with reuse water, resulting in considerably lower TGP. At facilities that use oxygen on a first-pass basis, high TGP could result in gas bubble disease even though the excess gas pressure is due to DO. In a recent study, rainbow trout *Oncorhynchus mykiss* reared in water having 200% DO, 90% DN, and 113% TGP, suffered low-level losses and poor growth (B.L., unpublished data). Bubbles were apparent between the fin rays, and tissue on the fins was heavily eroded. By reducing the DO to 150%, the TGP was reduced to 110% and there were no further problems.

Future plans are to alter the intake structure at Summerland Trout Hatchery and test other means of introducing the oxygen gas, such as ceramic tiles or spargers. The smaller bubbles produced by these devices will likely improve efficiency of DO addition and, possibly, of DN stripping. This will be useful where the contact time for oxygen in the supply line is limited.

The long-term effects on blood physiology of rearing trout in hyperoxygenated water need to be investigated. Some preliminary work with native rainbow trout has indicated that fish reared in hyperoxygenated water have lower hematocrit levels (B.L., unpublished data). The effect of this on survival of cultured fish once they are released is unknown.

Acknowledgments

We gratefully acknowledge the efforts of D. Larson and L. Lemke in providing facilities and staff to assist in the work, of D. Stanton in collecting the field data at Summerland Trout Hatchery, and of B. McLean, K. Ashley, D. Clough, and two unknown reviewers for reviewing the manuscript. Figures were prepared by G. Sutcliffe.

References

APHA (American Public Health Association), American Water Works Association, and Water Pollution Control Federation. 1975. Standard methods for the examination of water and wastewater, 14th edition. APHA, Washington, D.C.

Keenan, C. W., and J. H. Wood. 1971. General college chemistry. Harper and Row, New York.

Speece, R. E. 1981. Management of dissolved oxygen and nitrogen in fish hatchery waters. Pages 53–62 *in* L. J. Allen and E. C. Kinney, editors. Proceedings of the bio-engineering symposium for fish culture. American Fisheries Society, Fish Culture Section, Bethesda, Maryland.

American Fisheries Society Symposium 10:445–449, 1991

Design and Operation of Sealed Columns to Remove Nitrogen and Add Oxygen

HARRY WESTERS

Michigan Department of Natural Resources, Fisheries Division
Post Office Box 30028, Lansing, Michigan 48909, USA

GARY BOERSEN

Michigan Department of Natural Resources, Surface Water Quality Division
Post Office Box 30028, Lansing, Michigan 48909, USA

VERNON BENNETT

Michigan Department of Natural Resources
Harrietta State Fish Hatchery, Harrietta, Michigan 49638, USA

Abstract.—The State of Michigan equipped five state fish hatcheries with pressure swing absorption oxygen generators and sealed columns for aeration and degassing. Sealed columns can increase dissolved oxygen levels to far above saturation without exceeding total gas pressures. Many variables affect column performance, and best efficiency in performance is not necessarily the best economically. Column design and operational characteristics have been delineated but not firmly established. Formulas are presented to determine absorption efficiencies, oxygen flow requirements, and capacities of pure oxygen systems for facilities. Fine-tuning of the existing design is needed along with continued development of systems to introduce pure oxygen into solution under a variety of aquaculture operations.

The Michigan Department of Natural Resources, Fisheries Division, augmented or replaced spring or creek sources with well water during a hatchery renovation program. Packed columns (Owsley 1981) were provided for aeration and nitrogen removal. Dissolved oxygen (DO) levels were adequate (>90% saturation) and total gas pressure (TGP) was reduced to a low of 101–103% saturation. Despite these apparently favorable conditions, chronic losses of salmonid fingerlings plagued the new facilities, and eventually it was determined that these were caused by low-level gas supersaturation. Typical gas bubble disease symptoms were absent, but microbubbles could be found in the capillaries of the gills, eventually resulting in necrotic tissue.

Experiments with high-purity oxygen, added to modified packed columns, demonstrated that TGP could be reduced to less than 100% with the added benefit of increased dissolved oxygen levels (Boersen and Chesney 1986).

Column Design

A general column design is shown in Figure 1. From experiments it was discovered that a sealed column gave the best results. Compared with a normal packed column with the discharge above the water surface, the discharge is lowered into the water 15–20 cm below the surface. The column top is sealed hermetically. Oxygen and water are fed through the cover into the column, or the oxygen is injected directly into the water delivery line.

Packing media, such as pall rings, may be used to break up the water where clean, nutrient-free water is used. Debris-laden or nutrient-rich water precludes the use of packing because of plugging and slime growth. In such situations, nonplugging spray heads have been used. It has been demonstrated that media can improve column performance (Dwyer et al. 1991, this volume).

To simultaneously degas and oxygenate the water, it has been determined that a partial vacuum is advantageous. Flow rates are based on a column cross-sectional area of 1.0–2.0 cm^2/(L/min) flow (244–122 ft^2/[gal/min]). A relationship of 2.0 cm^2/(L/min) gave optimum performance for a 10-cm-diameter sealed column (Dwyer et al. 1991). Constricting the column outlet to half the diameter of the column itself makes it function as a flow-limiting device.

The column diameter can be determined with equation (1).

Column diameter

$$= 2\sqrt{\frac{[\mathrm{cm}^2/(\mathrm{L/min})] \times [\mathrm{L/min\ H_2O}]}{\pi}}. \qquad (1)$$

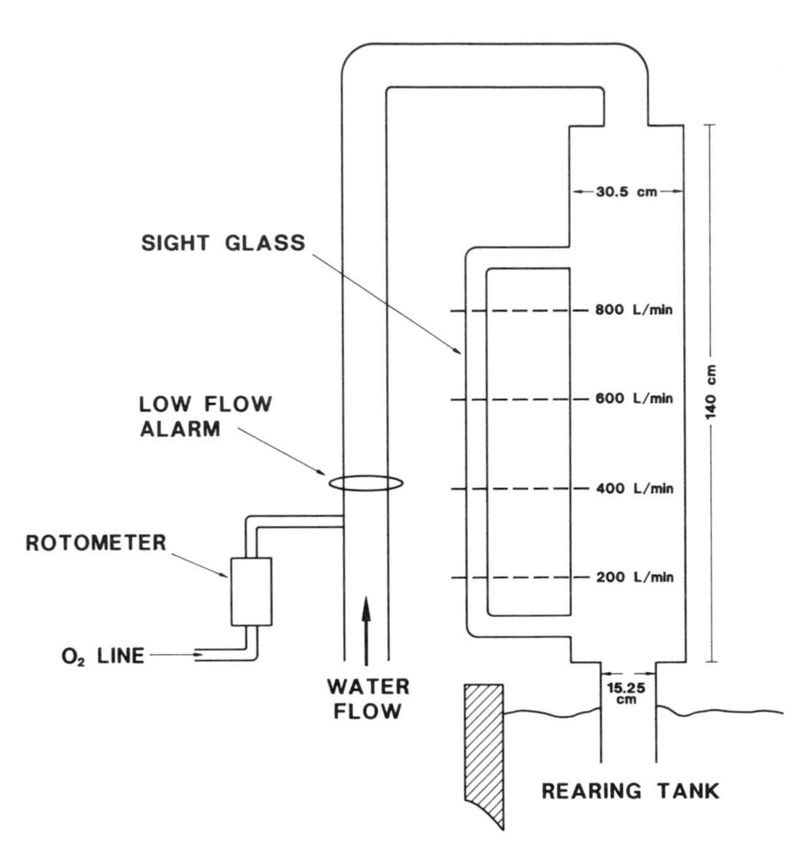

FIGURE 1.—Design of a sealed column.

The partial vacuum is created through venturi action and may range from zero to as much as 200 mm Hg, depending on flow rate, cross-sectional area, and the flow rate of oxygen added.

Many combinations of variables in design and operational modes of packed columns can result in the same oxygen gain while also maintaining TGP at 100%. The objective should be to achieve these results at the lowest possible cost (Nirmalakhandan et al. 1988). High absorption efficiencies can be achieved at low oxygen flow rates relative to the flow rate of water, but at a higher cost because a small amount of oxygen is added relative to the energy required and the capital investment. This is reflected in equation (2) developed by Nirmalakhandan et al. (1988).

Cost per kg O_2 absorbed

$$= \frac{(\text{amortization cost} + \text{energy cost} + O_2 \text{ cost})}{(\text{water quantity}) \times (\text{DO gain})}. \tag{2}$$

Wherever high-purity oxygen is used in aquaculture, it is important to determine the absorption efficiency of the system under different modes of operation. Equations (3)–(5) are offered for this purpose.

% absorption efficiency

$$= \frac{(H_2O \text{ flow in L/min}) \times 100}{1.43 \times 1{,}000 \times O_2 \text{ flow in L/min}} \times \text{mg/L DO increase}; \tag{3}$$

1.43 g is weight of 1 L of oxygen at standard temperature and pressure (20°C and 1.0 atmosphere), and 1,000 converts miligrams (in mg/L O_2) to grams. To further simplify the equation, 100/(1.43 × 1,000) is changed to 14.3, and

% absorption efficiency

$$= \frac{\text{L/min } H_2O}{14.3 \times \text{L/min } O_2} \times \text{mg/L DO increase}. \tag{4}$$

This equation is for 100% purity oxygen. In the case of oxygen generated by pressure swing absorption (PSA) equipment, the purity varies from 90 to 96%. This alters equation (4) to (5).

% absorption efficiency (PSA)

$$= \frac{\text{L/min } H_2O \times 100}{\text{\% purity } (14.3 \times \text{L/min } O_2)} \times \text{mg/L DO increase.} \quad (5)$$

From the equations it is apparent that the accuracy depends on the correct measurements of both the water flow and the flow rate of the oxygen added.

The transfer efficiency (kg O_2/kWh) is more difficult to determine because it includes the power requirements of all pumps and compressors. For the system described here, which includes pumping of all of the water, it is approximately 0.5 kg O_2/kWh. Colt and Watten (1991, this volume) report that the transfer efficiency for pure oxygen systems using a liquid oxygen source ranges from 0.20 to 4.0 kg/kWh.

Because any increased dissolved oxygen is available to the fish, even a 1.0-mg/L increase could easily represent a 20–25% increase in production capacity. When naturally incoming dissolved oxygen levels approach saturation, at best 50% of the oxygen will be available to the fish, because the effluent (salmonids) should contain from 5.0 to 7.0 mg/L DO. The effluent levels depend on the rearing-water temperature. At 15°C, 5.0 mg/L in the effluent is acceptable, but at 5°C this should be about 7.0 mg/L. In either case, with incoming DO near saturation, about 5.0 mg/L is available to the fish. Adding, therefore, just 1.0 mg/L results in 20% more available oxygen and production can be increased accordingly.

Once it has been decided by how much to increase the dissolved oxygen level, equation (6) can be applied to determine the flow of oxygen to be added.

$$\text{L/min } O_2 = \frac{\text{L/min } H_2O}{14.3 \times \text{expected \% AE}} \times \text{desired mg/L DO increase;} \quad (6)$$

AE is absorption efficiency.

To determine the minimum capacity of an oxygen system for a fish production facility, one must decide on how much the dissolved oxygen level is to be increased relative to the total flow of water to be treated. If on-site PSA-generated oxygen is the choice, the required capacity of the system can be determined with equation (7):

Cubic meters O_2 per hour

$$= \frac{\text{L/min } H_2O \times 60 \times 100}{\text{\% purity} \times 14.3 \times 1{,}000 \times \text{\% AE}} \times \text{mg/L DO increase.} \quad (7)$$

Because the capacity of PSA oxygen generators often is expressed in cubic feet per hour, equation (8) can be used.

Cubic feet O_2 per hour

$$= \frac{\text{L/min } H_2O \times 6{,}000}{\text{\% purity} \times 14.3 \times 28.3 \times \text{\% AE}} \times \text{mg/L DO increase.} \quad (8)$$

In the case of liquid oxygen, the size of the storage tank must be determined.

One gallon of liquid oxygen represented 115 ft^3 of gaseous oxygen at standard temperature and pressure. Assume that for each 100-ft^3h requirement (equation 8), 720 gal liquid are required per month. For a 6-month supply, a 4,320-gal tank is needed. To maintain proper pressure and temperature, approximately 0.25% of the liquid oxygen is vaporized off daily, if the tank is off-line. To prevent this loss, the minimum daily gas use rate should exceed the off-line vent rate.

Column Operations and Performance

The State of Michigan has equipped five state fish hatcheries with on-site PSA oxygen generators. All use sealed columns for aeration and degassing. Column design characteristics vary somewhat, and internally there may be differences also, such as the use of packing or nozzles to break up the water.

The Harrietta State Fish Hatchery operates nine 12-m^3 indoor rearing tanks, each furnished with a 30.5-cm-diameter column for maximum water flows of 800 L/min. Aeration chambers for the 12 outdoor raceways operate with 94.0-cm-diameter columns, rated for flows up to 4,000 L/min. Many operational data have been collected and analyzed and are summarized for the indoor column in Table 1.

Figure 1 shows the design of the 30.5-cm-diameter sealed column used for the indoor rearing tanks at the Harrietta State Fish Hatchery. This column receives aerated well water with a constant temperature of 8.2°C. The dissolved oxygen is 10.7 mg/L at 92% saturation. Dissolved nitrogen varies from 101 to 105% saturation. The sight glass functions as a flowmeter.

TABLE 1.—Operation and performance characteristics of a sealed packed column at the Harrietta State Fish Hatchery (Figure 1). Column diameter is 30.5 cm and discharge pipe diameter is 15.25 cm. Water characteristics: temperature, 8.2°C; dissolved oxygen (DO), 10.7 mg/L; O_2 saturation, 92%; N_2 saturation, 101–105%.

Measure	Water flow in L/min			
	200	400	600	800
Cross-sectional area-to-flow relationship (cm^2/[L/min])	3.65	1.80	1.20	0.91
DO gain in mg/L for O_2 inflows of				
1.0 L/min	2.5	1.5	1.0	0.9
1.5 L/min	3.9	2.3	1.8	1.3
2.0 L/min	4.7	3.7	2.1	1.8
O_2 absorption efficiency in percent for O_2 inflows of				
1.0 L/min	35	42	42	50
1.5 L/min	36	43	50	48
2.0 L/min	33	52	44	50
% gas:liquid ratios for O_2 inflows of				
1.0 L/min	0.50	0.25	0.17	0.13
1.5 l/min	0.75	0.37	0.25	0.19
2.0 L/min	1.00	0.50	0.33	0.25
Vacuum in mm Hg	36	54	74	91
%O_2 saturation for O_2 inflow of 1.5 L/min^2	121	108	104	103
%N_2 saturation for O_2 inflow of 1.5 L/min^2	94	98	99	99.5

TABLE 2.—Relationship of column wall (cm^2) to flow (L/min) for four column diameters at a flow of 1.0(L/min)/cm^2. Column length is 150 cm.

Measure	Column diameter (cm)			
	10	30	50	100
Column cross-sectional area (cm^2)	78.5	706.8	1,963.5	7,854
Column wall (cm^2)	4,710	14,145	23,565	47,100
Wall-to-flow relation (cm^2/[L/min])	60	20	12	6

The primary objective was to eliminate low-level nitrogen gas supersaturation and not to exceed a TGP over 100%. Starting with dissolved oxygen levels near saturation contributes to relatively poor absorption efficiencies, which ranged from 33 to 52%. Kindschi et al. (1988) found the best oxygen absorption efficiencies when dissolved oxygen levels were lowest. A decrease of only 1.0 mg/L DO increased absorption efficiency by 25.5%. Absorption efficiencies calculated by Watten et al. (1991, this volume) were 100% higher when DO started at 0% saturation than when it began at 100%.

Because of the initial high oxygen level (92%) at the Harrietta hatchery, the percent gas: liquid ratios are very small, ranging from 0.13 to 1.00%. The lower ratios are associated with better absorption efficiencies than those in the higher range. The same trend was observed by Nirmalakhandan et al. (1988): efficiency was better when the ratio was 1.00% than when it was higher, up to 3.80%. However, the cost per kilogram of oxygen was higher at a 1.00% ratio ($0.258) than at a ratio of 1.80% ($0.228). Dwyer et al. (1991), using a 10-cm-diameter sealed column, found better absorption efficiency (78%) at an oxygen-to-water flow ratio of 0.87% than with ratios of 0.10, 0.30, and 0.50%.

At Harrietta, somewhat better absorption efficiencies were obtained at the lowest column cross-sectional area-to-flow relationship—0.91 cm^2/(L/min)—but Nirmalakhandan et al. (1988) found little to no difference in absorption efficiencies when column cross-sectional area-to-flow varied from 1.4 cm^2/(L/min) to as high as 4.2 cm^2/(L/min). Table 2 shows that when a cross-sectional area-to-flow relationship is kept constant, the column wall area-to-flow relationship varies with column diameter. Less wall surface area per unit of flow could result in a reduced water-to-gas interface, especially in the case of spray nozzles. Therefore, larger columns may function better with less flow per cross-sectional area. Nevertheless, in studies conducted by Dwyer et al. (1991), the efficiency of a 15-cm-diameter column could be predicted accurately on the basis of the performance of a 10-cm-diameter column, at least at the water and oxygen flow rates tested.

Absorption efficiencies are also influenced by dissolved nitrogen (DN) gas levels (Watten et al. 1991). Elevated DN levels result in poorer oxygen absorption than low DN levels. This is reflected in the oxygen-plus-energy cost, which ranged from $0.24 to $0.52 per kilogram oxygen transferred (Watten et al. 1991). At 100% saturation of DN and zero DO, the cost was $0.24, whereas at 100% saturation DO the cost was $0.30. At 200% DN saturation and zero DO, the cost was $0.35, whereas at 100% saturation DO it was $0.52.

Discussion

The application of commercial oxygen in aquaculture has demonstrated economic and fish health benefits. Carrying capacity can be enhanced with minimal increases in cost. Capital costs for conventional hatchery construction may range from US$100 to 150 per kilogram fish production capacity per year. Increasing the production capacity by means of high-purity oxygen may cost no more than $1 per kilogram fish production capacity (Nirmalakhandan et al. 1988).

There is the need for more research and development efforts. For instance, the technology and methodology requires fine-tuning of existing approaches, and new ways to introduce the oxygen into the water must be found under a wide range of aquaculture operations, such as pond culture of channel catfish *Ictalurus punctatus*, net-pen culture, recirculation systems, and raceway culture. Design must also be optimized for a variety of objectives and water quality variables such as degassing of nitrogen and reaaeration in serial reuse.

The effects of a hyperoxic environment on fish and other aquatic organisms must also be determined. Some early investigations indicated fish tolerance of relatively high oxygen levels of 180% saturation followed by readjustments to normal levels (Kindschi et al. 1988). However, Colt et al. (1991) pointed out that a hyperoxic toxic threshold exists for all organisms.

It is also important to identify the limits to which oxygen can be made available before metabolic by-products such as ammonia, carbon dioxide, and solids, in combination or individually, limit production (Westers and Pratt 1977; Meade 1985; Colt and Watten 1988).

The application of high-purity oxygen has entered the aquaculture industry and its biological and economic potential has been recognized. We must now move on with the actual demonstration, data collection, research, and development aspects of this technology.

References

Boersen, G., and J. Chesney. 1986. Engineering considerations in supplemental oxygen. Pages 17–24 *in* G. Bouck, editor. Papers on the use of supplemental oxygen to increase hatchery rearing capacity in the Pacific Northwest. Bonneville Power Administration, Portland, Oregon.

Colt, J., K. Orwicz, and G. Bouck. 1991. Water quality considerations and criteria for high-density fish culture with supplemental oxygen. American Fisheries Society Symposium 10:372–385.

Colt, J., and B. Watten. 1988. Applications of pure oxygen in fish culture. Aquacultural Engineering 7:397–441.

Dwyer, W. P., G. A. Kindschi, and C. E. Smith. 1991. Evaluation of high- and low-pressure oxygen injection techniques. American Fisheries Society Symposium 10:428–436.

Kindschi, G. A., C. E. Smith, and S. K. Doulos. 1988. Use of the Aquatector oxygenation system for improving the quality of fish rearing water. U.S. Fish and Wildlife Service, Bozeman Information Leaflet 44, Bozeman, Montana.

Meade, J. W. 1985. Allowable ammonia for fish culture. Progressive Fish-Culturist 47:135–145.

Nirmalakhandan, N., Y. H. Lee, and R. E. Speece. 1988. Optimizing oxygen absorption and nitrogen desorption in packed towers. Aquacultural Engineering 7:221–234.

Owsley, D. E. 1981. Nitrogen gas removal using packed columns. Pages 71–82 *in* L. J. Allen and E. C. Kinney, editors. Proceedings of the bio-engineering symposium for fish culture. American Fisheries Society, Fish Culture Section, Bethesda, Maryland.

Watten, B., J. Colt, and C. E. Boyd. 1991. Modeling the effect of dissolved nitrogen and carbon dioxide on the performance of pure oxygen absorption systems. American Fisheries Society Symposium 10: 474–481.

Westers, H., and K. M. Pratt. 1977. Rational design of hatcheries for intensive salmonid culture, based on metabolic characteristics. Progressive Fish-Culturist 39:157–165.

American Fisheries Society Symposium 10:450–458, 1991

Bulk Box Aerators: Advantages and Design

DON D. MACKINLAY

Department of Fisheries and Oceans
555 West Hastings Street, Vancouver, British Columbia V6B 5G3, Canada

Abstract.—Many of the problems associated with packed-column aerators in salmon hatcheries (splashing, icing, wall flow, channeling, blasting, uneven flow control, fouling, and wasted space) can be reduced or eliminated by using a container with a large surface area (bulk box) for the media, rather than the usual battery of small (300-mm-diameter) columns. The recommended bulk box design includes layers of 25-mm-diameter pall media rings arranged on their sides, a media bed area sized for a hydraulic loading of 1,000–3,000 L/(min · m^2), bed depth calculated with a stage efficiency equation, a plate distributor designed according to an orifice flow equation, a stilling box to minimize inflow turbulence, and a modular box structure with louvered sides for ventilation. A bulk box can be used to increase aeration performance (oxygen in, nitrogen out) and decrease the costs of construction and operation of a packed column aeration system.

Canada's Salmonid Enhancement Program (SEP) operates many salmon and anadromous trout hatcheries and rearing facilities in the Pacific region, either directly or through contracts to private firms or community groups. Over 50 of these are supplied with groundwater (pumped, artesian, or gravity flow), which must be aerated to increase dissolved oxygen and decrease dissolved nitrogen concentrations. A few old hatcheries have wooden slat or aspirator aerators, but most facilities use minor modifications of the packed column described by Owsley (1981) to bring dissolved gas concentrations to acceptable levels.

A typical packed-column aerator (Figure 1) includes a pipe manifold distribution header supplying a battery of stacks of 300-mm-diameter by 300-mm-long column segments loaded with packing media, all mounted over the head tank for the hatchery water supply system. The packing medium is usually 38-mm-diameter pall rings, although 25- and 75-mm-diameter rings and different sizes of saddles, toroids, balls, and gratings have been used. The first, or distribution, column segment is usually 600 mm long and contains screens rather than rings, for the purpose of breaking up the inflow from a small-diameter supply pipe and spreading it over the entire cross-sectional area of the column. Each column receives between 500–700 L of water per minute, corresponding to a hydraulic loading of 7,000–10,000 L/(min · m^2).

Several problems with this kind of packed column have been identified in unpublished tests by our staff and consultants over the years. A review of the basic principles of aerator design and evaluation of the operational problems with batteries of small-diameter columns has led to the development of a large-surface-area packed column that we call a bulk box (Figure 2). The major difference offered by the bulk box is that a single, large, media bed replaces the battery of small beds used in the old design. Other features include a calibrated-orifice distribution plate that ensures an even initial distribution to the entire media bed, deeper segments (0.6–1.0 m versus 0.3 m for the old design) between ventilation gaps, a media bed of 25–50-mm-diameter pall rings laid on their sides in rows (rather than randomly packed), and louvered (rather than solid) side walls to allow maximum ventilation. Recommended flow rates are 1,000–3,000 L/(min · m^2).

The purpose of this paper is to encourage designers of packed-column aerators to build single, large units rather than batteries of small units. I discuss the advantages of the bulk-style over the battery-style packed-column aerators, considering both theoretical and practical grounds, and give a simple procedure for the design of a bulk box. The practical advantages are discussed first to give the reader background on problems in the operation of battery-style aerators.

Practical Advantages

Splashing

High flow rates cause water to splash out from the column at the top and between segments, either by bouncing off the media pieces, blowing sideways in the wind, or catching the bottom lip of the column segment and being deflected outward by surface tension. Splashed water is less well aerated because it plunges directly into the head

FIGURE 1.—Battery of segmented packed-column aerators at the Chehalis River Hatchery.

tank without being broken up by the packing (Hackney and Colt 1982). Design alterations to reduce splashing from small-diameter columns have included fluted segment tops to catch water from a wider area than the column diameter, funneled segment bottoms to direct the flow inward from the edges, and partial or complete wind shields around the battery of columns.

The bulk box reduces the opportunity for splashing because the total media bed perimeter is smaller than for the battery-style columns. The sum of the perimeters for ten 0.3-m-diameter columns is 9.4 m, compared with only 3.4 m for a square or 3.0 m for a round bulk box that would contain the same surface area of media bed.

Icing

In extreme cold, splashing leads to ice buildup, which poses both an operational problem by restricting or (if left unattended) blocking ventilation to the falling water, and a safety problem for staff who have to climb the tower and break off and clear away the ice (MacKinlay et al. 1987). Usually ice forms only at the margins of the aerator, because the latent heat in large water flows is so great that only a very slight drop in water temperature occurs between the supply and drain of an aeration tower, even in extremely cold air temperatures (−40°C).

Again, the smaller perimeter of a bulk box enclosing the same area of media bed reduces the opportunity for icing and the amount of ice that will form.

Wall Flow

Water flowing down the walls of a packed column, where the media pieces do not touch the sides, is less well aerated than water flowing through, and broken up by, the media bed (Colt and Bouck 1984). Once water hits the wall, surface tension makes it difficult for the water to be redistributed back into the media bed.

These effects are less significant in large-diameter columns because the proportion of the water that is flowing adjacent to the walls is smaller. The louvers forming the end walls of the bulk box direct water back into the media bed, minimizing wall flow.

Channeling

One of the most marked effects seen with the operation of packed columns is the decrease in aeration efficiency from segment to segment down the length of the column. The apparent cause of this trend is the coalescence of the flow into streams within confined areas of the column cross section. This reduces performance by flooding the parts of the aerator that have a stream moving through them and wasting the parts that have no water. Channeling can be worsened by structural components that concentrate or direct water flow, such as fluting, funnels, or screen supports. Channeling also tends to increase with the age of a packed column, as noted by decreased performance of aeration towers over time. This can be explained by movements of randomly packed media pieces in a column as water pummels them from above. Eventually, an orientation is reached for each piece that provides the least resistance to water flow, encouraging channeling and decreasing aeration efficiency.

The bulk box reduces channeling in two ways. First, the much more even distribution provided by a plate distributor, compared with manifold headers, is an important deterrent to channeling. This is especially true for the relatively small media sizes used, which tend to maintain an even water distribution if provided with one. Second, the layered packing does not have the ability to shift around as much as random packing, further reducing the opportunity for channeling. Layering of rings also allows for easier horizontal movement than vertical movement of water, which

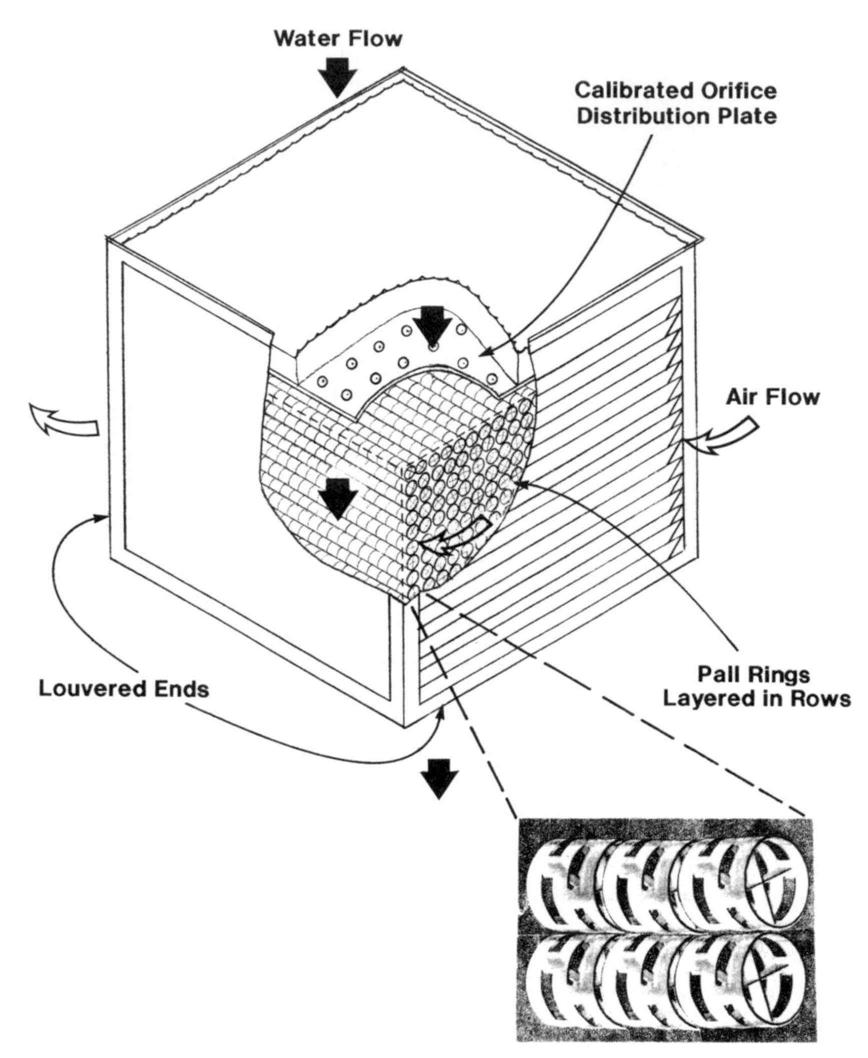

FIGURE 2.—Module of the prototype bulk box packed-column aerator.

makes each layer act as a kind of redistribution plate.

Blasting

The common pipe-manifold supply header operates at a back pressure to ensure equal flows to each column. This results in high velocity for the water entering the column, as well as a loss of hydraulic head in the supply system. When this water "blasts" into the distribution segment screens, it is broken up and mixed with the surrounding air into a kind of froth, which then falls through the screens and into the following, media-filled segments. This blasting may account for the increase in gas pressures reported by some workers in the initial segment(s) of packed-column aerators (Shrimpton 1987). Blasting creates above-ambient pressures in the microenvironment, which can actually increase the oxygen exchange rate, at the expense of nitrogen degassing. The loss of hydraulic head, however, could more beneficially be used to raise the water to the top of a taller column.

The plate distributor of the bulk box requires about a third (0.3 m versus 0.9 m) of the additional operating head required by a manifold-and-segment distributor combination, while providing a much more even flow distribution. Water entering the bulk box media bed has a lower velocity, resulting in closer-to-ambient pressures, and therefore provides better degassing than the manifold distributor.

Uneven Flow Control

Even with the use of individual valves at each column supply, it is often difficult to provide an equal flow of water to each column because of the need to accommodate different ranges of flow from different sources entering the manifold at different locations. Adjustments at one valve or pump alter the pressure distribution within the manifold and change the flow to the other columns in the system.

The distribution plate of the bulk box operates with a certain depth of water over the plate at the maximum design flow for the aerator. At lower flows, this depth is reduced, providing a lower flow through each orifice and a lower hydraulic loading to the media bed, thereby improving the aeration performance. Although the entire media bed may not be wetted at very low flow rates to the bulk box, the hydraulic loading is so low under these conditions that aeration performance remains high.

Fouling

Two types of fouling occur in packed column aerators: mineral and algal. Mineral fouling usually comes from anoxic water supplies that are rich in minerals, such as iron or manganese, that are insoluble in the oxidized state. When the water is aerated, the minerals precipitate, staining the aeration media and settling out in downstream structures. However, water with a mineral content high enough to clog an aerator would not be suitable for fish culture, because the precipitate would suffocate eggs or abrade fish gills (MacKinlay et al. 1987). Algal fouling is caused by mats of filamentous algae and fungi growing on the columns and media. The extent of this fouling depends upon nutrient availability in the supply water. Only the top layer of media pieces within a segment generally becomes fouled, indicating a light limitation to this phenomenon. Towers that are not enclosed usually require annual cleaning unless the water has such low nutrient content that substantial fouling does not occur.

The top of the bulk box media bed is covered by the distribution plate and is therefore not exposed, limiting light and thus algal fouling. The extent of potential algal fouling on media pieces near the louvered sides has not been evaluated.

Wasted Space

Because of the metalwork needed to support the columns and the requirement for access around each column, the media bed surface area in a battery of small-diameter packed columns occupies only a small proportion of the total plan area of the aeration tower structure, ranging from 7 to 20% in our 10 most recently constructed hatcheries. At the Kitimat Hatchery, where only 8.5% of the 81 m^2 of head tank area is occupied by the active surface area of the 95 aeration columns, a very large building was built to house a relatively small amount of aeration capacity.

FIGURE 3.—Tenderfoot Creek Hatchery aeration tower before renovation. Note algal fouling hanging from segments.

The aeration segment support system is one of the major costs in tower construction. At the Tenderfoot Creek Hatchery, expansion of the water supply from 12,000 to 20,000 L/min was to have included the construction of a new aeration tower at an estimated cost of CAN$60,000 (I. Ross, personal communication). Instead, a bulk box system was installed for less than CAN$25,000 in the space taken up by half of the old battery of aeration columns (Figures 3–5), providing a lower hydraulic loading and higher aeration performance than the old system.

Theoretical Advantages

Several factors control the efficiency of gas transfer in an aerator. For its dissolved gases to reach equilibrium with the surrounding air, water should flow over a large surface area in a thin film at ambient atmospheric pressure for an extended period of time, with thorough mixing and adequate ventilation.

Area of Contact

Gas transfer in an aerator is directly proportional to the wetted area, that is, the amount of air-to-water surface exposure. Maximum transfer

FIGURE 4.—Tenderfoot Creek Hatchery aeration tower after renovation to a bulk box. Note stilling box on left.

occurs when the water is spread out over the greatest possible area. The primary method for increasing the exposure surface in cascade aerators is to fill the column with a high-surface-area packing, such as pall rings. The bulk- and battery-style packed columns both use such packing. However, the denser packing obtained by layering the media pieces and the even water distribution obtained from the orifice plate maximize the wetted area of the bulk box compared with the battery-style aerator.

Film Thickness

Simply providing a large surface area within the aerator is not enough. The water must be made to flow over that area in a thin film, surrounded by air. Gas transfer is directly related to the amount of the water that is close to the surface, because

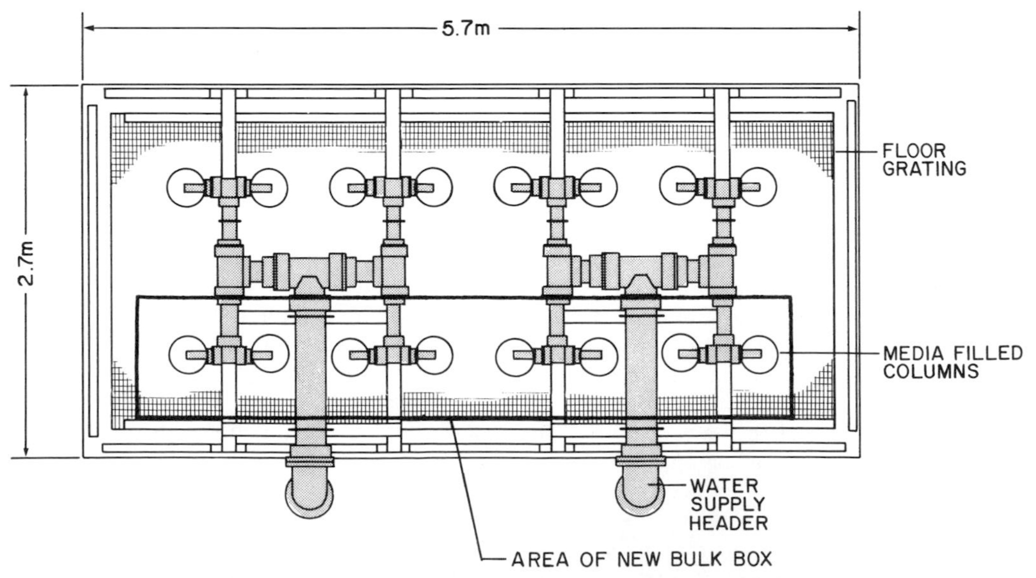

FIGURE 5.—Plan view of Tenderfoot Creek Hatchery aeration tower. The new bulk box area is shown as a heavy rectangle superimposed over locations of old column batteries.

the molecules closest to the surface will be most rapidly transferred across the air–water interface. Most studies of cascade aerators show that gas transfer efficiency decreases as flow to the aerator increases because, at a fixed contact surface area, higher flows must be in a thicker layer and thus are less exposed to the surrounding air (Hackney and Colt 1982; Bouck et al. 1984). The bulk box operates at a lower hydraulic loading than the battery-style columns, which results in a thinner film over the wetted area and consequent higher aeration efficiency.

Pressure Differential

One of the two functions of cascade aerators is degassing, or the reduction of nitrogen supersaturation. The rate at which a gas will transfer from one medium to another is a function of the difference in its partial pressures in the two phases, so it is important to keep the atmospheric pressure at or below ambient for degassing. This means avoiding high pressures, such as those that occur when the water is blasted into the media bed. However, below-ambient pressures reduce oxygen uptake, the other function of cascade aerators. The usual solution is to keep pressures at ambient throughout the media bed, although providing an oxygen-enriched atmosphere can increase both oxygen uptake and nitrogen degassing (Colt and Watten 1988).

The bulk box does not induce higher-than-ambient pressures through blasting, as do the battery columns.

Ventilation

In the battery column aerator, air is drawn down through the media pieces by the movement of the water. The reason for segmenting the columns was to allow the introduction of fresh air at intervals down the length of the column. In many packed columns in the chemical industry, fans force the gas up through the falling liquid to obtain thorough, countercurrent mixing (Coulson et al. 1978). The bulk box described here takes a unique approach to ventilation by allowing the air to flow across the media bed while the water flows down though it. Fans could assist this crosscurrent ventilation and would require less power than fans that have to force the air up through the media bed. The denser packing obtained by layering the media pieces in the bulk box, however, can cause it to flood at a lower flow rate than randomly packed media.

Turbulence

Severe turbulence caused by blasting is detrimental to degassing and a waste of hydraulic head. However, mild turbulence helps gas exchange by mixing the flow as it passes through the media bed, such that the molecules that are at the surface of the water film are being constantly replaced. A major advantage of pall rings over other media is that they promote this kind of turbulent mixing as the water is deflected from one piece to another. The layered structure of the media pieces in the bulk box maximizes such mixing without causing the high pressures produced in the initial segments of a battery-style aerator.

Exposure Time

The longer water is exposed to an atmosphere, the closer it will come to equilibrium with that atmosphere. Even a body of water several centimeters deep will reach equilibrium with the air if exposed for several hours. Exposure time is normally increased by increasing the height of the aerator, which increases the capital cost and the operational cost of pumping water to the top of a higher column. Inhibiting downward flow by the selection and arrangement of media is another alternative to increasing column height to cause a longer exposure time.

In general, smaller media provide a slower passage of water through an aerator segment because a falling droplet or stream will have a smaller distance to fall before encountering the next, lower, media piece. Orientation of individual media pieces can play a role in this function because some orientations provide more open area when viewed from above and thus less resistance to vertical flow than others. The classic example is the pall ring, which is quite closed to vertical water flow when it lies on its side but open when it stands on end.

The layering of rings on their sides in the bulk box slows the passage of water by providing a greater density of obstacles to vertical movement than does random packing.

Design of the Bulk Box

The preceding discussion of the advantages of the bulk-type over the battery-style packed column aerator has been based mainly on anecdotal and theoretical arguments. Much of the research on packed-column performance done by our group and others is of questionable applicability to bulk box design because of such factors as interference from blasting and uneven distribution and the uniqueness of the crosscurrent airflow attained by a layered-ring medium. Although experimental evidence to support many of the evaluations listed above is preliminary, the advantages of bulk boxes are consistent with the basic principles of packed-column design.

This section gives a simple procedure for designing the important components of a bulk box, although the design parameters have not been optimized under experimental conditions. Different designs of bulk boxes will produce differences in aeration performance, and this section is meant as an introduction to bulk box design and not as the final word. Designers are advised to use the following procedure to build a prototype, which they should test, modify, and optimize for the performance they require.

The design components include media type, bed area, bed depth, and structural details.

Media Type

Design textbooks recommend shaped media, such as rings and saddles, for use in packed columns because of the very high surface area that they provide for the liquid–gas interface (Treybal 1980). I have found that 25-mm-diameter pall rings preform better than rings, saddles, and balls of various sizes, and than screens, corrugated plates, or no media by 5–15% per meter and are particularly suited for laying in rows to provide high vertical resistance to water flow and minimal lateral resistance to airflow. Smaller media perform better than larger ones at the low hydraulic loadings desirable for fish culture aeration, but they are more expensive per unit volume. Sinnott (1983) recommended 50-mm-diameter rings as the cutoff at which the cost–performance factors are most advantageous. I have found no advantage using media smaller than 25 mm in diameter and therefore recommend a 25-mm-diameter pall ring media.

Bed Area

Lower hydraulic loadings give better performance but also cost more by requiring a larger media bed than higher loadings. This relationship has yet to be adequately modeled and optimized for the bulk box described here. Performance graphs distributed by media manufacturers show a decline in efficiency at high and low flows, but the decline at low flows is caused by a drop-off in wetted area, meaning that some of the media is not being used even though it is included in the evaluation. Hackney and Colt (1982) showed that the optimum performance for small rings was achieved at about 1,700 L/(min · m^2) in terms of weight of oxygen transferred per amount of energy used. As an initial recommendation, I suggest a design hydraulic loading of between 1,000 and 3,000 L/(min · m^2) to achieve good aeration performance at an economic cost.

To calculate the area required to treat a particular water supply, simply divide the total flow by the hydraulic loading.

Bed Depth

To calculate the height of tower required, the stage efficiency for the particular aeration unit must be known. Stage efficiency is a measure of the effectiveness of the aerator. It is calculated by comparing the actual change in gas pressure with the saturation value, which is the maximum change that could occur.

$$E_s = \frac{C_i - C_o}{C_i - 100} \times 100;$$

E_s = stage efficiency, expressed as percent;
C_i = inflow gas concentration (% saturation);
C_o = outflow gas concentration (% saturation).

Gas saturation values can be calculated from formulae and tables given in Colt (1984). The efficiency is a measure of how close the aerator moved the dissolved gas concentration towards equilibrium.

To calculate the outflow concentration of a gas after water passes through the aerator, a modification of the compound interest equation can be used (MacKinlay 1987):

$$C_o = 100 - (100 - C_i)\left(1 - \frac{E_s}{100}\right)^s;$$

s = the number of stages in the aerator. If the test stage, from which E_s is derived, were 1.0 m deep, E_s would be the aeration efficiency per meter of aerator and s would be the height in meters.

This equation can be rearranged to give the depth of media bed required to meet a certain outflow criterion:

$$s = \frac{\log_e \dfrac{(C_o - 100)}{(C_i - 100)}}{\log_e\left(1 - \dfrac{E_s}{100}\right)}.$$

As an example, to reduce nitrogen supersaturation from 150% to 103% with 25-mm-diameter pall rings at a flow loading of 3,000 L/(min · m^2), which should give at least 70% efficiency per meter of bed depth, would require an aerator 2.34 m high. It would require another 0.9 m to reduce supersaturation to 101%.

A tower with louvered sides would not require segmentation to provide ventilation. However, if the media were randomly packed, there should be periodic redistribution (with an orifice plate) at 1-m intervals to counteract potential channeling.

Structural Details

Distribution plate.—Very even initial distribution of water is critical to the performance of the bulk box, as it is with any packed column (Sinnott 1983). By using the Bernoulli equation for fluids (Merkel 1983), the flow (Q) through a 1.0-m^2 area of the distribution plate can be calculated from

$$Q = n\,(\pi r^2)\sqrt{\frac{2g\,h}{(K + 1)}} \times 60{,}000;$$

n = the number of holes in the 1.0-m^2 area,
r = the radius of each hole (m),
g = gravity constant (9.8 m/s^2),
h = head of water above the plate (m),
K = minor loss discharge coefficient,

60,000 is the conversion of m^3/s to L/min.

For a pattern of 5-mm-diameter (r = 0.0025 m) holes with square edges (K = 0.5) placed 25 mm apart (1,600 holes over 1.0 m^2), the flows at different depths in the head tank would be:

Depth (m)	Flow (L/min)
0.05	1,524
0.1	2,155
0.2	3,047
0.3	3,632
0.4	4,309

It should be noted that all commercial screens and perforated plates investigated were too porous to serve as distribution plates, so our plates were custom-made by drilling a pattern of holes in a metal sheet. Potential clogging of these small holes by sand or debris can be averted by suspending a fine-mesh screen or small-orifice perforated plate above the distribution plate.

Stilling box.—Another consideration in designing the header tank is how the supply water should be introduced. The best approach we have used to date involves connecting the supply pipe(s) to a separate stilling chamber from which the water overflows onto the distribution plate (Figure 4). A relatively deep chamber, with water inflow at the bottom, will reduce the turbulence and velocity of the water so that a gentle, even flow is provided over the distribution plate.

Box structure.—The original bulk boxes tested were built as trays placed in a framework of rails, such that they could be slid out for cleaning. The improved design for the boxes (Figure 2) stemmed from an attempt to use plastic milk carton baskets for holding the media on flat grids in an open plan. The baskets of media splashed terribly but the baskets could be stacked to fill as much three-dimensional space as was required. The presently favored design is made up of large boxes of media with built-in redistribution plates and louvered sides, which can be stacked under a master distribution plate. The only structural support required is for the distribution head tank, because the stacks of boxes are built with sufficient structural integrity to handle their own weight and the weight of the water passing through them.

Because the box is completely enclosed, little biological fouling is expected. Any cleaning or other maintenance can be done by unstacking the boxes. For this purpose, the boxes should be made small enough for a pair of workers to handle and valves should be provided to shut off the flow to each stack. If a single distribution plate is used in the head tank, portions of it can be turned off by simply laying some nonporous material over the holes in that section. This is necessary during maintenance because boxes with water running through them are a great deal heavier than dry boxes.

Future Work

As with the original battery-type packed-column aerator concept described by Owsley (1981), there are many details to work out in evaluating and improving the design of the bulk box aerator.

Ventilation

No decrease in performance has been detected due to lack of ventilation within a 1.0-m^3 bulk box media bed, although larger beds might be expected to suffer from some suffocation. Ventilation within large beds should be investigated both with and without fans or oxygen supplementation.

Media Type

The cost of small pall rings is very high per unit of volume, and the bulk box, with its layered packing and low hydraulic loading, requires more packing than the conventional batteries of segmented columns. The search should continue for a cheaper alternative packing medium, perhaps small-diameter plastic pipe sections lacerated down their length, or stacks of corrugated, porous sheets.

Bed Area and Depth

The performance of various combinations of bed depth and hydraulic loading needs to be quantified and modeled for design purposes. As with most packed columns (Sinnott 1983), design parameters will probably only be relevant for systems very similar to the tested prototype.

Conclusions

Use of the bulk box concept for packed-column aerator design for fish culture is actually a return to the simple principles used in packed-column design in other industries. Excessive concern over the restriction of ventilation in aeration columns led to the fragmentation of the flow into many small columns and the separation of those columns into segments. This led to many operational problems, which are mitigated by returning to the simpler, bulk box design.

In addition, the retrofit of a bulk box can increase aeration performance (nitrogen out, oxygen in) or increase the flow capacity of an existing aeration tower. The decreased space required for a bulk box reduces the size of superstructure and amount of supporting metalwork required to support a battery of small columns, offsetting the cost of extra media.

Acknowledgments

I sincerely acknowledge the contributions made by the SEP biologists, engineers, students, and consultants who have worked on defining and reducing the problems associated with packed-column aerators: B. Anderson, G. Berezay, A. Boreham, L. Ferriss, L. Fidler, D. Harding, G. Labinsky, W. McLean, V. McLeod, I. Ross, A. Rowland, and M. Shrimpton. Thanks to L. Fidler, J. Colt, and two anonymous reviewers for their thoughtful comments on the manuscript.

References

Bouck, G. R., R. E. King, and G. Bouck-Schmidt. 1984. Comparative removal of gas supersaturation by plunges, screens and packed columns. Aquacultural Engineering 3:159–176.

Colt, J. E. 1984. Computation of dissolved gas concentrations in water as functions of temperature, salinity, and pressure. American Fisheries Society Special Publication 14.

Colt, J. E., and G. R. Bouck. 1984. Design of packed columns for degassing. Aquacultural Engineering 3:251–273.

Colt, J. E., and B. Watten. 1988. Application of pure oxygen in fish culture. Aquacultural Engineering 7:397–441.

Coulson, J. M., J. F. Richardson, J. R. Blackhurst, and J. H. Harker. 1978. Chemical engineering, volume 2, 3rd edition. Pergamon Press, Oxford, UK.

Hackney, G., and J. E. Colt. 1982. The performance and design of packed column aeration systems for aquaculture. Aquacultural Engineering 1:275–295.

MacKinlay, D. D. 1987. Aeration: how much is too much? Canadian Aquaculture Magazine 3(2):21–23.

MacKinlay, D. D., D. D. MacDonald, M. V. D. Johnson, and R. F. Fielden. 1987. Culture of chinook salmon (*Oncorhynchus tshawytscha*) in iron rich groundwater: Stuart Pilot hatchery experiences. Canadian Manuscript Report of Fisheries and Aquatic Sciences 1944.

Merkel, J. A. 1983. Basic engineering principles, 2nd edition. AVI Press, Westport, Connecticut.

Owsley, D. E. 1981. Nitrogen removal using packed columns. Pages 71–82 *in* L. J. Allen and E. C. Kinney, editors. Proceedings of the bio-engineering symposium for fish culture. American Fisheries Society, Fish Culture Section, Bethesda, Maryland.

Sinnott, R. K. 1983. Chemical engineering: design, volume 6. Pergamon Press, Oxford, UK.

Shrimpton, J. M. 1987. Assessment of substrate performance and column geometry in aeration of hatchery water supplies. Canadian Manuscript Report of Fisheries and Aquatic Sciences 1886.

Treybal, R. E. 1980. Mass-transfer operations, 3rd edition. McGraw-Hill, Toronto.

American Fisheries Society Symposium 10:459–464, 1991

Use of Oxygen-Filled Enclosures for Aeration and Degassing

JAMES W. MEADE,[1] BARNABY J. WATTEN, WILLIAM F. KRISE, AND PAUL W. HALLIBURTON

U.S. Fish and Wildlife Service, National Fishery Research and Development Laboratory Rural Delivery 4, Box 63, Wellsboro, Pennsylvania 16901, USA

Abstract.—Surface aerators are used as temporary and permanent systems to add oxygen to and remove nitrogen gas from hatchery waters. Rates of gas transfer across a water–gas boundary are greater for an oxygen-enriched atmosphere than for air. We compared an enclosed agitator, a pumped spray, and diffusers (airstones) with respect to oxygen absorption efficiency and nitrogen stripping over a range of influent gas (oxygen or air) flow rates. Results indicate the enclosed agitator, receiving oxygen, can be used as a low-cost, portable oxygenation and degassing system.

An important aspect of oxygenation in fish culture is the method by which gas and water are brought into contact so that gas exchanges are effective and efficient. Hatchery managers who need to supplement dissolved oxygen may not be able to afford the capital and construction costs of installing commercially available systems and thus may have to work within the constraints of existing facilities and equipment. In this paper we summarize our tests of simple, conventional aeration devices that can suffice in such circumstances if they are used in an atmosphere of oxygen instead of air.

Operation of conventional aeration equipment in an oxygen-enriched atmosphere results in a substantial increase in oxygen transfer. This response becomes apparent with inspection of the following differential equation describing gas transfer rates (Lewis and Whitman 1924):

$$\frac{dc}{dt} = K_L a(C_s - C); \qquad (1)$$

dc/dt = gas transfer rate,
$K_L a$ = overall mass transfer coefficient (h^{-1});
C_s = saturation concentration of a dissolved gas (mg/L);
C = ambient concentration of a dissolved gas (mg/L).

Note that the rate of gas transfer depends in a first-order manner, on the extent of the dissolved gas deficit (i.e., the difference between C_s and C). Values of C_s, in turn, are related to gas phase composition as defined by Henry's Law:

$$C_s = B \cdot X \cdot P_T; \qquad (2)$$

B = Bunsen's coefficient at a given temperature and salinity;
X = mole fraction of a gas;
P_T = total pressure (mm Hg).

As the concentration of oxygen in water approaches air saturation (C_s), gas transfer rates approach zero. In a 90% oxygen atmosphere, however, C_s is increased approximately fourfold by the change in Xo_2 from 0.21 (in an air atmosphere) to 0.9 (equation 2). Therefore the dissolved gas deficit, and hence the oxygen transfer rate (equation 1), will remain high even when the dissolved oxygen (DO) concentration reaches or surpasses C_s with respect to ambient air. An increase in the mole fraction of oxygen also acts to increase the rate at which nitrogen is stripped from solution. The net effect on total dissolved gas pressures will vary with inlet dissolved gas concentrations, oxygen feed rate, and system operating pressure (Watten et al. 1991, this volume).

Methods

We compared the performances of an agitator (Figure 1), a pump spray system (Figure 2), and two airstones when operated with equal gas flow rates. Gas rates are expressed here in terms of a volumetric gas:liquid ratio (L gas:L water). The agitator was a 37.3-W Mino-saver[2] (model JA2P175N) with a shaft speed of 1,550 rounds/min. The pump spray system used a Little Giant centrifugal pump (model 501003) to force 54.0

[1]Present address: Tunison Laboratory of Fish Nutrition, 3075 Gracie Road, Cortland, New York 13045, USA.

[2]References to trade names or manufacturers do not imply government endorsement of commercial products.

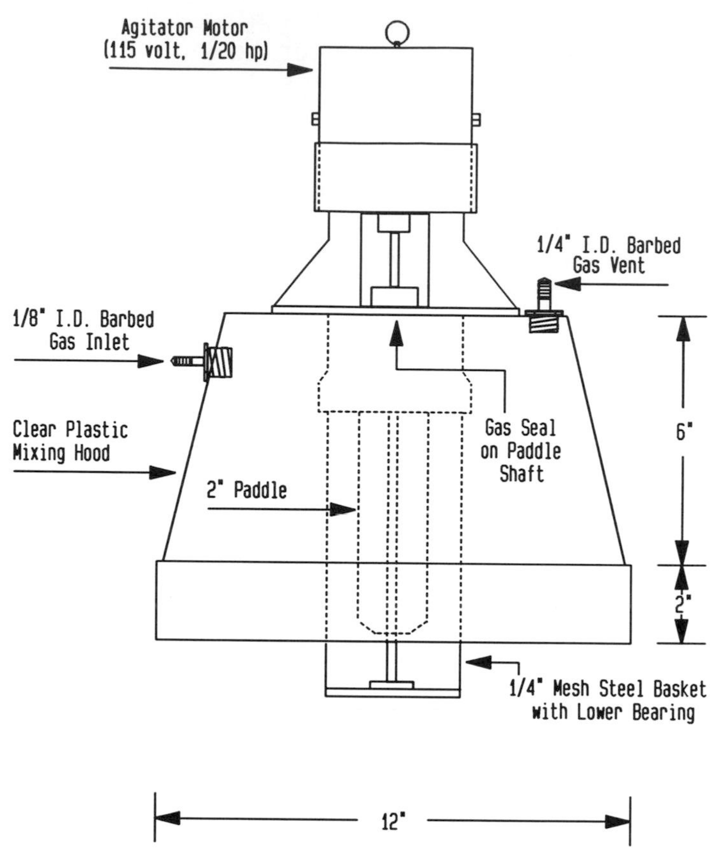

FIGURE 1.—Side view of the enclosed agitator used during air–water and oxygen–water contact tests. I.D. is inside diameter; hp is horsepower.

L/min through a 1.9-cm plastic cap perforated with six 2.87-mm holes. One of the two airstone types tested was 30 cm long and made of sandstone (commonly available at pet stores—Fritz Aquaculture, Mesquite, Texas, catalog number F1516); the other was a fine-bubble oxygen airstone (Aquatic Ecosystems, Apopka, Florida). Both airstone types were placed on the bottom of a fiberglass tub at a depth of 33 cm.

We metered air or oxygen through a research-grade rotometer (Cole Parmer, model 3216-45) to one of the modified aerator devices. The agitator and the pump spray head aeration equipment were enclosed in a 12-L, clear, polycarbonate tub. The airstones were placed beneath a Plexiglas sheet (26.6 × 41.9 cm) supported by a urethane float. Each of the four enclosures was placed in a 60-L fiberglass tub measuring 30 × 48 × 58 cm for testing. Gas leaving the enclosure outlets was directed through a monitor (Bio-Tek, model 74223) for direct reading of composition as percentage oxygen.

Tank water flow during all tests was 29 L/min. Influent and effluent DO concentrations were determined with a polarographic oxygen meter (YSI model 57). Total dissolved gas pressure was determined with a gasometer, functionally equivalent to that described by Bouck (1982). For each set of operating conditions tested, the overall mass transfer coefficient was calculated with the expression (Boyd and Watten 1989)

$$(K_L a)_{20°C} = \frac{\text{OTR}}{(C_s - C)V \times 10^{-3}(1.024^{T-20°})}; \qquad (3)$$

OTR = measured oxygen transfer rate (kg/h);
T = temperature (°C);
V = test tank volume (m^3).

Additionally, oxygen absorption efficiency was calculated based on the ratio of mass oxygen absorbed to mass applied:

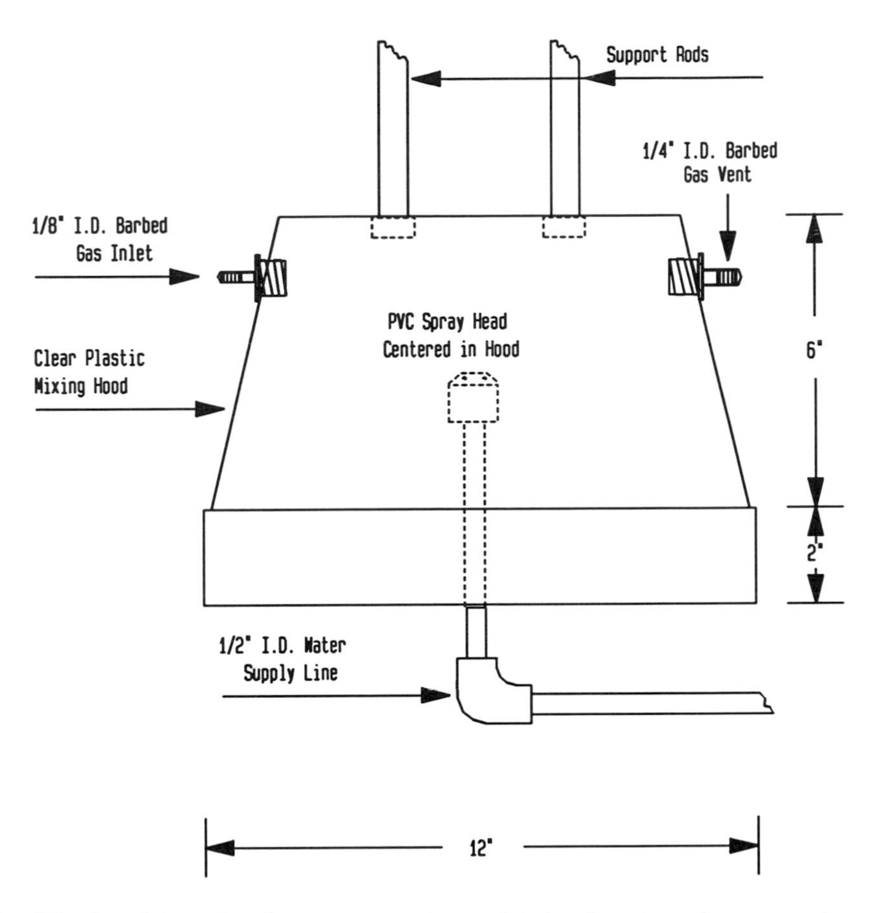

FIGURE 2.—Side view of the enclosed pump spray system used during air–water and oxygen–water contact tests. PVC is polyvinyl chloride; I.D. is inside diameter.

$$AE = \frac{Q_L(DO_{out} - DO_{in})10^{-3}}{Q_g Xo_2 Po_2} 100; \qquad (4)$$

AE = oxygen absorption efficiency (%);
Q_L = water flow rate (m^3/h);
Q_g = gas flow rate (m^3/h);
Po_2 = mass density of oxygen (kg/m^3).

Safety was an immediate problem with the electric agitator. Unless an electric motor is explosionproof, it should not be enclosed in an oxygen environment. To avoid oxygen contact with the motor, we sealed the agitator unit below the motor by removing the blade and fitting the shaft with a flexible rubber gasket. The pump spray system was first evaluated for optimum spray head depth.

Results

The optimum depth for the pump spray head was 4 cm (Figure 3). Spray heads at depths greater or less than 4 cm yielded lower oxygen transfer rates (OTR) and lower absorption efficiencies (AE). In the comparative tests, we used the 4-cm depth. Figures 4 and 5 compare contactor performances with air and oxygen applications. As gas flows were elevated, OTR generally increased, while AE decreased. It is also clear that the use of oxygen (versus air) resulted in highly improved OTR and AE values.

With oxygen, AE for the pump (Figure 5) was high only at extremely low oxygen flow rates, a condition that resulted in a relatively low OTR. However, at about 35% oxygen absorption, the pump system achieved almost 60% of the potential OTR. This would correspond to a DO_{out} of 19.2 mg/L with an oxygen:water ratio of 0.075—

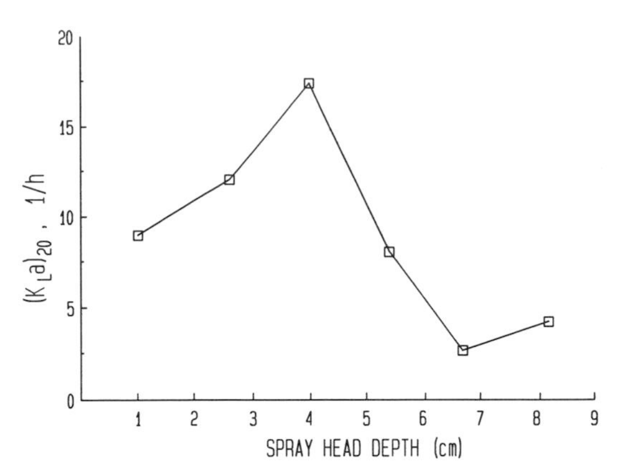

FIGURE 3.—Effect of spray head submergence on the overall mass transfer coefficient $(K_La)_{20°C}$.

more than sufficient for most direct-use (not side-stream and blending) fish culture needs. Figure 6 summarizes changes in DO with respect to absorption equipment, gas type (air or oxygen) and gas flow. Oxygen provided increases in DO exceeding 20 mg/L, whereas air provided DO increases of less than 1.5 mg/L. Overall, airstone performance was below that of either the pump or agitator at the same operating gas:liquid ratios. Changes in total gas pressure (TGP) across each

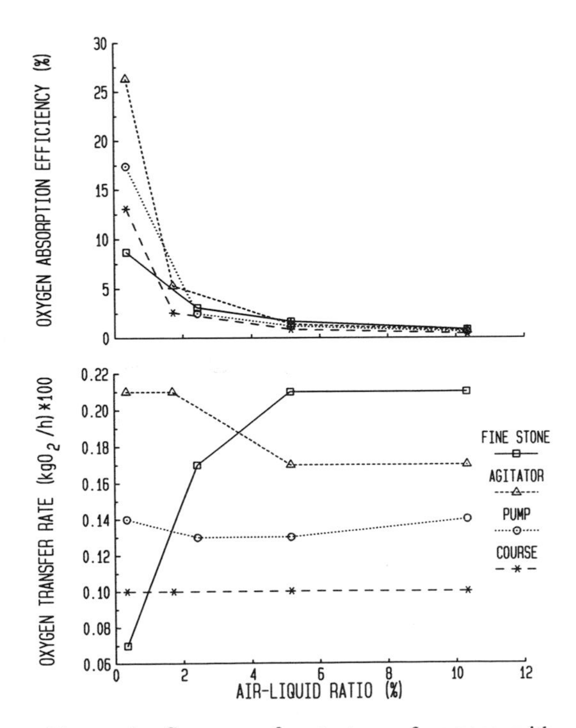

FIGURE 4.—Summary of contactor performances with air injections.

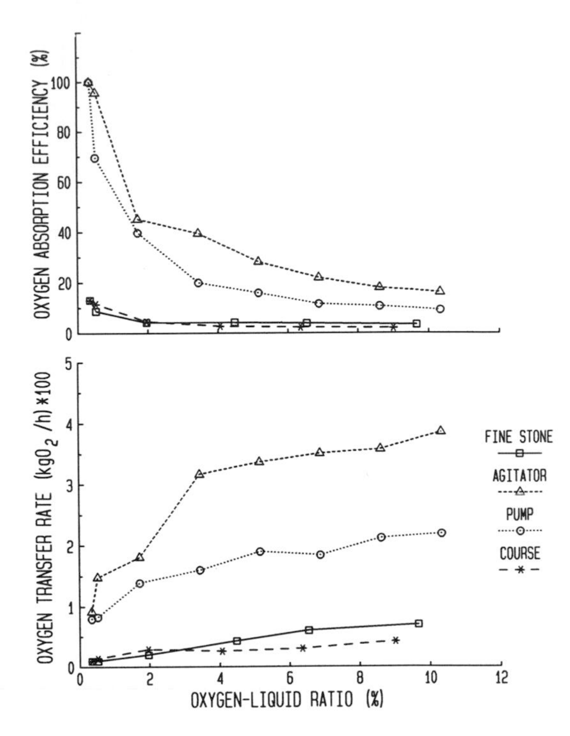

FIGURE 5.—Summary of contactor performances with oxygen injections.

contactor with air or oxygen are summarized in Figure 7. Corresponding changes in dissolved nitrogen are given in Figure 8. Contactor operation with air generally provided lower effluent TGPs than operation with oxygen, although nitrogen desorption was greatest when oxygen was applied.

Discussion

The four devices tested showed generally similar patterns of change in DO and TGP with increases in oxygen flow. The agitator system increased DO as much as 22 mg/L at the highest gas flow tested, compared with 12.6 mg/L for the pump and 2.6 and 4.2 mg/L for the coarse and fine airstones. The agitator was best at reducing overall TGP for all oxygen flow rates. The airstones reduced TGP in all but the 3.0-L/min oxygen flow test with the fine stone. The pump system increased DO fairly well but failed to decrease dissolved nitrogen; consequently, TGP actually increased at oxygen flows of 1.0 L/min or higher. Partly because of the shallow basin depth, the performance of the covered airstones was only about one-tenth as efficient as that of the covered agitator. Both airstones showed similar low gas

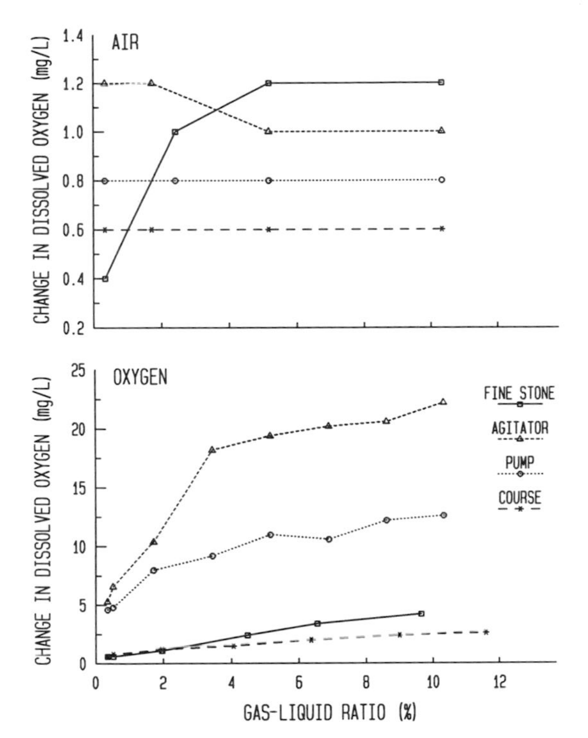

FIGURE 6.—Summary of changes in dissolved oxygen concentration with respect to contactor type, gas type (air or oxygen), and gas flow.

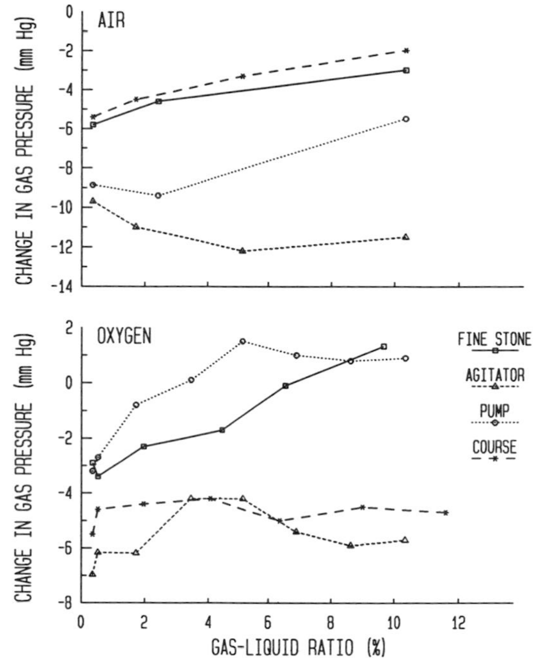

FIGURE 7.—Summary of changes in total gas pressure with respect to contactor type, gas type (air or oxygen), and gas flow.

transfer rates (0.004 and 0.007 kg/h at about 3.0-L/min oxygen flow), the fine-bubble stone having slightly higher values. Absorption efficiencies ranged from 13% at the lowest oxygen flow to less than 2% at the highest flow rate. In comparison, efficiency for the pump and agitator dropped from nearly 100% to about 10% over the range of oxygen flows tested. The efficacy of the airstones tested for dissolved nitrogen removal was extremely poor compared with that of the pump and agitator.

Of the four systems tested, the covered agitator provided the greatest increase in DO (22.2 mg/L at 3.0-L/min oxygen flow) and the highest transfer rate (up to 0.089 kg/h) and absorption efficiency (100%), while effecting the largest decrease in dissolved nitrogen (−53.9% at 3.0 L/min). Typically, oxygen absorption efficiency decreased as oxygen flow increased. However, as oxygen flow increased, the transfer rate (kg/h) increased. At the same gas:liquid ratio, the agitator provided more efficient absorption of oxygen than the pump, which was far more efficient than the airstones.

At high oxygen flow rates the oxygen absorption efficiency of these systems is much less than that of production-scale injection systems. Vacuum columns usually operate at about 40–50% efficiency (Boerson and Chesney 1987; Brock 1987), sidestream systems are documented at 65–93% oxygen absorption efficiency (Laks and Godfriaux 1981; Schutte 1986), and packed towers and U-tubes usually operate at 60–90% efficiency (Watten and Beck 1985; Nirmalakhandan et al. 1988). Absorption efficiencies for the systems tested here were low because the gas–liquid contact took place at relatively low pressures, incoming DO levels were near saturation, and the vented gas was not recycled (Watten and Beck 1985).

In this study our purpose was to show the potential use of simple reaeration techniques for emergency or retrofit oxygen contactor systems. We conclude that there may be extensive, small-scale application. For instance, the covered agitator provided a simple method of increasing DO and decreasing dissolved nitrogen. When operated at a gas:liquid ratio of 0.017, the agitator provided a 45% oxygen absorption efficiency, increased DO to 21.6 mg/L, and decreased dis-

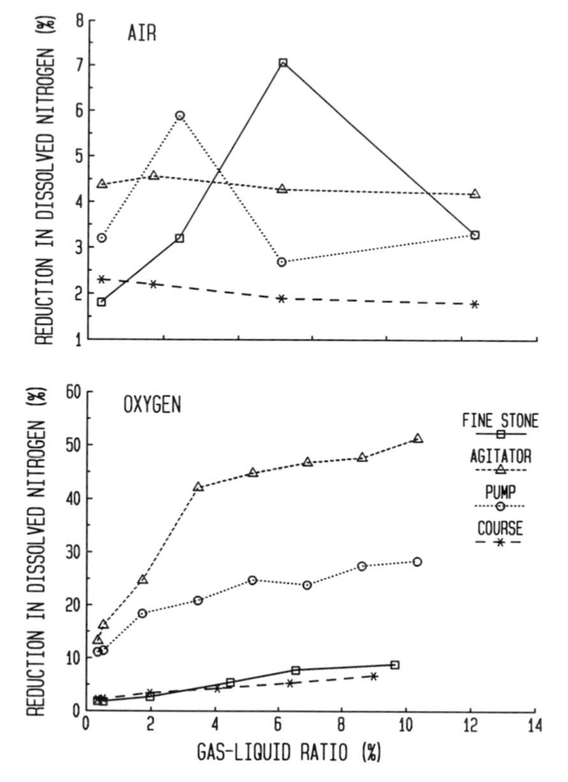

FIGURE 8.—Summary of changes in dissolved nitrogen with respect to contactor type, gas type (air or oxygen), and gas flow.

solved nitrogen by 26%. Thus the primary and essential equipment for making a low-fixed-cost, portable, oxygenation–degassing system is already available in most production facilities in the form of agitators and submersible pumps.

References

Boerson, G., and J. Chesney. 1987. Engineering considerations in supplemental oxygen. *In* L. Visscher and W. Godby, editors. Oxygen supplementation a new technology in fish culture. U.S. Fish and Wildlife Service, Denver Regional Office, Information Bulletin 1, Denver.

Bouck, G. R. 1982. Gasometer: an inexpensive device for continuous monitoring of dissolved gases and supersaturation. Transactions of the American Fisheries Society 111:505–516.

Boyd, C. E., and B. J. Watten. 1989. Aeration systems in aquaculture. Critical Reviews in Aquatic Sciences 1:425–472.

Brock, I. R. 1987. Use of pure oxygen in fish culture. *In* L. Visscher and W. Godby, editors. Oxygen supplementation a new technology in fish culture. U.S. Fish and Wildlife Service, Denver Regional Office, Information Bulletin 1, Denver.

Laks, R., and B. Godfriaux. 1981. Growing trout in waste heated water appears both practical, economical. Aquaculture Magazine (July–August):22–25.

Lewis, W. K., and W. C. Whitman. 1924. Principles of gas adsorption. Journal of Industrial and Engineering Chemistry 16:1215–1220.

Nirmalakhandan, N., Y. H. Lee, and R. E. Speece. 1988. Optimizing oxygen absorption and nitrogen desorption in packed towers. Aquacultural Engineering 7:221–234.

Schutte, A. R. 1986. Evaluation of the Aquatector, an aeration system for intensive fish culture. U.S. Fish and Wildlife Service, Bozeman Information Leaflet 42, Bozeman, Montana.

Watten, B. J., and L. T. Beck. 1985. Modeling gas transfer in a U-tube oxygen absorption system: effects of off-gas recycling. Aquacultural Engineering 4:271–297.

Watten, B. J., J. Colt, and C. E. Boyd. 1991. Modeling the effect of dissolved nitrogen and carbon dioxide on the performance of pure oxygen absorbtion systems. American Fisheries Society Symposium 10:474–481.

American Fisheries Society Symposium 10:465–473, 1991

Experiences in Evaluating Surface and Diffused-Air Aerators

Claude E. Boyd

Department of Fisheries and Allied Aquacultures
Alabama Agricultural Experiment Station, Auburn University, Alabama 36849, USA

Abstract.—Vertical pump, pump sprayer, propeller-aspirator-pump, paddle wheel, and diffused-air aerators powered by electric motors or by takeoffs from farm tractors are used widely to aerate aquaculture ponds. Performance tests suggested that electric paddle wheel aerators are the most efficient in transferring oxygen and circulating pond water. A design for a highly efficient paddle wheel aerator is described. Aeration can improve dissolved oxygen concentrations, enhance the efficiency of feed utilization, and increase aquacultural production and profits. Aerator placement and the use of aerator controllers are discussed.

Aeration has become an important management tool in pond aquaculture. Several basic types of aerators are available for use in ponds, and practical aquaculturists are confused about the relative efficiencies and performance characteristics of the different aerators. Researchers at Auburn University have worked with aerator manufacturers for several years to improve the design and performance of specific aerators. Production model aerators that are sold to aquaculturists also have been tested. The purpose of this report is to describe the basic types of aerators, to discuss aerator performance tests, to summarize performance data, and to make comments on the practical application of aerators in ponds.

Types of Aerators

The most common types of aerators used in aquaculture ponds are described briefly.

A vertical pump aerator consists of a submersible, electric motor with an impeller attached to its shaft. The motor is suspended by floats, and the impeller jets water into the air to effect aeration (Figure 1). These aerators are manufactured in sizes ranging from less than 1 kW to more than 100 kW, but units for aquaculture are seldom larger than 3 kW. Vertical pump aerators for wastewater treatment are classified as low speed (impeller speed, <200 rpm) or high speed (impeller speed, >800 rpm). Units for aquaculture have high-speed impellers—usually 1,730 or 3,450 rpm.

A pump sprayer aerator consists of a high-pressure pump that discharges water at high velocity through one or more orifices to effect aeration (Figure 2). Many designs have been used for the discharge orifices. The simplest procedure is to discharge the water directly from the pump outlet. The most complex design is to discharge the water from small orifices in a manifold that is attached to the pump outlet. Aerator sizes range from 7.5 to 15 kW, and the impeller speeds are from 500 to 1,000 rpm.

The primary aerator parts of a propeller-aspirator-pump aerator are an electric motor, a hollow shaft that rotates at 3,450 rpm, a hollow housing inside which the rotating shaft fits, a diffuser, and an impeller attached to the end of the rotating shaft (Figure 3). In operation the impeller accelerates water to a velocity high enough to cause a drop in pressure within the hollow, rotating shaft. Air is forced down the hollow shaft by atmospheric pressure, and fine bubbles of air exit the diffuser and enter the turbulent water around the impeller.

The rotating paddle wheel of paddle wheel aerators splashes water into the air to effect aeration. A floating, electric paddle wheel aerator is illustrated in Figure 4. The device consists of floats, a frame, motor, speed reduction mechanism, coupling, paddle wheel, and bearings. There is considerable variation in the design of the paddle wheel and in the mechanism for reducing the speed of the motor output shaft. Additional information on paddle wheel aerator design is provided later.

A diffused-air aerator consists of an air compressor or an air blower to provide air to diffusers positioned on the pond bottom or suspended in the water (Figure 5). Various types of diffusers have been used including ceramic dome diffusers, porous ceramic tubing, porous paper tubing, perforated rubber tubing, perforated plastic pipe, packed columns, and carborundum airstones. Most diffused-air aerators release a large volume of air at low pressure. The minimum permissible pressure increases with increasing depth of water above diffusers, because enough pressure must be available to force air from the diffuser against the

FIGURE 1.—Vertical pump aerator. Upper: top view; lower: aerator in operation.

total pressure at the discharge point (atmospheric pressure plus hydrostatic pressure).

Tractor-powered aerators may be used for emergency aeration in large ponds when fish are in severe stress from low dissolved oxygen (DO) concentrations. Two types are most common: paddle wheel aerators and pump sprayer aerators. The aerators are mounted on trailers and powered by the tractor takeoff. A paddle wheel aerator usually is constructed with a truck differential for a speed reduction mechanism and the axles for aerator shafts (Figure 6). Normally, the tractor takeoff is operated at 540 rpm and the paddle wheel revolves at 100–120 rpm. A pump sprayer aerator consists of a centrifugal pump and a discharge manifold or tube (Figure 7). Pumps are operated at 540 or 1,000 rpm.

Aerator Tests

Oxygen Transfer

The standard procedure for testing aerators for wastewater treatment applications can be used for evaluating the oxygen transfer efficiencies of aerators for aquaculture applications. A brief description of the aerator-testing procedure is given below; details of the test procedure can be found in Stuckenburg et al. (1977), American Society of Civil Engineers (1984), and Ahmad and Boyd (1988). In aerator tests, a known volume of clear water in a basin is deoxygenated with sodium sulfite and cobalt chloride (a catalyst), and DO concentrations are measured during reaeration with an aerator. Data on rate of change in the DO deficit, water temperature, and water volume permit calculation of the standard oxygen transfer rate (SOTR). The SOTR usually is reported as kilograms of oxygen per kilowatt-hour for standard conditions (20°C, 0 mg DO/L, and clear water). Power consumption of the aerator may be measured, and SOTR divided by power gives the standard aeration efficiency (SAE). The SAE normally is reported in terms of kilograms of oxygen

FIGURE 2.—Pump sprayer aerator.

FIGURE 3.—Propeller-aspirator-pump aerator.

FIGURE 4.—Paddle wheel aerator.

per kilowatt-hour. However, there are two ways of reporting SAE. Power may be measured at the aerator shaft with a torque sensor or estimated from electrical measurements and equations, and SAE is given in terms of brake power (Ahmad and Boyd 1988). Alternatively, power consumption by the motor may be measured with a watt meter or a watt-hour meter, and SAE is presented in terms of wire power. The later convention facilitates calculations of aerator operating costs. In the USA, motors are rated in horsepower (hp), but electricity consumption is measured in kilowatts or kilowatt-hours; 1 kW = 1.34 hp. Also, some aerator manufacturers in the USA report SOTR in pounds oxygen per horsepower or pounds oxygen per horsepower-hour; 1 kg O_2/h = 2.205 lb O_2/h and 1 kg O_2/kWh = 1.65 lb O_2/hph.

The efficiency of an aerator under actual operating conditions can be calculated for SOTR or SAE as follows:

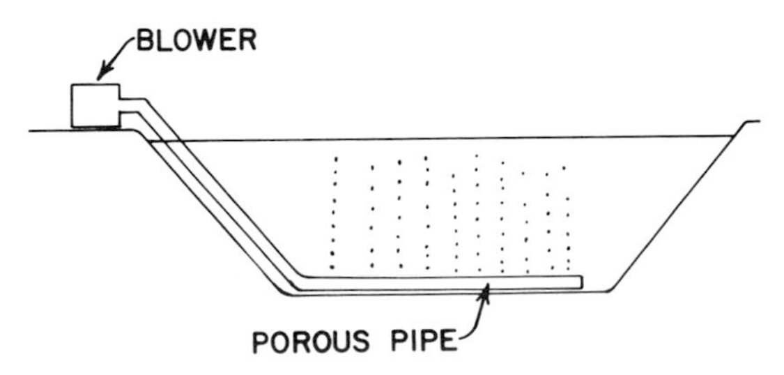

FIGURE 5.—Schematic representation of a diffused-air aeration system.

FIGURE 6.—Tractor-powered paddle wheel aerator.

$$OTR_f = SOTR\left[\frac{\beta C_s - C_p}{9.09} \times 1.024^{T-20} \times \alpha\right];$$

$$FAE = SAE\left[\frac{\beta C_s - C_p}{9.09} \times 1.024^{T-20} \times \alpha\right];$$

OTR_f = actual oxygen transfer rate (kg O_2/h);
FAE = actual aeration efficiency (kg O_2/kWh);
β = salinity correction (dimensionless);
C_s = DO concentration at saturation (mg/L);
C_p = DO concentration in pond (mg/L);
T = water temperature (°C);
α = correction for surfactants (dimensionless).

In the above equation, β is a factor obtained as follows:

$$\beta = \frac{C_s \text{ pond water}}{C_s \text{ clear fresh water}}.$$

Pond water and clear fresh water C_s values must be at the same temperature. The values for C_s may be made by saturating samples of pond water and clear fresh water at 20°C. However, for practical purposes it is easiest to measure the salinity of the pond water and obtain the value of C_s for the appropriate salinity from a table of DO

FIGURE 7.—Tractor-powered pump sprayer aerator.

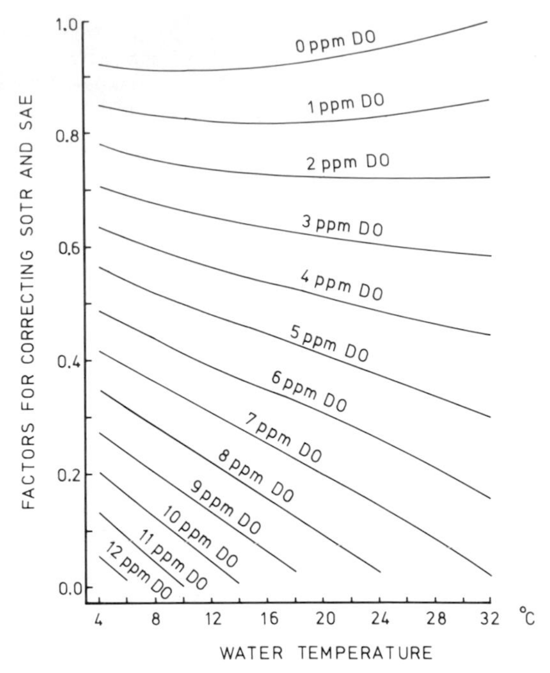

FIGURE 8.—Nomograph for correcting standard oxygen transfer rate (SOTR) and standard aeration efficiency (SAE) for actual conditions of water temperature and dissolved oxygen (DO) concentration.

concentrations at saturation for different temperatures and salinities (Boyd and Daniels 1987). If this is done, β may be omitted from the equation. Of course, in fresh water β will be approximately 1.0. The value for α in the above equation may be obtained by conducting aeration tests on pond water and clear fresh water at the same temperature and determining the oxygen transfer coefficients (K_La_T):

$$\alpha = \frac{K_La_T \text{ pond water}}{K_La_T \text{ clear fresh water}}.$$

For practical purposes, it usually is sufficient to assume an α value of 0.92 (Shelton and Boyd 1983).

To facilitate the estimation of actual aeration performance (OTR_f or FAE), Figure 8 was prepared. A correction factor corresponding to existing water temperature and DO concentration may be taken from Figure 8 and multiplied by SOTR or SAE to give OTR_f or FAE. In preparing Figure 8, $\alpha = 0.92$ and $\beta = 1.0$ were assumed. Boyd and Daniels (1987) showed that salinity does not affect SOTR and SAE for aerators that splash water into the air if all DO concentrations for saturation used in calculations of these two variables were for the actual salinity of the water used in aerator tests. Therefore, Figure 8 is appropriate for aerators operated in waters of any salinity. However, propeller-aspirator-pump aerators and diffused-air aeration systems operate more efficiently in brackish water than in fresh water (Ruttanagosrigit et al., in press), and Figure 8 would only apply when these aerators are used in fresh water.

Water Circulation

Two water circulation tests were developed by Boyd and Martinson (1984). In the dye test an intensely colored dye is poured in front of the aerator and the time required for the aerator to spread the dye over the entire pond surface is measured. Aerial photographs made from an airplane or helicopter are the best means of recording a dye test. The salt test may be used to determine the time required for an aerator to mix the entire volume of water in a pond. Enough sodium chloride to raise the conductivity of a pond by 100 to 200 μS is dissolved in a large container and poured around the aerator. Specific conductance is measured at intervals at several places and depths until all specific conductance values are essentially equal. The mixing rate may be estimated as

$$MR = \frac{(A)(D)}{(P)(T)};$$

MR = mixing rate (m^3/kWh);
A = pond area (m^2);
D = pond depth (m);
P = power consumption by aerator (kW);
T = time for complete mixing (h).

The water-circulating capabilities of aerators also may be estimated from the volumes of water discharged or from the velocities of water discharged. The pumping rate of aerators that discharge water in a well-defined stream or jet can be estimated from pump curves or from measurements with special weirs. The velocity of water at some distance from an aerator may be measured with a current meter.

TABLE 1.—Summary of standard oxygen transfer rate (SOTR, kg O_2/h) and standard aeration efficiency (SAE, kg O_2/kWh) values for electric aerators used in aquaculture. Values for SAE are in terms of power applied to the aerator shaft (brake power).

Type of aerator	Number of aerators	Range of SOTR	SAE	
			Average	Range
Paddle wheel	24	2.5–23.2	2.2	1.1–3.0
Propeller–aspirator–pump	11	0.1–24.4	1.6	1.3–1.8
Vertical pump	15	0.3–10.9	1.4	0.7–1.8
Pump sprayer	3	11.9–14.5	1.3	0.9–1.9
Diffused air	5	0.6–3.9	0.9	0.7–1.2

Aerator Performance

Boyd and Ahmad (1987) and Boyd (unpublished data) tested many electric aerators for oxygen transfer efficiency. Values for SOTR and SAE are summarized in Table 1. The reason that SOTR values had a wider range than SAE values was that aerators varied in size. These data demonstrated that paddle wheel aerators, in general, were more efficient than other types of aerators. Of course, some paddle wheel aerators were not as efficient as individual aerators of other types. Paddle wheel aerators constructed according to, or similar to, a design by Ahmad and Boyd (1988) had the highest SOTR and SAE values. Six manufacturers used this basic design for aerators. Values for SOTR ranged from 17.4 to 23.2 kg O_2/h, and values for SAE (based on brake power) ranged from 2.6 to 3.0 kg O_2/kWh. The average SAE (based on wire power) for these aerators was 2.2 kg O_2/kWh.

Boyd and Ahmad (1987) and Boyd and Stone (1988) tested several tractor-powered pump sprayer and paddle wheel aerators for SOTR. Values ranged from 7.8 to 73.8. The power applied to the aerator shaft was not measured, but a larger tractor was required for pump sprayer aerators than for paddle wheel aerators. For example, one pump sprayer aerator required a 60-kW tractor and had a SOTR of 21.2 kg O_2/h, whereas one paddle wheel aerator required a 50-kW tractor and had a SOTR of 29.8 kg O_2/h. In general, paddle wheel aerators performed better than pump sprayer aerators.

Although tests have been developed for evaluating water circulation by aerators, few data have been collected. Propeller-aspirator-pump aerators are much more efficient in mixing pond water than vertical pump aerators. A 1.5-kW propeller-aspirator-pump aerator spread dye over the surface of a 0.4-hectare pond in 32 min. After 32 min of operation in a 0.4-hectare pond, a 2.25-kW vertical pump aerator had spread dye over only one-fifth of the pond surface (Boyd and Martinson 1984). A 2.25-kW paddle wheel aerator spread dye over a 0.4-hectare pond in 28 min (Boyd and Watten 1989). A better comparison of water-mixing capabilities of surface aerators is afforded by salt-mixing tests, because these tests evaluate mixing of the entire pond volume rather than just surface water. The mixing rates for a propeller-aspirator-pump aerator and a vertical pump aerator were 1,778 and 305 m^3/kWh, respectively. A paddle wheel aerator had a mixing rate of 3,235 m^3/kWh.

These results indicate that paddle wheel aerators are more efficient in transferring oxygen and circulating water than other types of aerators commonly used in aquaculture. For aerators of 1 kW and larger size, paddle wheel aerators are equal or less in cost than other types of aerators. However, small paddle wheel aerators are more expensive than other types, because gear motors required for small paddle wheel aerators are very expensive. For this reason, vertical pump aerators, propeller-aspirator-pump aerators, and diffused-air aerators may be more suitable for small ponds than paddle wheel aerators even though paddle wheel aerators are more efficient.

Design for Paddle Wheel Aerators

An electric paddle wheel aerator consists of a motor, a speed reduction mechanism, and paddle wheel mounted on a trailer or on a flotation device. Floating electric paddle wheel aerators are much more common than trailer-mounted ones. The oxygen transfer efficiency of a paddle wheel aerator depends upon design and operating

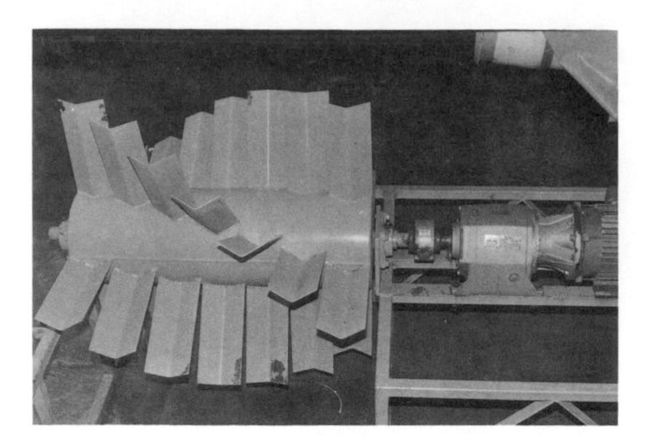

FIGURE 9.—A 2-kW aerator. Note the spiral arrangement of paddles on the hub.

FIGURE 10.—A 0.75-kW paddle wheel aerator fabricated of polyvinyl chloride pipe and polyurethane.

characteristics of the paddle wheel. Thus, paddle wheel fabrication specifications are rigid, whereas design of the flotation system is flexible.

The highest oxygen transfer efficiency was achieved with a paddle wheel 91 cm in diameter with triangular paddles (120–135° interior angle) spiraled on the hub (Ahmad and Boyd 1988). The most efficient of the electric paddle wheel aerators tested had paddles that extended 9–11 cm into the water and paddle wheels that rotated at 75–80 rpm. The optimum brake power was about 1 kW for each 40 cm of paddle wheel length. This design works well for 2-kW and larger aerators. A 2-kW paddle wheel aerator is shown in Figure 9. If either paddle submergence or paddle wheel speed is increased, power requirement will increase and oxygen transfer efficiency will decline. The spiral arrangement of paddles on the hub (Figure 9) allowed a fairly constant area of paddle surface to move continuously through the water, reducing vibration and wear.

For small paddle wheel aerators (<2 kW), the paddle wheel diameter should be 60 cm. A speed of 80 rpm also is sufficient for small aerators. In order to load a 0.75-kW motor properly when the paddle depth is 9 cm and speed is 80 rpm, a 60-cm-diameter paddle wheel must be about 1 m long.

Paddle wheel shafts are fitted with bearings and mounted on a metal frame that is floated with steel boxes, Styrofoam blocks, or plastic or metal tanks. Some means of raising and lowering both ends of the paddle wheel are provided so that minor adjustments in depth may be made once the aerator is installed in a pond. Takeup bearings provide one convenient way of adjusting paddle wheel elevation and paddle depth. Aerators usually are anchored in ponds by the aid of two metal stabilizer bars attached to each end of the floating frame and to metal bars driven into the pond bank.

A gear reducer is the simplest way to reduce motor output shaft speed to 75 or 80 rpm. A gearmotor has the gear reducer built onto the motor. The output shaft of a gearmotor can be connected to the input shaft of the paddle wheel with a flexible coupling. Alternatively, a gearbox may be connected on one side to the output shaft of the motor with sheaves and cog belts and coupled on the other side to the input shaft of the aerator.

For brackish water application, aerators may be fabricated of mild steel and hot-dip galvanized. After aging for a few months, the galvanized surfaces should be painted with epoxy resin. This procedure will greatly reduce the rate of corrosion. Alternatively, paddle wheel aerators may be constructed of stainless steel or plastic. The small aerator shown in Figure 10 is constructed of polyvinyl chloride pipe (floats) and polyurethane (hubs, paddles, and bearings).

After the aerator is installed in a pond, the electrical current used by the motor can be measured with an ammeter. If the current is higher than the rated current for full load of the motor, the paddle wheel should be raised. If current use is low, the paddle wheel should be lowered. For best service life, the motor should not be loaded beyond 90–95% of rated current.

Use of Aerators in Ponds

Effect on Production

Some research on aeration of ponds used to raise channel catfish *Ictalurus punctatus* is discussed. These findings probably are applicable to most other types of pond aquaculture. Tucker et al. (1979) demonstrated that at feeding rates below 30 kg/hectare · d, aeration usually would not be necessary. At this feeding rate, an annual channel catfish production of 3,000 kg/hectare can be achieved. At feeding rates between 30 and 50 kg/hectare · d, emergency aeration must be applied occasionally to most ponds or low DO will stress or kill fish. If electricity is unavailable at ponds, tractor-powered aerators are suitable for emergency aeration of ponds with feeding rates of 30–50 kg/hectare · d. At this feeding rate, an annual catfish production of 4,000–4,500 kg/hectare is normal. If feeding rates exceed 50 kg/hectare · d, aeration will be needed frequently in nearly all ponds during warm weather and electrically powered aerators are more efficient than tractor-powered ones. Lai-fa and Boyd (1988) suggested using nightly aeration (midnight to dawn) with aerators controlled by timers. At identical stocking and feeding rates, higher feed conversion efficiency and greater production of catfish were achieved in ponds with nightly aeration than in ponds where aeration was used only in emergencies. This resulted because DO concentrations were consistently higher in the ponds aerated nightly.

Channel catfish were stocked in ponds at rates of 1,200, 4,300, 8,600, 17,300, 26,000, and 34,600/hectare, and maximum daily feeding rates of 0, 28, 56, 84, 112, 168, and 224 kg/hectare, respectively, were established (Cole and Boyd 1986). Aeration was applied with vertical pump aerators at 6.1 kW/hectare when DO concentration was expected to fall below 2 mg/L. Aerators were seldom operated in ponds with feeding rates of 0–56 kg/hectare · d. Aeration was applied almost constantly at night in ponds with feeding rates of 112 kg/hectare · d and above. Even though aeration prevented extremely low DO concentrations in all ponds, net fish production increased with feeding rate only up to 112 kg/hectare · d of feed. Feed conversion values, calculated as amount feed applied divided by net fish production, were between 1.6 and 1.8 for daily feeding rates of 0–112 kg/hectare. Feed conversion values were 2.5 and 16.5 for daily feeding rates of 168 and 224 kg/hectare, respectively. Maximum net production of 6,000 kg/hectare was achieved at a feeding rate of 112 kg/hectare · d. Ammonia-nitrogen accumulated in ponds, and high ammonia concentrations apparently limited production at high feeding rates. This experiment shows that if economic considerations are ignored and enough aeration is applied to prevent DO depletion at high feeding rates, production still cannot be increased without limit. Some other water quality variable, probably ammonia, will impose limits on production even though there is adequate DO.

The experiment by Cole and Boyd (1986) and practical experience by commercial producers indicate that about 6,000 kg/hectare is the maximum production of channel catfish possible in static-water ponds. Production can be increased by water exchange to remove ammonia and toxic metabolites. Plemmons and Avault (1980) used continuous aeration with vertical pump aerators at 5.5 kW/hectare, emergency aeration when DO concentrations were low, and water exchange in channel catfish ponds. Net production averaged 12,060 kg/hectare. The maximum daily feeding rate was not reported. N. Parker (U.S. Fish and Wildlife Service, unpublished data) produced 15,800 kg of channel catfish per hectare in small ponds (200 m^2) with aeration and water exchange.

In Israel, fish are produced in intensive culture in small, concrete ponds (50 m^2) with center drains. One or two 0.75-kW paddle wheel aerators are operated continuously in each pond. Circular water movement produced by the aerator causes waste solids to accumulate in the center of the pond, and these wastes are removed twice per day through the center drain. In addition to aeration and solids removal, up to four volumes of water are flushed through ponds daily. The water is recycled through a large, earthen pond to effect waste treatment.

In trout raceways, water is clean and exchange times are short. Therefore, fish consume essentially all of the oxygen lost from raceway waters, and photosynthesis is not a significant source of oxygen. It is known that each kilogram of feed provided to trout in raceways requires about 0.2 kg of oxygen. Therefore, if the amount of oxygen in water entering raceways, the amount of oxygen supplied by aeration, and the average feed conversion efficiency of fish are known, one can compute the potential for trout production in a given raceway. In ponds, phytoplankton produces oxygen, and plankton, bacteria, benthos, insects, and fish use oxygen; fish are often a minor component in the DO budget for a pond. Methods for determining the proportion of the DO added to a pond by aeration that can be used by fish are not available. It is not possible to calculate how much added production is possible from a given amount

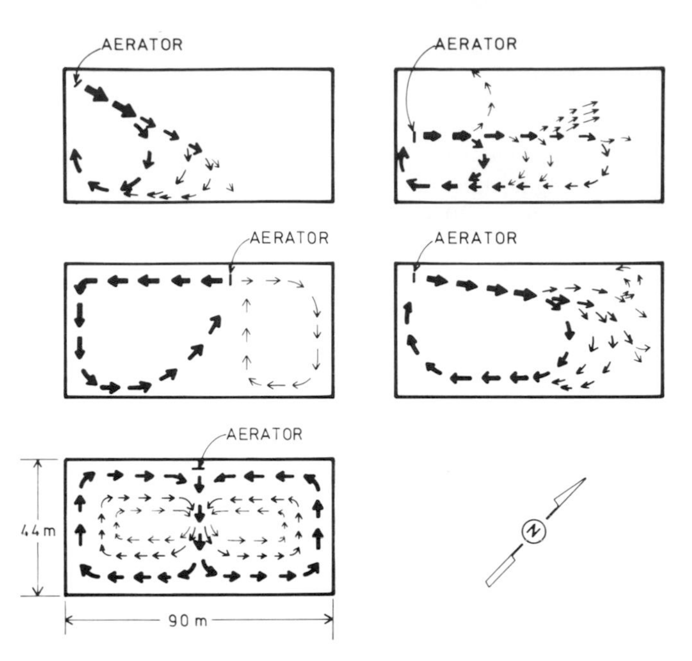

FIGURE 11.—Water circulation patterns for different aerator locations. Arrow thickness indicates strength of water currents; the thicker the arrow the stronger the current.

of aeration in a pond. Research in this area is badly needed.

Aerator Placement

Fish become conditioned to high DO concentrations around an aerator, and they come to this area when low DO concentrations occur in other parts of a pond. Some farmers feed fish in the area around an aerator to encourage fish to frequent this area. If aerators are permanently installed, there must be access to a power source and easy access to the aerator from an all-weather road for maintenance purposes. If an aerator fails, the replacement aerator should be placed near the failed unit so that fish do not have to swim through DO-deficient water to reach the oxygenated zone around the replacement. If mobile aerators are used for emergency aeration, an attempt should be made to locate the area where fish are concentrated and position the aerator in that place. But, if the mobile aerator is used frequently, it should be used in the same place night after night so that the fish train to it. Of course, for emergency aeration, an aerator that discharges water over a large area is desirable, for this increases the probability that fish or other aquatic organisms can find areas with adequate DO.

Water circulation tests show that aerators mix water throughout ponds, and paddle wheel aerators are especially efficient in circulating pond water. A recent study was conducted to ascertain the best position for a paddle wheel aerator in a rectangular pond. Sixty 1-L plastic bottles were filled with water so that they floated with only their caps above the water. These bottles were released at once in front of a 2.25-kW aerator in a 0.4-hectare pond, and the paths of the bottles were recorded (Figure 11). The best place to mount a paddle wheel aerator is at the middle of one of the long sides of the pond. The aerator should direct water parallel to the short sides of the pond. Of course, this study was for a specific aerator and pond. Different results no doubt would be obtained for larger ponds.

In intensive shrimp production, four paddle wheel aerators often are placed in a single pond. An aerator is placed in each corner of a pond, and the aerators all direct water in the same direction to provide a circular flow pattern. I also have seen single levees built across the centers of ponds, but with gaps between the ends of the center levees and the side levees. One or more paddle wheel aerators operated on one side of such a pond or in the gaps at the ends of the center levee will cause a circular movement of water.

Aerator Controllers

Several companies are trying to sell devices that automatically turn on aerators when DO concentrations fall below a certain level and turn off aerators when DO concentrations rise above a certain level. The DO concentration at which the aerator is turned on or off can be set according to the desires of the pond manager. However, such devices currently are expensive and unreliable. The biggest problem involves biological growths that foul the DO probe. This results in inaccurate DO measurements, and the aerator is activated at improper times. As technology improves, less expensive and more reliable devices for automatically controlling aerators in response to DO concentration can no doubt be developed. However, the usefulness of such devices is questionable. It is probably better to use inexpensive timers to turn aerators on and off. A device could be attached to the aerator circuit to notify the pond manager if the aerator motor shuts off because of malfunction. This device would not require a DO probe and the electrical equipment necessary to activate the aerator in response to DO concentration.

References

Ahmad, T., and C. E. Boyd. 1988. Design and performance of paddle wheel aerators. Aquacultural Engineering 7:39–62.

American Society of Civil Engineers. 1984. A standard for the measurement of oxygen transfer in clean water. American Society of Civil Engineers, New York.

Boyd, C. E., and T. Ahmad. 1987. Evaluation of aerators for channel catfish farming. Alabama Agricultural Experiment Station, Auburn University Bulletin 584.

Boyd, C. E., and H. V. Daniels. 1987. Performance of surface aerators in saline pond waters. Progressive Fish-Culturist 49:306–308.

Boyd, C. E., and D. J. Martinson. 1984. Evaluation of propeller–aspirator–pump aerators. Aquaculture 36:283–292.

Boyd, C. E., and N. Stone. 1988. Evaluation of aeration attachments for a Crisafulli pump. Alabama Agricultural Experiment Station, Auburn University, Circular 293.

Boyd, C. E., and B. J. Watten. 1989. Aeration systems in aquaculture. Critical Reviews in Aquatic Sciences 1:425–472.

Cole, B. A., and C. E. Boyd. 1986. Feeding rate, water quality, and channel catfish production in ponds. Progressive Fish-Culturist 48:25–29.

Lai-fa, Z., and C. E. Boyd. 1988. Nightly aeration to increase the efficiency of channel catfish production. Progressive Fish-Culturist 50:237–242.

Plemmons, B., and J. W. Avault, Jr. 1980. Six tons of catfish per acre with constant aeration. Louisiana Agriculture 23:6–9.

Ruttanagosrigit, W., Y. Musig, C. E. Boyd, and L. Sukchareon. In press. Effect of salinity on oxygen transfer by propeller-aspirator-pump and paddle wheel aerators used in shrimp farming. Aquacultural Engineering 10.

Shelton, J. L., Jr., and C. E. Boyd. 1983. Correction factors for calculating oxygen-transfer rates of pond aerators. Transactions of the American Fisheries Society 112:120–122.

Stuckenburg, J. R., V. H. Wahbeh, and R. E. McKinney. 1977. Experiences in evaluating and specifying aeration equipment. Journal of the Water Pollution Control Federation 49:66–82.

Tucker, L., C. E. Boyd, and E. W. McCoy. 1979. Effects of feeding rate on water quality, production of channel catfish, and economic returns. Transactions of the American Fisheries Society 108:389–396.

American Fisheries Society Symposium 10:474–481, 1991

Modeling the Effect of Dissolved Nitrogen and Carbon Dioxide on the Performance of Pure Oxygen Absorption Systems

BARNABY J. WATTEN

U.S. Fish and Wildlife Service, National Fishery Research and Development Laboratory Rural Delivery 4, Box 63, Wellsboro, Pennsylvania 16901, USA

JOHN COLT

James M. Montgomery, Consulting Engineers 2375 130th Avenue NE, Suite 200, Bellevue, Washington 98005, USA

CLAUDE E. BOYD

Department of Fisheries and Allied Aquacultures, Auburn University, Alabama 36830, USA

Abstract.—The effects of dissolved nitrogen and carbon dioxide on the performance of a pure oxygen packed column were evaluated with a multicomponent gas-transfer model. Simulation data indicated that the high solubility of carbon dioxide precludes significant desorption under typical column operating conditions. The control of carbon dioxide buildup in intensive culture systems will require use of conventional aeration equipment or chemical treatment. In contrast to the high solubility of carbon dioxide, the solubility of nitrogen (N_2) is low. Consequently, dissolved nitrogen can be stripped from solution during the oxygenation process. The transfer of nitrogen from the liquid phase into the gas phase within the contact enclosure substantially reduces oxygen absorption efficiency (mass oxygen absorbed/mass oxygen supplied) and transfer efficiency (mass oxygen absorbed/power input). Nitrogen stripping also acts to control the allowable change in dissolved oxygen across the absorber when regulated by criteria for contactor effluent total dissolved gas pressure. The extent of nitrogen desorption is determined by the concentration of dissolved gases in the influent, column height, oxygen flow rate, and column pressure.

Recent aquacultural applications of pure oxygen contact systems for nitrogen desorption or oxygen addition have shown favorable operational and economic characteristics (Collins et al. 1984; Gowan 1987; Severson et al. 1987). The design of such systems depends on selected water quality criteria, the layout of the culture system, and site-specific operating conditions that include influent concentrations of dissolved carbon dioxide (DC) and nitrogen (DN) (Colt and Watten 1988). Influent gas concentrations vary with temperature, water source, and previous treatment or use. Many of the system performance parameters, including concentrations of dissolved gas in the effluent, oxygen absorption efficiency (mass oxygen absorbed/mass oxygen supplied), and transfer efficiency (mass oxygen absorbed/power input), are interrelated; thus the selection of the actual operating point involves trade-offs between the various parameters. The effect of these trade-offs on performance may be substantial when a single contact system is used for both the addition of oxygen and the desorption of nitrogen.

Our purpose is to establish the effects of influent dissolved carbon dioxide and nitrogen on the performance of the packed column, a pure oxygen contactor widely used in fish culture (Boerson and Chesney 1987; Severson et al. 1987). After a multicomponent gas-transfer model is developed, simulation data are used to predict the operational characteristics of the packed column under typical culture conditions.

Methods

A multicomponent gas-transfer model was developed by the finite-difference method. The gas phase within the column was considered homogeneous (Watten 1989). Gas transfer was estimated with the algorithms developed by Hackney and Colt (1982) and Colt and Bouck (1984). Gas solubility was based on values given by Benson and Krause (1984) for oxygen, by Weiss (1970) for argon and nitrogen, and by Weiss (1974) for carbon dioxide. Nitrogen gas in this model is taken as the sum of nitrogen and argon gas, and the molecular weight, mole fractions, Bunsen coefficients, and other gas constants have been adjusted accordingly (Colt 1984). This procedure was used because the mole fraction of argon is small and the most common method of gas analysis determines the sum of nitrogen and argon. Reactions of DC in the liquid phase were not considered.

The model directs the following computational

procedure to establish steady-state performance. (1) Set the gas composition within the column equal to inlet gas values and compute the moles of gas present in the column. (2) Compute the moles of oxygen, nitrogen, and carbon dioxide transferred into and out of the water per time step. (3) Calculate the moles of gases added to the column enclosure from the enriched oxygen flow per time step. (4) Recompute the working composition of the gas within the column through steps (1) to (4). (5) Check the gas composition for convergence. If the gas composition does not converge, repeat the calculation sequence starting at step (2).

After convergence, the computer program computes steady-state effluent DN, DC, and dissolved oxygen (DO) concentrations. Dissolved gas pressures are then calculated by using methods recommended by Colt (1984) along with the standard performance parameters such as oxygen absorption efficiency, transfer efficiency, and economy. Unless otherwise stated, the following environmental conditions were used in the simulation runs: water temperature, 15°C; barometric pressure, 760 mm Hg; salinity, 0 g/kg; influent DO, saturation; influent DN, 105% saturation; influent DC, saturation; alpha factor, 1; beta factor, 1. The alpha and beta factors represent the field-water-to-clean-water ratios of mass transfer coefficients and saturation concentrations, respectively:

$$\alpha = \frac{G \text{ (field water)}}{G \text{ (clean water)}};$$

$$\beta = \frac{C_s \text{ (field water)}}{C_s \text{ (clean water)}};$$

$G = K_d + KZ$, overall mass transfer coefficient (dimensionless);
K_d = distribution plate mass transfer coefficient (dimensionless);
K = specific packing mass transfer coefficient (m^{-1});
Z = packed bed depth (m);
C_s = dissolved gas saturation concentration (mg/L).

The physical characteristics of the packed column evaluated, unless noted otherwise, are given in Table 1. Further, pure oxygen gas flows are presented as a volumetric ratio of gas to liquid, expressed as a percentage (100 · G/L). The cost of oxygen and electricity were set at U.S.$0.16/kg and $0.07/kWh. An additional 1.82 m of head loss was included in the computation of total dynamic head. Pump and vacuum power requirements were based on an overall motor and pump efficiency of 70%.

TABLE 1.—Physical characteristics of the packed column used in simulation runs.

Variable	Value
Packing type	3.81-cm pall rings
Packing void fraction	94%
K	1.71/m at 20°C
K_d	0.30 at 20°C
$G_{20°C}$ for oxygen	2.00
$G_{20°C}$ for nitrogen	1.88
$G_{20°C}$ for carbon dioxide	2.00
Column height	0.994 m
Column diameter	0.558 m
Oxygen purity	100.0%
Hydraulic loading	245 $m^3/m^2 \cdot h$

Gas Transfer

Gas transfer depends on the magnitude and sign of the partial pressure gradient between the gas phase and the liquid phase (Lewis and Whitman 1924):

$$\frac{dP_i}{dt} = K_La\left[X_i(P_c - Pwv) - C_i\frac{A_i}{B_i}\right]; \quad (1)$$

dP_i/dt = change in partial pressure of ith gas species with respect to time (mm Hg/h);
K_La = overall gas transfer coefficient (h^{-1});
X_i = gas phase mole fraction of gas species i (dimensionless);
P_c = column pressure (absolute, mm Hg);
Pwv = water vapor pressure (mm Hg);
C_i = concentration of gas species i (mg/L);
A_i = 760/100K_i, K_i being the ratio of molecular weight to volume for gas species i;
B_i = Bunsen coefficient of gas species i (L gas/[L water · 760 mm Hg pressure]).

A positive value of the gradient results in transfer from the gas phase into the liquid phase. Conversely, a negative value results in transfer from the liquid into the gas phase. No transfer of gas occurs when the gradient is zero.

Typically, within a packed column (Figure 1), oxygen is absorbed by the liquid phase whereas nitrogen and carbon dioxide are stripped from solution and enter the gas phase. This concurrent movement of gas into and out of the gas phase acts to lower X_{O_2} and increase X_{N_2} and X_{CO_2}. These changes, in turn, reduce the rates of gas transfer to levels below the level expected with a pure oxygen environment (equation 1). The extent of the gas phase contamination, under steady-

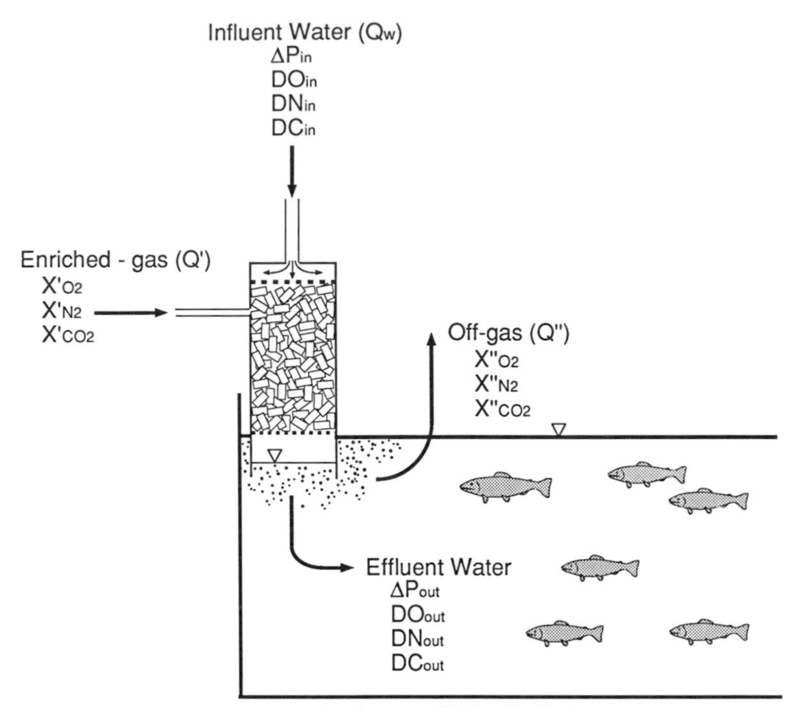

FIGURE 1.—Definition sketch for a packed column oxygenation system. The ΔP is the difference between the total gas pressure and barometric pressure; DO, DN, and DC refer to the concentration of dissolved oxygen, nitrogen + argon, and carbon dioxide, respectively; X is the mole fraction of the gas; and Q is the volumetric flow rate of the gas. The subscripts "in" and "out" refer to the influent and effluent waters; the single primes (′) refer to the influent gases and the double primes (″) to the effluent gases.

state conditions, is determined by column height, oxygen flow rate, influent dissolved gas concentrations, and column pressure. It is important to note that as influent DC and DN increase, so do steady-state values of X_{N_2} and X_{CO_2}. Carbon dioxide and nitrogen buildup is regulated by venting gas (off-gas) from the absorber unit, a condition that reduces oxygen absorption efficiency.

In applications in which nitrogen stripping is not required, complete oxygen absorption can be attained by regulating column pressure. That is, by increasing column pressure to a point at which the partial pressure of the gas phase is at least equal to the partial pressure of the liquid phase:

$$X_i\,(P_c - Pwv) = C_i\,\frac{A_i}{B_i}. \tag{2}$$

The mole fraction, X_i, of nitrogen and carbon dioxide that results in no stripping is shown in Figure 2 as a function of concentration and column pressure. Operation of the system above the curve results in an increase in dissolved gas concentration; operation below the curve results in gas stripping. Below atmospheric pressure, gas stripping can occur with air; above atmospheric pressure, however, the mole fraction of gas must be reduced below the atmospheric composition. Although this simple analysis can determine the direction of gas transfer, the multicomponent transfer model is needed to compute the steady-state mole fraction of the gases within the column.

Multicomponent Gas Transfer Modeling

Carbon Dioxide

The solubility of carbon dioxide is relatively high compared with that of oxygen and nitrogen (Colt and Watten 1988). Thus only a slight buildup of carbon dioxide in the gas phase dramatically reduces carbon dioxide stripping (Figure 2). For example, a DC concentration equal to 1,000% saturation (6.77 mg/L at 15°C) represents an equilibrium X_{CO_2} of just 3.5×10^{-3}. Over typical pure oxygen G/L ratios, the removal of carbon dioxide is small (Figure 3). Changes in column height, as indicated by G, had only a slight effect on DC stripping. Additional simulation runs indicated

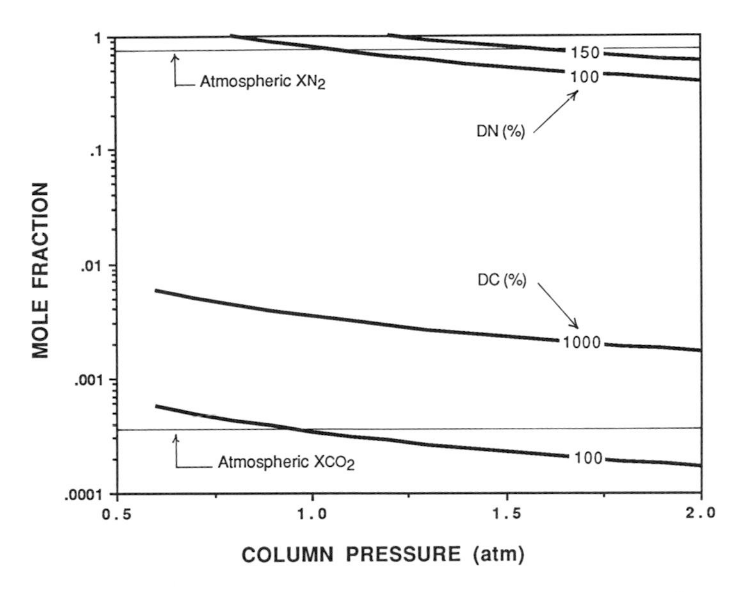

FIGURE 2.—Lines of no transfer of carbon dioxide or nitrogen between water or gas phases as functions of gas phase mole fraction (X), column pressure, and influent dissolved gas concentration (water temperature, 15°C; influent dissolved nitrogen [DN], 1.0 and 1.5 times saturation; dissolved carbon dioxide [DC], 1.0 and 10.0 times saturation). Atm is atmospheres.

that DC stripping was also insensitive to changes in column pressure. Performance indicators for the packed column are given in Figure 4 as a function of influent DC. As shown, an increase in DC up to 100 mg/L results in less than a 7% change in oxygen absorption efficiency (AE), transfer efficiency (TE), and variable costs.

Although the data presented here show that

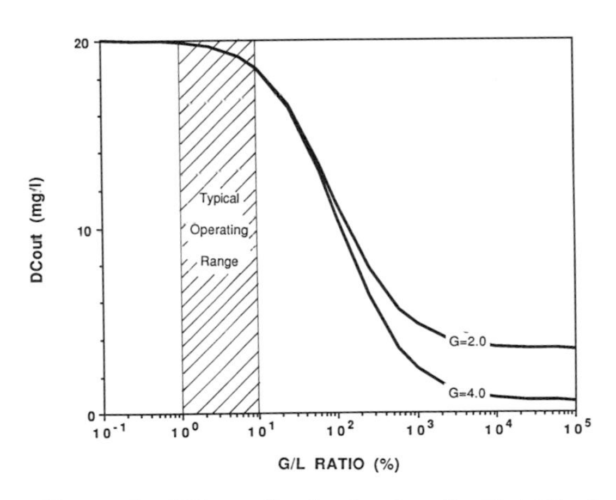

FIGURE 3.—Effluent dissolved carbon dioxide (DC_{out}) as a function of volumetric gas-to-liquid (G/L) ratio and overall column mass transfer coefficient (G_{oxygen}, dimensionless; water temperature, 15°C; influent DC, 20 mg/L; influent dissolved oxygen, 100% saturation; dissolved nitrogen, 105% saturation).

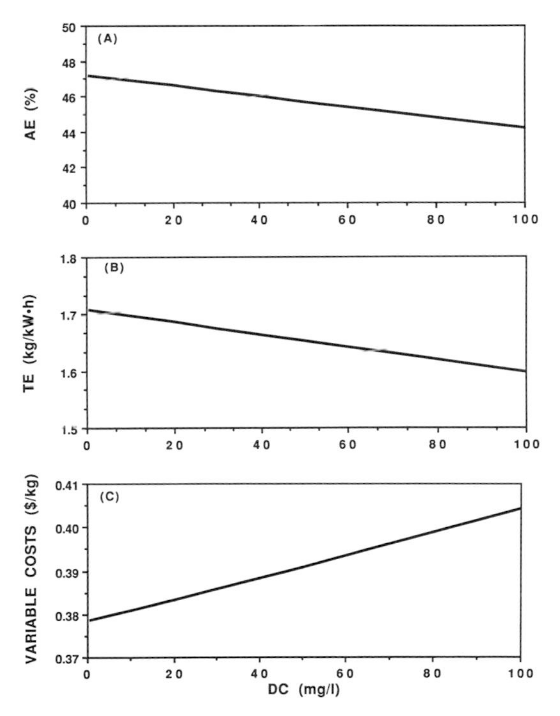

FIGURE 4.—Effects of influent dissolved carbon dioxide (DC) on **(A)** oxygen absorption efficiency (AE), **(B)** transfer efficiency (TE), and **(C)** variable costs (water temperature, 15°C; influent dissolved oxygen, 100% saturation; dissolved nitrogen, 105% saturation). The G/L ratio was set at 3% and G_{oxygen} at 2.0.

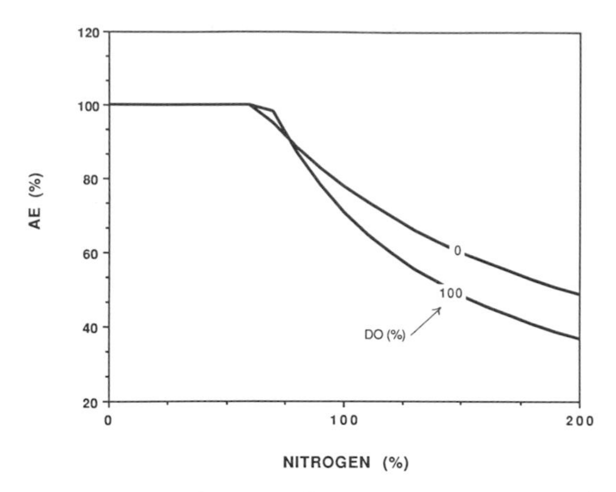

FIGURE 5.—Effects of influent dissolved nitrogen (% saturation) and influent dissolved oxygen (DO, % saturation) on oxygen absorption efficiency (AE) (water temperature, 15°C; effluent DO, 20 mg/L; column pressure, 760 mm Hg). The value for G_{oxygen} was set at 2.0.

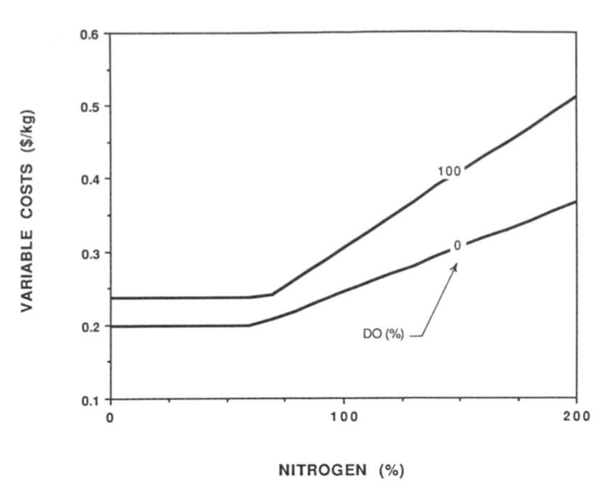

FIGURE 6.—Effects of influent dissolved nitrogen (% saturation) and dissolved oxygen (DO, % saturation) on variable costs (water temperature, 15°C; effluent DO, 20 mg/L; column pressure = 760 mm Hg). The value of G_{oxygen} was set at 2.0.

influent DC has no significant effect on column performance, carbon dioxide must not be ignored in culture facilities employing pure oxygen systems. Colt et al. (1991, this volume) show that under conditions typical of trout and salmonid culture (alkalinities < 1.0 meq/L), the total cumulative oxygen consumption is limited to about 14 mg/L based on a free carbon dioxide criterion of 20 mg/L. Under high-intensity culture conditions, surface aerators or packed column aerators operated at G/L ratios above 100 may be needed to control the buildup of carbon dioxide. Alternatively, carbon dioxide may be removed with calcium hydroxide or sodium carbonate treatments.

Nitrogen

In contrast to the solubility of carbon dioxide, that of nitrogen is relatively low. This difference is evident in the critical mole fraction values plotted in Figure 2. Note that X_{N_2} may approach 0.78 (column pressure, 1 atmosphere; C_i, saturation) whereas the corresponding X_{CO_2} is limited to 0.00032. Packed column operation with an elevated X_{N_2} can result in substantial DN stripping, but at the cost of reduced oxygen transfer (equation 1). Figure 5 gives the effects of influent DN and DO on AE with a required DO_{out} of 20 mg/L. The required DO_{out} was achieved by adjusting the G/L ratio. Below a DN of about 70%, oxygen absorption is complete (100%). Increasing the influent DN above 70% results in a substantial increase in the required G/L ratio. For example, at an influent DO of 100%, increasing the influent DN from 100 to 150% increases the required G/L ratio from 1.05 to 1.54%. Higher influent DO reduces AE by decreasing the partial pressure gradient (equation 2). Because, in this example, a constant DO_{out} was required, the TE was constant for all DN concentrations but did depend on the DO_{in} concentrations. The TE was equal to 0.898 kg/kWh at a DO_{in} of 100% and 1.811 kg/kWh at a DO_{in} of 0%. This difference in TE is due to the increased partial pressure gradient at lower DO concentrations. Total variable costs (the sum of oxygen and energy costs) range between about $0.20 and 0.52 per kilogram of oxygen transferred (Figure 6). The rise in variable costs above a DN of 70% reflects the reduction in oxygen absorption that occurs within that operating region. Figure 6 also illustrates the sensitivity of variable costs to changes in influent DO. As influent DO increases, so do variable costs.

In the 100% oxygen absorption region, the mole fraction of oxygen inside the column is higher than typically achieved because of reduced nitrogen stripping. At low G/L ratios, conditions can occur where not enough oxygen gas is supplied to maintain the column pressure. For example, for influent DO and DN = 0%, the final column pressure is only 387 mm Hg. As the column pressure falls, the water level in the column rises, if permitted, and reduces the column performance. If the G/L is increased, the DO_{out} exceeds the required DO, and problems with oxygen toxicity or gas supersaturation may occur. Under these conditions, a side-stream contact system

(Colt and Watten 1988) may be needed. With such a system, only a fraction of the total process water is passed through the absorber and then combined with the untreated water.

In addition to regulating AE, TE, and variable costs, nitrogen transfer controls the operating ΔDODN stripping ratio, which is defined as the change in DO per change in DN, the change in both gas species being expressed as mg/L. Applying the general packed column performance equations (Hackney and Colt 1982), one can calculate values of ΔDODN as follows:

$$\Delta \text{DODN} = \frac{[C^*(O_2) - \text{DO}_{in}]}{[C^*(N_2) - \text{DN}_{in}]} \cdot \frac{(1 - e^{-G})}{(1 - e^{-\phi G})}; \quad (3)$$

C^* = equilibrium gas concentration inside the column (mg/L);
DO_{in} = influent DO concentration (mg/L);
DN_{in} = influent DN concentration (mg/L);
ϕ = relative G for nitrogen gas (0.89–0.94).

The second term in equation (3) depends only on the physical characteristics of the column, such as media type, height, and distribution plate design. Note also that the C^* terms in equation (3) depend on column gas composition, column pressure, and water temperature and therefore cannot be simply computed. The stripping ratio (ΔDODN) is important because it largely determines the allowable change in DO between the absorber and the rearing unit when regulated by ΔP criteria for the fish being held. For a plug flow-type rearing vessel, the allowable ΔDO as a function of the stripping ratio, dilution ratio, absorber influent, and rearing unit influent ΔP is as follows (Colt and Watten 1988):

$$\Delta \text{DO} = \frac{\Delta P_{out} - \Delta P_{in}}{F(O_2) + [F(N_2)/\Delta \text{DODN}]}; \quad (4)$$

F = A_i/B_i for nitrogen or oxygen gas;
ΔDO = allowable change in DO across the system (mg/L);
ΔP_{out} = maximum allowable ΔP in the rearing unit influent (mm Hg);
ΔP_{in} = absorption unit influent ΔP (mm Hg).

The allowable change in DO across the absorption unit can also be established, given the water flow rate through the absorber and the total flow entering the rearing unit:

$$\Delta \text{DO}_{AB} = \Delta \text{DO} \frac{Q_{total}}{Q_{lox}}; \quad (5)$$

ΔDO_{AB} = allowable change in DO across the absorber (mg/L);
Q_{total} = total flow to the rearing unit (m^3/s);
Q_{lox} = flow through pure oxygen absorber unit (m^3/s).

As the allowable ΔDO increases, the volume of water that must be treated to satisfy a given oxygen demand is reduced. The allowable ΔDO, based on equation (4), is plotted in Figure 7 as a function of ΔP_{out} and the stripping ratio ΔDODN. At large values of ΔDODN, the allowable ΔDO approaches a horizontal asymptote that depends only on ($\Delta P_{out} - \Delta P_{in}$). The allowable ΔDO also approaches a vertical asymptote at a ΔDODN ratio of −2.24 (15°C). This asymptotic value, defined here as the critical stripping ratio, identifies the operating point that results in no net change in total gas pressure:

$$\Delta \text{DODN}_{critical} = -F(N_2)/F(O_2). \quad (6)$$

The ratio $-F(N_2)/F(O_2)$ varies only slightly with temperature (e.g., it is −2.26 at 10°C and −2.18 at 30°C). The maximum value of ΔDO given in Figure 7 is limited due to considerations of conservation of mass for nitrogen gas. At the maximum ΔDO, all of the nitrogen gas has been stripped out and replaced with oxygen gas. The curves plot to the left of the vertical asymptote when $\Delta P_{out} > \Delta P_{in}$, and to the right for when $\Delta P_{out} < \Delta P_{in}$. It is important to note that for each

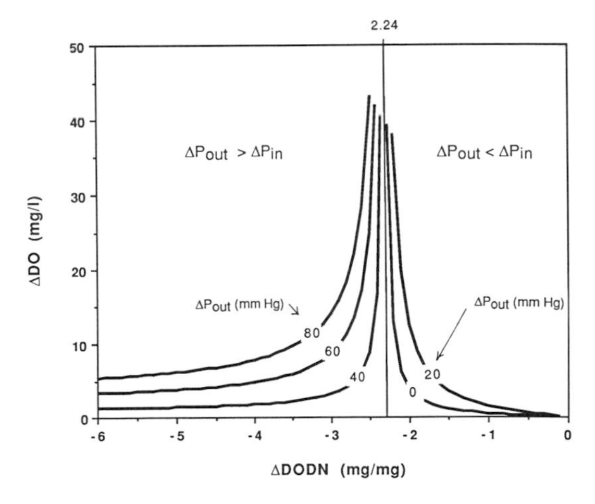

FIGURE 7.—Limitations in the allowable change in dissolved oxygen (ΔDO) across a packed column system serving a raceway as a function of the ΔDODN stripping ratio and ΔP_{out} (water temperature, 15°C; influent DO, saturation; influent dissolved nitrogen, 105% saturation; ΔP_{in}, 28.8 mm Hg; column pressure, 760 mm Hg).

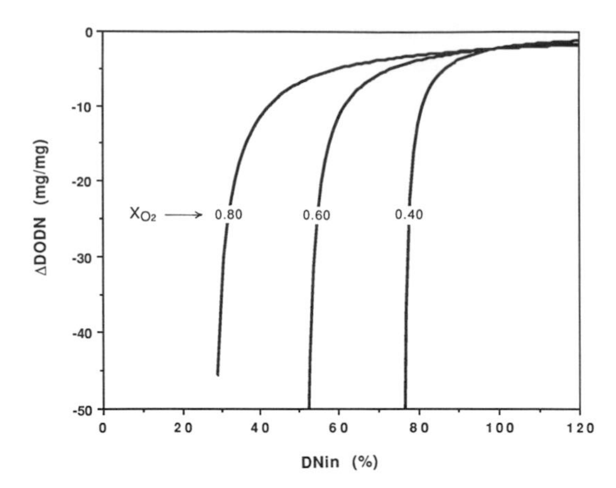

FIGURE 8.—Effect of influent dissolved nitrogen (DN_{in}, % saturation) on the ΔDODN stripping ratio as a function of the mole fraction of oxygen (X_{O_2}) in the gas phase (water temperature, 15°C; influent dissolved oxygen, saturation). The value for G_{oxygen} was set at 2.0.

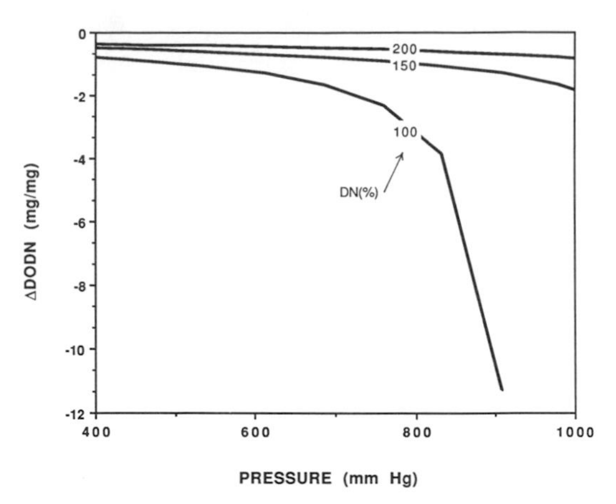

FIGURE 9.—Effect of pressure on the ΔDODN stripping ratio as a function of influent dissolved nitrogen concentration (DN, % saturation) (water temperature, 15°C; effluent dissolved oxygen, 20 mg/L; influent dissolved oxygen, saturation). The value for G_{oxygen} was set at 2.0.

curve, only the portion resulting in positive ΔDOs has been plotted. For $\Delta P_{out} > \Delta P_{in}$, operation to the right of the vertical asymptote results in oxygen stripping. As indicated in equation (5), the allowable ΔDO across the absorption unit can be increased by passing only a portion of the flow through the absorption unit and then mixing this flow back with untreated influent water. However, bubble formation during mixing will also, at some point, limit the ΔDO across the absorption system.

Allowable ΔDO is sensitive to changes in ΔDODN, particularly between values of −2 and −3 (Figure 7). The effects of DN_{in} on this ratio are illustrated in Figure 8. For a given DO_{out}, increasing the concentration of DN_{in} requires higher values of ΔDODN. These higher values can be achieved by operating the column at reduced pressures (Figure 9). For a given maximum ΔP_{out} of 40 mm Hg and a DO_{in} of 100%, the computed ΔDODNs (equation 4) for DN_{in} values of 100% and 200% are −3.10 and −0.50, respectively. The column pressures required to attain these ΔDODNs are approximately 802 mm Hg for a DN_{in} of 100% and 690 mm Hg for a DN_{in} of 200%. However, lower values of column pressure (vacuum operation) result in increased operating costs (Figure 10), due primarily to the higher G/L ratios required.

It should be noted that in a well-mixed system (circular or mixed-flow tanks), the allowable ΔDO is based on the dissolved gas pressures in the effluent from the rearing unit. Therefore, the ΔDO can be several times larger than it can for a raceway because of the rapid mixing of absorber effluent with the process water. The maximum allowable ΔDO from the absorber unit for a circular tank depends primarily on the ΔP and DO criteria and on the water flow rate through the absorber unit(s). Typically, reducing the influent nitrogen concentration to less than 103% eliminates the need for further nitrogen stripping within the oxygen absorber (Colt and Watten 1988).

Thus far the effects of DC and DN have been evaluated on the basis of a fixed bed depth in the

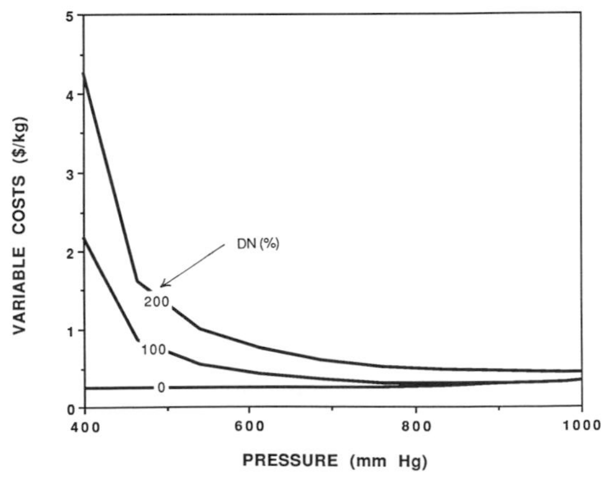

FIGURE 10.—Effect of pressure on variable costs ($/kg O_2 absorbed) as a function of the ΔDODN stripping ratio and influent dissolved nitrogen (DN, % saturation) (water temperature, 15°C; effluent dissolved oxygen, 20 mg/L; influent dissolved oxygen, saturation). The value for G_{oxygen} was set at 2.0.

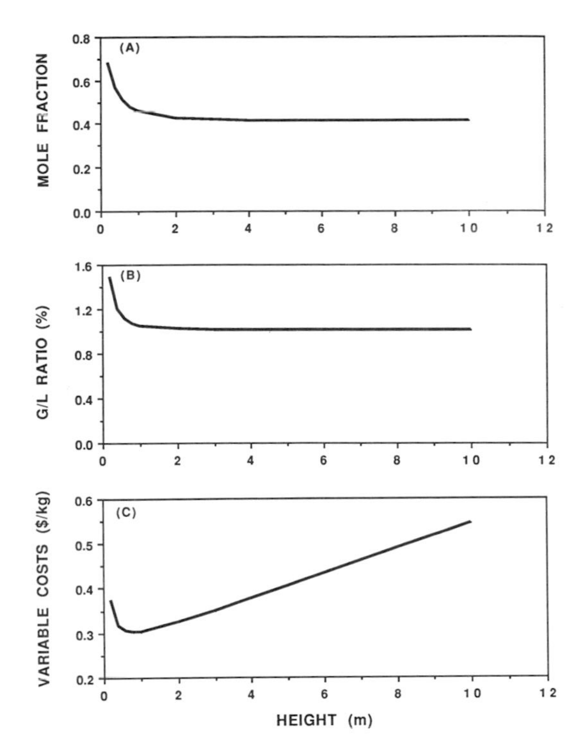

FIGURE 11.—Effects of column height on (A) the mole fraction of oxygen in the gas phase, (B) variable costs ($/kg O_2 absorbed), and (C) the required volumetric oxygen-to-water (G/L) ratio (water temperature, 15°C; influent dissolved oxygen and nitrogen, saturation; effluent dissolved oxygen, 20 mg/L; column pressure, 760 mm Hg).

packed column. The variations of oxygen mole fraction, G/L ratio, and variable costs are presented in Figure 11 as a function of bed depth for a DO_{out} of 20 mg/L, a DO_{in} of 100% saturation, and a DN_{in} of 100% saturation. As the bed depth is reduced, the G/L ratio (and therefore the oxygen mole fraction) must be increased to maintain the required DO_{out}. Under the conditions of this analysis, variable costs are lowest at a column height of 0.80 m. However, operation over the bed depth range of 0.40 to 1.7 m results in less than a 6% change in variable costs. Below a column height of 0.3 m, variable costs increase rapidly. Thus, it appears that column height is a less important design parameter for systems receiving pure oxygen than for those receiving air.

Acknowledgment

This project was funded in part by the Department of Fisheries and Allied Aquacultures, Auburn University, and the Fish Factory (J. Colt), Davis, California. We thank L. Benefield, B. Ducan, and J. Morgan for reviewing the manuscript.

References

Benson, B. B., and D. Krause. 1984. The concentration and isotopic fraction of oxygen dissolved in freshwater and seawater in equilibrium with the atmosphere. Limnology and Oceanography 29:620–632.

Boersen, G., and J. Chesney. 1987. Engineering considerations in supplemental oxygen. Pages 17–24 *in* G. Bouck, editor. Papers on the use of supplemental oxygen to increase hatchery rearing capacity in the Pacific Northwest. Bonneville Power Administration, Portland, Oregon.

Collins, C. M., G. L. Burton, and R. L. Schweinforth. 1984. Evaluation of liquid oxygen to increase channel catfish production in heated water raceways. Tennessee Valley Authority, TVA/ONRED/AWR-84/8, Gallatin, Tennessee.

Colt, J. 1984. Computation of dissolved gas concentrations in water as functions of temperature, salinity, and pressure. American Fisheries Society Special Publication 14.

Colt, J., and G. R. Bouck. 1984. Design of packed columns for degassing. Aquacultural Engineering 3:251–273.

Colt, J., K. Orwicz, and G. Bouck. 1991. Water quality considerations and criteria for high-density fish culture with supplemental oxygen. American Fisheries Society Symposium 10:372–385.

Colt, J., and B. J. Watten. 1988. Application of pure oxygen in fish culture. Aquacultural Engineering 7:397–441.

Gowan, R. 1987. Use of supplemental oxygen to rear chinook in seawater. Pages 35–39 *in* G. Bouck, editor. Papers on the use of supplemental oxygen to increase hatchery rearing capacity in the Pacific Northwest. Bonneville Power Administration, Portland, Oregon.

Hackney, G. E., and J. E. Colt. 1982. The performance and design of packed column aeration systems for aquaculture. Aquacultural Engineering 1:275–295.

Lewis, W. K., and W. C. Whitman. 1924. Principles of gas adsorption. Journal of Industrial and Chemical Engineering 16:1215–1220.

Severson, R. F., J. L. Stark, and L. M. Poole. 1987. Use of oxygen to commercially rear coho salmon. Pages 25–34 *in* G. Bouck, editor. Papers on the use of supplemental oxygen to increase hatchery rearing capacity in the Pacific Northwest. Bonneville Power Administration, Portland, Oregon.

Watten, B. J. 1989. Simulation of gas transfer within a multi-stage packed column oxygen absorber: numerical model development and application. Doctoral dissertation. Auburn University, Auburn, Alabama.

Weiss, R. F. 1970. The solubility of nitrogen, oxygen, and argon in water and seawater. Deep-Sea Research 17:721–735.

Weiss, R. F. 1974. Carbon dioxide in water and seawater: the solubility of a non-ideal gas. Marine Chemistry 2:203–215.

American Fisheries Society Symposium 10:482–486, 1991

Effect of Oxygen versus Air on Fish in Serial Production

JAMES W. MEADE[1] AND WILLIAM F. KRISE
U.S. Fish and Wildlife Service, National Fishery Research and Development Laboratory
Rural Delivery 4, Box 63, Wellsboro, Pennsylvania 16901, USA

DAVE ERDAHL AND JILL ORTIZ
U.S. Fish and Wildlife Service, Iron River National Fish Hatchery
HCR Box 44, Iron River, Wisconsin 54847, USA

ANTHONY J. BECKER, JR.
Mansfield University, Grant Science Center
Mansfield, Pennsylvania 16933, USA

Abstract.—The effects of air versus oxygen addition in a serial water use rearing system were assessed in terms of water quality and fish growth, hematology, and kidney condition. Replicate series of five rearing units, stocked with lake trout *Salvelinus namaycush* at sites in Wisconsin and Pennsylvania, were provided with either aeration or oxygenation chambers, and the water and fish were monitored for 2 months. As expected, ammonia concentration and water conductivity increased, while mean dissolved oxygen (DO) concentration, total gas pressure, and fish growth were reduced as water was used serially. Initially, hematocrit, serum protein, and hemoglobin levels fluctuated, but there were no significant differences due to level of water reuse (metabolic loading) after 2 months. Oxygen consumption was relatively constant, regardless of the serial reuse level. There were differences in total gas pressure between the aerated and oxygenated water, because the nitrogen-stripping effect of oxygenation resulted in much lower total gas pressure by the end of the series compared with water in similar aerated units. Thus the use of oxygen can be far more effective than use of air for control of dissolved nitrogen and total gas pressure. The advantage of oxygen over air is in its ability to supersaturate water with DO while removing dissolved nitrogen. However, we found no differences in growth or in health indices between fish reared in oxygenated water versus aerated water. We conclude that the use of oxygen rather than air for control of dissolved oxygen did not cause physiological problems, but neither did it enhance health or increase production.

Aeration and oxygenation are often used to increase production capacity and production intensification in fish-rearing facilities. Currently there is a trend toward the use of oxygenation rather than aeration. To make a rational choice, once dissolved oxygen (DO) management objectives have been established, the manager needs to address an important question in addition to costs: What are the effects of aeration and oxygenation on the water, the fish, and ultimately on production?

One must first define the objectives for the use of a dissolved gas management system. The transfer efficiency and the energy cost of the oxygen transferred or of dissolved nitrogen (DN) removed depend on both the starting (untreated) level of the gas pressure and the level to be reached. As dissolved gases approach equilibrium, the efficiency of aeration in an air atmosphere decreases, and the potential for the cost-effective use of oxygen (versus air) increases. If the manager wants to increase DO beyond saturation or to decrease DN or total dissolved gas pressure below 100%, simple aeration in an air atmosphere is not an option.

Economic comparisons must include (1) cost of additional investment (capital required for a new or different system), (2) operating costs, (3) maintenance (including periodic replacement) costs, and (4) the cost of risk (e.g., insurance) or system failure. Total costs will also depend on the facility design and equipment, the geographic location, and the oxygen production source. For instance, electricity is relatively inexpensive in the northwestern USA. Liquid oxygen is more available in parts of the northeast than in some other regions; and the higher cost (per unit volume) of liquid oxygen may be exacerbated by the hidden costs of maintenance and the increased risk associated with oxygen generation (pressure swing adsorber) systems.

In addition to determining the costs for a sys-

[1]Present address: Tunison Laboratory of Fish Nutrition, 3075 Gracie Road, Cortland, New York 13045, USA.

TABLE 1.—Source water quality at Iron River and Wellsboro. All values are in mg/L unless otherwise noted.

Determination	Iron River	Wellsboro
Alkalinity, total as $CaCO_3$	65	24
Calcium	19	11
Chloride	0.0	4.7
Hardness, total as $CaCO_3$	68	35.1
Iron	0.6	<0.01
Lead (μg/L)	9.8	0.3
Magnesium	5.6	1.8
Nitrate-N	0.03	0.85
Nitrite-N	0.00	0.01
pH (units)	7.7	7.1
Sodium	1.8	1.4
Specific conductance (μS/cm)	124	89
Sulfate	4.0	8.0
Total dissolved gas pressure (% saturation)	101.4	104.5
Dissolved nitrogen (% saturation)	105.0	107.8
Total dissolved solids	78	45
Zinc (μg/L)	0.0	5.0

tem specified to meet the objectives, there remains the other important question: Will the use of oxygen rather than air affect the fish in some significant way? That is, will the use of oxygen increase production, enhance health, or otherwise have a positive or a negative effect? The purpose of this study was to determine differences in water quality and in growth, oxygen consumption, hematocrit, hemoglobin, serum protein, and kidney histology of fish reared in serial use water in which the DO concentration was controlled either oxygen or air.

Methods

The study was conducted at the Iron River (Wisconsin) National Fish Hatchery and at the National Fishery Research and Development Laboratory, Wellsboro, Pennsylvania. Water quality profiles for the locations are shown in Table 1. Five rearing units of roughly 60-L capacity were arranged in a series of decreasing heights so that four of the units used the effluent water from the unit immediately above. Water flow to each rearing unit, including source water to the first rearing unit, entered a container that held a gas (air or oxygen) diffuser; the water then flowed from the aeration or oxygenation container into the fish-rearing container. We used three replicates of the five-level series (3 × 5) for air and three for oxygen treatments (30 rearing units) at Iron River. At Wellsboro we used three replicates (3 × 5) for oxygen and one series for the air treatment as a reference for previous and nearly identical studies conducted there.

Healthy lake trout *Salvelinus namaycush* were stocked at a rate sufficient to reduce the inflow water DO in each unit by roughly 2.5 mg/L. Fish were fed GR-6 (U.S. Fish and Wildlife Service formulation) pelleted diet at the rate of 0.5–0.7% of body weight per day. Aeration or oxygenation was adjusted at each rearing unit to bring DO of incoming water to roughly 90% of saturation concentration. The stocking and test start-up procedure followed that of Meade (1988).

Temperature and DO were measured twice daily. Dissolved gas levels were measured with a gasometer (Bouck 1982) as total gas pressure in mm Hg, and DN measurements were calculated according to Dawson (1986). Total ammonia-nitrogen (APHA et al. 1985) and conductivity were measured weekly at Wellsboro. Specific growth rate was calculated as $100(\log_e[\text{weight at end}] - \log_e[\text{weight at start}])/(\text{number of days})$. We measured hematocrit as percent and hemoglobin as methemoglobin (Wedemeyer and Yasutake 1977). Serum protein levels were measured with a refractometer. We used two-way analysis of variance (Neter et al. 1985) to determine differences, and Tukey's multiple comparisons test to compare means at the 0.05 level of significance.

Results

Water Quality

Mean water temperature was similar at Iron River and Wellsboro (Tables 2, 3). There was an

TABLE 2.—Water temperature and dissolved oxygen (DO) consumption at Iron River during 2-month study of effects of air versus oxygen on lake trout in five serial water use tubs.

	Air				Oxygen			
	Tub 1		Tub 5		Tub 1		Tub 5	
Variable	Mean	SD	Mean	SD	Mean	SD	Mean	SD
Temperature (°C)	7.9	1.3	8.8	1.0	8.0	1.4	8.8	1.5
DO in (mg/L)	11.4	0.4	10.3	0.6	11.7	0.6	10.4	2.0
DO out (mg/L)	9.0	0.6	7.9	1.0	9.5	0.5	8.0	1.8
DO consumption (mg/L)	2.4	0.5	2.4	0.7	2.8	2.2	2.9	1.8

TABLE 3.—Mean treatment water quality at Wellsboro during a 2-month study of effects of air versus oxygen on lake trout in five serial water use tubs.

	Air				Oxygen			
	Tub 1		Tub 5		Tub 1		Tub 5	
Variable	Mean	SD	Mean	SD	Mean	SD	Mean	SD
Temperature (°C)	8.3	1.6	8.9	1.7	8.4	1.3	8.9	1.4
DO in (mg/L)	10.4	2.0	9.1	1.7	12.2	2.0	11.1	2.0
DO out (mg/L)	8.5	1.6	7.1	1.4	9.9	1.6	9.1	1.8
DO consumption (mg/L)	1.9	0.5	2.0	0.6	2.3	0.6	2.0	0.6
Changes in total gas pressure (ΔP as mm Hg)			−14.7	5.3	−15.5	2.6	−100.7	16.5
Total ammonia nitrogen after feed (mg/L)	0.2	0.1	0.8	0.4	0.6	0.2	0.7	0.2
Conductivity (μS/cm)	81.5	0.5	86.8	1.8	81.4	1.1	86.5	1.8

increase in water temperature through successive water uses at Iron River (0.8–0.9°C) and at Wellsboro (0.5–0.6°C); thus fish in lower levels of serial reuse water were exposed to higher temperatures. The temperature of source water at Iron River increased slightly during the test (Figure 1). Mean outflow DO levels were somewhat reduced as the water passed through the serial units (Table 2), but DO consumption was not affected by serial use level. Variability of DO consumption was greater in the oxygenated tanks than in the aerated units, and the apparent consumption (ΔDO) tended to be greater among fish in oxygenated water. However, the variability in DO consumption and any increased oxygen consumption was probably an artifact of measuring incoming DO in

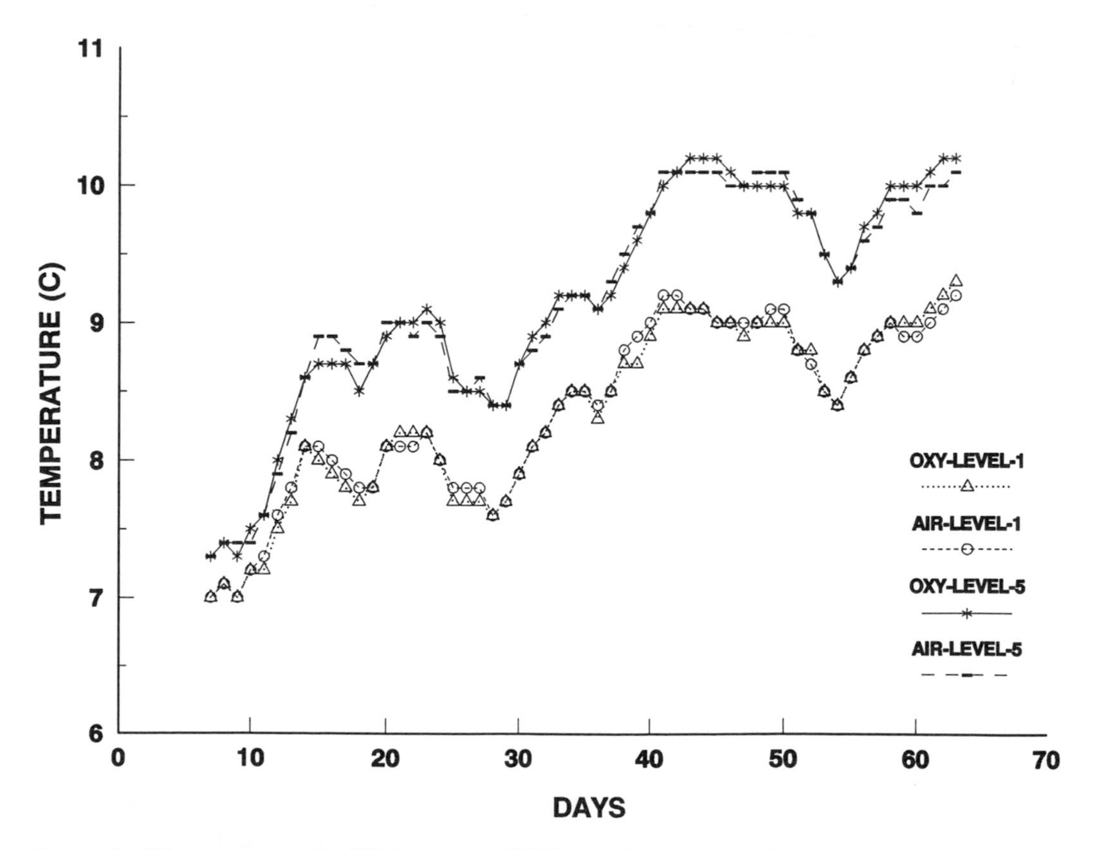

FIGURE 1.—Water temperatures (°C) in oxygen- (OXY) and air-treated serial rearing unit water at Iron River. Levels 1 and 5 refer to tubs 1 and 5 in series.

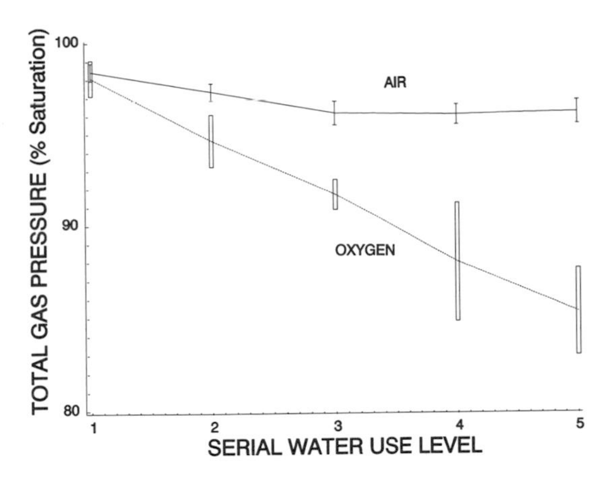

FIGURE 2.—Total dissolved gas pressure in oxygen- and air-treated serial rearing unit water at Wellsboro. Bars indicate two standard deviations.

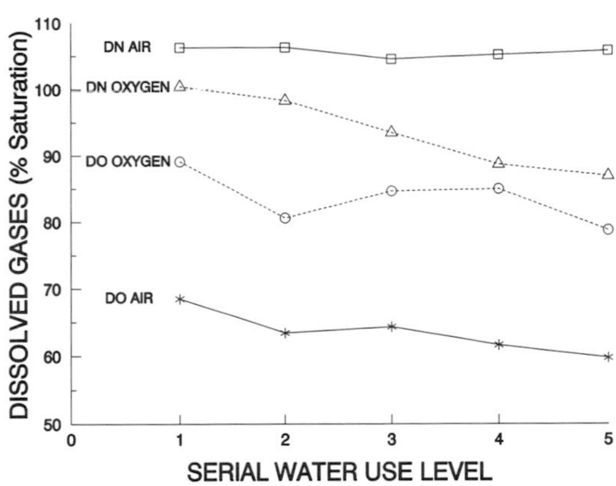

FIGURE 3.—Dissolved nitrogen (DN) and dissolved oxygen (DO) concentrations in oxygen- and air-treated serial rearing unit water at Wellsboro.

the oxygenation contact container. Patterns of DO consumption were similar between test sites and among treatments.

Trends in ammonia, conductivity, and total dissolved gas pressure were evident between the first and fifth water use (Table 3). Water samples taken before feeding showed that ambient ammonia levels at Wellsboro increased from a mean of 0.13 mg/L in the first water use to 0.48 mg/L in the fifth water use. Conductivity increased with increasing water use, and total gas pressure decreased as water was reused (Table 3). Total gas pressures and DN concentrations were lower in the oxygenated water than in the aerated water (Figures 2, 3).

Specific Growth Rate

At both sites, specific growth rate did not differ between fish from air and oxygen treatments. At Wellsboro, growth tended to be reduced with water use (Figure 4) but was not significantly different among fish from different levels of water reuse. Overall, fish in the first and second water uses outgrew fish from each of the other water uses. Growth increased in the first month and oscillated thereafter. Growth decreased as the study progressed at Iron River. As with the Wellsboro fish, there was a trend for growth to be reduced with serial use in the oxygen-treated water (Figure 4), but that trend was reversed in the aerated water. Overall, growth differences were not significant between fish reared at Iron River and Wellsboro.

Hematology and Kidney and Gill Histology

We measured hematocrit and serum protein levels at both locations and hemoglobin at Wellsboro. Mean levels of hematocrit, serum protein, and hemoglobin of fish among the serial use levels and between oxygen and aeration treatments were not significantly different at the end of the test from those at the start, or from one another. Although the gills were not different (they were equally poor among fish from the first and fifth water reuse levels, indicating less than optimal prestudy rearing conditions), the glomerular capillaries of the kidneys were more dilated in fish reared in fifth reuse water (both oxygenated and aerated) than in those reared in first reuse water.

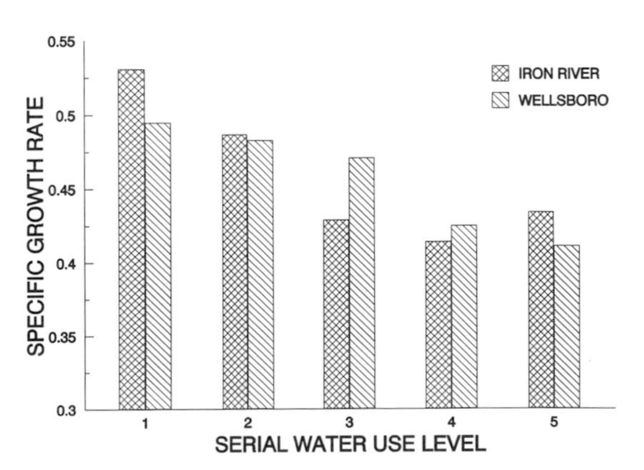

FIGURE 4.—Mean specific growth rate of lake trout in oxygen-treated serial reuse water for 56 d at Iron River and at Wellsboro.

Discussion

The trend of growth reduction with time at Iron River and the absence of significant differences in growth of fish among the serial treatments may indicate an insufficiently high feeding rate. Studies similar to this one should probably use ad libitum or excess feeding. In this study, there were extreme reductions in total gas pressure when oxygen was used, but perhaps the most notable outcome of the study was that there were no differences in growth between fish reared in reoxygenated versus reaerated water.

Although un-ionized ammonia concentrations were well below the levels reported by Smith and Piper (1975) to cause kidney damage, the kidney condition was similar to the water quality-related damage reported by Meade and Herman (1986). No other health implications were evident; however, the use of oxygen at supersaturated levels may have health implications not addressed in this study. The absence of differences in blood or other tissues between fish reared in oxygen versus air-treated water supports a hypothesis, based on growth, that there was no qualitative difference between aeration and oxygenation. On the other hand, the reversed growth trend of fish in the air-treated water at Iron River may indicate that conditioning of the water occurred with retention, aeration, or fish use. Oxygenated water did not impede growth, which suggests an inherent qualitative difference between aeration and oxygenation. Differences were not substantial but indicate the need for further investigation of limiting factors at Iron River.

In terms of only DO management, there is no reason to prefer oxygen over air based on this study, and the choice of DO control systems for production should probably be based on cost, convenience, and reliability. Thus, the application of results of this study may be primarily in supporting cost-effective choices of DO control systems. In addition to addressing DO maintenance systems, the results can be used to help address the question of economical reduction of DN and total gas pressure through oxygenation. Until other information becomes available, it seems prudent to assume that the use of oxygen for DO regulation near the saturation concentration will not cause physiological problems. But neither will oxygen significantly enhance health or production over that achieved through DO regulation with air, provided DO is maintained near the level of saturation in air, and influent DN is at acceptable concentrations.

Acknowledgments

Roger Herman of the National Fish Health Laboratory, Kearneysville, West Virginia, provided the histological analysis. Paul Halliburton and Lori Redell were responsible for fish care and data taking at Wellsboro. We thank Henn Gruenthal and Jerry McClain for their guidance and support, Chris Hanson for fish care at Iron River, Lori Redell and Les Mengel for assistance in data preparation, and Barnaby Watten for manuscript review and advice.

References

APHA (American Public Health Association), American Water Works Association, and Water Pollution Control Federation. 1985. Standard methods for the examination of water and wastewater, 16th edition. APHA, Washington, D.C.

Bouck, G. R. 1982. Gasometer: an inexpensive device for continuous monitoring of dissolved gases and supersaturation. Transactions of the American Fisheries Society 111:505–516.

Dawson, V. K. 1986. Computer program calculation of gas supersaturation in water. Progressive Fish-Culturist 48:142–146.

Meade, J. 1988. A bioassay for production capacity assessment. Aquacultural Engineering 7:139–146.

Meade, J. W., and R. W. Herman. 1986. Histological changes in cultured lake trout *Salvelinus namaycush*, subjected to cumulative loading in a water reuse system. Canadian Journal of Fisheries and Aquatic Sciences 43:228–231.

Neter, J., W. Wasserman, and M. Kutner. 1985. Applied linear statistical models, 2nd edition. Irwin, Homewood, Illinois.

Smith, C. E., and R. G. Piper. 1975. Lesions associated with chronic exposure to ammonia. Pages 497–514 *in* W. E. Ribelin and G. Migaki, editors. The pathology of fishes. University of Wisconsin Press, Madison.

Wedemeyer, G. A., and W. T. Yasutake. 1977. Clinical methods for assessment of the effects of environmental stress on fish health. U.S. Fish and Wildlife Service Technical Paper 89.

American Fisheries Society Symposium 10:487–494, 1991

Management Techniques to Minimize the Loss of Genetic Variability in Hatchery Fish Populations

RAYMOND C. SIMON[1]

National Fish Health Research Laboratory, Immunogenetics Section
Box 700, Kearneysville, West Virginia 25430, USA

Abstract.—Levels of inbreeding in hatchery fish populations may be greater than is generally recognized, and the consequent loss of genetic variability may be degrading the ability of fish to adapt to future environmental changes. Causes of inadvertent inbreeding include variations in family size, inequality of male and female numbers within generations and of breeding numbers among generations, and seemingly trivial degrees of brood-stock selection. All these factors reduce the effective population size (N_e) for breeding programs. Seven ways to increase N_e are suggested: (1) mate each available female with a single male; (2) include both early- and late-maturing fish in matings and avoid selection for other traits; (3) import fertilized eggs from areas within the hatchery's drainage or from adjacent drainages; (4) import sperm from wild or adjacent hatchery populations; (5) rear about 20 independent families that can be tagged to estimate family-size variance; (6) store sperm from in-house males cryogenically for later out-crosses; and (7) ensure that progeny of all matings are represented among released fish and replacement brood stock. Practical considerations determine which of these suggestions apply to particular hatchery operations, but the success of each depends on sustained commitment by management, careful monitoring, and objective evaluation.

Who, upon viewing hundreds or even thousands of adult anadromous fish entering the weir of a hatchery retaining pen each day, would hazard a statement of need for management changes to sustain fish quality in these populations? Such statements of need are certainly counter to intuition during times of resource abundance, but so was the need for investor caution during the "abundance" in the U.S. stock markets in early 1929 and 1987. Although financial and biological systems have little in common, neither monetary markets nor salmon abundance are indicative of long-term stability. In either system, the most effective management adjustments must be made during times of sufficiency because delay can result in declines to levels from which recovery is difficult or impossible.

Many will be skeptical about the need for improvement in the management of hatchery fish until clear reasons are provided to justify change. Are not hatchery personnel already providing their best effort (often with little staffing) that precludes undertaking any additional burden? Do not many fish hatcheries comfortably exceed the 500 or so breeding adults that several geneticists have warned is a minimum? Doesn't the historical production of many hatcheries show that traditional operations have done rather well? What, then, is the issue? First, one must agree that for protracted survival of any breeding population, genetic variability must be present to permit adaptations to inevitable changes of environmental conditions. Second, one must agree that the subject of genetic variability has not emerged to any appreciable extent in guidance issued for fish hatchery management. If these two postulates can be accepted without serious reservation, then a path is open for further discussion.

The objective of this report is to provide justification for the belief that management changes can be beneficial in many instances and to offer suggestions for their implementation. First, a brief outline of fundamentals that apply to genetic variability in populations is provided, followed by illustrations based on relevant published data. In certain respects to be considered later, the salmon data of Simon et al. (1986) are unique in allowing definition of management impacts on genetic factors. Their data are used extensively, but the present treatment differs substantially because it involves different calculation methods, emphasizes corrective actions, and explains the need for such actions.

The Problem

In the simplified or idealized population discussed in textbooks, the number of breeding individuals that transmit genes to the next generation remains constant from generation to gener-

[1]Present address: 5868 Old Highway 14, Box 341, Beulah, Wyoming 82712, USA.

ation. Further, there is no exchange of breeding individuals between populations, the population is "large," generations do not overlap, sexes are equal in number, mating takes place randomly, and selection is absent at all life stages. Probably no real population meets all of these simplifications. Departures from the ideal do not mean that the genetics of real populations are intractable, but rather than the effects of such departures on genetic variability must be understood. Inbreeding is no doubt the most widely understood factor that contributes to reduced genetic variability. Nearly everyone of adult age (and many who are not) knows that inbred lines are noted for their uniformity of individuals. Fewer persons realize that inbreeding may be present in some large populations, and fewer still understand that inbreeding is cumulative, so that small increases over many generations may reach significant levels. Inbreeding results from mating of close relatives, and close relatives are often the consequence of matings among adults of a small population. Although inbreeding is difficult to measure directly unless pedigree data are available, its levels can be conveniently estimated by knowing effective population size. According to Ralls and Ballou (1986), "Reality and theory are linked by the concept of effective population size. Any real population will have the same rate of change in genetic diversity or degree of inbreeding as an ideal population of some size. The size of this equivalent ideal population is called the effective population size of the real population. The effective population size of a real population is usually less than its actual size."

Little work has been published on the breeding genetics of hatchery populations of Pacific salmon, although much effort has been devoted to other aspects of fish population genetics. Many electrophoretic studies have defined the degree of genetic difference among populations and the amount of genetic variation within populations. Reviews of this progress include those of de Ligny (1969), Utter et al. (1974), Allendorf and Utter (1979), and Altukhov and Salmenkova (1981). In several electrophoretic studies, the allele frequencies in hatchery stocks differed from those in the wild populations from which these stocks were derived (e.g., Allendorf and Phelps 1980; Ryman and Stahl 1980; Vuorinen 1984; Allendorf and Ryman 1987; Verspoor 1988). These differences were suspected to have been caused by genetic drift, due to the founder effect (the use of progeny from small numbers of parents—that is, nonrepresentative samples—to establish the original hatchery population).

Even though inbreeding might be tolerated in some hatchery populations that would not exist without human assistance, its introduction into anadromous species that spend extended periods outside the shelter of the hatchery can adversely affect growth or survival. Ryman (1970) noted significantly lower return rates among inbred Atlantic salmon *Salmo salar* and concluded that inbreeding should not be encouraged in hatchery populations intended for release into natural environments. An additional factor leading to genetic change involves fitness, or survivorship of matings. Fitness is a measure of the numbers of progeny that survive to adulthood and successfully reproduce. Unequal reproductive success can be due to variable potency of males, which may also contribute to reduction in genetic variability (Gharrett and Shirley 1985), as well as to differential survival at any preadult life stage. If survivorship contains a significant genetic component, genetic variability will be further reduced in proportion to the heritability of survival (Nei and Murata 1966). Variation in reproductive success can also be due to differences in fecundity, or to repeated use of some males in the fertilization process. Regardless of the causes of variation in reproductive success, the variance in survivorship (usually called variance in family size, or V_k) can be measured and used to estimate effects on effective population size (N_e). One estimator of N_e derived by Hill (1979),

$$N_e = \frac{8NL}{V_{kf} + V_{km} + 4}, \tag{1}$$

emphasizes the importance of family size variance; V_{kf} is the variance in family size of female parents, V_{km} is the variance in male parents' family size, N is the sum of male and female parent numbers—assumed to be equal—and L is the generation interval in years. An estimate of $V_{kf} = V_{km} = 6.0$ was given by Gall (1987) for rainbow trout *Oncorhynchus mykiss* for which, presumably, $L = 2$. For this rainbow trout example, terms in the denominator of equation (1) cancel the numerator product $8L$; hence, N_e equals the census number (N) *for one generation*, provided that numbers of both sexes are equal. Equality of sex numbers enables the use of equation (1) directly. If numbers of the sexes are unequal, the N in equation (1) should be replaced by N':

$$N' = \frac{4N_f N_m}{N_f + N_m}. \quad (2)$$

The only other estimates of variance in family size for fish appear to be those of Simon et al. (1986) for a hatchery population of coho salmon *Oncorhynchus kisutch*. The largest V_k estimates from that study were for the 1976 year-class for which the calculated V_{kf} of about 81 was surprisingly large as judged from wire-tagging data (mean family size for that year-class was about 22). Data taken from Tables 1 and 2 of that study were used to calculate Table 1 of the present report, "lines" being defined as reproductively isolated breeding units. Adults from each line return every 3 years (the generation interval), two other lines occupying the intervening years.

I calculated family size variances from the data on average family size by using the regression equation of Simon et al. (1986):

$$V_{kf} = 3.99X - 6.551; \quad (3)$$

X is the mean family size for females. In that study, males were known to have been used to fertilize eggs from more than one female, but specific records were lacking. For purposes of illustration, N_e is shown in Table 1 as calculated from equations (1), (2), and (3) for matings of one to five females per male and a generation interval (L) of 3 years. Note that using two females per male is assumed to double the male family size, with an attendant increase in V_{km} according to equation (3); using three females per male triples the male family size; and so on.

Inequality of breeding numbers from generation to generation is a second factor that influences N_e in present and future generations. The exact solution to this problem is given by the equation

$$N_e = \frac{1}{2\left\{1 - \left[P\left(1 - \frac{1}{2N_i}\right)\right]^{1/t}\right\}} \quad (4)$$

(J. Felsenstein, University of Washington, personal communication); P refers to products of terms in parentheses to the right, and t is the number of generations for which measures of N or N' are available. Equation (4) was applied to the data in Table 1, the results of which are shown in Table 2.

Selection—an obvious additional factor leading to genetic change and inbreeding—is generally assumed to be trivial in most salmon hatcheries, except where broodstock culling is practiced (Mulcahy 1983) to reduce mortality from infectious hematopoietic necrosis (IHN) virus in the susceptible species sockeye salmon *O. nerka* and chinook salmon *O. tshawytscha*.

Numbers of adult coho salmon females returning to Big Creek Hatchery (near Knappa, Oregon) are not a reliable index of the numbers ultimately contributing to the smolt production (Simon et al.

TABLE 1.—Calculated effective population sizes (N_e) for Big Creek coho salmon, from data of Simon et al. (1986), and equations (1)–(3). Consequences of mating multiple females per male are shown. Each line is a reproductively isolated population unit that returns every 3 years to breed.

Females per male	Generation	N_e Line I	Line II	Line III
1	1	376	131	646
	2	295	429	1,010
	3	294	178	215
	4	222	84	81
	5	276	622	425
	6	704	352	718
	7	121	50	63
	8	110	120	574
	9	349		
2	1	157	56	262
	2	126	181	416
	3	128	78	94
	4	96	37	35
	5	117	330	173
	6	443	146	478
	7	53	22	28
	8	47	51	235
	9	147		
3	1	86	31	141
	2	70	100	225
	3	71	43	52
	4	53	20	20
	5	64	179	94
	6	245	79	357
	7	29	12	15
	8	26	28	125
	9	80		
4	1	54	20	88
	2	44	63	141
	3	45	27	33
	4	34	13	12
	5	40	112	58
	6	156	50	239
	7	18	8	10
	8	17	18	77
	9	51		
5	1	37	14	60
	2	30	43	97
	3	31	19	23
	4	23	9	9
	5	28	77	40
	6	107	34	167
	7	13	5	7
	8	11	12	53
	9	35		

TABLE 2.—Data from Table 1 used in equation (4) to take into account varying numbers per generation on present N_e.[a,b]

Females per male	N_e over time		
	Line I	Line II	Line III
1	203.94	132.59	195.34
2	88.56	57.75	85.49
3	48.73	31.50	47.09
4	30.93	20.41	29.74
5	21.14	13.53	21.10

[a]If three more generations of 50,000 each with equal sex numbers are added to line II (originally, mating was three females per male), N_e increases from 31.50 to 43.21, showing that the effects of small numbers are not efficiently removed by even dramatic population expansion.

[b]If all available females (Figure 1) were mated to an equal number of males, N_e for columns I, II, and III would have been 3,648, 3,326, and 1,962, respectively, by equation (4), provided that each mating was represented in smolts released.

1986). The numbers of females used for smolt production remain relatively constant (governed by the hatchery capacity) when compared with the total number returning and are generally much smaller than the number of females available (Figure 1). This reduction in number is a kind of nondirectional selection and is the consequence of setting aside egg lots (baskets) that were representative of few females (compared with the numbers returning). From this viewpoint, selection might be considerably more prevalent than one might have imagined, especially where the exportation of eggs is practiced. This sort of unpurposeful selection is particularly insidious because it is hidden behind the impression that "all is well" as judged from the number of adults available. Parallel situations exist for many fish species and are assuredly most distinct for species such as the striped bass *Morone saxatilis* in which the fecundity is so large that eggs from a few females can provide hundreds of thousands of progeny—enough to saturate the rearing space for fingerlings in even the largest operations.

A fourth major category of factors leading to genetic change is that of immigration or straying. This process, unlike the others described, tends to remove or reduce the consequences of inbreeding. Homing fidelity of coho salmon is not equal to that of some other species, (e.g., sockeye salmon), and some leakage of genes between populations occurs. The difficulty of assessing this factor, however, along with the other factors noted, is difficult.

My objective here is to define courses of action that will increase and maintain effective population size. These enhancements of fish quality can be undertaken at minimal cost with minor changes in existing hatchery practices. The remedies suggested have been partly stated elsewhere (Simon et al. 1986; Allendorf and Ryman 1987), and at least one (crossbreeding with outside populations) is not without risk (Altukhov 1981).

Discussion

Data provided in Table 1 reveal the probable consequences of two kinds of breeding factors that were previously not obvious (family size variance and nondirectional selection). A third factor, mating of each male to more than one female, can also greatly influence effective population size—and thus inbreeding levels (Table 2). Footnote calculations in Table 2 clarify two

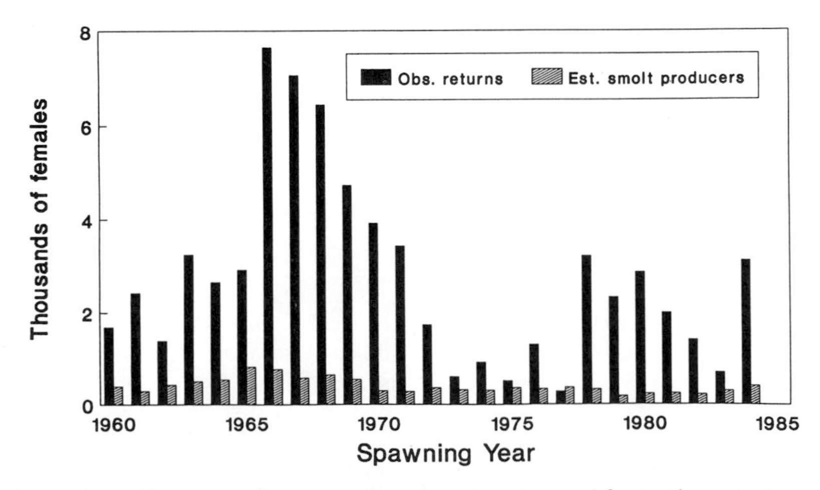

FIGURE 1.—Proportion of Big Creek Hatchery coho salmon females used for smolt production, compared with numbers available (observed returns).

points: (1) drastic reduction of fishing intensity to permit increase in breeding adult escapement will do little to reverse existing circumstances, and (2) decline in N_e to its present levels (whatever those exact values might be) could have been avoided and need not be further reduced in the future. If all factors affecting N_e must be defined by research studies before clear courses of action can be taken, one possible result will be a study of some populations to their extinction (given that few of the factors listed will be constant over time). I do not wish to be pessimistic in making this statement, because monitoring and other studies will be needed even if improvements are undertaken immediately. Furthermore, I do not discount the wisdom of Gall's (1972) plea that both the biology and genetics of a managed species be understood before one attempts to tamper with its genetic variability. I am simply trying to say that unintentional tampering has been commonplace, and that improvements can be made now without ultimate perfection of the issue.

Seven Avenues for Improvement

A simple list of remedies (not in priority order) can be given at the outset.

(1) Mate each female available with a single male, resorting to importation of sperm from nearby populations if too few males are available.

(2) Be certain that early- and late-maturing adults are included in matings, and avoid selection for any other traits.

(3) Import fertilized eggs from areas within the drainage or from adjacent drainages.

(4) Import sperm from wild populations or adjacent hatchery populations to replace that from in-house males.

(5) Rear about 20 families (each family separated from the others and representing different parents) to receive coded tags that will enable estimation of V_k.

(6) Store sperm from in-house males cryogenically to permit later crossing of lines from different generations and to assist with sperm imports or exports.

(7) To meet requirements for replacement brood stock or smolt production, (a) set aside a sample of fertilized eggs from every female as eggs are taken, (b) rear fingerlings from each mating separately until an equal number can be sampled from each family, or (c) sample mixed fingerlings from all females if exportation of small fish is desired and retain the required number. The intent in each of these three approaches is to obtain representation from all matings.

What is a Sufficient Number of Breeders to Sustain a Viable Population?

Published recommendations on minimal breeding numbers span a considerable range, partly because some apply to short-term adequacy and others are intended to apply far into the future. The minimum viable population concept is affected not only by genetic factors, but also by variable stochastic factors such as pollution, drought, competition, predators, and disease (Shaffer 1981). Because these factors are sporadically important and unpredictable, most estimates of minimal adequacy in number are probably too small if survival well into the future is desired and if genetic considerations alone form the basis for the recommendations (as has generally been the case). Furthermore, many sufficiency estimates are given in terms of actual numbers of adults rather than in effective numbers—a point fully appreciated by Ryman and Stahl (1980) but often missed when their recommendations are quoted (or misquoted, e.g., Hynes et al. 1981).

Some examples of recommended numbers illustrate their range. Franklin (1980) suggested a minimum effective size of about 50 for short-term avoidance of inbreeding consequences. He further suggested that a minimum effective size of about 500 would allow sufficient genetic variation for adjustments to changing environmental conditions. Ryman and Ståhl (1980) recommended that "no stock should be founded or perpetuated using less than 30 males and 30 females chosen at random from the stock in any generation." Soule (1980) suggested that substantially more than 500 would be needed to meet long-term evolutionary potential of a species. Kincaid (1976a) recommended more than 100 breeding pairs to minimize inbreeding, and later suggested not less than 500 adults (Kincaid 1979). Tave (1986) defined N_e levels between 263 and 344 as sufficient for populations of food fish and baitfish and between 424 and 685 for populations to be released into natural waters. Finally, a sufficiency estimate of 1,000 adults for hatchery fish populations was derived from computer simulation results (National Fish Health Research Laboratory 1984).

The estimates given above could all be adequate for some populations, at least for the short term, and at the same time all could be inadequate for other populations, given that environments, stochastic factors, and V_k will be different for different populations except for chance similarities. The practice of stating population requirements in terms of N_e, rather than in terms of numbers of adults (e.g., Tave 1986), is sensible and should be mandatory if recommendations are to be meaningful. This suggestion implies that N_e must be known if one wishes to know whether a given population is of sufficient size to avoid significant inbreeding. For hatchery populations, the application of equation (1) will often be appropriate because in a successful hatchery V_k can be expected to be large.

Hatcheries are well known for their ability to greatly enhance the survival of propagated species by affording them protection during early life stages, when mortality is greatest in natural populations. If two reproducing individuals survive from each parental mating in a natural population (or any population for that matter), that survival level will sustain a constant population number from one generation to the next. Average survival to adulthood in hatchery populations often far exceeds two per mating. The consequence of this surplus is that the capacity of the hatchery is exceeded, either in terms of space for adults (if they can be spawned repeatedly), or in terms of space to grow offspring to a size required in their management. Exportation of excess adults, fertilized eggs, or small fish is a sign of a successful hatchery, but at the same time this can be a source of compromise in population quality if V_k is large or if the individuals retained do not represent large numbers of parents.

TABLE 3.—Inbreeding increments per generation ($\Delta F = 1/[2N_e]$) in three lines of coho salmon, calculated from Table 1 for three females per male.

Generation	Line I	Line II	Line III
1	0.0058	0.0161	0.0035
2	0.0071	0.0050	0.0022
3	0.0070	0.0116	0.0096
4	0.0094	0.0250	0.0250
5	0.0078	0.0028	0.0053
6	0.0020	0.0063	0.0014
7	0.0172	0.0417	0.0333
8	0.0192	0.0625	0.0040
9	0.0063		
Total	0.0818	0.1710	0.0843

How Much Inbreeding Exists in the Big Creek Coho Salmon Example (Table 1), and What Does It Mean?

Three uncertainties prevent an accurate answer to the question of inbreeding level among Big Creek coho salmon. First, neither the number of parents that contributed to the founders of the hatchery population nor the inbreeding level that existed in the founder population is known. Second, the actual numbers of females mated per male have not been recorded, but fewer age-2 males than females have generally returned (requiring that some males be used more than once), and thus certain males are sometimes used repeatedly. The true average number of females per male is probably not as small as one nor as large as five; thus the situation involving three females per male has been chosen here for illustration (with the disclaimer that this number could be too small). Third, the depression of growth, vigor, and survival has not been estimated for various inbreeding levels in coho salmon.

The increment in inbreeding for each generation (ΔF) was estimated as shown in Table 3. If the assumption that, on average, three females were mated per male is reasonably accurate, inbreeding levels appear to be high enough to cause a depression in growth and survival of roughly 10–25%, by analogy with Kincaid's (1976a, 1976b) work. This estimate might be too low because Kincaid's estimates of depression referred to an an initial inbreeding level of zero, from which depression was measured for defined inbreeding increments. In the present case, the initial level was surely not zero, but is unknown; thus the increments shown in Table 3 represent only recent additions to the undefined prior inbreeding level. Because the fishery takes about three or four age-2 adults for each one returning to Big Creek Hatchery, small depressions caused by inbreeding are not small—either monetarily or in impact on the population over time. Would one dare to add up production losses for coho salmon hatcheries generally? I do not wish to make such conjecture here, given the paucity of required information listed thus far and the lack of N_e data for other hatcheries. However, perhaps someone should, as a reminder that efforts taken to change breeding practices are worthwhile. It seems more constructive to estimate the probable improvement to be made and to test the degree to which the estimate is realized.

Constraints and Priorities in Recommendations

All of the above recommendations are intended either to increase N_e or to assist in defining N_e.

None should be applied until possible adverse consequences have been examined.

Recommendations 1–7 are not arranged in order of priority because the appropriate order must be judged for each population. Priorities can also be governed by several practicalities such as cost, complexity, disease status of the population(s) in question, and availability of personnel. For example, the deliberate mixing of eggs infected with IHN virus in susceptible species such as chinook salmon would be ill-advised, even though N_e could be substantially improved by this action. Other diseases that are transmitted through the egg should also be absent in donor populations that are considered for fertilized egg sources. Perhaps the most important need is to know the genetic status of the population both before and after intervention in the breeding program. Unless some populations can be monitored to show that intervention has measurable results, there will be justifiable doubt that adjustments in breeding methodology have added anything more than greater complexity to the hatchery operation. In terms of present technology, monitoring entails estimating allele frequencies by electrophoresis, contributions of hatchery production to the fishery, percentages of smolts released that return to the hatchery, survival during the hatchery phase, and effective population size.

Although the suggestions for changes in breeding methods are biologically meaningful, they will be effective only if they receive high-level management support for their long-term continuance. Without such commitment, many hatchery populations can be expected to lose the genetic variability that would have permitted genetic adjustment to inevitable changes in environments of the future. Perhaps even the environments of the present are beyond the adjustment capacity of some populations that are in serious decline. Even so, some of these populations can still be revitalized, provided their plight does not continue to go unrecognized.

Acknowledgments

For reading the manuscript and making helpful suggestions, I thank Graham Bullock, National Fish Health Research Laboratory, Leetown, West Virginia; James Geiger and Anne Henderson-Arzapalo, Fish Culture Research Laboratory, Leetown; Joyce Mann, National Fisheries Research Center, Leetown; and Douglas Tave, Auburn University, Auburn, Alabama. I also thank Joseph Felsenstein, University of Washington, Seattle, and William Hill, University of Edinburgh, for assisting in the development of formulae.

References

Allendorf, F. W., and S. R. Phelps. 1980. Loss of genetic variation in a hatchery stock of cutthroat trout. Transactions of the American Fisheries Society 109:537–543.

Allendorf, F. W., and N. Ryman. 1987. Genetic management of hatchery stocks. Pages 141–159 *in* N. Ryman, and F. Utter, editors. Population genetics and fishery management. University of Washington Press, Seattle.

Allendorf, F. W., and F. M. Utter. 1979. Population genetics. Pages 407–454 *in* W. S. Hoar, D. J. Randall, and J. R. Brett, editors. Fish physiology, volume 8. Academic Press, New York.

Altukhov, Yu. P. 1981. The stock concept from the viewpoint of population genetics. Canadian Journal of Fisheries and Aquatic Sciences 38:1523–1538.

Altukhov, Yu. P., and E. A. Salmenkova. 1981. Applications of the stock concept to fish populations in the USSR. Canadian Journal of Fisheries and Aquatic Sciences 38:1591–1600.

de Ligny, W. 1969. Serological and biochemical studies of fish populations. Oceanography and Marine Biology. An Annual Review 7:411–513.

Franklin, I. R. 1980. Evolutionary change in small populations. Pages 135–149 *in* M. E. Soule and B. A. Wilcox, editors. Conservation biology: an evolutionary–ecological perspective. Sinauer, Sunderland, Massachusetts.

Gall, G. A. E. 1972. Phenotypic and genetic components of body size and spawning performance. University of Washington Publications in Fisheries, New Series 5:159–163.

Gall, G. A. E. 1987. Inbreeding. Pages 47–87 *in* N. Ryman and F. Utter, editors. Population genetics & fishery management. University of Washington Press, Seattle.

Gharrett, A. J., and S. M. Shirley. 1985. A genetic examination of spawning methodology in a salmon hatchery. Aquaculture 47:245–256.

Hill, W. G. 1979. A note on effective population size with overlapping generations. Genetics 92:317–322.

Hynes, J. D., E. H. Brown, Jr., J. H. Helle, N. Ryman, and D. A. Webster. 1981. Guidelines for the culture of fish stocks for resource management. Canadian Journal of Fisheries and Aquatic Sciences 38:1867–1876.

Kincaid, H. L. 1976a. Effects of inbreeding on rainbow trout populations. Transactions of the American Fisheries Society 105:273–280.

Kincaid, H. L. 1976b. Inbreeding in rainbow trout (*Salmo gairdneri*). Journal of the Fisheries Research Board of Canada 33:2420–2426.

Kincaid, H. L. 1979. Development of standard reference lines of rainbow trout. Transactions of the American Fisheries Society 108:457–461.

Mulcahy, D. 1983. Control of mortality caused by

infectious hematopoietic necrosis virus. Pages 51–75 *in* J. C. Leong and T. Y. Barila, editors. Proceedings: workshop on viral diseases of salmonid fishes in the Columbia River basin. Bonneville Power Administration, Portland, Oregon.

National Fish Health Research Laboratory. 1984. Minimum number of parents needed to protect genetic stability in fish brood stocks. Page 79 *in* Fisheries and wildlife research and development 1983. U.S. Fish and Wildlife Service, Denver.

Nei, M., and M. Murata. 1966. Effective population size when fertility is inherited. Genetical Research 8:257–260.

Ralls, K., and J. Ballou. 1986. Captive breeding programs for populations with a small number of founders. Trends in Ecology and Evolution 1:19–22.

Ryman, N. 1970. A genetic analysis of recapture frequencies of released young of salmon (*Salmo salar* L.). Hereditas 655:159–160.

Ryman, N., and G. Ståhl. 1980. Genetic changes in hatchery stocks of brown trout (*Salmo trutta*). Canadian Journal of Fisheries and Aquatic Sciences 37:82–87.

Shaffer, M. L. 1981. Minimum population sizes for species conservation. BioScience 31:131–134.

Simon, R. C., J. D. McIntyre, and A. R. Hemmingsen. 1986. Family size and effective population size in a hatchery stock of coho salmon (*Oncorhynchus kisutch*). Canadian Journal of Fisheries and Aquatic Sciences 43:2434–2442.

Soule, M. E. 1980. Thresholds for survival: maintaining fitness and evolutionary potential. Pages 151–169 *in* M. E. Soule and B. A. Wilcox, editors. Conservation biology: an evolutionary–ecological perspective. Sinauer, Sunderland, Massachusetts.

Tave, D. 1986. Genetics for fish hatchery managers. AVI, Westport, Connecticut.

Utter, F. M., H. O. Hodgins, and F. W. Allendorf. 1974. Biochemical genetic studies of fishes: potentialities and limitations. Pages 213–238 *in* D. C. Malins and J. R. Sargent, editors. Biochemical and biophysical perspectives in marine biology, volume 1. Academic Press, New York.

Verspoor, E. 1988. Reduced genetic variability in first-generation hatchery populations of Atlantic salmon (*Salmo salar*). Canadian Journal of Fisheries and Aquatic Sciences 45:1686–1690.

Vuorinen, J. 1984. Reduction of genetic variability in a hatchery stock of brown trout, *Salmo trutta* L. Journal of Fish Biology 24:339–348.

American Fisheries Society Symposium 10:495–506, 1991

System for Continuously Monitoring Dissolved Gas in Multiple Water Sources

WILLIAM F. KRISE, WILLIAM J. RIDGE, AND LESLIE J. MENGEL
U.S. Fish and Wildlife Service, National Fishery Research and Development Laboratory
Rural Delivery 4, Box 63, Wellsboro, Pennsylvania 16901, USA

Abstract.—Dissolved gas pressures were continuously monitored with gasometers, which were connected to a Commodore-64 computer that recorded and displayed supersaturation levels on a monitor. A line printer provided a hard copy of the data. The system was protected from power outages by an emergency backup power supply. Power to commercially available pressure transducers was supplied from an external dual-voltage power supply. Calibration lines were fitted by a built-in linear regression calibration program. Each gasometer was fitted with a valve system for purging moisture from the silastic tubing. The gasometer core was protected from moisture and abnormally low pressure readings were eliminated.

Continuous monitoring of dissolved gas pressures in hatchery water supplies is advantageous for facilities that have gas supersaturation or varying gas pressures, or that carry out fish culture research. The system we describe is used for monitoring supersaturation for research, but it can also monitor incoming water for a production hatchery. This system has three major components: (1) gasometers (Bouck 1982), used for measuring the pressure difference between ambient air and the fish culture water; (2) pressure transducers, which measure pressure differences (ΔP, mm Hg) within the silastic tubing of the gasometers; and (3) the computer system, which programs, displays, and prints the data. We describe the purpose, construction, and maintenance of the system and its components.

Measuring Dissolved Gases

Gasometers, constructed according to Bouck (1982), were used to measure total gas pressure in flowing water. The mercury manometer used by Bouck was replaced with an electronic pressure transducer to measure gas pressure readings and transfer measures to the computer. Differential gas pressure (ΔP, mm Hg above equilibrium) was obtained by pumping water across the encased silastic tubing. Fifty feet of 0.012-in (internal diameter) tubing was wrapped around a polyvinyl chloride (PVC) assembly to hold the tubing. The system accommodates several gasometers (we commonly used 10) operating simultaneously.

Pressure Transducers

A gauge-type pressure transducer (Table 1) was mounted in an electrical box (Figure 1) on the gasometer core housing. The disadvantage of this placement is the potential it creates for the transducers to become damaged by water or moisture; however, this location provides rapid equilibrium of gas pressures for sensing by the pressure transducer. We used silicone adhesive to seal the housing of the electrical box and reduce the possibility of moisture damage. We tested three pressure transducers (Table 1) that covered a range of gas measurements, including a transducer for 0–1 lb/in^2 (1 lb/in^2 = 52 mm Hg) for low supersaturation, and two models of a transducer of 0–5 lb/in^2 for water with higher supersaturation. Two of the three transducers did not function in negative gas pressures ($\Delta P < 0$). The third—and the most versatile—measured 0–5 lb/in^2, with the zero measurement at the midvoltage range to allow for positive and negative measurement of ΔP in a range of plus and minus 130 mm Hg.

Pressure transducers must first be calibrated in a water supply (Appendix 1) with a gasometer or other calibrated meter. The two numeric values, the machine value and the gasometer ΔP, are used as Y and X variables, respectively. A simple linear regression supplies the computer with a method of converting the pressure signal from the transducer to the ΔP reading shown on the monitor screen.

Circuit Description

We used electrical pressure transducers with a 1–6-V direct current (DC) output. Because our analog-to-digital (A–D) conversion board (Table 2) handled a maximum of +5 V DC, we used an LM 1458 operational amplifier (op-amp; Figure 2) to offset the pressure transducer's output by 1.0 V. All pin number connections refer to the op-amp.

TABLE 1.—Ratings of pressure transducers.

Microswitch identification number[a]	Pressure rating		Voltage out (V, DC)
	lb/in^2	mm Hg	
142PC01G	1	50	1–6
142PC05G	5	250	1–6
143PC05G	±5	±130	1–6 2.5

[a]Reference to trade names or manufacturers does not imply government endorsement of commercial products.

The system circuit required an external power supply with ±8-V output; the +8 V was connected to pin 8 and the −8 V to pin 4. Output from the transducer was connected to the inverting input of the first stage of the amplifier circuit. Offsetting was effected by a 1-MΩ potentiometer (R2). Resistance values were computed according to Noll (1984). The second-stage circuit was used to invert the now negative voltage to a positive voltage, and the output pin 7 of the op-amp was wired to the A–D conversion board.

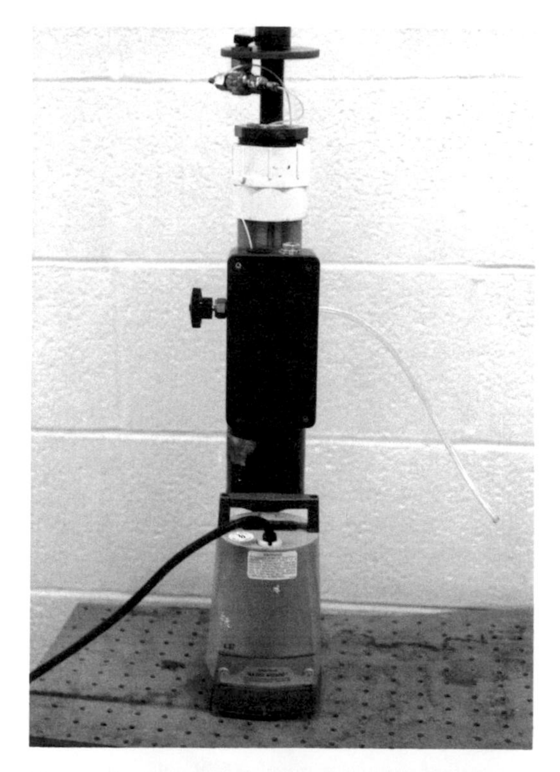

FIGURE 1.—Photograph of the gasometer unit showing, from the bottom, the pump, the silastic core housing with purge valve, the electrical box containing the pressure transducer, and the pressure equilibration valve.

TABLE 2.—Parts list for electronic components.[a]

Description	Supplier	Catalog number
Plastic utility box	Radio Shack	270-223
Chassis mount socket, 4 pin	Radio Shack	274-002
Plug, 4 pin	Radio Shack	274-001
LM 1458 Operational amplifier	Electronics supplier	
Low profile IC socket, 8 pin	Electronics supplier	
Resistor: 4,700 Ω, 0.25 W	Electronics supplier	
Resistor: 10,000 Ω, 0.25 W	Electronics supplier	
General purpose input–output board (analog to digital), mw-611	Micro World Electronics, Inc.	
Three way valve Whitey (¼-in swagelok fitting)	Air-line fitting supplier	B-1K54-X
Swagelok fitting, ¼–1⁄16-in	Air-line fitting supplier	B-100-R-4, B1G54
Backup power supply	Tripp Lite	BC 200
Adjustable power supply +8 V and −8 V	Jameco-Electronics	JE 215
Pressure transducers (microswitch)	Honeywell Distributor	

[a]Reference to trade names or manufacturers does not imply government endorsement of commercial products.

Costs

Materials cost US$200 (1987 prices) for each gasometer, half of which was for a high-volume pump and the remainder of the cost for the pressure transducer, housing, and valves. The power supply, wiring, A–D conversion board, and electronic materials cost $850. The costs of the computer–printer system (about $1,500) must be added to these costs. The cost of the 10-gasometer system totaled $4,350.

System Operation and Maintenance

The monitor system required one dedicated computer with a disk drive and a printer wired through a power-routing board to the gasometers (Figure 3). The printer enabled the recording of overnight measurements. A temperature recorder, an oxygen meter, and a barometer were required to program calculation of percent total gas pressure, percent nitrogen, and percent oxygen from the instrument recordings. The Commodore programming system required two programs, one for calibration (Appendix 1) and one for operation (Appendix 2). An IBM-compatible program was also written (Appendix 3). A backup

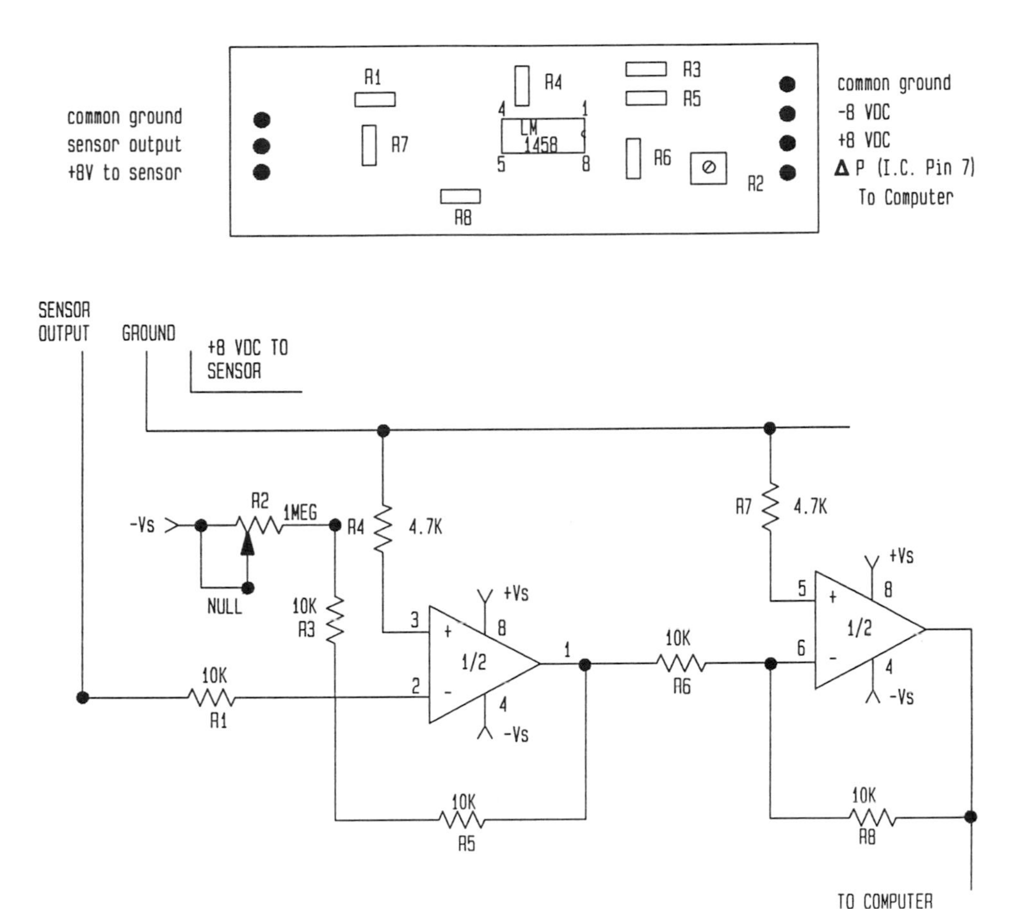

FIGURE 2.—Circuit wiring diagram for pressure transducers.

power supply (Table 2) was used to prevent loss of the operating program during interruptions of station power.

System operation can begin once the portable gasometers are calibrated. One simply unplugs the power to the water-circulating pump, drains the piping surrounding the silastic tubing, and opens the valve on the pressure transducer electrical box to equilibrate pressure in the silastic tubing to zero. Next, the transducer valve is closed, the meter is moved to the new location, and the pump is plugged in to begin operation. This procedure can be performed quickly enough that the operating-system voltage is not disturbed. Our gasometers attained an equilibrium reading in 20 min. Most operational difficulties occur because power is interrupted (Table 3) or the gasometers are placed in gas pressures too high or too low to accommodate the pressure transducer capacity (Table 4). If interruptions occur, one should check the amplifier box voltages (Figure 4), using procedures shown in Appendix 4. In enriched water, algae and other materials sometimes become attached to the silastic tubing; the tubing core should then be removed and lightly brushed. The gasometer is disinfected by pumping a solution of Roccal through it. We recommend placing the gasometers in head tanks free of fish when monitoring continues for long periods.

An air purge system keeps the silastic tubing clear of moisture. This system includes a three-way valve (Figure 1), which opens an air passage to the silastic tubing, limits total pressure, and provides access for a tubing hookup with the silastic tubing of the meter. This purge system eliminates the need for periodic changing of silastic tubing cores and is used to test the silastic tubing for leaks. To purge, we used small aquarium air pumps attached to the purge line. Purging (done monthly) usually required about 30 min/unit.

The purging air pump must not exceed the

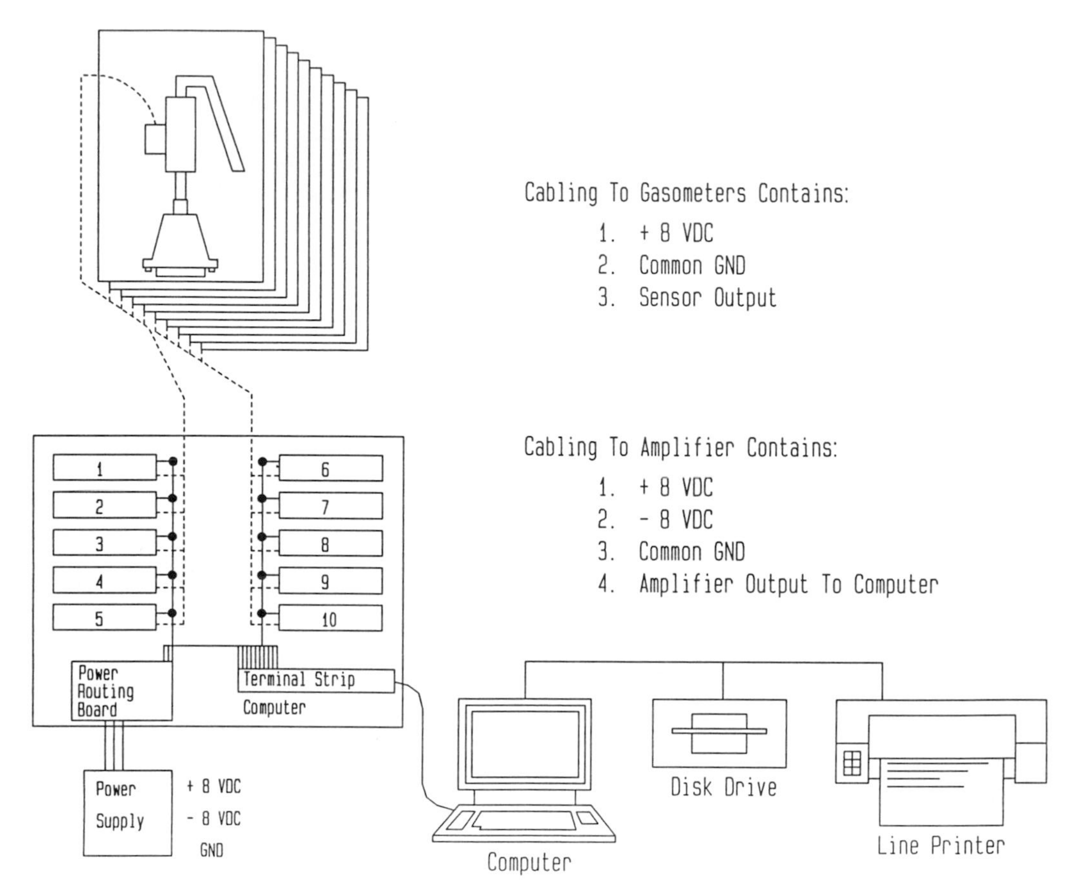

FIGURE 3.—Components of the monitoring system, indicating voltages in direct current (VDC), grounds (GND), and routing of power through the system.

pressure transducer capacity or pressure rating (Table 1). Exceeding the capacity rating can permanently damage the transducer. In measuring the pump output, either a pressure gauge is attached to the pump to obtain a pressure reading or a mercury manometer is attached to obtain a displacement reading. We do not recommend exceeding 50 mm Hg displacement on the 1-lb/in^2 transducer or 190 mm displacement on a 5-lb/in^2 transducer.

TABLE 3.—Troubleshooting guide for computer malfunctions.

Symptom	Probable cause	Solution
Time clock not running; Commodore 64 Basic V2 message; monitor not updating	Program not running	Load program; restart computer
Extremely low readings displayed on monitor	No 110 V AC to pumps	Check breakers on AC supply; reset if necessary
	DC power supply failure	Check power supply voltages; replace power supply if necessary
Abnormal readings displayed for all gasometers	DC power supply voltages have drifted	Check voltages; reset if necessary
	One gasometer is overpressured, causing an overload to the computer	Check voltages on computer cable terminal strip, looking for a voltage over +5.0 V DC or a negative voltage; unhook wire causing such readings; system should return to normal operation

TABLE 4.—Troubleshooting guide for gasometers.[a]

Symptom	Probable cause	Solution
High reading (low gas)	Exceeded capacity of the pressure transducer	Check assigned tank ΔP
Abnormal low readings	Core has accumulated moisture and needs purging	Purge
	Purge valve open	Close purge valve
	Hand (release) valve open	Check transducer with air for normal operation; replace if necessary
	Hole in silastic tubing	Remove core; use water tub and air to find hole; repair if possible; rewind or replace core
Low or high reading	Transducer not operating	Check transducer voltage input and output; replace transducer if needed
Negative reading on display	Loose or corroded contacts on connector	Spread and clean contacts or replace connector

[a]When a gasometer is removed from the system by removing the connector on the transducer box, the corresponding wire on the computer cable terminal strip must be disconnected to avoid a negative voltage to the computer. Negative voltage will distort all pressure readings.

FIGURE 4.—Troubleshooting voltages on the amplifier box.

This system provided reliable 24-h measures of ΔP, giving printouts at any chosen time interval. Fluctuations in ΔP ranged from less than 1.0 ΔP to 2.0 ΔP in a tank over a 24-h period. Maintenance and operation were relatively simple and routine over a 10-month period of use. Improvements can be made to this system by adding probes to determine other analytical parameters such as temperature, barometric pressure, and dissolved oxygen. Also, it is possible to add calculations for percent nitrogen and total gas pressure to the computer program.

Acknowledgment

We thank Joe Fuss, who designed the prototype gasometer system and wrote the first Commodore 64 programs.

References

Bouck, G. R. 1982. Gasometer: an inexpensive device for continuous monitoring of dissolved gases and supersaturation. Transactions of the American Fisheries Society 111:505–516.

Noll, E. M. 1984. Linear circuits. Heath Company, Benton Harbor, Michigan.

Appendixes follow

Appendix 1: Commodore 64 Gasometer Calibration Program

```
1 REM THIS IS A PROGRAM TO CALIBRATE THE PORTS ON THE A/D BOARD
5 POKE53280,0:POKE53281,2
10 PRINT"{SC}{HM}":PRINT
20 INPUT"NUMBER OF DEVICE TO CALIBRATE";N
30 IFN>100RN<1THEN10
40 A=57087+N
45 T=0
50 FORX=1TO100:POKEA,0:T=T+PEEK(A):NEXT
60 PRINTT/100:GOTO45
```

Appendix 2: Commodore 64 GAs Monitor System Operation Program

```
 1 PRINT"{WH}{HM}{SC}":PRINT:PRINT
 2 POKE53280,4:POKE53281,0:DIM T(16)
 3 REM INSERT CLOCK & CALENDAR HERE
 4 DEF FN T(X)=(INT(X/16)*10)+INT(A-(INT(X/16)*16))
 5 PRINT"{SC}{HM}":PRINT
 6 INPUT"HOW MANY GASOMETERS ARE CONNECTED";NM:PRINT"{SC}{HM}":PRINT
10 GOSUB8000:SYS65418
11 FORX=1TO12:READM$:NEXT
12 FORI=49152TO49296:READJ:POKEI,J:NEXT
13 DATAJAN,FEB,MAR,APR,MAY,JUN,JLY,AUG,SEP,OCT,NOV,DEC
14 DATA 76,30,192,120,173,20,3,141,28,192,169,45,141,20,3,173,21
15 DATA 3,141,29,192,169,192,141,21,3,88,96,49,234,120,173,28,192
16 DATA 141,20,3,173,29,192,141,21,3,88,96,173,24,208,41,240,74
17 DATA 74,133,254,169,28,133,253,160,0,173,11,220,72,41,127,162,186
18 DATA 32,120,192,173,10,220,32,120,192,173,9,220,162,174,32,120,192
19 DATA 173,8,220,32,137,192,104,16,3,169,144,44,169,129,32,141,192
20 DATA 169,141,145,253,169,216,133,254,169,1,145,253,136,16,251,108,28
21 DATA 192,72,32,133,192,104,32,137,192,138,32,141,192,96,74,74,74
22 DATA 74,41,15,9,176,145,253,200,96
23 SYS49155
24 X$="":REM CLOCK SETTING ROUTINE
25 PRINT"DO YOU NEED TO SET THE CLOCK?"
26 INPUT"('Y' OR 'RETURN')";X$
27 IFLEFT$(X$,1)="N"ORX$=""GOTO50
28 REM: CLOCK SETTING ROUTINE
29 PRINT"{SC}{HM}{CU}{RV}SET THE CLOCK ":PRINT
30 POKE56335,PEEK(56335)AND127
31 INPUT"AM OR PM";B$
32 A=128:IFLEFT$(B$,1)="A"THENA=0
33 INPUT"HOUR";A$:IFLEN(A$)>2THENPRINT"ERROR":GOTO33
34 IFLEFT$(B$,1)="P"ANDA$="12"THENA=0
35 GOSUB90
36 IFN>18THENPRINT"ERROR":GOTO33
37 POKE56331,A+N
38 INPUT"MINUTES";A$:IFLEN(A$)>2THENPRINT"ERROR":GOTO38
39 GOSUB90
40 IFN>89THENPRINT"ERROR":GOTO38
41 POKE56330,N
```

Appendix 2.—Continued.

```
 42 INPUT"SECONDS";A$:IFLEN(A$)>2THENPRINT"ERROR":GOTO42
 43 GOSUB90:IFN>89THENPRINT"ERROR":GOTO42
 44 POKE56329,N
 45 PRINT"WHEN YOU ARE READY TO START THE CLOCK,"
 46 PRINT"PRESS ANY KEY"
 47 GETA$:IFA$=""THEN47
 48 POKE56328,0:REM STARTS THE CLOCK
 50 PRINT"{SC}{HM}":Z$="0":IFPEEK(6Z3)<>11GOTO56
 51 GOSUB80:PRINT"PRESENT DATE IS ";C$:PRINT
 52 PRINT"DO YOU NEED TO RESET THE CALENDAR?"
 53 INPUT"('Y' OR 'RETURN')";Z$
 54 IFZ$="Y"GOTO56
 55 GOTO75
 56 PRINT:INPUT"WHAT YEAR IS IT";Y
 57 IFY<1986GOTO56
 58 Y$=RIGHT$(STR$(Y),4)
 59 POKE681,VAL(RIGHT$(Y$,2)):POKE682,VAL(LEFT$(Y$,2))
 60 INPUT"WHAT NUMBER MONTH IS IT";M:IFM>12GOTO60
 61 POKE680,M:RESTORE:FORX=1TOM:READM$:NEXT
 62 INPUT"WHAT DATE IS IT";D:D=INT(D):IFD<1GOTO62
 63 IFM<>2GOTO66
 64 IFY/A=INT(Y/A)ANDD>29GOTO62
 65 IFY/4<>INT(Y/4)ANDD>28GOTO62
 66 IFM=4ANDD>30GOTO66
 67 IFM=6ANDD>30GOTO66
 68 IFM=9ANDD>30GOTO66
 69 IFM=11ANDD>30GOTO66
 70 IFD>31GOTO66
 71 POKE827,D:D$=RIGHT$(STR$(D),2)
 72 IFD<10THEND$="0"+RIGHT$(STR$(D),1)
 73 POKE683,11:REM SET DATE FLAG
 74 C$=M$+" "+D$+", "+Y$
 75 V=1:PRINT"{SZ}{HM}";Z$="":PRINT
 76 PRINT"DO YOU WISH DATA TO BE PRINTED OUT?":INPUT"('N' OR 'RETURN')";
    Z$
 77 IFLEFT$(Z$,1)="N"THENV=0:PRINT"{SC}{HM}":PRINT:PRINT:PRINT:PRINT:
    PRINT:PRINT" SYSTEM OPERATING":GOTO160
 78 OPEN1,4:PRINT#1,CHR$(12):PRINT#1,C$:PRINT#1,:CLOSE1
 79 GOTO5000
 80 M=PEEK(680):RESTORE:FORX=1TOM:READM$:NEXT:REM DATE STRING GENERATOR
 81 D=PEEK(827):D$=RIGHTZ(STR$(D),2)
 82 IFD<10THEND$="0"+RIGHT$(D$,1)
 83 Y$=RIGHT$(STR$(PEEK(682)),2)+RIGHT$(STR$(PEEK(681)),2)
 84 IFPEEK(681)=0THENY$=LEFT$(Y$,2)+"00"
 85 C$=M$+" "+D$+", "+Y$:RETURN
 90 IFLEN(A$)=1THENT=0:GOTO92
 91 T=VAL(LEFT$(A$,1))
 92 U=VAL(RIGHT$(A$,1))
 93 N=16*T+U:RETURN
 99 REM
100 GOSUB6000:REM CHECKS CLOCK FOR CURRENT CALENDAR UPDATE
110 IFV=0THEN160
112 GOSUB7000
```

Appendix 2.—Continued.

```
114 IFVAL(MN$)<>SMTHENPF=0:GOTO160
115 IFVAL(HR$)<>SHTHENPF=0:GOTO160
116 IFPF=1GOTO160
117 PF=1:OPEN1,4
118 PRINT#1,:PRINT#1,TM$
120 FORX=1TONM:PRINT#1,X;":  ";T(X):NEXT
122 CLOSE1
130 GOSUB7000
132 SM=VAL(MN$)+F:SH=VAL(HR$)
133 IFSM>59THENSH=SH+1:SM=SM-60:GOTO133
134 IFSH>12THENSH=SH-12
160 FORX=1TONM:T(X)=0
162 FORY=1TO100
170 POKE57087+X,0:T(X)=T(X)+PEEK(57087+X)
180 NEXT:T(X)=T(X)/100:NEXT
200 FORX=1TONM:T$=STR$(T(X))
202 T(X)=VAL(LEFTZ(T$,6))
204 IFT(X)<10THENT(X)=VAL(LEFT$(T$,5))
205 NEXT
210 PRINT"{SC}{HM}"
220 FORX=1TONM
221 L$=STR$(X):IFX<10THENL$="  "+L$
222 L$=L$+":"
230 ONXGOTO231,232,233,234,235,236,237,238,239,240
231 T(X)=FNT1(T(X)):GOTO250
232 T(X)=FNT2(T(X)):GOTO250
233 T(X)=FNT3(T(X)):GOTO250
234 T(X)=FNT4(T(X)):GOTO250
235 T(X)=FNT5(T(X)):GOTO250
236 T(X)=FNT6(T(X)):GOTO250
237 T(X)=FNT7(T(X)):GOTO250
238 T(X)=FNT8(T(X)):GOTO250
239 T(X)=FNT9(T(X)):GOTO250
240 T(Z)=FNT0(T(X))
250 PRINT"{RV}"L$"{RO}";T(X)
251 NEXT
998 IFV=0THEN160
999 GOTO100
5000 REM FREQUENCY OF DATA PRINTOUTS
5010 PRINT"{SC}{HM}":PRINT:PRINT"HOW OFTEN, IN MINUTES, WOULD YOU LIKE";
5011 INPUT"DATA PRINTED OUT";F
5012 PRINT"{SC}{HM}":PRINT:PRINT:PRINT:PRINT:PRINT:PRINT:PRINT"
     SYSTEM OPERATING":GOTO130
6000 REM UPDATE CALENDAR -- FL = UPDATE FLAG
6010 IFPEEK(56331)<>18THENFL=0:GOTO6280
6012 IFFL=1GOTO6280
6020 IFPEEK(56331)<>18GOTO6280
6030 FL=1
6110 D=PEEK(827):M=PEEK(680)
6112 Y=INT(100*PEEK(682)+PEEK(681))
6120 IFD<28THENPOKE827,D+1:GOTO6260
6130 IFM=4ANDD=30GOTO6250
6140 IFM=6ANDD=30GOIQ6250
```

Appendix 2.—Continued.

```
6150 IFM=9ANDD=30GOTO6250
6160 IFM=11ANDD=30GOTO6250
6170 IFM<>2GOTO6200
6180 IFY/4=INT(Y/4)ANDD=29GOTO6250
6190 IFY/4<>INT(Y/4)ANDD=28GOTO6250
6200 IFM=12ANDD=31THENPOKE827,1:M=0POKE681,PEEK(681)+1:GOTO6220
6210 POKE827,D+1:IFD+1>31THENPOKE827,1:POKE680,M+1
6212 GOTO6260
6220 X=PEEK(681)
6222 IFX=100THENPOKE682,PEEK(682)+1:POKE683,0:POKE680,1:GOTO6260
6250 POKE827,1:POKE680,M+1
6260 GOSUB80
6270 IFV=1THENOPEN1,4:PRINT#1,:PRINT#1,:PRINT#1,C$:PRINT#1,:CLOSE1
6280 RETURN
7000 REM: TIME STRING (TM$) GENERATOR
7100 A=PEEK(56331):N$="AM"
7120 IFA>128THENN$="PM":A=A-128
7130 HR$=STR$(FNT(A)):A=PEEK(56330)
7150 MN$=RIGHT$(STR$(FNT(A)),2)
7151 IFVAL(MN$)<10THENMN$="0"+RIGHT$(MN$,1)
7160 A=PEEK(56329):SC$=RIGHT$(STR$(FNT(A)),2)
7171 IFVAL(SC$)<10THENSC$="0"+RIGHT$(SC$,1)
7180 TM$=HR$+":"+MN$+N$:REM SECONDS LEFT OFF
7190 RETURN
8000 REM CALIBRATION FUNCTIONS
8001 DEF FN T1(X)=.826*X+.928
8002 DEF FN T2(X)=.217*X-3.47
8003 DEF FN T3(X)=1.06*X-.66
8004 DEF FN T4(X)=.267*X+3.06
8005 DEF FN T5(X)=.988xX+2.02
8006 DEF FN T6(X)=.242xX-1.11
8007 DEF FN T7(X)=.26*X-4.95
8008 DEF FN T8(X)=1.00xX+2.30
8009 DEF FN T9(X)=1.0*X-1.0
8010 DEF FN T0(X)=1.03*X-1.50
8011 RETURN
```

Appendix 3: IBM-Compatible Program for Calibration and Operation

```
10 KEY OFF : WIDTH 40
20 DIM T(16), TT(1000)
30 ON ERROR GOTO 810
31 REM
32 REM  Definitions for converting analog reading to PPM,  Enter you
   linear
33 REM  regression here in the form: FNTx (N) =  slope * T + y
   intercept.
34 REM
40 DEF FNT1  (N) = .979 * T + .0522
50 DEF FNT2  (N) = .979 * T + .0522
60 DEF FNT3  (N) = .979 * T + .0522
```

Appendix 3.—Continued.

```
 70 DEF FNT4  (N) = .979 * T + .0522
 80 DEF FNT5  (N) = .979 * T + .0522
 90 DEF FNT6  (N) = .979 * T + .0522
100 DEF FNT7  (N) = .979 * T + .0522
110 DEF FNT8  (N) = .979 * T + .0522
120 DEF FNT9  (N) = .979 * T + .0522
130 DEF FNT10 (N) = .979 * T + .0522
140 CLS : LOCATE 10, 2
141 REM
142 REM  Input number of gasometers that are installed and if you want
143 REM  to correct the date , time , or print out the readings.
144 REM
150 INPUT "HOW MANY GASOMETERS ARE CONNECTED"; NM
160 IF NM < 1 OR NM > 10 THEN CLS : GOTO 140
170 CLS
180 X$ = "": REM Clock setting routine
190 LOCATE 10, 13: PRINT "TIME: "; TIME$
200 LOCATE 20, 5: PRINT "DO YOU NEED TO SET THE CLOCK?"
205 LOCATE 22, 12 : PRINT "('Y' OR 'RETURN')": X$ = INKEY$
210 IF X$ = "" GOTO 190 ELSE IF X$ <> "y" AND X$ <> "Y" THEN 230
220 CLS : SHELL "TIME"
230 X$ = "": REM Date setting routine
240 CLS
250 LOCATE 10, 13: PRINT "DATE="; DATE$
260 LOCATE 20, 4: PRINT "DO YOU NEED TO RE-SET THE DATE?"
265 LOCATE 22, 12: PRINT "('Y' OR 'RETURN')": X$ = INKEY$
270 IF X$ = "" THEN GOTO 250 ELSE IF X$ <> "y" AND X$ <> "Y" THEN 290
280 CLS : SHELL "DATE"
290 X$ = "": REM Printer setting routine
300 CLS : LOCATE 10, 3
310 PRINT "DO YOU WISH DATA TO BE PRINTED OUT?"
315 LOCATE 12 , 12 : PRINT "('Y' OR 'RETURN')";
320 Z$ = INKEY$
330 IF Z$ = "" THEN GOTO 320 ELSE IF Z$ <> "y" AND Z$ <> "Y" THEN 380
    ELSE Z$ ="Y"
340 CLS : LOCATE 10, 10: INPUT "HOW OFTEN IN MINUTES"; TM
350 CLS : LOCATE 10, 12: PRINT "TURN ON PRINTER.": LOCATE 13, 8: PRINT
    "PRESS ANY KEY WHEN READY"
360 A$ = INKEY$: IF A$ = "" THEN 360
370 OPEN "LPT1" FOR OUTPUT AS 1
380 REM READS PORTS AND CALCULATES READINGS
390 WIDTH 40
400 CLS
410 GOSUB 650
420 FOR N = 1 TO NM
430 GOSUB 650
440 A = 768 + N
441 REM
442 REM  The loop at line 450, is used to slow down the screen printout to
443 REM  make it readable.  Increase the value of 10 in lines 450 and 460
444 REM  to slow the screen printout or decrease it to speed it up.
445 REM
450 FOR X = 1 TO 10: OUT A, 0: TT(X) = INP(A): TT = TT + TT(X): NEXT
```

Appendix 3.—Continued.

```
460 T = TT / 10: TT = 0
470 ON N GOTO 480, 490, 500, 510, 520, 530, 540, 550, 560, 570
480 T(N) = FNT1(T(N)): GOTO 580
490 T(N) = FNT2(T(N)): GOTO 580
500 T(N) = FNT3(T(N)): GOTO 580
510 T(N) = FNT4(T(N)): GOTO 580
520 T(N) = FNT5(T(N)): GOTO 580
530 T(N) = FNT6(T(N)): GOTO 580
540 T(N) = FNT7(T(N)): GOTO 580
550 T(N) = FNT8(T(N)): GOTO 580
560 T(N) = FNT9(T(N)): GOTO 580
570 T(N) = FNT10(T(N))
580 LOCATE 10 + N, 1: PRINT "
    Channel "; : PRINT USING "##"; N; : PRINT " s reading";
590 PRINT USING "####.##"; T(N)
600 LOCATE 24, 10: PRINT "Press any key to End "; : A$ = INKEY$
610 IF A$ <> "" THEN 680
620 NEXT
630 IF Z$ <> "Y" THEN 640 ELSE IF TMR = 0 OR TMR = VAL(MID$(TIME$, 4,
    2)) - TM THEN GOSUB 730
640 GOTO 420
645 REM Prints date on printout
650 LOCATE 1, 31: PRINT DATE$
660 LOCATE 1, 1: PRINT TIME$: A$ = INKEY$
670 RETURN
680 CLS : LOCATE 12, 1: PRINT "(S)top, (R)estart, or (M)ake a change?"
690 A$ = INKEY$
700 IF A$ = "s" OR A$ = "S" THEN CLOSE #1: END
710 IF A$ = "r" OR A$ = "R" THEN CLS : GOTO 420
720 IF A$ = "m" OR A$ = "M" THEN CLOSE #1: WIDTH 80: GOTO 140 ELSE 690
729 REM Print to printer routine.
730 IF DAT$ <> DATE$ THEN PRINT #1, DATE$: DAT$ = DATE$
740 PRINT #1, TIME$;
750 FOR N = 1 TO NM: PRINT #1, USING " ##"; N;
760 PRINT #1, USING " ####.##"; T(N);
770 NEXT
780 PRINT #1,
790 TMR = VAL(MID$(TIME$, 4, 2))
800 RETURN
810 COLOR 21, 1: CLS
819 REM Error routine
820 IF ERR = 27 OR ERR = 24 THEN LOCATE 12, 4 ELSE 840
830 PRINT "PRINTER IS OUT OF PAPER or NOT ON": GOTO 850
840 LOCATE 12, 17: PRINT "ERROR = "; ERR; " ON LINE "; ERL
850 COLOR 0, 1: LOCATE 24, 8: PRINT "PRESS ANY KEY TO STOP ALARM"
860 A$ = INKEY$: IF A$ = "" THEN BEEP: GOTO 860
870 COLOR 15, 0: GOTO 680
```

Appendix 4: Simplified Troubleshooting Guide for the Amplifier Box

Each circuit board has been preset for a 1-V offset to reduce troubleshooting. To determine the correct electronic voltages check the following with a multimeter:

- Pin 8 of the IC +8 V DC
- Pin 4 of the IC −8 V DC
- Pin 7 of the IC op-amp output
- Green lead from transducer cable
- Pin 7 of op-amp output should be 1 V DC less than the green lead, if no difference, change the IC. Do not try to adjust the potentiometer until after the IC has been replaced.

American Fisheries Society Symposium 10:507–515, 1991

Automated Water Quality Analysis and Control[1]

G. Ellen Kaiser, Fred W. Wheaton, and Brian Wortman
Department of Agricultural Engineering, University of Maryland
College Park, Maryland 20742, USA

Abstract.—An automated water quality monitoring and control system, designed for a closed-cycle biofilter research system, used a personal computer as the central processing unit. The system was developed to monitor and control ammonia concentration, pH, temperature, and flow rate. The methods by which certain variables were measured and controlled, the problems encountered, and the accuracies achieved are presented. Potentially, the system can be applied to many water-related operations such as waste treatment and aquaculture.

A considerable market exists for aquaculture products in the USA and abroad, and U.S. culturists will have to increase domestic production if the USA is to increase or even maintain its share of the market. Per capita consumption of fisheries products in the USA was higher in 1987 than ever before, having increased for five consecutive years, and represented a 5% increase over 1986. At present, the USA imports far more fisheries products than it exports. The total trade imbalance for fisheries products in 1987, when edible and nonedible products were combined, was $7.2 billion (Office of Natural Resources and Rural Development 1988). As the U.S. demand increases, U.S. imports will also increase unless more cost-effective, domestic aquaculture production systems are developed.

Other industries such as chemical, wastewater, and various manufacturing industries have demonstrated that automation provides one important way to improve cost-effectiveness. The emergence of factory automation has been fueled by increased international competition, inflation, rapidly changing price factors, the high cost of capital, the decreasing availability of skilled labor, and the increasing emphasis on quality. Manufacturers avoiding factory automation are expected to find survival almost impossible in the 1990s (Hordeski 1987). It is reasonable to conclude, therefore, that automation should improve the performance of intensive aquaculture operations, lower costs, and perhaps reduce energy consumption.

The direction of automation for aquaculture must be carefully planned in order to maximize the rate of development and minimize costly errors. Failures resulting from overzealous application of automation may inhibit the growth of automation and instrumentation in the aquaculture industry.

Valuable lessons from the closely related wastewater treatment industry can assist in the development of a strategy for automating aquaculture. Garber and Anderson (1985) cited some specific cases in wastewater treatment where expensive computer systems (from $700,000 to $3.1 million) have deteriorated to junk while others remained in operation. Among the problems they observed, complex, specifically designed systems became obsolete and repair parts became unavailable. Only specifically trained experts could maintain and repair them. Thus, personnel at the operational level never accepted computer systems as tools they could use. The personnel could not program or troubleshoot, and quickly found that suspect data were possible or probable. Consequently, the systems were not used to capacity. The designers and programmers who had worked on the system quickly became unavailable so that major problems or questions concerning programs and computer system changes could not be readily answered. Dallimore and Thatcher (1985) also indicated that users' lack of understanding of how to use and maintain computer systems was a significant problem.

On the brighter side, Garber and Anderson (1985) also cited cases in which automation in wastewater treatment plants using programmable controllers or personal computers (PCs) were successful. For example, they cited a consulting firm that took over contract operation of several small treatment works and installed PCs of a popular make to provide some measure of operational control. The operators at first resisted the idea of computer control, but the consulting firm

[1]This is scientific article A-4841, contribution 7867, of the Maryland Agricultural Experiment Station. Partial funding for this project was provided by the Maryland Sea Grant Program.

suddenly found that operators were calling for more memory for their PC units. The operators had found it was easy to program the PC, that some available software could be used to aid their tasks, and that they could maintain their computers or have them easily maintained by others. As a result of their many observations in the wastewater treatment industry, Garber and Anderson (1985) recommended using PCs and plug-in interface boards for data acquisition and control. If a plug-in unit breaks down, another can quickly be plugged into place, and the original one sent out for repairs. Also, because of the thousands of PC owners, PC-based systems become outmoded less rapidly than other parts of the fast-moving computer field.

Microcomputers offer a new dimension in instrumentation because they can be programmed to perform the tasks of complex electronic circuits and the program can be readily modified until it does precisely what is wanted. The versatility and power of the system comes from the ability to use software to manipulate signals in an infinitely extendable variety of ways. Furthermore, almost anyone with a secondary school education can quickly master the elements of programming a microcomputer with the most common language, BASIC (Malcolm-Lawes 1984).

The Florida Department of Natural Resources' Bureau of Marine Research in St. Petersburg, installed a microcomputer-based system of environmental monitoring and control that has been controlling photoperiod and temperature for 3 years. It also monitors critical support equipment and notifies key personnel by telephone in the event of equipment failure. One of the major reasons cited for choice of a microcomputer-based system was economic—this system was much less expensive than conventional instrumentation. Moreover, the conventional system could not perform some of the functions available on the microcomputer-based system (Plaia 1987).

Microcomputer-based automation with plug-in data acquisition controllers appears to be the best approach for aquaculture for the same reasons that it is the best approach for the wastewater treatment industry (Garber and Anderson 1985). Among its benefits (Garber and Anderson 1985), cost of installation and maintenance is moderate; operators, not specialists, can maintain the computer system; operators can make program changes or updates and add functions; and the system will probably be long lived and the problem of obsolescence will be avoided if PCs and plug-in boards are chosen from manufacturers likely to survive the competition and swift development in the field.

The research system described in this paper is a PC-based data acquisition and control system. It has been developed to monitor and control various water quality variables for the purpose of conducting research on biological filters. Although development of the computer-automated system was not the higher goal of this research project, our findings should contribute to the development of such systems for intensive aquaculture.

Equipment

Figure 1 is a schematic of the closed-cycle biofilter research system and the water quality monitoring and control systems. The core of the biofilter system is presented first. Next, the equipment and instrumentation involved in extracting water samples from selected locations and measuring the quality of those samples are described. Then, the programmable controller and computer equipment are described. Finally, the equipment involved in the development of the control of ammonia concentration, pH, and temperature are presented briefly.

Core biofilter system.—The core of the biofilter system consisted of a mixing tank, a pump, five identical biofilters, and a settling basin. Water and chemicals that formed the nutrient solution to be fed to the biofilters in later studies were measured into and stored in the mixing tank. Submersible pumps thoroughly agitated water in the tank to facilitate rapid mixing. The nutrient solution was then pumped through five magnetic flowmeters (Fisher and Porter, Warminster, Pennsylvania[2]), which measured the water flow rate into each of the five biofilters. Water returned from the filters to the mixing tank via a settling basin where solids settled out of solution.

Sample extraction and measurement.—In order to study the response of the filters to various environmental conditions, the system was designed to automatically measure selected water quality variables both before and after each filter. Prefilter samples were obtained by opening a solenoid valve, which allowed solution to flow from the mixing tank to a sample chamber.

The sample chamber (bottom center of Figure

[2]Use of trade or vendor names does not imply endorsement by the University of Maryland.

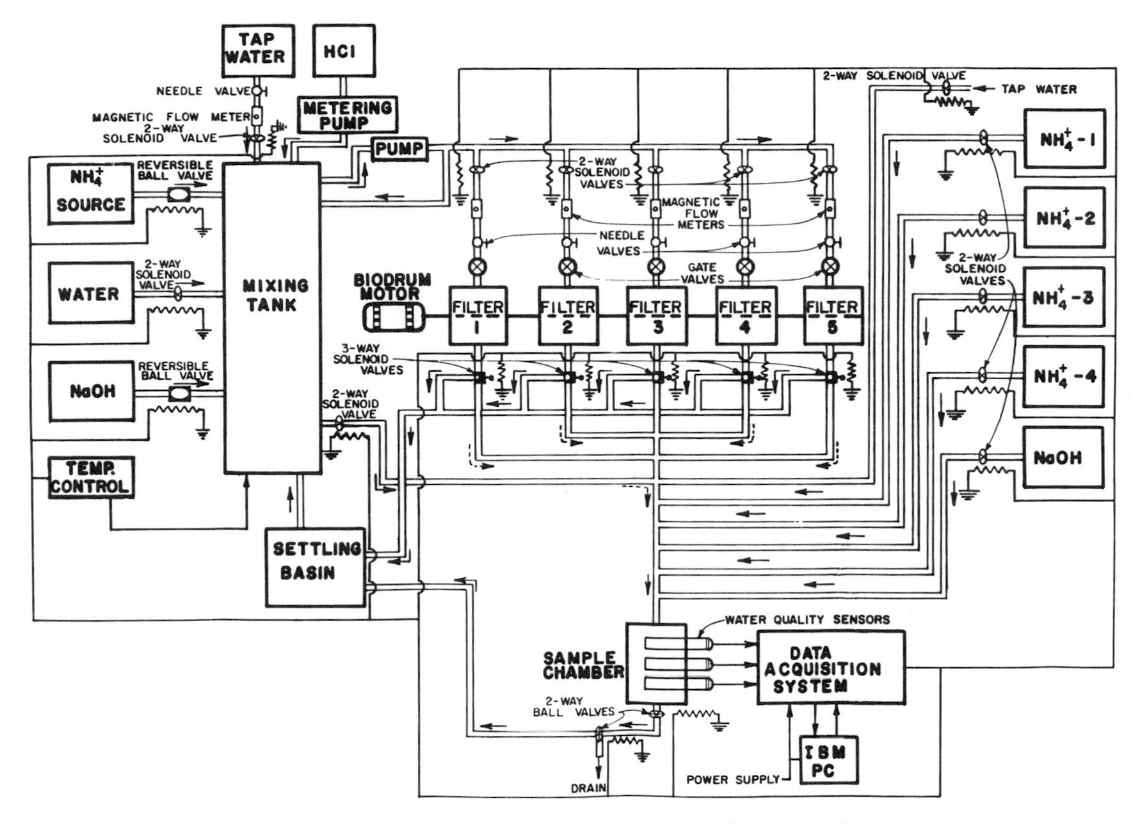

FIGURE 1.—Schematic representation of the biological filter research system.

1) was approximately 500 mL in volume and consisted of a clear, acrylic plastic pipe into which were mounted a Fisher (Silver Spring, Maryland) Accu-pHast pH sensing electrode, an Orion (Cambridge, Massachusetts) model 95-12 ammonia gas sensing electrode, and a copper-constantan thermocouple. The ammonia electrode was the only sensor calibrated automatically because it drifted rapidly and needed hourly calibration. Ammonium chloride (NH_4Cl) standards 1–4 (tanks shown on right of Figure 1) were sequentially directed to the sample chamber as the appropriate solenoid valves were opened. The sodium hydroxide (NaOH) tank supplied a strong base to raise the pH of the ammonia samples in the sample chamber so all ammonia was in the un-ionized form, NH_3 (the form measured by the ammonia electrode).

A three-way electrically actuated valve was placed after each filter so that the effluent from each could be tested automatically. The three-way valves allowed water from each filter to be directed either to the settling basin for recycling or to the sample chamber for testing.

Computer equipment.—The computer equipment consisted of an IBM-PC XT and an ISAAC 2000 (Cyborg, Newton, Massachusetts) data acquisition and control system. The IBM-PC had 384-kilobyte memory, an internal clock, a 20-megabyte hard disk, two RS-232 serial ports, and one parallel port. The ISAAC 2000, interfaced with the PC through one of the RS-232 serial ports, is a modular data acquisition and control system consisting of an I-130 module, two I-140 modules, two I-160 modules, and two I-120 modules. The I-160 modules provided multiplexing so that the 4-channel I-140s were expanded to 16 channels each, making 32 input channels available. All the sensors (pH and ammonia electrodes, thermocouples, and five magnetic flowmeters) were connected to the I-160 modules. The 4-channel I-140 modules, among other functions, provided amplification of the low–medium-level input signals acquired from the sensors through the I-160s. The I-130 module was a 12-bit analog input module that converted analog signals received from an I-140 into a digital number between 0 and 4095, which could, in turn, be sent to the PC. The I-120 modules were binary input–output (I–O) modules that allowed on–off computer control of devices such as solenoid valves, pumps, electrically actuated ball valves, and lights. Each I-120 interfaced with 16 solid-state relays mounted on a board, all of which were off-the-shelf items. This board was mounted in a metal box constructed with typical duplex receptacles on the front panel and recessed power supply plugs on the side panel. Each relay was wired so that when the I-120 switched a relay on, 120 V AC (up to 3 A maximum) were delivered to a receptacle, and when it switched the relay off, no voltage was available. Therefore, in this system, up to 32 devices such as solenoid and electrically actuated valves could be turned on or off independently after each was plugged into a separate socket. The I–O modules were controlled from the PC through a BASICA program. Readings from any input channel could be acquired on demand or any combination of relays could be switched on or off through the program.

Control equipment.—For the purpose of conducting controlled experimental trials on the biofilters, the pH, ammonium concentration, and temperature of the nutrient solution in the mixing tank were to be maintained at selected levels. Submersible heaters with a combined rating of 2,100 W were selected for temperature control. The HCl, ammonium source, water, and NaOH tanks (top left in Figure 1) contained chemicals to adjust the other variables. Flow from the ammonium and base sources was controlled by reversible ball valves and gravity flow, and flow from the HCl tank was to be controlled by a peristaltic pump. Submersible heaters with a combined rating of 2,100 W were used for temperature control.

Procedures

Automatic pH, temperature, and ammonia concentration measurements have been performed. Automated ammonia control has been developed and tested; pH and temperature control procedures had been developed but not tested by late 1988. In addition to testing the pH and temperature control procedures, we planned to incorporate dissolved oxygen, conductivity, and nitrite–nitrate electrodes into the system.

In the following sections, sensor calibration procedures are presented. Next, the overall system operation is described. Finally, the control procedures for ammonia concentration, pH, and temperature are presented.

Sensor Calibration

Selecting voltage scales.—The system included ammonia and pH sensors, thermocouples, and magnetic flowmeters. Each sensor generated an electric potential that varied as a function of the variable being measured. The ammonia-sensing electrode, for example, generated a voltage that varied between −100 mV and +100 mV, corresponding to an ammonia concentration range between 10^{-5} and 10^{-2} M NH_3 (0.17 and 170 mg NH_3/L). The pH electrode, on the other hand, generated a voltage that varied between −300 and +300 mV, corresponding to a pH range between 3 and 10. The I-140 input module had switches that were set to choose one of the following full-scale ranges: ±50 mV, ±100 mV, ±2.5 V, and ±5 V. Thus, each sensor was connected to a channel in which the proper voltage range was selected.

The voltage range of the sensor should fit the full-scale setting on the I-140 as closely as possible for the best resolution and accuracy of measurement. The full-scale resolution of the data acquisition system for, say, the ammonia measurement was 200 mV (±100 mV scale)/4,095 = 0.049 mV/digital number (i.e., each 0.049-mV change changed the ISAAC 2000 output by one digital number). In terms of concentration, because (170 − 0.17 mg NH_3/L)/200 mV = 0.85 mg/(L · mV), and (0.049 mV/digital number) × (0.85 mg/[L · mV]) = 0.04 mg NH_3/L per digital number, about 0.04 mg NH_3/L or greater change in concentration was detectable by the data acquisition machine. On the other hand, if the ±2.5-V scale was selected to enclose the ±300 mV range generated by the pH electrode, the resolution was not as good. Because 5,000 (±2,500-mV scale)/4,095 = 1.22 mV/digital number, each 1.22-mV change could have been detected, which was about 0.04 the sensitivity of the ammonia measurement. If ±300 mV corresponded to a pH range from 3 to 10, the electrode output was approximately (10 − 3 pH units)/600 mV = 0.01167 pH units/mV. If 1.22 mV was the smallest unit of change detected, 1.22 mV detected × 0.01167 pH units/mV = 0.014 detectable pH change. Thus, a pH change of 0.014 or greater could have been detected by the machine. However, if a voltage divider was used so that only ±100 mV of the pH electrode response was measured, the ±100-mV scale could be selected. Then, (7 pH units/200 mV) × (0.049 mV/digital number) = 0.0017, or a pH change of 0.0017 could be detected, which is about eight times more sensitive than in the previous case.

A calibration relationship must be established for each sensor (e.g., the relationship between the digital numbers received by the computer and the flow rates, or between the digital numbers and ammonia concentration). The following sections describe the various calibration procedures.

Ammonia electrode calibration.—The ammonia gas-sensing electrode interfaced with the ISAAC's I-160 module by means of a field effect transistor operational amplifier (FET op-amp) circuit (Kaiser and Wheaton 1988) and was calibrated automatically by the procedure described by Kaiser and Wheaton (1988). The calibration curve obtained at the end of each hour was used to predict the concentration of all the samples measured over the following hour. The ammonia concentration must be measured as accurately as possible to minimize the ammonia control error. For this reason, hourly calibration was necessary to obtain 2% reproducibility (Orion Research 1986).

pH electrode calibration.—The pH electrode interfaced with the ISAAC 2000 by means of an FET op-amp circuit identical to that used with the ammonia electrode. The pH-sensing electrode was calibrated manually before the two control runs with two pH buffers (pH 4 and 10). As the pH electrode was inserted into each buffer, a BASICA program displayed the digital number for each. Regression analysis was used to develop a linear calibration equation relating pH to digital number. A third buffer (pH = 7) was used to compare actual and predicted values to estimate the error of the calibration relationship.

After a 2-min stabilization period, five consecutive readings were taken in about 2 s and an average was calculated. This procedure reduced difficulties experienced with the electrode's sluggish and noisy response, which had developed over time.

Thermocouple calibration.—Thermocouples were connected directly to the ISAAC's I-140 card without preamplification because their internal resistance was small compared with the input impedance of the I-140. Eleven thermocouples were used to monitor the system temperature at different locations. To calibrate the thermocouples, all of the thermocouples were inserted into a common water bath along with a thermocouple calibrated by the National Bureau of Standards (NBS). A potentiometer was used to measure the voltage of the NBS-calibrated thermocouple. The temperature of the water bath was slowly raised from 0 to 35°C in 22 steps. At each step, the voltage from the NBS thermocouple was recorded along with the digital numbers from each of the 11 thermocouples. The voltage from the NBS thermocouple was translated into °C so that relationships between °C and digital number could be obtained. During a typical temperature measurement, the temperature was calculated from the digital number with a "look-up table" approach. All of the 22 data pairs of temperature and corresponding digital number were stored in the computer. When a digital number was read, the two points that bracketed the digital number were located and linear regression was performed on these two points. The resulting equation was then used to calculate the temperature predicted by this digital number. The error of a temperature measurement with this method was determined to be ±0.5°C. The contributors to this error included the error of the potentiometer, the NBS-calibrated thermocouple error, computer resolution, and the regression equation error.

Flowmeter calibration.—For future studies on the effect of ammonium loading rate on filter performance, the flow rate through each of the five filters was to be carefully controlled. The flow rate of the liquid to each of the five filters was regulated by manual needle valves, which were in line with the magnetic flowmeters. Three meter sizes, 4, 6, and 10 mm, have been chosen to cover the flow range of future experiments. The magnetic flowmeters produce a 4- to 20-mA analog output signal that varied in direct proportion to the flow rate through the meter. This linear output was converted to a 1–5-V linear voltage by connecting a 250-Ω, 3-W, ±0.1% resistor in parallel with the input of ISAAC's I-140 module.

Each flowmeter was calibrated with the aid of a BASICA program that opened a solenoid valve in line with the meter for a specified time, then closed it. The water volume collected during this time was measured and the flow rate (volume collected/valve opening time) was calculated. The digital number was also read for each flow rate.

Linear regression analysis was performed on the digital reading versus flow rate for three flow rates to determine a calibration curve. The other flow rates were compared with the digital number predicted by the regression equation for those flow rates to estimate the error of the calibration relationship.

System operation.—The various sensor limitations and the PC's ability to perform only one task at a time produced the requirement for careful coordination of sensor calibrations, water quality measurements, and control procedures. This time-sharing constraint affects how accurately various water quality variables can be controlled. For example, after a thermocouple is calibrated, it takes only a few seconds to read the temperature of a sample, compare it with an ideal setting, and turn on, say, a heater to bring the water to the desired temperature. This task can be slipped into a control program frequently and the temperature can be closely monitored in this way. The ammonia electrode, however, is not as simple to use. The hourly calibration, the time required to flush the sample chamber and inject the sample for testing, and the time required for the electrode to stabilize after immersion in a new solution add up to about 20 min in this system. Consequently, the control actions are separated by longer periods of time. These considerations have led to the following control schedule that has been translated into a BASICA program to intertwine the ammonia electrode calibration with the control process.

(1) Measure ammonium concentration in standards 1 through 4 (start-up step).

(2) Calibrate the relationship of ammonia concentration to digital number from ISAAC.

(3) Measure water quality in the mixing tank and perform any needed control action.

(4) Measure ammonium concentration in standards 1–4, measuring water quality and performing necessary control actions between each standard measurement.

(5) Calculate the new calibration curve.

(6) Repeat steps (3) through (5).

An automatic flushing routine was called between each measurement to prepare the sample chamber for the next sample. The chamber was first flushed with tap water for 1 min, then flushed with the sample solution, and then the sample was collected. Initially, the entire calibration procedure must be completed to measure the samples of unknown ammonium concentration. Obviously, if the ammonia electrode calibration were performed all at once every time, the mixing tank water would only be sampled about every 80 min, followed by a control action. It would be difficult at best to establish any control with such a long feedback time. Notice that the first calibration curve would be used until step (5) was completed, then the second calibration curve would be used starting after step (5) until the next calibration curve was completed, and so forth.

Although not mentioned in the stepwise procedure described above, pH, temperature, flow rate, and any other desired variables would also be measured at step (4). Most of these measurements require very little time (usually less than a second) and can be easily incorporated into the procedure.

Control Procedures

Ammonia control.—During future experimental trials, it is anticipated that nitrifying bacteria in the biofilters will convert ammonium in the nutrient solution originating from the mixing tank into nitrate via nitrification. Because most of the filters' effluent will be recycled to the mixing tank, this will tend to lower the ammonium concentration in the mixing tank. If the biofilters removed 100% of the ammonium from the nutrient solution before it was recycled to the mixing tank, then, in effect, the mixing tank solution would be diluted the same amount as if pure water were added at the same flow rate as that of the filters' effluent. This concept was used to develop the ammonia control before the filters were operating. The maximum flow rate from the filters in future experiments will be 4.09 L/min. Thus, to simulate 100% removal, pure water must be added to the system at a rate of 4.09 L/min. To simulate 10% removal, the flow rate of pure water added will be 0.409 L/min, etc. In this study, 46% removal (water flow rate = 1.9 L/min) and 77% removal (water flow rate = 3.14 L/min) were simulated to aid in the development of the ammonia control procedure.

The ammonia concentration in the mixing tank was controlled from the BASICA program by first measuring the ammonia concentration in the mixing tank, then comparing this with the selected level of 8.0 mg NH_3/L (to be used in future experiments). Next, the amount of control standard (9,432 mg NH_4Cl/L or 3,000 mg NH_3/L) that was required to raise the ammonia concentration from the measured value to the selected level in the 1,205-L mixing tank was calculated by the computer. The addition was accomplished by the computer opening a reversible ball valve at the base of the NH_4Cl tank (located above the mixing tank) to a particular position for a specific time, then closing it again. Flow through the ball valve was calibrated prior to the test so that the relationship between milliliters added, valve opening position, and time was determined. This calibration was incorporated into the control software. The control process was repeated after every measurement of mixing tank ammonia concentration.

pH control.—Nitrification tends to lower culture system pH, so the control system was designed only to raise the pH. To aid in the pH control, a buffer composed of potassium phosphate monobasic (KH_2PO_4) and potassium phosphate dibasic (K_2HPO_4) was added to the system. In a manner similar to the ammonia control procedure, after the mixing tank pH is measured and compared with the selected level, the computer computes the volume of 1.0 N NaOH needed to raise the 1,205 L in the mixing tank to the proper pH. A volume slightly less than that calculated was then added (to avoid overshoot) by opening a reversible ball valve at the base of the NaOH control tank to a calculated position for a calculated time, then closing it again.

Temperature control.—A computer-regulated temperature feedback subsystem was still in the process of being tested to most effectively maintain the desired water temperatures in the biofilter system. System temperature information is received by the computer from thermocouples and analyzed by a BASICA program. When the system temperature falls below the set temperature, the computer switches a series of relays, thereby activating submersible water heaters. Similarly, the computer deactivates the heaters when the water temperature exceeds the set-point temperature. The room temperature is maintained below the set water temperature by a room air conditioner. Preliminary temperature control trials indicate that the temperature can be maintained within approximately ±1°C.

Experimental constraints and thermodynamic considerations were employed in determining the size of the heaters required for the system. The total system water volume is approximately 1.6 m^3, 74% of which is in the large mixing or control tank. If heat loss to the room is ignored, about 2,100 W of power are required to raise the system water 1°C in 1 h. This heat is provided by three Electrothermal (Gillette, New Jersey) red rod immersion heaters (500 W, 120 V AC) and a Sethco (Freeport, New York) heater (600 W, 120 V AC).

The computer is only able to switch solid-state relays that have a maximum current rating of 3 A. This is sufficient for the various valves and motors in this system, but not for the heaters. For this reason, a solid-state relay is used to switch two Potter and Brumfield (Princeton, Indiana) mechanical relays (120 VA C, 15 A maximum), which deliver power to the heaters.

TABLE 1.—Minimum detectable limits and error ranges observed for sensors in the water quality monitoring system, based on experimental measurement and calibration procedures.

Water quality variable	Smallest detectable change by data acquisition system	Estimated error of reading
Ammonia concentration	0.04 mg NH_3/L	7.7–11.3%
pH	0.002 units	1.83%
Temperature	0.16°C	<±0.5°C
Flow rate range (L/min)		
0.35–4.4	3.8 mL/min	<0.5%
0.75–6.0	3.8 mL/min	<1.0%
14.9–17.5	10.0 mL/min	<1.0%
0.17–1.12	1.0 mL/min	<0.05%

Results and Discussion

Measurement results.—Table 1 shows the minimum detectable limits and the error ranges observed for each of the four variables measured with the calibration procedures and sensing system described. The accuracy of the sensor measurement sets the upper limit for the control accuracy of the variable measured.

Ammonia control.—The ammonia concentration was maintained near the ideal level of 8.0 mg NH_3/L (Figure 2). However, the process still needs to be refined. During the first trial, an average of 8.05 mg NH_3/L (SD, 0.30) was maintained after the initial start-up of 60 min. In the second trial, an average of 7.99 mg NH_3/L (SD, 0.34) was maintained. The control error, which ranged between 7.9 and 16.4%, mainly consisted of the ammonia electrode reproducibility limitations, the temperature difference between the standards and the sample, the measurement error of the four NH_4Cl standards, electrode age, the volumetric measurement error of the NH_4Cl control solution as it was added by the reversible ball valve, and possible sample contamination in the plumbing system. It is anticipated that some control improvement will be possible by reducing some of these errors in the future.

Because the ammonia electrode mentioned can measure below 0.07 mg NH_3/L (Orion Research 1986), there is no reason to believe that control at these levels would not also be possible. In the control trials, the measurement error was a greater contributor to the control error than the

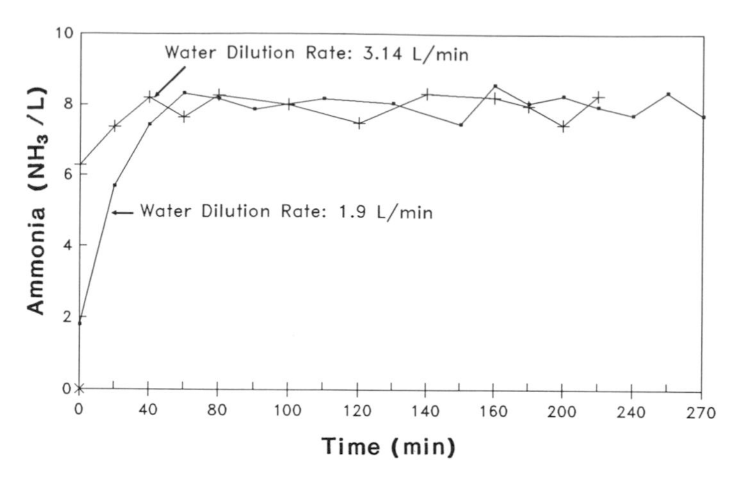

FIGURE 2.—Ammonia control response with dilution water added at 1.9 and 3.1 L/min.

error due to the actual compensating action (which consists of chemical metering error). In any case, as long as a variable can be measured accurately and the compensating chemical metered accurately, accurate control is possible.

Flow rate.—It appears that any flow rate between 0.17 and 17.5 L/min can be measured in this system with less than 1% error (Table 1). Manufacturer's specifications indicate the flowmeter is capable of an accuracy of ±1% of the flow rate plus ±0.01% of full scale for any 10:1 flow range between the meter's stated flow extremes (Fisher and Porter 1981). Thus, the system error appears no worse than the intrinsic error of the metering system.

pH system.—The pH was read during the ammonia control trials, but no control response was tested then. The pH of the system during those trials ranged from 8.6 to 8.9. There is a problem with the concept of inserting all of the electrodes in one central sample chamber and pumping samples from various points in the system to the chamber. In order to test the ammonia concentration in the chamber, a base was added to the sample to shift the pH to between 11 and 14 so that most of the ammonium ion was converted to ammonia gas, the ammonia form sensed by the electrode. The pH of the water sample was read before the base was added. However, prolonged and frequent immersions of the pH electrode in strong alkali solutions cause surface leaching of the glass membrane, resulting in extremely noisy or sluggish responses (Fisher 1984). Such responses were observed after about 3 months of intermittent use in this automated system. With a new electrode, a single reading could be taken within a few seconds after immersion in a solution with only about ±0.05% error. After 3 months of use, however, the error increased to 2%.

Further considerations.—Rogers and Klemetson (1985) noted the importance of proper management of the biofilter system in improving filter efficiency. Their study indicated that supplemental addition of ammonium salts to maintain a constant level of NH_4^+ in the biofilter influent could enhance the nitrification process because, as they showed, ammonia removals were the highest when steady state ammonia levels were maintained. This could be accomplished with a system similar to the one described in this paper.

The measurement and control trials described were conducted with relatively clean water with little or no organic contaminants. Sensors used with operating fish culture systems, however, have to withstand considerable organic loads. This is not a new problem to wastewater treatment engineers, and some lessons may be gleaned from that industry. In some cases, ultrafilters are employed in water quality measuring equipment to remove all algae and bacteria without changing the ionic species and dissolved gases in the sample (Meredith 1985). In another case, Tonegawa et al. (1985) examined the cleaning effect of ozone on sensors used in water quality measurement equipment and proved that slime adhesion was completely prevented. They used high-concentration ozone to clean the sensors and the water sampling piping.

Engineers are needed to help refine computer-automated systems that aquaculturists and biologists can easily learn to operate, maintain, and update. A PC-based system with central sample chambers containing various water quality sensing electrodes has many advantages, as described earlier. Moreover, there is only one set of electrodes in one central location to calibrate and maintain. In this case, plumbing to the remote measurement locations replaces wiring that would be required if separate sensors and microprocessor systems were located at each sample location.

In any event, if there is an agreement on an automation strategy for aquaculture, and if compatible systems are developed, not only researchers and engineers but every owner and operator would be able to contribute and thus speed the development and improve the efficiency of automation in the aquaculture industry.

Conclusions

At this stage of system development, automated ammonia measurements with less than 9% error, temperature measurements within ±0.5°C, and pH measurements with less than 2% error can be made. Ammonia concentration has been controlled at 8.0 mg NH_3/L with a maximum error range of 7.9–16.4% during preliminary trials. Marked improvements in accuracy are expected with continued refining of the control process.

References

Dallimore, J. M., and K. Thatcher. 1985. Distributed microprocessor based control systems in the automation of wastewater pumping stations. Pages 191–198 *in* R. A. R. Drake, editor. Instrumentation and control of water and wastewater treatment in transport systems. Pergamon Press, New York.

Fisher Scientific. 1984. Accu-pHast electrode instruction manual. First issue (4-0326-04/2229). Fisher Scientific, Pittsburgh, Pennsylvania.

Fisher and Porter. 1981. Instruction bulletin for model 10D1475 Mini-Mag Flowmeter. Fisher Scientific, Warminster, Pennsylvania.

Garber, W. F., and J. J. Anderson. 1985. From the standpoint of an operator—What is really needed in the automation of a wastewater treatment plant. Pages 429–442 *in* R. A. R. Drake, editor. Instrumentation and control of water and wastewater treatment in transport systems. Pergamon Press, New York.

Hordeski, M. F. 1987. Transducers for automation. Van Nostrand Reinhold, New York.

Kaiser, G. E., and F. W. Wheaton. 1988. Computerized rapid measurement of ammonia concentration in aquaculture systems. Marine Technology Society, Pre-Prints for Oceans '88, Washington, D.C.

Malcolme-Lawes, D. J. 1984. Microcomputers and laboratory instrumentation. Plenum Press, New York.

Meredith, W. D. 1985. Using micro-electronics to improve sensor reliability. Pages 547–549 *in* R. A. R. Drake, editor. Instrumentation and control of water and wastewater treatment in transport systems. Pergamon Press, New York.

Office of Natural Resources and Rural Development. 1988. Aquaculture Update 3(2). U.S. Department of Agriculture, Washington, D.C.

Orion Research. 1986. Model 95-12 ammonia electrode instruction manual. Orion Research, Boston.

Plaia, W. C. 1987. A computerized environmental monitoring and control system for use in aquaculture. Aquacultural Engineering 6:27–37.

Rogers, G. L., and S. L. Klemetson. 1985. Ammonia removal in selected aquaculture water reuse biofilters. Aquacultural Engineering 4:135–154.

Tonegawa, H., H. Yotsumoto, I. Nakahori, and T. Miyauchi. 1985. Water quality monitoring system using ozone cleaning and its application in wastewater treatment process control. *In* Proceedings of the 4th International Association on Water Pollution Research and Control (IAWPRC) workshop. Pergamon Press, New York.

American Fisheries Society Symposium 10:516–528, 1991

Microcomputer Models for Salmonid Hatcheries

WILLIAM E. MCLEAN

Department of Fisheries and Oceans, Enhancement Operations Branch
Quinsam Hatchery, Box 467, Campbell River, British Columbia V9W 5C1, Canada

JOHN O. T. JENSEN

Department of Fisheries and Oceans, Biological Sciences Branch
Pacific Biological Station, Nanaimo, British Columbia V9R 5K6, Canada

PETER J. ROMBOUGH

Zoology Department, Brandon University
Brandon, Manitoba R7A 6A9, Canada

Abstract.—Three microcomputer models are presented for salmonid fish culture use. They are based on current information compiled from recent literature and unpublished data. They deal with (1) the effect of temperature on egg and larval development rates and oxygen requirements of five species of Pacific salmon (chinook *Oncorhynchus tshawytscha*, chum *O. keta*, coho *O. kisutch*, pink *O. gorbuscha*, sockeye *O. nerka*) and steelhead *O. mykiss*; (2) the influence of temperature, fish size, and ration level on oxygen consumption rate and hence on the rearing capacity of a water supply; and (3) gas supersaturation and its effect on juvenile salmonid survival. The biological responses for each program are presented, the data sources are cited, and the mathematical relations are described. This is followed by a description of program inputs and outputs. Finally, illustrative examples are generated for each of the three microcomputer programs, based on commercial spreadsheet software.

In Canada, microcomputers were originally installed in federal hatcheries to improve financial and fish culture record keeping. The goal was to gain more complete understanding of costs and benefits so that a facility could be assessed objectively. Programs were designed so that data could be entered, processed, and transformed into useful information. This ongoing task is being accomplished increasingly by the use of commercial spreadsheet and data-base software. High-quality commercial software can be used to construct templates suited to almost any problem. The template is analogous to a blank data form. However, it also contains instructions and reminders for using the form, formulas for performing calculations, and capabilities allowing one to sort, search, and summarize the data files. Commercial software has evolved to the point at which templates can be constructed quickly and easily by fish culturists. Custom-built data forms are encouraged at federal hatcheries because they can take into account site-specific differences and are better accepted by hatchery staff.

Although data storage, organization, and retrieval are still the main functions of microcomputers in these hatcheries, a set of predictive models has been developed to assist hatchery workers with a wide range of fish culture problems. The programs focus on incubation, freshwater rearing, water quality, and fish health. They are fundamentally different from the data management programs because they predict how fish may interact with the culture environment.

Evolution of useful predictive models occurs as a result of challenging existing models with real data. Discrepancies must be highlighted and rationalized. Small differences may lead to minor alterations in the mathematical model, whereas major discrepancies might lead to a totally new conceptual model. The search for discrepancies is essential in defining the boundaries of an existing model. Access to microcomputers has greatly accelerated this model-building process. Not only can simulation runs be generated for the most complex mathematical models but data management systems have also made comparative information from fish culture facilities readily available.

Predictive microcomputer programs at federal hatcheries were originally custom-programmed in BASIC and were presented to fish culture workers as a package of utility programs (Kling et al. 1983). Upgrading of these programs is currently under way. As with the data-processing programs, there will be an increasing trend to use commercial software. Fish culture workers are familiar with the spreadsheet or data-base format, and predictive models are easily presented within this context.

This report describes the biological models on which three of these programs are based. They deal with (1) the effect of temperature and ambient dissolved oxygen supply on egg and larval development rates of different species of Pacific salmon *Oncorhynchus* spp.; (2) the influence of temperature, fish size, and ration level on the rearing capacity of a water supply; and (3) gas supersaturation and its effects on fish health.

Incubation Model

The possibility of constructing an incubation model capable of predicting egg development rates, oxygen requirements, and oxygen consumption rates for different species in production hatcheries was discussed by McLean and Lim (1985) and by Rombough (1986). These papers described a conceptual framework for such an undertaking but could not provide detailed relationships required for a quantitative model. Since that time much information has become available (Heming 1982; Rombough 1985; Velsen 1987; Murray and McPhail 1988), and it now seems possible to build a mathematical model useful to hatchery operators.

The operation of an incubation facility is centered around several key stages in development. The eyed stage, for instance, determines when eggs can be handled on a large scale, and the hatching stage marks a completely new phase in incubation with more complex environmental requirements. The attainment of neutral buoyancy and maximum alevin weight marks the transition to the fry stage with its requirement for external feeding (Heming 1982). Coping with millions of eggs from different species, fertilized at different times and incubated under a variety of temperature regimes, requires an ability to predict these major events. The task becomes more demanding and the need for predictive models more critical when temperature control is available and a specific timetable is required.

The rates at which eggs or alevins consume oxygen and produce ammonia and carbon dioxide also depend on developmental stage, water temperature, and species. Water flowing through the incubator counters the effects of metabolism—excreted ammonia and carbon dioxide are diluted while the oxygen concentration is maintained. The goal of the hatchery operator is to set a flow regime in which the requirements of the fish are satisfied and the water usage is minimized. This is critical where the water supply is limited or where water costs are high due to heating, pumping, or treatment.

Development Rates

The accumulated temperature units (ATUs) to a specific developmental stage are often considered to be a constant for a species:

$$\text{ATU} = T \cdot D = k = \text{constant}; \qquad (1)$$

D = number of days to a particular stage, T = average temperature to that stage, and k = constant for a species. The time to hatch for any temperature regime can then be predicted from k/T. This model is reasonably accurate at temperatures greater than 5°C but increasingly overestimates incubation times at lower temperatures. To overcome this difficulty two models were used to model egg and larval development rates based on data compiled by Velsen (1987), Murray and McPhail (1988), and Rombough (unpublished data). The first model, often called the Belehradek model, is described by the equation

$$D = a/(T - c)^b, \qquad (2)$$

which rearranges to yield ATU as follows:

$$\text{ATU} = D \cdot T = T\,[a/(T - c)^b]; \qquad (3)$$

a, b, and c are constants characteristic of a species and stage of development. This model was used to predict time–temperature relations for two embryonic stages described by Ballard (1968) based on the data from Velsen (1987). It was also used to model time to maximum alevin wet weight (MAWW), or emergence, shown by Heming (1982) and Rombough (1985) to occur at approximately the same time, with data (Table 1) obtained from Heming (1982), Murray and McPhail (1988), Rombough (1986, unpublished data), and F. P. J. Velsen (Department of Fisheries and Oceans, unpublished data). Rombough (1985) also found for chinook salmon eggs that initial egg size significantly ($P < 0.01$) influenced the time to MAWW. However, this influence accounted for only 2% of the overall variability, whereas temperature accounted for 96.4%. Therefore, only temperature has been used here to model time to MAWW.

A second model, originally proposed by Schnute (1981) to describe growth curves for fish, was found to give a better fit to the time–temperature data for salmonid egg and larval development. However, this model is more complex and

TABLE 1.—Times to maximum alevin wet weight and emergence at various temperatures, used to develop models for six species of salmonids.

Temperature (°C)	Time (d)	Temperature (°C)	Time (d)	Temperature (°C)	Time (d)	Temperature (°C)	Time (d)
			Chinook salmon *Oncorhynchus tshawytscha*				
2.3	316.0[a]	7.3	136.9[b]	10.2	88.3[b]	12.5	62.2[b]
5.1	191.0[a]	7.3	142.3[b]	10.2	94.5[b]	5.9	190.0[c]
8.1	115.0[a]	7.3	143.3[b]	10.0	98.7[b]	7.9	141.0[c]
11.2	84.0[a]	10.0	84.3[b]	10.2	94.5[b]	9.8	107.0[c]
13.9	63.0[a]	10.2	90.8[b]	12.5	64.8[b]	11.8	85.0[c]
5.0	200.5[b]	10.0	89.8[b]	12.5	58.4[b]		
7.3	135.9[b]	10.2	95.7[b]	12.5	64.9[b]		
			Chum salmon *O. keta*				
5.1	160.7[a]	13.9	86.0[a]	8.1	122.9[d]	12.0	84.0[d]
8.1	124.2[a]	3.9	209.8[d]	8.1	119.4[d]	12.0	92.5[d]
11.2	98.0[a]	8.1	117.8[d]	10.0	115.1[d]		
			Coho salmon *O. kisutch*				
2.3	228.0[a]	13.9	61.2[a]	8.0	94.2[d]	11.9	61.4[d]
5.1	138.5[a]	3.9	183.1[d]	8.0	98.7[d]	12.0	63.6[d]
8.1	109.1[a]	3.9	190.8[d]	8.0	99.5[d]	12.0	64.0[d]
11.2	73.7[a]	3.9	200.2[d]	8.0	109.7[d]	12.0	65.0[d]
			Pink salmon *O. gorbuscha*				
5.1	173.2[a]	11.2	90.8[a]	3.0	215.8[d]	8.0	130.4[d]
8.1	120.2[a]	13.9	72.1[a]	4.0	200.0[d]		
			Sockeye salmon *O. nerka*				
2.3	282.2[a]	2.1	267.4[d]	4.0	225.1[d]	8.0	131.9[d]
5.1	173.0[a]	2.3	269.8[e]	4.1	223.9[d]	8.0	144.0[d]
8.1	120.5[a]	3.1	238.6[e]	4.1	215.7[d]	11.9	85.3[d]
11.2	90.3[a]	3.1	222.6[e]	4.1	202.0[d]	12.0	89.1[d]
13.9	72.0[a]	3.1	210.1[e]	4.1	198.1[d]	12.0	89.4[d]
1.1	301.9[e]	3.1	213.6[e]	7.9	128.9[d]	12.0	90.3[d]
1.3	304.5[e]	4.0	236.2[d]	8.0	146.6[d]		
2.1	282.9[e]	4.0	235.9[d]	8.0	131.9[d]		
			Steelhead *O. mykiss*				
6.0	110.5[f]	9.1	68.6[f]	15.1	38.0[f]		

[a]Murray and McPhail (1988).
[b]Rombough (1985).
[c]Heming (1982).
[d]Rombough (unpublished data).
[e]Velsen (unpublished data).
[f]Rombough (1986).

requires larger data sets than the Belehradek model. Therefore, it was only used to model time to hatch. The model, with constants a and b not equal to zero, is described by the equation

$$D = \left[D_1^b + \frac{(D_2^b - D_1^b)(1 - e^{-a(T-T_1)})}{(1 - e^{-a(T_2-T_1)})} \right]^{1/b}; \qquad (4)$$

T_1 and T_2 are the minimum and maximum incubation temperatures (°C) from the data set used, and D_1 and D_2 are the predicted development times (days) at those temperatures. Calculation of ATUs was then simply made by multiplying the predicted number of days to hatch (D) by the average temperature (T).

Jensen (1988) described an incubation program with equations describing 25 stages of development for six species. However, for the purposes of the incubation computer program in this paper, only four developmental stages are included. They are (1) completion of epiboly or yolk plug closure (stage 14), (2) the eyed stage (stage 20), (3) median hatch, and (4) MAWW or emergence. The model coefficients for these stages are listed in Table 2.

Oxygen Consumption Rates

Oxygen consumption rates have been measured under laboratory conditions (Rombough 1986) for steelhead and chinook, coho, pink, chum, and sockeye salmon. The experimental procedures used to obtain these measurements have been fully described by Rombough (1986). Measurements were performed on eggs incubated at constant temperatures. Therefore, the original oxygen consumption rate equations had to be modified to take into account the variable temperature regimes encountered in production hatcher-

TABLE 2.—Coefficients and development times used in text equations (2) and (4)[a] for four development stages modeled for each of the six salmonid species, and the number of data points (N).

Development stage	Coefficient			Development time (d)		
	a	b	c	D_1	D_2	N
Chinook salmon						
Epiboly	207.1222814	1.143230343	−1.001756111			7
Eyed egg	883.2385275	1.414574147	−2.457463872			7
Hatch	0.098319	−0.0371603		239.1	27.8	295
MAWW	5209077.4	3.92184502	−19.07204758			26
Chum salmon						
Epiboly	510.6446646	1.503134190	−1.325714686			5
Eyed egg	2378.325397	1.761904580	−3.947781414			5
Hatch	0.1362313	0.7193096		177.8	25.8	101
MAWW	314.41470	0.51820254	1.6820473			11
Coho salmon						
Epiboly	1303048.207	3.436728916	−19.21278684			5
Eyed egg	16661.25802	2.255270212	−8.618379214			5
Hatch	0.0964044	0.0790329		213.2	21.6	108
MAWW	923367.454	2.8995311	−15.02834151			16
Pink salmon						
Epiboly	704.6378875	1.567007981	−1.593069000			6
Eyed egg	390.088759	1.209049777	−0.343034298			6
Hatch	0.5672965	4.4479176		180.4	49.7	54
MAWW	977660.461	2.7231174	−18.83916220			7
Sockeye salmon						
Epiboly	1071672210	5.109288900	−21.5645848			3
Eyed egg	1097342643	4.801222359	−29.07423777			3
Hatch	0.1730653	1.0483823		212.4	36.4	81
MAWW	34186861.2	3.56109377	−24.96801241			31
Steelhead						
Epiboly	394.1655832	1.500516000	−2.941684794			4
Eyed egg	3223106353	5.173971576	−29.75685594			4
Hatch	0.4084140	2.3613821		139.3	18.3	127
MAWW	922049.740	3.00725581	−14.19575994			3

[a]Equation (2) was used to describe time to completion of epiboly, the eyed stage, and maximum alevin wet weight (MAWW). Equation (4) was used to describe time to hatching. Auxiliary parameters for all species were set at T_1 = 1°C and T_2 = 20°C.

ies. This was accomplished by using the ATU model (Rombough 1986). The oxygen consumption rate (R_O) equations, modified for a variable temperatures regime, show that R_O depends on how far development has progressed (i.e., the ATUs) and also on the ambient temperature T.

Oxygen consumption rate equations have been derived for two incubation phases: phase 1, from fertilization to completion of epiboly, and phase 2, from completion of epiboly to MAWW. Equations for R_O in units of milligrams of oxygen per 1,000 eggs per hour (mg O_2/[1,000 eggs · h] or μg O_2/[egg · h]) are summarized in Table 3.

Oxygen Concentration Requirements

As the ambient oxygen concentration drops, a point is reached at which the oxygen consumption rate of the egg or alevin becomes dependent on the supply; below this concentration, respiration is suppressed. This level is termed the critical oxygen concentration, P_c (Rombough 1986), and is suggested as a lower limit for production hatcheries.

In reality, minimum oxygen criteria based on P_c must be used very carefully. Large incubators often have poor water flow distributions; as a result they may have localized areas of low oxygen and high ammonia concentration. Oxygen concentrations will be much lower in these areas than at the outflow of the incubator. If the outflow oxygen is allowed to drop to P_c in these incubators, large numbers of eggs may already be suffering the effects of low dissolved oxygen. Thus P_c must be looked on as a guideline rather than as a target concentration. Equations for P_c are presented in Table 4 for three incubation phases:

TABLE 3.—Oxygen consumption rates (R_O, mg O_2/[1,000 eggs · h]) for six salmonids as a function of accumulated temperature units (ATU) and ambient temperature (T) for two phases of incubation: (1) fertilization to epiboly; and (2) epiboly to MAWW. The R^2% value (where sufficient data exist) and the number of data points (N) are also shown.

Phase	Equation
	Chinook salmon
(1)	$\log_e(R_O) = -2.780 + 0.944\log_e(T)$; $R^2 = 99.8$, $N = 5$
(2)	$\log_e(R_O) = -16.897 + 2.873\log_e(\text{ATU}/T) + 3.840\log_e(T)$; $R^2 = 96.3$, $N = 75$
	Coho salmon
(1)	$R_O = -0.240 + 0.364\log_e(T)$; $R^2 = 90.2$, $N = 5$
(2)	$\log_e(R_O) = -15.417 + 2.746\log_e(\text{ATU}/T) + 3.528\log_e(T)$; $R^2 = 96.3$, $N = 74$
	Chum salmon
(1)	$R_O = 0.60$; $N = 2$
(2)	$\log_e(R_O) = -15.437 + 2774\log_e(\text{ATU}/T) + 3.212\log_e(T)$; $R^2 = 95.7$, $N = 23$
	Pink salmon
(1)	$R_O = 0.60$; $N = 2$
(2)	$\log_e(R_O) = -14.603 + 2.668\log_e(\text{ATU}/T) + 2.749\log_e(T)$; $R^2 = 99.0$, $N = 28$
	Sockeye salmon
(1)	$R_O = 0.30$; $N = 2$
(2)	$\log_e(R_O) = -14.032 + 2.383\log_e(\text{ATU}/T) + 2.866\log_e(T)$; $R^2 = 98.5$, $N = 30$
	Steelhead
(1)	$\log_e(R_O) = -2.977 + 0.1969T$; $R^2 = 98.3$, $N = 4$
(2)	$\log_e(R_O) = -18.139 + 2.950\log_e(\text{ATU}/T) + 4.549\log_e(T)$; $R^2 = 97.9$, $N = 28$

phase 1 is fertilization to completion of epiboly, phase 2 is epiboly to hatch, and phase 3 is hatch to maximum alevin wet weight.

Prediction of Oxygen Concentrations

In a single-pass incubator, oxygen is the most important factor limiting egg or larval carrying capacity of the water supply (McLean and Lim 1985). If the inflow water is the only source of oxygen and the eggs are the only major consumers, the oxygen concentration at the outflow, C_f, is given by

$$C_f = C_i - [N \cdot R_O/(Q \cdot 60{,}000)]; \qquad (5)$$

C_i = oxygen concentration at inflow (mg/L), R_O = oxygen consumption rate (mg O_2/[1,000 eggs · h]), Q = water flow (L/min) to the incubator, and N = number of eggs in incubator. The R_O values from Table 3 can be substituted into equation (5) to obtain a predictive model for C_f.

Reaeration is significant in vertically stacked Heath trays and so equation (5) was modified for this type of incubator (McLean and Lim 1985):

$$C_f = C_s - (C_s - C_i)e^{-kn} + \frac{R_O \cdot N_t(e^{-kn} - 1)}{60{,}000 \cdot Q \cdot k}; \qquad (6)$$

C_s = oxygen saturation concentration (mg/L), n = number of Heath trays, N_t = number of eggs per tray, Q = water flow (L/min), and k = reaeration coefficient for a Heath tray. The reaeration coefficient for Heath trays operated at a flow rate of 12 L/min and a temperature of 10°C was 0.0555 per tray (McLean and Lim 1985).

Model Inputs, Outputs, and Examples

The model was programmed with commercial spreadsheet software so that a report is generated for each week of incubation. The oxygen consumption rate, oxygen concentration at the inflow and outflow of the incubator, and the critical oxygen level are calculated. Also major events (completion of epiboly, eyed stage, hatch, and maximum alevin wet weight) are forecast. Inputs required are average weekly temperature, number of eggs in the incubator, water flow rate, and the percentage oxygen saturation of the water supply. If Heath tray incubation is indicated, the program includes the effects of reaeration.

A simulation of chinook salmon incubation is shown in Figure 1. Values were predicted over the 19-week incubation period extending from fertilization to maximum alevin weight (week 19). In this simulation, the incubator contained 52,000 eggs, the inflow dissolved oxygen was 95% of saturation, and the water flow was 12 L/min until hatch. After hatch, the flow was increased to 15 L/min. With no reaeration, the model predicted an oxygen concentration of 6.0 mg/L at the outflow

TABLE 4.—Critical oxygen concentration P_c (mg/L) as a function of accumulated temperature units (ATU), the accumulated temperature units from hatch (ATUa), and the ambient temperature (T) for three phases of incubation: (1) fertilization to epiboly, (2) epiboly to hatch, and (3) hatch to maximum alevin wet weight. The R^2% and N (where sufficient data exist) are also shown.

Phase	Equation
	Chinook salmon
(1)	$P_c = 2.0$[a]
(2)	$\log_e(P_c) = -3.302 + 0.976\log_e(\text{ATU}/T) + 0.161T$; $R^2 = 91.5$, $N = 40$
(3)	$P_c = 4.4$[b]; mean = 3.98, SD = 0.41, $N = 46$
	Coho salmon
(1)	$P_c = 2.0$[a]
(2)	$\log_e(P_c) = -2.141 + 0.799\log_e(\text{ATU}/T) + 0.128T$; $R^2 = 88.7$, $N = 41$
(3)	$P_c = 5.586 - 0.664\log_e(\text{ATUa}/T)$; $R^2 = 62.8$, $N = 38$
	Chum salmon
(1)	$P_c = 2.0$[a]
(2)	$\log_e(P_c) = -3.634 + 0.976\log_e(\text{ATU}/T) + 0.193T$; $R^2 = 88.9$, $N = 12$
(3)	$P_c = 4.942 - 0.021\log_e(\text{ATUa}/T)$; $R^2 = 49.9$, $N = 13$
	Pink salmon
(1)	$P_c = 2.0$[a]
(2)	$\log_e(P_c) = -3.261 + 1.017\log_e(\text{ATU}/T) + 0.131T$; $R^2 = 93.2$, $N = 14$
(3)	$P_c = 4.2$[b]; mean = 3.65, SD = 0.55, $N = 15$
	Sockeye salmon
(1)	$P_c = 2.0$[a]
(2)	$\log_e(P_c) = -2.214 + 0.693\log_e(\text{ATU}/T) + 0.141T$; $R^2 = 90.1$, $N = 16$
(3)	$P_c = 3.575 - 0.013(\text{ATUa}/T)$; $R^2 = 42.1$, $N = 12$
	Steelhead
(1), (2)	$\log_e(P_c) = -4.50 + 0.949\log_e(\text{ATU}/T) + 1.45\log_e(T)$; $R^2 = 97.1$, $N = 20$
(3)	$P_c = 5.56 - 1.04\log_e(\text{ATUa}/T) + 0.131T$; $R^2 = 79.5$, $N = 20$

[a]Data suggest that P_c is 2.0 mg from fertilization to completion of epiboly.

[b]No significant relationship exists with temperature; therefore, the SD was added to the mean to give a conservative estimate of P_c.

at week 19. With reaeration, and with an assumption that the eggs had been distributed equally in eight vertically stacked Heath trays, the predicted oxygen concentration was 7.4 mg/L. Figure 2 shows how the accumulated temperature units to hatch vary with the incubation temperature for various species. In this simulation, temperature was constant throughout development.

Rearing-Pond Oxygen Model

In a typical single-pass rearing pond with good-quality water and sufficient rearing space, dissolved oxygen is the most important factor limiting the carrying capacity of the water supply. A simple oxygen model was developed under the assumption that the inflow water is the only source of oxygen and that the fish are the only consumers. This is valid for most fish culture containers. However for long shallow channels, reaeration within the channel may be significant and should be incorporated into a pond oxygen model (McLean 1979).

If reaeration is neglected, the oxygen concentration at the outflow of a pond is related to the biomass of fish in the pond, their oxygen consumption rate, the oxygen concentration at the inflow of the pond, and the water flow rate (McLean 1979):

$$C_f = C_i - (R_O \cdot B/60 \cdot Q) = C_i - (R_O \cdot L_r/60); \quad (7)$$

C_f = oxygen concentration at the outflow (mg/L), C_i = oxygen concentration at the inflow (mg/L), R_O = oxygen consumption rate (mg/[kg · h]), B = biomass (kg), Q = flow (L/min), and L_r = load rate or ratio of biomass to flow (kg/[L/min]). If the minimum acceptable oxygen concentration at the outflow is specified (C_f), equation (7) can be rearranged and used to predict the maximum allowable load rate or carrying capacity of the water supply:

$$L_r = (C_i - C_f)60/R_O. \quad (8)$$

To make use of these equations, knowledge of the oxygen consumption rate (R_O) is required.

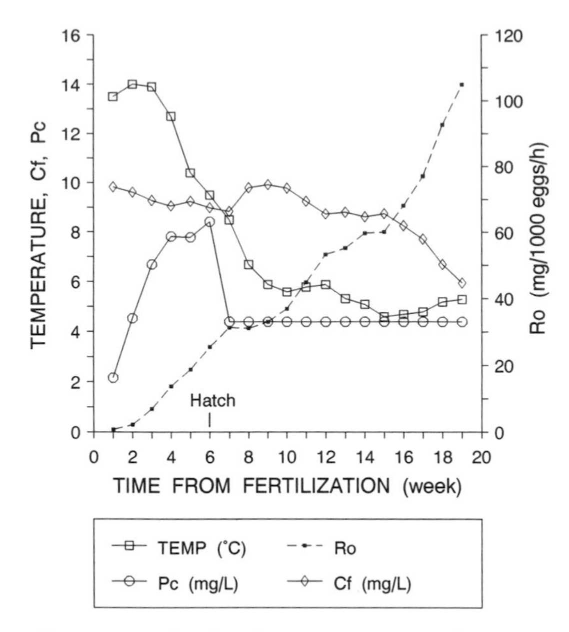

FIGURE 1.—Predicted oxygen consumption rates (R_O), critical oxygen levels (P_c), and outflow oxygen concentrations (C_f) for 52,000 chinook salmon eggs incubated over a 19-week period. The simulation was carried out from fertilization to maximum alevin wet weight (ponding) over a decreasing temperature regime. Flow was increased from 12 L/min to 15 L/min at hatch, inflow oxygen concentration was 95% of saturation, and reaeration within the incubator was assumed to be insignificant.

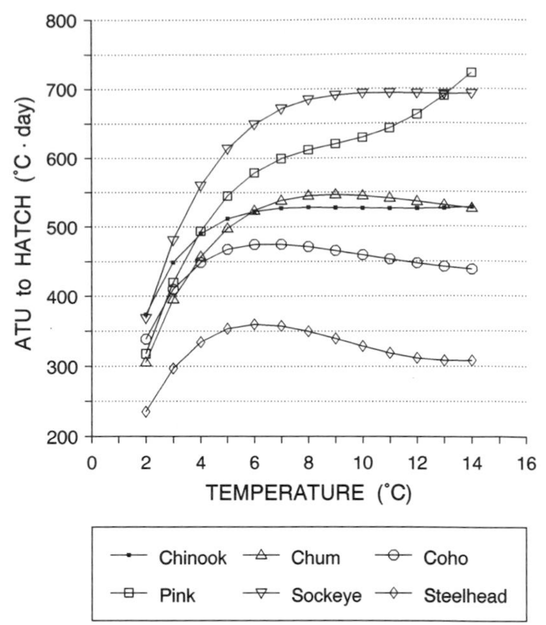

FIGURE 2.—Accumulated temperature units (ATU) to hatch for eggs of steelhead and several species of salmon incubated at a constant temperature. Simulations were carried out between 2 and 14°C.

Oxygen Consumption Rate Model

A conceptual model for R_O was based on research performed by Brett (1964, 1965, 1976) and Brett and Glass (1973) into the effects of temperature, swimming speed, fish weight, and ration level on the metabolic rate of sockeye salmon. Their work underlined the theoretical problems inherent in modeling R_O. It was shown that metabolic rate, as well as being a function of temperature, weight, and ration level, was also influenced by many other factors such as excitement, activity level, and feeding strategy and could vary rapidly within wide limits. The difference between standard and active metabolic rates illustrates the problem of predicting oxygen consumption. Approximate standard and active metabolic rates for 10-g sockeye salmon were 70 and 640 mg/(kg · h), whereas the limits for 1,000-g fish were 40 and 590 mg/(kg · h). These theoretical limits on metabolism highlight the problems of modeling R_O when many of the factors that contribute to increased swimming speed are unknown.

From these studies it was predicted that the oxygen consumption rate of juvenile fish would be highly variable over the day—depending to a large extent on the feeding intensity and the level of disturbance. Because of this hour-to-hour variation, it was decided to build the pond oxygen model around the average daily oxygen consumption rate rather than on the instantaneous rate. This parameter would be much less variable and therefore easier to model. However, it was recognized that the peak consumption rates would also have to be taken into account in a practical pond oxygen model.

To deal with this problem, a safety factor was derived by examining the ratio between the peak and average daily oxygen consumption rates. This safety factor, when multiplied by the average daily consumption rate, yielded a rate that was expected to exceed the peak rate most of time. This "safe" rate could be used to predict the carrying capacity of the water supply or the lowest daily oxygen level in the pond. It should be noted that the active metabolic rate of salmon is so high that even in well-planned rearing programs, there is always the possibility that extreme bursts of activity could result in oxygen levels dropping below the predicted minimums. Setting the carrying capacity in anticipation of such an unlikely event is not practical; hence, the safe R_O value was chosen to include most of the rates likely to be encountered during typical rearing operations.

Mathematical Model for R_O

A multiple curvilinear regression model for R_O was derived by the method of least squares:

$$R_O = A_0 + A_1F + A_2T + A_3W + A_4F^2 + A_5T^2 + A_6W^2 + A_7F \cdot T + A_8F \cdot W + A_9W \cdot T; \quad (9)$$

R_O = average daily oxygen consumption rate (mg/[kg · h]); F = ration level (g dry food per 100-g fish per day or %/d); T = average daily temperature (°C); W = fish weight (g); $F \cdot T$, $F \cdot W$, and $W \cdot T$ = products of the main variables; and A_0, A_1, A_2, . . . , A_9 = regression coefficients.

The regression coefficients were derived by using a combination of measured and theoretical values of R_O. Oxygen was measured continuously so that average and peak daily values could be calculated. Ninety measurements of the average daily R_O (mean, 241 mg/[kg · h]; SD = 75.3) were made for feeding juvenile coho, chum, and sockeye salmon. These measurements showed that the ratio of peak to average daily R_O values was 1.22 with a SD of 0.13 (McLean 1988).

Temperatures, weights, and ration levels ranged from 5 to 20°C, 0.65 to 33.3 g, and 0 to 3.1%/d, respectively. In this data set there was a shortage of values at low and high ration levels, and the set was augmented by addition of several theoretical values derived from Brett and Glass (1973). The regression model derived from this augmented data set is referred to as a semi-empirical model. The least-squares regression coefficients for this model were A_0 = 50.9205, A_1 = 76.1692, A_2 = 5.7331, A_3 = −2.6280, A_4 = −10.9408, A_5 = −0.0042, A_6 = 0.0420, A_7 = 3.9391, A_8 = 1.6212, and A_9 = −0.0219.

The regression model accounted for 87% of the variation in R_O. The variable with the greatest influence on R_O was the product of ration level and temperature ($F \cdot T$)—this variable alone accounted for 71% of the variation. Ration level (F) was also a good predictor but could only account for 49% of the variability in R_O. The 90 measured R_O values have been plotted against $F \cdot T$ in Figure 3.

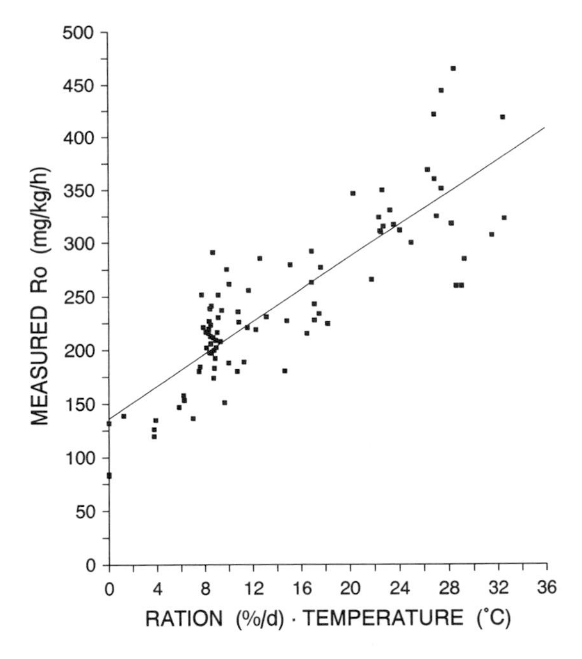

FIGURE 3.—Measured average daily oxygen consumption rates (R_O) for juvenile salmonids versus the product of ration level and temperature. The least-squares line relating R_O to $F \cdot T$ is shown: R_O = 136.2 + 7.5 FT; R^2 = 0.71, N = 90.

A comparison of model predictions with the 90 measured values showed that the average difference between the measured and predicted values was 0 (i.e., unbiased) and that the standard deviation of the differences was 38.05. Although not sensitive to species differences, this equation is a reliable predictor of R_O for juvenile salmon (W < 40 g) in a rearing environment where temperatures are between 5 and 20°C.

Safety Factor for R_O

The R_O values generated by the predictive model are average daily rates. The peak daily consumption rate was found by multiplying the average daily rate by the peak-to-average R_O ratio (mean, 1.22; SD, 0.13). The uncertainty in this calculated value is a reflection of the uncertainties in both the peak-to-average ratio and the predicted average daily R_O. The standard deviations of these quantities were combined mathematically (Baird 1963) to estimate the standard deviation associated with the peak daily R_O values (SD_p):

$$SD_p = 1.22\,R_O\,[(SD_1/1.22)^2 + (SD_2/R_O)^2]^{0.5}; \quad (10)$$

R_O = average daily R_O value predicted by equation (9), SD_1 = SD of peak-to-average ratio (0.13), and SD_2 = SD of difference between measured and predicted R_O (38.05). The safe oxygen consumption rate was defined as one standard deviation above the predicted peak daily rate: 1.22 R_O + SD_p.

Maximum Load Rates

To predict the maximum load rate (L_r) with a margin of safety, the safe R_O value (1.22 R_O + SD_p) was substituted into equation (8):

$$L_r = \frac{(C_i - C_f)60}{1.22\,R_O + SD_p}. \quad (11)$$

This load rate was considered safe because the minimum daily oxygen concentration specified at the pond outflow (C_f) would be satisfied most of the time. In theory, if the peak R_O values are distributed normally, the specified outflow oxygen concentration is satisfied 84% of the time.

In federal rearing operations in British Columbia, minimum oxygen concentration at the outflow is based on criteria developed by Davis (1975), typically 6 mg/L. The choice of a minimum oxygen level is complicated if other stressors are present or if the fish have suffered gill damage; in such cases higher oxygen levels would be required to protect the population.

Program Inputs, Outputs, and Examples

The oxygen model (equations 7–11) has been programmed with commercial spreadsheet software. The required inputs are fish weight, ration level, temperature, salinity, barometric pressure, percentage oxygen saturation of inflow water, and minimum oxygen concentration at the pond outflow. The program calculates the average daily oxygen consumption rate, maximum allowable load rate, the inflow oxygen concentration, and the maximum ration.

The inflow oxygen level is based on the solubility relationships of Hitchman (1978) and the maximum ration is derived from Stauffer (1973). If the load rate (rather than the minimum oxygen concentration) is specified, the program predicts the average and minimum daily oxygen concentration at the outflow. The spreadsheet format allows several simulation runs to be performed simultaneously and displays the results in tabular form. Graphical output is also possible.

The results of two simulation runs are shown in Figures 4 and 5. Figure 4 shows the effects of

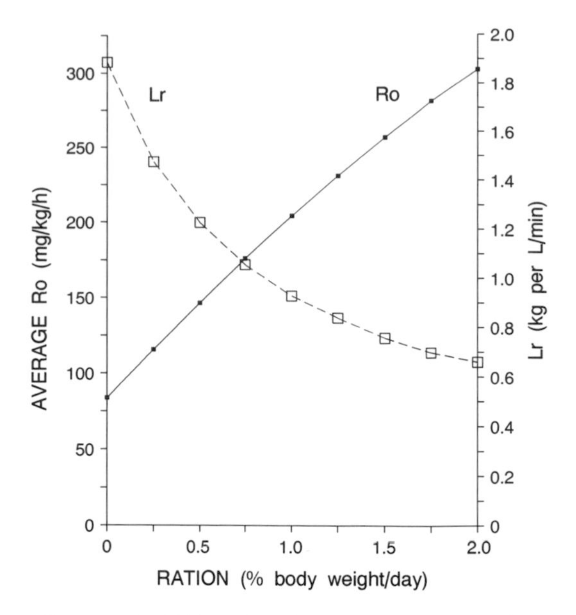

FIGURE 4.—Effect of ration on the average daily oxygen consumption rate (R_o) for juvenile salmonids and the load rate (L_r). In this simulation, temperature, fish weight, and the inflow oxygen concentration were 10°C, 10 g, and 95%, respectively, and the outflow dissolved oxygen was set at 6 mg/L.

increasing ration on the average daily oxygen consumption rate (R_O) and the maximum allowable load rate (L_r). In this simulation, the ration level was increased from 0 to 2%/d for 10-g fish at a water temperature of 10°C while the outflow oxygen concentration was set at 6 mg/L and the inflow level was assumed to be 95% of saturation. Figure 5 shows the effects of temperature on R_O and L_r for 1-, 10- and 30-g fish. In this trial, ration level has been set at the maximum level and temperature has been increased from 4 to 16°C. As in the first simulation, the inflow and outflow oxygen levels were set at 95% of saturation and 6 mg/L, respectively.

Gas Supersaturation Model

Excess gas tension can pose a serious threat to fish health and is routinely measured at fish culture facilities. Because the effects of supersaturation are more severe when the oxygen-to-nitrogen gas ratio is low (Jensen et al. 1986), estimating the contribution of the individual gas components is required.

Calculation of Gas Components

Characterizing the gas components in a hatchery water supply is carried out by measuring the

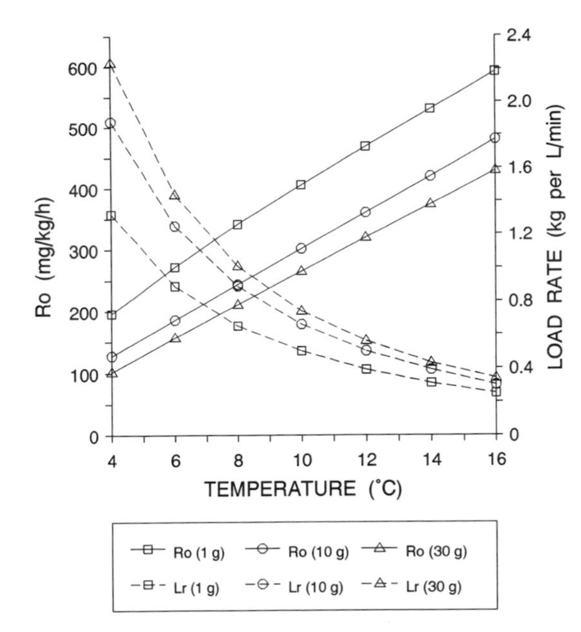

FIGURE 5.—Effect of temperature on the average daily oxygen consumption rate (R_O) for juvenile salmonids and load rate (L_r). Simulations were carried out for 1-g, 10-g, and 30-g fish; the temperature and inflow and outflow oxygen concentrations were 10°C, 95%, and 6 mg/L, respectively.

differential gas pressure (ΔP), oxygen concentration, temperature, and barometric pressure (BP) (McLean and Boreham 1980). These values are used to calculate the partial pressures and percentage saturation levels of oxygen and nitrogen and also the total gas pressure (TGP).

In the calculations that follow, nitrogen denotes argon plus nitrogen. Also, it was assumed that the carbon dioxide contribution to the total gas pressure was negligible.

Differential gas pressure, as commonly measured in hatcheries by means of a portable saturometer, is the total gas pressure exerted by all the gas components relative to the barometric pressure: ΔP = TGP − BP. Therefore, TGP in units of mm Hg is simply ΔP + BP. However, this parameter is traditionally expressed as a percentage of the barometric pressure (TGP%) and is calculated from

$$TGP\% = [(\Delta P + BP)/BP] \times 100. \qquad (12)$$

In this model, the oxygen and nitrogen components were expressed as a percentage of the saturation values. Percentage oxygen ($O_2\%$) was calculated by expressing the measured oxygen concentration as a percentage of the saturation value C_s, which was calculated from the temperature, barometric pressure, water vapour pressure, and salinity with a relationship derived by Hitchman 1978. The water vapor pressure was calculated from the temperature (N. Shrimpton, British Columbia Ministry of the Environment, unpublished program) whereas the other variables in the equation were measured quantities. Percentage nitrogen saturation ($N_2\%$) was found by first calculating the nitrogen partial pressure and then expressing it as a percent of the partial pressure of nitrogen in air-equilibrated water (Colt 1984).

Biological Response to Supersaturation

Jensen et al. (1986) modeled time to 50% mortality (ET50) of juvenile salmonids in response to gas supersaturation, using data compiled from the literature (Jensen et al. 1985). The data, primarily from studies conducted to determine lethal thresholds for water supersaturated as a result of hydroelectric dams, indicated that ET50 was related exponentially to excess TGP. This relation was further influenced by ancillary variables such as water depth, proportions of oxygen and nitrogen, and fish size. Attempts to model the modifying effects of temperature and barometric pressure were less successful due to limitations in the data. Nevertheless, reasonable predictions of expected 50% mortality for salmonids are now possible by using some of the models developed.

Two models (models 7 and 13) have been incorporated into a spreadsheet-formatted computer program. Model 7 (Jensen et al. 1986) predicts ET50 in response to TGP(%), water depth (m), and fish size (mm). This model is based on the larger of the two data sets modeled, but does not include information on the effects of oxygen or nitrogen. A smaller data set, with less predictive power, was used to generate several models that included oxygen and nitrogen as well as TGP as predictors of ET50. From these, model 13 (Jensen et al. 1986) was chosen for this paper. Model 13 predicts ET50 in response to TGP% and $O_2\%$. Model coefficients were developed by applying general multivariate dose–response model-fitting procedures (Schnute and Jensen 1986) (Table 5).

Criteria for Safe Exposure

Jensen et al. (1986) discussed the definition of a "safe" level of gas supersaturation and used 95% confidence levels to estimate minimum thresholds. However, a large uncertainty associated

TABLE 5.—Effect of gas supersaturation on juvenile salmonids. Multivariate dose-response model[a] parameter estimates for ET50 models 7 and 13 (Jensen et al. 1986). Model 7 factors are 1 = TGP (%), 2 = depth (m), and 3 = fish size (mm). Model 13 factors are 1 = TGP (%) and 4 = O_2 (%). ET50 is time (h) to 50% mortality; TGP is total gas pressure (% of saturation).

	Parameter	Model 7	Model 13
Auxiliary parameters			
	v (minimum ET50 h)	1.5	1.5
	V (maximum ET50 h)	1,182.0	1,182.0
Factor 1	u_1 (minimum TGP, %)	111.0	111.0
	U_1 (maximum TGP, %)	141.4	141.4
Factor 2	u_2 (minimum depth, m)	0.14	
	U_2 (maximum depth, m)	1.00	
Factor 3	u_3 (minimum size, mm)	45.0	
	U_3 (maximum size, mm)	209.0	
Factor 4	u_4 (minimum O_2, %)		40.5
	U_4 (maximum O_2, %)		307.9
Model parameters			
	α_1	−1.5	1.8
	α_2	1.0	
	α_3	−3.6	
	α_4		1.2
	γ	−0.041	−0.170
	p	0.29	0.14
	q_1	−1.01	−1.00
	q_2	0.24	
	q_3	−0.30	
	q_4		0.36

[a]The model is defined by the following equations:

(a) x_i-variable transformation, $E_i = (2x_i^{\alpha i} - U_i^{\alpha i} - u_i^{\alpha i})/(U_i^{\alpha i} - u_i^{\alpha i})$;

(b) y-variable transformation, $N = (2y^\gamma - V^\gamma - v^\gamma)/(V^\gamma - v^\gamma)$;

(c) dose–response equation, Model 7: $N = p + q_1E_1 + q_2E_2 + q_3E_3$;
Model 13: $N = p + q_1E_1 + q_4E_4$.

with ET50 response at lower levels of gas supersaturation resulted in very broad confidence limits. Hence, the decision regarding the ''safe'' level of gas supersaturation cannot be based solely on a mathematical expression derived from the existing data. The approach we have taken herein is to calculate the expected ET50. The ET50s from which the models were derived ranged from less than 1 d to 50 d. Hence, gas supersaturation conditions that cause 50% mortality in 10 d (i.e., ET50 = 10 d) or less would be considered acutely lethal, whereas an ET50 of 50 d would be caused by chronically lethal gas supersaturation. Fish culturists can then use the appropriate TGP model to decide whether or not a particular set of conditions of TGP, water depth, fish size, and oxygen concentration are putting their fish at risk.

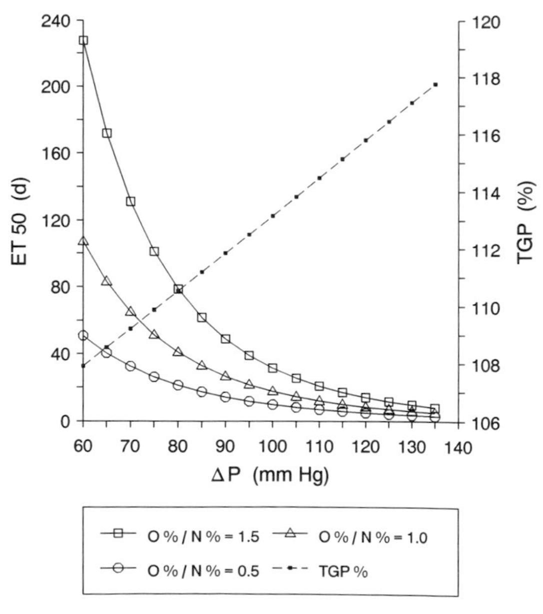

FIGURE 6.—Predicted times to 50% mortality (ET50) for juvenile salmonids as a function of differential gas pressure (ΔP). The response is predicted for O_2/N_2 ratios of 0.5, 1.0, and 1.5. Simulation was performed for a temperature of 15°C, a barometric pressure of 760 mm Hg, and a water depth of 0 m.

Program Inputs, Outputs, and Examples

As with other models, the predictive equations have been programmed with commercial software. The inputs required are differential gas pressure (mm Hg), temperature (°C), dissolved oxygen (mg/L), available water depth (m), and barometric pressure (mm Hg). The program predicts the ET50 and also calculates the TGP% and

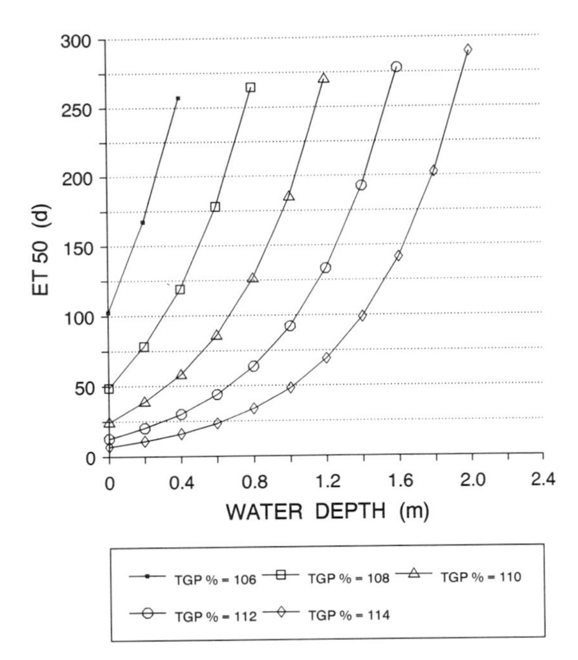

FIGURE 7.—Effect of water depth on the time to 50% mortality (ET50) of juvenile salmonids for total gas pressure (TGP) levels between 106 and 114%. These responses were calculated for a temperature of 10°C, a barometric pressure of 760 mm Hg, and an O_2/N_2 ratio of 1 (i.e., air-supersaturated water).

the percentage saturation levels of nitrogen and oxygen. Predicted ET50 values were generated as the differential gas pressure increased from 60 to 140 mm Hg (Figure 6). The ET50 values are shown for three distributions of oxygen and nitrogen, simulating groundwater ($O_2\%/N_2\%$ = 0.5), water below dams ($O_2\%/N_2\%$ = 1.0), and water supplemented with pure oxygen ($O_2\%/N_2\%$ = 1.5). The most severe case (lowest ET50 values) involved low oxygen and high nitrogen values; the least harmful case (highest ET50 values) involved high oxygen and low nitrogen values. These simulations were generated for a temperature of 15°C, a barometric pressure of 760 mm Hg, and a water depth of 0 m. Finally, the sparing effect of water depth on ET50 is shown in Figure 7. For this simulation, it was assumed that the water was supersaturated with air ($O_2\%/N_2\%$ = 1).

Acknowledgments

We gratefully acknowledge D. F. Alderdice for his invaluable contributions to this project. It was his interest and guidance in the experimental studies on embryonic and larval development that enabled the authors and F. P. J. Velsen to obtain the data necessary for the models reported here. We are indebted to J. R. Brett for his advice and helpful suggestions in the area of fish metabolism and energetics. Erick Groot is gratefully acknowledged for his preparation of the final figures. Finally, we thank Bruce Shepherd for his review of the manuscript.

References

Baird, D. C. 1963. Experimentation. Prentice-Hall, London.

Ballard, W. W. 1968. History of the hypoblast in *Salmo*. Journal of Experimental Zoology 168:257–272.

Brett, J. R. 1964. The respiratory metabolism and swimming performance of young sockeye salmon. Journal of the Fisheries Research Board of Canada. 21:1183–1226.

Brett, J. R. 1965. The relation of size to rate of oxygen consumption and sustained swimming speed of sockeye salmon (*Oncorhynchus nerka*). Journal of the Fisheries Research Board of Canada 22:1491–1501.

Brett, J. R. 1976. Feeding metabolic rate of young sockeye salmon, *Oncorhynchus nerka*, in relation to ration level and temperature. Canada Fisheries and Marine Service, Technical Report 675.

Brett, J. R., and N. R. Glass. 1973. Metabolic rates and critical swimming speeds of sockeye salmon (*Oncorhynchus nerka*) in relation to size and temperature. Journal of the Fisheries Research Board of Canada 30:379–387.

Colt, J. 1984. Computation of dissolved gas concentrations in water as functions of temperature, salinity and pressure. American Fisheries Society Special Publication 14.

Davis, J. C. 1975. Minimal dissolved oxygen requirements of aquatic life with emphasis on Canadian species: a review. Journal of the Fisheries Research Board of Canada 32:2295–2332.

Heming, T. A. 1982. Effects of temperature on utilization of yolk by chinook salmon (*Oncorhynchus tshawytscha*) eggs and alevins. Canadian Journal of Fisheries and Aquatic Sciences 39:184–190.

Hitchman, M. L. 1978. Measurement of dissolved oxygen. Wiley-Interscience, New York.

Jensen, J. O. T. 1988. A microcomputer program for predicting embryonic development in Pacific salmon and steelhead trout. Journal of the World Aquaculture Society 19:80–81.

Jensen, J. O. T., A. N. Hally, and J. Schnute. 1985. Literature data on salmonid response to gas supersaturation and ancillary factors. Canadian Data Report of Fisheries and Aquatic Sciences 501.

Jensen, J. O. T., J. Schnute, and D. F. Alderdice. 1986. Assessing juvenile salmonid response to gas supersaturation using a general multivariate dose–response model. Canadian Journal of Fisheries and Aquatic Sciences 43:1694–1709.

Kling, A., C. Cross, W. E. McLean, and C. J. West. 1983. A guide to the use of the Fish Culture Information System for Hatcheries (FISH) utility programs and Apple II$^+$ microcomputer. Canada Department of Fisheries and Oceans, Ottawa.

McLean, W. E. 1979. A rearing model for salmonids. Master's thesis. University of British Columbia, Vancouver.

McLean, W. E. 1988. Oxygen consumption rates and water flow requirements for salmon in the fish culture environment. Canada Department of Fisheries and Oceans, Campbell River, British Columbia.

McLean, W. E., and A. L. Boreham. 1980. The design and assessment of aeration towers. Canadian Department of Fisheries and Oceans, Vancouver.

McLean, W. E., and P. G. Lim. 1985. Some effects of water flow rate and alevin density in vertical incubation trays on early development of chinook salmon (*Oncorhynchus tshawytscha*). Canadian Technical Report of Fisheries and Aquatic Sciences 1357.

Murray, C. B., and J. D. McPhail. 1988. Effect of incubation temperature on the development of five species of Pacific salmon (*Oncorhynchus*) embryos and alevins. Canadian Journal of Zoology 66:266–273.

Rombough, P. J. 1985. Initial egg weight, time to maximum alevin wet weight, and optimal ponding times for chinook salmon (*Oncorhynchus tshawytscha*). Canadian Journal of Fisheries and Aquatic Sciences 42:287–291.

Rombough, P. J. 1986. Mathematical model for predicting the dissolved oxygen requirements of steelhead (*Salmo gairdneri*) embryos and alevins in hatchery incubators. Aquaculture 59:119–137.

Schnute, J. 1981. A versatile growth model with statistically stable parameters. Canadian Journal of Fisheries and Aquatic Sciences 38:1128–1140.

Schnute, J., and J. O. T. Jensen. 1986. A general multivariate dose–response model. Canadian Journal of Fisheries and Aquatic Sciences 43:1684–1693.

Stauffer, G. D. 1973. A growth model for salmonids reared in hatchery environments. Doctoral dissertation. University of Washington, Seattle.

Velsen, F. P. J. 1987. Temperature and incubation in Pacific salmon and rainbow trout: compilation of data on median hatching time, mortality, and embryonic staging. Canadian Data Report of Fisheries and Aquatic Sciences 626.

American Fisheries Society Symposium 10:529–538, 1991

The Effect of Moist Air Incubation Conditions and Temperature on Chinook Salmon Egg Survival

JOHN O. T. JENSEN AND ERICK P. GROOT

*Department of Fisheries and Oceans, Pacific Biological Station
Nanaimo, British Columbia V9R 5K6 Canada*

Abstract.—Eggs of chinook salmon *Oncorhynchus tshawytscha* were incubated in water or in moist air for 6-, 12-, or 24-h intervals between 0.5 h of flooding with water at temperatures of 10.2–20.2°C. Egg mortality was very rapid at high temperatures. Total mortality occurred before hatching at temperatures greater than 17.2°C, regardless of whether eggs were incubated in water or moist air. Significant ($P < 0.05$) egg losses occurred at temperatures greater than 14°C. Hatching rates appeared to be affected primarily by temperature. However, the 12- and 24-h moist air treatments showed a slight delay in hatching compared with eggs incubated at similar temperatures in 6-h moist air treatments or in water. Alevin losses occurred at the higher temperatures but were not associated with exposure to moist air. Response surface analysis of mortality indicated that, below the lethal threshold of 14°C, incubation in moist air with 4–8% water exposure per day (about 6–10-h intervals between flooding) should produce greater egg survival than if eggs were incubated entirely in water.

Salmon eggs can survive out of water for periods of time. This tolerance has facilitated transport of developing eggs to almost anywhere in the world. The basic requirements for transport arc that the eggs remain cool and moist. Salmonid eggs generally survive such conditions without problem (Affleck 1953). Hence, little research has been conducted to determine conditions affecting short-term egg exposure to a dewatered incubation environment during transport or in hatchery incubators. Instead, research has focused on survival of eggs in dewatered redds under field conditions. Such research often has arisen from concerns about streamflow fluctuation and reduction, and about increased sediment loads and other physical changes occurring in rivers and streams as a result of logging, mining, and hydroelectric flow management (Reiser and White 1981a, 1981b, 1983; Becker et al. 1982, 1986; Neitzel and Becker 1985).

Interest in egg survival in a moist air environment has renewed with the recent increase in commercial salmon farming. The potential benefits of incubating salmon eggs in moist air include (1) reduced water requirements, hence reduced cost; (2) cheaper heating costs to accelerate egg development; and (3) more effective therapeutic treatment of disease, particularly when a drug is readily washed off the egg.

These benefits are already being realized: several government hatcheries and commercial fish farms in British Columbia now incubate eggs in moist air. The moist air incubators being used vary in design and method of keeping eggs moist. One method, termed the flood-type, has eggs in a series of stacked trays in a boxed enclosure that is flooded with water periodically (M. Foy, Department of Fisheries and Oceans, personal communication). Another method, mist-type, uses a fogger nozzle to create a fine mist that is timer-controlled to spray at regular intervals to ensure that eggs remain moist (M. Wolfe, Department of Fisheries and Oceans, personal communication). The latter method has been used to incubate eggs even to the point of hatching in air; the hatched larvae then fall through the suspending screen into water. These methods generally result in normal egg survival (i.e., >90% survival).

In this paper we present results from incubating eggs of chinook salmon *Oncorhynchus tshawytscha* in moist air (i.e., dewatered) for various periods between flooding and over a range of temperatures. This information will allow fish culturists to decide between the potential advantages of moist air incubation and the hazards of increased temperature and prolonged exposure to dewatering.

Methods

Egg source and fertilization procedure.—Eggs and milt were obtained from five female and five male chinook salmon at the Big Qualicum hatchery, and were transported to the Pacific Biological Station, Nanaimo, British Columbia. These gametes were pooled, mixed, and activated (i.e., water uptake was started) in 10°C water. Immedi-

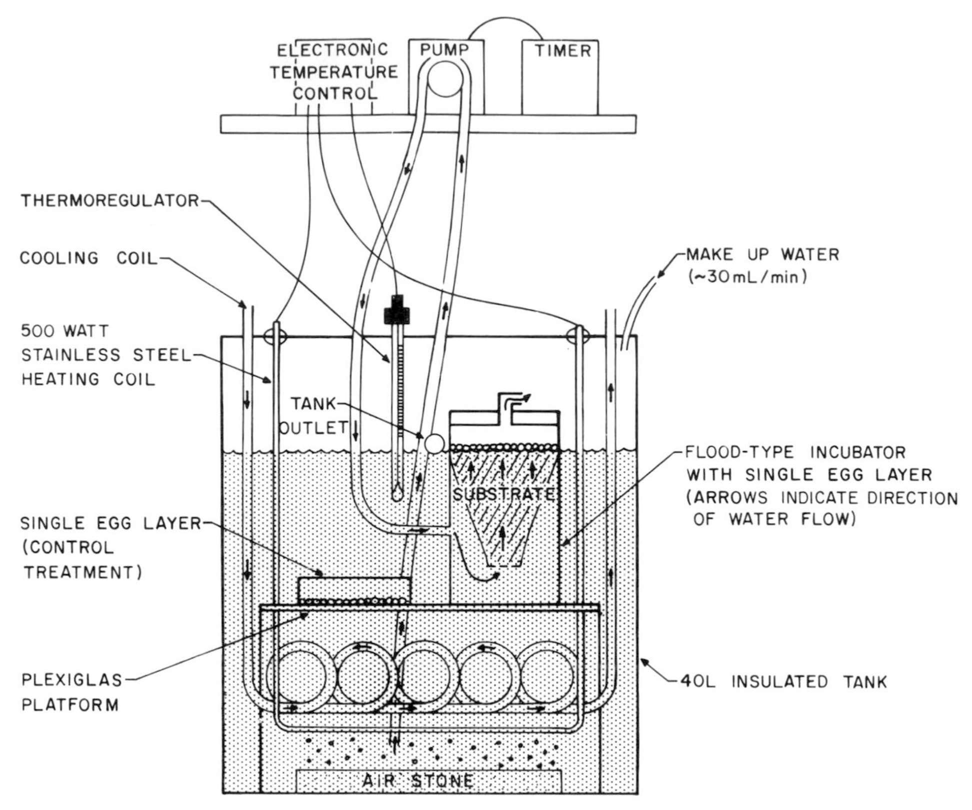

FIGURE 1.—Schematic diagram of the test tanks with an experimental incubator and egg basket.

ately after water-activation, small groups of eggs (about 30 eggs per group, two groups per treatment) were placed into incubation baskets in a single layer. These baskets were placed either into experimental flood-type moist air incubators (Figure 1) or directly into the test water. Total delay from activation to completion of egg allocation into test treatments was 9 min. The eggs were left in water for 1.5 h to allow complete water uptake, after which exposure to moist air was begun.

Experimental design.—The experimental design consisted of 20 treatments, each replicated once (i.e., two groups of eggs per treatment). Water temperatures of 10, 12, 14, 16, 18, and 20°C were chosen to determine the upper lethal temperature and to allow modeling of mortality rates at higher temperatures. Three periods of exposure to moist air were tested by using programmed timers that activated pumps at 6-, 12-, and 24-h intervals for a 0.5-h flooding period. Incubation baskets placed directly in test water served as controls. Hence, the proportional times of exposure to water during incubation were 100% (control), 8.3% (6-h interval between flooding), 4.2% (12-h interval between flooding), and 2.1% (24-h interval between flooding). Although a full factorial design was not possible because of space and equipment limitations, 20 treatment combinations were tested.

Eggs in the controls and 6-h-interval moist air treatments were incubated together in the same test tank at each of six test temperatures, whereas eggs in the 12- and 24-h treatments were incubated together at four of the test temperatures. Hence, treatment temperatures varied as a result of differences in water temperature in addition to the influence of air temperature and the number of flooding cycles. Each tank was covered by 2-cm-thick neoprene insulation, which kept air temperature near that of water temperature.

Responses monitored and data analyses.—The test tanks had a volume of 40 L and water temperatures were maintained (±0.1°C) by electronic temperature controllers. Temperature recordings were made for each moist air treatment by placing temperature probes directly on the egg incubation screens. In addition, each treatment was monitored at 5-min intervals for up to 24 h to

TABLE 1.—Summary of measured temperature conditions in test tanks and in incubators.

Test conditions				Temperature (°C)		
					Incubator	
Test number	Nominal temperature (°C)	Interval between flooding	Water exposure (%)	Tank, mean (SD)	Mean (SD)	Range
1	10	Control	100.0	10.2 (0.006)	10.2 (0.006)	10.1–10.3
2	10	6 h	8.3	10.2 (0.006)	11.7 (0.5)	10.5–12.2
3	10	12 h	4.2	10.1 (0.005)	11.7 (0.5)	10.3–12.4
4	10	24 h	2.1	10.1 (0.005)	11.9 (0.5)	10.3–12.8
5	12	Control	100.0	11.7 (0.004)	11.7 (0.004)	11.6–11.8
6	12	6 h	8.3	11.7 (0.004)	12.5 (0.3)	11.8–12.8
7	12	12 h	4.2	11.6 (0.003)	13.2 (0.4)	11.8–13.8
8	12	24 h	2.1	11.6 (0.003)	13.3 (0.5)	11.6–14.0
9	14	Control	100.0	14.0 (0.009)	14.0 (0.009)	13.9–14.1
10	14	6 h	8.3	14.0 (0.009)	14.3 (0.3)	13.6–14.5
11	16	Control	100.0	16.4 (0.06)	16.4 (0.06)	16.3–16.5
12	16	6 h	8.3	16.4 (0.06)	16.3 (0.1)	16.1–16.5
13	16	12 h	4.2	16.3 (0.006)	16.9 (0.3)	16.1–17.5
14	16	24 h	2.1	16.3 (0.006)	17.2 (0.3)	16.1–17.5
15	18	Control	100.0	18.0 (0.004)	18.0 (0.004)	17.9–18.1
16	18	6 h	8.3	18.0 (0.004)	17.4 (0.2)	17.1–17.9
17	20	Control	100.0	20.2 (0.006)	20.2 (0.006)	20.1–20.3
18	20	6 h	8.3	20.2 (0.006)	19.4 (0.2)	19.2–20.0
19	20	12 h	4.2	20.3 (0.006)	20.0 (0.1)	19.8–20.2
20	20	24 h	2.1	20.3 (0.006)	20.1 (0.1)	19.8–20.2

obtain a typical temperature pattern for each treatment.

Water was circulated in the test tanks by aeration turbulence at the back of each tank. Hence, both temperature and dissolved oxygen (DO) levels were uniform in each tank. One series of DO measurements was made at each test temperature to ensure that oxygen concentration would not influence egg survival. Levels of DO ranged from 95 to 101% of saturation, higher saturation levels occurring with increasing temperature. A small amount of make-up water (about 30 mL/min or approximately one full exchange per day) was metered into each tank to prevent buildup of metabolites.

The eggs in moist air treatments were flooded for 0.5 h at appropriate times with water withdrawn by peristaltic pumps from the test tank in which each incubator was held. The water was forced up through a cone filled with porcelain substrate to ensure even flow past the single layer of eggs (Figure 1). The experimental setup required that flows for the 6-h moist air treatments be split between two incubators. Therefore, the flow rates in those cases (treatments 2, 6, 10, 12, 16, and 18) were considerably lower than for 12- and 24-h intervals (treatments 3, 4, 7, 8, 13, 14, 19, and 20). Mean flow rates ranged from 151.4 (for 6-h dewatering intervals) to 384.8 mL/min (for 12- and 24-h dewatering intervals), which translates to apparent bulk velocities (cross-sectional area = 102 cm^2) of 257.4–654.2 cm/h.

Each treatment tank was monitored daily for temperature and pump operation. Dead eggs were removed and counted during incubation. Eggs incubated in moist air were transferred to water just before hatching. Hatching rates were determined. Alevin mortality then was monitored until alevins had reached the swim-up or button-up stage (i.e., yolk sac no longer visible) or until total mortality occurred.

Analysis of variance of combined or total egg and alevin mortality was conducted to determine the incubation conditions that caused significant increases in mortality. Finally, response surface analysis (Schnute and McKinnell 1984) was employed to model egg mortality and combined or total mortality; the latter response was analysed to test for delayed effects of egg exposure to the dewatering treatments.

Results

Several differences in temperature regimes during the experiments were observed and are summarized in Table 1. The control treatments experienced very stable temperatures that varied by

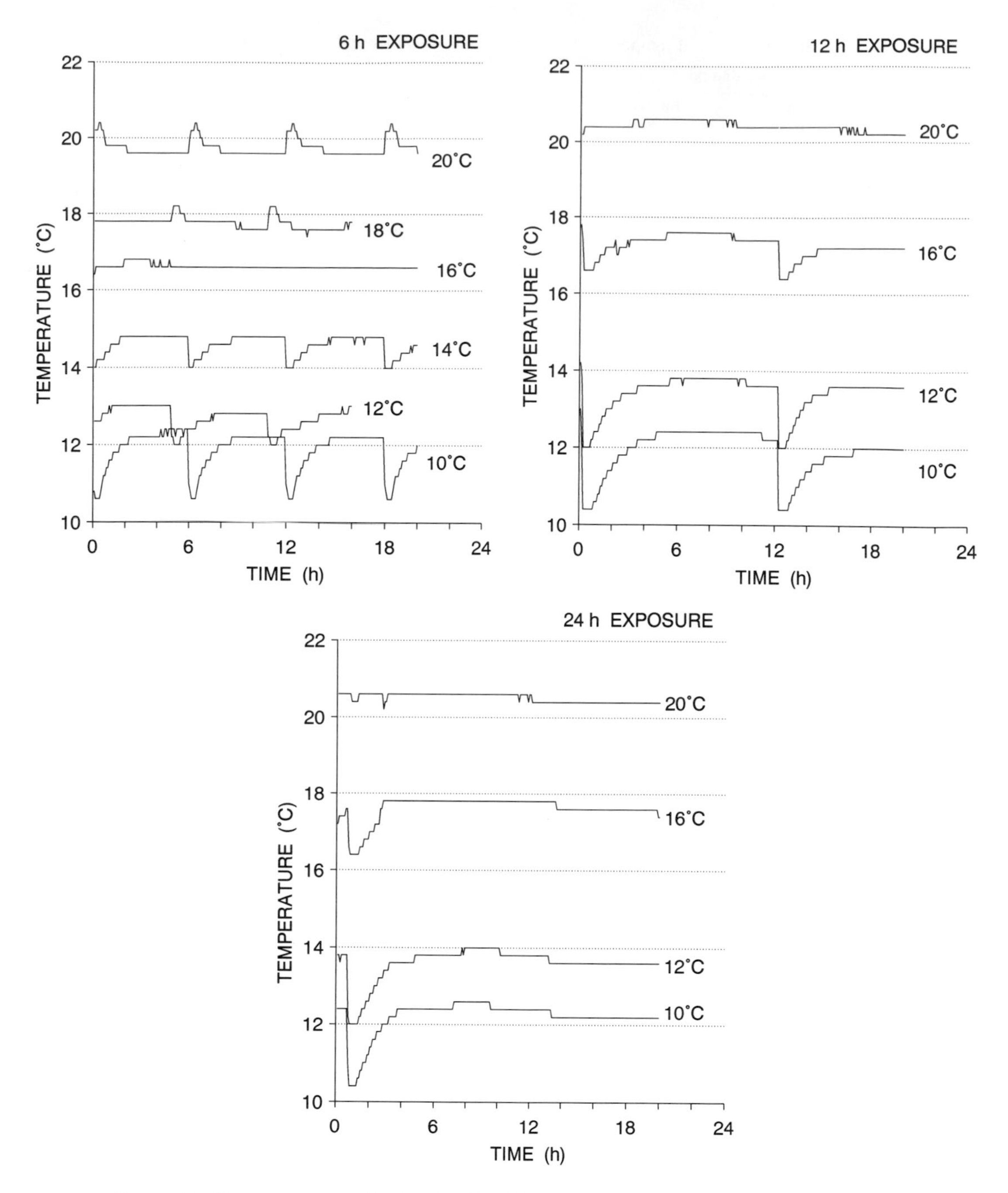

FIGURE 2.—Temperature profiles for the 6-h, 12-h, and 24-h moist air incubation treatments.

only 0.1°C around the mean. Temperatures during the moist air treatments fluctuated. The magnitude of this fluctuation around the mean depended on the intervals between flooding and the difference between the water and air temperature. For example, moist air treatments with 10 and 12°C water (treatments 2, 3, 4, 6, 7, and 8) averaged 1.5 to 1.7°C higher than the control treatments (1 and 5). In contrast, at 18 and 20°C, the 6-h moist air treatments (16 and 18) were actually lower in temperature than their respective control treatments (15 and 17). To illustrate these differences

TABLE 2.—Summary of egg mortality, time to 50% hatch, alevin mortality, combined mortality for each test condition (two replicates plus mean values until no eggs survived to hatch), and time to end of experiment for the various conditions. Test means are marked with asterisks.

Test number	Mean temperature (°C)	Water exposure (%)	Initial number of eggs	Egg mortality (%)	Time to 50% hatch (d)	Alevin mortality (%)	Combined mortality (%)	Time to end of experiment (d)
1	10.2	100.0	32	25.0	51.2	0.0	25.0	
			34	17.6	51.2	0.0	17.6	
				21.3*	51.2*	0.0*	21.3*	86.7
2	11.7	8.3	31	6.5	48.2	0.0	6.5	
			29	6.9	47.3	0.0	6.9	
				6.7*	47.8*	0.0*	6.7*	86.7
3	11.7	4.2	30	20.0	46.1	0.0	20.0	
			35	22.9	45.9	0.0	22.9	
				21.4*	46.0*	0.0*	21.4*	87.8
4	11.9	2.1	27	29.6	45.9	0.0	29.6	
			33	12.1	45.3	0.0	12.1	
				20.9*	45.6*	0.0*	20.9*	87.8
5	11.7	100.0	29	24.1	43.6	0.0	24.1	
			27	33.3	42.7	0.0	33.3	
				28.7*	43.1*	0.0*	28.7*	70.5
6	12.5	8.3	30	10.0	40.8	3.7	13.3	
			29	10.3	40.0	0.0	10.3	
				10.2*	40.4*	1.9*	11.8*	70.5
7	13.2	4.2	34	11.8	40.7	0.0	11.8	
			31	12.9	40.1	3.3	16.1	
				12.3*	40.4*	1.7*	13.9*	65.8
8	13.3	2.1	26	11.5	40.2	8.7	19.2	
			30	33.3	40.0	25.0	53.3	
				22.4*	40.1*	16.8*	36.3*	65.8
9	14.0	100.0	32	21.9	35.8	4.0	25.0	
			29	20.7	35.6	3.6	24.1	
				21.3*	35.7*	3.8*	24.6*	62.8
10	14.3	8.3	32	28.1	35.0	21.7	43.8	
			28	28.6	33.8	7.7	35.7	
				28.3*	34.4*	14.7*	39.7*	62.8
11	16.4	100.0	33	57.6	31.5	100.0	100.0	
			31	71.0	32.8	100.0	100.0	
				64.3*	32.1*	100.0*	100.0*	62.8
12	16.3	8.3	31	74.2	32.3	100.0	100.0	
			32	78.1	32.3	100.0	100.0	
				76.2*	32.3*	100.0*	100.0*	62.8
13	16.9	4.2	29	75.9	31.7	28.6	82.8	
			30	73.3	31.4	100.0	100.0	
				74.6*	31.5*	66.7*	91.4*	63.0
14	17.2	2.1	35	97.1	33.4	100.0	100.0	
			30	90.0	32.3	100.0	100.0	
				93.6*	32.9*	100.0*	100.0*	44.8
15	18.0	100.0	33	100.0			100.0	32.0
			31	100.0			100.0	32.0
16	17.7	8.3	24	100.0			100.0	32.0
			15	100.0			100.0	32.0
17	20.2	100.0	30	100.0			100.0	17.8
			33	100.0			100.0	17.8
18	19.4	8.3	21	100.0			100.0	17.8
			26	100.0			100.0	17.8
19	20.0	4.2	36	100.0			100.0	10.8
			26	100.0			100.0	10.8
20	20.1	2.1	35	100.0			100.0	10.8
			36	100.0			100.0	10.8

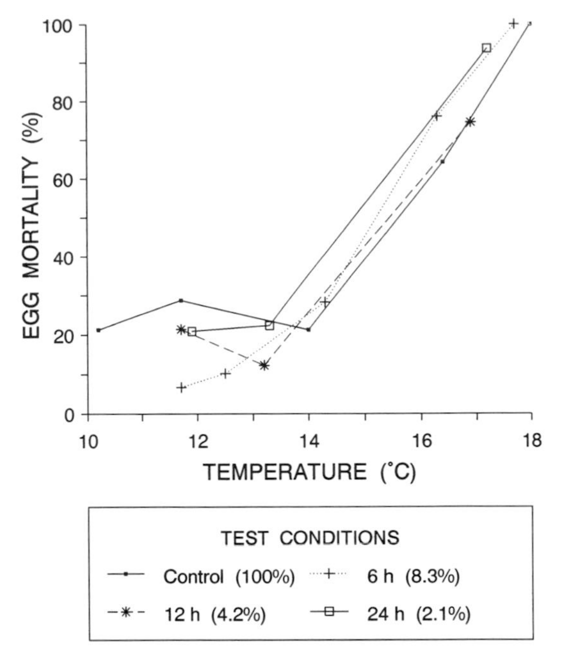

FIGURE 3.—Mean egg mortality (%) for eggs incubated in water (control, 100% of the time spent in water) or in moist air (6-, 12-, or 24-h intervals between flooding) at various temperatures.

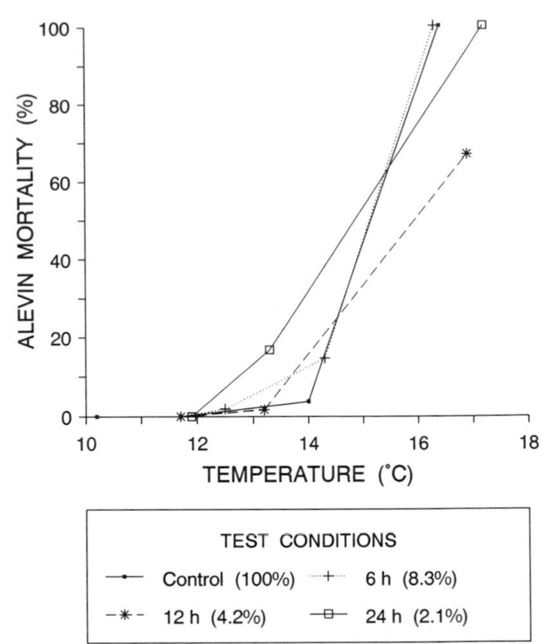

FIGURE 5.—Mean alevin mortality (%) occurring after eggs were incubated in water (controls) or in moist air (6-, 12-, or 24-h intervals between flooding) at various temperatures.

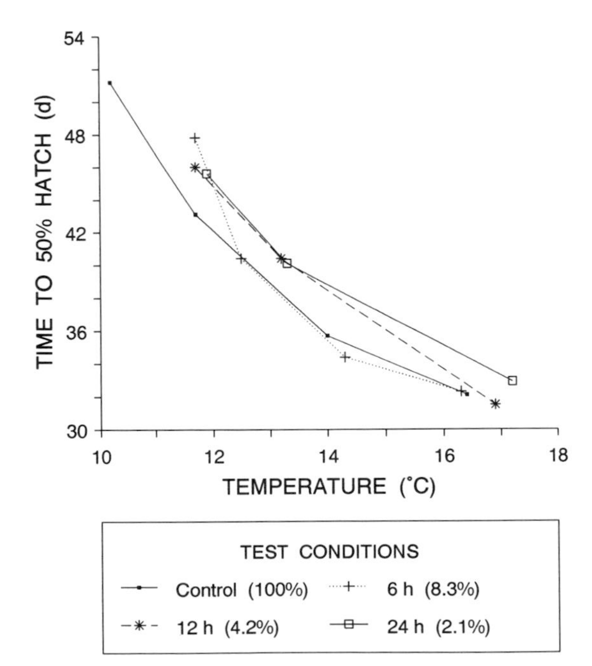

FIGURE 4.—Time (days) to 50% hatch for eggs incubated in water (control) or in moist air (6-, 12-, or 24-h intervals between flooding) at various temperatures.

in temperature exposure, temperature profiles for the various dewatered treatments are presented in Figure 2.

Egg mortality occurred quickly at high temperatures. All eggs died before hatching at temperatures greater than 17.2°C, regardless of whether they were incubated in water or moist air (Table 2). At temperatures of 14.3°C and less, egg mortality ranged from 6.5 to 33.3%; mortality increased markedly at temperatures of 16.3°C and greater (Figure 3). Analysis of variance of arcsine-transformed data indicated that significant ($P < 0.05$) mortality started at temperatures greater the 14.3°C.

Hatching rates (i.e., time to 50% hatch) were determined for treatments 1–14 (Table 2). Hatching trends were similar for eggs incubated in water and in moist air, but eggs dewatered 12 and 24 h hatched slightly later than control eggs and eggs dewatered 6 h (Figure 4).

Alevin mortality continued after hatch at the higher temperatures (Table 2; Figure 5). At 13.3°C mortality of alevins from the 24-h moist air treatment (16.8%) was substantially greater than it was for alevins from other treatments (0.0–1.9%). At higher temperatures of 14.0 and 14.3°C, control

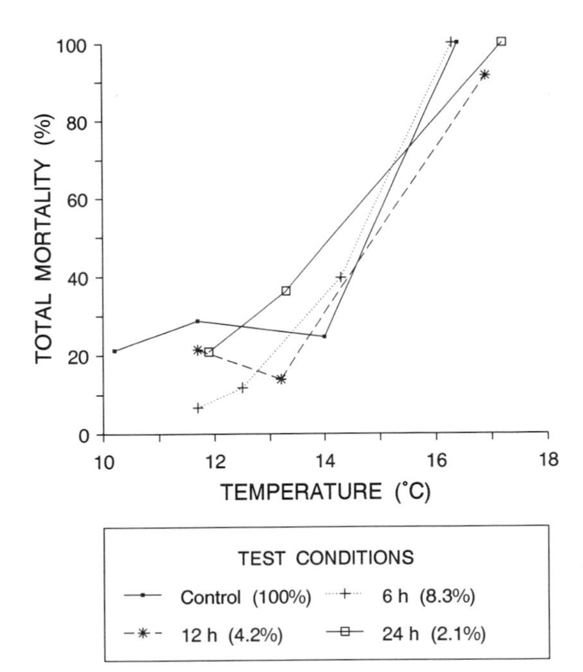

FIGURE 6.—Combined egg and alevin mortality (%) to the point of alevin swim-up for eggs that had been incubated in water (controls) or in moist air (6-, 12-, or 24-h intervals between flooding) at various temperatures.

alevins and those from the 6-h moist air treatments showed mortality increases of 3.8 and 14.7%, respectively.

Combined or total mortality trends, recorded when alevins had reached swim-up (Table 2; Figure 6), were similar to those for egg mortality. Two possible exceptions were observed in the 24-h moist air treatment at 13.3°C (36.3% mortality) and in the 6-h treatment at 14.3°C (39.7% mortality). However, mortality was not statistically significant ($P > 0.05$) at or below 14.3°C, regardless of whether eggs had been incubated in water or moist air.

Egg mortality and combined or total egg and alevin mortality were modeled in response to temperature and percent water exposure. The resultant model coefficients are listed in Table 3. Contour plots (Figures 7, 8) illustrate how incubation temperature was the predominant cause of mortality. The contours are vertical and increase in magnitude with increasing temperature. The influence of exposure to moist air on egg mortality is evident by the sharp rightward shift in contours (i.e., less mortality at the same temperature) occurring at about 4–8% exposure to water, followed by a similar sharp leftward change of all contours as water exposure approaches 0% (Figure 7). The combined egg and alevin mortality contour plot (Figure 8) illustrates the total or final mortality resulting from moist air egg incubation

TABLE 3.—Parameter estimates for response surface models[a] generated for egg and combined mortality (data listed in Table 1); factor x_1 = temperature (°C) and x_2 = water exposure (%).

Parameter	Egg mortality (model 1)	Combined mortality (model 2)
Auxiliary parameters		
v (minimum % mortality)	6.7	6.7
V (maximum % mortality)	100.0	100.0
u_1 (minimum temperature, °C)	10.2	10.2
U_1 (maximum temperature, °C)	20.2	20.2
u_2 (minimum % water)	2.1	2.1
U_2 (maximum % water)	100.0	100.0
Model parameters		
α_1	7.079155	5.182744
α_2	−5.304747	−0.2228003
τ	1.138138	0.8990445
p	0.5821490	1.031628
q_1	0.9051015	0.9052239
r_{11}	−1.156074	−1.185902
r_{12}	−2.7773662E−03	−3.1452909E−02
q_2	−4.0508866E−02	3.9937634E−04
r_{22}	0.5751039	0.1813477

[a] The model is defined by the following equations (Schnute and McKinnell 1984):

(a) x_i-variable transformation, $E_i = (2x_i^{\alpha i} - U_i^{\alpha i} - u_i^{\alpha i})/U_i^{\alpha i} - u_i^{\alpha i})$;

(b) y-variable transformation, $N = (2y^\tau - V^\tau - v^\tau)/(V^\tau - v^\tau)$;

(c) response surface equation, $N = p + q_1E_1 + r_{11}E_1^2 + r_{12}E_1E_2 + q_2E_2 + r_{22}E_2^2$.

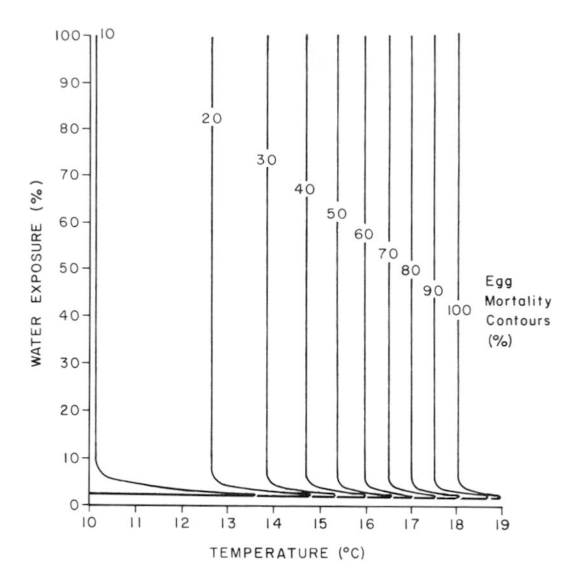

FIGURE 7.—Two-factor contour plot of egg mortality (%) at various temperatures and in three moist air treatments.

and subsequent exposure to the various tank temperatures. The contours lean to the right to a point of about 8% water exposure (i.e., mortality is least at a given temperature when water exposure was about 8%) and then make a sharp turn to the left (i.e., mortality increases at a given temperature) as water exposure approaches zero.

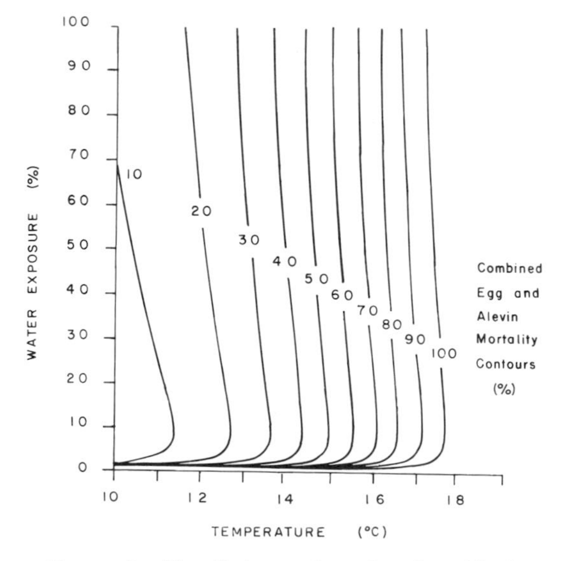

FIGURE 8.—Two-factor contour plot of combined egg and alevin mortality (%) at various temperatures and in three moist air treatments.

Discussion

Chinook salmon eggs incubated at temperatures below 14°C, either in water or in moist air with flooding intervals of up to 24 h, showed no statistically significant ($P > 0.05$) difference in mortality. This generally agrees with the findings of Reiser and White (1981a, 1981b), who observed 81–91% egg survival to hatch when chinook salmon eggs were incubated under moist cotton cloth (kept moist by a wick siphon at a flow rate of 315 mL/h). Water temperature ranged from 8.5 to 17.0°C with a mean temperature of 11.2°C. Neitzel and Becker (1985) and Becker et al. (1986) also found that eggs of chinook salmon and rainbow trout *Oncorhynchus mykiss* held at 10°C exhibited high survival as long as humidity was maintained near 100%. Significant mortality of the eggs occurred after a single 8-h exposure to 75% (chinook salmon) or 90% (rainbow trout) humidity.

Our study indicates that egg mortality is primarily caused by exposure to temperatures greater than 14°C (Figure 3). The effect of egg incubation in moist air is also apparent but less pronounced in part because the moist air maintained the eggs in 100% humidity. Therefore, the periods of exposure to air proved to be beneficial rather than causing egg mortality. The contour plot of egg mortality (Figure 7) illustrates the combined effects of these two factors. The sharp rightward shift in mortality near 8.3% water exposure (6-h flooding intervals) indicates that eggs have a greater tolerance of higher temperature if they are incubated in moist air with frequent flooding.

Hatching rates of eggs were primarily affected by incubation temperature and not by exposure to moist air (Figure 4). However, eggs incubated in 12 and 24 h moist air treatments generally took longer to hatch than eggs incubated at the same temperature but in water or in 6-h moist air treatments. This hatching delay may be a result of longer exposure to air, which provides greater access to oxygen. Rombough (1988) stated that low oxygen levels initiate release of the hatching enzyme. Therefore, eggs in air may experience less of a stimulus to commence hatching.

In general, alevin mortality showed no association with prior egg exposure to moist air. Alevin mortality occurred over a narrower temperature range than did egg mortality (Table 1; Figure 5). No alevins died at temperatures below 12.5°C and only a small percentage of alevins died below 14°C—except mortality was 16.8% among alevins

TABLE 4.—Total mortality estimates expected when chinook salmon eggs are incubated in various moist air conditions (% water exposure) and temperatures (°C). Estimates were calculated with model 2 (Table 3).

Water exposure (%)	Temperature (°C)	Mortality (%)
100	10	12.0
10	10	3.7
1	10	21.0
100	12	23.4
10	12	14.4
1	12	33.2
100	14	45.4
10	14	36.0
1	14	56.5
100	16	79.5
10	16	69.9
1	16	92.4

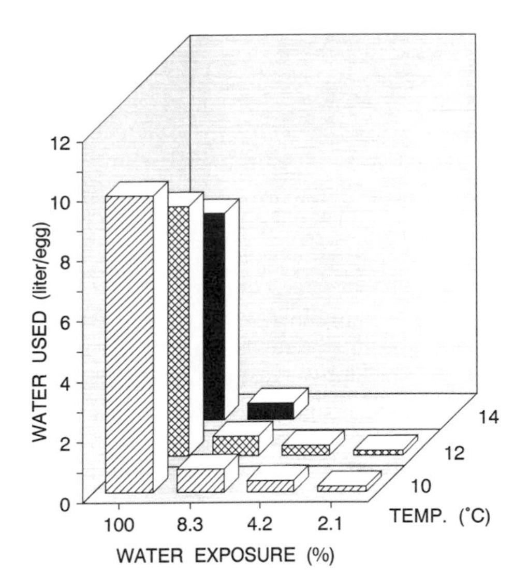

FIGURE 9.—Comparative bar graph illustrating the amount of water used (liters of water used per egg to 50% hatch, based on a standard 12-L/min flow per 90,000 eggs) at three temperatures for eggs incubated in water (control or 100%) or in three moist air treatments (8.3%, 4.2%, and 2.1% water exposure).

from the 24-h (2.1% water exposure) moist air treatment at 13.3°C. Perhaps this was a delayed effect of longer exposure of eggs to moist air at temperatures near the lethal threshold of 14°C. However, only one moist air treatment (test 10, a 6-h treatment) was conducted at the nominal temperature of 14°C, so we are unable to draw further conclusions here.

The combined egg and alevin mortality to the swim-up stage indicates the feasibility of moist air incubation for the aquaculture industry. It takes into account egg losses, which are a direct result of the incubation environment, plus losses that occur after hatching, which may be associated with the conditions the eggs were exposed to. As discussed above, it appears from this experiment that alevin mortality is not associated with prior incubation conditions. Therefore, the combined total mortality picture (Figure 6) looks much the same as that of egg mortality. Neither temperature nor exposure to moist air affected egg survival below 14°C, with the exception of an increase in mortality at 13.3°C and 2.1% water exposure. However, at 14.3°C and 8.3% water exposure, combined mortality increased to 39.7%, and it reached 91–100% for all incubation temperatures greater than 14.3°C, regardless of water exposure. These results indicate that constant incubation temperatures greater than 14°C should be avoided for chinook salmon eggs. In addition, if moist air techniques with prolonged intervals between flooding (e.g., 24 h) are to be used, then temperatures should be maintained below 13°C.

Response surface analysis of combined or total mortality (Figure 8) provides further insight into the interaction between the two incubation factors of temperature and exposure to water. Clearly, the main factor influencing egg and alevin survival is temperature. It appears from the mortality contours that moist air incubation may result in greater egg-to-fry survival if chinook salmon eggs are incubated in moist air with intervals between flooding of 6–10 h. A table of mortality estimates (Table 4) was developed with model-2 parameters in Table 3. The greater survival of eggs in moist air (with 10% water exposure) is evident over the temperature range of 10–12°C. However, at 14–16°C, all treatments show increases in mortality. We caution that these estimates are presented to illustrate trends. There is considerable uncertainty as to the magnitude of mortality; recall that no statistically significant ($P > 0.05$) differences were noted in treatments below 14.3°C.

A significant saving of water can be achieved when eggs are incubated by flood-type moist air incubation. Figure 9 compares calculated water usage (i.e., liters of water used per egg to 50% hatch) for eggs incubated in water (i.e., 100% water exposure) and in the three moist air treatments of 6-, 12-, and 24-h intervals between flooding (i.e., 8.3, 4.2, and 2.1% water exposure, respectively) at temperatures of 10, 12, and 14°C.

The effect of a temperature increase is to accelerate development to hatch and therefore to reduce the water requirement. This effect is most obvious for eggs incubated in water, for which water usage would decline from 9.83 L/egg at 10°C to 6.85 L/egg at 14°C, a reduction of 30%. In comparison, moist air treatments with 24-h intervals between flooding reduced the water usage by 97.9%. A further reduction in water requirement is also realized if the flooding water is recycled. However, the amount of recycling will be limited by ammonia excretion. Further tests are needed to determine safe limits of water recycling when coupled with moist air incubation.

Responses other than mortality have not been dealt with in this paper, nor have the underlying physiological processes likely affected by the incubation conditions reported herein. Other methods of moist air incubation, such as the mist-type, also should be tested for their effectiveness. This type of incubation requires more frequent intervals of water exposure but for shorter duration. Perhaps some of the temperature fluctuations observed in this experiment can be avoided by using this technique.

The objective of our study was to obtain information on the lethal limits of the combined effects of exposure to elevated temperature and various periods in moist air for chinook salmon eggs. This objective has been achieved and a first predictive model (Figure 8; Table 4) is now available. Hence, fish culturists can now estimate incubation mortality, using various combinations of temperature and water exposure, which then can be balanced against the advantages of embryonic development when eggs are incubated at higher temperatures and in moist air. These predictions, based solely on the results of this single experiment, should be verified with more tests, particularly with more test temperatures in the range of 12–16°C and with a complete range of moist air conditions at all test temperatures. Nevertheless, moist air incubation appears to have high potential for use on a commercial scale as long as temperatures and durations of dewatering are kept below lethal levels.

Acknowledgments

We are grateful to Matt Foy, Rheal Finnigan, and Mel Sheng (Department of Fisheries, Vancouver, British Columbia) and Mike Wolfe (Department of Fisheries, Robertson Creek hatchery, Port Alberni, British Columbia) for their information about moist air incubation methods.

References

Affleck, R. J. 1953. The stability of the vitelline membrane and the requirements of developing trout ova. Australian Journal of Marine and Freshwater Research 4:82–95.

Becker, C. D., D. A. Neitzel, and D. W. Carlile. 1986. Survival data for dewatered rainbow trout (*Salmo gairdneri* Rich.) eggs and alevins. Journal of Applied Ichthyology 3:102–110.

Becker, C. D., D. A. Neitzel, and D. H. Fickeisen. 1982. Effects of dewatering on chinook salmon redds: tolerance of four developmental phases to daily dewaterings. Transactions of the American Fisheries Society 111:624–637.

Neitzel, D. A., and C. D. Becker. 1985. Tolerance of eggs, embryos, and alevins of chinook salmon to temperature changes and reduced humidity in dewatered redds. Transactions of the American Fisheries Society 114:267–273.

Reiser, D. W., and R. G. White. 1981a. Incubation of steelhead trout and spring chinook salmon eggs in a moist environment. Progressive Fish-Culturist 43: 131–134.

Reiser, D. W., and R. G. White. 1981b. Influence of streamflow reductions on salmonid embryo development and fry quality. Idaho Water and Energy Resources Research Institute, Research Technical Completion Report, Moscow.

Reiser, D. W., and R. G. White. 1983. Effects of complete redd dewatering on salmonid egg-hatching success and development of juveniles. Transactions of the American Fisheries Society 112:532–540.

Rombough, P. J. 1988. Respiratory gas exchange, aerobic metabolism, and effects of hypoxia during early life. Pages 59–161 *in* W. S. Hoar, D. J. Randall, and J. R. Brett, editors. Fish physiology, volume 11, part A. Academic Press, New York.

Schnute, J., and S. McKinnell. 1984. A biologically meaningful approach to response surface analysis. Canadian Journal of Fisheries and Aquatic Sciences 41:936–953.

American Fisheries Society Symposium 10:539–547, 1991

Pen Rearing of Juvenile Fall Chinook Salmon in the Columbia River: Alternative Rearing Scenarios

JERRY F. NOVOTNY[1]

U.S. Fish and Wildlife Service, Columbia River Field Station
Cook, Washington 98605, USA

THOMAS L. MACY

U.S. Fish and Wildlife Service, Vancouver Fisheries Assistance Office
Vancouver, Washington 98665, USA

Abstract.—The upriver bright strain of fall chinook salmon *Oncorhynchus tshawytscha* was successfully reared in net pens at three locations along the Columbia River during March, April, and May in 1985, 1986, and 1987. Juvenile fish (0.6–1.5 g) were transferred from the hatchery of origin and reared in net pens at rearing densities of 0.43–4.20 kg/m^3; they were released after 4–9 weeks at an average weight of about 4.5 g. Mortality was low among fish reared in the pens, and physiological development and growth were faster than in control groups reared in the hatchery. The costs of rearing fish at most of the densities tested were lower in the net pens than in concrete hatchery raceways. Off-station rearing facilities, such as net pens, offer diversity to fishery managers for stocking juvenile salmon and may prove to be an efficient method of increasing hatchery production and the return of adults to areas adversely affected by environmental alterations.

Rearing fish in cages or net pens is a relatively new technology in terms of worldwide application. However, in parts of southeast Asia, where the concept is thought to have originated, these methods have been used since the beginning of the 20th century (Beveridge 1984) and probably much longer (Pantulu 1976). Many types of cages and enclosures have been fabricated to culture commercially important tropical species; net pens are now used to rear various species of fish in temperate waters as well. The variety and design of rearing enclosures are as diverse as the geographical locations and environments in which they have been used (Pillay and Dill 1976; Bulleid 1980; Beveridge 1984). Because pen culture involves the use of already available water, requires a relatively low initial capital investment, and is based on simple technology, the method has provided fish culturists with an efficient alternative to traditional culture techniques.

More recently, net pens have been used in the cool waters of the Pacific Northwest, where the rising costs of more traditional methods of culturing anadromous Pacific salmonids led to the investigation of alternative, more economical techniques that would be applicable for commercial use, as a hatchery supplement, or in restoration projects (Novotny 1975, 1980; Mahnken and Waknitz 1979). In most of these rearing situations, the net pens have been installed in marine waters, in either estuaries or protected bays. However, anadromous salmonids have also been successfully reared in and released from net pens that were installed in lentic freshwater environments by the Washington Department of Wildlife and the Quinault Indian Tribe (R. Paulsen, Washington Department of Wildlife and M. Figg, Quinault Lake Hatchery, personal communications).

Anadromous fish populations were severely affected by the construction and operation of hydroelectric dams on the Columbia River, USA (Blumm 1981). Much of the spawning and nursery habitat was destroyed and the movement of adults and juveniles through or over the dams resulted in high mortalities. These losses were partly mitigated by increased hatchery releases, but fish stocks from the lower river hatcheries where the fish were reared and released failed to provide enough fish for the severely impacted reaches of the mid and upper Columbia River basin for which they were intended.

The Pacific Northwest Electric Power Planning and Conservation Act of 1980 authorized the mitigation, protection, and enhancement of Columbia River basin fish and wildlife resources that had been adversely affected by hydroelectric development (Blumm 1985). The process involved formulation of a comprehensive plan, the Colum-

[1]Present address: U.S. Fish and Wildlife Service, Fisheries and Federal Aid, 911 Northeast 11th Avenue, Portland, Oregon 97232-4181, USA.

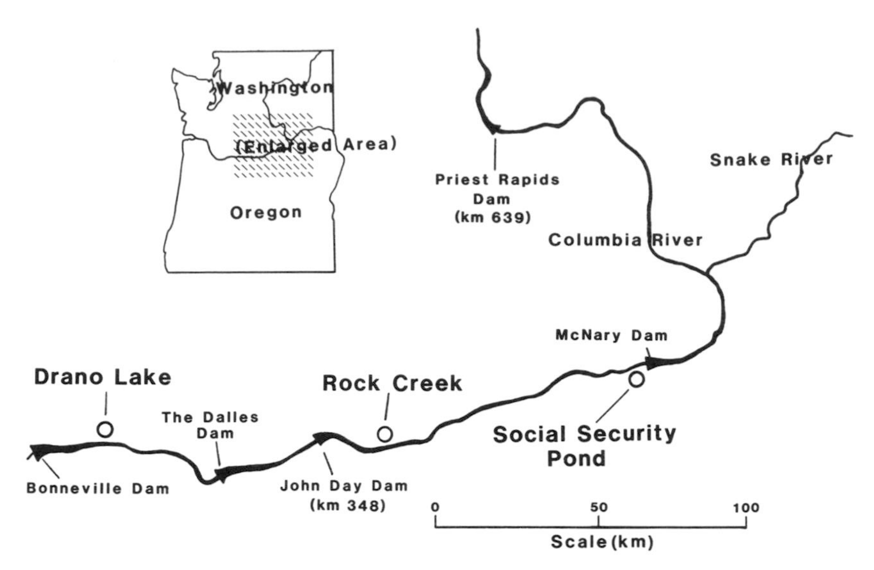

FIGURE 1.—Reach of the Columbia River from Priest Rapids Dam to Bonneville Dam, showing locations of main-stem dams and reservoirs in relation to the three rearing areas, Drano Lake, Rock Creek, and Social Security Pond.

bia Basin Fish and Wildlife Program (Northwest Power Planning Council 1982), which incorporated the recommendations from various state and federal agencies, Indian tribes, and public interests. The plan identified various measures that would enable mitigating, enhancing, or protecting anadromous fish populations (and other identified fish and wildlife in the basin), including increased hatchery production, transportation of juvenile migrants around the lower dams, and the installation of facilities to allow juveniles to pass the dams. A part of this plan involved developing and evaluating economic methods of reestablishing anadromous fish stocks that were most severely affected by habitat changes caused by hydropower development.

The upriver bright strain of fall run chinook salmon *Oncorhynchus tshawytscha* has been identified as one of the more severely affected anadromous stocks in the system, due to the inundation of spawning habitat after the construction of dams in the main stem of the Columbia and Snake rivers (Fulton 1968). This stock was selected to evaluate pen-rearing methods because the life history characteristics of the fish were compatible with short-term rearing, and because their production would compensate for fish with similar characteristics that were lost because of habitat destruction.

The objective of this study was to evaluate the feasibility of rearing and acclimating age-0 fall chinook salmon "off-station" in temporary rearing facilities. Rearing success, costs, and procedures could then be compared with traditional hatchery rearing and release practices. In future years, success of the rearing phase of the study will be judged partly by the survival of the fish produced by these methods, as indicated by the number of adults captured in ocean and in-river fisheries, and recaptured at the respective rearing sites.

Methods

Accessible study sites for rearing fish in net pens were evaluated in 1983 in the section of the Columbia River from John Day Dam (river km 348 from the mouth) to Priest Rapids Dam (river km 639) (Figure 1). Sites were evaluated according to physical characteristics, accessibility, water level and temperature fluctuations, public use, river access, and water quality and source (Novotny et al. 1984).

Sites selected for net-pen installation, starting in 1984, included Rock Creek (RC), a backwater at river km 367, and Social Security Pond (SSP), which is adjacent to the Columbia River at river km 468. Both sites are in John Day Reservoir (Figure 1). In 1987, after it was discovered that the stock had been exposed to infectious hematopoietic necrosis (IHN) virus in the hatchery where fish for the study had been spawned and hatched (Novotny et al. 1987), the entire project was transferred to Drano Lake (DL), a backwater in Bonneville Reservoir at river km 261. Various combinations of rearing densities and feeding rations given preliminary tests in 1984 (Novotny et al. 1985) were subsequently used to rear and release large numbers of fish in 1985–1987 (Novotny et al. 1986a, 1986b).

Net pens selected for the study were chosen for ease of construction, comparatively low cost, and expected durability. Pen framing, a modular unit consisting of a frame of 5.1-cm aluminum tubing (outside diameter), 6.1 m square and supported by eight floats (one at each corner and one in the middle of each side), was obtained from a commercial supplier.

The dimensions of the net deployed within the framing were 6.1 m square by 3.0 m deep; 0.9 m of the net extended above the water line, where it was attached to the framing. Two mesh sizes—3.2 and 4.8 mm—were used over the rearing period, depending on size of the fish. Total usable rearing area within the net pens was 79 m^3. Each of the pens was covered with a nylon-mesh wind screen, attached firmly around the perimeter of the frame, that excluded predacious birds.

Pens were anchored individually in water depths of at least 5 m, leaving a minimum of 3–4 m between the bottom of the enclosure and pond bottom at the rearing site. A 5- to 6-m space between the pens was maintained throughout the study. The units could be easily detached from the anchoring system and moved if necessary, allowing relocation of the pens while they were stocked with fish.

Minimum stocking densities of 0.47 kg/m^3 ("regular") were established in 1984 rearing trials (Novotny et al. 1985) as the densities to be used in standard rearing and release groups over the succeeding 3 years. The number of rearing densities tested was increased over the course of the study to a maximum of four in 1987. The additional, higher densities—double, or 0.95 kg/m^3; triple, 1.40 kg/m^3; and quadruple, 1.90 kg/m^3—were incorporated into the study in 1986 and 1987 (Novotny et al. 1986b, 1987).

Stocking rates for all four densities were determined for fish averaging 2.0 g in weight. Although sizes and numbers of fish stocked in each pen were controlled as closely as possible, absolute densities differed somewhat among pens and treatments because of slight differences in sizes of fish and in fish inventory when transferred from the hatchery.

Twelve pens were stocked at regular density at both SSP and RC in 1985 and 1986, but only at DL in 1987. Double and triple densities (two pens of each) were tested at RC in 1986 and at DL in 1987; in addition, quadruple densities (two pens) were tested at DL in 1987. All groups reared were fed a ration of dry, commercial fish meal (Abernathy Dry preparation) at a rate of 3–4% of body weight per day. Food was dispensed periodically throughout the daylight hours by battery-powered, automatic feeders installed in each of the pens.

Control groups were reared in raceways in two U.S. Fish and Wildlife Service National Fish Hatcheries (NFH)—Little White Salmon NFH and Spring Creek NFH—by using standard hatchery procedures and fish cultural techniques. Hatchery fish were hand-fed rations of Oregon moist pellets at a rate of 3–4% of body weight per day. The hatcheries had refrigerated storage facilities, which enabled them to feed the moist ration, whereas off-station storage was limited, making dry feed most practical. Studies with coho salmon *O. kisutch* have shown that in a controlled environment, dry and moist rations yielded similar growth responses (J. Westgate, J. P. Lagasse, and W. Fairgrieve, Oregon Department of Fish and Wildlife, unpublished report).

In 1985, fish were transferred to pens from the hatchery during the second week of April at a weight of 1.3–1.5 g; in 1986 and 1987, fish were taken from the hatchery in the second week of March at a much smaller size (weight, <1.0 g). The larger fish in 1985 were stocked directly into pens at the specified density, but the smaller fish in 1986 and 1987 were held in pens for 3–4 weeks before they were redistributed (at an average weight of about 2.0 g) for the various treatments.

Fish were held at the hatchery in 1985 until April to facilitate coded wire tagging of the groups. In 1986 and 1987, tagging was completed at the rearing sites, which enabled transfer of the fish from the hatchery at a smaller size.

Growth and survival of fish reared in net pens, as well as of those reared as controls in the hatchery, were monitored during alternate weeks from stocking through release. Differences in growth among the various treatments were compared by using a one-way analysis of variance procedure for balanced and unbalanced samples (Zar 1984). Significant differences were then compared by using a Newman–Keuls multiple-comparison test.

Gill Na^+,K^+-ATPase activities (micromoles inorganic phosphate per milligram protein per hour:

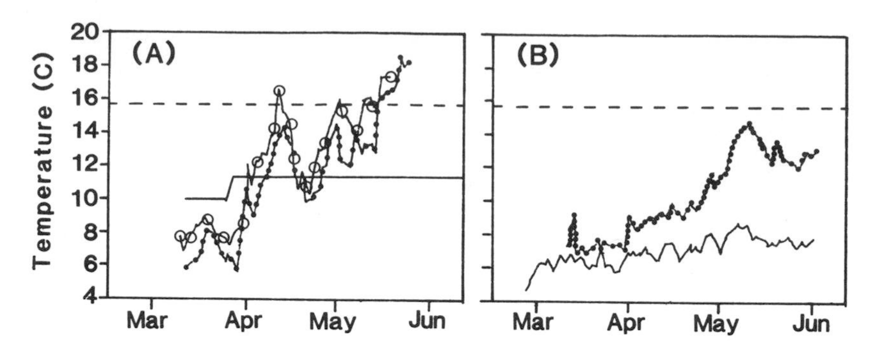

FIGURE 2.—Representative water temperatures for each of the rearing areas: (A) Rock Creek (●—●), Social Security Pond (○—○), and Spring Creek NFH (——) in 1985; and (B) Drano Lake (●—●) and Little White Salmon NFH (——) in 1987. Horizontal broken line denotes temperature of 15.6°C.

μmol P_i/[mg protein · h]) were monitored every other week in all groups as an indication of smoltification and readiness for outmigration (Folmar and Dickoff 1981). Gills were clipped, fixed, and analyzed according to the methods of Zaugg (1982). Routine disease checks were completed throughout the rearing at a certified fish disease laboratory.

We used the "present value theory" technique of Senn et al. (1984) to compare the various rearing densities and procedures. This tool of basic economics incorporates the different life expectancies and replacement costs of the various facilities to develop a "hatchery efficiency ratio" (HER) for each of the treatments. Fixed initial expenses, life expectancy of net pens (frames, nets, covers, anchors, and feeders), and estimated costs of construction of a concrete hatchery raceway (plus plumbing and water expense; Senn et al. 1984) were standardized for the rearing of 100 kg of fish. These costs reflected the original expenditures for purchasing each of the various facilities. Costs subject to variation in an annual expenditure (feed, operation and maintenance, and labor) were calculated separately because changes in operation would first appear in these expenditures.

Rearing Review

Typically, mean water temperatures rose from about 5°C in the second week of March to more than 15°C by mid-May at RC and SSP (Figure 2). Periods of increased water temperatures in mid to late April were common, but the higher temperatures were not sustained. Temporary changes in temperature of as much as 3°C in a 3- or 4-d period, at least in the upper strata of the water columns, were not unusual at either site. Temperature increases in DL were delayed in comparison with those in either RC or SSP and remained lower throughout the 1987 study period than temperatures previously observed at the other off-station sites. Temperatures in the hatcheries were much cooler and more stable, especially at Spring Creek NFH, throughout the period of rearing (5–11.5°C).

Preliminary rearing trials in 1984 indicated that rearing fish off-station in pens at temperatures above 15.6°C became increasingly difficult because of disease problems that developed during the later stages of the rearing period (Novotny et al. 1984). Therefore, in later trials, releases were made when water temperatures were projected to exceed 15.6°C (Figure 2).

Fish were taken from Little White Salmon NFH in 1985 during the second and third weeks of April at about 1.5 g, and reared for a maximum of 4 weeks; they were released on May 15 and 16. In 1986 and 1987, fish were taken from the hatchery during the second and third weeks of March at 0.6–0.8 g and released during the second and third weeks of May, after 6–8 weeks of rearing. Although mean size at stocking varied between brood years, fish were distributed among the treatments at an average weight of about 2.0 g, regardless of size when transferred from the hatchery to the off-station sites; all fish were released at a minimum weight of 4.5 g. Transfer of the fish from the hatchery at an earlier date made it possible to increase the rearing period by as much as 4 weeks.

Mean numbers of fish stocked per pen ranged from 18,758 at regular density to more than 75,000 at quadruple density; stocking densities ranged

TABLE 1.—Summary means for all chinook salmon broods combined, including numbers of fish stocked and released, densities (kg/m^3) at stocking and at release, mortalities, and biomass produced (kg) for each treatment during net-pen studies, 1985–1987. Estimates were based on a stocking size of about 2.0 g and a release size of 4.5 g.

Stocking density	Numbers of fish		Density (kg/m^3)		Mortality[a] (%)	Biomass (kg)		
	Stocked	Released	Stocked	Released		Stocked	Released	Increase (%)
Regular	18,758	18,464	0.47	1.05	0.8	38.1	83.3	119
Double	37,606	37,282	0.95	2.12	0.3	75.3	168.1	123
Triple	55,156	54,605	1.40	3.10	0.5	110.2	245.8	123
Quadruple	75,202	74,358	1.90	4.22	0.6	150.7	334.6	122

[a]On average, sampling mortality accounted for an additional 0.5% of fish from each treatment.

from 0.47 to 1.90 kg/m^3 (Table 1). Release numbers ranged from 18,464 to 74,358 fish/pen and densities at release from 1.05 kg/m^3 at the regular density to 4.22 kg/m^3 for the quadruple density. Percent increase in fish biomass produced was similar among the four treatments, and differences in total biomass produced were predictably related to the density at which the fish were reared, quadruple density producing the most fish.

Mortalities were low, ranging from 0.3 to 0.8% among all pen-reared groups in 1985–1987. Holding fish longer had resulted in high mortalities (up to 14%) during preliminary rearing trials in 1984 (Novotny et al. 1985). These mortalities were attributed primarily to enteric redmouth disease, which developed with higher water temperatures; after treatment with medicated feed, survival through release improved. Hatchery mortality ranged from 1.3 to 5.5% in 1985–1987. In 1987, however, the entire upriver bright stock reared at Little White Salmon NFH was destroyed after IHN was diagnosed in hatchery raceways; IHN was not diagnosed in fish reared in the net pens (Novotny et al. 1987).

Sizes of pen-reared fish typically increased at a faster rate than sizes of control fish in the hatchery in all years—for example, 1986 (Figure 3). Growth in DL in 1987 was slightly slower than that observed at either of the upstream sites in 1985 and 1986, but because of the lower water temperatures, it was possible to rear the fish for a longer period. Although pen-reared fish were significantly larger than hatchery fish at the time of release ($P < 0.05$), there was no apparent difference in the size of fish reared at different densities in the pens (Table 2). Hatchery fish were scheduled for release when they approached 4.5–5.0 g, normally during the last week of May or first week of June.

Gill ATPase activity was generally similar between hatchery and pen-reared fish during March and early April—for example in 1986 (Figure 3). However, values for pen-reared fish began to increase about the first week in May, whereas those for hatchery fish remained low through the time of release. Final comparisons before release indicated that most ATPase values for pen-reared fish were significantly higher than values for hatchery fish ($P < 0.05$), regardless of rearing density (Table 2). However, there was no significant difference in ATPase values among fish reared at different densities in the pens.

Densities tested during these trials were low in comparison with densities at which hatchery fish were reared, or with the maximum densities rec-

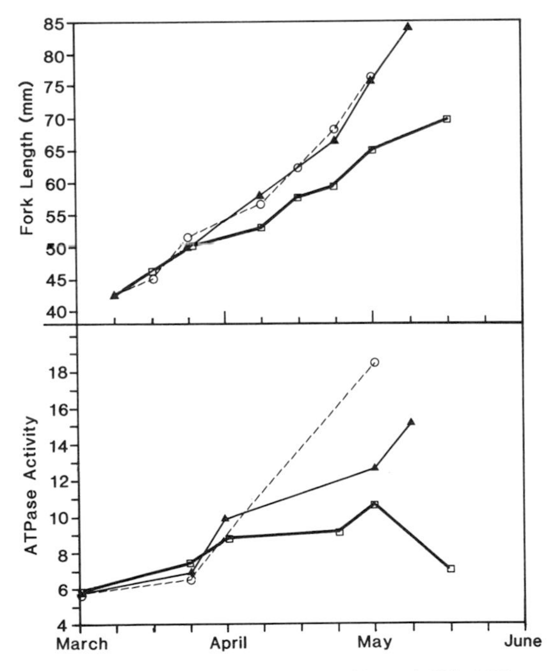

FIGURE 3.—Fork lengths and gill Na^+,K^+-ATPase activity (μmol P_i/[mg protein · h]) of fish stocked at "regular" density (0.47 kg/m^3) at Social Security Pond (○----○) and Rock Creek (▲——▲), and of control fish in the hatchery (□——□), 1986.

TABLE 2.—Summaries of mean fork lengths (FL, mm) and ATPase values (μmol P_i/[mg protein · h]) of chinook salmon at release, and results of analyses of variance completed for each of the respective densities and locations, 1985–1987. Values within a column with a letter in common were not significantly different ($P > 0.05$). Treatment key: SSP/REG—Social Security Pond/regular density; REG—regular density (1985, 1986, Rock Creek; 1987, Drano Lake); DBL—double density (1986, Rock Creek; 1987, Drano Lake); TRPL—triple density (1986, Rock Creek; 1987, Drano Lake); QDPL—quadruple density (1987, Drano Lake); HTC—hatchery control.

Treatment or statistic	1985		1986		1987	
	FL	ATPase	FL	ATPase	FL	ATPase
SSP/REG	70.9 x	24.5 x	76.4 x	18.3 x		
REG	71.9 x	16.4 y	75.6 xy	16.6 xy	72.0 xy	15.7 x
DBL			73.9 xy	20.7 x	73.8 x	16.9 x
TRPL			72.3 y	15.1 xy	70.4 y	18.8 x
QDPL					71.1 y	16.4 x
HTC	65.5(SSP) y 69.2(RC) y	9.1 z	65.0 z	11.1 y	[a]	[a]
F-value	5.07(SSP) 5.07(RC)	39.8	24.40	4.20	3.18	0.48

[a]Hatchery comparisons not available at release, however, 10 d before release, both variables were significantly lower than observed in any off-station treatment (Novotny et al. 1987).

ommended, both in terms of space and in terms of the density index commonly used in national fish hatcheries (Table 3). However, flow is a primary factor in determining the number of fish that can be reared in a facility. Therefore, an additional expression of density, a flow index, which incorporated available rearing space, size of the fish, and flow rate, was also used to compare the off-station treatments with hatchery controls (Wedemeyer et al. 1981). The estimates of water exchange in net pens in DL of 1–4 h (Novotny et al. 1987) were assumed to be similar to exchanges at other rearing sites. We believe that fish reared at densities equivalent to the double or triple treatments tested may have exceeded the maximum densities recommended for situations when water exchange was limited. Manifestations of poor water exchange or quality were never observed, however—indicating that our estimates of exchange were either conservative, or that the periods of limited exchange were too brief to observe undesirable effects in fish.

TABLE 3.—Summary of density comparisons (expressed in terms of space, density indices, and flow indices) between chinook salmon reared off- and on-station, and maximum recommended for each respective density expression. Mean size of fish for these calculations was 4.5 g.

Number of fish per pen or raceway	Spatial density (kg/m³)	Density index[a]	Flow index[b]
Off-station			
18,500	1.05	0.0226	0.08–0.15
37,000	2.10	0.0456	0.16–0.67
55,500	3.15	0.0683	0.25–1.01
74,000	4.20	0.0923	0.34–1.36
On-station			
250,000 (SC)[c]	10.30	0.2265	0.54
250,000 (LWS)[d]	26.48	0.5818	0.86
Maximum	14.30[e]	0.35–0.40[f]	0.56–0.80

[a]Piper et al. (1982): weight/(length × volume).

[b]Wedemeyer et al. (1981), includes a range of temperatures from 9 to 14°C and is based on a water exchange in net pens of 1–4 h: 100 · kg/(length × L/min).

[c]Spring Creek National Fish Hatchery.

[d]Little White Salmon National Fish Hatchery.

[e]Banks et al. (1979).

[f]S. Leek, U.S. Fish and Wildlife Service, personal communication.

Highest HERs (in terms of U.S.$/kg) were in the lower density treatments; costs were obviously progressively lower as densities increased (Table 4). Hatchery expenses were higher than three of the four densities tested off-station, and were nearly twice as high as estimated expenses for quadruple treatments.

Rearing expenses reflected the costs of rearing fish in aluminum-framed pens, used in this study because they were the least-expensive framing available. Other types of framing may be more cost-effective over an extended period, depending on location and needs (Novotny et al. 1987).

Fish Culture Implications

Currently, hatcheries in the Columbia River basin are commonly able to spawn and hatch more fish than they can rear through release. Culturists collect excess eggs to replace those lost to mortality and to ensure adequate production, but crowding that occurs with growth generally results in the release of the excess fish at a smaller than desirable size. Survival of these "thinning"

TABLE 4.—Costs of rearing 100 kg of juvenile chinook salmon in net pens at regular (REG), double (DBL), triple (TRP), and quadruple (QDPL) densities, and of rearing fish in a concrete hatchery raceway (HTC); includes fixed expenses and variable, annual expenses. Costs are based on a stocking size of 2.0 g and a release size of 4.5–5.0 g. Equipment life expectancy in years (y) is in parentheses; net-pen rearing costs are based on 1987 U.S. dollars.

Expense item[a]	REG	DBL	TRP	QDPL	HTC
Number of pens required	2.2	1.1	0.7	0.5	
Fixed, initial					
Frames, etc. (20 y)	$4,347	$2,129	$1,464	$1,065	
Feeders (20 y)	896	439	302	219	
Nets, covers (7 y)	1,241	608	418	304	
Raceway (50 y)					$1,419
Plumbing (25 y)					1,322
Water					138
Total	$6,484	$3,176	$2,184	$1,588	$2,879
Variable, annual					
Feed costs	$121	$99	$88	$93	$132
Operations, maintenance	648	317	218	159	518
Labor	809	396	273	198	184
Total	$1,578	$812	$579	$450	$834
Hatchery efficiency ratio ($/kg)[b]	$19.76	$12.29	$8.63	$6.59	$12.62

[a]Expenses include: (1) frames, anchors, etc.—$2,014; (2) feeders—$415; (3) nets and covers—$575; (4) Abernathy dry feed (off-station)—$0.99/kg; Oregon moist pellets (hatchery)—$1.32/kg; (5) operations and maintenance figured at 10% off-station and $5.18/kg in hatchery (1986 dollars, J. Bodle, Little White Salmon National Fish Hatchery, personal communication); (6) labor figured as 0.015 person-year per pen at $25,000/year.

[b]Senn et al. (1984).

releases is generally considered poor and the fish are essentially lost to production.

Hatchery fish are now released directly from the hatchery or hauled to and released at preselected locations along the river. These scenarios normally do not take into account the physiological state of the fish, but are related to a predetermined size and general release date. Previous studies have suggested that maximum gill Na^+,K^+-ATPase activity is usually not achieved in fish reared in hatcheries (Dickoff et al. 1985). Therefore, most hatchery fish are probably not fully developed smolts when released, regardless of size. After release, these fish undergo a period of acclimation, which may simulate natural migration, and which may be required for the fish to complete metamorphosis from parr to smolts (Zaugg et al. 1985). Natural migration of juvenile salmonids as they pass from fresh water to salt water is an important developmental period in which the smolts prepare for seawater entry (Dickoff et al. 1985; Zaugg et al. 1985; Rondorf et al. 1988).

The integration of a period of acclimation could accelerate growth and physiological development of smolts, similar to that observed in the current study prior to both thinning and regular hatchery releases. Used as an adjunct to existing hatchery production, net pens could provide flexible, efficient use of existing facilities, allow incremental on-site release of hatchery fish—which would moderate their effects on wild fish already present in the system—alter migration times (fish reared off-station would migrate earlier than wild fish), and probably increase survival (Mullan 1987). Release of fish could, therefore, be coordinated with physiological indicators, such as rising gill-ATPase activities, after acclimation in the relatively natural surroundings possible with off-station net-pen rearing.

Pen-rearing facilities also provide the option of rearing fish economically, even at relatively low densities such as those tested in our study. Rearing fish at lower densities may present several advantages, including better survival as juveniles and greater contribution to the fishery as adults (Soderberg and Krise 1986; Sandercock 1987). A "higher quality" smolt, larger and more advanced physiologically than now reared in hatcheries, would thus be economically achievable, and rearing strategies could be designed that would emphasize the quality rather than quantity of fish produced. Mullan (1987) suggested that some combination of hatchery and reservoir rearing would improve the quality of smolts released and would relieve hatchery stresses that often lead to disease problems.

Freezing, flooding, and warm water can be

primary concerns when fish are reared in off-station areas (Senn et al. 1984). For example, emplacing the net pens in 1985 was delayed until surface ice had thawed. On the other hand, high water temperatures limited the length of the rearing period in May. In some sections of the Columbia River, ice in January and February may delay construction of facilities and stocking of fish, and water temperatures may warm to above 15.6°C by the first or second week of May, leaving a rearing "window" of only 8–10 weeks.

Rearing periods could be extended by modifying the pens and methods used during this study. By increasing the depth of the nets as much as the specific rearing area allows, total available rearing space could be increased, and access to cooler water present in the deeper strata of most backwaters would be provided. In addition, high and fluctuating temperatures could be moderated by placing pens within the reservoirs. Precautions would be required, such as selection of protected rearing locations on the lee side of obstacles to prevailing winds, heavy anchoring systems, and the use of breakwaters. These types of structures have been designed to protect net-pen arrays in other areas (Kato et al. 1976), and it is likely that they could be used in main-stem reservoirs as well. Certainly, some type of modification would be required to provide rearing space in the main bodies of the Columbia or Snake river reservoirs.

Expenses presented here are a useful means of comparing various off-station rearing procedures with those of a typical Pacific Northwest salmon hatchery. Our costs were not presented, nor intended, as an expenditure goal that fish culturists should strive to achieve for pen rearing; actual management scenarios may be much different, depending on the types of facilities purchased and the labor costs required to maintain them. Rearing fish in low-cost, off-station facilities may prove to be an appealing method of improving current hatchery production and of returning adults to reaches of the Columbia River basin where environmental alterations have affected their populations. If properly sited and planned, rearing of salmonids in net pens would provide fishery managers in the Columbia River basin with diverse alternatives to traditional rearing and release strategies.

Growth and development of pen-reared salmonids could be expected to be better than for the same stock of fish reared in the hatchery, depending on water temperature differences and rearing scenarios for each situation, even over short periods of time. This is not to say that off-station rearing outperforms traditional hatchery practices. However, our studies revealed definite advantages in the transfer and rearing of fall chinook salmon off station. Similar results for the rearing of other Pacific salmonids may be possible, provided that the rearing scenarios tested in this study, and practical adjustments on a site-by-site basis for each potential species, are incorporated into the planning phase and site selection.

Acknowledgments

This study was funded by the ratepayers in the Pacific Northwest and administered through the Bonneville Power Administration. W. Nelson, C. Burley, and D. Rondorf wrote and submitted the original study proposals and lent counsel throughout the study. Hatchery personnel at Spring Creek and Little White Salmon national fish hatcheries allowed us to use their facilities. We especially thank the persons who constructed the pens, cared for the fish, and assisted in laboratory analysis: D. Allard, J. Beeman, R. Burkhardt, A. Ecklund, G. Evans, M. Faler, J. Gardenier, M. Gross, M. Kohn, K. Pohlod, F. Stein, P. Veliz, and S. Woods.

References

Banks, J. L., W. G. Taylor, and S. L. Leek. 1979. Carrying capacity recommendations for Olympia area national fish hatcheries. Abernathy Salmon Cultural Development Center, Longview, Washington. (Mimeo).

Beveridge, M. C. M. 1984. Cage and pen fish farming: carrying capacity models and environmental impact. FAO (Food and Agriculture Organization of the United Nations) Fisheries Technical Paper 255.

Blumm, M. C. 1981. Hydropower vs. salmon: the struggle of the northwest's anadromous fish runs for a peaceful coexistence with the federal Columbia River system. Environmental Law 11:211–300.

Blumm, M. C. 1985. Restoring Columbia Basin salmon under the Northwest Power Act. Marine Recreational Fisheries 10:161–171.

Bulleid, M. J., chairman. 1980. Cage fish rearing symposium. Reading University, Institute of Fisheries Management, Reading, UK.

Dickoff, W. W., C. Sullivan, and C. V. W. Mahnken. 1985. Methods of measuring and controlling the parr to smolt transformation (smoltification) of juvenile salmon. Pages 5–9 *in* C. J. Sindermann, editor. Proceedings of 11th U.S.–Japan meeting on aquaculture, salmon enhancement.

Folmar, L. C., and W. W. Dickoff. 1981. Evaluation of some physiological parameters as predictive indices of smoltification. Aquaculture 23:309–324.

Fulton, L. A. 1968. Spawning areas and abundance of

chinook salmon (*Oncorhynchus tshawytscha*) in the Columbia River basin—past and present. U.S. Fish and Wildlife Service Special Scientific Report—Fisheries 571.

Kato, J., T. Noma, and Y. Vekita. 1976. Design of floating breakwaters. Pages 458–466 *in* T. V. R. Pillay and W. A. Dill, editors. Advances in aquaculture. Fishing News Books, Farnham, UK.

Mahnken, C. V. W., and F. W. Waknitz. 1979. Factors affecting growth and survival of coho salmon (*Oncoryhnchus kisutch*) and chinook salmon (*O. tshawytscha*) in saltwater net-pens in Puget Sound. Proceedings of the World Mariculture Society 10: 280–305.

Mullan, J. W. 1987. Status and propagation of chinook salmon in the mid-Columbia River through 1985. U.S. Fish and Wildlife Service Biological Report 87(3).

Northwest Power Planning Council. 1982. Columbia River basin fish and wildlife program. Northwest Power Planning Council, Portland, Oregon.

Novotny, A. J. 1975. Net-pen culture of Pacific salmon in marine waters. U.S. National Marine Fisheries Service Marine Fisheries Review 37(1):36–47.

Novotny, A. J. 1980. Cage culture of salmonids in the United States. Pages 3–14 *in* M. J. Bulleid, chairman. Cage fish rearing symposium. Reading University, Institute of Fisheries Management, Reading, UK.

Novotny, J. F., T. L. Macy, M. P. Faler, and J. W. Beeman. 1987. Pen rearing and imprinting of fall chinook salmon: annual report 1987. Report of U.S. Fish and Wildlife Service, National Fishery Research Center to Bonneville Power Administration, Portland, Oregon.

Novotny, J. F., T. L. Macy, and J. T. Gardenier. 1984. Pen rearing and imprinting of fall chinook salmon: annual report 1983. Report of U.S. Fish and Wildlife Service, National Fishery Research Center to Bonneville Power Administration, Portland, Oregon.

Novotny, J. F., T. L. Macy, and J. T. Gardenier. 1985. Pen rearing and imprinting of fall chinook salmon: annual report 1984. Report of U.S. Fish and Wildlife Service, National Fishery Research Center to Bonneville Power Administration, Portland, Oregon.

Novotny, J. F., T. L. Macy, and J. T. Gardenier. 1986a. Pen rearing and imprinting of fall chinook salmon: annual report 1985. Report of U.S. Fish and Wildlife Service, National Fishery Research Center to Bonneville Power Administration, Portland, Oregon.

Novotny, J. F., T. L. Macy, J. T. Gardenier, and J. W. Beeman. 1986b. Pen rearing and imprinting of fall chinook salmon: annual report 1986. Report of U.S. Fish and Wildlife Service, National Fishery Research Center to Bonneville Power Administration, Portland, Oregon.

Pantulu, V. R. 1976. Floating cage culture of fish in the lower Mekong Basin. Pages 423–427 *in* T. V. R. Pillay and W. A. Dill, editors. Advances in aquaculture. Fishing News Books, Farnham, UK.

Pillay, T. V. R., and W. A. Dill, editors. 1976. Advances in aquaculture. Fishing News Books, Farnham, UK.

Piper, R. G., I. B. McElwain, L. E. Orme, J. P. McCraren, L. G. Fowler, and J. R. Leonard. 1982. Fish hatchery management. U.S. Fish and Wildlife Service, Washington, D.C.

Rondorf, D. W., J. W. Beeman, M. B. E. Free, and D. E. Liljegren. 1988. Correlation of biological characteristics of smolts with survival and travel time: annual report 1988. Report of U.S. Fish and Wildlife Service, National Fishery Research Center to Bonneville Power Administration, Portland, Oregon.

Sandercock, F. K. 1987. Hatchery loading and flow. Pages 183–194 *in* G. R. Bouck, editor. Proceedings of workshop: improving hatchery effectiveness as related to smoltification. Bonneville Power Administration, Portland, Oregon.

Senn, H., J. Mack, and L. Rothfus. 1984. Compendium of low-cost Pacific salmon and steelhead trout production facilities and practices in the Pacific Northwest. Bonneville Power Administration, Division of Fish and Wildlife, Portland, Oregon.

Soderberg, R. W., and W. F. Krise. 1986. Effects of rearing density on growth and survival of lake trout. Progressive Fish-Culturist. 48:30–32.

Wedemeyer, G. H., R. L. Saunders, and W. C. Clarke. 1981. The hatchery environment required to optimize smoltification in the artificial propagation of anadromous salmonids. Pages 6–20 *in* L. J. Allen and E. C. Kinney, editors. Proceedings of the bio-engineering symposium for fish culture. American Fisheries Society, Fish Culture Section, Bethesda, Maryland.

Zar, J. G. 1984. Biostatistical analysis. Prentice-Hall, Englewood Cliffs, N.J.

Zaugg, W. S. 1982. A simplified preparation for adenosine triphosphatase (ATPase) determined in gill tissue. Canadian Journal of Fisheries and Aquatic Sciences 39:215–217.

Zaugg, W. S., E. F. Prentice, and F. W. Waknitz. 1985. Importance of river migration to the development of seawater tolerance in Columbia River anadromous salmonids. Aquaculture 51:33–47.

American Fisheries Society Symposium 10:548–553, 1991

Offshore Release of Salmon Smolts

WILLIAM J. MCNEIL
Hatfield Marine Science Center
Oregon State University, Newport, Oregon 97365, USA

RON GOWAN
Paradise Bay Sea Farms, Ltd.
Box 162, Heriot Bay, British Columbia V0P 1H0, Canada

RICHARD SEVERSON
8303 Thurston Road, Springfield, Oregon 97477, USA

Abstract.—Bird predators appear to exhibit functional and numerical responses to the availability of salmon smolts. Although the mortality rate associated with smolt predation in the vicinity of Yaquina and Coos bays, Oregon, is not known, the potential for predation by at least one bird species, the common murre *Uria aalge*, appears to be very high. One strategy to reduce mortality of salmon smolts is to transport them in a vehicle for release offshore where predators are less likely to concentrate. Offshore releases of Atlantic salmon *Salmo salar* in Sweden and Norway yielded consistently higher returns than onshore releases. Preliminary results with offshore release of coho salmon *Oncorhynchus kisutch* and chinook salmon *O. tshawytscha* in Oregon are mixed, but handling and transport methods are suspected to play an important role in survival of smolts released offshore. Smolts released offshore from a towed net pen exhibited a much higher return rate than smolts released from a tank aboard a fishing vessel or from a towed barge. Straying was less of a problem when smolts were released from a towed net pen in comparison with a towed barge or deck tank.

Processes controlling marine survival of salmon involve many interactions between fish and their environment. Past investigations have tended to place emphasis on the roles of health, nutrition, physiology, and genetics of hatchery stocks and hatchery practices in ocean survival. Research is continuing in these areas, but more attention is also being paid to behavior, release strategies, and interactions between salmon and the marine ecosystem.

Of the several factors that can be controlled during the hatchery production cycle, size and date at release have shown an association with marine survival of smolts of coho salmon *Oncorhynchus kisutch* (Gowan and McNeil 1984). Smolts released at a relatively large size (>40 g) in August have experienced higher survival than smolts released earlier at a smaller size. This result, combined with other information, has drawn attention to questions about predation and how to minimize its impact.

Based on the following four suppositions, offshore transport and release of salmon smolts has the potential to increase marine survival. (1) Ocean mortality is highest and most variable during and shortly after smolts enter the sea. (2) Ocean mortality is due largely to predation, but the ability of smolts to osmoregulate, grow, and resist disease can affect their vulnerability to predation. (3) Ocean mortality from predation is density dependent, but functional responses between mortality and smolt abundance are not well understood. (4) Ocean mortality from predation can be reduced by modifying release strategy and procedures.

Juvenile salmon are forage fish for a wide variety of marine fishes, birds, and mammals. Predation has two characteristics that potentially are important to an offshore release strategy: a functional relationship between prey consumption per predator and prey density; and a numerical response of predators to prey.

Several functional relationships have been described in the literature. All are based on assumptions that individual predators eventually become satiated and that percent prey mortality per predator declines as prey density increases above a certain level. At lower prey densities, some models of functional relationships show increased consumption of prey per predator up to a prey population level at which consumption declines as satiation is approached. Peterman and Gatto (1978) described some theoretical relationships of food consumption per predator to prey abundance.

Numerical response is related to numbers of predators preying on a prey population. Highly mobile predators have the capacity to respond to an increased prey population by increasing their population as well. Observations at Coos and Yaquina bays, Oregon, indicate that at least one important sea bird predator exhibits both functional and numerical responses to salmon smolts.

Sea birds have been observed to alter their feeding behavior and concentrate in the vicinity of Coos and Yaquina bays during periods of smolt release. The common murre *Uria aalge* was the most prevalent bird predator; Bayer and Varoujean (1984) estimated daily numbers as high as 9,600 adult birds in the vicinity of Yaquina Bay. These authors also estimated that 22,000 common murres nest within 50 km of Yaquina Bay. Foraging range of common murres is believed to extend 150 km from nesting colonies (Varoujean and Mathews 1983). Other birds observed to concentrate at Yaquina Bay when smolts were present included brown pelicans *Pelecanus occidentalis*, gulls *Larus* spp. and cormorants *Phalacrocorax* spp., but these species were much less numerous than murres.

According to Scott (1973), the common murre feeds largely within 8 km of shore in spring and summer. Scott further reported that common murres can pursue prey to depths exceeding 100 m and can remain underwater for 4 min. Each bird requires about 250 g/d of fish and invertebrates. Common murres normally do not eat juvenile salmon (Livingston 1980; Mathews 1983), but Matthews (1983) observed them consuming large numbers of salmon smolts released from private salmon ranching sites in the Coos Bay area. The incidence of smolts in stomach contents of common murres increased about 30-fold in 1981, when 12.8 million smolts were released. A similar high incidence of salmon smolts in common murre stomach contents was also observed at Yaquina Bay in 1982, when 20.9 million smolts were released. However, 1982 was the only year in which stomach contents of common murres were evaluated at Yaquina Bay, and baseline data for years of low smolt numbers were not available.

The high incidences of salmon smolts in common murre stomachs at Yaquina and Coos bays were evidence of a functional response of predators to increased prey abundance. There was also a simultaneous numerical response: densities of common murres feeding in vicinity of Coos and Yaquina bays increased (Mathews 1983; Bayer and Varoujean 1984).

Offshore release of tagged groups of salmon smolts was undertaken to facilitate their dispersal at sea and to reduce mortality from birds and other predators. Information on adult returns and straying is assessed in this report to compare three methods of offshore transport of smolts.

Methods

Oregon private salmon ranching companies began to evaluate offshore release of salmon smolts in 1982. Cooperating companies were Oregon AquaFoods, Inc., Springfield, Oregon, and Anadromous, Inc., Corvallis, Oregon. Three modes of transport were tried—(1) deck tank aboard a fishing vessel, (2) towed net pen, and (3) towed barge. Tagged coho salmon smolts were released from a deck tank at distances of 8 and 24 km offshore in 1982 and 1983, from a towed net pen at distances of 8 and 24 km offshore in 1983 and 1984, and from a towed barge at distances ranging from 6 to 32 km offshore in 1985 and 1986. Tagged smolts of chinook salmon *O. tshawytscha* were also released from a towed barge in 1985 and 1986, and control groups were released from shore.

Smolts were transported by tanker truck from freshwater hatcheries to saltwater acclimation sites at Yaquina and Coos bays on the central Oregon coast 10–30 d prior to release. All groups of smolts released offshore and from shore were imprinted to return as maturing adults to their acclimation ponds, which consisted of raceways and tanks receiving seawater pumped from estuaries. Effluent water is discharged through a fish ladder so that returning adults have ready access to acclimation ponds, where they are collected and examined for tags.

Deck tank.—Groups of coho salmon were transported by truck from the acclimation site to a fishing vessel moored approximately 1 km from the acclimation site. The fish were transferred to a deck tank holding approximately 4 m^3 of aerated seawater. Ten groups of approximately 15,000 coho salmon smolts tagged with coded wire microtags were released offshore by draining the tank overboard.

Net pen.—The net pen had a rigid external iron frame to which the net was attached to minimize billowing while under tow. Its volume was approximately 100 m^3. Flotation was provided by airtight plastic cylinders secured to the frame. A seam in the net was opened at sea by a diver to allow fish to escape.

Five groups of 40,000 to 50,000 coho salmon smolts tagged with coded microwires were re-

TABLE 1.—Ratios of percent return of coho and chinook salmon smolts released offshore to those released from shore at Yaquina and Coos bays, Oregon, 1982–1987.

	Type of transport system							
					Towed barge			
	Deck tank		Towed net pen		1985		1986	
Ratio	Samples	%	Samples	%	Samples	%	Samples	%
>1.2	4	40	4	80	2	5	7	35
0.8–1.2	0	0	1	20	13	35	5	25
<0.8	6	60	0	0	22	60	8	40
Total	10		5		37		20	

leased from the net pen. The pen was towed at slow speed (<2 km/h) in relatively calm seas.

Barge.—The barge was designed to transport smolts in rougher seas and at higher speeds than the net pen. Constructed of aluminum, the barge contained about 230 m^3 of seawater. It could be towed at speeds in excess of 8 km/h under most sea conditions. It was supported by two flotation chambers extending its 12-m length. Both bow and stern structures were porous to allow a continuous exchange of seawater while under tow. Fish were released by opening a stern gate.

Forty-seven groups of coho salmon and seven groups of chinook salmon smolts were released from the barge. Numbers of tagged smolts per group ranged from 3,400 to 58,000.

Statistical analysis.—Statistical tests for significant differences in response of adult salmon to release method have not been attempted because observations were few, especially for groups released from the deck tank and net pen, and because the different transport techniques were tried in different years. Simple deterministic models are used here to develop hypotheses about effects of release method on adult return and straying.

Experiments designed for more rigorous statistical comparisons are desirable but costly to implement. Perhaps relationships hypothesized here about relationships between release method and adult return and straying will prove useful in future research.

Adult Returns

The percentage of maturing adults returning to acclimation sites from each group released offshore was compared with the percentage returning from a control group released from shore at the same size. A ratio was calculated by dividing return rate of the offshore test group by the control group. A ratio greater than 1.0 indicates a higher rate of return from offshore than from control releases, whereas, a ratio less than 1.0 indicates the reverse.

Ratios were recorded in three categories to compare returns from the three offshore transport treatments. Ratios in the range of 0.8–1.2 are considered to indicate similar return rates for control and offshore release groups. Ratios above 1.2 are considered to indicate a higher return rate for offshore groups; ratios below 0.8 are considered to indicate a lower return rate for offshore groups. Results are summarized in Table 1 for the three transport methods. Data from the towed barge are separated into two release years (1985 and 1986) because sea conditions at the time of offshore release tended to favor higher survival in 1986 than in 1985. A much higher percentage of releases were made in rough seas in 1985 than in 1986, based on unpublished reports by personnel accompanying the barge. Furthermore, the speed of towing was reduced somewhat in 1986 to reduce further stress on fish during transport.

The five marked groups from the net pen yielded an average ratio of 2.3 (i.e., 2.3 times greater return for groups released offshore than for control groups). The 10 observations from the deck tank yielded an average ratio of 0.7, and the 57 observations from the barge yielded an average ratio of 0.8 for the 2 years combined. By separating the 1985 and 1986 barge results, we calculated average ratios of 0.7 for 1985 (37 observations) and 1.1 for 1986 (20 observations). The improved performance in 1986 is thought to arise from reduced stress on fish as a consequence of slower towing speed and selection of release dates when sea conditions were relatively calm.

The towed net pen produced the most favorable returns to the acclimation sites. Although returns were less favorable for groups released offshore from the deck tank and barge than for net-pen groups, it does not necessarily follow that marine survival was reduced proportionately. The possibility that straying or catchability may have been higher for groups released offshore from the deck tank and towed barge than for the net-pen groups cannot be ignored, because these factors individually or collectively might explain differences in return.

The results reinforce the conclusion by Gowan (1988) that return rate is greatly affected by transport and handling methods. Gowan mentioned the probable importance of handling and transport techniques in developing, implementing, and evaluating an offshore release program designed to reduce mortality from predation without a commensurate increase in mortality from handling. The net pen, which produced a two- to threefold higher return rate than the deck tank or barge, is intended for use only when seas are nearly calm. Both the deck tank and barge, on the other hand, can be operated in rough and calm seas.

Adult Straying

A high percentage of the tagged salmon surviving natural marine mortality are harvested by sport and commercial fishers. The remainder return to saltwater acclimation sites or stray to natural streams. Evaluations in the previous section are based entirely on tagged fish returning to the acclimation sites. Evaluation of straying will make use of recovered tagged fish reported from sport and commercial fisheries as well as those returning to acclimation sites.

Government agencies routinely sample sport and commercial catches of salmon for tagged fish. Even though a small and variable portion of the total catch is sampled, tagged fish reported from the fishery can be used along with tagged fish recovered at acclimation sites to evaluate straying.

The number of adult salmon returning to an acclimation site can be described by the equation

$$R_i = \left(\frac{k_i}{f_i} - k_i\right)X_i - S_i; \tag{1}$$

R_i = number of the ith group of tagged smolts returning to the acclimation site;
X_i = number of the ith group reported from sport and commercial fisheries;
S_i = number of the ith group returning to locations other than the acclimation site (i.e., strays);
k_i = a coefficient to expand number of tagged fish in the ith group recovered by sampling sport and commercial fisheries to the total number caught;
f_i = proportion of the population taken by the fishery for the ith group.

Equation (1) states that the number of tagged fish returning to the acclimation site is equal to the total number of fish surviving natural mortality ($[k_i/f_i]X_i$), minus the number caught in sport and commercial fisheries (k_iX_i) and the number straying (S_i). The expression for total number of fish surviving natural mortality can be derived as follows.

Let N_i = total number of tagged fish in group i surviving natural mortality,
k_iX_i = total number of tagged fish in group i caught in sport and commercial fisheries, and
E_i = total number of tagged fish escaping sport and commercial fisheries.

The following relationships are defined:

$$N_i = k_iX_i + E_i, \tag{2}$$

and

$$E_i = (1 - f_i)N_i. \tag{3}$$

It follows, therefore, that

$$N_i = \frac{k_i}{f_i}X_i. \tag{4}$$

A model to estimate straying results from rearrangement of equation (1):

$$S_i = \left(\frac{k_i}{f_i} - k_i\right)X_i - R_i. \tag{5}$$

Solution of equation (5) requires estimates of k_i and f_i, which are beyond the scope of this analysis. However, if $([k_i/f_i] - k_i)$ is assumed to be equal for the ith groups of fish released from shore and offshore, then R_i/X_i may be used to compare stray rates. We rearrange equation (5) to obtain

$$\frac{S_i}{X_i} = \left(\frac{k_i}{f_i} - k_i\right) - \frac{R_i}{X_i}, \tag{6}$$

and use the ratio R_i/X_i to compare straying between groups released from acclimation sites and offshore. Higher values of R_i/X_i for control groups than for offshore groups indicate a higher rate of straying for the offshore groups under the assumption that the fraction harvested by sport and commercial fishers is the same for the offshore

TABLE 2.—Comparisons of average values of R_i/X_i for tagged groups of salmon released into Yaquina Bay from shore (control release) and offshore, 1982–1987. R_i is the number of fish in the ith group returning to the acclimation site; X_i is the number reported from sport and commercial fisheries.

Type	R_i/X_i	
	Control	Offshore
Net pen	10	8
Barge	7	4
Deck tank	13	5

and control groups being compared. Values of R_i/X_i for tagged groups of salmon from the Yaquina Bay acclimation site are compared in Table 2 for net pen, barge, and deck tank.

The R_i/X_i values were consistently higher for the control groups, suggesting that straying of adults was greater for offshore than for control groups. Control groups had R_i/X_i values that were 2.6 times higher than for offshore groups released from the deck tank, 1.8 times higher than for fish released from the barge, and 1.2 times higher for groups released from the net pen. Equal values of R_i/X_i for control and offshore groups would indicate that the rate of straying was unaffected by offshore release. It appears, therefore, that transport and handling methods can exert a profound influence on rate of straying. The net pen had the least influence and the deck tank the greatest.

Conclusions and Recommendations

Conditional results presented here suggest that straying of tagged groups of salmon smolts released offshore from the net-pen was much less than straying of groups released from the barge and, especially, from the deck tank. The small difference in values of R_i/X_i calculated for groups released from shore and offshore from the net pen suggests the possibility that straying will be less affected by offshore release from a net pen than from a deck tank or barge.

An important question that merits further analysis is that of ocean survival. Return rates to acclimation sites were much higher for groups released offshore from a net pen than for fish released from either the deck tank or barge. On the other hand, rate of straying appeared to be higher for groups released from the deck tank and barge than for groups from the net pen. It may be premature, therefore, to conclude that marine survival of groups released offshore from the deck tank and barge was generally lower than survival of groups released from acclimation sites.

Data on offshore releases of smolts of Atlantic salmon *Salmo salar* in Europe (Hansen 1982; Larsson 1982) support the hypothesis that marine survival will be enhanced through reduction of predation by releasing smolts offshore. The hypothesis has a theoretical basis, but more research is needed to develop techniques and criteria that will maximize survival and minimize straying. Considerably more information must be gathered on transport and handling procedures before an offshore release program can be implemented with consistent success. It now appears that a transport system incorporating net pens should receive priority attention. Investigators might well be advised to place emphasis on stress-related interactions with predator avoidance and possibly disease resistance.

Straying also merits priority attention for at least two reasons: straying increases interbreeding of hatchery and naturally produced salmon, and it decreases economic returns to salmon ranches. However, it must be recognized that any increase in marine survival and return of ranched salmon implies increased numbers of fish straying into natural streams, even if the rate of straying is unaffected. Techniques to minimize interbreeding of stray ranched adults with naturally produced fish merit further work. Examples include selection of ranched stocks that mature earlier or later than natural stocks and use of sterile triploid salmon (including hybrids) for hatchery production of market fish.

References

Bayer, R. D., and D. Varoujean. 1984. El Niño and bird numbers near a salmon hatchery at Yaquina estuary, Oregon. Weyerhaeuser Co., Research Report, Springfield, Oregon.

Gowan, R. 1988. Release strategies for coho and chinook salmon released into Coos Bay, Oregon. Pages 75–80 *in* W. J. McNeil, editor. Salmon production, management and allocation. Oregon State University Press, Corvallis.

Gowan, R., and W. McNeil. 1984. Factors associated with mortality of coho salmon (*Oncorhynchus kisutch*) from saltwater release facilities in Oregon. Pages 3–18 *in* W. G. Pearcy, editor. The influence of ocean conditions on the production of salmonids in the North Pacific. Oregon State University, Sea Grant Publication ORESU-W-83-001, Corvallis.

Hansen, L. P. 1982. Salmon ranching in Norway. Pages 95–108 *in* C. Ericksson, M. P. Ferranti, and P. O. Larsson, editors. Sea ranching of Atlantic salmon. Commission of the European Communities, COST 46/4 Workshop, Luxembourg.

Larsson, P. O. 1982. Salmon ranching in Sweden. Pages 127–137 *in* C. Ericksson, M. P. Ferranti, and P. O. Larsson, editors. Sea ranching of Atlantic salmon. Commission of the European Communities, COST 46/4 Workshop, Luxembourg.

Livingston, P. 1980. Marine bird information synthesis. National Marine Fisheries Service, Northwest and Alaska Fisheries Center, Processed Report 80-2, Seattle.

Mathews, D. R. 1983. Feeding ecology of the common murre, *Uria aalge*, off the Oregon coast. Master's thesis. University of Oregon, Eugene.

Peterman, R. M., and M. Gatto. 1978. Estimation of functional responses of predators on juvenile salmon. Journal of the Fisheries Research Board of Canada 35:797–808.

Scott, J. M. 1973. Resource allocation in four syntopic species of marine diving birds. Doctoral dissertation. Oregon State University, Corvallis.

Varoujean, D. H., and D. R. Mathews. 1983. Distribution, abundance, and feeding habits of seabirds off the Columbia River, May–June, 1982. Oregon Institute of Marine Biology, Report OIMB 83-1, Charleston.

American Fisheries Society Symposium 10:554–561, 1991

Fecundity and Egg Size of Brook Trout and Brown Trout Brood Stocks in Heated, Recirculated Water

NOEL C. FRASER[1]

Fish and Wildlife Division, Alberta Department of Forestry, Lands and Wildlife
9945-108 Street, Edmonton, Alberta T5K 2G6, Canada

CLYDE W. PARKE[2]

Allison Creek Brood Trout Station
Fish and Wildlife Division, Alberta Department of Forestry, Lands and Wildlife
Post Office Box 394, Coleman, Alberta T0K 0M0, Canada

Abstract.—Heating and recirculating water at a brood-stock station resulted in increases in individual fecundities, egg sizes, and volume of eggs produced per kilogram of fish for brown trout *Salmo trutta* and brook trout *Salvelinus fontinalis*. Relative fecundities increased in brook trout but did not vary predictably in brown trout. In the original station, mean weekly water temperatures ranged from 0.0°C in winter to 15.0°C in summer. Since 1984, water temperatures have been kept at 8.0–10.0°C year-round with 80% of the water being recirculated each cycle. After renovations, brook trout fecundities increased from 556 to 1,789 eggs/female (2,298 to 3,806 eggs/kg) for 2-year olds, 1,455 to 4,071 eggs/female (2,778 to 3,232 eggs/kg) for 3-year olds, and 4,514 to 5,920 eggs/female (3,224 to 3,947 eggs/kg) for 4-year olds. The size of brook trout eggs increased from 26,340 to 16,974 eggs/L, 21,370 to 13,423 eggs/L and 15,351 to 13,796 eggs/L for 2-, 3-, and 4-year olds, respectively. The volume of eggs produced per kilogram of female rose from 0.087 to 0.22 L/kg for 2-year-olds, 0.13 to 0.24 L/kg for 3-year-olds, and 0.21 to 0.29 L/kg for 4-year-olds. Individual fecundities of brown trout increased from 1,279 to 1,798 eggs/female for 3-year-olds, and from 1,695 to 3,629 eggs/female for 4-year-olds. Within age-3 fish, the volume of eggs produced per kilogram of female was 0.20 L/kg both before and after renovations, whereas the value for 4-year-olds doubled from 0.14 to 0.28 L/kg after renovations. The size of brown trout eggs increased from 16,027 to 13,255 eggs/L and from 16,027 to 9,230 eggs/L for 3- and 4-year-olds, respectively. These improvements in fecundities and egg sizes were attributed to the heavier mean weights and more efficient egg production of the brood trout held in constant-temperature water. These results demonstrate the feasibility of successfully rearing trout brood stocks in heated, recirculated water.

Economics of fish production tend to limit the use of recirculated-water systems to early rearing stages, laboratory systems, and public display systems (Muir 1982). However, where suitable water sources are rare and sources of disease-free trout eggs are limited, it may be practical to rear trout brood stocks in recirculated water.

The Province of Alberta has few available sources of high-quality water with adequate flow rates and suitable temperature regimes in which to rear trout (T. W. McFadden, Alberta Fish Culture Section, personal communication). To reduce the risk of introducing new fish diseases from outside the province, the provincial authority for fisheries management adopted a policy that discourages importing eggs to provincial hatcheries. As a result, the Allison Creek Brood Trout Station at Coleman, Alberta, Canada (50°N, 114°W), was commissioned in 1982.

During the station's first 2 years of operation, low winter water temperatures were accompanied by slow fish growth and unsatisfactory egg production by brood stocks of brook trout *Salvelinus fontinalis* and brown trout *Salmo trutta*. Water temperature is one of the main regulators of fish growth and egg development (Ursin 1979; Leitritz and Lewis 1980; Piper et al. 1982; Billard 1985). Consequently, to improve brood-stock performance the station was converted to a heated, recirculated-water system in 1984.

This paper reports on the feasibility of rearing trout brood stocks in heated, recirculated water. It also describes changes in egg size and brood-stock fecundity that were observed after the station was renovated.

Prerenovation Conditions

The Allison Creek Brood Trout Station initially used up to 68 L/s of water from Allison Creek as

[1]Authors contributed equally to this paper.
[2]To whom correspondence should be sent.

TABLE 1.—Comparison of water quality and trout brood-stock densities at the Allison Creek Brood Trout Station before and after renovation in 1984. Water quality was measured at the outlet of rearing units.

Condition	Before renovation (single-pass creek water)	After renovation (heated, recirculated groundwater)
Water pH	8.1–8.4	7.2–8.4
Total hardness (mg/L as $CaCO_3$)	76–99	231
Alkalinity (mg/L as $CaCO_3$)	88–127	180
Total dissolved solids (mg/L)	89–132	250–290
Un-ionized ammonia (mg/L)	Negligible	0.0007–0.0120
Brood-stock densities (kg/m^3)	64–96	32–66
Temperature (°C)	0.0–15.0	8.0–10.0

its main water source (Table 1). Water quality was good, although alkalinity and total hardness were below the optimum range of 120–400 mg/L prescribed by Piper et al. (1982). When the silt load was excessive during spring runoff, or flow rate in the creek was inadequate, water from Chinook Lake was used.

At the station, dissolved gases in the water were equilibrated by cascading the water through a 1.5-m tower of stacked troughs. Water then flowed down through two mechanical filters (diameter, 6.2 m) that consisted of a 30-cm layer of silica sand covered by a 30-cm layer of anthracite coal. The filters were designed for a filtration rate of 1.4 L/(s · m^2) of filter area. Finally, water flowing to incubation and early rearing units, and to brood-stock ponds, was irradiated with ultraviolet light (Figure 1).

In the first 2 years of operation, performance of the brood stocks was unsatisfactory: growth was slow, fecundity was low, eggs were small, and age of first maturity was delayed. The winter water supply was colder and the period of low temperatures was longer than expected (Figure 2). Diurnal fluctuations of as much as 9°C were recorded. These factors appeared to be the main constraints on brood-stock performance. To overcome these deficiencies, the station was renovated in 1984.

Postrenovation Conditions

Groundwater at 6–7°C is now the primary source of water. It is pumped to the station from a depth of about 26 m at a mean rate of 13 L/s. A fraction of this water is heated to 82°C by a natural gas boiler and heat exchanger system. This hot water is mixed back with the cold groundwater to achieve the desired rearing temperature. Since 1984, the station has operated within a range of 8.0–10.0°C. After being heated, the fresh water is passed through columns (height, 2.0 m; diameters 20.3 and 25.4 cm) packed with 3.8-cm-diameter Koch rings to equilibrate the dissolved gases.

BEFORE STATION RENOVATION

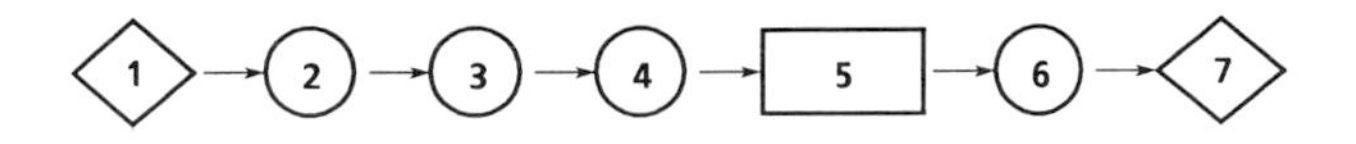

AFTER STATION RENOVATION

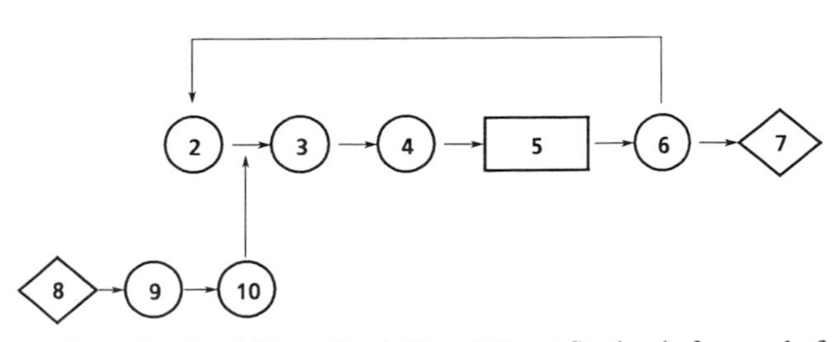

FIGURE 1.—Water flow schematics for Allison Creek Brood Trout Station before and after renovations in 1984. (1) creek water, gravity flow to station; (2) tower of stacked troughs for gas equilibration; (3) mechanical filters before renovation, mechanical–biological filters after renovation; (4) ultraviolet light sterilizers; (5) rearing units; (6) settling pond; (7) outlet; (8) groundwater, pumped to station; (9) boilers to heat water; (10) packed columns for gas equilibration.

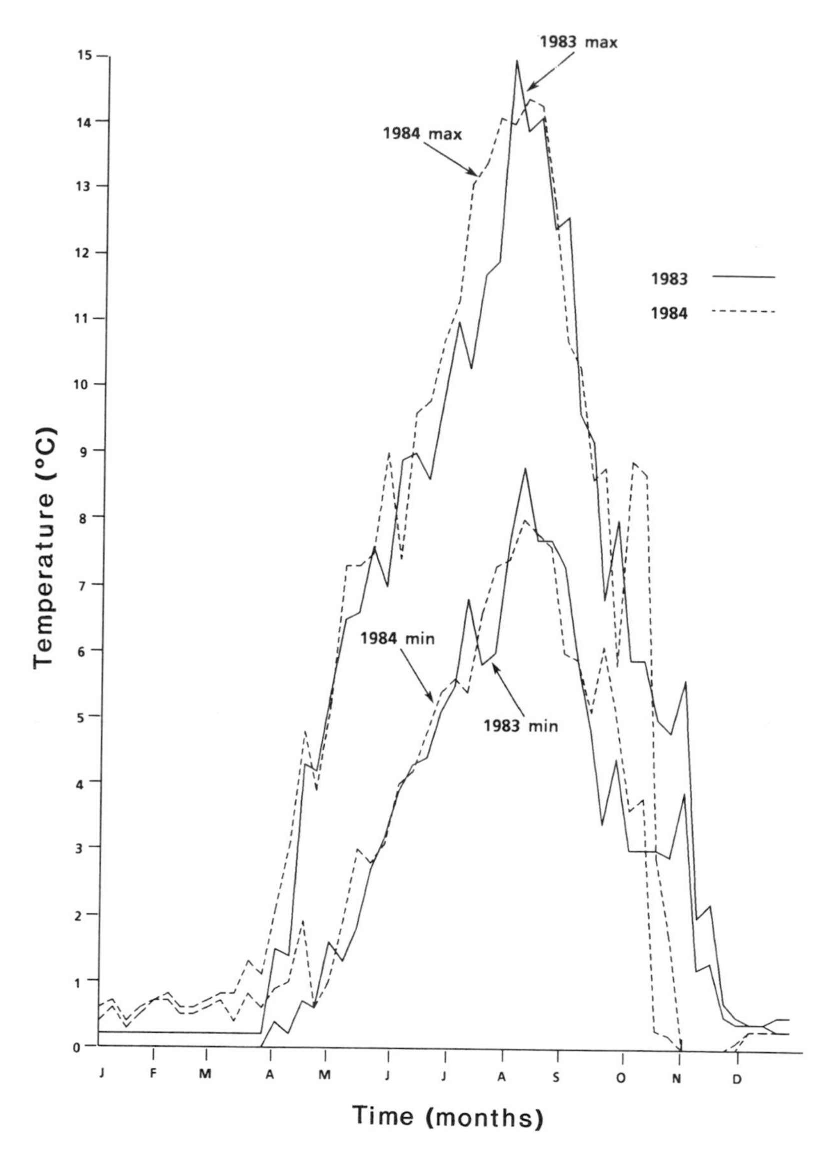

FIGURE 2.—Mean maximum and minimum temperatures of the surface water supply at the Allison Creek Brood Trout Station, 1983–1984. Measurements were taken daily; means were calculated on a weekly basis. The abscissa scale is in months (Jan, Feb, Mar, etc.).

To reduce heating costs, 80% of the water is recirculated each turnover cycle. It is pumped from an effluent settling pond (volume, 91.5 m^3; inlet flow, 61 L/s) up 11.3 m to the tower of stacked troughs for gas equilibration. After dissolved gas equilibration, the recirculated water mixes with the fresh water in a head tank before it is filtered.

The mixed water flows down through the submerged filters, which now operate as both biological and mechanical filters. The filter media were replaced and upgraded in 1988; the media now consist of 10 cm of filter gravel (3.2-cm × 1.9-cm rock), 10 cm of garnet sand, 23 cm of silica sand, and 30 cm of anthracite coal.

The final water treatment, ultraviolet light sterilization, remains unchanged (Figure 1). One water exchange is completed in 2.5 h.

Water quality (Table 1) remains within the optimum range described by Piper et al. (1982). Brood-stock densities were reduced (Table 1) in anticipation of the increased fecundities.

TABLE 2.—Mean fecundity, egg size, and weight of female brook trout at Allison Creek Brood Trout Station, and the history of fish exposure to constant and seasonally variable water temperatures.

Age (years)	Year-class	Year spawned	% of time with variable water temperature	Volume of eggs per kg female (L/kg)	Number of females	Eggs per female	Eggs per kg	Egg size: number per		Female weight (kg)
								Liter	Ounce	
2	1981	1982–1983	9	0.13	235	1,141	3,126	24,041	711	0.37
2	1982	1983–1984	25	0.087	416	556	2,298	26,340	779	0.24
2	1984	1985–1986	0	0.22	884	1,393	3,806	16,974	502	0.37
2	1986	1987–1988	0	0.20	699	1,789	3,727	18,766	555	0.48
3	1981	1983–1984	35	0.13	1,824	2,153	2,778	21,370	632	0.78
3	1982	1984–1985	35	0.13	1,460	1,455	2,787	16,061	475	0.52
3	1984	1986–1987	0	0.23	773	3,555	3,232	14,201	420	1.1
3	1986	1988–1989	0	0.24	145	4,071	3,180	13,423	397	1.3
4	1982	1985–1986	27	0.21	892	4,514	3,224	15,357	454	1.4
4	1984	1987–1988	0	0.29	225	5,920	3,947	13,796	408	1.5

Methods

Spawning.—The dry fertilization method (Piper et al. 1982) was used to strip and mix gametes from females and males. The number and size of eggs spawned were estimated by the Von Bayer method (Leitritz and Lewis 1980). The mean individual fecundity was calculated as B/C; B is the total number of eggs expressed, and C is the total number of females spawned. The relative fecundity was calculated as A/W; A is the mean number of eggs per female, and W is the mean weight (kg) of females spawned.

Eggs left in the gonads after stripping were not included in the fecundity estimates. Fish from representative samples of each lot were weighed individually before they were stripped of eggs.

Brood stocks.—Until 1986, brood-stock recruits were reared initially at the Sam Livingston Fish Hatchery. Located in Calgary, Alberta, this station heats groundwater to 8–12°C and recirculates 90% of its supply. Since 1986, brood stocks have been spawned, incubated, and reared entirely at the Allison Creek Brood Trout Station.

The brown trout brood stock was started from fertilized eggs collected from a wild population in the Bow River in 1980 and 1981. This population was originally established in 1925, probably from Loch Leven stock (Paetz and Nelson 1970). The brook trout stocks were derived from eggs purchased from Beity's Resort in Valley, Washington, in 1978.

Results

Brook Trout

Mean individual fecundity, mean relative fecundity, and egg size increased with the relative duration of exposure to constant-temperature water within all three age-groups (Table 2). In addition, eggs per female, fish weight, and egg size increased with age from ages 2 to 4. Because egg size and fecundity are often negatively correlated (Bagenal 1978), we calculated the volume of eggs produced per kilogram of female fish (Buss and McCreary 1960; Springate and Bromage 1984). Within each age-group this measure increased substantially for fish in constant-temperature water (Table 2).

At age 2, the mean weight of the 1984 year-class was lower than that of the 1986 year-class; this would account for the lower individual fecundity of the 1984 year-class despite the same exposure to constant-temperature water. The relative fecundities and the volumes of eggs per kilogram of fish of the two year-classes were about the same.

At age 3, the 1981 and 1982 year-classes had nearly identical relative fecundities and volumes of eggs per kilogram of fish; they were also exposed to a similar period of variable water temperatures. However, the station switched to heated groundwater 2–3 months before the 1982 year-class spawned, whereas the 1981 year-class was spawned in cold (<2°C) creek water. This could account for the larger egg size of fish in the 1982 year-class despite the smaller mean weight of females.

Brown Trout

Within age-groups 3 and 4, the mean individual fecundity increased as the relative duration of exposure to variable-temperature water decreased (Table 3). As well, the mean individual fecundities increased with age. The mean relative fecundity and the number of eggs per liter de-

TABLE 3.—Mean fecundity, egg size, and weight of female brown trout at Allison Creek Brood Trout Station, and the history of fish exposure to constant and seasonally variable water temperatures.

Age (years)	Year-class	Year spawned	% of time with variable water temperature	Volume of eggs per kg female (L/kg)	Number of females	Eggs per female	Eggs per kg	Egg size: number per		Female weight (kg)
								Liter	Ounce	
3	1980	1983–1984	24	0.20	84	1,279	3,166	16,027	474	0.40
3	1984	1987–1988	0	0.20	911	1,798	2,684	13,255	392	0.67
4	1980A	1984–1985	46	0.14	61	1,695	2,322	16,027	474	0.73
4	1980B	1984–1985	24	0.18	280	2,147	1,804	10,076	298	1.2
4	1981	1985–1986	0	0.23	424	2,827	2,827	12,409	367	1.0
4	1984	1988–1989	0	0.28	410	3,629	2,592	9,230	273	1.4
5	1981	1986–1987	0	0.24	328	4,439	2,466	10,178	301	1.8
6	1981	1987–1988	0	0.30	169	5,260	2,023	6,796	201	2.6

clined with age, although they did not appear to vary predictably within age-groups 3 and 4. Within age-group 3, the volume of eggs per kilogram of fish was the same for both year-classes. Within age-group 4, however, the volume of eggs per kilogram of fish increased substantially as the relative exposure to variable water temperatures decreased.

At age 3, the mean weight of females in constant-temperature, recirculated water was greater than that of females exposed to variable-temperature water. This could account for the increase in individual fecundity despite the larger eggs produced.

At age 4, the increasing individual fecundities and volume of eggs per kilogram of female were most closely related to the declining exposure to variable-temperature water. Egg size increased with mean weight but was not related to the mean relative fecundity or volume of eggs per kilogram of female. The mean weight of the 1981 year-class of females was intermediate between those of the two 1980 year-class lots. The 1981 year-class was spawned once at the Sam Livingston Fish Hatchery before being transferred to the Allison Creek Brood Trout Station. This year-class was stressed through overcrowding before its second spawning. This may have contributed to the unexpectedly small egg size of 12,409 eggs/L (367/oz) at age 4; when first spawned at age 3, egg size was 11,192 eggs/L (331/oz) (W. Schenk, Sam Livingston Fish Hatchery, unpublished data).

Discussion

Individual fecundity of both brook trout and brown trout increased as the fishes' exposure to variable water temperatures declined. Mean weight of fish also increased and could account for the increase in individual fecundity (Rounsefell 1957; Blaxter 1969; Bagenal 1978; Wootton 1979, 1982). For brook trout, relative fecundity, egg size, and volume of eggs per kilogram of female increased as well. Thus it was not only the larger size of brook trout that resulted in higher fecundities but also an improved efficiency of egg production, in terms of both larger numbers and volumes of eggs per kilogram of female and larger sizes of eggs. For brown trout, there was no apparent pattern to the changes in relative fecundity within age-groups 3 and 4. However, 4-year-olds showed an increase in egg size and volume of eggs per kilogram of female as exposure to variable-temperature water decreased. From this we conclude that the brown trout brood stocks also performed better under constant temperature conditions.

Many authors have shown that larger eggs produce larger fry (see reviews in Bagenal 1978; Springate and Bromage 1985), but debate continues over whether larger fry survive at higher rates under the controlled conditions of hatcheries (Springate and Bromage 1984, 1985). Although many other factors undoubtedly influence growth and survival of fry in our hatchery, the larger eggs from brood stocks in constant-temperature water produced larger fry with higher survival rates than smaller eggs from brood stocks in variable-temperature water (J. Enns, Sam Livingston Fish Hatchery, unpublished data).

We conclude that change in water-temperature regime was the most probable cause of the improved brood-stock performance. Piper et al. (1982) recommended spawning temperatures of 7.2–12.8°C and 8.9–12.8°C for brook trout and

TABLE 4.—Mean fecundity, egg size, and weight of brook and brown trout brood stocks reared in selected flow-through water systems in North America compared with data for brood stocks reared in recirculated, constant-temperature water at the Allison Creek Brood Trout Station.

Age (years)	Water temperature (°C)	Number of eggs per		Egg size: number per		Female weight (kg)	Strain[a]	Fish culture station[b]
		Female	Kilogram	Liter	Ounce			
Brook trout								
2	2–7	850	2,576	16,195	479	0.33	GR	WS-MB
2	0–17	2,600–3,000	2,364–2,727	12,172–13,186	360–390	1.1	WV	WH-CA
2	3–20	2,175	3,107	18,765	555	0.7	WH	CL-CA
2		1,811	3,321			0.55	HY	BS-PA
2	4–14	1,161–1,347	3,414–3,962	21,740–23,261	643–688	0.34	OW	SR-WY
2	8–10	1,591	3,767	17,885	529	0.43	BE	AL-AB
2	12	2,700–2,800	3,500–3,857	13,524	400	0.7–0.8	OW	WS-WV
3	2–7	1,300	1,182–1,444	14,302	423	0.9–1.1	GR	WS-MB
3	4–11	1,060–1,282	1,514–1,603	12,138–13,558	359–401	0.7–0.8	LN	DO-ON
3	0–17	4,000	2,222	8,453–9,467	250–280	1.8	WV	WH-CA
3	8–10	3,813	3,206	13,828	409	1.2	BE	AL-AB
3		4,584	3,879			1.2	HY	BS-PA
3	4–14	2,817	4,942–6,260	19,880–20,827	588–616	0.45–0.57	OW	SR-WY
4	2–7	2,200	1,375–1,571	12,848	380	1.4–1.6	GR	WS-MB
4	4–11	1,700–1,800	1,545–1,636	11,867–12,476	351–369	1.1	LN	DO-ON
4	4–14	3,350–3,810	3,350–4,233	17,852–18,866	528–558	0.9–1.0	OW	SR-WY
4	8–10	5,920	3,947	13,794	408	1.5	BE	AL-AB
5	4–11	2,250	1,607	11,867	351	1.4	LN	DO-ON
5	2–7	3,800	1,520–1,900	16,195	479	2.0–2.5	GR	WS-MB
Brown trout								
2	0–17	1,800–1,900	2,000–2,111	14,031–14,538	415–430	0.9	WH	WH-CA
2	3–11	2,000	2,222	11,834–12,307	350–364	0.9	MS	MS-CA
2		2,170	3,672			0.59	DO	BS-PA
3	9	3,000	1,875	7,607	225	1.6	SC	SC-MT
3	12	3,500–3,700	2,313–2,500	<8,453	<250	1.4–1.6	PR	WS-WV
3	2–7	1,800	2,571	11,834–13,524	350–400	0.7	GM	WS-MB
3	8–10	1,798	2,684	13,254	392	0.67	BR	AL-AB
3		4,374–4,762	2,601–3,380			1.4–1.7	DO	BS-PA
3	1–13	2,475–3,382	2,727–3,780	15,958–18,562	472–549	0.74–1.2	GN	CT-ON
3	4–14	2,160–2,360	3,789–5,244	20,387–21,740	603–643	0.45–0.57	PR	SR-WY
4	2–7	2,200	1,467–2,200	12,848	380	1.0–1.5	GM	WS-MB
4		6,183	1,916			3.2	DO	BS-PA
4	8–10	3,228	2,710	10,819	320	1.2	BR	AL-AB
4	1–13	3,084–4,792	1,945–2,958	11,597–12,645	343–374	1.5–1.7	GN	CT-ON
4	4–14	3,197–3,399	2,829–3,008	15,620–16,398	462–485	1.13	PR	SR-WY
5	2–7	3,400	1,360–1,700	12,172	360	2.0–2.5	GM	WS-MB
5	4–14	4,509	2,254–2,505	12,780	378	1.8–2.0	PR	SR-WY
5	8–10	4,439	2,466	10,177	301	1.8	BR	AL-AB

[a]Strain codes:
BE Beity
BR Bow River
DO Domestic
GM German Brown (Massachusetts)
GN Ganaraska River
GR God's River
HY Unknown hybrids
LN Lake Nipigon
MS Mount Shasta
OW Owhi
PR Plymouth Rock Domestic
SC Spring Creek
WH Whitney
WV West Virginia

[b]Fish culture station codes and references for unpublished data:
AL-AB Allison Creek, Alberta
BS-PA Benner Spring Research, Pennsylvania (Buss and McCreary 1960)
CT-ON Codrington, Ontario (G. Raine, Ontario Ministry Natural Resources)
CL-CA Crystal Lake, California (M. McCormack, The Resources Agency, California Department Fish and Game)
DO-ON Dorion, Ontario (P. Richard, Ontario Ministry Natural Resources)
MS-CA Mt. Shasta, California (T. Nevison, The Resources Agency, California Department Fish and Game)
SC-MT Spring Creek, Montana (private operation)
SR-WY Saratoga, Wyoming (J. Hammer, U.S. Fish and Wildlife Service)
WS-MB White Shell, Manitoba (J. Ziemanski, Manitoba Department of Natural Resources)
WS-WV White Sulphur Springs, West Virginia (W. Eubank, U.S. Fish and Wildlife Service)
WH-CA Whitney, California (J. Riley, The Resources Agency, California Department Fish and Game)

brown trout, respectively. Hokanson et al. (1973) found that to stimulate successful spawning by brook trout, water temperature should not exceed 12°C during the normal breeding season; mean water temperatures below 9°C were required for optimal spawning activity and gamete viability, though the lower temperature limit was not determined. Before the Allison Creek station was renovated, the spawning period for both species peaked between early November and mid-February when median water temperatures were less than 1.8°C. After renovation, the brood stocks were spawning in 8.0–10.0°C water, which is much closer to the temperatures recommended by Hokanson et al. (1973) and Piper et al. (1982).

The brook trout and brown trout brood stocks did not shift their spawning season in response to the constant-temperature environment. They did, however, reduce the period of spawning from as high as 4.5 months to as low as 2.0 months (C.W.P., unpublished data). Our results agree with those of Henderson (1963) and Hokanson et al. (1973), who concluded that temperature had little influence on the timing of the spawning season for brook trout.

Most factors other than temperature remained relatively constant before and after the renovation. We assumed the factors that did change were minor relative to the change in temperature regime. The higher ammonia levels of the recirculated water probably increased stress on the fish (Colt and Armstrong 1981). Spawning techniques, feeding rates, and diet were largely unchanged. Brood-stock recruits were selected to be representative of the parent stock; fecundity and egg size were not selection criteria. Lower fish densities may have contributed to faster growth of fish. Brood stocks were no longer exposed to high silt loads in the spring because the water source was converted from surface water to groundwater; this improvement in water quality probably also helped the fish.

The quality of the brook trout and brown trout brood stocks in recirculated water at the Allison Creek Brood Trout Station is comparable with the standard of brood stocks reared in many flow-through systems in North America (Table 4). Factors such as water temperature, feeding rate, nutrition, size of fish, and brood-stock strain at the other stations probably account for the range in values observed. We consider our brown trout stock to be "wild" and do not expect their growth rate to be as high as those of domesticated stocks.

In Alberta, where sources of high-quality water with adequate flow rates and suitable temperatures are rare, we have shown that heating and recirculating water is a biologically viable approach to rearing brook trout and brown trout brood stocks. When this alternative is considered, its higher fixed costs, higher energy consumption, greater technological complexity, and greater inherent risk should be weighed against the benefits of increased fecundity, egg size, and reduced numbers of brood stock required to meet production targets.

References

Bagenal, T. B. 1978. Aspects of fish fecundity. Pages 75–101 *in* S. D. Gerking, editor. Ecology of freshwater fish production. Blackwell Scientific Publications, London.

Billard, R. 1985. Environmental factors in salmonid culture and the control of reproduction. Pages 70–87 *in* R. N. Iwamoto and S. Sawer, editors. Salmonid reproduction, an international symposium. Washington Sea Grant Program, University of Washington, Seattle.

Blaxter, J. H. S. 1969. Development: eggs and larvae. Pages 177–252 *in* W. S. Hoar and D. J. Randall, editors. Fish physiology, volume 3. Academic Press, New York.

Buss, K., and R. McCreary. 1960. A comparison of egg production of hatchery-reared brook, brown, and rainbow trout. Progressive Fish-Culturist 22:7–10.

Colt, J. E., and D. A. Armstrong. 1981. Nitrogen toxicity to crustaceans, fish, and molluscs. Pages 34–47 *in* L. J. Allen and E. C. Kinney, editors. Proceedings of the bio-engineering symposium for fish culture. American Fisheries Society, Fish Culture Section, Bethesda, Maryland.

Henderson, N. E. 1963. Influence of light and temperature on the reproductive cycle of the eastern brook trout, *Salvelinus fontinalis* (Mitchill). Journal of the Fisheries Research Board of Canada 20:859–897.

Hokanson, K. E., J. H. McCormick, B. R. Jones, and J. H. Tucker. 1973. Thermal requirements for maturation, spawning, and embryo survival of the brook trout, *Salvelinus fontinalis*. Journal of the Fisheries Research Board of Canada 30:975–984.

Leitritz, E., and R. C. Lewis. 1980. Trout and salmon culture. California Department of Fish and Game, Fish Bulletin 164.

Muir, J. F. 1982. Recirculated water systems in aquaculture. Pages 357–446 *in* J. F. Muir and R. J. Roberts, editors. Recent advances in aquaculture. Croom Helm, London.

Paetz, M. J., and J. S. Nelson. 1970. The fishes of Alberta. Queen's Printer, Edmonton, Alberta.

Piper, G. P., I. B. McElwain, L. E. Orme, J. P. McCraren, L. G. Fowler, and J. R. Leonard. 1982. Fish hatchery management. U.S. Fish and Wildlife Service, Washington, D.C.

Rounsefell, G. A. 1957. Fecundity of North American

salmonidae. U.S. Fish and Wildlife Service Fishery Bulletin 57:451–468.

Springate, J. R. C., and N. R. Bromage. 1984. Egg size and number—it's a "trade-off". Fish Farmer (July): 12–14.

Springate, J. R. C., and N. R. Bromage. 1985. Effects of egg size on early growth and survival in rainbow trout (*Salmo gairdneri* Richardson). Aquaculture 47:163–172.

Ursin, E. 1979. Principles of growth in fishes. Symposia of the Zoological Society of London 44:63–87.

Wootton, R. J. 1979. Energy costs of egg production and environmental determinants of fecundity in teleost fishes. Symposia of the Zoological Society of London 44:133–159.

Wootton, R. J. 1982. Environmental factors in fish reproduction. Pages 210–219 *in* C. J. J. Richter and H. J. Th. Goos, compilers. Proceedings of the international symposium on reproductive physiology of fish. Centre for Agricultural Publishing and Documentation, Wageningen, Netherlands.

American Fisheries Society Symposium 10:562–566, 1991

Sea Releases of Baltic Salmon: Increased Survival with a Delayed-Release Technique

T. ERIKSSON

*Department of Aquaculture, Swedish University of Agricultural Sciences
S-90183 Umeå, Sweden*

Abstract.—Delayed releases of Baltic (Atlantic) salmon *Salmo salar* into the sea resulted in enhanced total recapture rates compared with river releases. The recapture rates achieved from sea releases at six sites in the mid and southern Baltic Sea varied between 8.5 and 39% compared with a mean recapture rate of about 10.5% for river-released fish. The increased recapture rates in this study are supported by results from a series of experiments in 1980–1982 in which Baltic salmon smolts were released in the northern part of the Bothnian Sea. The survival rate in that study was improved three to five times by using a release technique including delayed sea releases. In the Baltic Sea, delayed sea release of Baltic salmon has generally improved survival rates.

The anadromous Baltic (Atlantic) salmon *Salmo salar* exhibits two major habitat shifts during its life cycle. Adults enter the rivers in autumn to spawn. After 1–5 years in fresh water, the juveniles leave the riverine environment and migrate to their feeding areas in the central Baltic Sea. Before entering the seaward migratory phase, the parr undergo smoltification, preparing for life in the sea (Hoar 1976; Wedemeyer et al. 1980).

A positive relationship between smolt size and survival after release has been reported for Baltic salmon by several authors (e.g., Österdahl 1964; Carlin 1969; Eriksson 1988; Lundqvist et al. 1988). By keeping the smolts beyond the normal time of migration—that is, by delaying the times of release—the size of the fish will increase before release. With the intention of achieving increased survival, such delayed-release experiments have been conducted in rivers. These experiments have mainly resulted in lower recapture rates for fish released beyond their normal migration period, despite enhanced size of the fish before release (Peterson 1973; Larsson 1979).

When these results are interpreted, it is important to be aware that smoltificating itself is a dynamic process. Eriksson and Lundqvist (1982) and Lundqvist and Eriksson (1985) demonstrated the seasonality of smoltification among Baltic salmon, expressed in the smoltification–desmoltification cycle. They stated that smoltification by Baltic salmon is basically a seasonal, reversible, and annually reoccurring process. Thus parr released in the river during autumn are assumed to act as stationary, freshwater-adapted fish; they remain and face a high predatory risk.

Based on knowledge regarding the annual dynamic process of Baltic salmon smoltification, an alternative release technique with nonriver-based delayed releases has been developed (Eriksson and Eriksson 1985; Eriksson 1988). This technique includes a transfer of Baltic salmon smolts to the sea at the normal time of migration. Fish that reach the sea at this period follow a "sea-phase track," in which they behave as migratory fish and allocate energetic resources to growth (Eriksson et al. 1987).

In a series of experiments from 1980 to 1982, the sea-ranching technique was tested (Eriksson and Eriksson 1985). Baltic salmon smolts were kept in net pens after transfer to the sea and released from the pens into the northern part of the Bothnian Sea after a varying period of delay. The survival rate was improved three to five times by this technique (Eriksson 1988).

The technique of delayed releases at sea is under further evaluation in a joint national research program organized by the Salmon Research Institute in cooperation with the Department of Aquaculture, Swedish University of Agricultural Sciences, the National Board of Fisheries, and the fishery administration in Mörrum. In this paper I present the results from this extended experiment, including releases at six places in the central and southern Baltic Sea, and evaluate the effect on survival of delayed sea releases of postsmolts in the southern and middle Baltic Sea.

Methods

The releases of fish were conducted at six release sites in three main areas: (1) the Stockholm archipelago, (2) the Bråviken archipelago, and (3) the Mörrumå River mouth.

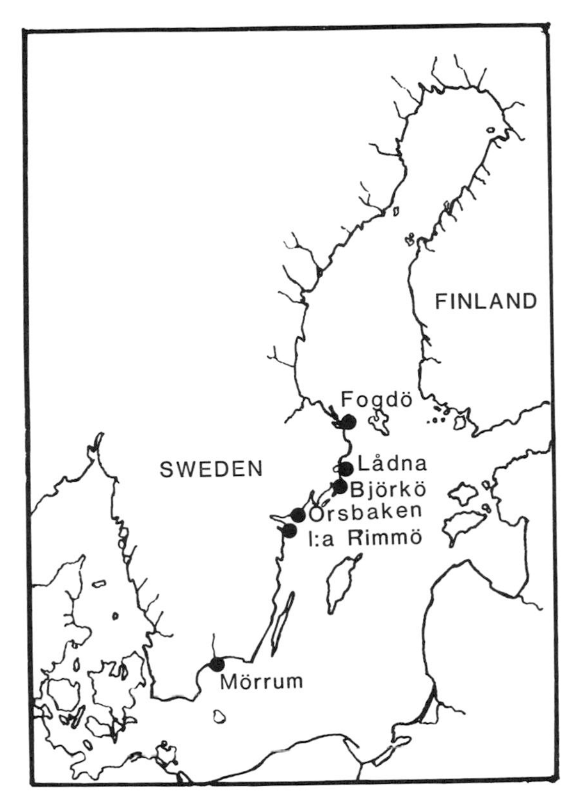

FIGURE 1.—Map of the Baltic Sea showing the six release sites used in the study.

In the Stockholm archipelago, fish were released at three locations: Fogdö (60°10′N, 18°45′E), Lådna (59°25′N, 19°45′E), and Björkösund (59°05′N, 19°10′E) (Figure 1). The smolts released were randomly selected from 1-year-old fish originating from the Lule River salmon stock. They had been reared under natural conditions of water temperature and photoperiod at the Salmon Research Institute at Dalälven River (60°35′N, 17°30′E). At the normal time of the smolt run (early June), the fish were transferred to the sea and kept in net pens until being released on September 4. Before release, about 1,000 fish at each of the three release places were individually tagged with Carlin tags (Carlin 1955) and measured (fork length) to the nearest 0.5 cm.

In the Bråviken area fish were released at two places: Örsbaken (58°40′N, 17°10′E) and Lilla Rimmö (58°25′N, 16°45′N) (Figure 1). The smolts released were randomly selected from 2-year-old fish originating from both the Lule River salmon stock and the Dalälven River stock. They had been reared under natural conditions of water temperature and photoperiod at the Salmon Research Institute on the Dalälven River (60°35′N, 17°30′E). At the normal time of the smolt run (early June), the fish were transferred to the sea and kept in net pens until they were released on September 3 and 5. Before release, about 2,000 fish at each release place were individually tagged with Carlin tags and measured (fork length, nearest 0.5 cm).

Mörrumå River releases were made in the spring of 1985. About 1,500 1-year-old and 1,500 2-year-old smolts were randomly selected at the Mörrum hatchery (56°07′N, 14°45′E) (Figure 1). The fish originated from wild parents from the Mörrumå River and were reared under ambient conditions. They were individually tagged with external Carlin tags on May 22–23. At the normal time of releases of 1-year-old smolts (in late May), the smolts were transferred directly to the sea, where they were kept in net pens about 5 km south of the river mouth. A control group of fish from the same stock was released in the river on May 31. In total, two groups were used, each consisting of 500 1-year-old and 500 2-year-old smolts. After the fish had been acclimatized in the net pens for about 10 d, the first group was released (June 11). Release of the second group was delayed until September 20. Each group of tagged fish released from the net pens consisted of about 500 1-year-old and 500 2-year-old postsmolts.

In this analysis I used all recaptures of fish reported until August 1, 1988. Recaptures reported from the open sea fishery, the coastal fishery, and the riverine fishery were used to estimate survival. The fishing pressure on salmon in the Baltic is high and Eriksson (1988) suggested that fishing mortality allows a good estimate of survival rate. The recapture data from tagged fish were collected in computer files at the Salmon Research Institute, previously described by Carlin (1971).

To compare the rate of survival between fish of different sizes released from the Mörrumå River, fish were graded in size-groups so each group had at least five recaptured fish. Data were subjected to regression analysis. Tests of proportions (5% level) were used to discover differences in percentage recoveries of salmon between Mörrumå River release groups.

Results

The delayed sea releases of Baltic salmon postsmolts generally showed high recapture rates compared with river-released fish (Table 1). In the

TABLE 1.—Recapture rates for Baltic salmon released at six sites in the Baltic Sea. Date of release, mean fork length, and age of smolts and numbers of tagged fish are indicated.

Release site	Date of release (1985)	Mean length (mm)	Age of fish (years)	Tagged fish released (number)	Recapture rate (%)
Fogdö	Apr 9	245	1	999	24.3
Lådna	Apr 9	238	1	998	32.2
Björkö	Apr 9	249	1	998	39.1
Örsbaken	Apr 9	236	2	1,994	24.8
Lilla Rimmö	Sep 4	233	2	1,997	21.9
Mörrum coast	Sep 20	246	1	426	8.5
Mörrum coast	Sep 20	298	2	435	24.6

Stockholm archipelago area, the recapture rate of two-summer-old fish released in September varied between 24 and 39%. The three-summer-old fish released in the Bråviken area showed a recapture rate of 22–25%. The corresponding mean recapture rate of river-released smolts was 8.7% for recaptures until January 1988. The recaptures reported for the delayed releases include reports until July, and the recaptures reported from the river releases should be raised by about 2% to be comparable.

When comparing release groups of Mörrumå River smolts, I found an increased rate of recapture for the delayed sea release technique (Figure 2). The same general pattern appeared for both 1-year-old and 2-year-old smolts.

The river-released fish showed recapture rates of 4.0 and 7.7% for 1- and 2-year-old smolts, respectively. A release after transfer to net pens in the sea and 10 d of acclimatization increased the survival rates to 5.2 and 14.2%, respectively. When delayed for about 4 months before release, fish showed a further increase in survival, giving recapture rates of 7.3 and 22.8%, respectively, for two- and three-summer fish.

A comparison of length at release and recapture rate within the groups showed a positive size-related relationship. Correlation coefficients were between 0.92 and 0.98 for 1-year-old fish and between 0.54 and 0.91 for 2-year-old fish (Table 2).

Discussion

The delayed sea releases of Baltic salmon resulted in high recapture rates compared with recaptures from river-released fish. The increased recapture rates achieved at experimental releases at Ulvön (Eriksson 1988) agree with the results

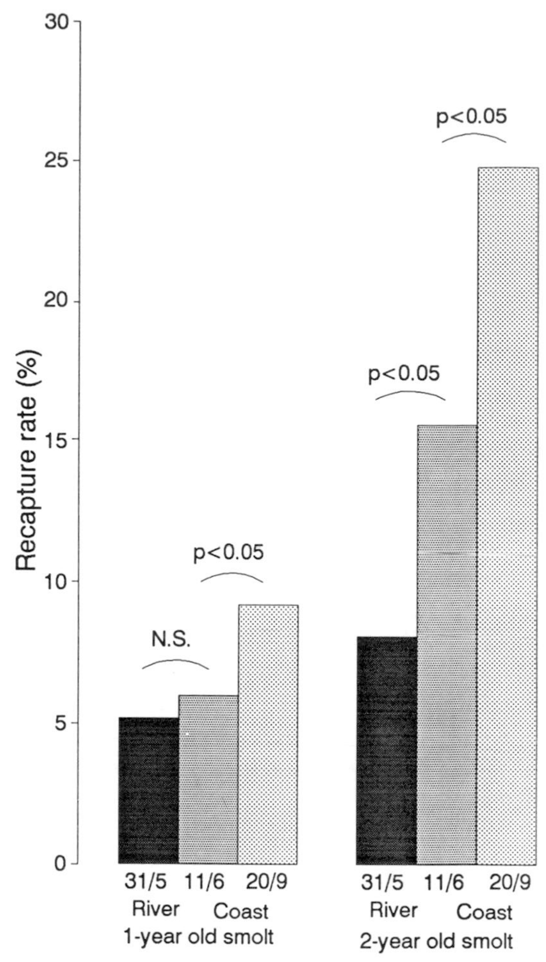

FIGURE 2.—Recapture rates reported for Mörrumå River Baltic salmon released in 1985. For each 1- and 2-year-old smolt group, fish were released in the river in May (31/5) or transferred to the sea and kept in cages before release in June (11/6) or September (20/9).

TABLE 2.—Correlation coefficient between length at release and recapture rate within each group of Mörrumå River Baltic salmon released in 1985.

Release group	Correlation coefficient	P-value
Age-1 fish		
River	0.92	0.25
Jun	0.98	0.12
Sep	0.97	0.17
Age-2 fish		
River	0.54	0.46
Jun	0.81	0.05
Sep	0.91	0.005

reported in this study, including releases of fish at the six release sites. In both studies a high number of releases ($N \sim 20$) were performed and consistently showed increased recapture rates. It therefore seems reasonable to state that in the Baltic Sea delayed release of Baltic salmon directly into the sea generally improves survival rate.

Similar experiences with delayed releases have been reported for Pacific salmon *Oncorhynchus* spp. (Novotny 1980) and Atlantic salmon (Saunders 1977). In a more recent study, Gunneröd et al. (1988) showed that open sea releases of Atlantic salmon on the Norwegian coast increased mean recapture rate by 111% compared with river releases. However, other Norwegian experimental releases have not shown any significant increase in survival rate for delayed sea releases (Hansen and Jonsson 1986).

The mortality risk for juvenile salmon during migration has been suggested to be very high (Ricker 1976; Larsson 1984). Eriksson (1988) found that mortality rates peaked during the downstream migration and entry into the sea. The mortality during the first 2 weeks was estimated to be between 25 and 30% per week. In a study of Mörrumå and Emå river smolt migrations, Larsson (1984) calculated a mortality rate in the rivers of about 50%. The sea releases of Mörrumå River smolts in the present study further confirm these earlier results. After protected transfer of smolts to sea and acclimatization of 10–14 d before release, the recapture rates improved, indicating a reduced mortality rate, because size did not significantly increase during this time.

These studies indicate that there is a strong positive size-dependent survival among Baltic salmon released in the Baltic Sea. In the experiments reported in this study, the increased size before release seems to explain the major part of the increased recapture rates of the delayed releases of fish, compared with sea releases at the normal time of smolt migration. A similar relationship has been presented for Pacific salmon. Mathews and Buckley (1976) presented a model for natural mortality during 18 months of marine life of different-sized coho salmon *Oncorhynchus kisutch*. They found a high degree of agreement between the hypothesized inverse weight relationship of mortality. Ricker (1976) also concluded that the assumption that mortality is inversely proportional to weight is realistic though unproven.

Acknowledgments

I thank C. Eriksson and L. Karlsson at the Salmon Research Institute, who kindly supplied information during preparation of this paper. N. Johansson and H. Lundqvist gave valuable comments on an earlier draft of the manuscript. C. Johansson and his staff in Mörrum are greatly appreciated for their valuable assistance during the experimental releases. The experiments were funded by charges imposed on the commercial fishery.

References

Carlin, B. 1955. Tagging of salmon smolts in the River Lagan. Institute of Freshwater Research, Annual Report 1954, Drottningholm, Sweden.

Carlin, B. 1969. Salmon tagging experiments. Swedish Salmon Research Institute, Report 3/1969, Älvkarleby, Sweden.

Carlin, B. 1971. Data processing in Swedish salmon tagging experiments. Swedish Salmon Research Institute, Report 3/1971, Älvkarleby, Sweden.

Eriksson, L.-O., and T. Eriksson. 1985. Non river based sea-ranching experiments and net-pen rearing of Baltic salmon (*Salmo salar* L.) in the Bothnian Sea. Pages 108–129 *in* Preliminary report of the salmonid workshop on biological and economical optimization of smolt production. Japan Ministry of Agriculture, Forestry, and Fisheries, Tokyo.

Eriksson, L.-O., and H. Lundqvist. 1982. Circannual rhythms and photoperiod regulation of growth and smolting in Baltic salmon (*Salmo salar* L.) parr. Aquaculture 28:113–121.

Eriksson, T. 1988. Migratory behaviour of Baltic salmon (*Salmo salar* L.); adaptive significance of annual cycles. Doctoral dissertation. University of Umeå, Umeå, Sweden.

Eriksson, T., L.-O. Eriksson, and H. Lundqvist. 1987. Adaptive flexibility in life-history tactics of mature male Baltic salmon parr in relation to body size and environment. American Fisheries Society Symposium 1:236–243.

Gunneröd, T. B., N. A. Hvidsten, and T. G. Heggberget. 1988. Open sea releases of Atlantic salmon, *Salmo salar*, in central Norway, 1973–83. Canadian Journal of Fisheries and Aquatic Sciences 45:1340–1345.

Hansen, L. P., and B. Jonsson. 1986. Salmon ranching experiments in the River Imsa: effects of day and night release and of sea-water adaption on recapture rates of adults. Institute of Freshwater Research Drottningholm Report 63:47–51.

Hoar, W. S. 1976. Smolt transformation: evolution, behaviour and physiology. Journal of the Fisheries Research Board of Canada 33:1234–1252.

Larsson, P.-O. 1979. The impact of water temperature at release on the subsequent return rates of hatchery reared Atlantic salmon smolts. Swedish Salmon Research Institute, Report 3/79 Älvkarleby, Sweden.

Larsson, P.-O. 1984. Some characteristics of the Baltic salmon, *Salmo salar* L., population. Doctoral dissertation. University of Stockholm, Stockholm.

Lundqvist, H., W. C. Clarke, and H. Johansson. 1988. The influence of precocious sexual maturation on survival to adulthood of river stocked Baltic salmon (*Salmo salar*) smolts. Holarctic Ecology 11:60–69.

Lundqvist, H., and L.-O. Eriksson. 1985. Annual rhythms of swimming behaviour and seawater adaption in young Baltic salmon, *Salmo salar*, associated with smolting. Environmental Biology of Fishes 14:259–267.

Mathews, S. B., and R. Buckley. 1976. Marine mortality of Puget Sound coho salmon (*Oncorhynchus kisutch*). Journal of the Fisheries Research Board of Canada 33:1677–1684.

Novotny, A. J. 1980. Delayed release of salmon. Pages 356–369 *in* J. E. Thorpe, editor. Salmon ranching. Academic Press, New York.

Österdahl, L. 1964. Smolt investigations in the river Rickleån. Swedish Salmon Research Institute, Report 8/64, Älvkarleby, Sweden.

Petersson, H. H. 1973. Adult returns to date from hatchery-reared one-year old smolts. Pages 219–226 *in* M. V. Smith and W. M. Carter, editors. International Atlantic Salmon Symposium. Atlantic Salmon Foundation, New York.

Ricker, W. E. 1976. Review of the rate of growth and mortality of Pacific salmon in salt water, and noncatch mortality caused by fishing. Journal of the Fisheries Research Board of Canada 33:1483–1542.

Saunders, R. L. 1977. Salmon ranching—a promising way to enhance populations of Atlantic salmon for angling and commercial fisheries. International Atlantic Salmon Foundation, Special Publication Series 7:17–24.

Wedemeyer, G. A., R. L. Saunders, and W. C. Clarke. 1980. Environmental factors affecting smoltification and early marine survival of anadromous salmonids. U.S. National Marine Fisheries Service Marine Fisheries Review 42(6):1–14.

LAST POST

OTHER BOOKS BY THE AUTHOR

Into India
India Discovered
Eccentric Travellers
Explorers Extraordinary
Highland Drove
The Honourable Company: A History of the English East India Company
Indonesia: From Sabang to Merauke
Explorers of the Western Himalayas, 1820–1895
(first published as *When Men and Mountains Meet* and *The Gilgit Game*)

The Royal Geographical Society's History of World Exploration
(general editor)
The Robinson Book of Exploration (editor)
Collins Encyclopaedia of Scotland (co-editor with Julia Keay)

LAST POST

The End of Empire in the Far East

JOHN KEAY

JOHN MURRAY
Albemarle Street, London

First published in 1997
by John Murray (Publishers) Ltd.,
50 Albemarle Street, London W1X 4BD

A catalogue record for this book is available from the British Library

ISBN 0-7195-5346 6

Typeset in 12.25/13.5 Monotype Garamond by Servis Filmsetting, Manchester

Printed and bound in Great Britain by
the University Press, Cambridge

making a nonsense of the civilising mission sometimes assumed by the colonial powers, gave to these countries a ready-made focus of identity, and hence a sense of nationhood.

At the height of the Vietnam War American generals sometimes talked of 'bombing North Vietnam back to the Stone Age'. They were unaware that going back to the Stone Age was official policy in Hanoi. Improbably, North Vietnam, though experiencing desperate shortages, subject to the heaviest bombardment in history, and taking 10,000 casualties a month, still found the resources and personnel to fund an ambitious programme of archaeological research. Professor Cao Xuan Pho was then excavating the pre-historic Hoa Binh culture. His work, occasionally helped but more often hindered by US carpet bombing, went on regardless. As he recalled, even Ho Chi Minh acknowledged the importance of archaeological study.

> We were telling our people that they were going into war with 4,000 years of history behind them. What does this mean? It is a matter of tradition. We are poor, OK, desperate; but we have to divert some expenditure into archaeology to prove to our people that once we were an independent nation with a social organisation of our own and that after that we were oppressed by foreigners.

History in the case of Indonesia and geography in the case of Malaya and the Philippines served much the same purpose as archaeology in Vietnam. In varying degrees most of the peoples of the Far East had some pre-colonial claim to a nucleus of national pride and cohesion which awaited only a catalyst – economic collapse, world war, revolution, all of which were waiting in the wings – to excite mass support and become a righteous demand for national liberation.

No less important than nationalism in the anti-colonial equation, according to the Hanoi professor, was another feature shared by most of the peoples of the Far East. They all ate rice, a lot of it and not much else; and nearly all the rice was grown on irrigated land. Until the third millennium BC Vietnam's Stone Age people had lived in upland caves and knew nothing of irrigation.

> The sea then came up to Dien Bien Phu and the delta was inundated. But when the sea receded, the people came down to the plain and had to close ranks, to support one another and to work together to build an irrigation system for wet rice cultivation. So the first characteristic of our people is solidarity, the solidarity learnt from working together to clear, and make fertile with water, the land.

Constructing and operating a system of water management makes exceptional demands on the collective will of a community. The consequent tradition of peasant 'solidarity' is seen as a distinctive feature of most east and south-east Asian societies and, in Vietnam, as the bed-rock of resistance to both the French and the Americans. The enemy had more sophisticated weapons but, in Cao Xuan Pho's opinion, the Vietnamese had the more cohesive society 'which, in a people's war, is what counts'.

It was also, of course, a Communist society. Whether or not Marx had anything to teach the Vietnamese about solidarity, and whether or not one subscribes to the idea that Communist ideology, originally tailored to an industrial proletariat in Europe, found an Asian equivalent in the equally organised and exploited masses in the rice fields, the fact remains that in the Far East Communism was as much a part of the colonial challenge as nationalism.

This complicated the end of empire considerably. Scrutinising the fabric of anti-colonialism so as to distinguish strands of Communist ideology from those of nationalist resurgence would so obsess the imperial powers that their final hour seemed as much about Cold War containment as about sponsoring successor states. Initially in China and Malaya, and eventually in Indonesia, nationalists came out against the Communist challenge, but in Indo-China the strands proved indistinguishable and the US found itself in the anomalous position of waging a colonial war in defence of the free world.

The vigour of nationalism, then, and the wide appeal of Communism distinguished the process of decolonisation in the Far East from that elsewhere and considerably hastened empire's end. Additionally and crucially, the colonies of the Far East, though totally dissimilar and firmly rooted in the very different imperial cultures of France, the Netherlands, Great Britain and the USA, nevertheless underwent a common historical experience in the mid-twentieth century. All suffered from the worldwide depression of the 1930s, and all witnessed the defeat of their colonial overlords by the Japanese in the 1940s. The myth of colonial prosperity was wrecked by the depression; that of imperial invincibility was exploded by the war. Within weeks of Pearl Harbor the erstwhile master races were to be seen being marched off to prison camps in the midday sun, their bags and bundles slung about them, without a servant in sight.

Regaining their colonies would be as much about restoring pride